THEORY OF ELECTROMAGNETIC WELL LOGGING

THEORY OF ELECTROMAGNETIC WELL LOGGING

C. RICHARD LIU
Bitswave, Inc.

elsevier.com

Elsevier
Radarweg 29, PO Box 211, 1000 AE Amsterdam, Netherlands
The Boulevard, Langford Lane, Kidlington, Oxford OX5 1GB, United Kingdom
50 Hampshire Street, 5th Floor, Cambridge, MA 02139, United States

Copyright © 2017 Elsevier Inc. All rights reserved.

No part of this publication may be reproduced or transmitted in any form or by any means, electronic or mechanical, including photocopying, recording, or any information storage and retrieval system, without permission in writing from the publisher. Details on how to seek permission, further information about the Publisher's permissions policies and our arrangements with organizations such as the Copyright Clearance Center and the Copyright Licensing Agency, can be found at our website: www.elsevier.com/permissions.

This book and the individual contributions contained in it are protected under copyright by the Publisher (other than as may be noted herein).

Notices

Knowledge and best practice in this field are constantly changing. As new research and experience broaden our understanding, changes in research methods, professional practices, or medical treatment may become necessary.

Practitioners and researchers must always rely on their own experience and knowledge in evaluating and using any information, methods, compounds, or experiments described herein. In using such information or methods they should be mindful of their own safety and the safety of others, including parties for whom they have a professional responsibility.

To the fullest extent of the law, neither the Publisher nor the authors, contributors, or editors, assume any liability for any injury and/or damage to persons or property as a matter of products liability, negligence or otherwise, or from any use or operation of any methods, products, instructions, or ideas contained in the material herein.

British Library Cataloguing-in-Publication Data
A catalogue record for this book is available from the British Library

Library of Congress Cataloging-in-Publication Data
A catalog record for this book is available from the Library of Congress

ISBN: 978-0-12-804008-9

For Information on all Elsevier publications
visit our website at https://www.elsevier.com/books-and-journals

Publisher: Candice Janco
Acquisition Editor: Marisa LaFleur
Editorial Project Manager: Marisa LaFleur
Production Project Manager: Anitha Sivaraj
Designer: Victoria Pearson Esser

Typeset by MPS Limited, Chennai, India

This book is dedicated to
my mentor, a pioneer in the EM well logging theory, and my colleague,
Professor
Liang Chi Shen

CONTENTS

Preface   *xiii*
Acknowledgment   *xv*

1. Introduction to Well Logging   1
   1.1   Oil and Gas Exploration   1
   1.2   Well Logging Methods   2
   1.3   Nuclear Logging   6
   1.4   Sonic Logging   8
   1.5   Nuclear Magnetic Resonance Logging   9
   1.6   Dielectric Logging   11
   1.7   Wireline Logging and Logging While Drilling   12
   1.8   Geosteering   13
   1.9   Summary of Electromagnetic Logging Tools   15
   References   15

2. Fundamentals of Electromagnetic Fields and Induction Logging Tools   17
   2.1   Maxwell's Equations   17
   2.2   Complex Permittivity   20
   2.3   Sources $\overline{J}_s, \overline{P}_s$ and $\overline{M}_s$ 21
   2.4   Hertz Potential $\overline{\Pi}_m$ 21
   2.5   Electromagnetic Fields Due to a Magnetic Dipole in a Homogeneous Medium   23
   2.6   Induced Electromotive Force (EMF) in the Receiving Coil and the Use of Bucking Coil   26
   2.7   Quasistatic Approximations and Skin Depth   31
   2.8   Apparent Conductivity   34
   2.9   Tool Constant and Skin-Effect Correction   35
   2.10   Direct Inversion of Induction Logging Data   37
   2.11   Spectrum Domain Solutions and Two-Coil Induction Tools in Layered Media   37
   2.12   Induction Arrays   45
   References   49

3. Electrical Properties of Sediment Rocks: Mixing Laws and Measurement Methods   51
   3.1   Resistivity and Dielectric Constant of Rocks   52
   3.2   Archie's Law   55
   3.3   Mixing Laws   55
   3.4   Frequency Dispersion of the Dielectric Constant   64
   3.5   Frequency Dispersion of the Conductivity   67
   3.6   Measurement Methods of Electrical Properties of Rocks   69

3.7	TM_{010} Resonant Cavity Technique	95
	References	110
	Appendix A E Field Analysis of the Circuit Model of the Parallel-Disk Sample Holder	112
	Appendix B Equipment Calibration Synopsis	117

4. Triaxial Induction and Logging-While-Drilling Resistivity Tool Response in Homogeneous Anisotropic Formations — 121

4.1	Magnetic Dipole in Homogeneous Lossy Media	122
4.2	Finite Coil in Homogeneous Formation	127
4.3	LWD Tool Response in Homogeneous Formation	127
4.4	Triaxial Induction Logging Tool Response in Biaxial Anisotropic Homogeneous Formation	136
	References	160
	Appendix A Derivation of Parameters a and b in Eqs. (4.53)–(4.55)	161

5. Triaxial Induction Tool and Logging-While-Drilling Tool Response in a Transverse Isotropic-Layered Formation — 163

5.1	Introduction	163
5.2	Summary of a Magnetic Dipole Source in a Transverse Isotropic Homogeneous Formation	165
5.3	Magnetic Dipole in a Layered Formation	166
5.4	Convergence Algorithm	172
5.5	Simulation Results and Analysis	172
5.6	Analysis of Anisotropy Impact to the Resistivity LWD Tool	175
	References	178
	Appendix A Derivation of Hertz Vector Potential in Multiple Layer Formation	179

6. Triaxial Induction and Logging-While-Drilling Logging Tool Response in a Biaxial Anisotropic-Layered Formation — 187

6.1	Spectral-Domain Solution to Maxwell's Equations in a Homogeneous Biaxial Anisotropic Medium	188
6.2	Propagation in Unbounded Medium	193
6.3	Propagation in Layered Medium	193
6.4	Computation of the Double Integrals	196
6.5	Numerical Examples	197
	References	202
	Appendix A Derivation of Matrix A	202

7. Induction and LWD Tool Response in a Cylindrically Layered Isotropic Formation — 205

7.1	Introduction	205
7.2	Induction and LWD Tool Response in a Four-Layer Cylindrical Medium	206

7.3	Response of Induction and LWD Tools in Arbitrary Cylindrically Layered Media	217
7.4	Conclusions	241
	References	242
	Appendix A Derivation for the Magnetic Fields in Spectral Domain	243
	Appendix B Derivation for the Expression of Electrical Field for the Homogeneous Formation in Spectral Domain	245
	Appendix C Derivation for the Expression of Electrical Field for Arbitrary Cylindrical Layered Formations in Spectral Domain	247

8. Induction and Logging-While-Drilling Resistivity Tool Response in a Two-Dimensional Isotropic Formation — 251

8.1	Introduction	251
8.2	Formulations	252
8.3	Numerical Consideration	263
8.4	Verifications	265
8.5	Array Induction Logs	270
8.6	Measurement-While-Drilling Logs	273
8.7	Simulation of Effects of Mandrel Grooves on MWD Conductivity Logs	281
8.8	Summary	292
	References	293

9. Theory of Inversion for Triaxial Induction and Logging-While-Drilling Logging Data in One- and Two-Dimensional Formations — 295

9.1	Introduction	296
9.2	Gauss–Newton Algorithm	297
9.3	Cholesky Factorization	299
9.4	Line Search	301
9.5	Jacobian Matrix	302
9.6	Constraints	303
9.7	Initial Values	304
9.8	Inversion Results and Analysis	310
9.9	Inversion of Induction Logs in a Two-Dimensional Formation	330
9.10	Summary	347
	References	348

10. The Application of Image Theory in Geosteering — 351

10.1	Introduction	351
10.2	Theory of Forward Modeling Using Image Theory	353
10.3	Simulation Results and Discussions	361
10.4	Boundary Distance Inversion	391
10.5	Conclusion	404
	References	405

11. Ahead-of-the-Bit Tools and Far Detection Electromagnetic Tools — 407

- 11.1 Introduction — 407
- 11.2 Ahead-of-the-Bit Field Distribution of LWD Tools — 410
- 11.3 Toroidal Transmitter — 415
- 11.4 Boundary Detection Using Orthogonal Antennas — 420
- 11.5 Deep-Looking Directional Resistivity Tool — 424
- 11.6 Distance Inversion Based on the Gauss–Newton Algorithm — 435
- 11.7 Conclusions — 444
- References — 445

12. Principle of Dielectric Logging Tools — 447

- 12.1 Introduction — 447
- 12.2 History of Dielectric Tool Study — 449
- 12.3 Frequency Selection of a Dielectric Tool — 451
- 12.4 Antenna Spacing — 451
- 12.5 Sensitivity Analysis — 454
- 12.6 Sensitivity Analysis in Anisotropic Formation — 459
- 12.7 Dielectric Logging Tool Design and Modeling Using Three-Dimensional Numerical Modeling Software Package — 463
- 12.8 Cavity-Backed Slot Antenna — 467
- 12.9 Effects of the Pad — 470
- 12.10 Borehole Mud Influence — 475
- 12.11 Vertical Resolution — 479
- 12.12 Mud Cake and Invasion — 483
- 12.13 Depth of Investigation — 491
- 12.14 Applications of Dielectric Tools — 495
- 12.15 Summary — 496
- References — 497
- Appendix — 498

13. Finite Element Method for Solving Electrical Logging Problems in Axially Symmetrical Formations — 503

- 13.1 Overview of the Numerical Simulation Methods for Well Logging Problems — 504
- 13.2 Finite Element Method Based on Magnetic Field — 505
- 13.3 Analysis of Transverse Electric Mode and Transverse Magnetic Mode — 507
- 13.4 Vector Matrix Equation of Magnetic Field and Impedance Matrix — 509
- 13.5 The Basis Functions — 511
- 13.6 Evaluation of Impedance Element Matrix for Rectangular Element Based on H_ϕ — 515
- 13.7 Evaluation of Impedance Element Matrix for Rectangular Element Based on ρH_ϕ — 518
- 13.8 Evaluation of Impedance Element Matrix for Triangular Element Based on H_ϕ — 520
- 13.9 FEM Based on Electric Field — 532

13.10	Evaluation of Triangular Element Matrix Based on E_ϕ (TE Mode)	535
13.11	FEM Model of Sources	537
References		545
Appendix A Vector Analysis in Cylindrical Coordinates		546
Appendix B Computation Method for Matrix Assembling Rule		547
Appendix C Computation Method of Element Matrix for Rectangular Element Based on H_ϕ (Section 13.6)		548
Appendix D Computation Method of Element Matrix for Rectangular Element Based on ρH_ϕ (Section 13.7)		550
Appendix E Term 4 (Section 13.8.3.9)		552
Appendix F Computation Method of Element Matrix for Triangular Element Based on H_ϕ (Section 2.7.4)		554
Appendix G Computation Method of Element Matrix for Triangular Element Based on E_ϕ (Section 3.3)		556

14. Resistivity Imaging Tools — 559

14.1	Introduction	559
14.2	Water-Based Mud Resistivity Imaging Tool	561
14.3	The Oil-Based Mud Resistivity Imager	562
14.4	Conclusions	576
References		578

15. Laterolog Tools and Array Laterolog Tools — 579

15.1	Introduction	579
15.2	Basics of Electrode Type of Logging Tools	580
15.3	The Laterolog Focusing Principle and the Model of Dual Laterolog Tool	583
15.4	Application of Finite Element Method on Alternating Current Dual Laterolog Tool	590
15.5	Validation of the Computational Method	594
15.6	Simulation Result	597
15.7	Array Laterolog Tool	603
References		619
Appendix A Computation Method of Source Model for Alternating Current Dual Laterolog Tool		620

16. Theory of the Through-Casing Resistivity Logging Tool — 625

16.1	Introduction	625
16.2	Through-Casing Resistivity Measurement Procedure	626
16.3	Circuit Model of the Through-Casing Resistivity Tool	628
16.4	Finite Element Method Simulation of the Through-Casing Resistivity Logging Tool	632
16.5	Through Casing Logs from a Toroidal Antenna	637
References		643

17. Electromagnetic Telemetry System and Electromagnetic Short Hop Telemetry in a Logging-While-Drilling/Measuring-While-Drilling Tool — **645**

- 17.1 Introduction to Logging-While-Drilling/Measuring-While-Drilling Uplink and Downlink Technologies — 645
- 17.2 The Numerical Model of Electromagnetic Telemetry System — 649
- 17.3 Application of Finite Element Method on Electromagnetic Telemetry Systems — 652
- 17.4 Validation of the Computation Algorithm in a Cased Borehole — 659
- 17.5 Simulation Result Without Casing — 661
- 17.6 Short Hop Electromagnetic Telemetry Used in a Near Bit Logging-While-Drilling Sensor — 680
- 17.7 Conclusions — 687
- References — 688
- Appendix A Computation Method of Source Model for EM Telemetry System (Section 17.3.2) — 688

Appendix A: LogSimulator User Manual — *691*
Index — *703*

PREFACE

Well logging is a very important field in petrophysical exploration. It is also a relatively small sector in the oil and gas industry. Its purpose is to locate oil and gas in a formation along a borehole using physical measurements. Unlike seismic exploration, which has a scope measured in kilometers, well logging tools map formations locally on a much smaller scale. Therefore, well logging methods have relatively high spatial resolution and can provide very detailed formation information to petrophysicists. On the other hand, they are indirect measurements that are based on the physical and chemical characteristics of the formation; the major ones include electrical, sonic, nuclear, and nuclear magnetic resonance. Among them, electrical parameters (resistivity and dielectric constant) account for approximately 70% of the entire logging data.

Methods used in logging that are related to the electrical and electromagnetic (EM) characterization have been studied and reported by many researchers. Research and textbooks that have contents pertaining to logging tools can also be found in large amounts. Many of them are designed as a part of the physical background from a petrophysicist's point of view. However, a complete description of various electrical and electromagnetic logging tools is not available. This book intends to include the analytical and numerical methods for most electrical and electromagnetic logging tools that are used in the logging industry. It is designed to serve as a textbook for the undergraduate and graduate level courses in the well logging area. Students are assumed to have basic knowledge of electrical and electromagnetics.

This book is based on my 26 years' experience in teaching and research in the Well Logging Laboratory at the University of Houston. It is essentially a distillation of my research and teaching notes, as well as dissertations and theses of graduate students. It covers most areas of electrical and electromagnetic logging tools from a tool designer's point of view. I have tried to update the contents so that most recent developments in the area are also addressed. The contents are intended to give students solid analytic background in EM logging. Therefore, many basic mathematical and physical process are described and derived in detail. The purpose of the analytical methodology is to offer the reader a better understanding of the tool physics and tool performances from a designer's point of view, since analytical methods can make the physics clear, which will help readers in improving existing tool performances and creating new tools. In recent years, numerical methods have become both more available and more mature in analyzing logging tools. It provides the students and researchers handy tools to obtain quick results. By fully understanding the analytical methods, the numerical methods can be more meaningful.

The performance of a logging tool is related to the tool structure and formation geometry and characteristics. Formations may be homogenous, cylindrically layered, vertically layered, isotropic, transverse isotropic (TI), and biaxial anisotropic. Tool structures can also vary in complexity in practical applications depending on the purpose of the analysis. For tool design, minute structural details such as antenna groves and even the distance between antenna wires must be considered. However, for analysis of tool performance in petrophysical applications, tool structure can be simplified.

This book uses analytical methods for cases wherever possible, otherwise numerical analysis is employed. From Chapter 1 to Chapter 10, analytical methods are used. In Chapters 11–17, numerical analysis is used.

The first three chapters serve as the basics of the book in terms of EM background and rock physics including rock measurement methods for electrical properties. Chapters 4–6 provide detailed forward modeling methods for coil type of logging tools such as induction & LWD tools in various homogenous and vertically layered formations with anisotropy. Chapter 7 provides a forward modeling method for induction and LWD tools in isotropic and cylindrically layered formations. Chapter 8 handles tool responses in a vertically 2D formation, which has both radial and vertical layers. Chapter 9 deals with data inversion methods for induction and LWD tools in a vertically layered TI formation and isotropic formations with both cylindrical and vertical layers. Chapter 10 gives an approximation of the analytical method and can be used in Geosteering cases. In Chapter 11, we explore the possible ways to implement a "look ahead" tool and far detection logging tools to help drilling process. Chapter 12 discusses dielectric logging tools using numerical analysis. Chapter 13 covers the fundamentals of finite element analysis for well logging tools. Chapters 14–17 give detailed tool physics and tool performance for resistivity imaging tools, laterolog tools, through casing tools, and EM telemetry tools using numerical methods, respectively.

ACKNOWLEDGMENT

I would like to first thank my mentor, a pioneer in EM well logging, and colleague, late Professor Liang Chi Shen. I am grateful to my family for their constant support and understanding. I would also like to thank my colleagues at Bitswave, Inc. Dr. Jing Li, and Dr. Zhong Wang for their support and help. The efforts from colleague Dr. Suming Wu in developing the software interface for this book is also acknowledged. I would also like to thank my colleagues at the University of Houston: Professors Donald Wilton, David Jackson, Jack Wolf, Haluk Ogmen, and Stuart Long for their involvement in the well logging program and their support. I am very grateful to my students and post docs who worked in the well logging areas and contributed their work to this book. They include: Dr. Ning Yuan, Dr. Gong Bo, Dr. Jing Wang, Dr. Yinxi Zhang, Dr. Chen Guo, Dr. Zhili He, Dr. Jinjuan Zhou, Dr. Hamid Nasari, Dr. Mark Collings, Dr. Xiang Tian, Dr. Zhijuan Zhang, Dr. Miao Luo, Dr. Hanming Wang, Dr. Shanjun Li, Jing Li, El Emir Fouad Shehab, David Navarro, and Li Zhong. I would also like to acknowledge the sponsorship to the Well Logging Lab at the University of Houston from many oil and service companies: BP, BakerHughes, Chevron, CNPC, Exxon Mobile, Great Wall Drilling Company, Haliburton, Saudi Aramco, Schlumberger, Shell, and Weatherford.

CHAPTER 1

Introduction to Well Logging

Contents

1.1 Oil and Gas Exploration	1
1.2 Well Logging Methods	2
1.2.1 Basic resistivity logging methods	2
1.2.2 Basic induction logging tool	4
1.2.3 Basic propagation logging method	5
1.2.4 Basic laterolog	6
1.3 Nuclear Logging	6
1.3.1 Gamma ray log	7
1.3.2 Neutron log	7
1.3.3 Density log	8
1.4 Sonic Logging	8
1.5 Nuclear Magnetic Resonance Logging	9
1.6 Dielectric Logging	11
1.7 Wireline Logging and Logging While Drilling	12
1.8 Geosteering	13
1.9 Summary of Electromagnetic Logging Tools	15
References	15

1.1 OIL AND GAS EXPLORATION

Oil and gas has been discovered and used in our ordinary life for thousands of years. The word "petra" means "stone/rock" in Latin, and "-oleum" means "oil" (http://wiki.answers.com/Q/Where_does_the_word_petroleum_come_from#ixzz1JJoSo7Cb). Early Chinese scientist Shen, Kuo (1031–95) once recorded early discovery of oil in his *Dream Pool Essays*. As a summary in his writing, he predicted that "...this thing will have a great future." According to Wikipedia (http://en.wikipedia.org/wiki/History_of_petroleum), "The earliest known oil wells were drilled in China in 347 AD or earlier. They had depths of up to about 800 feet (240 m) and were drilled using bits attached to bamboo poles. The oil was burned to evaporate brine and produce salt. By the 10th century, extensive bamboo pipelines connected oil wells with salt springs. .... In his book Dream Pool Essays written in 1088, the polymathic scientist and statesman Shen Kuo of the Song Dynasty coined the word 石油 (*Shíyóu*, literally "rock oil") for petroleum, which remains the term used in contemporary Chinese."

Modern oil and gas explorations are more sophisticated. In general, the oil and gas exploration can be divided into three steps. The first step is to find an oil/gas bearing reservoir. To do so, geology and geophysical methods such as seismic imaging are used. The scale of seismic image is in the range of kilometers with a resolution of hundreds of meters. Once the oil reservoir is located, pilot drilling will be done and the geophysical properties of the earth formations will be carefully studied based on drilling samples and logging data. Logging is a closer look of the physical properties of the formation using various indirect measurements. Logging can be done while drilling, or after drilling. When the measurement equipment is directly attached to a drilling bit, the measurement is performed during the drilling. This direct measurement method is called logging while drilling (LWD) or measuring while drilling (MWD). When the logging is performed after drilling activity in a borehole, the logging tools used are wireline tools. The name wireline comes from the cables that send power to the logging device and carry logging signals to the surface. The logging data gives a more detailed geophysical description of the formation surrounding the well drilled. For most logging tools, the depth of investigation is in the range of a few centimeters to about 30 ft. The resolution can be as high as millimeter to about a few feet. The results of the logging process will give petrophysicists a quantitative measure of the formation parameters so that the production rate, formation quality, depth of production zone, and productivity can be evaluated.

1.2 WELL LOGGING METHODS
1.2.1 Basic resistivity logging methods

Resistivity of the formation indicates the capability of the materials contained in the formation to resist the flow of electric current. Generally speaking, dry rocks, oil and gas are good insulators and cannot conduct electricity, but the mineralized water contained in the pores of the rocks makes it feasible to measure finite resistivities for different formations. Since oil and gas are much more resistive than most formation waters, the resistivity logs can greatly help in determining the fluid content of the reservoir. The unit of resistivity used in well logging is ohm-meter or ohm-m. The reverse of the resistivity, named conductivity, describes the ability of a matter to conduct electric current flow.

Electrical logging, or electrical survey, is considered as the earliest resistivity logging method. The tool was invented by the Schlumberger brothers in 1927 [1], and was eventually replaced by induction logging and laterolog after 1960s [2].

The principle of electrical logging is simply based on Ohm's Law. As we know, if a point current source is surrounded by an infinite, homogeneous, and isotropic medium, the equipotential surfaces will be perfect spheres. Assume if such conditions could be simulated by placing an emitting electrode in the borehole, the formation resistivity would be ready to obtain if the potential difference between any two equipotential spheres is given. Consequently, electrical survey can be taken only in uncased wells and with conductive muds, such as water-based mud and oil-emulsion mud, otherwise the measurement would be greatly influenced by the mud resistivity.

Practical resistivity logging devices utilize multiple electrodes of various configurations and dimensions to serve different needs. The normal device and the lateral device are two frequently used electrode arrangements.

The principle of the normal device is shown in Fig. 1.1. A point electrode, A, is connected to a current source with an intensity of I. Two other electrodes, M and N, are also placed in the hole. M is near A, while N is far away enough to be approximately seen as infinite distance. Assume the formation is uniform, its resistivity R can thus be expressed by

$$R = K \frac{\Delta V}{I}$$

where ΔV is the potential difference between M and N. K is a coefficient that depends on the distances between the electrodes.

The lateral device is illustrated in Fig. 1.2. The principle is quite similar; only the source is connected with two electrodes, A and B, and both of them are placed in the

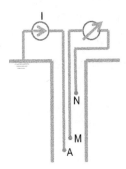

Figure 1.1 The normal electrode device.

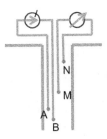

Figure 1.2 The lateral device.

hole. Sometimes the roles of A—B and M—N are interchanged, with M as the emitting electrode, and the potential difference between A and B being measured.

The resistivity calculated by the equation above is called apparent resistivity, R_a. In practice, it reflects the average resistivity of the formation, and is affected by the borehole. The apparent resistivity is the actual resistivity of the formation assuming the formation is isotropic, homogeneous, and no borehole. Resistivity inversion models and digital processing are further needed to obtain the true resistivity R_t if the above conditions are not satisfied, which is often the case. Therefore the apparent resistivity obtained from any resistivity tool is not actual formation resistivity without data inversion.

In general, the properties of a device depend on the relative positions of electrodes. Different electrode distributions may affect the performance of defining bed boundaries, estimation of fluid content, or showing thinner/thicker layers. Therefore a combination of two or more different devices is often used to provide sufficient information.

1.2.2 Basic induction logging tool

Induction logging measures the formation conductivity instead of its reciprocal, resistivity. It is best used in highly resistive drilling fluids, e.g., oil-based mud, air, etc., and makes more accurate measurements than conventional resistivity logging tools. Therefore after its introduction in 1940s, induction logging soon became widespread and dominated the entire resistivity survey market [3].

A basic induction logging tool using two coils is shown in Fig. 1.3. Both coils are mounted coaxially on an insulating mandrel. The transmitter coil is fed by an alternating current source, and induces eddy current loops by generating magnetic field in the formations. These current loops in turn induce currents in the receiver coil in the same way, and the induced voltage is directly proportional to the formation conductivity, which can be expressed by

$$V = K\sigma_a \tag{1.1}$$

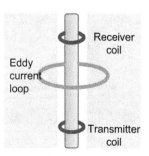

Figure 1.3 Induction logging tool.

where K is a calibration factor that relates to the tool geometry and transmitting current, and σ_a is the apparent conductivity. Here the voltage V is complex, since a phase shift exists between V and the transmitter current I_T. As a result, σ_a is also complex. Assume the formation surrounded the tool is infinite and homogeneous, σ_a can be seen as an integration over the whole space, given by

$$\sigma_a = \sigma_R + j\sigma_X = \int_{-\infty}^{+\infty}\int_{0}^{+\infty} g(\rho, z, \sigma) \times \sigma(\rho, z) \mathrm{d}\rho \mathrm{d}z \qquad (1.2)$$

where $g(\rho, z, \sigma)$ represents the contribution of each specific eddy current loop to the total conductivity, or the sensitivity of the tool at (ρ, z).

Based on the two-coil tool, multiple-coil tools are developed to focus the measuring signal, reduce the borehole influence, and improve the vertical resolution. Such tools include dual induction tool and array induction tool. After 1990s the Triaxial tool has been introduced to the market, in which the transmitter and receiver coils are mounted orthogonally to obtain measurements in all three directions. The operation frequency of induction tool is usually at tens of kilohertz.

1.2.3 Basic propagation logging method

This resistivity logging method taking advantage of electromagnetic (EM) wave propagation properties was proposed in 1986 [4]. This type of tools also adopts coils as transmitter(s) and receivers, but instead of generating low-frequency fields in the formations as in induction logging, it uses high frequency (typically around 2 MHz), propagating EM wave to measure the resistivity.

The basic configuration of a propagation logging tool is shown in Fig. 1.4. The tool contains a single transmitter, T1, and two receivers, R1 and R2, located at distances z_1 and z_2 from the transmitting coil, respectively. By computing the amplitude ratio and phase difference between the signals received by R1 and R2, the information

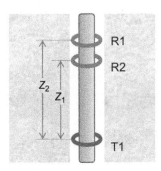

Figure 1.4 Propagation logging tool.

of formation conductivity can be obtained. In addition, the depths of investigation of the two measurements are different, which is useful in detecting invasion.

Some tools incorporate a second transmitter above the receiver pair symmetrically [5,6]. Two transmitters send signal alternately, and an average is taken to improve the reliability.

Propagation logging can be conducted in all types of muds, and exhibits better vertical resolution than induction tool.

1.2.4 Basic laterolog

Laterolog was designed to act complementarily to induction logging, which has severe borehole effect in wells with highly conductive mud. Also, laterolog makes more accurate definition of the bed boundaries despite of the bed thickness.

Two basic laterolog tool configurations are illustrated in Fig. 1.5. In addition to the measuring current sent from central electrode A_0, there are also auxiliary currents fed through A_1, A_2, A_1', and A_2'. These currents are adjusted so that a zero potential difference can be maintained at the planes shown by the *dash lines*. Therefore the measuring current is focused as a sheet between the "guarded" planes and forced horizontally into the formation. This leads to a much higher depth of investigation than the unguarded electrode tools, and improves the measurement of the invaded zones.

The Guard electrode tool shares the same design as the Point electrode tool, only the point electrodes are replaced by elongated bar electrodes. Some Dual laterolog tools combine the two arrangements together by adding bar electrodes above and below the point electrodes. In this way, the current can be further focused so as to obtain more accurate measurements.

1.3 NUCLEAR LOGGING

Nuclear logging, or radiation logging, is in use to determine the formation properties by detecting the radioactivity. Such radioactivity may be either naturally emitted by the formation substances, or reflected from the formation induced by a Gamma or

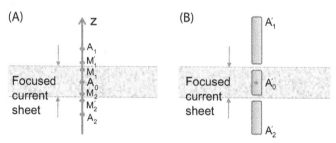

Figure 1.5 Two configurations of laterolog: (A) Point electrode; (B) Guard electrode.

neutron source. These two cases lead to Gamma Ray logging and Neutron logging, respectively. Nuclear logging can be conducted in open or cased holes, and can be used with any type of borehole liquid.

1.3.1 Gamma ray log

Gamma rays are basically short bursts of high-frequency EM waves emitted by the atomic nuclei. Such emission may happen when a nucleus is collided by another particle, or naturally unstable. Some elements contained in the earth spontaneously emit gamma rays: potassium-40, thorium, uranium, and the radioactive families of the last two. These elements mainly exist in shales, which are thus more radioactive than any other formations. Therefore the formation radioactivity is detected by the Gamma Ray logging tool, and used for bed definition and correlation.

Two types of gamma ray detectors are used in the logging tools: the Geiger–Mueller (G-M) counter and the Scintillation counter. The G-M counter consists of a gas chamber and a power-fed electrode, and detects the voltage pulses caused by the gas ionization when a gamma ray enters. The scintillation counter uses a sodium iodide crystal, which gives off a tiny flash of light whenever penetrated by a gamma ray. Such flashes are then converted into electrical pulses by a photomultiplier tube. Generally, the Scintillation counter has a superior sensitivity, and often preferred in modern logging tools. However, it is usually more expensively made, and cannot stand very high temperature as well as the G-M counter.

In addition to the total gamma rays, the tools today can also record the gamma ray spectrum emitted by different minerals, and quantitatively analyze the contributions of each element. This method can be used in clay type identification, or evaluating the source rock potential.

1.3.2 Neutron log

Neutron logging is also based on the detection of radiation. However, such radiation is not naturally produced by the substances in formation, but results from the bombardment of a neutron source.

The neutron logging tool consists of a source of fast neutrons and a proper radiation detector. The neutrons emitted from the source will gradually slow down in the formation because of the collisions with hydrogen atoms, until they are finally captured while emitting secondary gamma rays. This process can be evaluated in two different ways: by detecting the capture gamma rays; or by counting the slowed neutrons. If the surrounding formation contains a large concentration of hydrogen, most neutrons will be frequently collided and soon captured near the source, leading to a low counting rate at the detector. On the contrary, if the concentration of hydrogen decreases, the neutrons will go further without collision, and more likely to be detected by the receiver.

In unshaly zones, hydrogen primarily exists in water, oil, and gas, which are contained in the pores of formations. Therefore the counting rate, which shows the concentration of hydrogen, closely relates to the rock porosity. Shaly formations can also cause high counting rate because of the water bounded in the pores, but since shale is practically impervious, it does not contribute to the effective porosity. Consequently, the Neutron log needs to be compared with other logs (e.g., gamma ray) to determine the real porous zones.

1.3.3 Density log

The density log, or photoelectric (Pe) log, measures the formation density around the borehole. It is also used to derive a value for the overall porosity.

The tool includes a gamma ray source and two or more detectors. Medium-energy gamma rays are emitted through the formation, and lose energy from time to time due to the Compton scattering effect (a formation electron may be ejected out of its orbit when collided by a gamma ray). By detecting the remained low-energy gamma rays, one can estimate the number of Compton scattering collisions, which directly relates to the electron density of the formation.

If the type of the formation rock and that of the fluid it contains are known, the porosity is given by

$$\varnothing = \frac{\rho_{\text{ma}} - \rho_{\text{b}}}{\rho_{\text{ma}} - \rho_{\text{f}}} \tag{1.3}$$

where ρ_{ma} and ρ_{f} represent the densities of the rock matrix and the contained fluid, respectively. The average bulk density of the formation ρ_{b} can be derived from the measured electron density.

1.4 SONIC LOGGING

Sonic log, or acoustic log, determines the fluid content or porosity of the formations by measuring the speed of sound waves that travel through the earth. Generally, sonic wave travels much slower in liquids than in solid materials. In oil and water, the average speed of sound is around 4300–5300 ft/s; while in rock materials, it ranges from 6000 ft/s (shales) to 26,500 ft/s (dolomites). Therefore a continuous record of sonic velocity with respect to depth forms a porosity indication, as well as a reliable reference of lithology variation.

In practice, the sonic velocity is indirectly obtained by recording the traveling time of a sound wave through a constant distance through formations. As shown in

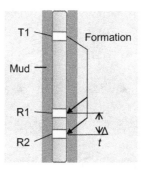

Figure 1.6 Sonic logging.

Fig. 1.6, a sonic pulse emitted by the transmitter T1 passes through the mud and enters the formations, where it propagates in all directions. A small fraction of it may "bend" back, penetrate the mud again and reach the receivers. The time difference between the pulse arriving at two receivers, Δt, is recorded, so that the round-trip time through the mud can be eliminated. Note that the logging tool body must be made from low-velocity materials (e.g., rubber), in order to minimize the energy loss.

One problem of the tool design in Fig. 1.6 is that if the tool is not parallel to the borehole, or the borehole size changes fast, and Δt is too vague to determine, the dual-receiver system will not give satisfying results [7]. This can be solved by the Borehole Compensating tool, which incorporates two transmitters and four receivers, aligning symmetrically on the tool. The two transmitters emit sound pulses alternately from both ends of the tool, and the signals captured at the central receivers are averaged to compensate for the tool misalignment.

1.5 NUCLEAR MAGNETIC RESONANCE LOGGING

The Nuclear Magnetic Resonance (NMR) effect was first successfully measured in 1946. After that, it rapidly became a very powerful tool in multiple disciplines, such as biology, physics, chemistry, and medicine, assisting in various analyses at a molecular level. For the oil and gas industry, NMR logging is also considered as a big breakthrough in recent history. The early measurement attempts were performed since the 1960s, but it took several decades till the first modern logging tool was brought to the market in 1991. NMR responds only to fluids, so the logs usually give more accurate indications to the fluid quantity, fluid properties, and formation porosity than any other logging methods [8–10].

The NMR measurement is based on the intrinsic magnetic moment of protons and neutrons. Some atoms, such as ^{1}H, ^{13}C, and ^{23}Na, have an odd number of protons and/or neutrons, the spinning of which forms a net magnetic moment that can contribute to the macroscopic magnetization signal detected by the NMR logging tool. Most of the existing NMR logging tools are designed for hydrogen detection, since hydrogen produces a strong signal, and is naturally abundant in water and hydrocarbons. When no external magnetic field exists, the hydrogen atoms (protons) are randomly aligned in the formations.

When a magnetic nucleus is placed in a static magnetic field B_0, a perpendicular torque is applied to its spinning axis, making it process around the direction of B_0 with a frequency v. For hydrogen, the nucleus can align either with or against B_0, as shown in Fig. 1.7. Since the alignment with B_0 is in the low-energy state, this direction is preferred by most nuclei, so the macroscopic magnetization M is parallel to B_0. This alignment process is called polarization. However, if an oscillating magnetic field B_1 is applied perpendicular to B_0, and the frequency of B_1 is exactly equal to v, the low-energy state nuclei may absorb energy and jump to the high-energy state. Consequently, the direction of M is gradually tipped while B_1 lasts.

In NMR logging, the static field B_0 is generated by a permanent magnet, and the oscillating field B_1 is transmitted from an antenna around the magnet. Such radiofrequency (RF) energy is in the form of precisely timed bursts: First, a 90-degree pulse is applied to the polarized protons to change the precessional direction to the transverse plane, generating the first resonance signal on the plane; then a series of 180-degree pulses follows, reproducing the resonances, or spin echoes, by reversing the magnetization vectors on the transverse plane. As a result, a decaying signal pulse series is detected by the antenna on the tool, and this is the raw data measured by the NMR tool, containing most of the logging information. The pulse train causing spin echoes is called a CPMG sequence, which is illustrated in Fig. 1.8.

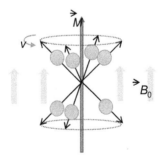

Figure 1.7 Polarization.

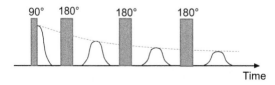

Figure 1.8 CPMG sequence and spin-echo trains.

Since the detected magnetic intensity M is proportional to the density of hydrogen atoms in formations, and the quantity of hydrogen contained in water is certain, the NMR measurements can be converted to an apparent water-filled porosity. Moreover, additional information can be extracted from the variation of the formation magnetization with respect to time. For example, the exponentially decreasing envelope of the spin echoes shown in Fig. 1.8 has a time constant T_2, called transverse relaxation time, which plays an important role in the determination of fluid types and properties. Also, based on the different behaviors of NMR tools in bulk fluids and fluids contained in pores, the pore size of formations can be calculated in the further place.

Because of its unique capability of accurate fluid logging, NMR data are independent of rock matrix, and do not need to be calibrated to lithology. This fundamentally distinguishes NMR from other logging tools. In addition, the abundant information contained in NMR measurements makes it possible to analyze the formation fluid properties in detail. For example, light oil, medium-viscosity oil, and heavy oil can be distinguished from each other. Conventional logging tools are not able to provide such measurements.

1.6 DIELECTRIC LOGGING

Dielectric logging is developed to solve the problems met by resistivity tools in flooded zones, where the difference between the resistivities of oil-bearing zone and the fresh water-bearing zone is difficult to detect. While conventional resistivity tools are greatly affected in such formations, the contrast between relative dielectric constants of hydrocarbons and water is quite high, as listed in Table 1.1. Also, the permittivity of water is less sensitive to the salinity variation than resistivity is. This makes the dielectric logs particularly useful when the water salinities are unknown [11].

The principle of dielectric logging is based on the propagation characteristics of electromagnetic waves traveling in the formations. The complex wave number k can be expressed by

$$k \equiv \beta + j\alpha = \omega\sqrt{\mu\varepsilon_c}$$

Table 1.1 Relative dielectric constant ε_r of subsurface fluids [12]

Gas	1
Oil	2
20 ohm-m water	79
1 ohm-m water	77
0.1 ohm-m water	59

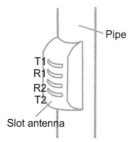

Figure 1.9 Pad-type dielectric logging tool.

where β and α are called phase constant and attenuation constant, respectively, shown by the phase shift and attenuation ratio of the transmitting waves, which are directly measured by the logging tool.

ε_c is the complex permittivity, given by

$$\varepsilon_c = \varepsilon' - j\left(\frac{\sigma}{\omega} + \varepsilon''\right)$$

where the real part $\varepsilon' = \varepsilon_0 \varepsilon_r$, and the imaginary part involves the effects of both conductivity σ and the dielectric loss ε''. When the frequency is low, the imaginary part is dominated by σ, and is much greater than the real part. To obtain relatively accurate calculation of ε', the operation frequency must be chosen so high (>15 MHz) that the real and imaginary parts of ε_c are comparable.

Two types of dielectric logging tools are used in practice: coil-type and pad-type. The coil-type tool uses coils as antennas, similar with the propagation tool shown in Fig. 1.4, only the operating frequency is higher (30–50 MHz). The pad-type tool has a conductive pad with slot antennas acting as a current sheet in a cavity, as illustrated in Fig. 1.9. During the measurement, the pad is pushed against the borehole by a mechanical control system. Pad-type tools operate at hundreds to thousands of megahertz.

Quantitatively, there is a relationship between the dielectric constant, porosity, and water saturation. Therefore the last two can also be calculated from the dielectric measurements. Moreover, the measurement of permittivity also assists in other formation evaluation methods (e.g., crosswell radar investigation), which may be greatly affected by the variation of permittivity.

1.7 WIRELINE LOGGING AND LOGGING WHILE DRILLING

Early logs were run with the logging tools connected by a multiple-conductor wireline. After a well is drilled, the tools are lowered to the open borehole, and perform measurements while being pulled upward. The tension and speed of the wireline is monitored so that any stuck can be detected immediately. The logging data are either stored in the memory of the tools and retrieved on the surface, or directly transmitted up through the wireline. However, when the borehole is highly deviated or horizontal, it becomes difficult for the tools to enter by gravity. Sometimes the tools must be pumped into the wellbore to obtain measurements, which makes the logging procedure extremely time-consuming.

LWD offers an alternative way to solve this problem. Instead of logging after drilling, the LWD tools are integrated to the drill pipe as part of the Bore Hole Assembly, and conduct continuous measurements while the drilling proceeds. Downhole power is supplied by specially designed batteries or mud turbines, which leverages the flow energy of the drilling fluid. Data are still recorded to the memory for download afterwards, or the tools can send part of the information to a telemetry system, which communicates with the surface in real time.

Several types of telemetry methods are used in practice. The most widespread one is the mud pulse system, invented in the 1960s [13], which utilizes a valve to adjust the mud pressure so that it can represent different digital bits. Some companies provide EM telemetry or acoustic telemetry system, using low-frequency electric source or sonic source as downhole wireless transmitters. In the past decade, a wired pipe system enters the market, offers a new possibility for the telemetry approaches. All of these methods have advantages and drawbacks in different environments, and the practice selection depends on the well depth, formation properties, cost, and so on.

The advent of LWD not only decreases the down time, it also helps in optimizing the drilling operation. With the real-time logging information, the operator can respond quickly to improve the efficiency and productivity, as well as avoid potential accidents and tool loss. Also, with the logging tools installed near the bit, the logs are obtained as the hole is freshly drilled. As a result, the measurements are less affected by the mud invasion, and hence more accurate. Nowadays, wireline logging is gradually replaced by LWD, and mostly used for benchmark purposes.

1.8 GEOSTEERING

Geosteering is developed for the purpose of directional drilling. With a mud motor and a bent subconnected, the orientation and inclination of the well can be changed without pulling out the drill pipe.

Before drilling a well, the well paths are planned in order to meet specific requirements, such as maximizing the productivity, or reducing the expense. To follow this trajectory accurately, MWD techniques must be applied. Based on the real-time information gathered by MWD tools, the borehole position and bit conditions are continuously updated. The information includes inclination, azimuth, weight-on-bit, tool

Table 1.2 Summary of electrical and electromagnetic logging tools

	Tool name	Major applications	Characteristics	Major specifications	Operating frequency	Preferred borehole mud	Maximum DOI	Minimum vertical resolution
1	Induction/Array induction	Wireline resistivity for formation evaluation	Accurate for low resistivity formations	Resistivity range 0.1–500 ohm-m	10k–100 kHz	Higher resistivity mud	90"	12"
2	Triaxial induction	Wireline resistivity for formation evaluation	Azimuth resistivity, bed boundary determination, true dip, cross bedding, unconventional oil and gas	Resistivity range 0.1–500 ohm-m	10k–100 kHz	Higher resistivity mud	90"	12"
3	Laterolog (Dual laterolog, Array laterolog)	Wireline resistivity for formation evaluation	High resistivity measurements	Resistivity range 0.2–10k ohm-m	10–400 Hz	Low resistivity mud	50"	12"
4	LWD propagation resistivity	LWD geosteering, resistivity for formation evaluation	Measuring while drilling, EM propagation, phase and amplitude resistivity	Resistivity range 0.2–500 ohm-m (phase resistivity); 0.2–300 ohm-m (amplitude resistivity)	100k–2 MHz	High resistivity mud	78"	6"
5	LWD directional propagation resistivity	Geosteering, boundary distance and direction detection and distance to boundary measurement, resistivity for formation evaluation	Measuring while drilling, EM propagation, phase and amplitude resistivity, azimuth resistivity and cross component measurement	Resistivity range 0.2–500 ohm-m (phase resistivity); 0.2–300 ohm-m (amplitude resistivity)	100k–2 MHz	High resistivity mud	78"	6"
6	Electromagnetic telemetry for LWD data transmission	Wireless downhole and surface data communication	Measuring while drilling, low frequency	Transmission distance: 3000 m, bit rate is 1–10 bps	1–50 Hz	Medium resistivity	NA	NA
7	Dielectric constant tool	Water saturation and unconventional	Wireline tool, high frequency	Both dielectric constant and conductivity	20 MHz–1 GHz	High resistivity	1–4"	1"
8	Far boundary detection tool	Extremely far boundary detection	LWD tool, very low frequency, azimuth antennas	Boundary detection distance can reach 100 ft	100 Hz–5 kHz			
9	Resistivity imaging tools	High resolution and shallow DOI	Both LWD and wireline	Near borehole resistivity imaging Resistivity range: 0.2–2000 ohm-m	10 Hz–10 kHz	Both oil- and water-based mud	4"	1"
10	Through casing resistivity tool	Resistivity measurement through metal casing	Wireline	0.1–200 ohm-m	1–5 Hz	NA		

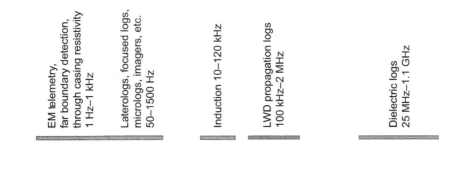
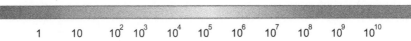

Figure 1.10 Spectrum of electrical and electromagnetic logging tools.

face, torque, etc. Correspondingly, the drilling operation can be instantaneously adjusted if necessary.

LWD also plays an important role in geosteering. Resistivity, neutron, and density logs are often referred to as a basic indication of the lithology, in case the well is not being drilled toward or within the anticipated zone.

1.9 SUMMARY OF ELECTROMAGNETIC LOGGING TOOLS

The EM logging tools are widely used in the formation evaluation and geosteering activities. In this book, we will discuss mostly used EM logging tools including induction, propagation resistivity, dielectric, boundary detection tools, laterolog tools, and variations of these tools. Due to the differences in operation principles, applications of these tools are different. Table 1.2 summarizes the characteristics of these tools. Fig. 1.10 shows a spectrum of frequencies used in the EM logging tools.

REFERENCES

[1] S.W.S. Corporation, Introduction to Schlumberger Well Logging, Schlumberger Well Surveying Corp, 1958.
[2] W.C. Lyons, G.J. Plisga, Standard Handbook of Petroleum and Natural Gas Engineering, Elsevier, New York, 2005.
[3] B. Anderson, Modeling and inversion methods for the interpretation of resistivity logging tool response, DUP Science Publication, Delft, 2001.
[4] P.F. Rodney, M.M. Wisler, Electromagnetic wave resistivity MWD Tool, SPE Drill. Eng. 1 (5) (1986) 33701/01/1986.
[5] P.D. Fredericks, et al., Formation evaluation while drilling with a dual propagation resistivity tool, in: Presented at the SPE Annual Technical Conference and Exhibition, San Antonio, Texas, 1989.

[6] B. Clark, et al., Electromagnetic propagation logging while drilling: theory and experiment, SPE Form. Eval. 5 (3) (1990) pp 263–271, 01/01/1990.
[7] P. Glover, The sonic or acoustic log. Available: <http://www2.ggl.ulaval.ca/personnel/paglover/CD%20Contents/GGL-66565%20Petrophysics%20English/Chapter%2016.PDF>.
[8] G.R. Coates, et al., NMR Logging: Principles and Applications, Haliburton Energy Services, 1999.
[9] M. Appel, Nuclear magnetic resonance and formation porosity, PetroPhysics 45 (2004) 296–307 01/01/2004.
[10] B.L. Hou, G.R. Coates, Nuclear magnetic resonance logging methods for fluid typing, in: Presented at the SPE International Oil and Gas Conference and Exhibition in China, Beijing, China, 1998.
[11] R.A. Meador, P.T. Cox, Dielectric constant logging, a salinity independent estimation of formation water volume, in: Presented at the Fall Meeting of the Society of Petroleum Engineers of AIME, Dallas, Texas, 1975.
[12] L.C. Shen, Problems in dielectric-constant logging and possible routes to their solution, Log Anal. 26 (6) (1985) 01/01/1985.
[13] W. Gravley, Review of downhole measurement-while-drilling systems, SPE J. Pet. Technol. 35 (1983) 1439–1448.

CHAPTER 2

Fundamentals of Electromagnetic Fields and Induction Logging Tools

Contents

2.1	Maxwell's Equations	17
2.2	Complex Permittivity	20
2.3	Sources $\overline{\mathbf{J}}_s, \overline{\mathbf{P}}_s$ and $\overline{\mathbf{M}}_s$	21
2.4	Hertz Potential $\overline{\mathbf{\Pi}}_m$	21
2.5	Electromagnetic Fields Due to a Magnetic Dipole in a Homogeneous Medium	23
2.6	Induced Electromotive Force (EMF) in the Receiving Coil and the Use of Bucking Coil	26
2.7	Quasistatic Approximations and Skin Depth	31
2.8	Apparent Conductivity	34
2.9	Tool Constant and Skin-Effect Correction	35
2.10	Direct Inversion of Induction Logging Data	37
2.11	Spectrum Domain Solutions and Two-Coil Induction Tools in Layered Media	37
2.12	Induction Arrays	45
References		49

2.1 MAXWELL'S EQUATIONS

Maxwell's Equations are fundamentals for solving electrical and electromagnetic (EM) logging problems. As described in Chapter 1, Introduction to Well Logging, resistivity is one of the most important parameters of the formations. To measure the resistivity of the formations, one or more transmitters are used to generate electromagnetic fields in the formation. A few receivers are placed along the axel of the logging tool away from the transmitters. The received signals by the receivers are functions of the formation resistivity. By processing the received signals, the formation resistivity can be obtained. The relations between transmitted signal and received signals can be well described by Maxwell's equations. Due to the geometric limits of the borehole environments, the transmitter and receiver antennas are either coils (induction and measuring while drilling (MWD) tools) or electrodes (laterolog tools). Depending on the applications, the operating frequencies are in the range of kilohertz to megahertz. For example, induction logging tools mostly use 20 kHz frequencies while MWD tools use 400 kHz and 2 MHz. In an induction logging tool, the transmitters induce eddy current in the formation, the eddy current in the formation induces an

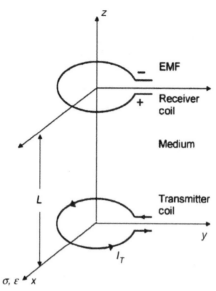

Figure 2.1 Transmitting and receiving coils of an induction sonde.

electromagnetic motive force (EMF) in the receiver coils, which is sometimes called secondary field, which will be picked up by the receiver circuits for processing. Fig. 2.1 shows a basic logging tool configuration, in which one transmitter and one receiver are used.

The problem becomes the solution of the received EMF induced by the transmitter through the formation. To solve this problem, Maxwell's equations must be introduced [1–3]:

$$\nabla \times \overline{\mathbf{E}} = -\frac{\partial \overline{\mathbf{B}}}{\partial t} \qquad (2.1)$$

$$\nabla \times \overline{\mathbf{H}} = \overline{\mathbf{J}} + \frac{\partial \overline{\mathbf{D}}}{\partial t} \qquad (2.2)$$

$$\nabla \cdot \overline{\mathbf{B}} = 0 \qquad (2.3)$$

$$\nabla \cdot \overline{\mathbf{D}} = \rho_v \qquad (2.4)$$

where
$\overline{\mathbf{E}}$ = electric field strength (V/m)
$\overline{\mathbf{D}}$ = electric flux density (C/m^2)
$\overline{\mathbf{H}}$ = magnetic field strength (A/m)
$\overline{\mathbf{B}}$ = magnetic flux density (Wb/m^2) or (T)
$\overline{\mathbf{J}}$ = electric current density (A/m^2)
ρ_v = electric charge density (C/m^3)

Note that all the field quantities are vectors in the space. The time domain expression of Maxwell's equations can be simplified when applying time-harmonic conditions, which means when the signals are time harmonic (sinusoidal), which is the case in both induction and logging while drilling (LWD) tools.

The Maxwell's equations (2.1)–(2.4) are a series of partial differential equations in vector form. The variables are space parameters and time. According to the theory of partial differential equations, the right-hand side of the equations represents the source of the field. Therefore Eqs. (2.1) and (2.4) mean the source of the electrical field **E** is the varying magnetic field with the time. Similarly, the sources of magnetic field are both changing electric field and static current flow **J** as shown in Eq. (2.2).

The electromagnetic field vectors and the electric current and charge densities discussed in Eqs. (2.1)–(2.4) are real functions of space and time. In the case of time-harmonic fields at a single frequency, it is assumed that all field vectors and current and charge densities vary sinusoidally with time at a single angular frequency ω due to the fact of the single frequency excitation. Then we can write, e.g., the x-component of the real $\overline{\mathbf{E}}$ vector, $\overline{\mathbf{E}}_x$ in the following form:

$$\overline{\mathbf{E}}_x(x, y, z, t) = \overline{\mathbf{E}}_1(x, y, z)\cos(\omega t + \phi_1) = \mathbf{Re}\{[\overline{\mathbf{E}}_1(x, y, z)\exp(j\phi_1)]\exp(j\omega t)\} \qquad (2.5)$$

Now let

$$\overline{\mathbf{E}}_x(x, y, z, t) = \overline{\mathbf{E}}_1(x, y, z)\exp(j\phi_1) \qquad (2.6)$$

The complex function $\overline{\mathbf{E}}_1(x, y, z)$, together with knowledge of the frequency ω, contains all necessary information about the original real function $\overline{\mathbf{E}}_x(x, y, z, t)$.

We call $\overline{\mathbf{E}}_1(x, y, z)$ the complex representation of $\overline{\mathbf{E}}_x(x, y, z, t)$.

Same reasoning process applies to the y and z components of the $\overline{\mathbf{E}}$ vector. In summary, we can write

$$\overline{\mathbf{E}}(x, y, y, t) \leftrightarrow \overline{\mathbf{E}}(x, y, z) \qquad (2.7)$$

The vector is now a complex vector, each component of which possesses both real and imaginary parts. Similar expressions apply to all other field vectors. It can be shown that the time derivatives can be represented by $j\omega$. Thus

$$\frac{\partial \overline{\mathbf{B}}(x, y, z, t)}{\partial t} \leftrightarrow j\omega \overline{\mathbf{B}}(x, y, z) \qquad (2.8a)$$

and similarly for D. With these complex notations for the time-harmonic quantities, Maxwell's equations in (2.1)–(2.4) become

$$\nabla \times \overline{\mathbf{E}} = -j\omega \overline{\mathbf{B}} \qquad (2.8b)$$

$$\nabla \times \overline{\mathbf{H}} = \overline{\mathbf{J}} + j\omega \overline{\mathbf{D}} \qquad (2.9)$$

$$\nabla \cdot \overline{\mathbf{B}} = 0 \quad (2.10)$$

$$\nabla \cdot \overline{\mathbf{D}} = \rho_v \quad (2.11)$$

Notice that all field vectors are now complex quantities independent of time. To obtain the real space-time expression for quantity given in the complex notation, we merely have to multiply the complex quantity by $e^{j\omega t}$ and then take the real part.

2.2 COMPLEX PERMITTIVITY

In Maxwell's Equations, the following constitutive relations exist

$$\overline{\mathbf{D}} = \varepsilon \overline{\mathbf{E}} \quad (2.12)$$

$$\overline{\mathbf{B}} = \mu \overline{\mathbf{H}} \quad (2.13)$$

$$\overline{\mathbf{J}} = \sigma \overline{\mathbf{E}} \quad (2.14)$$

where
- ε: dielectric constant or permittivity (F/m)
- μ: magnetic permeability (H/m)
- σ: conductivity (mho/m), inverse of resistivity
- ε_0: free-space permittivity $= 8.85 \times 10^{-12}$ F/m
- μ_0: free-space permeability $4\pi \times 10^{-7}$ H/m

and
- $\varepsilon = \varepsilon_r \varepsilon_0$
- $\mu = \mu_r \mu_0$

ε_r and μ_r are relative dielectric constant and relative magnetic permeability. Therefore we can rewrite Eqs. (2.8a,b) and (2.9) in the following format:

$$\begin{aligned}\nabla \times \overline{\mathbf{H}} &= j\omega \overline{\mathbf{D}} + \overline{\mathbf{J}} = j\omega\varepsilon\overline{\mathbf{E}} + \sigma\overline{\mathbf{E}} + \overline{\mathbf{J}}_{\text{source}} \\ &= j\omega(\varepsilon - j\sigma/\omega)\overline{\mathbf{E}} + \overline{\mathbf{J}}_{\text{source}} \\ &= j\omega\varepsilon^*\overline{\mathbf{E}} + \overline{\mathbf{J}}_{\text{source}}\end{aligned} \quad (2.15)$$

$$\varepsilon^* = \varepsilon - j\sigma/\omega = \varepsilon' - j\varepsilon'' \quad (2.16)$$

ε^* is the complex permittivity, sometimes it is also called complex dielectric constant. ε' is the real part of the complex permittivity, and $\varepsilon'' = \sigma/\omega$ is the imaginary part. It is clear that the complex permittivity ε^* is a function of both dielectric permittivity ε' and conductivity σ. It is also a function of angular frequency ω. Greater σ means higher loss to the electromagnetic fields. From the definitions of the complex permittivity ε^*, we can derive that at low frequencies, the imaginary of the complex permittivity dominates, while at high frequencies, real dielectric permittivity plays a major rule.

2.3 SOURCES $\overline{J}_s, \overline{P}_s$ AND $\overline{M}_s$

In practice, there are three major sources that can generate electromagnetic fields other than varying electrical and magnetic fields. In Maxwell's equations (2.1)–(2.4) and (2.15), we can identify current density J is a source of the fields. Using subscript s to denote the word "source," we can see that

$$\nabla \times \overline{E} = -j\mu\omega\overline{H}$$

$$\nabla \times \overline{H} = \overline{J}_s + j\omega\varepsilon\overline{E}$$

The current source J_s can also be implemented using a dipole charge P_s varying in time:

$$J_s = \frac{d}{dt}P_s \tag{2.17a}$$

Assuming the dipole charge is a time-harmonic source varying at an angular frequency ω,

$$J_s = \frac{d}{dt}P_s = j\omega P_s \tag{2.17b}$$

We can obtain

$$\nabla \times \overline{E} = -j\mu\omega\overline{H} \tag{2.18a}$$

$$\nabla \times \overline{H} = j\omega\overline{P}_s + j\omega\varepsilon\overline{E} \tag{2.18b}$$

Eq. (2.18a,b) is now in the form of electric dipole format. Following the idea of electric dipole, we can image that there is a magnetic dipole source $\overline{M}_s$ which is part of the source of electric field in Eq. (2.18a), therefore, (Eq. 2.18a,b) can be written as

$$\nabla \times \overline{E} = -j\mu\omega\overline{H} - j\omega\mu\overline{M}_s \tag{2.19a}$$

$$\nabla \times \overline{H} = j\omega\varepsilon\overline{E} \tag{2.19b}$$

Note that in Eq. (2.18a,b), only magnetic dipole source is considered.

In a well logging problem, the above three sources can be used to analyze the logging tools. Fig. 2.2 are examples of these sources.

2.4 HERTZ POTENTIAL $\overline{\Pi}_m$

The Maxwell's equations discussed in the previous sections are not easy to solve. To simplify the solution procedure, it is necessary to analyze the properties of the field

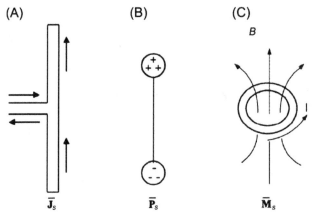

Figure 2.2 Sources of electromagnetic waves. (A) Oscillating electric current, (B) oscillating electric dipole, and (C) oscillating magnetic dipole (the circle is a conducting loop with current flow of *I*).

quantities. In induction and LWD cases, the antennas are loop antennas, which can be described by using a magnetic dipole source shown in Fig. 2.2C. In this section we will introduce a Hertz potential to solve the field problem with a magnetic dipole source $\overline{\mathbf{M}}_s$.

Consider the case only magnetic dipole source exists, no other sources at present. According to Eq. (2.11), when $\rho_v = 0$,

$$\nabla \cdot \overline{\mathbf{D}} = \nabla \cdot \varepsilon \overline{E} = \varepsilon \nabla \cdot \overline{E} = 0 \qquad (2.20)$$

Since ε is a nonzero constant, we can conclude that

$$\nabla \cdot \overline{E} = 0 \qquad (2.21)$$

From the theory of vector analysis, we know that if a vector has a zero divergence, it can be expressed as a rotation of an arbitrary vector since for any vector $\overline{A}$, and a constant α the following relation is always true:

$$\nabla \cdot \nabla \times \alpha \overline{A} = 0 \qquad (2.22)$$

Therefore we can define a new vector $\overline{\Pi}_m$ and let

$$\overline{E} = -j\omega\mu \nabla \times \overline{\Pi}_m \qquad (2.23)$$

$\overline{\Pi}_m$ is the Hertz potential or vector potential of the field. The reason it is called potential since the derivative of $\overline{\Pi}_m$ with respect to the space is the E field, similar to the scalar potential. We can obtain the following relation between the Hertz potential $\overline{\Pi}_m$ and H field [1],

$$\overline{H} = \nabla \nabla \cdot \overline{\Pi}_m + k^2 \overline{\Pi}_m \qquad (2.24)$$

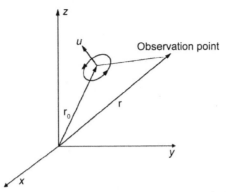

Figure 2.3 An oscillating magnetic dipole is located at $\mathbf{r}_0$ oriented in $\hat{u}$ direction.

where

$$k = \omega\sqrt{\mu\varepsilon^*} \qquad (2.25)$$

$$\nabla^2 \overline{\Pi}_m + k^2 \overline{\Pi}_m = -\overline{\mathbf{M}}_s \qquad (2.26)$$

Solutions to the above Helmholtz equation in a spherical coordinate is

$$\overline{\Pi}_m = \frac{1}{4\pi} \int \frac{\overline{\mathbf{M}}_s(\bar{\mathbf{r}}') e^{-jk|\bar{\mathbf{r}}-\bar{\mathbf{r}}'|}}{|\bar{\mathbf{r}} - \bar{\mathbf{r}}'|} d\bar{\mathbf{r}}' \qquad (2.27)$$

In an induction or LWD logging problem, consider a coil antenna with N_T turns, a current excitation of I_T, and a coil area of A_T in the direction of $\hat{u}$, located at $\bar{\mathbf{r}}_0$ in a homogeneous formation (Fig. 2.3), the magnetic dipole moment $\overline{\mathbf{M}}_s$ can be defined as

$$\overline{\mathbf{M}}_s = I_T A_T N_T (\bar{\mathbf{r}} - \bar{\mathbf{r}}_0) \hat{u} \qquad (2.28)$$

Substituting Eq. (2.28) into (2.27), we can obtain the solution of $\overline{\Pi}_m$ in a homogeneous formation

$$\overline{\Pi}_m = \frac{I_T A_T N_T e^{-jk|\bar{\mathbf{r}}-\bar{\mathbf{r}}'|}}{4\pi |\bar{\mathbf{r}} - \bar{\mathbf{r}}'|} \hat{u} \qquad (2.29)$$

Fig. 2.3 defines quantities appear in Eqs. (2.27)–(2.29).

2.5 ELECTROMAGNETIC FIELDS DUE TO A MAGNETIC DIPOLE IN A HOMOGENEOUS MEDIUM

In Section 2.4, we obtained the solution of Hertz potential $\overline{\Pi}_m$ due to a magnetic dipole source (or an equivalent coil antenna). In this section, we will derive the electromagnetic fields using the solution of the vector potential. Referring to Fig. 2.3, we assume the magnetic dipole is lined up with z axis of the formation (usually this is the direction of the borehole axis), therefore, $\hat{u} = \hat{z}$. Due to the fact that the logging tool moves only along the borehole, therefore, a cylindrical coordinates is more appropriate.

Eq. (2.29) is simplified to

$$\overline{\Pi}_m = \frac{I_T A_T N_T e^{-jk\sqrt{\rho^2+z^2}}}{4\pi\sqrt{\rho^2+z^2}} \hat{z} \qquad (2.30)$$

Hence, $\overline{\Pi}_m = \Pi_{mz}\hat{z}$

where $\Pi_{mz} = \dfrac{I_T A_T N_T e^{-jk\sqrt{\rho^2+z^2}}}{4\pi\sqrt{\rho^2+z^2}}$.

Note that the Hertz potential $\overline{\Pi}_m$ is a function of ρ and z. It is not a function of ϕ due to the symmetry of the system. We also noticed that $\overline{\Pi}_m$ only has z component. Using Eq. (2.23), referring Table 2.1, we can obtain the solution of the E and H fields

$$\overline{E} = -j\omega\mu\nabla\times(\Pi_{mz}\hat{z}) = -j\omega\mu(\nabla\Pi_{mz})\times\hat{z} = +j\omega\mu\frac{\partial\Pi_{mz}}{\partial\rho}\hat{\phi} \qquad (2.31)$$

$$\overline{H} = \nabla\frac{\partial\Pi_{mz}}{\partial z} + k^2\overline{\Pi}_m \qquad (2.32)$$

Writing the above vector equations into scalar format, we have

$$H_z = \left(\nabla\frac{\partial\Pi_{mz}}{\partial z}\right)_z + k^2\overline{\Pi}_m \qquad (2.33)$$

In Eq. (2.33), $\left(\nabla\frac{\partial\Pi_{mz}}{\partial z}\right)_z$ means the z component of the function inside the presences.

$$\mathbf{H}_\rho = \frac{\partial^2\Pi_{mz}}{\partial\rho\partial z} \qquad (2.34)$$

$$\mathbf{H}_\phi = 0 \qquad (2.35)$$

$$\mathbf{E}_\phi = \frac{-j\omega\mu I_T A_T N_T}{4\pi}\frac{\rho e^{-jk\sqrt{\rho^2+z^2}}\left(1+jk\sqrt{\rho^2+z^2}\right)}{(\rho^2+z^2)^{3/2}} \qquad (2.36)$$

Table 2.1 Coordinates and ∇ operators

	Cartesian	Cylindrical	Spherical
Unit length	dx, dy, dz	$d\rho, \rho d\phi, dz$	$dr, rd\theta, r\sin\theta\, d\phi$
∇f	$\hat{x}\dfrac{\partial f}{\partial x} + \hat{y}\dfrac{\partial f}{\partial y} + \hat{z}\dfrac{\partial f}{\partial z}$	$\hat{\rho}\dfrac{\partial f}{\partial \rho} + \hat{\phi}\dfrac{1}{\rho}\dfrac{\partial f}{\partial \phi} + \hat{z}\dfrac{\partial f}{\partial z}$	$\hat{r}\dfrac{\partial f}{\partial r} + \hat{\theta}\dfrac{1}{r}\dfrac{\partial f}{\partial \theta} + \hat{\phi}\dfrac{1}{r\sin\theta}\dfrac{\partial f}{\partial \phi}$
$\nabla^2 f$	$\dfrac{\partial^2 f}{\partial x^2} + \dfrac{\partial^2 f}{\partial y^2} + \dfrac{\partial^2 f}{\partial z^2}$	$\dfrac{1}{\rho}\dfrac{\partial}{\partial \rho}\left(\rho\dfrac{\partial f}{\partial \rho}\right) + \dfrac{1}{\rho^2}\dfrac{\partial^2 f}{\partial \phi^2} + \dfrac{\partial^2 f}{\partial z^2}$	$\dfrac{1}{r^2}\dfrac{\partial}{\partial r}\left(r^2\dfrac{\partial f}{\partial r}\right) + \dfrac{1}{r^2\sin\theta}\dfrac{\partial}{\partial \theta}\left(\sin\theta\dfrac{\partial f}{\partial \theta}\right) + \dfrac{1}{r^2\sin^2\theta}\dfrac{\partial^2 f}{\partial \phi^2}$
$\nabla \cdot \overline{A}$	$\dfrac{\partial A_x}{\partial x} + \dfrac{\partial A_y}{\partial y} + \dfrac{\partial A_z}{\partial z}$	$\dfrac{1}{\rho}\dfrac{\partial}{\partial \rho}(\rho A_\rho) + \dfrac{1}{\rho}\dfrac{\partial A_\phi}{\partial \phi} + \dfrac{\partial A_z}{\partial z}$	$\dfrac{1}{r^2}\dfrac{\partial}{\partial r}(r^2 A_r) + \dfrac{1}{r\sin\theta}\dfrac{\partial}{\partial \theta}(\sin\theta A_\theta) + \dfrac{1}{r\sin\theta}\dfrac{\partial A_\phi}{\partial \phi}$
$\nabla \times \overline{A}$	$\hat{x}\left(\dfrac{\partial A_z}{\partial y} - \dfrac{\partial A_y}{\partial z}\right)$ $+\hat{y}\left(\dfrac{\partial A_x}{\partial z} - \dfrac{\partial A_z}{\partial x}\right)$ $+\hat{z}\left(\dfrac{\partial A_y}{\partial x} - \dfrac{\partial A_x}{\partial y}\right)$	$\hat{\rho}\left(\dfrac{1}{\rho}\dfrac{\partial A_z}{\partial \phi} - \dfrac{\partial A_\phi}{\partial z}\right)$ $+\hat{\phi}\left(\dfrac{\partial A_\rho}{\partial z} - \dfrac{\partial A_z}{\partial \rho}\right)$ $+\hat{z}\left[\dfrac{1}{\rho}\dfrac{\partial}{\partial \rho}(\rho A_\phi) - \dfrac{1}{\rho}\dfrac{\partial A_\rho}{\partial \phi}\right]$	$\hat{r}\dfrac{1}{\sin\theta}\left[\dfrac{\partial}{\partial \theta}(\sin\theta A_\phi) - \dfrac{\partial A_\theta}{\partial \phi}\right]$ $+\hat{\theta}\dfrac{1}{r}\left[\dfrac{1}{\sin\theta}\dfrac{\partial A_r}{\partial \phi} - \dfrac{\partial}{\partial r}(rA_\phi)\right]$ $+\hat{\phi}\left[\dfrac{1}{r}\dfrac{\partial}{\partial r}(rA_\theta) - \dfrac{\partial A_r}{\partial \theta}\right]$

From Eqs. (2.33)–(2.36) we can see that the electromagnetic fields due to a magnetic dipole in the $\hat{z}$ direction in a homogeneous medium have the following nonzero components:

$$\mathbf{E}_\phi, \mathbf{H}_z, \mathbf{H}_\rho$$

and the following zero components

$$E_\rho, E_z, \text{ and } H_\phi$$

Note that the E field has only ϕ component, which means that the current flow in the formation is in circular direction. This means that the field generated by a magnetic dipole (or a small size coil) will induce an E field in the circular direction inside a conducting formation. According to Maxwell's Equation in Eq. (2.4), the circular E field will generate a circular current density **J**, which is also in the ϕ direction. Since the current is from the excitation of the magnetic dipole, we call the current as induced eddy current:

$$\mathbf{J}_\phi = \sigma \mathbf{E}_\phi \quad \text{(eddy current)} \tag{2.37}$$

Note that the strength of the induced eddy current is a linear function of the formation conductivity σ. Fig. 2.4 shows the E field distribution around the transmitter coil for the formation conductivity of 0.1 S/m when the frequency is 20 kHz. We can clearly see that the field decays very fast in both ρ and z directions.

2.6 INDUCED ELECTROMOTIVE FORCE (EMF) IN THE RECEIVING COIL AND THE USE OF BUCKING COIL

As a sensor of the formation conductivity, the purpose of the induction or LWD logging tools are to extract the value of σ in a formation. From Eqs. (2.36) and (2.37), we can obtain the induced E field and eddy current in the formation. If a receiver is placed in the formation a distance away from the transmitter, it is possible to sense the induced E field and therefore, obtain the formation conductivity value. Consider a two-coil system shown in Fig. 2.1, if the radius of the receiving coil is a, located on the z axis separated from the transmitter coil by a distance of L, the induced voltage in the receiving coil is called electromagnetic motive force [1,2,4–6] and it can be calculated as:

$$EMF = N_R \int_0^{2\pi} \overline{E} \cdot d\overline{l} \tag{2.38}$$

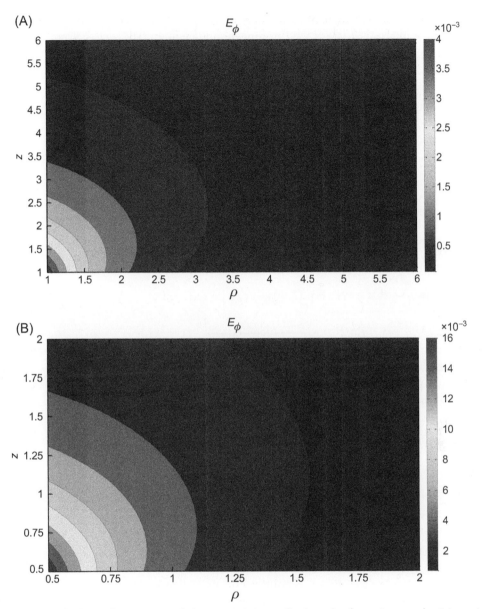

Figure 2.4 E field distribution around the transmitting coil when the formation conductivity σ is 0.1 s/m. The operating frequency is 20 kHz with antenna moment is 1. (A) Real part of E field and (B) Imaginary part of E field.

where N_R is the number of turns in the receiver coil and $\vec{dl}$ is in the direction of ϕ. Referring to Figure 2.1, Eq. (2.38) can be simplified as

$$EMF = 2\pi a N_R E_\phi(\rho = a, \ z = L) \tag{2.39}$$

Substituting Eq. (2.31) into (2.39) we have

$$EMF = \frac{-j\omega\mu I_T A_T N_T N_R \pi a^2}{2\pi(a^2+L^2)^{3/2}}(1+jk\sqrt{a^2+L^2})e^{-jk\sqrt{a^2+L^2}} \tag{2.40a}$$

$$EMF = -\frac{j\omega\mu I_T A_T N_T N_R A_R}{2\pi L^3\left[1+\left(\frac{a}{L}\right)^2\right]^{\frac{3}{2}}}\left(1+jkL\sqrt{1+\left(\frac{a}{L}\right)^2}\right)e^{-jkL\sqrt{1+\left(\frac{a}{L}\right)^2}} \tag{2.40b}$$

where $\pi a^2 = A_R$.

If we assume that $a \ll L$, Eq. (2.40a,b) can be reduced to

$$EMF = \frac{-j\omega\mu I_T A_T N_T N_R A_R}{2\pi L^3}(1+jkL)e^{-jkL} \tag{2.41}$$

Eq. (2.40a,b) is valid for general case, Eq. (2.41) is valid for the cases when $a \ll L$. Note that the received voltage in the receiver coil is a complex value. Figs. 2.5 and 2.6 show the real and imaginary parts of the received EMF as a function of formation conductivity. At the induction frequency, the real part changes with formation conductivity significantly whereas the imaginary does not change much with the

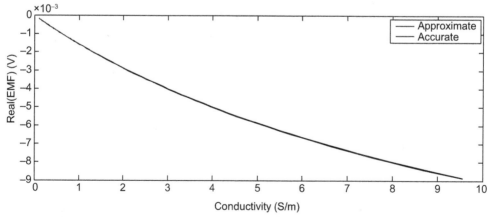

Figure 2.5 Comparing the real part of Eqs. (2.40a,b) and (2.41) at $f = 20$ kHz and $a/L = 0.09338$, for $L = 40''$.

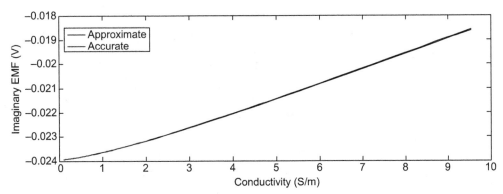

Figure 2.6 Comparing the imaginary part of Eqs. (2.40a,b) and (2.41) at $f = 20$ kHz and $a/L = 0.09338$, for $L = 40''$.

conductivity of the formation. This phenomenon indicates that the real part of the EMF reflects the conductivity of the formation and the imaginary part of the received EMF is largely a directly coupled signal from the transmitter. However, the existence of imaginary signal may affect the receiver operation. Therefore many modern induction tools are equipped with another antiwound coil to reduce the effects of the direct coupling, which is called "bucking coil."

Figs. 2.5 and 2.6 also show a comparison of the results computed by using accurate Eq. (2.40a,b) and approximate Eq. (2.41) at the induction frequency of 20 kHz and a/L value is 0.09338. It can be seen that the values from the two equations are very close.

Fig. 2.7 is the percentage errors when using Eq. (2.41) instead of Eq. (2.40a,b) when a/L increases. It can be seen that when a/L is less than 0.22, the errors of Eq. (2.41) is within 5%. In most of the cases, in logging tool design, this condition is satisfied. Therefore Eq. (2.41) is often used in the computation for simplicity.

Consider the case when $|kL| \ll 1$, Eq. (2.41) can be written as

$$EMF \cong \frac{j\omega\mu I_T A_T N_T N_R A_R}{2\pi L^3}(1 - k^2 L^2) \qquad (2.42)$$

Note that k is a complex number given by $k = \omega\sqrt{\mu\varepsilon^*} = \omega\sqrt{\mu\varepsilon(1 - \frac{j\sigma}{\omega\varepsilon})}$ in Eq. (2.16), therefore,

$$\mathrm{Re}(EMF) \cong \frac{\omega^2 \mu^2 \sigma I_T A_T N_T N_R A_R}{2\pi L^2} \cdots \qquad (2.43)$$

$$\mathrm{Im}(EMF) \cong \frac{\omega\mu I_T A_T N_T N_R A_R}{2\pi L^3}(1 + \omega^2 \mu\varepsilon L) \qquad (2.44)$$

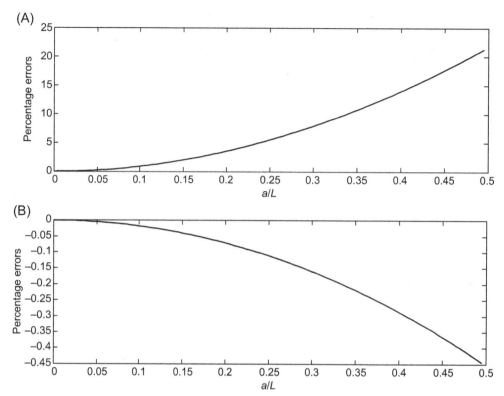

Figure 2.7 The percentage errors when using Eq. (2.41) and Eq. (2.40a,b). (A) Real Part of EMF and (B) Imaginary Part of EMF.

Eq. (2.44) can be further reduced to

$$\mathrm{Im}(EMF) \cong \frac{\omega\mu I_T A_T N_T N_R A_R}{2\pi L^3} \quad (2.45)$$

From Eqs. (2.43) and (2.45), we can also conclude that the received EMF has a real part that is proportional to the conductivity of the formation and the imaginary part is not a strong function of the formation conductivity. We also noticed that the real part of EMF decays with the transmitter–receiver distance L at its second power. However, the imaginary part of the EMF decays with L at its third power.

If we place another bucking coil in the system to reduce the tool influence from the directly coupled signal as shown in Fig. 2.8, assuming the current flow in the bucking coil has the same value as the main transmitter coil, we can cancel the directly coupled EMF (imaginary part in Eq. 2.46) by setting the number of turns in the bucking coil as

$$N_{Rb} = -\frac{L_b^3}{L^3} N_R \quad (2.46)$$

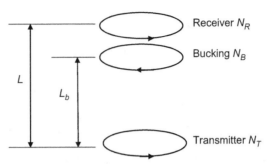

Figure 2.8 The use of bucking coil to reduce directly coupled signals to the receiver. Note that the bucking coil is antiwound with respect to the main receiver coil.

the real part of the received signal including the signal from the bucking coil can be found:

$$\mathrm{Re}(EMF) \cong \frac{\omega^2 \mu^2 \sigma I_T A_T N_T N_R A_R}{2\pi L^2}\left(1 - \frac{L_b}{L}\right) \qquad (2.47)$$

2.7 QUASISTATIC APPROXIMATIONS AND SKIN DEPTH

Assuming the operating frequency is relatively low, Eqs. (2.40a,b) and (2.41) can be further simplified using the quasistatic approximations. In EM theory, when the system frequency is low but not zero, we can approximately express the field quantities using quasistatic approximation. We can use this approach to obtain the characteristics of the received EMF.

Using Eq. (2.25) we know that $k = \omega\sqrt{\mu\varepsilon^*}$ and $\varepsilon^* = \varepsilon - j\sigma/\omega$

$$k = \omega\sqrt{\mu\varepsilon^*} = \omega\sqrt{\mu\varepsilon\left(1 - \frac{j\sigma}{\omega\varepsilon}\right)}$$

Consider the low frequency case we define the loss tangent (the tangent of the complex number in Eq. 2.42) as

$$\tan\delta = \frac{\sigma}{\omega\varepsilon} \gg 1 \qquad (2.48)$$

Therefore

$$k^2 \approx -j\omega\mu\sigma \qquad (2.49)$$

and

$$k = \frac{1-j}{\sqrt{2}}\sqrt{\omega\mu\sigma} = (1-j)\frac{1}{d_s} \qquad (2.50)$$

where d_s is the skin depth

$$d_s = \sqrt{\frac{2}{\omega\mu\sigma}} \qquad (2.51)$$

As seen from Eq. (2.40a,b), the attenuation of the EMF is largely due to the exponential term kL. Consider the exponential term in Eq. (2.40a,b), it can be expressed in terms of skin depth d_s:

$$e^{-jkL} = e^{-L/d_s}e^{-jL/d_s} \qquad (2.52)$$

The skin depth actually is a measure of the depth of the EM field inside the conducting media. One skin depth indicates the distance of the EM field reaches 36% of its maximum value. Fig. 2.9 shows the skin depth when the frequency is 20 kHz and 2 MHz in a media when $\mu = \mu_0$.

From Eq. (2.51) we conclude that the skin depth is a square root of both conductivity and frequency. If you would like to develop a logging tool that can "see" double the distance from an existing logging tool, such as common induction logging tool, you have to reduce the frequency by a factor of 4 if the other tool parameters are kept the same. Note that in each skin depth, the EM field will be attenuated by a factor of 0.36. In numerical simulation, we usually consider the field at 5 skin depth is generally vanished. In most service company specifications, we can find a parameter called "Depth of Investigation (DOI)." It is difficult to define DOI since the DOI of an EM logging tool is a function of conductivity. It is more meaningful to give the DOI together with the conductivity and operating frequency.

From Fig. 2.6, we know that the approximation in Eq. (2.41) is accurate enough when a/L is small. They can be used to qualitatively analyze the logging tool performance. However, for more accurate approximation, higher order terms must be used. With the understanding of skin depth and approximation of the EMF in an induction logging tool, we can improve the computation accuracy. Consider the complex term in Eq. (2.41), let $x = kL$ and expand the exponential term into a Taylor series, we have,

$$\begin{aligned}(1+jx)e^{-jx} &= (1+jx)(1-jx-x^2/2+jx^3/6+x^4/24-jx^5/120+\cdots) \\ &= 1+x^2/2-jx^3/3-x^4/8+jx^5/30\end{aligned} \qquad (2.53)$$

$$x = L/d_s(1-j) = \frac{\sqrt{2}L}{d_s}e^{-j\frac{\pi}{4}} = \Delta e^{-j\frac{\pi}{4}}\sqrt{2} \qquad (2.54)$$

where

$$\Delta = \frac{L}{d_s} = \sqrt{\pi f \mu \sigma} L \qquad (2.55)$$

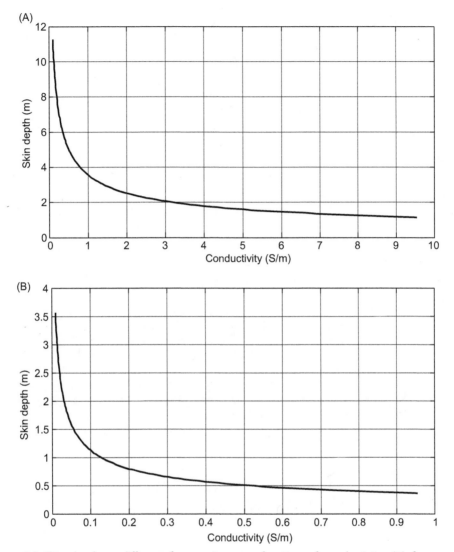

Figure 2.9 Skin depth at different frequencies as a function of conductivity: (A) frequency is 20 kHz; and (B) frequency is 2 MHz.

Therefore we can obtain

$$(1+jkL)e^{-jkL} \approx 1 - j\Delta^2 + \left(\frac{-1+j}{\sqrt{2}}\right)\frac{2\sqrt{2}\Delta^3}{3} + \frac{\Delta^4}{2} + \frac{-1-j}{\sqrt{2}}\frac{\Delta^5}{15}2\sqrt{2} + \cdots$$

$$EMF = \frac{\omega\mu I_T A_T N_T N_R A_R}{2\pi L^3}\left\{-j - \Delta^2 + (1+j)\frac{2\Delta^3}{3} - j\frac{\Delta^4}{2} + (-1+j)\frac{2\Delta^5}{15} + \cdots\right\}$$

(2.56)

2.8 APPARENT CONDUCTIVITY [6–8]

Take only $(-\Delta^2)$ term in EMF expression in Eq. (2.56), and we know that the real part of the received EMF is a strong function of the formation conductivity as shown in Fig. 2.16. The received real part of the EMF in Eq. (2.56) is actually the received voltage in the receiver antenna, we rename the real part of the EMF as V_R for simplicity, then

$$V_R = \frac{-\omega^2 \mu^2 I_T A_T N_T N_R A_R \sigma}{4\pi L} \quad (2.57)$$

From Eq. (2.57), we can see that the approximate expression of the received signal V_R is a linear function of the conductivity in a homogeneous formation. Therefore the conductivity of the formation can be simply "inverted" from Eq. (2.57),

$$\sigma_a = \frac{-4\pi L V_R}{\omega^2 \mu^2 I_T A_T N_T N_R A_R} \quad (2.58)$$

or

$$\sigma_a = \frac{-4\pi L}{\omega^2 \mu^2 I_T A_T N_T N_R A_R} \text{Re}(EMF) \quad (2.59)$$

The σ_a in Eqs. (2.58) and (2.59) are defined as "Apparent Conductivity" of the formation. The word "Apparent" means that this value is obtained from the measured voltage by the receiver; it is a conductivity "seen" by the logging tool. It may not be the "true" conductivity of the formation when the formation is not homogeneous. In logging industry, resistivity is mostly used instead of conductivity. Therefore the apparent resistivity of the formation can be obtained by inversing Eq. (2.58) or (2.59),

$$R_a = \frac{-\omega^2 \mu^2 I_T A_T N_T N_R A_R}{4\pi L \text{Re}(EMF)} = \frac{-\omega^2 \mu^2 I_T A_T N_T N_R A_R}{4\pi L V_R} \quad (2.60)$$

Therefore the apparent conductivity or the apparent resistivity is a linear approximation of the received signal with respect to the formation conductivity or resistivity assuming the formation is homogeneous. If the formation is layered, we still use the definition given in Eq. (2.59) or (2.60) to convert the received voltage into apparent conductivity or resistivity. However, in this case, the converted conductivity or resistivity may no longer be the true conductivity or resistivity of the formation. It will be a value related to the properties of the formations near the logging tool. To obtain the true conductivity or resistivity, the measured values must be processed using inversion algorithms, which will be discussed in the later chapters.

Now the apparent conductivity and resistivity are defined as shown in Eqs. (2.59) and (2.60), using the definition of EMF given in Eq. (2.42), we can obtain a general expression for the apparent conductivity:

$$\sigma_a = \frac{-2}{\omega \mu L^2} \text{Im}\{(1+jkL)e^{-jkL}\} \tag{2.61}$$

For $\Delta \ll 1$,

$$\sigma_{aa} = \sigma\left\{1 - \frac{2}{3}\Delta + \frac{2}{15}\Delta^3 + \cdots\right\} \tag{2.62}$$

where σ_{aa} is the approximate apparent conductivity.

2.9 TOOL CONSTANT AND SKIN-EFFECT CORRECTION

For a given tool parameters (transmitter and receiver parameters), the received EMF is a complex function of the conductivity of the formation. Using the concept of apparent conductivity or resistivity given in Eqs. (2.59) and (2.60), the received voltage is converted into apparent conductivity or apparent resistivity. However, from the analysis process in Section 2.8, we understand that the apparent conductivity is just a linear approximation of the complex function of the formation conductivity and received signal. In general cases, we can express the apparent conductivity approximately as a linear function of the received voltage:

$$R_a = f(1/V_R) \approx \alpha_T/V_R \tag{2.63}$$

where, α_T is called tool constant of the induction logging device. From Eq. (2.60), it can be seen that the tool constant for a two-coil induction logging tool is

$$\alpha_T = \frac{-\omega^2 \mu^2 I_T A_T N_T N_R A_R}{4\pi L} \tag{2.64}$$

From Eq. (2.64), we can see that the tool constant is a function of tool parameters. Once the hardware is determined, the parameters in the tool constant can be obtained. Due to the approximation, even in the homogeneous formation, the apparent conductivity or resistivity is not the true resistivity of the formation. To obtain the real conductivity or resistivity of the formation, we can use either higher order approximation in EMF or use an iterative algorithm to improve the accuracy of the apparent conductivity converted from the measured voltage using the definition in Eq. (2.59) in a homogeneous formation. In logging industry, this process is called skin depth correction. There are two approaches that are mostly used. The first method is based on the higher order of the expansion of the apparent conductivity in Eq. (2.62).

Therefore the skin depth correction means for a given measured $\sigma_m = \sigma_a$, to find the effective σ of the formation. This process sometime is also called homogeneous inversion. From Eq. (2.55), we have

$$\Delta = \frac{L}{d_s} = \sqrt{\pi f \mu \sigma} L \qquad (2.55)$$

Note that Δ is a function of formation conductivity σ. And we know that

$$\sigma_{aa} = \sigma\left\{1 - \frac{2}{3}\Delta + \frac{2}{15}\Delta^3 + \cdots\right\} \qquad (2.62)$$

Therefore we can use σ_m as σ in Eq. (2.55) to calculate an approximate Δ, and calculate σ using Eq. (2.62) with σ_{aa} being the measured apparent conductivity. The procedure can be summarized as follows:

1. let $\sigma = \sigma_m$ in Eq. (2.55) to calculate approximate Δ, marked as Δ_a,

$$\Delta_a = L\sqrt{\pi f \mu \sigma_m} \qquad (2.65)$$

2. let $\Delta = \Delta_a$ in Eq. (2.62) to calculate σ:

$$\sigma = \frac{\sigma_m}{\left\{1 - \frac{2}{3}\Delta_a + \frac{2}{15}\Delta_a^3\right\}} \qquad (2.66)$$

In most cases, this procedure can obtain good approximation if $\Delta_a < 0.1$.

The more accurate skin depth correction is to use an iterative algorithm. The following iterative procedure is one way to invert the formation conductivity from measured apparent conductivity σ_m:

1. let $\sigma = \sigma_m$ in Eq. (2.65) to calculate Δ_a

$$\Delta_a = L\sqrt{\pi f \mu \sigma_m} \qquad (2.65)$$

2. calculate σ using Eq. (2.66):

$$\sigma = \frac{\sigma_m}{\left\{1 - \frac{2}{3}\Delta_a + \frac{2}{15}\Delta_a^3\right\}} \qquad (2.66)$$

3. calculate k using Eq. (2.50) and the σ obtained in step (2)
4. calculate σ_α using Eq. (2.61)

$$\sigma_a = \frac{-2}{\omega \mu L^2} \text{Im}\left\{(1 + jkL)e^{-jkL}\right\} \qquad (2.61)$$

5. Is $|\sigma_a - \sigma_m| <$ tolerance?
   if yes, terminate the iteration;
   if no, go to step (6);
6. change the σ value to $\sigma(\sigma_m/\sigma_a)$; go to step (4).

This algorithm is the extension of the first method using accurate k and σ_a expression. In most cases, the tolerance is a predetermined value, such as 1 milliohm.

2.10 DIRECT INVERSION OF INDUCTION LOGGING DATA

Note that in Eq. (2.40a,b), the received EMF is a function of formation conductivity and the expression of Eq. (2.41) is accurate. However, the definition of apparent conductivity defined in Eqs. (2.58) and (2.59) is a first order approximation of Eq. (2.40a,b). Therefore the apparent conductivity is an approximate representation of formation conductivity. Another way to overcome this defect is to directly invert the formation conductivity from Eq. (2.40a,b). This can be mathematically expressed as

$$EMF(\sigma) = -\frac{j\omega\mu I_T A_T N_T N_R A_R}{2\pi L^3 \left[1+\left(\frac{a}{L}\right)^2\right]^{\frac{3}{2}}} \left(1+jkL\sqrt{1+\left(\frac{a}{L}\right)^2}\right) e^{-jkL\sqrt{1+\left(\frac{a}{L}\right)^2}} \quad (2.67)$$

$$\sigma = EMF^{-1}(\sigma) \quad (2.68)$$

Note that this is an inversion problem instead of forward. However, this inversion is a single variable process. In practice, Eq. (2.68) is solved by using precalculated table. The one-dimensional table is interpolated for a given value of EMF. The precalculated table serves as the plot in Figs. 2.6 and 2.7. For the values between data points in the table, interpolation method is used. Interpolation method can be linear, quadrature, or spline depending on the density of the table. In general, a 20-point logarithm based table with a linear interpolation is satisfactory in most of the cases covering conductivity range of 0.001–10.

This method can also be used when borehole and eccentricity exist. In this case, a four-dimensional table must be established and a corresponding interpolation method must be used.

2.11 SPECTRUM DOMAIN SOLUTIONS AND TWO-COIL INDUCTION TOOLS IN LAYERED MEDIA

In this section, we will discuss the use of spectrum domain solutions to solve the induction and LWD problems in a layered one-dimensional formation. There are

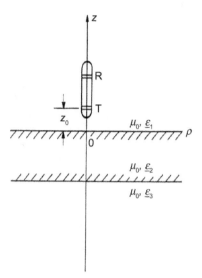

Figure 2.10 A two-coil induction sonde in a layered medium.

many different ways to approach this problem, such as the generic reflection and transmission matrix method given in Ref. [3]. Here, we will introduce a slightly different way for the same problem in a manner that is physically easier to understand using the knowledge given in previous sections. Our discussion is limited to two- and three-layer formations. For multilayer formations, the discussions will be given in the later chapters. Consider Fig. 2.10, we have a simple tool structure of two coils over a layered formation. From the previous sections, we know that the field of a coil in a homogeneous media can be obtained using Hertz potential for a z-directed magnetic dipole as given in Eq. (2.69):

$$\overline{\mathbf{M}}_s = -I_T A_T N_T \delta(\overline{\mathbf{r}} - \overline{\mathbf{r}}_0)\hat{z} \qquad (2.69)$$

From Eq. (2.26) we have

$$\nabla^2 \overline{\mathbf{\Pi}}_z + k^2 \overline{\mathbf{\Pi}}_z = -I_T A_T N_T \delta(\overline{\mathbf{r}} - \overline{\mathbf{r}}_0) \qquad (2.70)$$

Note that the Hertz potential is a function of both ρ and z. In a homogeneous formation, we can find a simple solution as shown in Eq. (2.29). However, in a layered formation, formation property changes in the z direction (has boundaries) but no boundaries in the ρ direction. In order to hand layers, we will use Fourier transform in z coordinate and solve the partial differential equation (2.70) in the spectrum domain. Since there is no change in the formation in the f direction, partial derivatives with respect to f is zero in reference to Table 2.1. From Table 2.1 and consider $\overline{\Pi}_z$ has only z component, Eq. (2.70) becomes

$$\frac{1}{\rho}\frac{\partial}{\partial \rho}\left(\rho\frac{\partial \Pi_z}{\partial \rho}\right) + \frac{1}{\rho^2}\frac{\partial^2 \Pi_z}{\partial \varphi^2} + \frac{\partial^2 \Pi_z}{\partial z^2} + k^2 \Pi_z = -I_T A_T N_T \delta(z-z_0) \qquad (2.71)$$

We know that from the theory of differential equations, solutions to Eq. (2.71) have two parts: a homogeneous solution and a particular integral [9]. The homogeneous solution is when the right-hand side of Eq. (2.71) is zero and the particular integral is one solution when the right-hand side is nonzero, i.e.,

$$\Pi_z = \Pi_{hz} + \Pi_{pz} \qquad (2.72)$$

Let's consider the homogenous solution since the particular integral is the dipole solution in a homogeneous media and has been given in Section 2.5:

$$\overline{\Pi}_{pz} = \frac{I_T A_T N_T e^{-jkR}}{4\pi R} \qquad (2.73)$$

where $R = \sqrt{\rho^2 + (z-z_0)^2}$.

The homogenous solution satisfies:

$$\frac{1}{\rho}\frac{\partial}{\partial \rho}\left(\rho\frac{\partial \Pi_{hz}}{\partial \rho}\right) + \frac{1}{\rho^2}\frac{\partial^2 \Pi_{hz}}{\partial \varphi^2} + \frac{\partial^2 \Pi_{hz}}{\partial z^2} + k^2 \Pi_{hz} = 0 \qquad (2.74)$$

Use the method of separation of variable and let

$$\Pi_{hz}(\rho, z) = R(\rho)Z(z) \qquad (2.75)$$

and apply Fourier transform to z to obtain spectrum domain expression:

$$\tilde{\Pi}_{hz}(\rho, \xi) = \int_{-\infty}^{\infty} \Pi_{hz}(\rho, z)e^{-j\xi z} dz \qquad (2.76a)$$

and

$$\Pi_{hz}(\rho, z) = \int_{-\infty}^{\infty} \tilde{\Pi}_{hz}(\rho, \xi)e^{j\xi z} d\xi \qquad (2.76b)$$

Substitute Eq. (2.76b) into (2.74), we can find the differential equation that $\tilde{\Pi}_{hz}(\rho, \xi)$ satisfies:

$$\rho^2 \frac{\partial \tilde{\Pi}_{hz}}{\partial \rho^2} + \rho \frac{\partial \tilde{\Pi}_{hz}}{\partial \rho} + (\lambda \rho)^2 \tilde{\Pi}_{hz} = 0 \qquad (2.77)$$

where

$$\lambda^2 = k^2 - \xi^2 \tag{2.78}$$

Using Eq. (2.75), we can obtain

$$\rho^2 \frac{\partial R(\rho)}{\partial \rho^2} + \rho \frac{\partial R(\rho)}{\partial \rho} + (\lambda\rho)^2 R(\rho) = 0 \tag{2.79}$$

And

$$Z(z) = e^{\pm z\sqrt{k^2 - \lambda^2}} \tag{2.80}$$

The solution to Eq. (2.79) is a zeroes order Bessel function of the first kind $J_0(\lambda\rho)$ [10]. Therefore from Eqs. (2.75) and (2.80) we have

$$\overline{\Pi}_{hz} = J_0(\lambda\rho) e^{\pm\sqrt{\lambda^2 - k^2} z} \tag{2.81}$$

From properties of Bessel functions [10], we have the following useful formulas:

$$\frac{e^{-jkR}}{R} = \int_0^\infty \frac{\lambda}{\xi} e^{-\xi|z-z_0|} J_0(\lambda\rho) d\lambda \tag{2.82}$$

$$\frac{\rho(1+jkR)e^{-jkR}}{R^3} = \int_0^\infty \frac{\lambda^2}{\xi} e^{-\xi|z-z_0|} J_1(\lambda\rho) d\lambda \tag{2.83}$$

$$\frac{2(1+jk|z-z_0|)e^{-jk|z-z_0|}}{(|z-z_0|)^3} = \int_0^\infty \frac{\lambda^3}{\xi} e^{-\xi|z-z_0|} d\lambda \tag{2.84}$$

Note that from Eq. (2.80) we can find spectrum domain variable ξ as:

$$\xi = (\lambda - k)^{1/2}(\lambda + k)^{1/2}$$

Consider in the complex domain, ξ has a branch cut as shown in Fig. 2.11:

$$-\pi/2 < \arg(\lambda - k) < 3\pi/2$$
$$-3\pi/2 < \arg(\lambda + k) < \pi/2$$

When the tool travels in the z direction, the wave will be expressed differently due to the reflections from the boundaries. When $z_0 > 0$, which means the tool is located in the top layer, besides the radiated wave from the transmitter, there is a

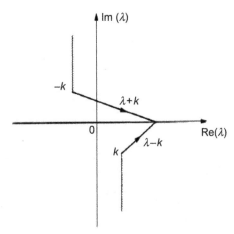

Figure 2.11 Branch cuts on the complex plane.

reflected wave traveling in the opposite direction with unknown magnitude. Therefore the Hertz potential can be expressed as

$$\overline{\Pi}_{1z} = \frac{\overline{\mathbf{M}}_T}{4\pi} \int_0^\infty \left[F_1(\lambda)e^{-\xi_1 z} + \frac{\lambda}{\xi_1} e^{-\xi_1|z-z_0|} \right] J_0(\lambda\rho)d\lambda, \quad z_0 > 0 \qquad (2.85)$$

$F_1(\lambda)$ is an unknown reflection coefficient at the boundary located at $z = 0$.

When the tool is located in the middle layer, the wave is composed of a positively going wave and a negatively going part, both with unknown magnitude:

$$\overline{\Pi}_{2z} = \frac{\overline{\mathbf{M}}_T}{4\pi} \int_0^\infty [F_2(\lambda)e^{-\xi_2 z} + G_2(\lambda)e^{-\xi_2 z}] J_0(\lambda\rho)d\lambda, \quad -h < z_0 < 0 \qquad (2.86)$$

where F_2 and G_2 are the reflection coefficients by the boundary at $z = 0$ and $z = -h$ respectively.

When $z_0 < -h$, the wave only has the component traveling in the negative z direction:

$$\overline{\Pi}_{3z} = \frac{\overline{\mathbf{M}}_T}{4\pi} \int_0^\infty G_3(\lambda)e^{\xi_3(z+h)} J_0(\lambda\rho)d\lambda, \quad z_0 < -h \qquad (2.87)$$

where $\xi_i = \sqrt{\lambda^2 - k_i^2}$ $(i = 1, 2, 3)$, and $\overline{\mathbf{M}}_T = \overline{\mathbf{I}}_T A_T N_T$.

We can use boundary conditions to determine the unknowns F_1, F_2, G_2, and G_3.

Let's consider the boundary conditions. From the definition of Hertz potential (2.23), we know that E field is perpendicular with the Hertz potential, which is in the ϕ direction, and H field has both z and ρ components. We can derive the boundary conditions for the Hertz potential:

1. $\overline{\Pi}_z$ continuous at $z = 0$ and at $z = -h$
2. $\frac{\partial \overline{\Pi}_z}{\partial z}$ continuous at $z = 0$ and $z = -h$

From Eq. (2.39) we have,

$$EMF = 2\pi a N_R E_\phi = j\omega\mu 2\pi \rho N_R \frac{\partial \pi}{\partial \rho} \tag{2.88}$$

Note that

$$\frac{\partial}{\partial \rho}[J_0(\lambda\rho)] = \lambda J_0'(\lambda\rho) = -\frac{1}{2}\lambda^2\rho \quad \text{for small } \rho$$

We have,

$$EMF = \frac{-j\omega\mu \overline{M}_T N_R A_R}{4\pi} \int_0^\infty \lambda^2 \left[F_1 e^{-\xi_1 z} + \frac{\lambda}{\xi_1} e^{-\xi_1|z-z_0|} \right] d\lambda \tag{2.89}$$

We can obtain apparent conductivity from Eq. (2.59),

$$\sigma_a = \frac{L}{\omega\mu} \text{Re}\left\{ j \int_0^\infty \lambda^2 \left[F_1(\lambda) e^{-\xi_1 z} + \frac{\lambda}{\xi_1} e^{-\xi_1|z-z_0|} \right] d\lambda \right\} \tag{2.90}$$

Therefore we can use the boundary conditions to solve for the unknown parameters and the apparent conductivity can be found to be:

for $z_0 > 0$

$$\sigma_a = \frac{|z-z_0|}{\omega\mu} \text{Re}[jI_m] \quad m = 1, 2, 3$$

$$I_1(z, z_0) = \int_0^\infty d\lambda \frac{\lambda^3[\xi_2(\xi_1 - \xi_3)\cosh(\xi_2 h) + (\xi_1\xi_3 - \xi_2^2)\sinh(\xi_2 h)]}{\xi_1[\xi_2(\xi_1 + \xi_3)\cosh(\xi_2 h) + (\xi_1\xi_3 + \xi_2^2)\sinh(\xi_2 h)]} e^{-\xi_1(z+z_0)}$$
$$+ \frac{2[1 + jk_1|z-z_0|]e^{-jk_1(z-z_0)}}{|z-z_0|^3} \tag{2.91a}$$

$$I_2(z, z_0) = \int_0^\infty d\lambda \frac{\lambda^3[(\xi_2 - \xi_3)e^{-\xi_2(h+z)} + (\xi_2 + \xi_3)e^{\xi_2(h+z)}]}{[\xi_2(\xi_1 + \xi_3)\cosh(\xi_2 h) + (\xi_1\xi_3 + \xi_2^2)\sinh(\xi_2 h)]} e^{-\xi_1 z_0} \tag{2.91b}$$

$$I_3(z, z_0) = \int_0^\infty \frac{(+2)\lambda^3 \xi_2 e^{-\xi_3(h+z)} e^{-\xi_1 z_0}}{[\xi_2(\xi_1 + \xi_3)\cosh(\xi_2 h) + (\xi_1\xi_3 + \xi_2^2)\sinh(\xi_2 h)]} d\lambda \quad (2.91c)$$

$\xi_m = \sqrt{\lambda^2 - k_m^2}$

for $-h < z_0 < 0$

$$I_1(z, z_0) = \int_0^\infty d\lambda \frac{\lambda^3[(\xi_2 - \xi_3)e^{-\xi_2(h+z)} + (\xi_2 + \xi_3)e^{\xi_2(h+z)}]}{[\xi_2(\xi_1 + \xi_3)\cosh(\xi_2 h) + (\xi_1\xi_3 + \xi_2^2)\sinh(\xi_2 h)]} e^{-\xi_1 z_0} \quad (2.92a)$$

$$I_2(z, z_0) = \frac{2[1 + jk_2|z - z_0|]e^{-jk_2|z-z_0|}}{|z - z_0|^3} + \int_0^\infty d\lambda \frac{\lambda^3}{2\xi_2 \Delta_0} \{(\xi_2 - \xi_3)[(\xi_2 + \xi_1)e^{-\xi_2 z_0}$$
$$+ (\xi_2 - \xi_1)e^{\xi_2 z_0}] + (\xi_1 - \xi_2)[(\xi_3 - \xi_2)e^{-\xi_2(h+z_0)} - (\xi_3 + \xi_2)e^{\xi_2(h+z_0)}]e^{-\xi_2(h+z_0)}\}$$
$$(2.92b)$$

$$I_3(z, z_0) = \int_0^\infty d\lambda \frac{\lambda^3[(\xi_2 - \xi_1)e^{\xi_2 z_0} + (\xi_2 + \xi_1)e^{-\xi_2 z_0}]}{\Delta_0} e^{-\xi_3(z+h)} \quad (2.92c)$$

$$\Delta_0 = \xi_2(\xi_1 + \xi_3)\cosh(\xi_2 h) + (\xi_1\xi_3 + \xi_2^2)\sinh(\xi_2 h) \quad (2.92d)$$

for $z_0 < -h$

$$I_1(z, z_0) = \int_0^\infty d\lambda \frac{2\lambda^3 \xi_2 e^{-\xi_1 z} e^{\xi_3(h+z_0)}]}{[\xi_2(\xi_1 + \xi_3)\cosh(\xi_2 h) + (\xi_1\xi_3 + \xi_2^2)\sinh(\xi_2 h)]} e^{-\xi_1 z_0} \quad (2.93a)$$

$$I_2(z, z_0) = \int_0^\infty d\lambda \frac{\lambda^3[(\xi_2 - \xi_1)e^{\xi_2 z_0} + (\xi_2 + \xi_1)e^{-\xi_2 z_0}]}{\Delta_0} e^{\xi_3(z_0 + h)} \quad (2.93b)$$

$$I_3(z, z_0) = \int_0^\infty d\lambda \frac{\lambda^3[\xi_2(\xi_3 - \xi_1)\cosh\xi_2 h + (\xi_3 \xi_1 - \xi_2^2)\sinh\xi_2 h]}{\xi_3 \Delta_0} e^{\xi_3(z_0 + 2h + z)}$$
$$+ \frac{2[1 + jk_3|z - z_0|]e^{-jk_3|z-z_0|}}{|z - z_0|^3} \quad (2.93c)$$

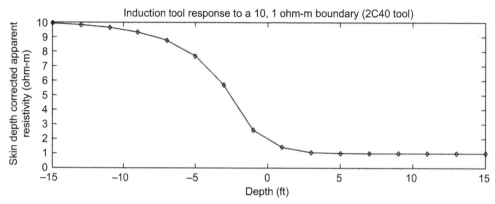

Figure 2.12 Computed apparent conductivity of an induction tool in a two-layer formation. The tool spacing is 40 in. and the formation conductivities are 0.1 and 1 S/m (resistivity of 10 and 1 ohm-m), respectively. The operating frequency of the tool is 20 kHz.

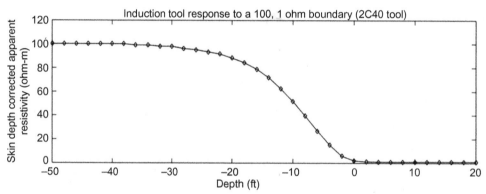

Figure 2.13 Computed apparent conductivity of an induction tool in a two-layer formation. The tool spacing is 40 in. and the formation conductivities are 0.01 and 1 S/m (resistivity of 100 and 1 ohm-m), respectively. The operating frequency of the tool is 20 kHz.

From the above equations, we can compute apparent conductivity for a given tool structure and formations. Fig. 2.12 shows the computed induction tool response in a two-layer formation with conductivity of 0.1 and 1 S/m (resistivity of 10 and 1 ohm-m). The induction tool has a transmitter–receiver separation of 40 in., and coil turns of 1 turn.

Fig. 2.13 shows the same tool and formation geometry but with a relatively higher conductivity contrast. Fig. 2.14 shows the tool response in a three-layer formation when the center layer is 400 in. (33.33 ft) and the shoulder and layer conductivities are 1 and 0.1 S/m. We can clearly see that the tool measures accurately in the center layer. The effect of shoulder conductivity is rather negligible. However, when the conductivity contrast goes higher, as shown in Fig. 2.15, the shoulder effects show up.

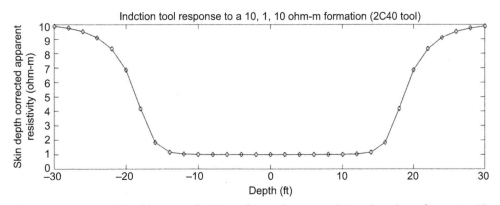

Figure 2.14 Computed apparent conductivity of an induction tool in a three-layer formation. The tool spacing is 40 in. and the formation conductivities are 1, 0.1, and 1 S/m, respectively. The center layer thickness is 32 ft. The operating frequency of the tool is 20 kHz.

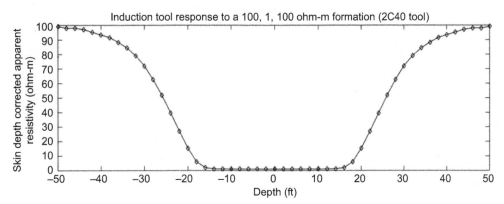

Figure 2.15 Computed apparent conductivity of an induction tool in a three-layer formation. The tool spacing is 40 in. and the formation conductivities are 1, 0.01, and 1 S/m (resistivity of 100, 1, 100 ohm-m), respectively. The center layer thickness is 32 ft. This log shows that the shoulder effect will affect the conductivity measurements when the conductivity contrast is high. The operating frequency of the tool is 20 kHz.

The apparent conductivity measured in the high-resistivity center layer has less reading than the true formation value due to the effects from the low shoulder resistivity. Figs. 2.16 and 2.17 shows that for the same tool, the tool response in a thinner layer formation sandwiched in a two-shoulder bed. We can see that the measured conductivity is greatly affected by the shoulders for a thinner bed.

2.12 INDUCTION ARRAYS

As seen from Figs. 2.14–2.17, the induction tool response to both thin layer and high-resistivity formations are greatly affected by the low-resistivity shoulder beds.

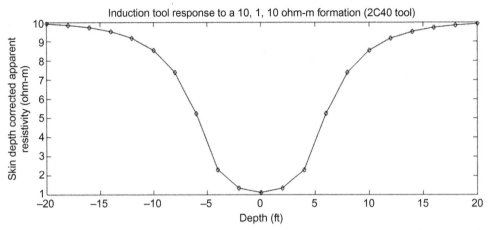

Figure 2.16 Computed apparent conductivity of an induction tool in a three-layer formation. The tool spacing is 40 in. and the formation conductivities are 1, 0.1, and 1 S/m (resistivity of 100, 1, 100 ohm-m), respectively. The center layer thickness is 80 in. The operating frequency of the tool is 20 kHz.

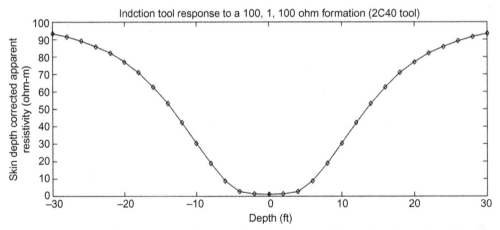

Figure 2.17 Computed apparent conductivity of an induction tool in a three-layer formation. The tool spacing is 40 in. and the formation conductivities are 1, 0.01, and 1 S/m (resistivity of 100, 1, 100 ohm-m), respectively. The center layer thickness is 80 in. The operating frequency of the tool is 20 kHz.

To reduce the shoulder impact to the high-resistivity measurements, more induction coils can be added to the tool. Therefore multicoil induction tools are practically used. The final measurement data is a superposition of the individual measurements. The superposition of the measurements by an array of induction coils with different spacing can greatly improve the tool performance. In this section, we will define the apparent conductivity of a multicoil induction tool and use the algorithms developed in Section 2.11 to investigate the tool response of a multicoil induction tool.

If an induction tool has m transmitters, and n receivers, we use the following notations to describe the tool:

N_{Ti}: number of turns of the ith transmitting coil, positive for forward winding, negative for reverse winding

N_{Rj}: number of turns of the jth receiving coil, positive or negative

L_{ij}: spacing between Ti and Rj

The apparent conductivity for a multicoil induction tool is defined as

$$\sigma_a = \frac{1}{\omega\mu \left(\sum_{i=1}^{m}\sum_{j=1}^{n} \frac{N_{Ti}N_{Rj}}{L_{ij}}\right)} \sum_{i=1}^{m}\sum_{j=1}^{n} N_{Ti}N_{Rj}\text{Re}\{j\text{Im}(Z_{Ti}, Z_{Ri})\} \qquad (2.94)$$

For homogeneous medium:

$$\sigma_a = \frac{2}{\omega\mu \left(\sum_{i=1}^{m}\sum_{j=1}^{n} \frac{N_{Ti}N_{Rj}}{L_{ij}}\right)} \sum_{i=1}^{m}\sum_{j=1}^{n} \frac{N_{Ti}N_{Rj}}{(Lij)^3} \text{Re}\{j(1+jkLij)e^{-jkL_{ij}}\} \qquad (2.95)$$

For homogeneous media and for $|kL_{ij}| \ll 1$ for all L_{ij}:

$$\sigma_a \approx \sigma \left[1 - \frac{2\sum_{i=1}^{m}\sum_{j=1}^{n}(N_{Ti}N_{Rj})}{3d_s \left(\sum_{i=1}^{m}\sum_{j=1}^{n} N_{Ti}N_{Rj}\right)}\right] \qquad (2.96)$$

Skin depth is given in Eq. (2.51) $d_s = \frac{1}{\sqrt{\pi f \mu \sigma}}$ and complex propagation constant is given in Eq. (2.50) $k = \frac{1-j}{d_s}$.

Figs. 2.18–2.20 are the tool response of a multicoil induction tool named 6FF40. The parameters of the tool are shown in Table 2.2 [11]. Comparing the results from 6FF40 and that of a two-coil induction tool, we can clearly see that the response of a multicoil induction tool is closer to the resistivity of the true formation.

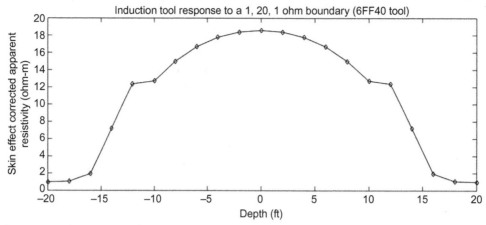

Figure 2.18 The response of a 6FF40 in a three-layer formation. The formation conductivities are 1, 0.05, and 1 S/m (resistivity of 1, 20, 1 ohm-m), respectively. The center layer thickness is 400 in. The operating frequency of the tool is 20 kHz.

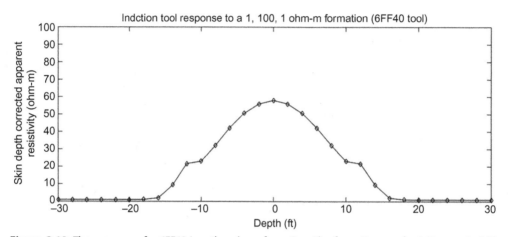

Figure 2.19 The response of a 6FF40 in a three-layer formation. The formation conductivities are 1, 0.01, and 1 S/m, respectively. The center layer thickness is 400 in. The operating frequency of the tool is 20 kHz.

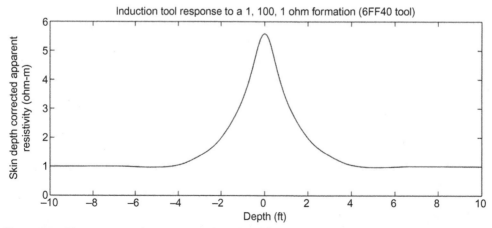

Figure 2.20 The response of a 6FF40 in a three-layer formation. The formation conductivities are 1, 0.01, and 1 S/m (resistivity of 1, 100, 1 ohm-m), respectively. The center layer thickness is 4 ft. The operating frequency of the tool is 20 kHz.

Table 2.2 The antenna parameters of 6FF40 induction logging tool

	Number of turns	Positions (in.)
Transmitters	60	20
	−15	10
	−4	−50
Receivers	60	−20
	−15	−10
	−4	50

REFERENCES

[1] L.C. Shen, J.A. Kong, Applied Electromagnetism, third ed., PWS Series in Engineering, 1995, ISBN-13: 978-0534947224
[2] R.F. Harrington, Time-Harmonic Electromagnetic Fields (McGraw-Hill Texts in Electrical Engineering), McGraw-Hill, New York, NY, 1961.
[3] W.C. Chew, Waves and Fields in Inhomogeneous Media, IEEE Press, New York, 1995.
[4] H.G. Doll, Introduction to induction logging and application to logging of wells drilled with oil base mud, AIME J. Pet. Technol. TP 2641, 1 (6) (1949) 148−162.
[5] W.C. Duesterhoeft Jr., Propagation effects in induction logging, Geophysics 26 (No. 2) (April 1961) 192−204.
[6] W.C. Duesterhoeft Jr., R.E. Hartline, H.S. Thomsen, The effect of coil design in the performance of the induction log, J. Pet. Technol. (1961) 1137−1150.
[7] W.C. Duesterhoeft Jr., Propagation effects on radial response in induction logging, Geophysics 27 (No. 4) (1962) 463−469.
[8] J.H. Morgan, K.S. Kunz, Basic theory of induction logging and application to study of two-coil sondes, Geophysics 27 (No. 6), Part 1 (1962) 829−858
[9] G.B. Afken, H.J. Weber, Mathematical Methods for Physicists, sixth ed., Elsevier, Amsterdam, 2005 (Chapter 9).
[10] G.B. Afken, H.J. Weber, Mathematical Methods for Physicists, sixth ed., Elsevier, Amsterdam, 2005 (Chapter 11).
[11] B.I. Anderson, Modeling and Inversion Methods for the Interpretation of Resistivity Logging Tool Response, DUP Science, Delft, 2001.

CHAPTER 3

Electrical Properties of Sediment Rocks: Mixing Laws and Measurement Methods

Contents

3.1	Resistivity and Dielectric Constant of Rocks	52
3.2	Archie's Law	55
3.3	Mixing Laws	55
	3.3.1 Background review	55
	3.3.2 Mixing formulas	57
3.4	Frequency Dispersion of the Dielectric Constant	64
3.5	Frequency Dispersion of the Conductivity	67
3.6	Measurement Methods of Electrical Properties of Rocks	69
	3.6.1 Background of rock measurements	70
	3.6.2 Parallel-disk measurement method	71
	3.6.3 The circuit model of the parallel-disk sample holder	72
	3.6.4 The circuit parameters of the parallel-disk sample holder	74
	3.6.5 Performance analysis at high frequencies	76
	3.6.6 Computation of the dielectric constant and the conductivity of test samples	77
	3.6.7 Analysis of dynamic range of the parallel-disk technique	78
	3.6.8 Automatic measurement system using the parallel-disk technique	80
	3.6.9 Experimental data and discussions	83
	3.6.10 Differences between the LF and the HF measurements	92
	3.6.11 Error analysis	95
3.7	TM_{010} Resonant Cavity Technique	95
	3.7.1 Theory of the TM_{010} resonant cavity technique	95
	3.7.2 Dynamic range of the resonant cavity technique	98
	3.7.3 Measurement system and experimental data of the resonant cavity technique	104
References		110
Appendix A E Field Analysis of the Circuit Model of the Parallel-Disk Sample Holder		112
Appendix B Equipment Calibration Synopsis		117

3.1 RESISTIVITY AND DIELECTRIC CONSTANT OF ROCKS

As discussed in Chapter 1, Introduction to Well Logging, and Chapter 2, Fundamentals of Electromagnetic Fields Induction Logging Tools, electrical logging is to find the resistivity and dielectric constant of the formation. However, finding these parameters is not the final goal of the oil and gas exploration and production. The purpose of the logging is to help in locating the oil and gas reservoir. To do so, it is essential to study the relations between the resistivity and the physical properties of the formation rocks. In this chapter, we will use analytical and empirical method to analyze the resistivity and dielectric constant of rocks. As we know, the rock is a mixture of rock frame and materials existing in its pore space. The materials in the pore space can be oil, water, and gas and the resistivity of the rock is directly related to the properties of these materials and geometry of the rock pores.

Oil and gas are contained inside various rocks, such as sandstone and carbonate. In general, most oil- and gas-bearing rocks are porous. The pore spaces in rocks are where the oil and gas stored. The greater the pore space is, the higher the possibility of oil and gas reserves. However, there might be water in the pore space. An important task of well logging tools is to measure the porosity of rocks. The logging measurements should include the information about how easily the liquid can flow in the formation, which is defined by the permeability of the rocks. When both porosity and permeability of a formation are known, the next step is classifying the liquid property, e.g., identifying water, oil, and/or gas.

Porosity φ of a rock is defined as the ratio of the pore space volume V_p and the total volume of the rock V_t:

$$\varphi = \frac{V_p}{V_t} \tag{3.1}$$

Rocks with higher porosity usually have a higher capability in bearing oil, water, or gas. To describe the fluidity in a rock formation, the concept of formation permeability has been used in fluid dynamics. The permeability of a rock formation is defined in a similar way that electrical conductivity is defined. Reference to Fig. 3.1, if a pressure ΔP is applied to a rock formation with the thickness of Δx, the fluid velocity in the formation is v, and the fluid viscosity is μ, the permeability k is

$$k = v\mu \frac{\Delta x}{\Delta P} \tag{3.2}$$

Currently, direct measurement of porosity and permeability in the downhole condition is rather difficult. However, using the electrical and other measurements,

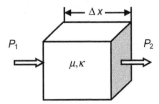

Figure 3.1 Definition of permeability of the formation given in Eq. (3.1), where $\Delta P = P_1 - P_2$.

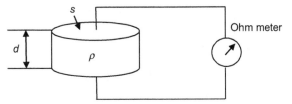

Figure 3.2 Definition of resistivity measurements using a simple parallel-plate pair.

these parameters can be determined indirectly. As we know, electrical resistivity is a parameter to measure the number of free moving electrons and their mobility in a material. Therefore the resistivity of a composite material such as sediment rock is largely dependent on the resistivity of each component in the rock such as the pore space and the liquid or gas that occupies the pore space. From a measurement point of view, the electric resistivity of a material can be simply measured by a pair of conducting plates. The resistivity of a bulk material is defined as follows:

$$\rho = \frac{RS}{d} \qquad (3.3)$$

where ρ is the resistivity of the material, R is the resistance measured by an ohm meter, S is the surface area of the electrodes, and d is the distance between two electrodes as shown in Fig. 3.2.

The measurement set up shown in Fig. 3.2 actually provides a useable method in the lab conditions where the rock samples with desired shapes and sizes are available. However, in a downhole situation, other methods must be employed to obtain the formation resistivity. We will discuss various logging methods to obtain the resistivity downhole in the later chapters. Since the purpose of logging is to derive the mechanical and physical properties from the electric property measurements, in this section, we will discuss the relations between the electric properties of rocks with their mechanical and other physical properties. The electric properties mentioned here mainly include resistivity and dielectric constant.

Reference to Chapter 2, Fundamentals of Electromagnetic Fields Induction Logging Tools, we know that electromagnetic (EM) fields, the dielectric constant, and the resistivity of a material are correlated by Maxwell equations as described by Eq. (2.15):

$$\nabla \times \overline{H} = \overline{J} + j\omega\varepsilon\, \overline{E}$$
$$= j\omega\left(\varepsilon - \frac{j\sigma}{\omega}\right)\overline{E}$$
$$= j\omega\varepsilon^*\overline{E}$$

where $\varepsilon^* = (\varepsilon - j\sigma/\omega)$ is referred to as the complex dielectric constant of the material, in which each of the dielectric constant and the conductivity takes a counterpart. In the case of high frequency (megahertz to gigahertz) or nonconductive media, the dielectric constant plays a major role in the complex dielectric constant. While in the low frequency case, the conductivity dominates. The electric logging tools are mainly designed to measure either the resistivity or the dielectric constant of downhole formations. Generally, the downhole formation is composed of different types of materials, including rock matrix, water, oil, and others. It is a mixture of various materials. Hence the concepts of effective dielectric constant and effective resistivity have been introduced to describe the overall dielectric constant and overall resistivity of mixed materials.

If give a thought to the resistivity of a formation, one can easily find that there are several factors that contribute to the effective resistivity of the rock formation such as the porosity, permeability, fluids, or gases in the formation, and the microstructure of the rocks. It has been one of the most interesting research subject in the well logging industry in the past a few decades to establish a mathematical correlation between these factors and the effective resistivity of the rock formation. The study of the resistivity of rocks is generally divided into two approaches. The first approach is based on experimental data to establish empirical formulas, and the second approach lies in building structural models of rock and theoretically calculating rock's resistivity. Early study on rock properties is limited to empirical methods [1,2] due to limited equipment capability. More recent research uses advanced CT image of the rocks (so-called digital rocks) [3] to facilitate the building of rock's structural models. Analytical methods for microscopic analysis of rock resistivity generally assume a periodic structure or random structure derived from crystals in solid physics [4].

Since the resistivity and the dielectric constant each composes a counterpart in the complex dielectric constant, the methods that are developed to determine the effective conductivity of a composite material can also be applied to calculate the effective dielectric constant of the material, and vice versa. Therefore in the following theoretical discussions, we may concentrate on one of the two parameters.

3.2 ARCHIE'S LAW

From a petrophysicist's point of view, the important issue is not the resistivity of the rock itself, rather the porosity, density, permeability, and contents of the rocks. Since the direct measurement of these physical parameters is difficult, the indirect measurement methods have been developed and well-accepted. The so-called indirect measurement methods usually include the following steps. The first step is to construct a mathematical model/correlation between rock's resistivity and its other physical properties of interest. The second step is the measurement of rock's resistivity. And in the third step the physical properties of rocks are extracted from the built mathematical model and the measured electric properties. One of the most popular mathematical models used in the well logging industry is Archie's Law [1]. Archie's Law is an empirical formula developed in early 1940s by Archie and Leverett [1,2]. The law explains that the resistivity of the rock is a function of the resistivity of water saturation, porosity, and other parameters such as the microscopic structure of the rock. Through numerous experiments, Archie summarized the relations as follows:

$$S_w^n = \frac{a}{\varphi^m} \frac{R_w}{R_t} \quad (3.4)$$

S_w is the water saturation; φ is the porosity; R_w is the resistivity of the water content in the rock; R_t is the resistivity of the rock; a is the tortuosity constant determined by the connectivity of the pore spaces of a rock; n is the saturation exponent; and m is the cementation factor. These factors are largely related to the microgeometry of the rock and are unknown constants to be determined for a given rock sample. For clean rock $m = n = 2$, $a = 1$. In general, m is in the range 1.3–2.6; a is in the range of 0.5–1.5, and n is close to 2. For unconsolidated sand, m is a number close to 1.3; n is close to $2m$; and a is about 1. These constants are dependent on the rock microstructure, clay contents, and rock itself.

Once the rock microstructure-related constants are obtained and the resistivities R_t and R_w are measured, the water-saturated porosity of the rock can be derived from Archie's Law. Detailed discussion of Eq. (3.4) can be found in Refs. [1,2].

3.3 MIXING LAWS

3.3.1 Background review

A mixing law is basically a formula or algorithm used to calculate the effective dielectric constant or conductivity of a multicomponent mixture based on the parameters of each component and microgeometry of the mixture. In most cases, an oil- or gas-bearing reservoir generally consists of three components: rock grain, oil, and water.

The effective dielectric constant or conductivity of the rock is a function of the dielectric constant or conductivity of each component, the microgeometry of the rock frame, and the volume fractions of the components. McPhedran and McKenzie [4] studied the effective conductivity of simple cubic, body-centered, and face-centered lattice when porosity is greater than 38%. The porosity of such a two-component material is defined as the volume fraction of the spaces not occupied by the inclusions and the total volume of the composite. Analytical methods are also described by Bergman [5] for calculating the dielectric constant of a two-phase composite consisting of a simple cubic array of identical spherical inclusions embedded in a homogeneous host. A spectral representation is derived for the effective dielectric constant and numerical results are presented for the poles and residues.

The dielectric response of water-saturated rocks based on a realistic model of the pore space was proposed by Sen et al. [6]. They presented data from glass beads for DC conductivity and the dielectric constant at 1.1 GHz. Garrouch and Sharma [7] discussed the influence of clay content, salinity, stress, and wettability on the dielectric constant of brine-saturated rocks in a frequency range of 10 Hz to 10 MHz.

Based on these experiments, the effect of rock/water interaction on the dielectric behavior of saturated sandstones was shown by Knight and Nur [8]; Knight and Endres [9]; and Knight and Abad [10]. In these studies, the dielectric constant of sandstone samples was measured as a function of water saturation in the frequency range of 60 kHz to 4 MHz. It was experimentally shown that rock/water interaction had significant effects on the effective dielectric constant. The experiments on hydrophobic rock samples confirmed that changes in the dielectric constant of sandstone at low saturation are caused by rock/water interaction. This study illustrated the importance of the chemical state of the rock surface in determining the dielectric behavior of sandstone. Shen et al. [11] discussed existing mixing formulas at UHF band using the measured data.

Numerical techniques for computation of the effective dielectric constant of a two-component rock have been developed by Shen et al. [12] and Liu and Shen [13]. They proposed a method to calculate the effective dielectric constant of two-component periodic composite material using a Fourier series expansion technique. Four types of microstructures are investigated: simple, body, face-centered cubic lattices (Fig. 3.3), and simple cubic lattice with hyperboloidal coating surfaces, with a porosity ranging from 0 to 100%.

Liu and Zou [14] presented a method to compute the effective dielectric constant of a two-component, three-dimensional mixture with geometric symmetry using a simplified Fourier series expansion. Liu and Wu [15] proposed a mixing formula using Bergman—Milton simple pole theory. The numerical methods show a satisfactory accuracy in predicting the effective dielectric constant for the simple cubic, body-centered, and face-centered lattices with simple formulas.

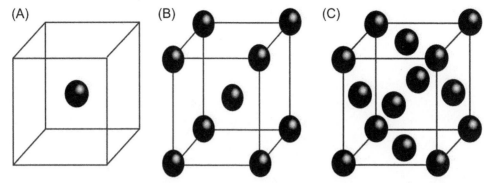

Figure 3.3 Theoretical rock models based on crystal structures used in solid-state physics. (A) Simple lattice structure, (B) body-centered structure, and (C) face-centered structure.

However an oil reservoir often consists of three different substances. For example, a sandstone reservoir may contain sand rock grain, oil, and water; each of which has very different dielectric constants.

To calculate the effective dielectric constant of a sediment rock, it is necessary to compute the effective dielectric constant of a three-component material. To model a real reservoir bearing oil, we assume that sandstone consists of three components: sand grain, oil, and water, and that the porosity ranges from 0 to 50% with water saturation ranging from 0 to 100%. A unit sandstone model is shown in Fig. 3.3. In the following discussions, the dielectric constant of the rock grain can be represented as ε_g with its value of 4.65. ε_h and ε_w are dielectric constants of oil and water with values of 2.2 and 78.5, respectively.

3.3.2 Mixing formulas

Many mixing formulas have been developed to calculate the effective dielectric constant of a mixture of materials with two or three components. Each formula is based on an assumption of the geometry of the mixture. These mixing formulas are discussed in the following sections.

3.3.2.1 Bruggeman–Hanai–Sen (BHS) formula

Under the Theory of Disordered Systems in solid-state physics and under the assumption of static fields, Sen et al. derived the following mixture formula for a two-component system:

$$\left(\frac{\varepsilon_{\mathit{eff}} - \varepsilon_{c2}}{\varepsilon_{c1} - \varepsilon_{c2}}\right)\left(\frac{\varepsilon_{c1}}{\varepsilon_{\mathit{eff}}}\right)^c = \varphi \qquad (3.5)$$

c is a constant determined by the geometrical shape of the rock grain. If the rock grain is spherical, $c = 1/3$; ε_{c1} and ε_{c2} are the dielectric constants of medium 1 and 2, respectively, ε_{eff} is the effective dielectric constant of the mixture with two components, and φ is the porosity of the sandstone and is defined as the volume ratio of medium 2 and the total volume of the rock.

To obtain the effective dielectric constant of a three-component mixture, Eq. (3.5) is used twice. At first, the dielectric constant of the mixture of oil and water is calculated, and then the sandstone with grain and the oil—water mixture is calculated.

For the mixture of water and oil, we must take the distribution of fluids into consideration. Generally, water or oil may exist in pore space in different forms. If oil exists as a dispersed phase in pore space, then water is the first medium ($\varepsilon_{c1} = \varepsilon_w$), and oil is the second medium ($\varepsilon_{c2} = \varepsilon_h$), then Eq. (3.5) can be written as Eq. (3.5a). However, if water exists as a dispersed phase in pore space, then oil is the first medium ($\varepsilon_{c1} = \varepsilon_h$), and water is the second medium ($\varepsilon_{c2} = \varepsilon_w$), then Eq. (3.5a′) is obtained.

When take the oil—water mixture as one component, the sandstone becomes a two-component system with grain and oil—water mixture. Eq. (3.5) can be written as Eq. (3.5b).

$$\left(\frac{\varepsilon_{wh} - \varepsilon_h}{\varepsilon_w - \varepsilon_h}\right)\left(\frac{\varepsilon_w}{\varepsilon_{wh}}\right)^c = S_w \tag{3.5a}$$

$$\left(\frac{\varepsilon_{wh} - \varepsilon_w}{\varepsilon_h - \varepsilon_w}\right)\left(\frac{\varepsilon_h}{\varepsilon_{wh}}\right)^c = 1 - S_w \tag{3.5a′}$$

$$\left(\frac{\varepsilon_{\text{eff}} - \varepsilon_g}{\varepsilon_{wh} - \varepsilon_g}\right)\left(\frac{\varepsilon_{wh}}{\varepsilon_{\text{eff}}}\right)^c = \varphi \tag{3.5b}$$

ε_{wh} is the dielectric constant of the mixture fluid with oil and water, ε_h is the dielectric constant of oil, ε_w is the dielectric constant of water, ε_g is the dielectric constant of rock grain, and S_w is water saturation—the ratio of water volume to the volume of total pore space.

Based on Eq. (3.5a) or (3.5a′), the dielectric constant of oil—water mixture can be obtained. According to Eq. (3.5b), we can calculate the effective dielectric constant of sandstone containing rock grain, oil, and water.

Fig. 3.4 is the result from Bruggeman—Hanai—Sen (BHS) formula. Fig. 3.4 shows the result from Eq. (3.5b), and it illustrates a linear relationship between ε_{eff}, ε_{wh}, and ε_{eff} increases smoothly with ε_{wh} at the same porosity.

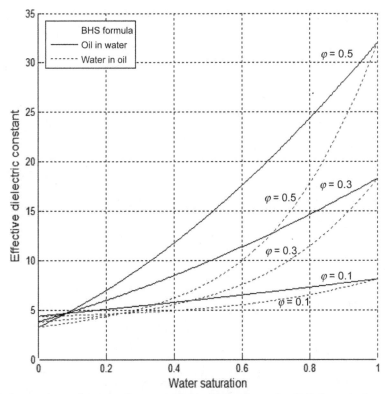

Figure 3.4 Comparison of results from two-fluid distributions: the BHS formula is used for the computation. The *solid line* is the case of oil in water; and *dashed line* for the case of water in oil. Porosity range is of 10–50%.

3.3.2.2 Lorentz–Lorenz, Clausius–Mossotti (LLCM) formula

Based on the dilute assumption in which the EM mutual couplings among rock grains are ignored, Lorentz–Lorenz and Clausius–Mossotti presented a formula to calculate the effective dielectric constant of a mixture of two components:

$$\frac{\varepsilon_{eff} - \varepsilon_{c1}}{\varepsilon_{eff} + 2\varepsilon_{c1}} = \frac{\varepsilon_{c2} - \varepsilon_{c1}}{\varepsilon_{c2} + 2\varepsilon_{c1}}(1 - \varphi) \tag{3.6}$$

where, ε_{c1} is the dielectric constant of pore material, ε_{c2} is that of rock frame, and ϕ is the porosity. Similarly, we use Eq. (3.6) twice to calculate the effective dielectric constant of a mixture with three components. Based on Eq. (3.6), Fig. 3.5 is obtained, and shows a linear relationship between the effective dielectric constant of sandstone (ε_{eff}) and that of oil–water mixture (ε_{wh}) in pore space. Fig. 3.5 shows the computed effective dielectric constant of sandstones when the water saturation changes using LLCM formula. The *solid line* is the case of oil existing in pore space as a dispersed phase while the *dashed line* is the case of water existing in pore space as a dispersed phase. An almost linear relationship between ε_{eff} and water saturation S_w is seen for the case of oil in water. However a different relationship between ε_{eff} and S_w is seen for the case of water in oil where ε_{eff} changes very little with S_w at low water saturation, but rises sharply with S_w at very high water saturation (>60%).

3.3.2.3 The complex refractive index method formula

The complex refractive index method (CRIM) assumes that the complex refractive index of the mixture, $\sqrt{\varepsilon_{eff}}$, is the volumetric sum of the complex refractive indexes of all components. For a three-component mixture consisting of water (ε_w), rock grain (ε_g), and oil (ε_h), the CRIM formula is as follows:

$$\sqrt{\varepsilon_{eff}} = \varphi S_w \sqrt{\varepsilon_w} + (1 - \varphi)\sqrt{\varepsilon_g} + \varphi(1 - S_w)\sqrt{\varepsilon_h} \qquad (3.7)$$

The CRIM formula can be readily derived from the assumption of two components in parallel. Consider a two-layer formation with a plane wave impinging from

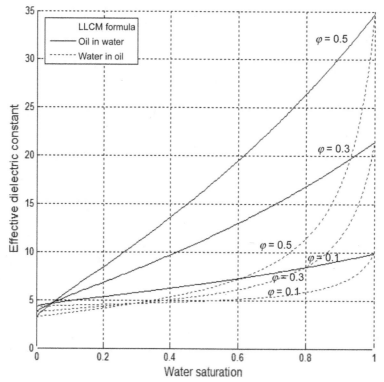

Figure 3.5 Comparison of results from two-fluid distributions. The LLCM formula is used for the computation. The *solid line* is the case of oil in water; and *dashed line* for the case of water in oil.

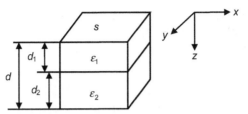

Figure 3.6 A two-layer model for the derivation of CRIM formula.

the top layer and propagating in the direction perpendicular to the layers as shown in Fig. 3.6. The incident E field can be written as

$$E_y(z) = E_0 e^{-jkz}$$

k is the wave number of the media.

At the interface between two layers, the E field is continuous and the field can be written as

$$E_y(d_1) = E_0 e^{-jk_1 d_1} \tag{3.8a}$$

At the exit of layer 2, the field is

$$E_y(d_2) = E_0 e^{-j(k_1 d_1 + k_2 d_2)} \tag{3.8b}$$

Consider the two-layer media as an effective media with an effective dielectric constant of ε_{eff} in the equivalent length d of wave propagation, then the field can be written as:

$$E_y(d) = E_0 e^{-jk_{eff} d} \tag{3.8c}$$

Compare Eqs. (3.8b) and (3.8c) we will have:

$$k_{eff} d = k_1 d_1 + k_2 d_2 \tag{3.8d}$$

Note that

$$k = \frac{2\pi}{\lambda} = \frac{2\pi}{\lambda_0} \sqrt{\varepsilon_r}$$

Apply the above relation to Eq. (3.8d), we have,

$$\sqrt{\varepsilon_{eff}} = \sqrt{\varepsilon_1} \frac{d_1 s}{ds} + \sqrt{\varepsilon_1} \frac{d_2 s}{ds}$$

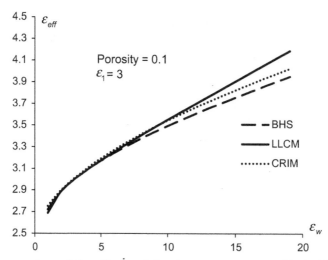

Figure 3.7 The comparison of the effective dielectric constant computed by LLCM, BHS, and CRIM formulas as a function of the dielectric constant of the pore material when the porosity of the material is 0.1.

Since $d_1 s/ds$ is equal to the volume ratio of the material 1; $d_2 s/ds$ is equal to the volume ratio of material 2, the above equation can be written as:

$$\sqrt{\varepsilon_{\text{eff}}} = \sqrt{\varepsilon_1}\frac{d_1 s}{ds} + \sqrt{\varepsilon_1}\frac{d_2 s}{ds} = \sqrt{\varepsilon_1}\varphi + \sqrt{\varepsilon_2}(1-\varphi) \qquad (3.8e)$$

where $\varphi = v_1/v$ is the volume ratio of the material 1.

Similarly, if the model shown in Fig. 3.6 is expanded into n layers, Eq. (3.8e) becomes:

$$\sqrt{\varepsilon_{\text{eff}}} = \sum_{n=1}^{N} \sqrt{\varepsilon_n}\varphi_n \qquad (3.8f)$$

where ε_n and φ_n are the dielectric constant and volume ratio of the material n.

From Eq. (3.8f), Eq. (3.9) can be readily obtained.

Figs. 3.7 and 3.8 show the comparison of the effective dielectric constants as a function of the dielectric constant of the material in the pore space when the porosity of the material is 0.1 and 0.3, respectively. From figure shown, we can conclude that the values of the effective dielectric constant from these three algorithms are very close to each other when the dielectric constant of the pore material is relatively small. However, they differ a bit from each other when the dielectric constant of the pore material increases. For a given porosity and dielectric constant of the rock frame, the LLCM mixing formula gives the highest effective dielectric constant, CRIM is the next and BHS gives the least.

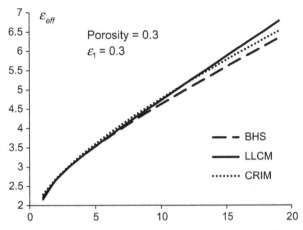

Figure 3.8 The comparison of the effective dielectric constant computed by LLCM, BHS, and CRIM formulas as a function of the dielectric constant of the pore material when the porosity of the material is 0.3.

3.3.2.4 Lichtnecker–Rother formula

The Lichtnecker–Rother (LR) formula is derived from a three-layer model when stratified geometry is assumed as discussed in CRIM formula. However, the square root of the dielectric constant is modified to an adjustable root value c and c is called cementation factor. Eq. (3.9) is the LR formula

$$\varepsilon_{\mathit{eff}}^{c} = \phi S_{w}\varepsilon_{w}^{c} + (1-\phi)\varepsilon_{g}^{c} + \phi(1-S_{w})\varepsilon_{h}^{c} \tag{3.9}$$

The exponent c is called the cementation factor. Eq. (3.9) reduces CRIM formula when $c = 0.5$. In most of the cases, the cementation factor $c = 1/3$.

All the formulas discussed earlier can provide reasonable approximations of the effective dielectric constant of the mixture with three components. Curves of $\varepsilon_{\mathit{eff}}$ versus S_w by different formulas are shown on Fig. 3.9. The relationships between $\varepsilon_{\mathit{eff}}$ and S_w from four algorithms are similar to each other as shown in Fig. 3.9A and B. In each case, the effective dielectric constant of sandstone increases with water saturation when the porosity maintains unchanged. Comparing Fig. 3.9A and B, it can be seen that the higher the formation porosity, the larger the effective dielectric constant could reach. An interesting result is obtained when ε_{c1} and ε_{c2} of Eq. (3.5a) or (3.5a′) are exchanged and used to calculate the dielectric constant of oil–water mixture. In this case, quite different relationships between $\varepsilon_{\mathit{eff}}$ and S_w are observed using BHS or LLCM formulas.

$\varepsilon_{c1} = \varepsilon_w$ and $\varepsilon_{c2} = \varepsilon_h$ mean that oil exists in dispersed form in pore space and the sandstone is water-wet. $\varepsilon_{c1} = \varepsilon_h$ and $\varepsilon_{c2} = \varepsilon_w$ mean that water exists in dispersed form in pore space, and that the sandstone is oil-wet.

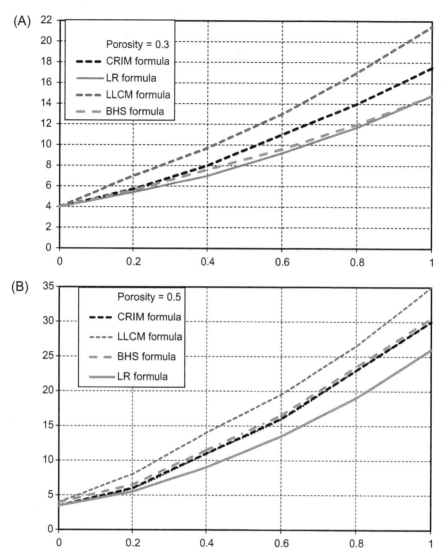

Figure 3.9 Comparison of ε_{eff} versus S_w by different by different algorithms. (A) Porosity = 0.3, (B) porosity = 0.5.

3.4 FREQUENCY DISPERSION OF THE DIELECTRIC CONSTANT

In general, any material will have dielectric constant and conductivity at the same time. In other words, the field applied to the material will generate both a conducting current and a displacement current. The conducting current is mainly determined by the conductivity, whereas the displacement current is controlled by dielectric constant of the material. We know that at low frequencies, the dielectric constant effect can be ignored since the displacement current is rather weak compared to the conductive current. The displacement current increases as frequency goes up. When frequency further increases, the dielectric constant will split into real part and imaginary part itself due to polarization effect [16].

The frequency dependency properties of dielectric constant of solid materials are studied by many researchers [17–19]. In this section, we discuss two popular theories: the Debye model and the Cole–Cole model. Debye model is expressed as follows:

$$\varepsilon^*(\omega) = \varepsilon' - j\varepsilon'' = \varepsilon_\infty + \frac{\varepsilon_s + \varepsilon_\infty}{1 + j\omega\tau} \qquad (3.10)$$

where ε' is the real part of the complex dielectric constant, ε'' is the imaginary part of the complex dielectric constant, ε_s is the dielectric constant at low frequencies, ε_∞ is the dielectric constant at very high frequencies, ω is the angular frequency, α is the relaxation constant, and τ is the time constant.

To better describe the frequency dependency of the dielectric constant of a material, Cole presented a slightly different model, the Cole–Cole model [20,21]. The Cole–Cole model is widely used since this model has more parameters and better fits to the measured data. The Cole–Cole model is given as:

$$\varepsilon^*(\omega) = \varepsilon' - j\varepsilon'' = \varepsilon_\infty + \frac{\varepsilon_s - \varepsilon_\infty}{1 + (j\omega\tau)^{1-\alpha}} \qquad (3.11)$$

where, α is called relaxation constant and the rest of the parameters are the same as in the Debye model (3.9).

To study the effects of these parameters, we can plot the effective dielectric constant as a function of the frequency. Fig. 3.10 shows the effective dielectric

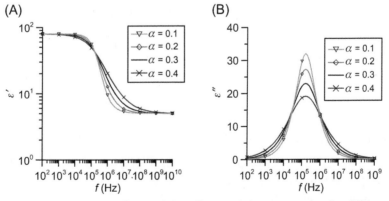

Figure 3.10 The frequency dependency of the effective dielectric constant for different values of α. (A) Real part of the effective dielectric constant and (B) imaginary part of the effective dielectric constant ($\varepsilon_\infty = 5$, $\varepsilon_0 = 80$, $\tau = 0.000001$ s).

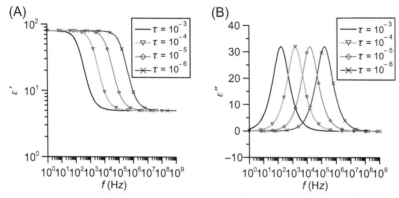

Figure 3.11 The frequency dependency of the effective dielectric constant for different values of τ. (A) Real part of the effective dielectric constant and (B) imaginary part of the effective dielectric constant ($\varepsilon_\infty = 5$, $\varepsilon_0 = 80$, $\alpha = 0.1$).

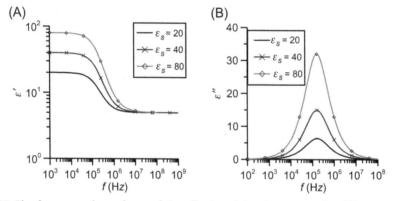

Figure 3.12 The frequency dependency of the effective dielectric constant for different values of ε_s. (A) Real part of the effective dielectric constant and (B) imaginary part of the effective dielectric constant ($\varepsilon_\infty = 5$, $\alpha = 0.1$, $\tau = 0.000001$ s).

constant at different frequencies when the relaxation constant changes. It is clearly shown that the value of α controls the bandwidth of the effective dielectric constant, e.g., a smaller α results in narrower bandwidth. The frequency dependency of the real part of the effective dielectric constant is more like a low-pass filter. However the imaginary part behaves more like a bandpass filter. Fig. 3.11 plots the effects of time constant τ. The value of t determines the cutoff frequency for the real part of the effective dielectric constant and the center frequency of the imaginary part. As τ reduces, the cutoff frequency of the real part of the effective dielectric constant increases, or equivalently, the center frequency of the imaginary part of the effective dielectric constant rises. In Fig. 3.12, the effect of the ε_s is depicted. We can see that

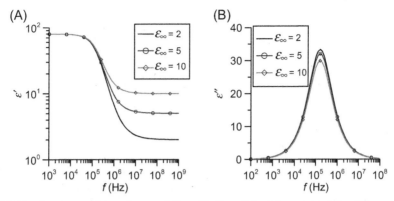

Figure 3.13 The frequency dependency of the effective dielectric constant for different values of ε_∞. (A) Real part of the effective dielectric constant and (B) imaginary part of the effective dielectric constant ($\varepsilon_0 = 80$, $\alpha = 0.1$, $\tau = 0.000001$ s).

the value of the ε_s gives the initial low frequency value of the real part of the dielectric constant. Similarly, as shown in Fig. 3.13, the value of ε_∞ determines the final value of the real part of the effective dielectric constant at high frequencies.

3.5 FREQUENCY DISPERSION OF THE CONDUCTIVITY

Pelton [22] developed a Cole–Cole model for the resistivity dispersion with frequency:

$$\rho^*(\omega) = \rho' + j\rho'' = \rho_0 \left[1 - m\left(1 - \frac{1}{1+(j\omega\tau)^c}\right)\right] \quad (3.12)$$

where ρ_0 is the resistivity at DC, τ is the average relaxation factor, c is the microgeometric factor, and m is the polarization constant, $j = \sqrt{-1}$. As for complex conductivity, we can obtain:

$$\sigma^*(\omega) = \sigma' + j\sigma'' = \sigma_0 \left[1 + m\left(\frac{(j\omega\tau)^c}{1+(j\omega\tau)^c(1-m)}\right)\right] \quad (3.13)$$

In Eq. (3.13), σ' is the real part of the conductivity, σ'' is the imaginary conductivity, σ_0 is the DC conductivity, τ is the average relaxation factor, c is the geometric factor, and m is the polarization constant.

To analyze the effective resistivity with respect to different parameters, we can plot the frequency dispersions of the effective resistivity or conductivity as we did for the effective dielectric constant in Figs. 3.10–3.13. These features are plotted in Figs. 3.14–3.17. We can see that the shapes of these figures are very much the same as in Figs. 3.10–3.13. For example, Fig. 3.14 shows the value of effective conductivity

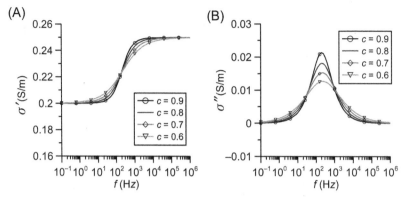

Figure 3.14 The effective conductivity as a function of microgeometric factor c. (A) Real part of effective conductivity and (B) imaginary part of the effective conductivity ($\sigma_0 = 0.2$ S/m, $m = 0.2$, $\tau = 0.001$ s).

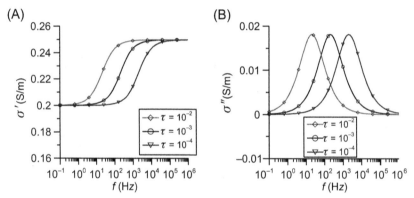

Figure 3.15 The effective conductivity as a function of relaxation factor τ. (A) Real part of effective conductivity and (B) imaginary part of the effective conductivity ($\sigma_0 = 0.2$ S/m, $m = 0.2$, $c = 0.8$).

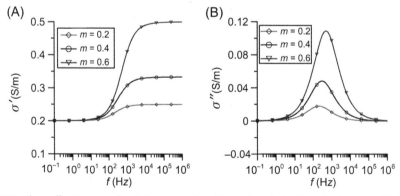

Figure 3.16 The effective conductivity as a function of polarization factor m. (A) Real part of effective conductivity and (B) imaginary part of the effective conductivity ($\sigma_0 = 0.2$ S/m, $\tau = 0.001$ s, $c = 0.8$).

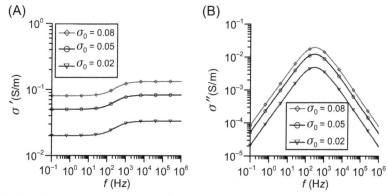

Figure 3.17 The effective conductivity as a function of DC conductivity σ_0. (A) Real part of effective conductivity and (B) imaginary part of the effective conductivity ($m = 0.4$, $\tau = 0.001$ s, $c = 0.8$).

as a function of frequency with different values of microgeometric factor c. We can see that the values of c affect the bandwidth. Figs. 3.15–3.17 illustrate the changes of the frequency dispersions with respect to τ, m, and σ_0, respectively.

From the above analysis, we can see that the frequency dependency of the rocks is largely related to the microstructure and the contents of the rocks. Using the frequency dependency characteristics, it is possible to develop new logging tools that can scan through a wide frequency band instead of a few discrete frequency points and obtain a spectrum information of the formation. Using the spectrum, more information of the rocks such as the microgeometry, water content, and materials in the rocks could be derived. Recent development in the logging tools, such as Schlumberger's Dielectric Scanner tool, offers multifrequency data at high frequency band up to 1 GHz and shows the trend to the spectrum EM measurement of formations in the logging industry.

3.6 MEASUREMENT METHODS OF ELECTRICAL PROPERTIES OF ROCKS

In the previous sections, we discussed the theoretical aspect of the electrical properties of rocks as well as the empirical correlations between rock's electrical properties and mechanical properties, which provide us with useful means to indirectly measure rock's mechanical properties from its electrical measurements. However, due to the variety of rocks, the electrical properties of rocks cannot be computed directly from the theory. To derive the mechanical properties of rocks, it is necessary to physically measure the dielectric constant and conductivity of rocks under different conditions. In this section, several measurement methods covering a frequency range of 10 kHz–1.1 GHz for electrical properties of rock samples will be discussed. Measurement examples of different rocks, such as sandstones with different porosities, limestone, asphalt, saline solutions with different salinities, and other samples will be presented.

3.6.1 Background of rock measurements

As we have discussed in the previous chapters, the main objective of well logging industry is to discover and to quantify hydrocarbon reservoirs. The use of the relative dielectric constant and the conductivity as diagnostic parameters in well logging has become a subject of great interest over the years. Several techniques have been developed to measure the electrical properties of rocks in the Lab, such as the coaxial line technique [23–26], the waveguide technique [27], the two-electrode method [28], the parallel-disk sample holder method [29–30], and the resonant cavity method.

In most cases, the coaxial line method uses a one-port coaxial-type device and measures the power reflection from the sample inside the device. Inversion of the measured data is usually done by approximating the impedance mismatch at the sample, and consequently relates the measured capacitance to the permittivity of the sample [23]. The coaxial line measurement method is reviewed by Stuchly [23], used by Poley [24] to measure sandstone samples up to 1.2 GHz and Tam [28] to measure nine different dry rocks from 150 Hz to 1 GHz.

The waveguide method used in the permittivity measurement employs a circular/rectangular waveguide in which the sample is placed. The sample is inserted into the waveguide and the scattering parameters of the rock sample in the waveguide can be measured. With appropriate calibration of the waveguide, the electric properties can be obtained from the measured scattering parameters. Roberts and Hippel [31] used this method to measure various materials in a frequency range of 500 MHz to 1.1 GHz.

The two-electrode technique uses two electrodes with platinum electrodes clamped onto the flat faces of the disk-shaped samples [28]. Electrode polarization effects have frequency limitations. Tam [28] used this method and measured eight different sandstone samples.

The parallel-disk sample holder method is based on parallel-capacitor principle. The sample holder consists of two parallel brass disks to sandwich the sample in the middle to be measured. The diameter of the disks must be much greater than the separation between them to reduce edge effect [32]. Good contact must be maintained between the conducting disks and the sample by applying proper pressure to the disks [29]. Compared to the coaxial line method, the parallel-disk holder method requires less precise machining of the sample. For many reservoir rocks, precise sample machining is difficult.

The resonant cavity method makes use of a cavity resonator that is actually a volume enclosed by metallic walls except for two small coaxial cable entrances, one for a probe to couple electromagnetic power into the volume and the other for a probe to detect the electromagnetic fields inside the cavity. The core sample inside its sample holder stands in the middle of the cavity. Compared to other methods, the resonant cavity technique requires a simple sample machining procedure. It can also measure unconsolidated materials, like fluids [33]. Similarly, good contact between the core sample and the cover of the cavity is critical for accurate measurement. One disadvantage is that the sample can only be measured at a single frequency, which is the resonant frequency of the cavity and it is determined by the physical size of the cavity [33].

To select proper measurement technique for your application, the following aspects need to be taken into account, such as limitation on frequency range, difficulties in sample machining, application of pressure and temperature, and wave polarization. If a sample is anisotropic, the polarization of the E field in the sample holder must also be considered. For example, the parallel-disk method has an E field applied in the sample's axial direction which is perpendicular to the surfaces of the two electrodes. Therefore the electrical properties along sample's axial direction will be measured. However, in the coaxial line method, the E field is in the radial direction, and the measured dielectric constant and conductivity are the sample properties in the radial direction. To understand deeper in the electrical property measurement technique, two typical methods will be discussed in detail below.

3.6.2 Parallel-disk measurement method

The parallel-disk technique is commonly used for measuring the complex dielectric permittivity of dissipative materials, such as reservoir rocks, soil, and saline solutions [32]. This parallel-disk sample holder consists of two parallel brass disks with the diameter much greater than the separation between them [32]. The sample is machined and inserted between the disks.

The advantages of this technique are
- Wide frequency range: according to current apparatus, it can work from 10 kHz to 1 GHz;
- Easy sample machining and preparation;
- High accuracy;
- Both the relative dielectric constant and the conductivity of the test sample can be determined in one measurement.

The parallel-disk sample holder was designed and constructed in the machine shop in the Electrical Engineering Department at the University of Houston [32]. The physical structure is shown in Fig. 3.18. The sample holder is designed to provide fixed parasitic parameters, which can be calibrated accurately. Apparently, thinner sample thickness will have less fringing effects and therefore, higher measurement frequency limit and accuracy. However, making sample too thin results in mechanical difficulties, especially, for rocks that are fragile such as shale. The simplified schematic diagram of the sample holder is shown in Fig. 3.18 [32].

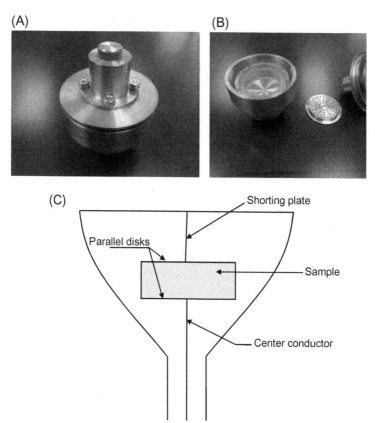

Figure 3.18 Parallel-plate sample holder and simplified schematic diagram. (A) Parallel-plate sample holder, (B) internal structure of the sample holder, and (C) simplified schematic diagram.

The test sample can be sandwiched between the two disks. The bottom disk is connected to the center connector of a coaxial cable; the upper one is connected to the electrical ground. The two disks and the outside hemispherical shell are all made of brass. The thickness of the test sample is 0.3125 in. and radius of the test sample is 1.000 in. [32].

3.6.3 The circuit model of the parallel-disk sample holder

Considering the effect of the metal shielding, the sample holder can be modeled by a π network as shown in Fig. 3.19 [32]. When the unknown test sample is placed into the sample holder, the admittance measured at the measurement plane is equal to the admittance of the unknown test sample plus the effect of the π network. Y_1, Y_2, and Y_3 are three coefficients of the π network, which can be determined using three standard samples (air, brass, and plexiglass) [32]. The parameters of the three standard samples are known.

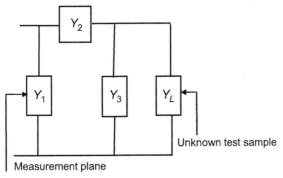

Figure 3.19 The π network circuit model.

The admittance measured at the measurement plane can be written as [32],

$$Y_m = Y_1 + (Y_L + Y_3)//Y_2 = Y_1 + \frac{Y_2 Y_3 + Y_2 Y_L}{Y_2 + Y_3 + Y_L} \quad (3.14)$$

where
  Y_L = the admittance due to the contribution of the sample between the two disks;
  Y_m = the measured values at the measurement plane;
  Y_1, Y_2, and Y_3 are three coefficients of the π network.

When the three standard samples (air, brass, and plexiglass) are respectively put into the sample holder, the measured admittances can be expressed as [32],

$$Y_{ma} = Y_1 + \frac{Y_2 Y_3 + Y_2 Y_a}{Y_2 + Y_3 + Y_a} \quad (3.15)$$

$$Y_{mb} = Y_1 + Y_2 \quad (3.16)$$

$$Y_{mp} = Y_1 + \frac{Y_2 Y_3 + Y_2 Y_p}{Y_2 + Y_3 + Y_p} \quad (3.17)$$

where
  Y_{ma} = the measured admittance of the air sample;
  Y_{mb} = the measured admittance of the brass sample;
  Y_{mp} = the measured admittance of the plexiglass sample;
  Y_a = the admittance of the air sample;
  Y_p = the admittance of the plexiglass sample.

The expressions of Y_1, Y_2, and Y_3 can be derived from Eqs. (3.15)–(3.17) [32],

$$Y_2 = \left[\frac{Y_{mp} - Y_{ma}}{Y_p - Y_a}(Y_{23} + Y_p)(Y_{23} + Y_a)\right]^{1/2} \tag{3.18}$$

$$Y_3 = Y_{23} - Y_2 \tag{3.19}$$

$$Y_1 = Y_{mb} - Y_2 \tag{3.20}$$

where

$$Y_{23} = \frac{Y_a(Y_{ma} - Y_{mb}) - Y_p(Y_{mp} - Y_{mb})}{Y_{mp} - Y_{ma}} \tag{3.21}$$

So far the only unknowns here are Y_a and Y_p. Because the relative dielectric constant and the conductivity of the air sample and the plexiglass sample are known, we can substitute them into the parallel-disk circuit model (it will be depicted in Section 3.6.4),

$$Y = \frac{\pi a^2}{d}(\sigma + jw\varepsilon)\frac{1 + \frac{j}{8}w\mu a^2(\sigma + jw\varepsilon)}{1 + \frac{j}{4}w\mu a^2(\sigma + jw\varepsilon)} \tag{3.22}$$

Hence, Y_a and Y_p can be solved at each frequency. Therefore Y_1, Y_2, and Y_3 are acquired and the π network is solved.

When the unknown test sample is put into the sample holder, the measured admittance satisfies [29],

$$Y_{mu} = Y_1 + \frac{Y_2 Y_3 + Y_2 Y_{lu}}{Y_2 + Y_3 + Y_{lu}} \tag{3.23}$$

where
Y_{lu} = the admittance of the unknown test sample;
Y_{mu} = the measured admittance of the unknown test sample.

3.6.4 The circuit parameters of the parallel-disk sample holder

The test sample and the two brass disks make up a parallel-disk capacitor as shown in Fig. 3.20 [32]. The relationship between the complex admittance and the complex dielectric constants of the test sample can be derived by solving the E fields between the two disks.

Because the distance between the parallel disks is relatively small, the E fields between them are independent of the Y axis. Considering the symmetry of the structure, the E fields are also independent of azimuthal angle φ [32].

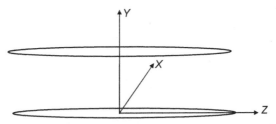

Figure 3.20 The coordinate setup in a parallel-disk capacitor.

At low frequencies, only the static fields need to be considered [29]. After solving the E fields, the relationship between the complex admittance and the dielectric constants of the test sample can be derived as,

$$Y_l = \frac{\pi a^2 (\sigma + jw\varepsilon)}{d} \quad (3.24)$$

where
a = the radius of the test sample;
d = the distance between the disks;
ε = the dielectric constant of the test sample between the disks;
σ = the conductivity of the test sample between the disks.

When the frequency goes higher than 20 MHz, the E fields cannot be expressed accurately by the static fields only; higher order corrections must be considered. The new relationship is derived as Eq. (3.25) [29], which has been proven valid when the frequency goes up to 250 MHz [30],

$$Y_l = \frac{\pi a^2}{d}(\sigma + jw\varepsilon)\frac{1 + \frac{j}{8}w\mu a^2(\sigma + jw\varepsilon)}{1 + \frac{j}{4}w\mu a^2(\sigma + jw\varepsilon)} \quad (3.25)$$

When the frequency goes even higher, more orders of the E fields must be added. The new relationship can be derived as,

$$Y_l = \frac{\pi a^2}{d}\frac{(\sigma + jw\varepsilon)\left[1 + \frac{j}{8}w\mu a^2(\sigma + jw\varepsilon)\right] - \frac{1}{192}w^2\mu^2\sigma^2 a^4(\sigma + j3w\varepsilon)}{1 + \frac{j}{4}w\mu a^2(\sigma + jw\varepsilon) - \frac{1}{64}w^2\mu^2\sigma a^4(\sigma + j2w\varepsilon)} \quad (3.26)$$

The derivations of Eqs. (3.24)–(3.26) are shown in Appendix A. The differences between Eqs. (3.25) and (3.26) are $-\frac{1}{192}w^2\mu^2\sigma^2 a^4(\sigma + j3w\varepsilon)$ in the numerator and $-\frac{1}{64}w^2\mu^2\sigma a^4(\sigma + j2w\varepsilon)$ in the denominator, respectively.

3.6.5 Performance analysis at high frequencies

To simplify the analysis, the components in Eq. (3.26) can be denoted by A, B, C, and D as,

$$A = (\sigma + jw\varepsilon)\left[1 + \frac{j}{8}w\mu a^2(\sigma + jw\varepsilon)\right] \qquad (3.27)$$

$$B = 1 + \frac{j}{4}w\mu a^2(\sigma + jw\varepsilon) \qquad (3.28)$$

$$C = -\frac{1}{192}w^2\mu^2\sigma^2 a^4(\sigma + j3w\varepsilon) \qquad (3.29)$$

$$D = -\frac{1}{64}w^2\mu^2\sigma a^4(\sigma + j2w\varepsilon) \qquad (3.30)$$

Substitute them into Eqs. (3.25) and (3.26). Eq. (3.25) is simplified as $Y_l = \frac{A}{B}$. Eq. (3.26) is simplified as $Y_l = \frac{A+C}{B+D}$.

Both $\frac{A}{B}$ and $\frac{A+C}{B+D}$ are harmonically increasing with frequencies changing from 1 MHz to 1 GHz, which are plotted in Fig. 3.21. Hence, only 1 GHz needs to be investigated in order to determine whether the working frequency can go to 1 GHz.

Substitute $\mu_0 = 4\pi \times 10^{-7}$ H/m, $\varepsilon_0 = \frac{1}{36\pi} \times 10^{-9}$ F/m, $a = 2.54 \times 10^{-2}$ m, and $f = 1$ GHz into Eqs. (3.27)–(3.30), we have Eqs. (3.31)–(3.34),

$$A = \sigma(1 - 0.1792 \times 10^{-2} \times \varepsilon_r) + j\left[\frac{\varepsilon_r}{36\pi} + 10.134 \times \sigma^2 - 7.9 \times 10^{-6} \times (\varepsilon_r)^2\right] \qquad (3.31)$$

$$B = 1 - 0.1792 \times 10^{-2} \times \varepsilon_r + j0.203 \times \sigma \qquad (3.32)$$

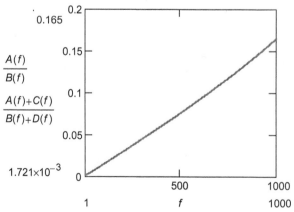

Figure 3.21 $\frac{A}{B}$ and $\frac{A+C}{B+D}$ versus frequency from 1 MHz to 1 GHz.

Table 3.1 Differences between Eqs. (3.25) and (3.26) for various samples

Materials	$\dfrac{A}{B}$	$\dfrac{A+C}{B+D}$ (MHz)	$\left|\dfrac{\frac{A+C}{B+D} - \frac{A}{B}}{\frac{A}{B}}\right| * 100\%$
Rock with oil	$3.940 \cdot 10^{-3} + j0.262$	$3.766 \cdot 10^{-3} + j0.262$	0.067
Sandstone	$7.721 \cdot 10^{-3} + j0.225$	$7.460 \cdot 10^{-3} + j0.225$	0.116
Limestone	$9.629 \cdot 10^{-3} + j0.301$	$9.092 \cdot 10^{-3} + j0.301$	0.179
Mortar	$4.134 \cdot 10^{-3} + j0.198$	$4.022 \cdot 10^{-3} + j0.198$	0.056
Nylon	$6.978 \cdot 10^{-3} + j0.205$	$6.779 \cdot 10^{-3} + j0.205$	0.098
Plexiglass	$2.158 \cdot 10^{-3} + j0.165$	$2.116 \cdot 10^{-3} + j0.165$	0.025

$$C = \sigma^2(-34.23 \times 10^{-4} \times \sigma - j0.908 \times 10^{-4} \times \varepsilon_r) \quad (3.33)$$

$$D = \sigma(-1.027 \times 10^{-2} \times \sigma - j1.816 \times 10^{-4} \times \varepsilon_r) \quad (3.34)$$

where

ε_r = the relative dielectric constant of the test sample;

σ = the conductivity of the test sample.

Usually the conductivity $\sigma \ll 1$ ohm/m (most of the time $\sigma < 0.1$ ohm/m), we can derive $\text{Re}(C) \ll \text{Re}(A)$, $\text{Im}(C) \ll \text{Im}(A)$, $\text{Re}(D) \ll \text{Re}(B)$, and $\text{Im}(D) \ll \text{Im}(B)$ from Eqs. (3.31)–(3.34). Therefore $A + C \approx A$ and $B + D \approx B$ at 1 GHz.

The differences between the values derived from Eqs. (3.25) and (3.26) are much less than 1% for various samples at 1 GHz as shown in Table 3.1. Even though higher order fields are considered, the fringing effects between the two disks do not interfere much at 1 GHz. Hence, Eq. (3.25) is still valid at 1 GHz.

3.6.6 Computation of the dielectric constant and the conductivity of test samples

When the unknown test sample is put into the sample holder, considering the π network, the measured admittance can be represented as,

$$Y_{mu} = Y_1 + \frac{Y_2 Y_3 + Y_2 Y_{lu}}{Y_2 + Y_3 + Y_{lu}} \quad (3.35)$$

Y_1, Y_2, and Y_3 are three coefficients of the π network, which are solved in Section 3.6.3. Measured admittance of the test sample Y_{mu} can be calculated from the measured admittance values. Hence, the admittance of the test sample Y_{lu} can be determined.

The expression of the dielectric constant ε and the conductivity σ of the test sample is Eq. (3.36), which is derived from Eq. (3.25),

$$\sigma + jw\varepsilon = \frac{Y_{lu}d}{\pi a^2} - \frac{4}{jw\mu a^2} \pm \sqrt{\frac{(Y_{lu}d)^2}{(\pi a^2)^2} + \left(\frac{4}{jw\mu a^2}\right)^2} \quad (3.36)$$

Substitute Y_{lu} and the dimensions of the test sample (a is the sample radius and d is the sample thickness) into Eq. (3.36). Two pairs of possible solutions of ε and σ are obtained. Only one pair is physically correct; the other pair should be discarded. It is easy to recognize the right pair just from the physical values. Therefore the dielectric constants ε and the conductivity σ of the test sample are determined at each frequency.

3.6.7 Analysis of dynamic range of the parallel-disk technique

Each measurement technique has its dynamic range for the sample measurement. For an ideal parallel-disk capacitor, the E field distribution between the two disks is uniform. For a finite size capacitor, the E field distribution is the strongest in the center and slightly decreases in the radial direction. The E field distributions between the disks will change with different materials inside the sample holder. Investigating the E field distribution between the disks is an efficient way to estimate the operating dynamic range of this measurement technique. The E field distribution can be modeled using numerical simulation software package such as High Frequency Electromagnetic Field (HFSS) with various sample materials. Considering the industry implementation as well as the working frequency range of the parallel-disk sample holder, 2 MHz and 1 GHz are chosen as the simulation frequencies to analyze the field distribution of the parallel-disk sample holder.

3.6.7.1 E field distribution inside the sample holder at 1 GHz

The dimensions of the simulation structures here are the same as shown previously by Fig. 3.18. The frequency range to be simulated starts from 200 MHz and ends at 1.5 GHz with 100 MHz increments. The solution frequency is 1 GHz, which is the current top frequency of this technique. The E fields inside the sample holder are simulated consecutively at the presence of different samples. Part of the simulated field patterns are shown in Figs. 3.22 and 3.23.

The simulated E field distribution between the disks can be considered as a parallel-disk capacitor working normally for the materials whose properties are from $\varepsilon_r = 1$ F/m, $\sigma = 0$ S/m to $\varepsilon_r = 3$, $\sigma = 1$ S/m, as shown in Fig. 3.22. The E fields become a little abnormal at the edge of the disks for the material with larger dielectric constant and conductivity such as $\varepsilon_r = 30$ and $\sigma = 2$ S/m as shown in Fig. 3.23. Due to the boundary condition, the E fields are abnormal for the material with $\varepsilon_r > 30$ or $\sigma > 1$ S/m. Therefore the dynamic ranges of this technique can be classified as 1−30 for ε_r and 0−1 S/m for σ at 1 GHz.

3.6.7.2 E field distribution inside the sample holder at 2 MHz

The frequency of 2 MHz is of great interest because the widely used measuring-while-drilling logging tools operate at 2 MHz. Similar to 1 GHz case, the E fields in

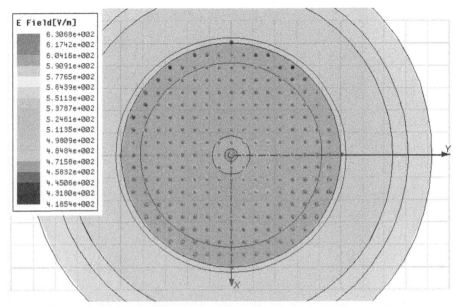

Figure 3.22 E fields between the disks for the air sample ($\varepsilon_r = 1$, $\sigma = 0$ S/m) at 1 GHz.

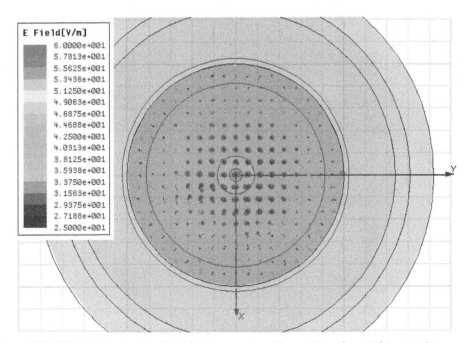

Figure 3.23 E fields between the disks for the sample with $\varepsilon_r = 20$ and $\sigma = 2$ S/m at 1 GHz.

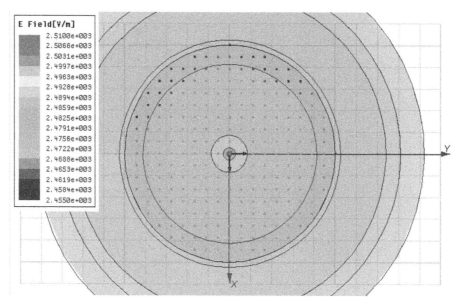

Figure 3.24 E fields between the disks for the sample with $\varepsilon_r = 80$ and $\sigma = 0$ S/m at 2 MHz.

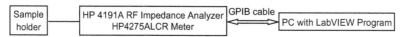

Figure 3.25 Block diagram of the automatic measurement system using the parallel-disk sample holder technique.

the parallel-disk sample holder at 2 MHz are modeled and simulation results are shown in Fig. 3.24. At 2 MHz, the parallel-disk sample holder is able to work for all the rock samples with different saturation levels ($\varepsilon_r = 1 - 80$ and $\sigma < 100$ S/m).

By the modeled E field distribution, the dynamic ranges of the parallel-disk sample holder technique have been verified to be 1−30 for ε_r and 0−1 S/m for σ at 1 GHz; and $\varepsilon_r = 1 - 80$ for ε_r and $\sigma < 100$ S/m for σ at 2 MHz.

3.6.8 Automatic measurement system using the parallel-disk technique

The automatic measurement system consists of the parallel-disk sample holder, the sample measurement equipment, and the LabVIEW computer control programs. Two instruments can be used in this technique. One is a HP4275A LCR meter for the low-frequency (LF) measurement, which works from 10 kHz to 10 MHz. The other is a HP4191A RF Impedance Analyzer for the high-frequency (HF) measurement, which works from 1 MHz to 1 GHz. The block diagram of the automatic measurement system is shown in Fig. 3.25. The flowchart of the sample measurement is shown in Fig. 3.26.

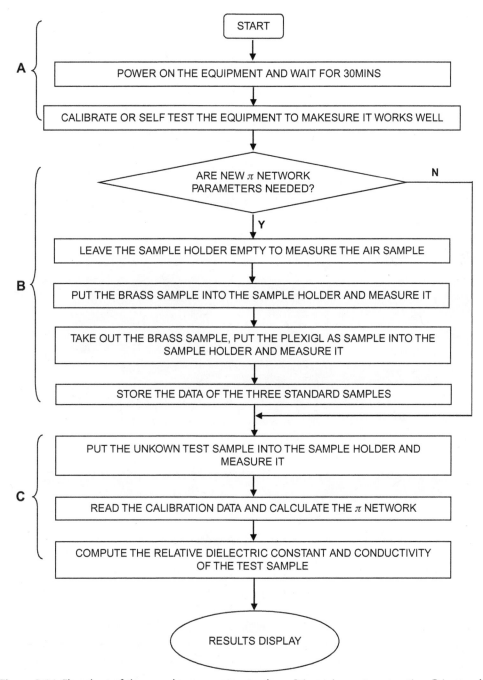

Figure 3.26 Flowchart of the sample measurement, where **A** is equipment preparation; **B** is sample holder calibration; **C** is test sample measurement.

The workstation for the LF measurement is shown in Fig. 3.27. A computer-controlled HP4275A LCR Meter is used to measure the capacitance (C) and the conductance (G). The measurement tolerance of the LCR Meter can be smaller than 1%. The self-diagnostic functions can be automatically performed or done whenever desire to confirm normal operation of the meter [28]; a HP16047A Calibration Standard Kit is used for the self-test.

The workstation for the HF measurement is shown in Fig. 3.28. A computer-controlled HP4191A RF Impedance Analyzer is used to measure the magnitude and the phase of the admittance at the APC-7 connector test port. This connector, which is located on the RF Impedance Analyzer, provides the means to connect and install a user-built test fixture (the sample holder) [31]. The measurement accuracy of

Figure 3.27 Workstation for the low-frequency measurement (the HP4275A LCR Meter).

Figure 3.28 Workstation for the high-frequency measurement (the HP4191A RF Impedance Analyzer).

the Impedance Analyzer is 1%. The calibration can be made by using the three reference terminals (open, short, and 50-ohm terminals).

The systematic calibration procedure of the HP4275A LCR Meter and the HP4191A RF Impedance Analyzer are included in Appendix B.

3.6.9 Experimental data and discussions

The test samples are shown in Fig. 3.29; the first row shows (from left to right): teflon, plexiglass, nylon, limestone, rock contaminated with oil; the second row (from left to right): two sandstones with different porosities, mortar, and asphalt. The sample measurement procedure is described in Section 3.4. The relative dielectric constants ε_r and the conductivities σ of the samples over 10 kHz–10 MHz are measured and shown in Tables 3.2–3.10. There are some negative values of the conductivity

Figure 3.29 Pictures of various test samples.

Table 3.2 Experimental data of the teflon sample for the low-frequency measurement

Frequency	ε_r (F/m)	σ (S/m)
10 kHz	1.7312	-1.3809×10^{-8}
20 kHz	1.7118	-4.9987×10^{-8}
40 kHz	1.7346	-8.9051×10^{-8}
100 kHz	1.7613	-1.8539×10^{-7}
200 kHz	1.7899	-3.7764×10^{-7}
400 kHz	1.7811	-3.2655×10^{-7}
1 MHz	1.8022	1.6142×10^{-7}
2 MHz	1.8143	3.4081×10^{-7}
4 MHz	1.8163	2.5951×10^{-5}
10 MHz	1.8464	9.0593×10^{-5}

Table 3.3 Experimental data of the plexiglass sample for the low-frequency measurement

Frequency	ε_r (F/m)	σ (S/m)
10 kHz	2.6188	1.4471×10^{-8}
20 kHz	2.5822	1.8222×10^{-8}
40 kHz	2.6141	1.2080×10^{-8}
100 kHz	2.6250	1.6708×10^{-8}
200 kHz	2.6111	1.6319×10^{-8}
400 kHz	2.5962	2.3420×10^{-8}
1 MHz	2.6022	1.1643×10^{-6}
2 MHz	2.6072	3.8963×10^{-6}
4 MHz	2.5954	1.9864×10^{-5}
10 MHz	2.6234	6.0592×10^{-5}

Table 3.4 Experimental data of the nylon sample for the low-frequency measurement

Frequency	ε_r (F/m)	σ (S/m)
10 kHz	4.3284	1.6499×10^{-7}
20 kHz	4.1242	4.1822×10^{-7}
40 kHz	3.9229	7.2136×10^{-7}
100 kHz	3.7927	1.5956×10^{-6}
200 kHz	3.6724	2.7497×10^{-6}
400 kHz	3.5445	4.5993×10^{-6}
1 MHz	3.4313	9.3843×10^{-6}
2 MHz	3.4043	1.6865×10^{-5}
4 MHz	3.3575	2.4460×10^{-5}
10 MHz	3.3055	4.7129×10^{-5}

Table 3.5 Experimental data of the rock contaminated with oil for the low-frequency measurement

Frequency	ε_r (F/m)	σ (S/m)
10 kHz	4.27805	4.9283×10^{-8}
20 kHz	4.2017	1.6679×10^{-8}
40 kHz	4.2571	1.0160×10^{-7}
100 kHz	4.2392	3.9831×10^{-7}
200 kHz	4.14465	8.0414×10^{-7}
400 kHz	4.1095	1.6317×10^{-6}
1 MHz	4.0786	3.9685×10^{-6}
2 MHz	4.0469	7.0487×10^{-6}
4 MHz	4.0062	1.4740×10^{-5}
10 MHz	4.0017	1.8206×10^{-5}

Table 3.6 Experimental data of the limestone sample for the low-frequency measurement

Frequency	ε_r (F/m)	σ (S/m)
10 kHz	6.9606	5.7263×10^{-7}
20 kHz	6.4772	1.1298×10^{-6}
40 kHz	6.0383	1.7916×10^{-6}
100 kHz	5.7006	3.5413×10^{-6}
200 kHz	5.4062	5.8890×10^{-6}
400 kHz	5.1475	9.8780×10^{-6}
1 MHz	4.9112	2.0654×10^{-5}
2 MHz	4.8544	3.8079×10^{-5}
4 MHz	4.6976	9.9707×10^{-5}
10 MHz	4.5437	3.1357×10^{-4}

Table 3.7 Experimental data of the sandstone No. 2 for the low-frequency measurement

Frequency	ε_r (F/m)	σ (S/m)
10 kHz	12.147	4.5762×10^{-6}
20 kHz	9.0937	6.6121×10^{-6}
40 kHz	7.3360	8.7938×10^{-6}
100 kHz	6.0473	1.360×10^{-5}
200 kHz	5.3217	1.9239×10^{-5}
400 kHz	4.7770	2.7128×10^{-5}
1 MHz	4.3029	4.5007×10^{-5}
2 MHz	4.1094	6.9655×10^{-5}
4 MHz	3.9024	9.9261×10^{-5}
10 MHz	3.6905	3.3244×10^{-4}

Table 3.8 Experimental data of the sandstone No. 3 for the low-frequency measurement

Frequency	ε_r (F/m)	σ (S/m)
10 kHz	11.3140	3.2528×10^{-6}
20 kHz	8.6980	5.1692×10^{-6}
40 kHz	7.0815	7.2343×10^{-6}
100 kHz	5.8850	1.1721×10^{-5}
200 kHz	5.2035	1.6922×10^{-5}
400 kHz	4.6924	2.4377×10^{-5}
1 MHz	4.2488	4.1445×10^{-5}
2 MHz	4.0682	6.5151×10^{-5}
4 MHz	3.8685	9.4507×10^{-5}
10 MHz	3.6569	3.2990×10^{-4}

Table 3.9 Experimental data of the dry mortar for the low-frequency measurement

Frequency	ε_r (F/m)	σ (S/m)
10 kHz	4.3546	5.7146×10^{-7}
20 kHz	3.9290	9.4058×10^{-7}
40 kHz	3.6386	1.3207×10^{-6}
100 kHz	3.4180	2.1622×10^{-6}
200 kHz	3.2210	3.0858×10^{-6}
400 kHz	3.1145	4.3679×10^{-6}
1 MHz	3.0349	7.2928×10^{-6}
2 MHz	3.0255	1.1381×10^{-5}
4 MHz	3.0021	1.1367×10^{-5}
10 MHz	2.9766	1.9274×10^{-5}

Table 3.10 Experimental data of the asphalt sample for the low-frequency measurement

Frequency	ε_r (F/m)	σ (S/m)
10 kHz	3.8260	-4.0740×10^{-8}
20 kHz	3.8177	1.6524×10^{-8}
40 kHz	3.7889	-3.8025×10^{-9}
100 kHz	3.8100	-2.7860×10^{-8}
200 kHz	3.7391	-9.7945×10^{-8}
400 kHz	3.7093	-1.1221×10^{-7}
1 MHz	3.7000	3.1932×10^{-7}
2 MHz	3.7477	1.4826×10^{-6}
4 MHz	3.7373	1.5786×10^{-6}
10 MHz	3.7327	1.2624×10^{-5}

because the conductivity values are too small and fall beyond the tolerance of the measurement system. The relative dielectric constants of the samples over 10 kHz—10 MHz are plotted in Figs. 3.30—3.33. The relative dielectric constants ε_r and the conductivities σ of various test samples over 10 MHz—1 GHz are shown in Figs. 3.34—3.37. Current experimental data are in good agreement with the values in the existing dielectric permittivity tables and the values obtained by another method developed in the Well Logging Lab previously.

The experimental data above illustrate reasonable correlations between the relative dielectric constants of various samples and the operating frequency. The relative dielectric constants of the teflon sample are harmonically increasing from 1.7340 to 1.9500 with the increasing frequency, and become stable around 1.9500 when the frequency is higher than 900 MHz. The relative dielectric constants of the plexiglass sample are around 2.6400 F/m. The relative dielectric constants of the nylon sample decrease from 4.3284 to 3.2200 over 10 kHz—720 MHz, stay around 3.2200 F/m,

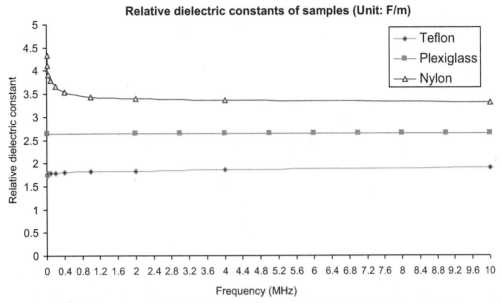

Figure 3.30 Plot of measured relative dielectric constants (Unit: F/m) of the teflon, plexiglass, and nylon samples for the low-frequency measurement.

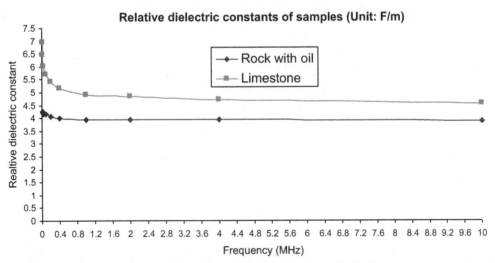

Figure 3.31 Plot of measured relative dielectric constants (Unit: F/m) of the rock contaminated with oil and the limestone for the low-frequency measurement.

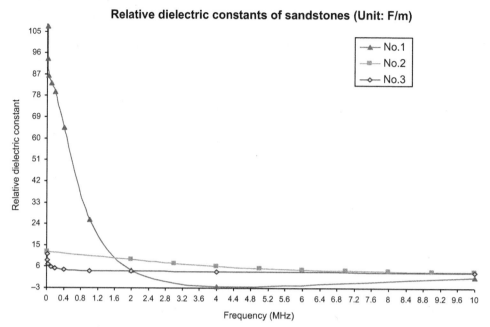

Figure 3.32 Plot of measured relative dielectric constant (Unit: F/m) of the sandstones with different porosities for the low-frequency measurement.

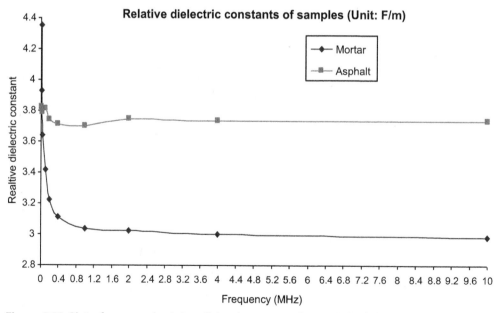

Figure 3.33 Plot of measured relative dielectric constant (Unit: F/m) of the dry mortar and the asphalt for the low-frequency measurement.

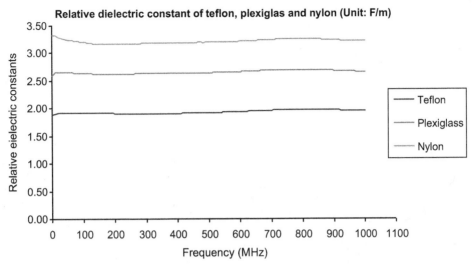

Figure 3.34 Plot of measured relative dielectric constant (Unit: F/m) of the teflon, plexiglass, and nylon for the high-frequency measurement.

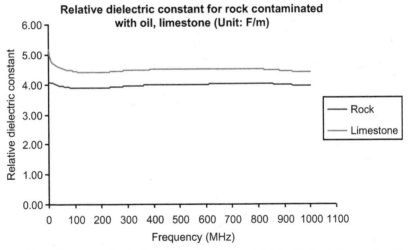

Figure 3.35 Plot of measured relative dielectric constant (Unit: F/m) of the rock with oil and limestone for the high-frequency measurement.

and then decrease a little when the frequency is higher than 920 MHz. The relative dielectric constants of the rock contaminated with oil go from 4.2694 to 4.0000 over 10 kHz–610 MHz; they are stable from 610 to 860 MHz and decay a few percent when the frequency is higher than 860 MHz. The relative dielectric constants of the limestone sample go from 6.9606 to 4.5500 over 10 kHz–390 MHz; they are

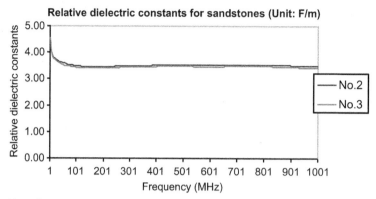

Figure 3.36 Plot of measured relative dielectric constant (Unit: F/m) of the sandstones with different porosities for the high-frequency measurement.

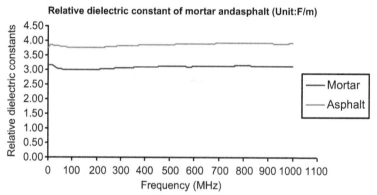

Figure 3.37 Plot of measured relative dielectric constant (Unit: F/m) of the dry mortar and the asphalt for the high-frequency measurement.

stable from 390 to 810 MHz and decay a few percent when the frequency is higher than 810 MHz. The relative dielectric constants of the sandstone No. 2 go from 12.1470 to 3.6000 over 10 kHz–390 MHz, and stay around 3.6000 when the frequency is higher than 390 MHz. The relative dielectric constants of the sandstone No. 3 go from 11.3140 to 3.4500 and stay around 3.4500. The relative dielectric constants of the dry mortar go from 4.3257 to 2.9500 over 10 kHz–200 MHz, increasing a little when the frequency goes up, and then stay around 3.0500 F/m. The relative dielectric constants of the asphalt sample go from 3.8260 to 3.2000 and then stay around 3.2000 F/m. The measured conductivities of all the samples are very small. There are a few measured conductivities with negative values, which are very small, around 10^{-8}. Those conductivities can be considered as zero. The different trends of the relative dielectric constants and the conductivities of the different samples are determined by their structure properties.

Multiple runs of sample measurements are carried in order to reduce the random measurement error. The upper and lower deviations of the measured relative dielectric constants of the rock contaminated with oil are calculated in both number and percentage format, which are shown in Table 3.11. The overall deviations are much less than 1%. That shows the relative dielectric constant measurements are consistent. Similarly, the upper and lower deviations of the measured conductivity of the oil-contaminated rock sample are calculated and shown in Table 3.12. The deviations are within 3%. From the above analysis, it proves that the sample measurements are accurate.

Table 3.11 Error bar for the measured relative dielectric constants of the rock contaminated with oil

Frequency	Measurement error of relative dielectric constant			
	Upper derivation		Lower derivation	
	Number	Percentage	Number	Percentage
10 kHz	0.000750	0.0222	0.000950	0.0222
20 kHz	0.000180	0.0003	0.004284	0.0076
40 kHz	0.001690	0.0007	0.039683	0.0167
100 kHz	0.000200	0.0003	0.004718	0.0071
200 kHz	0.000650	0.0003	0.015680	0.0060
400 kHz	0.000360	0.0005	0.008759	0.0131
1 MHz	0.000435	0.0097	0.010664	0.0237
2 MHz	0.003609	0.0020	0.089102	0.0492
4 MHz	0.000482	0.0004	0.012025	0.0104
10 MHz	0.002360	0.0009	0.058940	0.0235

Table 3.12 Error bar for the measured conductivity of the rock contaminated with oil

Frequency	Measurement error of conductivity			
	Upper derivation		Lower derivation	
	Number	Percentage	Number	Percentage
10 kHz	3.2930×10^{-10}	0.6638	4.7170×10^{-10}	0.9664
20 kHz	2.6830×10^{-10}	1.5832	4.7170×10^{-10}	2.9105
40 kHz	2.6850×10^{-10}	2.5748	1.1650×10^{-10}	1.1600
100 kHz	3.4122×10^{-10}	0.8494	4.2378×10^{-10}	1.0754
200 kHz	3.5580×10^{-10}	0.4405	4.2820×10^{-10}	0.5353
400 kHz	8.3600×10^{-10}	0.5097	3.1640×10^{-10}	1.9774
1 MHz	2.1950×10^{-10}	0.5501	2.1050×10^{-10}	0.5333
2 MHz	6.9255×10^{-10}	0.9730	6.5245×10^{-10}	0.9343
4 MHz	3.7645×10^{-10}	2.4904	4.1550×10^{-10}	0.2827
10 MHz	5.4380×10^{-10}	2.9003	3.2420×10^{-10}	1.8130

3.6.10 Differences between the LF and the HF measurements

The LF measurement, covering a frequency range from 10 kHz to 10 MHz, can be carried out by a HP4275A LCR Meter; the HF measurement can be completed by a HP4191A RF Impedance Analyzer in a higher frequency range from 1 MHz to 1 GHz. The overlapping frequency range of these two measurement system is from 1 MHz to 10 MHz. To determine the accuracy of the parallel-disk technique, the measurement differences of the LF and the HF in their overlapping frequency range will be analyzed. The relative dielectric constants of various samples measured by the HP4275A (LF) and the HP4191A (HF) in the overlapping frequency range are plotted in Figs. 3.38–3.45. It shows that the differences between the LF

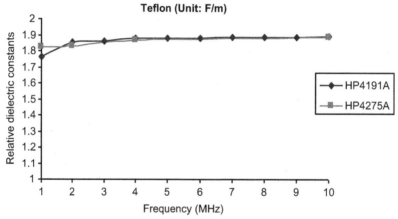

Figure 3.38 Plot of the measured relative dielectric constants (Unit: F/m) of the teflon sample for the overlapping frequencies.

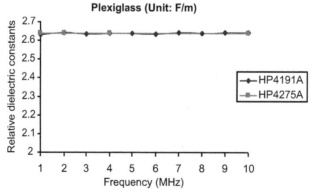

Figure 3.39 Plot of the measured relative dielectric constants (Unit: F/m) of the plexiglass sample for the overlapping frequencies.

Electrical Properties of Sediment Rocks: Mixing Laws and Measurement Methods 93

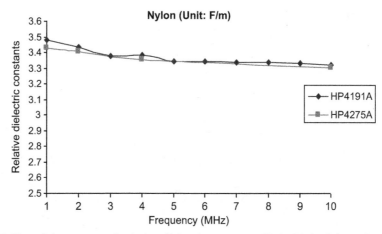

Figure 3.40 Plot of the measured relative dielectric constants (Unit: F/m) of the nylon sample for the overlapping frequencies.

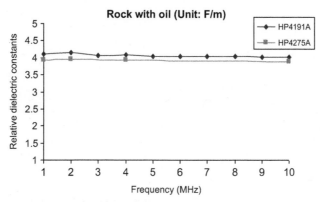

Figure 3.41 Plot of the measured relative dielectric constants (Unit: F/m) of the rock sample with oil for the overlapping frequencies.

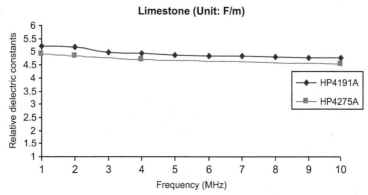

Figure 3.42 Plot of the measured relative dielectric constants (Unit: F/m) of the limestone sample for the overlapping frequencies.

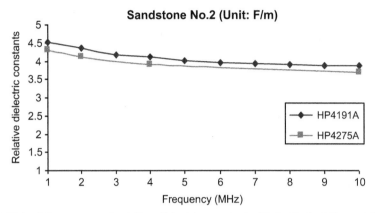

Figure 3.43 Plot of the measured relative dielectric constants (Unit: F/m) of the sandstone No. 2 for the overlapping frequencies.

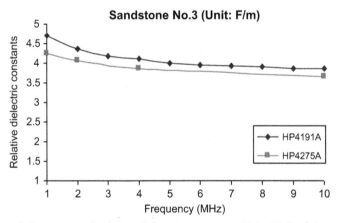

Figure 3.44 Plot of the measured relative dielectric constants (Unit: F/m) of the sandstone No. 3 for the overlapping frequencies.

and the HF are less than 5% for the test samples. That means the LF and HF measurements have good consistency and the parallel-disk measurement technique has satisfactory accuracy.

There are two possible contributions to the differences between the LF and the HF. Firstly, C and D factors in Eq. (3.26): the extra fringing effect may not interfere much compared to Eq. (3.25), but it still has some effect. Secondly, the overlapping frequencies are the "ending" frequencies for both instruments. The measurement accuracy might be less at these frequencies than at other frequencies. There may be other reasons also.

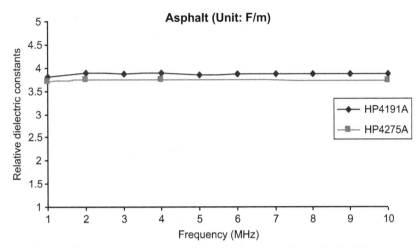

Figure 3.45 Plot of the measured relative dielectric constants (Unit: F/m) of the asphalt for the overlapping frequencies.

3.6.11 Error analysis

The system errors result from measurement limitations of instrumentation, sample preparation, and approximations in theoretical derivation. They directly influence accuracy and precision of the sample measurement and the data processing.

The instruments have measurement and test signal-level accuracy. Actual measurement error is the sum of the instrument error and the error peculiar to the test fixture (leads) used. When the test cables are used in the HF measurement for the HP4275A LCR Meter, the displayed test limitations and the calibration errors, which are dominated by the particular quality of the individual reference terminations.

For the parallel-disk technique, the spring is adjustable to make sure there is good contact between the test sample and the sample holder. However the test sample preparation still could cause different grain sizes on the sample surface and an unknown amount of gap between the sample and the sample holder.

The approximation in theory derivation can bring errors too. Eq. (3.25) is used in the data processing instead of Eq. (3.26); the approximation is another error source.

3.7 TM_{010} RESONANT CAVITY TECHNIQUE

3.7.1 Theory of the TM_{010} resonant cavity technique

A TM_{010} resonant cavity made of metal material can be used as a device to measure the complex dielectric permittivity of the core samples [33]. When a cylindrical sample is placed along the axis at the center of the cavity as shown in Fig. 3.46, the cross

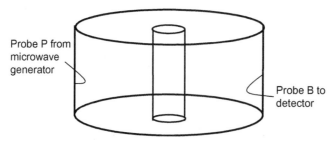

Figure 3.46 Cylindrical cavity in cylindrical coordinate with the core sample in the middle.

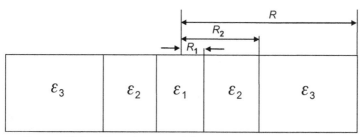

ε_1 = the permittivity of the unknown sample;
ε_2 = the permittivity of the sample holder;
ε_3 = the permittivity of the air sample;
R_1, R_2, and R_3 are the radii of the core sample, sample holder and the cavity, R_2 = 4.125

Figure 3.47 Cross section of the cavity with the core sample and the sample holder.

section of the cavity with the sample and the sample holder is shown in Fig. 3.47 [33]. The radii of the sample, the sample holder, and the resonant cavity are denoted by R_1, R_2, and R_3, respectively.

The E fields of this mode are axially symmetrical and reach the maximum magnitude at the center of the cavity. Cylindrical symmetry can therefore be maintained when a cylindrical core sample is placed at the center of the cavity. The TM_{010} mode assumes no variations in the Z axis. The resonant frequency of this mode depends only on the diameters of the cavity and the sample; it is independent of their lengths. The cavity resonator is enclosed completely by the metallic walls, except for a small probe P to couple electromagnetic power into the volume and another similar probe B to detect the electromagnetic fields in the cavity.

Probes P and B are optimized for best coupling. When electromagnetic power with an arbitrary frequency enters the cavity through Probe P, the cavity generally reflects almost all the incident power back to the source. Probe B is connected to a sensitive receiver, generally able to detect very weak signals. If the frequency of the incident power is close to the resonant frequency of the cavity, a comparatively large amount of power can be absorbed by the cavity, resulting in strong electromagnetic

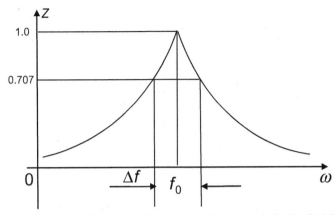

Figure 3.48 A resonant curve of the cavity. The resonant frequency is f_0, $Q = f_0/\Delta f$.

fields being established inside the cavity. The magnitude of the electromagnetic fields reaches their maximum when the frequency of the source is exactly equal to one of the resonant frequencies of the cavity [33].

When the electromagnetic fields detected by the receiver are plotted as a function of the frequency, a resonant curve can be obtained as shown in Fig. 3.48 [33].

The choice of the sample size represents a compromise among several competing effects. The sample must be small enough that the Q factor is not too small to observe [33]. The sizes are chosen to give an approximate Q factor between 50 and 100 [33]. The complex resonant frequency can be obtained from the resonant curve (the real resonant frequency and the Q factor), which is shown in Fig. 3.48 [33].

Applying the proper boundary conditions at R_1, R_2, and R_3 regions and analyzing the fields of R_1, R_2, and R_3, the formula to calculate the relative dielectric constant can be expressed as [33],

$$\varepsilon_1 = \frac{2\sqrt{\varepsilon_2}}{\omega\sqrt{\mu_0\varepsilon_0}R_1} A \tag{3.37}$$

where
 ε_1 = the relative dielectric constant of the test sample;
 ε_2 = the relative dielectric constant of the sample holder;
 w = the angular resonant frequency;
 R_1 = the radius of the test sample;
and A and B are two coefficients shown as,

$$A = \frac{(J_{121}Y_{022} - J_{022}Y_{121}) - B(J_{121}Y_{122} - J_{122}Y_{121})}{(J_{021}Y_{022} - J_{022}Y_{021}) - B(J_{021}Y_{122} - J_{122}Y_{021})} \tag{3.38}$$

$$B = \frac{k_2(J_{033}Y_{032} - J_{032}Y_{033})}{k_3(J_{033}Y_{132} - J_{132}Y_{033})} \quad (3.39)$$

where J_{mnp} and Y_{mnp} represent $J_m(k_n)R_p$ and $Y_m(k_n)R_p$, the Bessel functions of the first kind and the second kind, respectively, $m = 0$ or 1, $n = 1, 2, 3$, and $p = 1, 2, 3$.

3.7.2 Dynamic range of the resonant cavity technique

3.7.2.1 Simulation data

Similar to the parallel-disk technique, the dynamic range of the resonant cavity technique can be estimated from the simulation data by HFSS. The simulation structures, which are of the same physical size as the actual ones, are built up in HFSS as shown in Figs. 3.49 and 3.50.

The solution data to be found are the quality factor Q and the dominant resonant frequency f from Eqs. (3.40) and (3.41),

$$Q = \left| \frac{\text{Mag}(freq)}{2 \times \text{Im}(freq)} \right| \quad (3.40)$$

$$f = \frac{k_0 c}{2\pi} \quad (3.41)$$

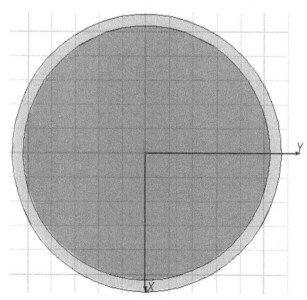

Figure 3.49 Simulation structure of the empty resonant cavity (front view).

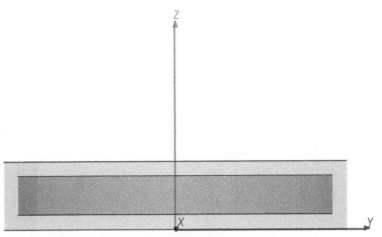

Figure 3.50 Simulation structure of the empty resonant cavity (side view).

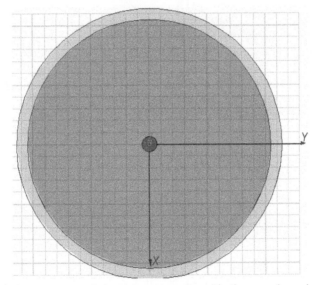

Figure 3.51 Simulation structure of the resonant cavity with the sample and the sample holder (front view).

where c is the light speed; Mag(*freq*) and Im(*freq*) represent the magnitude and imaginary part of the complex resonant frequency; f is the dominant resonant frequency.

The first simulation case is an empty cavity made of aluminum material ($\sigma = 3.8 \times 10^7$ S/m) as shown in Figs. 3.49 and 3.50. The simulation results are $f_0 = 1.0921 + j \times (6.5322 \times 10^{-5})$GHz and $Q = 8359.5$.

The second simulation structure consists of the aluminum resonant cavity, the plexiglass sample holder ($\varepsilon_r = 2.69$ F/m), and the core sample, as shown in Figs. 3.51 and 3.52.

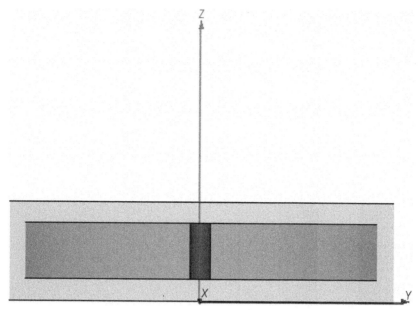

Figure 3.52 Simulation structure of the resonant cavity with the sample and the sample holder (side view).

Table 3.13 Simulation results for various water-based mud

Water-based mud ($\varepsilon_r = 80$)	Dominant resonant frequency f (GHz)	Frequency shift Δf (MHz)	Q
I: $\sigma = 0.1$ S/m	$1.0721 + j \times (1.7455 \times 10^{-4})$	20	3071
II: $\sigma = 0.5$ S/m	$1.0721 + j \times (5.1289 \times 10^{-3})$	20	1045.1
III: $\sigma = 1$ S/m	$1.0721 + j \times (9.3551 \times 10^{-3})$	20	572.99
IV: $\sigma = 1.5$ S/m	$1.0721 + j \times (1.3582 \times 10^{-2})$	20	394.68
V: $\sigma = 2$ S/m	$1.0721 + j \times (1.7808 \times 10^{-2})$	20	301.01

Various materials, such as water-based mud and oil-based mud, are used to build core samples in the simulation model. The diameter of the core sample is 1 mm. The diameter of the sample holder is 0.5 in.

The simulated data for various sample materials are shown in Tables 3.13–3.16. The resonant frequency of the empty cavity is f_0 (GHz). The resonant frequency of the cavity becomes f (GHz) when a core sample and its sample holder are present at the center of the cavity. The frequency shift is $\Delta f = f_0 - f$ (GHz).

For the common water-based mud, the relative dielectric constant is around 50 F/m; the conductivity varies from 0.1 to 2 S/m. Simulation results are shown in Table 3.13.

Table 3.14 Simulation results for various oil-based mud

Oil-based mud ($\varepsilon_r = 2.8$)	Dominant resonant frequency f (GHz)	Frequency shift Δf (MHz)	Q
I: $\sigma = 0.001$ S/m	$1.0756 + j \times (-5.9486 \times 10^{-5})$	16.5	5937.6
II: $\sigma = 0.01$ S/m	$1.0756 + j \times (9.7941 \times 10^{-5})$	16.5	5491.1
III: $\sigma = 0.1$ S/m	$1.0756 + j \times (1.7137 \times 10^{-4})$	16.5	3138.3

Table 3.15 Measured data versus simulated data using aluminum for the core samples

Samples	Measured data				Simulated data	
	ε_{rm}	σ_m (S/m)	Q_m	f_m (GHz)	Q	f (GHz)
Fresh water	81.39	0.01	143.20	1.068657	5445.10	1.0717
10 kppm saline solution	78.01	3.79	47.77	1.069403	162.95	1.0719
15 kppm saline solution	78.13	4.56	39.81	1.069403	136.07	1.0719

Table 3.16 Simulation results for various water-based mud after calibration process

Water-based mud ($\varepsilon_r = 80$)	Dominant resonant frequency f (GHz)	Frequency shift Δf (MHz)	Q	$\Delta Q = Q_0 - Q$
I: $\sigma = 0.1$ S/m	$1.0748 + j \times (3.7591 \times 10^{-3})$	13.6	142.96	3.16
II: $\sigma = 0.5$ S/m	$1.0748 + j \times (4.0913 \times 10^{-3})$	13.6	131.35	14.77
III: $\sigma = 1$ S/m	$1.0748 + j \times (4.5066 \times 10^{-3})$	13.6	119.25	26.87
IV: $\sigma = 1.5$ S/m	$1.0748 + j \times (4.9218 \times 10^{-3})$	13.6	109.19	36.93
V: $\sigma = 2$ S/m	$1.0748 + j \times (5.3368 \times 10^{-3})$	13.6	100.7	45.42

For the common oil-based mud, the relative dielectric constant is about 2.8; the conductivity varies from 0.001 to 0.1 S/m. Simulation results for various oil-based mud are shown in Table 3.14.

Measured values of the relative dielectric constant ε_{rm}, the conductivity σ_m, the quality factor Q_m, and the dominant resonant frequency f_m of some samples are shown in Table 3.15. Assign the measured relative dielectric constant and conductivity to the core sample in HFSS, and then analyze the simulation structure. Comparison of Q and f between the simulated and the measured data are shown in Table 3.15.

The simulated Q is much bigger than the measured quantity; this is because the surface of the aluminum cavity is oxidized, resulting in a higher surface impedance to the induced electric currents than the pure aluminum used in HFSS model. To get better results, it is necessary to "calibrate" the oxidized aluminum.

The calibration is described as follows: building a core sample of distilled water in HFSS simulation model since the relative dielectric constant and conductivity of distilled water are known; editing the conductivity of the cavity material used in the HFSS simulation model until the simulated Q value becomes close enough to the measured Q value of distilled water sample; recording the conductivity of the cavity material in the HFSS model of the time. The recorded conductivity in the HFSS model can be considered as the calibrated conductivity of the oxidized aluminum cavity. When the conductivity of the calibrated aluminum is 11,500 S/m, the simulated Q is 145.85, which is close enough to the measured value of 143.20. Hence, the "calibrated" aluminum with $\sigma = 11,500$ S/m can be assigned to the metallic wall of the cavity in HFSS to the rest of simulations. Simulation results are shown in Tables 3.16—3.18. The dominant resonant frequency of the empty cavity f_0 is 1.0884 GHz.

Measured values of the relative dielectric constant ε_{rm}, the conductivity σ_m, the quality factor Q_m, and the dominant resonant frequency f_m of some samples are shown in Table 3.18. Once the measured relative dielectric constant and conductivity of the core sample are assigned to the simulation sample in HFSS, the simulation structure of the resonant cavity can be analyzed. The simulated Q and f are shown in Table 3.18.

Table 3.17 Simulation results for various oil-based mud after calibration process

Oil-based mud ($\varepsilon_r = 2.8$)	Dominant resonant frequency f (GHz)	Frequency shift Δf (MHz)	Q	$\Delta Q = Q_0 - Q$
I: $\sigma = 0.001$ S/m	$1.0785 + j \times (3.6879 \times 10^{-3})$	9.9	146.23	0.11
II: $\sigma = 0.01$ S/m	$1.0785 + j \times (3.6953 \times 10^{-3})$	9.9	145.94	0.18
III: $\sigma = 0.1$ S/m	$1.0785 + j \times (3.7673 \times 10^{-3})$	9.9	143.15	2.97

Table 3.18 The measured data versus the simulated data for the core samples

Samples	Measured data				Simulated data		Relative error (%)	
	ε_{rm}	σ_m (S/m)	Q_m	f_m (GHz)	Q	f (GHz)	$\frac{\Delta Q}{Q}$	$\frac{\Delta f}{f}$
Fresh water	81.390	0.010	143.20	1.068657	145.85	1.074700	1.82	0.56
10 kppm saline solutions	77.930	1.981	57.77	1.069403	78.81	1.074900	26.70	0.51
15 kppm saline solutions	78.030	3.070	47.90	1.069403	72.06	1.074900	33.53	0.51

The differences of Q and f between the measured data and the simulated data are calculated as $\frac{\Delta Q}{Q} = \left|\frac{Q_m - Q}{Q}\right| \times 100\%$ and $\frac{\Delta f}{f} = \left|\frac{f_m - f}{f}\right| \times 100\%$, which are shown in Table 3.18.

3.7.2.2 Error analysis of simulation data

There is a big difference of Q between the measured data and simulated data. There may be several error contributions.

First, the simulation structure and the actual one are not identical. In the simulation structure, the entire cavity is made of aluminum; there is no other metal. In addition, there are no probes or screws. However the actual one is not the same, which is shown in Figs. 3.53 and 3.54. In the actual cavity, the probes coupling the energy for the actual cavity may cause energy loss. Part of the cavity cover is made of copper instead of aluminum. There are screws on the top of the cavity, which may cause energy leakage.

Secondly, sample heights are not identical. The thickness of the sample is exactly 1 in. in the simulation structure. However the actual sample height is hard to be exactly 1 in. It maybe a little bit more or less. If the actual height is less than 1 in., the capacitance of the air gap between the sample and the cavity cover can cause errors in the measurement results. If the actual height is more than 1 in., the extra volume may touch the cavity cover, which also causes errors.

There may be other possibilities. However the absolute values of Q are not the key point. In the resonant cavity algorithm, Q is calculated from the frequency shift Δf, which is shown in Fig. 3.48. The simulation data shows that the values of Δf are

Figure 3.53 The actual resonant cavity.

Figure 3.54 The opened resonant cavity.

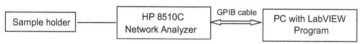

Figure 3.55 Block diagram of the resonant cavity measurement system.

in a reasonable range so that the calculated values of Q are big enough for the core sample measurement.

Considering the error analysis, the estimated dynamic range of the resonant cavity in HFSS is valid, which is $\sigma = 0.1 - 2.0$ S/m for saline solutions or water-based mud ($\varepsilon_r = 80$) and $\sigma = 0.001 - 0.1$ S/m for oil-based mud ($\varepsilon_r = 2.8$).

3.7.3 Measurement system and experimental data of the resonant cavity technique

3.7.3.1 Automatic measurement system of the resonant cavity technique

The apparatus used in the resonant cavity technique is a PC with a GPIB card, a LabVIEW software package, a HP8510C Network Analyzer and the TM_{010} resonant cavity developed by the Well Logging Lab. The block diagram of the automatic measurement system is provided in Fig. 3.55. The automatic measurement workstation is pictured in Fig. 3.56.

Like the parallel-disk technique, the computer program is developed in LabVIEW to control the HP8510C Network Analyzer, to measure the core sample, to process the measured data and to display the results. A HP8510C Network Analyzer is used to measure the S parameters of the resonant cavity. An user-friendly interface is shown in Fig. 3.57. The operator can set the test frequency, the dimensions of the cavity, the sample and its sample holder, the relative dielectric constant of the sample holder,

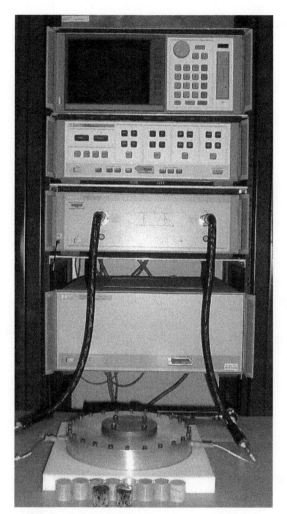

Figure 3.56 Workstation of the resonant cavity measurement system.

and then the program is ready to run. After the program finishes running, the relative dielectric constant and the conductivity of the core sample will be displayed by the user interface. The values of Q and $w_0 = 2\pi f_0$ will be displayed. The S parameter measured by the HP 8510C will be plotted and shown by the user interface.

3.7.3.2 Experimental data of saline solutions with different salinities

The relative dielectric constants ε_r and the conductivities σ of the saline solutions are measured with salinities ranging from 5 to 20 kppm at room temperature. Current experimental data and previous data [30] are shown in Table 3.19. The differences

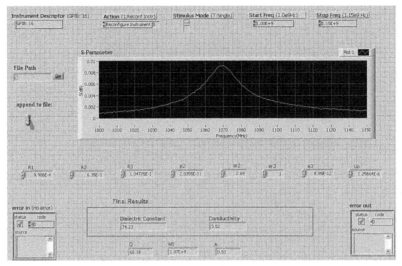

Figure 3.57 User interface of the resonant cavity measurement system.

Table 3.19 Experimental data of the saline solutions with different salinities

Saline solutions with different salinities (kppm)	Experimental data		Previous data		Relative error ε_r/σ (%)
	ε_r (F/m)	σ (S/m)	ε_r (F/m)	σ (S/m)	
5	80.2000	1.3020	79.1000	1.2460	1.3720/4.3010
8	76.6500	1.7470	76.5200	1.7190	0.1700/1.6030
10	77.9300	1.9810	77.0400	2.0760	1.1420/4.5760
13	78.8600	2.5360	76.7000	2.7160	2.7390/6.6270
15	78.0300	3.0700	75.3200	3.0600	3.4730/0.3260
20	76.1400	3.6030	73.0800	3.7860	4.0190/4.8340

between them are less than 5%. It shows that the measurement accuracy of the resonant cavity is good.

3.7.3.3 Experimental data of the asphalt and the mortar with different moisture contents

The asphalt sample and its sample holder are pictured in Fig. 3.58. The measurement data are $\varepsilon_r = 4.300$ and $\sigma = 0.006$ S/m at room temperature.

There are no available data in existing electrical property tables for the mortar samples. It is very useful to obtain them using the resonant cavity technique.

The mortar samples are prepared using Quikrete Mortar Mix No. 1102, a blend of masonry cement and graded sand. In the product direction, 27.2 kg mortar needs

Electrical Properties of Sediment Rocks: Mixing Laws and Measurement Methods 107

Figure 3.58 The asphalt sample and its sample holder.

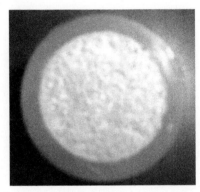

Figure 3.59 The mortar sample and its sample holder.

1 gallon water. In this chapter, 1 kg mortar is mixed with 150 mL water at room temperature (Fig. 3.59). Two kinds of measurements are performed according to different sample preparation procedures, which will be described in detail.

The first sample preparation is made immediately after the mortar and water are mixed well. The mix is put into the sample holder and compressed tightly. The sample became dry because of evaporation and cement hydration. The water left in the sample is free water and structural water, which is not sensitive to the dielectric constant measurement. The dielectric constant measurement may be more sensitive to the change of free water. Unfortunately, it is hard to determine how much free water is left in the mortar sample.

The first measurement is done 1 hour after the sample preparation; the second one is made in 2 hours, then in 3 hours, 4 hours, and so on. Increasing time roughly represents the decreasing moisture contents. Experimental data of the mortar sample with different moistures at room temperature is shown in Figs. 3.60 and 3.61.

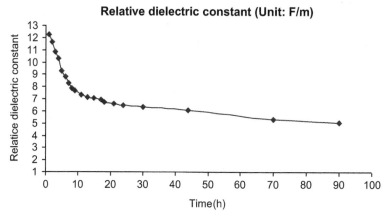

Figure 3.60 Relative dielectric constants (Unit: F/m) of the mortar sample with different moisture contents.

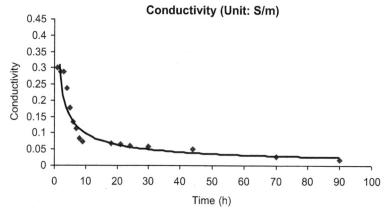

Figure 3.61 Conductivity (Unit: S/m) of the mortar sample with different moisture contents.

The second sample preparation is made some time after the mortar is mixed with water. The sample No. 1 is made in 6 hours. The mortar mix is half-solid at that time. The measurement data are $\varepsilon_r = 7.7000-8.1000$ and $\sigma = 0.0600-0.1000$ S/m. The sample No. 2 is prepared in 11 hours. The measurement data are $\varepsilon_r = 7.3000-7.7000$ and $\sigma = 0.0700-0.1000$ S/m. One side of the sample No. 2 is exposed to air for 24 hours; then the sample is tested. The measurement data are $\varepsilon_r = 5.3000$ and $\sigma = 0.0400$ S/m. The other side of the sample No. 2 is exposed to air for 24 hours; then the sample is tested. The measurement data are $\varepsilon_r = 4.5000$ and $\sigma = 0.0200$ S/m.

3.7.3.4 Error analysis

The system errors result from limitation of sample preparation, the physical size of the sample holder and the resonant cavity, and the measurement accuracy of the instrument, which can directly influence accuracy and precision of the experiment.

The resonant cavity must be in good condition before sample measurements. All the screws must be evenly tight in order to avoid RF leakage and maintain even pressure on the cavity. A regular check is recommended.

The physical size of the sample holder is critical for experimental results considering the small volume of the core sample. Accurate and careful manufacture and measurement can greatly reduce the error.

The limitation of the measurement and the test-signal frequency could also cause errors in the experiment. Instrument calibration is recommended before each experiment.

The operating errors are generated when the system is operated. Some of the operating errors are discussed in the following paragraphs.

Saline solution preparation has limited salinity accuracy. Tapes used to seal the sample holders for the asphalt and the mortar measurement can cause inaccuracy. The porosity of the mortar sample may change during the second type of sample preparation; it introduces errors.

The operating errors cannot be avoided. Only some errors can be reduced. For example, the volume error of the saline solutions can be reduced by using a syringe. The operator can inject liquid using a syringe into the sample holder carefully to avoid air bubbles. The water level must be little convex to make sure there is good contact. In addition, the operator can reduce the random errors by repeating the measurements.

The operating errors, which are much greater than the system errors, are considered as the main factor to determine the measurement accuracy.

3.7.3.5 Conclusions

Two automatic measurement systems are established to measure the electrical properties of rock samples and borehole fluids in this chapter. The systems could cover a wide frequency band, which is from 10 kHz to 1.1 GHz.

Theory analysis is completed to prove the capability of a parallel-disk measurement system to work from 10 kHz to 1 GHz according to current equipment. The E fields between the parallel disks are simulated in HFSS at 50 MHz. The E fields are the maximum at the center of the sample holder and decreased slightly as the distance moved further away. The dynamic range is determined by analyzing the E field distribution at 2 MHz and 1 GHz in HFSS. They are 1–30 for the relative dielectric constant ε_r and 0–1 S/m for the conductivity σ at 1 GHz, which is the stop frequency according to current equipment. It could work for almost all rock samples with different saturations and borehole fluids at 2 MHz. The sample inconsistency problem is fixed by adding a 0.055-in. plexiglass ring. The mathematical formulation is corrected. Hence, the errors in the measurement system are eliminated and the measurement system is ready to use.

The equipment used in the parallel-disk measurement system is the parallel-disks sample holder, a HP4275A LCR Meter, a HP4191A RF Impedance Analyzer, and two LabVIEW computer control programs. The LCR Meter is used over 10 kHz−10 MHz; the Impedance Analyzer is employed over 1 MHz−1 GHz.

The TM_{010} resonant cavity technique is operated at 1.1 GHz. The automatic measurement system is reestablished by using the resonant cavity, a HP8510C Network Analyzer, and a LabVIEW computer control program. Thorough maintenance of the cavity is completed to make sure the cavity worked well.

The operating software of the system is developed in LabVIEW for equipment control, sample measurement and data processing. Easy-to-use software interfaces are provided to the operators. By implementing the automatic measurement system, the procedure for the sample measurement and the data processing are simplified and became more efficient than previous.

Electrical property measurements of different samples with different structures and moisture contents are performed to verify the satisfactory measurement accuracy of these systems. The mortar samples with different moisture contents are measured. The experimental data, which are not available in previous existing electrical property tables, are presented in this chapter.

REFERENCES

[1] G.E. Archies, Electrical resistivity log as an aid in determining some reservoir characteristics, Trans. AIME 146 (1942) 1942.
[2] M.C. Leverett, Flow of oil-water mixtures through unconsolidated sands, Trans. AIME 132 (1938) 1938.
[3] H. Andrä, et al., Digital Rock Physics Benchmarks—Part I: Imaging and Segmentation Digital Rock, Elsevier, Amsterdam, 2012.
[4] R.C. Mc. Phedran, D.R. Mc. Kezie, The conductivity of lattices of spheres II: the body centered and face centered cubic lattices, Proc. R. Soc. A 363 (1978) 211−232.
[5] D.J. Bergman, The dielectric constant of a simple cubic array of identical spheres, J. Phys. C Solid State Phys. 12 (1979) 4947−4960.
[6] P.N. Sen, C. Scala, M.H. Cohen, A self-similar model for sedimentary rocks with application to the dielectric constant of fused glass beads, Geophysics 46 (5) (1981) 781−795.
[7] A.A. Garrouch, M.M. Sharma, The influence of clay content, salinity, stress and wettability on the dielectric properties of brine-saturated rocks: 10 Hz to 10 MHz, Geophysics 59 (6) (1994) 909−917.
[8] R. Knight, A. Nur, The dielectric constant of sandstones, 60 KHz to 4 MHz, Geophysics 52 (5) (1987) 644−654.
[9] R. Knight, A. Endres, A new concept in modeling the dielectric response of sandstones: defining a wetted rock and bulk water system, Geophysics 55 (5) (1990) 586−594.

[10] R. Knight, A. Abad, Rock/water interaction in dielectric properties: experiments with hydrophobic sandstones, Geophysics 60 (1995) 431–436.
[11] L.C. Shen, W.C. Skevre, J.M. Price, K. Athavale, Dielectric properties of reservoir rocks at ultrahigh frequencies, Geophysics 50 (1985) 692–704.
[12] L.C. Shen, C. Liu, J. Korringe, K.J. Dunn, Computation of conductivity and dielectric constant of periodic porous media, J. Appl. Phys. 67 (1990) 7071–7081.
[13] C. Liu, L.C. Shen, Dielectric constant of two-component, two-dimensional mixtures in terms of Bergman-Milton simple poles, J. Appl. Phys. 73 (1993) 1897–1903.
[14] C. Liu, Q. Zou, A simplified Fourier series expansion method for computing the effective dielectric constant of a two-component, three-dimensional composite material with geometric symmetry, Modelling Simul. Mater. Sci. Eng. 4 (1996) 55–71.
[15] C. Liu, H. Wu, Computation of the effective dielectric constant of two-component, three-dimensional mixtures using a simple pole expansion method, J. Appl. Phys. 82 (1997) 345–350.
[16] B. Scaife, Principles of Dielectrics (Monographs on the Physics & Chemistry of Materials), second ed., Oxford University Press, New York, NY, September 3, 1998. ISBN 978-0198565574.
[17] H. Fricke, A mathematical treatment of the electrical conductivity and capacity of disperse systems, Phys. Rev. 24 (1924) 575–587.
[18] T. Hanai, Electric properties of emulsions, in: P. Sherman (Ed.), Emulsion Science, Academic Press, New York, NY, 1968, pp. 353–478.
[19] C. Liu, L.C. Shen, Dispersion characteristics of two-component two-dimensional mixtures, Modeling Simul. Mater. Sci. Eng. 1 (5) (1993) 723–730.
[20] K.S. Cole, R.H. Cole, Dispersion and absorption in dielectrics—I Alternating current characteristics, J. Chem. Phys. 9 (1941) 341–352. Bibcode:1941JChPh...9..341C. http://dx.doi.org/10.1063/1.1750906.
[21] K.S. Cole, R.H. Cole, Dispersion and absorption in dielectrics—II Direct current characteristics, J. Chem. Phys. 10 (1942) 98–105. Bibcode:1942JChPh..10...98C. http://dx.doi.org/10.1063/1.1723677.
[22] W.H. Pelton, S.H. Ward, P.G. Hallof, et al., Mineral discrimination and removal of inductive coupling with multifrequency IP, Geophysics 43 (1978) 588–609.
[23] M.A. Stuchly, S.S. Stuchly, Coaxial line reflection methods for measuring dielectric properties of biological substrates at radio and microwave frequencies-a review, IEEE Trans. Instrum. Meas. IM-29 (1980) 176–183.
[24] J.P. Poley, J.J. Nooteboot, P.J. de Waal, Use of VHF dielectric measurements for borehole formation analysis, Log Anal. 19 (3) (1978) 8–30.
[25] R. Rau, R. Wharton, Measurement of core electrical parameters at ultrahigh and microwave frequencies, J. Pet. Technol. 34 (11) (1982) 2689–2700.
[26] L.C. Shen, A laboratory technique for measuring dielectric properties of core samples at ultrahigh frequencies, Soc. Pet. Eng. J. 25 (4) (1985) 502–514.
[27] M. Taherian, D. Yuen, T. Habashy, J. Kong, A coaxial-circular waveguide for dielectric measurement, IEEE Trans. Geosci. Remote Sens. 29 (2) (1991) 321–329.
[28] K.F. Tam, Dielectric Property Measurements of Rocks in the VHF-UHF Region (Ph.D. dissertation), Department of Geophysics, Texas A&M University, College Station, Texas, 1974.
[29] L.C. Shen, H. Marouni, Y.X. Zhang, X.D. Shi, Analysis of parallel disk sample holder for dielectric permittivity measurement, Well Log. Lab. Tech. Rep., 7, University of Houston, 1986, pp. 7.1–7.26.
[30] L.C. Shen, H. Marouni, Y.X. Zhang, Analysis of the parallel-disk sample holder for dielectric permittivity measurement, IEEE Trans. Geosci. Remote Sens. GE-25 (5) (1987) 534–540.
[31] S. Roberts, A. Von Hippel, New method for measuring dielectric constant and loss in the range of centimeter waves, J. Appl. Phys. 17 (1946) 610–616.
[32] H. Marouni, Dielectric Constant and Conductivity Measurement of Reservoir Rocks in the Range of 20–50 MHz (Master's thesis), Department of Electrical and Computer Engineering, University of Houston, 1985.
[33] X. Zhao, Measurement of Samples in a TM010 Cylindrical Cavity at 1.1 GHz (Master's thesis), Department of Electrical and Computer Engineering, University of Houston, 1991.

APPENDIX A E FIELD ANALYSIS OF THE CIRCUIT MODEL OF THE PARALLEL-DISK SAMPLE HOLDER

A parallel-disk capacitor is shown in Fig. A.1 [34]. Because the distance between the two disks is small, the E fields are independent of X and Y axis; also because of the symmetry of the structure, the E fields are independent of φ.

When the frequency is low, only the static fields are considered. The Gauss law is,

$$\oiint \underline{D} \cdot d\underline{s} = q \tag{A.1}$$

$$\oiint_s \hat{n} \cdot \underline{E}^{(0)} ds = \frac{1}{\varepsilon} \iiint_V \rho_V dv = \frac{1}{\varepsilon} \rho_S A \tag{A.2}$$

where A is the surface area of S1 and S2; S1 and S2 the upper and lower surface of the parallel disk as shown in Fig. A.1.

In this problem $\underline{E}$ is only in the $\hat{z}$ direction, it gives,

$$\iint_{S1} E_Z^{(0)} ds + \iint_{S2} E_Z^{(0)} ds = \frac{\rho_S A}{\varepsilon} \tag{A.3}$$

Because the surface S1 is shorted to the ground, it gives $E_Z^{(0)} = 0$ on S1.
Since $E_Z^{(0)}$ is constant on surface S2, it gives,

$$\iint_{S2} E_Z^{(0)} ds = E_Z^{(0)} A \tag{A.4}$$

Substituting Eq. (A.4) into (A.3) gives,

$$E_Z^{(0)} A = \frac{\rho_S A}{\varepsilon} \tag{A.5}$$

Eq. (A.5) is equal to Eq. (A.6),

$$E_Z^{(0)} = \frac{\rho_S}{\varepsilon} = E_0 \tag{A.6}$$

Figure A.1 A parallel-disk capacitor.

It is easy to have,

$$J_Z^{(0)} = \sigma E_Z^{(0)} = \sigma E_0 \tag{A.7}$$

One of the Maxwell equations is,

$$\nabla \times \underline{H} = \underline{J} + jw\varepsilon\underline{E} \tag{A.8}$$

In cylindrical coordinates, the curl is defined as [35],

$$\nabla \times \underline{A} = \hat{\rho}\left(\frac{1}{\rho}\frac{\partial A_z}{\partial \phi} - \frac{\partial A_\phi}{\partial z}\right) + \hat{\phi}\left(\frac{\partial A_\rho}{\partial z} - \frac{\partial A_z}{\partial \rho}\right) + \hat{z}\left(\frac{1}{\rho}\frac{\partial(\rho A_\varphi)}{\partial \rho} - \frac{1}{\rho}\frac{\partial A_\rho}{\partial \phi}\right) \tag{A.9}$$

Appling Eq. (A.9) into (A.8) gives,

$$\frac{1}{\rho}\frac{\partial[\rho(H_\phi)]}{\partial \rho} = J_Z + jw\varepsilon E_Z \tag{A.10}$$

Multiplying both sides of Eq. (A.10) by ρ gives,

$$\frac{\partial[\rho(H_\phi)]}{\partial \rho} = \rho(J_Z + jw\varepsilon E_Z) \tag{A.11}$$

For the zeroth order of field, $J_Z = \sigma E_0$, Eq. (A.11) becomes,

$$\frac{\partial[\rho(H_\phi^{(0)})]}{\partial \rho} = \rho\sigma E_0 \tag{A.12}$$

$$\rho(H_\phi^{(0)}) = \frac{1}{2}\rho^2\sigma E_0 \tag{A.13}$$

$$H_\phi^{(0)} = \frac{1}{2}\rho\sigma E_0 \tag{A.14}$$

When the frequency goes higher, the E fields cannot be represented accurately by the static fields only. Higher order fields must be added. One of the Maxwell equations is,

$$\nabla \times \underline{E} = -\frac{\partial \underline{B}}{\partial t} \tag{A.15}$$

$$\underline{B} = \mu\underline{H} \tag{A.16}$$

$$\nabla \times \underline{E} = -\mu\frac{\partial \underline{H}}{\partial t} = -jw\mu\underline{H} \tag{A.17}$$

Appling Eq. (A.9) into (A.17) gives,

$$-\hat{\phi}\frac{\partial E_Z}{\partial \rho} = \hat{\phi}(-jw\mu)H_\phi \tag{A.18}$$

$$\frac{\partial E_Z}{\partial \rho} = jw\mu H_\phi \tag{A.19}$$

One of the Maxwell equations is,

$$\nabla \times \underline{E}^{(1)} = -jw\mu \underline{H}^{(0)} \tag{A.20}$$

$$\frac{\partial E_Z^{(1)}}{\partial \rho} = jw\mu \left(\frac{1}{2}\sigma\rho E_0\right) \tag{A.21}$$

$$E_Z^{(1)} = \frac{j}{4}w\mu\sigma\rho^2 E_0 \tag{A.22}$$

Applying Eq. (A.22) into (A.11) gives,

$$\frac{\partial[\rho(H_\phi^{(1)})]}{\partial \rho} = \rho E_Z^{(1)}[\sigma + jw\varepsilon] = \rho(\sigma + jw\varepsilon)\frac{j}{4}w\mu\sigma\rho^2 E_0 \tag{A.23}$$

$$H_\phi^{(1)} = \frac{j}{2}w\varepsilon\rho E_0 + \frac{j}{16}w\mu\sigma^2\rho^3 E_0 \tag{A.24}$$

The first order of fields are solved. Now we need to solve the second-order fields.

$$\nabla \times \underline{E}^{(2)} = -jw\mu \underline{H}^{(1)} \tag{A.25}$$

$$\frac{\partial E_Z^{(2)}}{\partial \rho} = jw\mu H_\phi^{(1)} = (jw\mu)\left(\frac{j}{2}w\varepsilon\rho E_0 + \frac{j}{16}w\mu\sigma^2\rho^3 E_0\right) \tag{A.26}$$

$$E_Z^{(2)} = (jw\mu)\left(\frac{j}{4}w\varepsilon\rho^2 E_0 + \frac{j}{64}w\mu\sigma^2\rho^4 E_0\right) = -\frac{1}{4}w^2\mu E_0\left(\varepsilon\rho^2 + \frac{1}{16}\mu\sigma^2\rho^4\right) \tag{A.27}$$

Eq. (A.28) can be derived from Eq. (A.10),

$$\frac{\partial[\rho(H_\phi^{(2)})]}{\partial \rho} = \rho\left(J_Z^{(2)} + jw\varepsilon E_Z^{(1)}\right) = \rho\left(\sigma E_Z^{(2)} + jw\varepsilon E_Z^{(1)}\right) \tag{A.28}$$

Applying Eqs. (A.22) and (A.27) into Eq. (A.28) gives,

$$\frac{\partial[\rho(H_\phi^{(2)})]}{\partial \rho} = \rho\left[-\frac{1}{4}w^2\mu\sigma E_0\left(\varepsilon\rho^2 + \frac{1}{16}\mu\sigma^2\rho^4\right) + jw\varepsilon\left(\frac{j}{4}w\mu\sigma\rho^2 E_0\right)\right] \quad (A.29)$$

$$\rho(H_\phi^{(2)}) = -\frac{1}{8}E_0 w^2\mu\varepsilon\sigma\rho^4 - \frac{1}{384}E_0 w^2\mu^2\sigma^3\rho^6 \quad (A.30)$$

$$H_\phi^{(2)} = -\frac{1}{8}E_0 w^2\mu\varepsilon\sigma\rho^3 - \frac{1}{384}E_0 w^2\mu^2\sigma^3\rho^5 \quad (A.31)$$

The second order of fields are solved. Now we need to solve the third-order fields.

$$\nabla \times \underline{E}^{(3)} = -jw\mu\underline{H}^{(2)} \quad (A.32)$$

$$\frac{\partial E_Z^{(3)}}{\partial \rho} = jw\mu H_\phi^{(2)} = (jw\mu)\left(-\frac{1}{8}E_0 w^2\mu\varepsilon\sigma\rho^3 - \frac{1}{384}E_0 w^2\mu^2\sigma^3\rho^5\right) \quad (A.33)$$

$$\begin{aligned}E_Z^{(3)} &= (jw\mu)\left(-\frac{1}{32}E_0 w^2\mu\varepsilon\sigma\rho^4 - \frac{1}{384*6}E_0 w^2\mu^2\sigma^3\rho^6\right) \\ &= -\frac{j}{32}E_0 w^3\mu^2\varepsilon\sigma\rho^4 - \frac{j}{384*6}E_0 w^3\mu^3\sigma^3\rho^6\end{aligned} \quad (A.34)$$

$$\frac{\partial[\rho(H_\phi^{(3)})]}{\partial \rho} = \rho\left(J_Z^{(3)} + jw\varepsilon E_Z^{(2)}\right) = \rho\left(\sigma E_Z^{(3)} + jw\varepsilon E_Z^{(2)}\right) \quad (A.35)$$

$$\frac{\partial[\rho(H_\phi^{(3)})]}{\partial \rho} = \rho\bigg\{\sigma\left(-\frac{j}{32}E_0 w^3\mu^2\varepsilon\sigma\rho^4 - \frac{j}{384*6}E_0 w^3\mu^3\sigma^3\rho^6\right) \\ + jw\varepsilon\left[-\frac{1}{4}w^2\mu E_0\left(\varepsilon\rho^2 + \frac{1}{16}\mu\sigma^2\rho^4\right)\right]\bigg\} \quad (A.36)$$

$$\rho(H_\phi^{(3)}) = -\frac{j}{16}E_0 w^3\mu\varepsilon^2\rho^4 - \frac{j}{128}E_0 w^3\mu^2\varepsilon\sigma^2\rho^6 - \frac{j}{384*48}E_0 w^3\mu^3\sigma^4\rho^8 \quad (A.37)$$

$$H_\phi^{(3)} = -\frac{j}{16}E_0 w^3\mu\varepsilon^2\rho^3 - \frac{j}{128}E_0 w^3\mu^2\varepsilon\sigma^2\rho^5 - \frac{j}{384*48}E_0 w^3\mu^3\sigma^4\rho^7 \quad (A.38)$$

Considering the terms up to ρ^4 gives,

$$E_Z = E_Z^{(0)} + E_Z^{(1)} + E_Z^{(2)} + E_Z^{(3)} \text{ (terms up to } \rho^4) \tag{A.39}$$

$$E_Z = E_0 + \frac{j}{4} w\mu\sigma\rho^2 E_0 - \frac{1}{4} w^2 \mu E_0 \left(\varepsilon\rho^2 + \frac{1}{16}\mu\sigma^2\rho^4\right) - \frac{j}{32} E_0 w^3 \mu^2 \varepsilon\sigma\rho^4 \tag{A.40}$$

$$E_Z = E_0 \left[1 + \frac{j}{4} w\mu\rho^2(\sigma + jw\varepsilon) - \frac{1}{64} w^2 \mu^2 \sigma \rho^4 (\sigma + j2w\varepsilon)\right] \tag{A.41}$$

$$H_\phi = H_Z^{(0)} + H_Z^{(1)} + H_Z^{(2)} + H_Z^{(3)} \text{ (terms up to } \rho^5) \tag{A.42}$$

$$H_\phi = \frac{1}{2}\rho\sigma E_0 + \frac{j}{2} w\varepsilon\rho E_0 + \frac{j}{16} w\mu\sigma^2 \rho^3 E_0 + \left(-\frac{1}{8} E_0 w^2 \mu\varepsilon\sigma\rho^3 - \frac{1}{384} E_0 w^2 \mu^2 \sigma^3 \rho^5\right)$$
$$+ \left(-\frac{j}{16} E_0 w^3 \mu\varepsilon^2 \rho^3 - \frac{j}{128} E_0 w^3 \mu^2 \varepsilon\sigma^2 \rho^5\right)$$

$$\tag{A.43}$$

$$H_\phi = \frac{\rho E_0}{2} \left\{(\sigma + jw\varepsilon)\left[1 + \frac{j}{8} w\mu\rho^2 (\sigma + jw\varepsilon)\right] - \frac{1}{192} w^2 \mu^2 \sigma^2 \rho^4 (\sigma + j3w\varepsilon)\right\} \tag{A.44}$$

From the derivation of E and H fields, the admittance between the two parallel disks can be calculated as,

$$Y = \frac{I}{V} = \rho = a \frac{2\Pi a H_\phi}{dE_z}$$

$$= \frac{2\Pi a}{d} \cdot \left. \frac{\frac{\rho E_0}{2}\left\{(\sigma + jw\varepsilon)\left[1 + \frac{j}{8} w\mu\rho^2(\sigma + jw\varepsilon)\right] - w^2\mu^2\sigma^2\rho^4(\sigma + j3w\varepsilon)\frac{1}{192}\right\}}{E_0\left[1 + \frac{j}{4}w\mu\rho^2(\sigma + jw\varepsilon) - \frac{1}{64}w^2\mu^2\sigma\rho^4(\sigma + j2w\varepsilon)\right]} \right|_{\rho=a}$$

$$= \frac{\Pi a^2}{d} \cdot \left. \frac{(\sigma + jw\varepsilon)\left[1 + \frac{j}{8}w\mu a^2(\sigma + jw\varepsilon)\right] - \frac{1}{192}w^2\mu^2\sigma^2 a^4(\sigma + j3w\varepsilon)}{1 + \frac{j}{4}w\mu a^2(\sigma + jw\varepsilon) - \frac{1}{64}w^2\mu^2\sigma a^4(\sigma + j2w\varepsilon)} \right|_{\rho=a}$$

$$\tag{A.45}$$

APPENDIX B EQUIPMENT CALIBRATION SYNOPSIS

Most of the content in this appendix is from operation or service manuals of the equipment. The main purpose for including them in this chapter is to offer a complete reference to the reader when he/she tries to use this chapter as reference for the current automatic measurement systems.

B.1 HP4275A LCR Meter

Most of the content here is from the HP4275A Multi-Frequency LCR Meter Operating Manual. A HP4275A LCR Meter has self-diagnostic functions, which are automatically performed or can be done anytime to confirm the normal operation of the instrument.

Self-tests can be performed each time before sample measurement (before self-test, make sure the instrument is set to be LOCAL). The correct operating procedures for the self-test are: Display Test (the first step of SELF TEST) and Analog Circuit Test (SELF TEST). The Analog Circuit Test is divided into an "open" and a "short" test. 16047A Test Fixture is used for SELF TEST. The procedures are shown below.

For an "open" condition, nothing should be connected. Set the Display A as C. Press SELF TEST button on the front panel, Display A shows "OP." After 2 or 3 seconds, all the display windows show: "⊟ ⊟. . .." Then press the "ZERO OPEN" button and the "SELF TEST" button. Display A shows "CAL" and the calibration starts. The "open" test comprises 20 steps of the diagnostic tests. During the test, the Display A exhibits normal test results as "OP." After one "open" test, push the SELF TEST button and the instrument will be released. If an abnormal result occurs during the test, the number of the abnormal step is displayed in the window of Display A.

For a "short" condition, a low-impedance shorting strap is connected across the HIGH and LOW sides of the test fixture contact blocks. Set the Display A as L or R. Press SELF TEST button on the front panel and Display A shows "SH." After 2 or 3 seconds, all the display windows show: "⊟ ⊟. . .." Then press the "ZERO OPEN" button and the "SELF TEST" button and Display A shows "CAL" and the calibration starts. The "short" test comprises seven steps of the diagnostic tests. During the test, the Display A exhibits a normal test result as "SH." After one "short" test, push the "SELF TEST" button and the instrument will be released. If an abnormal result occurs during the test, the number of the abnormal step is displayed in the window of Display A.

SELF-TEST procedures are also talked about in the HP4275A Multi-Frequency LCR Meter Operating Manual.

B.2 HP4191A RF Impedance analyzer

Most of the content here is from the HP4191A RF Impedance Analyzer Operation and Service Manual. Autocalibration of HP4191A is performed by using reference termination under automatic settings of the measurement parameter (reflection coefficient) and of the test frequency as it sweeps the programmed frequency range. The detail of the procedure is shown below.

1. Press the LINE button to turn on the instrument, wait 10 minutes to warm up the instrument.
2. Press CALIBRATION key and the indicator lamps light. Concurrently "Conn" "0Ω" figures appear in the displays to indicate that 0Ω reference termination should be connected to UNKNOWN connector.
3. *0Ω calibration*: carefully couple 0Ω termination to the UNKNOWN connector, rotate 0Ω termination cap nut clockwise until it is firm. Press CALIBRATION START button. Test frequency display succeeding changes in a higher frequency direction and ends at 1000.0 MHz. "Conn" "0 S" figures appear in the displays to indicate that 0 S reference termination should be connected to UNKNOWN connector.
4. *0 S calibration*: carefully remove the 0Ω termination and couple 0 S termination to the UNKNOWN connector, rotate 0 S termination cap nut clockwise until it is firm. Press CALIBRATION START button. Test frequency display succeeding changes in a higher frequency direction and ends at 1000.0 MHz. "Conn" "50Ω" figures appear in the displays to indicate that 0 S reference termination should be connected to UNKNOWN connector.
5. *50Ω calibration*: carefully couple 0 S termination to the UNKNOWN connector, rotate 50Ω termination cap nut clockwise until it is firm. Press CALIBRATION START button. Test frequency displays succeeding changes in a higher frequency direction and ends at 1000.0 MHz. Carefully remove 50Ω reference termination from UNKNOWN connector. Calibration is finished and calibration data is stored in the internal memory.

The operator can also perform selective calibration.

- *0Ω, 0 S or 50Ω calibration*
    1. Connect the 0Ω reference termination to the UNKNOWN connector, set the measurement parameter to R-X, press CALIBRATION START button.
    2. Connect the 0 S reference termination to the UNKNOWN connector, set the measurement parameter to G-B, press CALIBRATION START button.
    3. Connect the 50Ω reference termination to the UNKNOWN connector, set the measurement parameter to $\Gamma_x - \Gamma_y$, press CALIBRATION START button.
- *Calibration on a defined frequency range*

    Set the start and stop frequencies, then the operator can perform calibration in the predetermined frequency range.

B.3 HP8510C Network analyzer

Most of the content here is from the HP8510C Network Analyzer Operation and Service Manual. Calibration greatly reduces repeatable systematic errors for your measurement. The procedure of system measurement calibration is shown sequentially.

1. Choose the selective calibration frequency range
   Set the start frequency and stop frequency from the input of the front panel
2. Calibration procedure
   - **2.1** Press *CAL* key on the front panel; choose *CAL 3.5 mm* or *CAL 2.4 mm* soft key to select the approximate calibration kit
   - **2.2** Press *FULL 2-PORT* soft key on the CRT Display
   - **2.3** Press *REFLECT'N* soft key on the CRT Display

     Connect standards (OPEN, SHORT, LOADS) at Port1, and press the appropriate soft key under (S11 :) on the CRT Display.

     Connect standards (OPEN, SHORT, LOADS) at Port2, and press the appropriate soft key under (S22 :) on the CRT Display.

     Press *REFLECT'N DONE* soft key on the CRT Display to store data.
   - **2.4** Press *TRANSMISSION* soft key on the CRT Display

     Connect Port 1 to Port 2 Thru.

     When the trace is correct, press *FWD. TRANS. THRU*. S21 frequency response is measured.

     Press *REV. MATCH. THRU*. S21 load match is measured.

     Press *REV. TRANS. THRU*. S21 frequency response is measured.

     Press *REV. MATCH. THRU*. S21 load match is measured.

     Press *TRANS. DONE* to store data.
   - **2.5** Press *ISOLATION* soft key on the CRT Display

     Connect a load at Port 1 and a load at Port 2.

     When the trace is correct, press *FWD. ISOL'N ISOL'N STD*. S21 noise floor is measured. Press *REV. ISOL'N ISOL'N STD*. S12 noise floor is measured.

     Press *ISOLATION DONE* to store data.
   - **2.6** Press *SAVE 2-PORT CAL*, and then select a Cal Set to save the calibration data. Cal menu is displayed with CORRECTION ON
3. Perform measurement considering calibration data
   - **3.1** Correct trace is displayed
   - **3.2** Connect the test device; set the stimulus frequency range. It must match those used during the calibration. Press any parameter key from S11, S12, S21, and S22 to display corrected data for that parameter.

CHAPTER 4

Triaxial Induction and Logging-While-Drilling Resistivity Tool Response in Homogeneous Anisotropic Formations

Contents

4.1 Magnetic Dipole in Homogeneous Lossy Media	122
4.1.1 Magnetic dipole in homogeneous isotropic lossy media	122
4.1.2 Magnetic dipole in homogeneous transverse isotropic lossy media	124
4.2 Finite Coil in Homogeneous Formation	127
4.3 LWD Tool Response in Homogeneous Formation	127
4.3.1 Introduction of commercial LWD/MWD tools	127
4.3.2 Dielectric constant model and conversion charts	136
4.4 Triaxial Induction Logging Tool Response in Biaxial Anisotropic Homogeneous Formation	136
4.4.1 Spectrum-domain solution to Maxwell's equations in a homogeneous biaxial anisotropic medium	140
4.4.2 Full magnetic field response of a triaxial induction sonde in a biaxial anisotropic medium	143
4.4.3 The full magnetic field response with arbitrary tool axis	148
4.4.4 Computation of the triple integrals	149
4.4.5 Numerical examples	152
References	160
Appendix A Derivation of Parameters a and b in Eqs. (4.53)–(4.55)	161

In this chapter, we will study the response of induction and logging-while-drilling (LWD) resistivity logging tools in homogeneous lossy formations. As we know, in induction and LWD resistivity logging tools, the transmitter and receiver coils are of finite dimensions. However, compared with the skin depth, these antennas are electrically small. To simplify the solutions, in the analysis below, we use a magnetic dipole to replace the antennas. Therefore the coil antennas can be simplified as equivalent magnetic dipoles without much loss of accuracy. Therefore the solutions to the response of induction or LWD logging tools in homogeneous lossy media is converted to the problem of the electromagnetic (EM) field generated by magnetic dipoles in homogeneous lossy media.

4.1 MAGNETIC DIPOLE IN HOMOGENEOUS LOSSY MEDIA

4.1.1 Magnetic dipole in homogeneous isotropic lossy media

According to its electrical property, a medium can be isotropic, transverse isotropic (TI), or biaxial anisotropic (fully anisotropic). The electrical conductivity of a medium is actually a positive-definite symmetric second-rank tensor [1,2]. In the principal axis coordinate system, the conductivity tensor diagonalizes

$$\hat{\sigma} = \begin{bmatrix} \sigma_x & 0 & 0 \\ 0 & \sigma_y & 0 \\ 0 & 0 & \sigma_z \end{bmatrix} \quad (4.1)$$

For an isotropic medium, the conductivity is a scalar, i.e., $\sigma_x = \sigma_y = \sigma_z = \sigma$; for a TI medium, two of the three principal conductivities are equal, i.e., $\sigma_x = \sigma_y$. A practical example of the TI situation is thin-bedded sequences of alternating high- and low-resistivity layers that occur in logging environments. If thin-laminated sequences have a fracture pattern that cuts across bedding, the conductivity of the medium is fully anisotropic. Full anisotropy is referred to as "biaxial" anisotropy in crystals. In this case, all three principal conductivities are different, representing the differences of pore-connectivity and conductivity in the vertical and lateral directions.

As a start, we consider the field generated by magnetic dipoles in homogeneous isotropic medium. Assuming the harmonic time dependence to be $e^{-i\omega t}$, Maxwell's equations for the electric and magnetic fields are

$$\nabla \times \boldsymbol{H}(\boldsymbol{r}) = (\sigma - i\omega\varepsilon)\boldsymbol{E}(\boldsymbol{r}) = \sigma'\boldsymbol{E}(\boldsymbol{r}) \quad (4.2a)$$

$$\nabla \times \boldsymbol{E}(\boldsymbol{r}) = i\omega\mu\boldsymbol{H}(\boldsymbol{r}) + i\omega\mu\boldsymbol{M}_s(\boldsymbol{r}) \quad (4.2b)$$

where $\boldsymbol{M}_s$ is the moment of the magnetic dipole, $\sigma' = \sigma - i\omega\varepsilon$ is the complex conductivity, ε is the dielectric permittivity, and σ is the conductivity of the medium.

In order to solve Eq. (4.2a,b), we introduce the Hertz vector potential $\boldsymbol{\Pi}$

$$\boldsymbol{E} = i\omega\mu\nabla \times \boldsymbol{\Pi} \quad (4.3a)$$

$$\boldsymbol{H} = \nabla \times \nabla \times \boldsymbol{\Pi} - \boldsymbol{M}_s \quad (4.3b)$$

Substituting Eq. (4.3a,b) into (4.2a,b), we can obtain the equation for the Hertz potential

$$\nabla^2 \boldsymbol{\Pi} + k^2 \boldsymbol{\Pi} = -\boldsymbol{M}_s \delta(\boldsymbol{r}) \quad (4.4)$$

where $k^2 = \omega^2 \mu \varepsilon$. Solving Eq. (4.4), we can obtain the Hertz potential and then the electric and magnetic field in homogeneous medium can be obtained from Eq. (4.3a,b).

For a z-directed magnetic dipole $\boldsymbol{M} = (0, 0, M_z)^T$, the Hertz potential is given by

$$\boldsymbol{\Pi} = \frac{M_z e^{ikr}}{4\pi r} \hat{z} \tag{4.5}$$

For a x-directed magnetic dipole $\boldsymbol{M} = (M_x, 0, 0)^T$, the Hertz potential is given by

$$\boldsymbol{\Pi} = \frac{M_x e^{ikr}}{4\pi r} \hat{x} \tag{4.6}$$

and for a y-directed magnetic dipole $\boldsymbol{M} = (0, M_y, 0)^T$, the Hertz potential is

$$\boldsymbol{\Pi} = \frac{M_y e^{ikr}}{4\pi r} \hat{y} \tag{4.7}$$

Substituting Eqs. (4.5–4.7) into (4.3b), we can obtain the magnetic field generated by a magnetic dipole $\boldsymbol{M} = (M_x, M_y, M_z)\delta(\boldsymbol{r})$ in a homogeneous isotropic medium.

$$H_{xx} = \frac{e^{ikr}}{4\pi} \left[\frac{k^2}{r} + \frac{ik}{r^2} - \frac{k^2 x^2 + 1}{r^3} - \frac{3ikx^2}{r^4} + \frac{3x^2}{r^5} \right] \tag{4.8}$$

$$H_{yx} = H_{xy} = -\frac{xy e^{ikr}}{4\pi} \left[\frac{k^2}{r^3} + \frac{3ik}{r^4} - \frac{3}{r^5} \right] \tag{4.9}$$

$$H_{zx} = H_{xz} = -xz \frac{e^{ikr}}{4\pi r^3} \left[k^2 + \frac{3ik}{r} - \frac{3}{r^2} \right] \tag{4.10}$$

$$H_{yy} = \frac{e^{ikr}}{4\pi} \left[\frac{k^2}{r} + \frac{ik}{r^2} - \frac{k^2 y^2 + 1}{r^3} - \frac{3iky^2}{r^4} + \frac{3y^2}{r^5} \right] \tag{4.11}$$

$$H_{zy} = H_{yz} = -yz \frac{e^{ikr}}{4\pi r^3} \left[k^2 + \frac{3ik}{r} - \frac{3}{r^2} \right] \tag{4.12}$$

$$H_{zz} = \frac{e^{ikr}}{4\pi r} \left[k^2 + \frac{ik}{r} - \frac{(k^2 z^2 + 1)}{r^2} - \frac{3ikz^2}{r^3} + \frac{3z^2}{r^4} \right] \tag{4.13}$$

4.1.2 Magnetic dipole in homogeneous transverse isotropic lossy media

Next, we consider the field generated by magnetic dipoles in homogeneous TI medium. The Maxwell's equations for the electric and magnetic fields in TI medium are:

$$\nabla \times \mathbf{H}(\mathbf{r}) = (\hat{\sigma} - i\omega\hat{\varepsilon})\mathbf{E}(\mathbf{r}) = \hat{\sigma}'\mathbf{E}(\mathbf{r}) \tag{4.14a}$$

$$\nabla \times \mathbf{E}(\mathbf{r}) = i\omega\mu \mathbf{H}(\mathbf{r}) + i\omega\mu \mathbf{M}_s(\mathbf{r}) \tag{4.14b}$$

In Eq. (4.14a), we define a complex conductivity tensor $\hat{\sigma}'$ for convenience

$$\hat{\sigma} = \begin{bmatrix} \sigma'_h & 0 & 0 \\ 0 & \sigma'_h & 0 \\ 0 & 0 & \sigma'_v \end{bmatrix} = \begin{bmatrix} \sigma_h - i\omega\varepsilon_h & 0 & 0 \\ 0 & \sigma_h - i\omega\varepsilon_h & 0 \\ 0 & 0 & \sigma_v - i\omega\varepsilon_v \end{bmatrix} \tag{4.15}$$

Following Moran and Gianzero [3,4] we can introduce the Hertz vector potential π and scalar potential Φ,

$$\hat{\sigma}' \cdot \mathbf{E}(\mathbf{r}) = i\omega\mu\sigma'_h \nabla \times \pi \tag{4.16a}$$

$$\mathbf{H}(\mathbf{r}) = i\omega\mu\sigma'_h \pi + \nabla\Phi \tag{4.16b}$$

Introduce a gauge condition for the scalar potential,

$$\nabla \cdot (\hat{\sigma}' \cdot \pi) = \sigma'_v \Phi \tag{4.17}$$

We will have

$$\mathbf{E}(\mathbf{r}) = i\omega\mu\sigma'_h \hat{\sigma}^{-1} \cdot \nabla \times \pi \tag{4.18}$$

and

$$\mathbf{H}(\mathbf{r}) = i\omega\mu\sigma'_h \pi + \nabla\left(\frac{\nabla \cdot (\hat{\sigma}' \cdot \pi)}{\sigma'_v}\right) \tag{4.19}$$

Following the procedure in Refs. [2,3], we can obtain the expression of the Hertz vector potential and scalar potential in a homogeneous medium for the x-, y- and z-directed magnetic dipoles.

For a x-directed magnetic dipole $\mathbf{M} = (M_x, 0, 0)^T$, the Hertz vector potential is given by

$$\pi = \pi_x \hat{x} + \pi_z \hat{z} \tag{4.20}$$

where

$$\pi_x = \frac{M_x}{4\pi\lambda} \frac{e^{ik_v s}}{s} \qquad (4.21)$$

$$\pi_z = \frac{M_x x}{4\pi\rho^2} \left(\lambda z \frac{e^{ik_v s}}{s} - z \frac{e^{ik_h r}}{r} \right) \qquad (4.22)$$

where

$$\lambda^2 = \sigma_h'/\sigma_v' \qquad (4.23)$$

$$k_h^2 = i\omega\mu\sigma_h' \qquad (4.24)$$

$$k_v^2 = i\omega\mu\sigma_v' \qquad (4.25)$$

$$r = \sqrt{x^2 + y^2 + z^2} \qquad (4.26)$$

$$\rho = \sqrt{x^2 + y^2} \qquad (4.27)$$

and

$$s = \sqrt{x^2 + y^2 + \lambda^2 z^2} \qquad (4.28)$$

For a y-directed magnetic dipole $\boldsymbol{M} = (0, M_y, 0)^T$, the Hertz vector potential is given by

$$\boldsymbol{\pi} = \pi_y \hat{y} + \pi_z \hat{z} \qquad (4.29)$$

where

$$\pi_y = \frac{M_y}{4\pi\lambda} \frac{e^{ik_v s}}{s} \qquad (4.30)$$

$$\pi_z = \frac{M_y y}{4\pi\rho^2} \left(\lambda z \frac{e^{ik_v s}}{s} - z \frac{e^{ik_h r}}{r} \right) \qquad (4.31)$$

For a z-directed magnetic dipole $\boldsymbol{M} = (0, 0, M_z)^T$, the Hertz vector potential has only z component

$$\boldsymbol{\pi} = \pi_z \hat{z} \qquad (4.32)$$

where

$$\pi_z = \frac{M_z}{4\pi} \frac{e^{ik_h r}}{r} \qquad (4.33)$$

Substituting Eqs. (4.20)–(4.33) into (4.19), we can obtain all the nine components of the magnetic field generated by a magnetic dipole $\boldsymbol{M} = (M_x, M_y, M_z)\delta(\boldsymbol{r})$ in a homogeneous TI medium,

$$H_{xx} = \frac{e^{ik_v s}}{4\pi}\left[\frac{k_h^2}{\lambda s} + \frac{ik_h s - k_h k_v x^2}{s\rho^2} - \frac{2ik_h x^2}{\rho^4}\right]$$
$$- \frac{e^{ik_h r}}{4\pi}\left[\frac{ik_h r - k_h^2 x^2}{r\rho^2} - \frac{2ik_h x^2}{\rho^4} - \frac{ik_h}{r^2} + \frac{(k_h^2 x^2 + 1)}{r^3} + \frac{3ik_h x^2}{r^4} - \frac{3x^2}{r^5}\right] \qquad (4.34)$$

$$H_{yx} = H_{xy} = xy\frac{e^{ik_v s}}{4\pi\rho^2}\left[-\frac{k_h k_v}{s} - \frac{2ik_h}{\rho^2}\right] - \frac{e^{ik_h r} xy}{4\pi}\left[-\frac{k_h^2}{r\rho^2} - \frac{2ik_h}{\rho^4} + \frac{k_h^2}{r^3} + \frac{3ik_h}{r^4} - \frac{3}{r^5}\right] \qquad (4.35)$$

$$H_{zx} = H_{xz} = -xz\frac{e^{ik_h r}}{4\pi r^3}\left[k_h^2 + \frac{3ik_h}{r} - \frac{3}{r^2}\right] \qquad (4.36)$$

$$H_{yy} = \frac{e^{ik_v s}}{4\pi}\left[\frac{k_h^2}{\lambda s} + \frac{ik_h s - k_h k_v y^2}{s\rho^2} - \frac{2ik_h y^2}{\rho^4}\right]$$
$$- \frac{e^{ik_h r}}{4\pi}\left[\frac{ik_h r - k_h^2 y^2}{r\rho^2} - \frac{2ik_h y^2}{\rho^4} - \frac{ik_h}{r^2} + \frac{(k_h^2 y^2 + 1)}{r^3} + \frac{3ik_h y^2}{r^4} - \frac{3y^2}{r^5}\right] \qquad (4.37)$$

$$H_{zy} = H_{yz} = -yz\frac{e^{ik_h r}}{4\pi r^3}\left[k_h^2 + \frac{3ik_h}{r} - \frac{3}{r^2}\right] \qquad (4.38)$$

$$H_{zz} = \frac{e^{ik_h r}}{4\pi r}\left[k_h^2 + \frac{ik_h}{r} - \frac{(k_h^2 z^2 + 1)}{r^2} - \frac{3ik_h z^2}{r^3} + \frac{3z^2}{r^4}\right] \qquad (4.39)$$

We can see that in homogeneous TI medium, all the cross-coupling components are the same.

4.2 FINITE COIL IN HOMOGENEOUS FORMATION

The magnetic field response of the induction tools in homogeneous formation can be obtained by the formulas of either the magnetic dipole model or the finite coil model in homogeneous formation. Then the apparent resistivity/conductivity can be converted from the magnetic field.

For multicoil sondes, as discussed in Chapter 2, Fundamentals of Electromagnetic Fields Induction Logging Tools, the tool response is the normalized summation of the individual two-coil responses, weighted by the appropriate number of windings and spacing, i.e.,

$$\sigma_{total}^a = \frac{\sum_{i,j} \frac{T_i R_j \sigma_{i,j}^a}{L_{ij}}}{\sum_{i,j} \frac{T_i R_j}{L_{ij}}} \quad (4.40)$$

where T and R are transmitter and receiver turns, respectively, L is the spacing between a transmitter–receiver coil pair, and σ^a is the apparent conductivity signal. The normalization factor in the denominator of the above equation is often referred to as the *sensitivity* of the sonde. The sensitivity is a meaningful quantity in itself, since it is too low, the signal level of a tool may be so small that the measurement is impractical.

4.3 LWD TOOL RESPONSE IN HOMOGENEOUS FORMATION

4.3.1 Introduction of commercial LWD/MWD tools [5]

MWD (measurement while drilling) and LWD (logging while drilling) are the most important and popular logging services in recent years. They deliver the real-time data at transmission rates quadruple the industry standard and acquire high-quality data for geosteering and formation evaluation. In Section 1.2.3, we have briefly introduced the principle and application of LWD tools. In this section, we will introduce the most popular, commercial LWD tools which are in use today. We will introduce the some of the popular tools in the sequence of the service companies' name. The data in this section is obtained in the public literature and can also be found in Ref. [5].

4.3.1.1 APS WPR Wave Propagation Resistivity Sub [5]

The Wave Propagation Resistivity (WPR) Sub is developed by APS (Advanced Products and Systems) Technology. It is a spatially compensated, dual frequency (400 kHz and 2 MHz), dual spacing device designed for wireline-equivalent logging-while-drilling (LWD) and measurements-after-drilling (MAD) services in all well types. WPR's symmetrical design, with centrally located receive antennas, provides real-time compensation, eliminates invasion effects due to measurement delays, and improves accuracy by canceling variations in receiver channels. WPR operates in all mud types including oil-based mud and salt-saturated water-based mud, and provides real-time resistivity with flexible transmission formats. High-resolution data are stored in downhole memory, which can be retrieved and processed during trips.

Fig. 4.1 shows a 3.5-in. WPR tool and the parameters of the tool such as the collar size, transmitter and receiver spacing, working frequencies are given in Table 4.1. It should be noted that 22.5 in. short spacing is not included in the 3.5-in. WPR. The two receivers are spaced at 8.5 in.

4.3.1.2 Multiple Propagation Resistivity [5]

Baker Hughes has developed various LWD tools including Multiple Propagation Resistivity (MPR), NaviGator, Deep Propagation Resistivity (DPR) tool, DeepTrak and AziTrack. MPR tool is also a symmetrical/compensated tool, as shown in Fig. 4.2. It has different collar sizes varying from 23/8, 31/8 (Ultra Slim MPR), 4.75, 6.75, 8.25 to 9.5 in (MPR 91/2). The distance between two receivers is 8 in. The physical parameters of the MPR tool are given in Fig. 4.2 and Table 4.2.

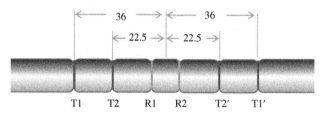

Figure 4.1 APS's WPR tool.

Table 4.1 Physical parameters of APS's WPR

Name	Frequency	Phase resistivity		Amplitude resistivity	
WPR	400 kHz	Long spacing	RPCECL	Long spacing	RACECL
		Short spacing	RPCECSL	Short spacing	RACECSL
	2 MHz	Long spacing	RPCECH	Long spacing	RACECH
		Short spacing	RPCECSH	Short spacing	RACECSH

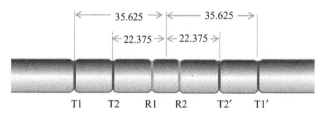

Figure 4.2 An illustration of MPR (Multiple Propagation Resistivity) tool by Baker Hughes.

Table 4.2 Physical parameters of MPR

Name	Frequency	Phase resistivity		Amplitude resistivity	
MPR	400 kHz	Long spacing	RPCL	Long spacing	RACL
		Short spacing	RPCSL	Short spacing	RACSL
	2 MHz	Long spacing	RPCH	Long spacing	RACH
		Short spacing	RPCSH	Short spacing	RACSH

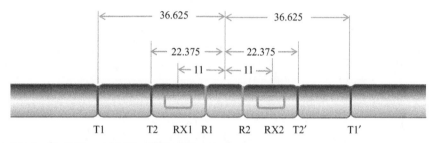

Figure 4.3 An illustration of AziTrak™ by Baker Hughes.

4.3.1.3 AziTrak Deep Azimuthal Resistivity [6,7]

AziTrak is Baker Hughes' most popular commercial tool, a fully integrated MWD/LWD suite, offering real-time Deep Azimuthal Resistivity, directional, azimuthal gamma ray, MPR, downhole pressure, and vibration measurements in a single sub. All the sensors are as close to the bit as possible for early detection of reservoir sections. Dual receiver/quadruple transmitter array, symmetrically oriented, can eliminate downhole temperature and pressure effects. AziTrak Deep Azimuthal Resistivity measurement tool detects reservoir boundaries, offering accurate formation evaluation to reduce wellbore position uncertainty. The physical parameters of the AziTrak tool are given in Fig. 4.3. From Figs. 4.2 and 4.3, we can see that the AziTrak tool is based on the MPR structure by adding two X-receivers RX1 and RX2. Other than the regular MPR measurements, AziTrak can also provide cross-component images and formation information, which can be used to invert distance to bed surface and bed-surface angles.

4.3.1.4 Centerfire and Compact Propagation Resistivity [8,9]

The main propagation tools developed by General Electrical (GE) Energy are the Centerfire system and the Compact Propagation Resistivity (CPR) Tool. The Centerfire has three probe sizes: 4.75, 6.75, and 8.25 in. (12, 17, and 21 cm). It provides multiple depths of investigation with 400 kHz and 2 MHz transmitting frequencies. The thin-bed resolution can be down to 6 in. (15 cm). The CPR Tool utilizes two frequencies (2 MHz and 400 kHz) and three transmitter−receiver spacings

Table 4.3 Physical parameters of GE Centerfire

Name	Frequency	Phase resistivity		Amplitude resistivity	
Centerfire	400 kHz	Long spacing	R41PLF	Long spacing	R41ALF
		Short spacing	R19PLF	Short spacing	R19ALF
	2 MHz	Long spacing	R41PHF	Long spacing	R41AHF
		Short spacing	R19PHF	Short spacing	R19AHF

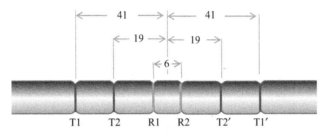

Figure 4.4 The Centerfire system developed by GE Energy.

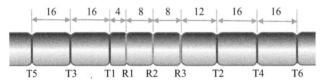

Figure 4.5 The antenna configuration of the EWR-M5 tool by Haliburton.

(18, 27, and 36 in.) to provide a total of 12 resistivity curves and 12 depths of investigation. It uses a borehole compensation (BHC) system for shorter overall tool length. The physical parameters of the Centerfire tool are given in Table 4.3 and the tool configurations are shown in Fig. 4.4.

4.3.1.5 Electromagnetic Wave Resistivity (EWR) and Azimuthal Deep Resistivity (ADR) [10,11]

The main LWD tools developed by Halliburton are the Azimuthal Deep Resistivity (ADR) sensor and the Electromagnetic Wave Resistivity (EWR) series: EWR-M5, EWR, EWR-Phase 4. The EWR-PHASE 4 LWD induction resistivity sensors measure both the phase shift and the attenuation for each of the four transmitter–receiver spacings. It can operate in water- and oil-based muds, as well as in air- and foam-drilled boreholes. It also can detect the horizontal and vertical resistivity in anisotropic formation. Fig. 4.5 shows an illustration of EWR-M5 and Table 4.4 shows the measurement quantities of the tool.

Table 4.4 Physical parameters of Haliburton EWR-M5

Name	Frequency	Phase resistivity		Amplitude resistivity	
EWR-M5	250 kHz	16" Spacing	RL16P	16" Spacing	RL16A
		24" Spacing	RL24P	24" Spacing	RL24A
		32" Spacing	RL32P	32" Spacing	RL32A
		40" Spacing	RL40P	40" Spacing	RL40A
		48" Spacing	RL48P	48" Spacing	RL48A
	500 kHz	16" Spacing	RM16P	16" Spacing	RM16A
		24" Spacing	RM24P	24" Spacing	RM24A
		32" Spacing	RM32P	32" Spacing	RM32A
		40" Spacing	RM40P	40" Spacing	RM40A
		48" Spacing	RM48P	48" Spacing	RM48A
	2 MHz	16" Spacing	RH16P	16" Spacing	RH16A
		24" Spacing	RH24P	24" Spacing	RH24A
		32" Spacing	RH32P	32" Spacing	RH32A
		40" Spacing	RH40P	40" Spacing	RH40A
		48" Spacing	RH48P	48" Spacing	RH48A

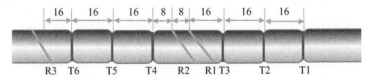

Figure 4.6 ADR Azimuthal Deep Resistivity Sensor.

The InSite ADR Azimuthal Deep Resistivity Sensor acquires measurements in 32 discrete directions and 14 different depths of investigation to determine distance and direction to multiple-bed boundaries. The azimuthal reading provides for derivation of anisotropy values and dip. An illustration of the ADR tool is shown in Fig. 4.6. The ADR tool has 6 z-directed transmitters and 3- to 45-degree oriented receivers.

4.3.1.6 Array Wave Resistivity [5]

The Array Wave Resistivity (AWR) tool is developed by PathFinder, which was acquired by Schlumberger in 2013. The physical parameters and curve mnemonics of the tool are given in Table 4.5. Fig. 4.7 shows the AWR tool configuration.

The PathFinder AWR array wave resistivity tool uses dual frequencies to provide up to 12 resistivity curves in all mud types. It has 12 diameters of investigation with three sensors at spacings of 15, 25, and 45 in. The spacings are optimized to provide robust quantitative data for invasion corrections and processing without dielectric assumptions or single-frequency results.

Table 4.5 Physical parameters of PathFinder AWR tool

Name	Frequency	Phase resistivity		Amplitude resistivity	
AWR	500 kHz	15″ Spacing	RSPL	15″ Spacing	RSAL
		25″ Spacing	RMPL	25″ Spacing	RMAL
		45″ Spacing	RDPL	45″ Spacing	RDAL
	2 MHz	15″ Spacing	RSPH	15″ Spacing	RSAH
		25″ Spacing	RMPH	25″ Spacing	RMAH
		45″ Spacing	RDPH	45″ Spacing	RDAH

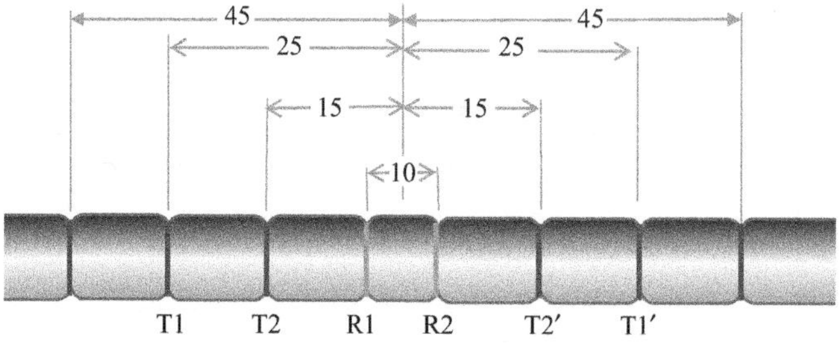

Figure 4.7 AWR tool by PathFinder.

4.3.1.7 Compensated Wave Resistivity

The PathFinder Compensated Wave Resistivity (CWR) tool offers four resistivity curves (two phase and two attenuation) and symmetrical receiver spacings at 25 and 55 in. for maximum depth of investigation. The balanced transmitter–receiver array provides repeatable measurements that are less susceptible to measurement noise induced by borehole rugosity and washouts, making accurate resistivity much easier to determine. PathFinder CWR measurements are made well beyond the invaded zone to give quick and accurate readings through porous and permeable lithology.

4.3.1.8 Slim Compensated Wave Resistivity

The PathFinder Slim Compensated Wave Resistivity (SCWR) is developed for boreholes as small as 55/8 in. It supplies the same high-quality resistivity measurement as the CWR tool, but with a limber design in a 43/4-in. collar. The SCWR tool resolves beds as thin as 6 in. and directly measures resistivity in beds thicker than 4 ft.

4.3.1.9 Compensated Dual Resistivity

Compensated Dual Resistivity (CDR) tool is a pioneer borehole-compensated tool developed by Schlumberger. It has upper and lower transmitters that fire alternately. The average of these phase shifts and attenuations for the upward and downward propagating waves provides a measurement with BHC similar in principle to that of

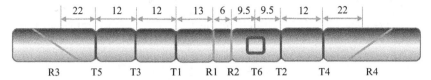

Figure 4.8 The tool configuration of PeriScope.

the Borehole-Compensated Sonic Tool (BHC). BHC reduces borehole effects in rugose holes, improves the vertical response, increases measurement accuracy, and provides quality control for the log. An electrical hole diameter is computed from the CDR data and is used as an input to hole size corrections.

4.3.1.10 PeriScope [12,13]

PeriScope bed boundary mapper was launched by Schlumberger in 2005 and has been extensively used around the world in environments such as heavy oil, coalbed methane, and tight gas, including horizontal shale gas drilling in the United States. PeriScope mapper makes 360-degree directional measurements that show the orientation of boundaries as far as 21 ft from the borehole. It uses a state-of-art tilted coil technology and multiple frequencies and spacings: 2 MHz, 400 kHz, and 100 kHz at 96, 84, 74, 44, 40, 34, 28, 22, and 16 in. A tool configuration is shown in Fig. 4.8.

4.3.1.11 Array Resistivity Compensated

The Array Resistivity Compensated (ARC) measurements provided by Schlumberger is called arcVISION LWD. It is available in a full tool size range from 31/8 to 9 in. and provides real-time resistivity, gamma ray, inclination, and annular pressure-while-drilling measurements that help produce and evaluate reservoirs. The tools can withstand a high sand content and high mud flow rates which ensure maximum power transfer.

The ARC tool combines the benefits of multispacing probes for formation evaluation with the advantages of BHC pioneered by the CDR tool. The response characteristics and the number of outputs of the ARC tool are purposely similar to those of recently developed wireline multispacing resistivity tools, allowing for a number of shared answer products.

Fig. 4.9 shows the ARC475 tool which is 21-ft long and 4.75-in. OD collar. The ARC475 antenna array consists of five transmitters and two receivers. Three transmitters are located above the midpoint of the receivers at spacings of 10, 22, and 34 in. Two transmitters are located below the midpoint of the receivers at spacings of 16 and 28 in. Each transmitter sequentially broadcasts a 2-MHz electromagnetic signal into the surrounding formation. For each transmitter, the phase shift and attenuation of the electromagnetic signals are measured between the receivers for a total of five raw phase shifts and five raw attenuations, which are shown in Table 4.6.

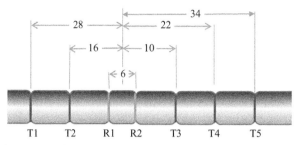

Figure 4.9 Tool configuration of ARC475.

Table 4.6 Physical parameters of Schlunberger's ARC tool

Name	Frequency	Phase resistivity		Amplitude resistivity	
ARC 4	400 kHz	10" Spacing	P10L	16" Spacing	A10L
		16" Spacing	P16L	24" Spacing	A16L
		22" Spacing	P22L	32" Spacing	A22L
		28" Spacing	P28L	40" Spacing	A28L
		34" Spacing	P34L	48" Spacing	A34L
	2 MHz	10" Spacing	P10H	16" Spacing	A10H
		16" Spacing	P16H	24" Spacing	A16H
		22" Spacing	P22H	32" Spacing	A22H
		28" Spacing	P28H	40" Spacing	A28H
		34" Spacing	P34H	48" Spacing	A34H

4.3.1.12 Multifrequency Resistivity High-Temperature LWD Sensor [5]

The Multifrequency Resistivity High-Temperature (MFR HT) sensor is designed by Weatherford. It operates in all mud types at 2 MHz and 400 kHz with transmitter−receiver spacing of 20, 30, and 46 in. These three independent transmitter−receiver antenna spacings and two operating frequencies provide accurate measurements over a wide range of drilling conditions. Any three compensated measurements can be combined to compute invasion diameter, flushed resistivity zone and true resistivity over a wide range of borehole conditions and resistivity contrasts. The tool configuration is shown in Fig. 4.10 and the measurements are shown in Table 4.7.

In the implementation of the antennas, in order to protect the antenna wires from abrasive mud and formation rocks, the antennas must be covered by metal. On the other hand, the EM field must not be shielded by the covers. Therefore slotted antenna structures are used. Fig. 4.11 shows one of the antenna implementation. Some of the antennas have ferrites installed under the antenna coil to increase receiver sensitivity. However, due to the temperature nonlinearity of the ferrites, the temperature performance will be compromised if ferrites are used. The opening is usually filled with high-strength epoxy. The radiation of the EM field from the antenna is from these opening as shown in Fig. 4.11B. The antenna efficiency is directly proportional to the width of each slot. The length of the slots is not as critical as the width

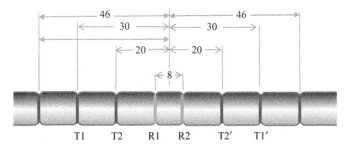

Figure 4.10 Weatherford MFR tool.

Table 4.7 Physical parameters of Weatherford AFR tool

Name	Frequency	Phase resistivity		Amplitude resistivity	
AWR	500 kHz	20" Spacing	RAD4	20" Spacing	RPD4
		30" Spacing	RAM4	30" Spacing	RPM4
		46" Spacing	RAS4	46" Spacing	RPS4
	2 MHz	20" Spacing	RAD2	20" Spacing	RPD2
		30" Spacing	RAM2	30" Spacing	RPM2
		46" Spacing	RAS2	46" Spacing	RPS2

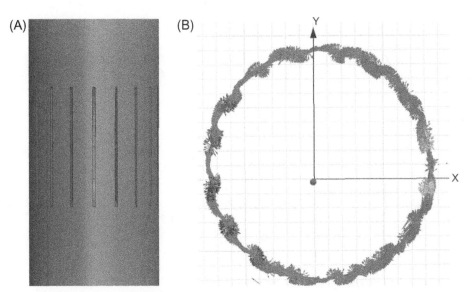

Figure 4.11 The slots around the collar with antenna loop (*red* (black in print versions) wire) in the middle of the slots (A) the antenna structure with slots and (B) E field distribution around the collar in the cross section of the antenna loop.

Table 4.8 Dielectric constant model

Company	Dielectric constant model
APS Technology	$\varepsilon_r = 210 \times R_t^{-0.42}$ (for 2 MHz)
	$\varepsilon_r = 480 \times R_t^{-0.49} + 8$ (for 400 kHz)
Baker Hughes INTEQ	$\varepsilon_r = 6.4 + 4.5255\sqrt{1 + \sqrt{1 + (2275/R_t)^2}}$ (for 2 MHZ)
	$\varepsilon_r = 6.4 + 4.5255\sqrt{1 + \sqrt{1 + (11375/R_t)^2}}$ (for 400 kHz)
GE Energy	$\varepsilon_r = 108.5 \times R_t^{-0.35} + 5$
Halliburton Sperry-Sun	$\varepsilon_r = 10$
PathFinder	$\varepsilon_r = 108.5 \times R_t^{-0.35} + 5$ (AWR)
	$\varepsilon_r = 10$ (CWR)
Schlumberger Anadrill	$\varepsilon_r = 108.5 \times R_t^{-0.35} + 5$ (for 2 MHz)
	$\varepsilon_r = 279.7 \times R_t^{-0.46} + 5$ (for 400 kHz)
Weatherford	$\varepsilon_r = 210 \times R_t^{-0.42}$ (for 2 MHz)
	$\varepsilon_r = 480 \times R_t^{-0.49} + 8$ (for 400 kHz)

of the slots. There are several ways to build the antenna slots, one of them is the use of two-halfs of slotted metal to cover antenna grove.

4.3.2 Dielectric constant model and conversion charts

In the simulation of the LWD tool responses, the effective dielectric constant should be used. Table 4.8 and Fig. 4.12 show the effective dielectric constant model for different service companies.

The magnetic field response of the LWD tools in homogeneous media can be obtained following the same procedure and formulas for the induction tools in Section 4.2. Different from the induction tool, the LWD tool measures the amplitude ratio and phase shift of the magnetic field received at the two receivers and convert the signals into apparent resistivity using conversion charts. The conversion chart is made for a given LWD tool and frequency in homogeneous medium. Figs. 4.13 and 4.14 show the conversion chart for ARC475 at 400 kHz and 2 MHz, respectively.

4.4 TRIAXIAL INDUCTION LOGGING TOOL RESPONSE IN BIAXIAL ANISOTROPIC HOMOGENEOUS FORMATION

Traditional induction tools have only coaxial transmitter–receiver coils and measure one magnetic field component at different receiver locations, thus have no sensitivity to formation anisotropy. To characterize the anisotropic conductivity/resistivity,

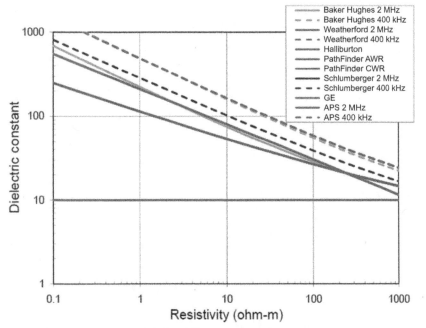

Figure 4.12 Dielectric constant models for different LWD tools.

conventional induction logging methods must be extended to provide additional information. Triaxial induction logging tools [14–21] are designed to obtain anisotropy of the formation. The rich information provided by triaxial induction measurement enables the determination of complex formations such as biaxial anisotropic media. A sketch of the standard triaxial tool is shown in Fig. 4.15A, which comprises three mutually orthogonal transmitters and three mutually orthogonal receivers. The transducer axes of the sonde can be arbitrarily oriented with respect to the principal axes of the conductivity tensor of the biaxial anisotropic medium. Triaxial tools can resolve conductivity anisotropy of earth formations, including reservoir rocks by making multicomponent electromagnetic measurements. The study of the impact of anisotropy on the tool response is of great importance for the correct interpretation of measurements. In this section, we will introduce how to simulate the response of a triaxial induction tool in a fully anisotropic homogeneous formation. Analytical solution is first derived [2] and the critical techniques for numerical evaluation are presented.

If a layered medium has a fracture pattern that cuts across bedding, the conductivity is fully anisotropic, as shown in Fig. 4.16. Full anisotropy is referred to as "biaxial" anisotropy in crystals. In this case, all three principal conductivities are different, representing the differences of pore-connectivity and conductivity in the vertical and lateral directions. Next, we will describe how to calculate the response of triaxial tools in homogeneous biaxial anisotropic medium.

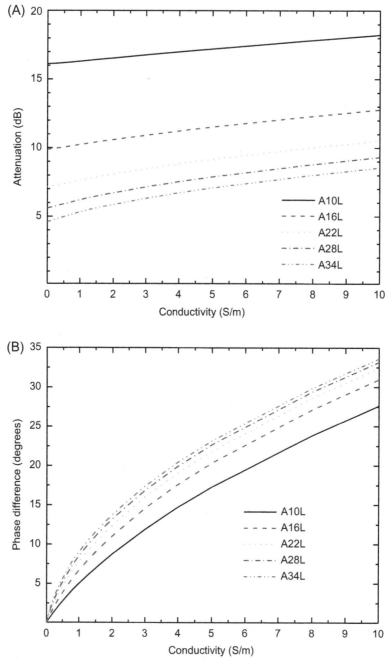

Figure 4.13 Conversion chart for ARC475 at 400 kHz. (A) Attenuation conversion chart, (B) phase shift conversion chart.

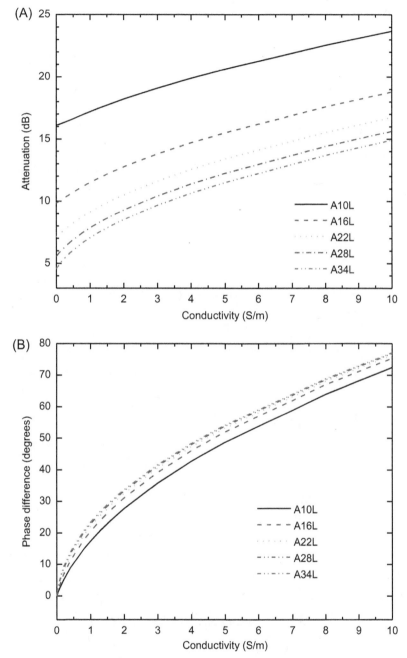

Figure 4.14 Conversion Chart for ARC475 at 2 MHz. (A) Attenuation conversion chart, (B) phase shift conversion chart.

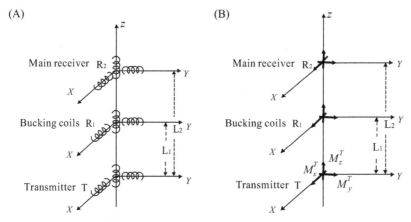

Figure 4.15 Basic structure of a triaxial induction tool and its equivalent dipole model. (A) Basic structure of a triaxial induction tool, (B) equivalent dipole model.

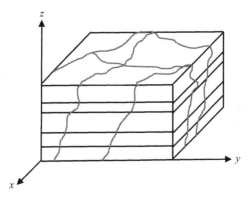

Figure 4.16 General anisotropy caused by fractured, layered medium [2].

4.4.1 Spectrum-domain solution to Maxwell's equations in a homogeneous biaxial anisotropic medium

A homogeneous, biaxial, unbounded medium can be characterized by the tensor conductivity defined in Eq. (4.1) (expressed in the principal axis system). Assuming the harmonic time dependence to be $e^{-i\omega t}$ (suppressed throughout the book), Maxwell's equations for the electric and magnetic fields are

$$\nabla \times \boldsymbol{H}(\boldsymbol{r}) = (\hat{\sigma} - i\omega\hat{\varepsilon})\boldsymbol{E}(\boldsymbol{r}) + \boldsymbol{J}_s(\boldsymbol{r}) \qquad (4.41\mathrm{a})$$

$$\nabla \times \boldsymbol{E}(\boldsymbol{r}) = i\omega\mu_0\boldsymbol{H}(\boldsymbol{r}) + i\omega\mu_0\boldsymbol{M}_s(\boldsymbol{r}) \qquad (4.41\mathrm{b})$$

where μ_0 is the magnetic permeability of the air, $\boldsymbol{r} = (x, y, z)$ is the position vector, $\hat{\varepsilon}$ is the dielectric constant tensor, $\boldsymbol{M}_s(\boldsymbol{r})$ is the magnetic-source flux density, and $\boldsymbol{J}_s(\boldsymbol{r})$ is the electric-source current density. For logging devices operating at relatively low frequencies and formations with conductivities greater than 10^{-4} S/m, we can assume that contributions from displacement current determined by $i\omega\hat{\varepsilon}$ can be ignored in comparison with $\hat{\sigma}$. For the induction logging problems which will be considered in this chapter, it is assumed that $J_s(r) = 0$, which means only magnetic dipoles are used to represent induction coils. Therefore Maxwell's equations are reduced to

$$\nabla \times H(r) = \hat{\sigma} E(r) \quad (4.42a)$$

$$\nabla \times E(r) = i\omega\mu_0 H(r) + i\omega\mu_0 M_s(r) \quad (4.42b)$$

The solution for Eq. (4.41a,b) in the space-domain can be expressed in terms of triple Fourier transforms of their spectral-domain counterparts $\tilde{E}(k)$ and $\tilde{H}(k)$

$$E(r), H(r) = \frac{1}{(2\pi)^3} \int_k dK e^{iK \cdot r} \tilde{E}(K), \tilde{H}(K) \quad (4.43)$$

where $K = (\xi, \eta, \varsigma)$ and

$$\tilde{E}(K), \tilde{H}(K) = \int_r dr e^{-iK \cdot r} E(r), H(r) \quad (4.44)$$

$$\int_k dK e^{iK \cdot r} = \int_{-\infty}^{\infty} \int_{-\infty}^{\infty} \int_{-\infty}^{\infty} d\xi d\eta d\varsigma e^{i(\xi x + \eta y + \varsigma z)} \quad (4.45)$$

Thus, for mathematical convenience, we will first obtain the solution to Eq. (4.42a,b) in the spectral domain and then use Eq. (4.43) to obtain the required space-domain solutions.

For the spectral-domain solutions $\tilde{E}(K)$ and $\tilde{H}(K)$, we can first eliminate the magnetic fields using electric fields in Eq. (4.42a) and solve for $\tilde{E}(K)$ from Eq. (4.42b) in the presence of the source $M_s(r)$. Then, since the source singularity has been totally accounted for in this solution, the magnetic fields can be determined from a homogeneous form of Maxwell's equations. Next, we will describe this procedure in detail.

Substituting Eq. (4.42a) into (4.42b) will result in the following vector wave equation

$$\nabla\nabla \cdot E(r) - \nabla^2 E(r) - \hat{k}^2 E(r) = i\omega\mu \nabla \times M_s(r) \quad (4.46)$$

where $\hat{k}^2 = i\omega\mu_0 \hat{\sigma}$. Applying the triple Fourier transform defined in Eq. (4.45) to (4.46), we obtain

$$\Omega(K) \cdot \tilde{E}(K) = -i\omega\mu \nabla \times M_s(K) \quad (4.47)$$

where the coefficient matrix Ω is given by

$$\Omega = \begin{bmatrix} k_x^2 - (\eta^2 + \varsigma^2) & \xi\eta & \xi\varsigma \\ \xi\eta & k_y^2 - (\xi^2 + \varsigma^2) & \eta\varsigma \\ \xi\varsigma & \eta\varsigma & k_z^2 - (\xi^2 + \eta^2) \end{bmatrix} \quad (4.48)$$

Then the solutions for the space-domain fields are

$$E(r) = -\frac{i\omega\mu}{(2\pi)^3} \int_K d\mathbf{K} e^{i\mathbf{K}\cdot\mathbf{r}} \Omega^{-1}(\mathbf{K}) \nabla \times M_s(\mathbf{K}) \quad (4.49)$$

The inverse matrix Ω^{-1} can be computed in terms of its adjoint and determinant as

$$\Omega^{-1} = \Lambda / \det \Omega \quad (4.50)$$

Let $\omega_{ij}(i = 1, 2, 3; j = 1, 2, 3)$ denote element (i, j) in the inverse matrix, these elements are found to be

$$\omega_{11} = \frac{[k_y^2 - (\xi^2 + \varsigma^2)][k_z^2 - (\xi^2 + \eta^2)] - \eta^2\varsigma^2}{\det \Omega} \quad (4.51a)$$

$$\omega_{12} = \omega_{21} = \frac{-\xi\eta[k_z^2 - (\xi^2 + \eta^2 + \varsigma^2)]}{\det \Omega} \quad (4.51b)$$

$$\omega_{13} = \omega_{31} = \frac{-\xi\varsigma[k_y^2 - (\xi^2 + \eta^2 + \varsigma^2)]}{\det \Omega} \quad (4.51c)$$

$$\omega_{22} = \frac{[k_x^2 - (\eta^2 + \varsigma^2)][k_z^2 - (\xi^2 + \eta^2)] - \xi^2\varsigma^2}{\det \Omega} \quad (4.51d)$$

$$\omega_{23} = \omega_{32} = \frac{-\eta\varsigma[k_x^2 - (\xi^2 + \eta^2 + \varsigma^2)]}{\det \Omega} \quad (4.51e)$$

$$\omega_{33} = \frac{[k_x^2 - (\eta^2 + \varsigma^2)][k_y^2 - (\xi^2 + \varsigma^2)] - \xi^2\eta^2}{\det \Omega} \quad (4.51f)$$

The determinant of the coefficient matrix can be written in the following factored form,

$$\det \Omega = k_z^2(\varsigma^2 - \varsigma_o^2)(\varsigma^2 - \varsigma_e^2) \quad (4.52)$$

where ς_o and ς_e are the axial wave numbers of the ordinary and extraordinary modes of propagation. The two distinct modes of propagation can be found to be (Appendix A)

$$\varsigma_{o,e}^2 = a \pm \sqrt{b}, \tag{4.53}$$

where

$$a = \frac{k_z^2(k_x^2 + k_y^2) - \xi^2(k_x^2 + k_z^2) - \eta^2(k_y^2 + k_z^2)}{2k_z^2} \tag{4.54}$$

$$b = a^2 - \frac{(\xi^2 + \eta^2 - k_z^2)(\xi^2 k_x^2 + \eta^2 k_y^2 - k_x^2 k_y^2)}{k_z^2} \tag{4.55}$$

The positive and negative square roots correspond to ς_e and ς_o, respectively. The derivation of the Eq. (4.55) followed Ref. [22] but modified all the major and minor typographical errors. Detailed derivation of parameters a and b can be found in Appendix A.

Once the space-domain electric field is obtained from Eq. (4.49), the corresponding magnetic field can be determined from the source-free Maxwell equations.

4.4.2 Full magnetic field response of a triaxial induction sonde in a biaxial anisotropic medium

In this section, we will derive the full magnetic field response of a triaxial induction sonde in a biaxial anisotropic medium. The basic structure of a triaxial tool is shown in Fig. 4.15A, consisting of one group of transmitter coils, one group of bucking coils, and one group of receiver coils. All the transmitter, bucking, and receiver coils are oriented in three mutually orthogonal directions. In the analysis, the coils are assumed to be sufficiently small and replaced by point magnetic dipoles in the modeling. Thus, the magnetic-source excitation of the triaxial tool can be expressed as $\boldsymbol{M} = (M_x, M_y, M_z)\delta(\boldsymbol{r})$, as shown in Fig. 4.15B.

For each component of the transmitter moments M_x, M_y, and M_z, there are in general three components of the induced field at each observation point in the medium. Thus there are nine field components at each receiver location. These field components can be expressed by a matrix representation of a dyadic $\hat{\boldsymbol{H}}$ as:

$$\hat{\boldsymbol{H}} = \begin{bmatrix} H_{xx} & H_{xy} & H_{xz} \\ H_{yx} & H_{yy} & H_{yz} \\ H_{zx} & H_{zy} & H_{zz} \end{bmatrix} \tag{4.56}$$

where the first subscript corresponds to the transmitter index and the second corresponds to the receiver index. Therefore H_{ij} denotes the magnetic field received by the

j-directed receiver coil excited by the i-directed transmitter coil. Next, we will derive the expressions for the nine magnetic field components in a homogeneous biaxial medium.

1. *The magnetic field components generated by a unit x-directed magnetic dipole $\boldsymbol{M} = (1,0,0)^T$*

   For an x-directed magnetic dipole $\boldsymbol{M} = (1,0,0)^T$ located at $\boldsymbol{r}' = (x', y', z')$, Eq. (4.47) can be rewritten as

$$\Omega(\boldsymbol{K}) \cdot \begin{bmatrix} \tilde{E}_{xx}(\boldsymbol{K}) \\ \tilde{E}_{xy}(\boldsymbol{K}) \\ \tilde{E}_{xz}(\boldsymbol{K}) \end{bmatrix} = \omega\mu_0 e^{-i(\xi x' + \eta y' + \varsigma z')} \begin{bmatrix} 0 \\ \varsigma \\ -\eta \end{bmatrix} \quad (4.57)$$

Solving Eq. (4.57), we can get

$$\begin{bmatrix} \tilde{E}_{xx}(\boldsymbol{K}) \\ \tilde{E}_{xy}(\boldsymbol{K}) \\ \tilde{E}_{xz}(\boldsymbol{K}) \end{bmatrix} = \omega\mu_0 e^{-i(\xi x' + \eta y' + \varsigma z')} \begin{bmatrix} \varsigma\omega_{12} - \eta\omega_{13} \\ \varsigma\omega_{22} - \eta\omega_{23} \\ \varsigma\omega_{32} - \eta\omega_{33} \end{bmatrix} \quad (4.58)$$

The corresponding components of the magnetic field can be determined from the source-free Maxwell equations:

$$\tilde{H}_x = \frac{1}{\omega\mu_0}(\eta\tilde{E}_z - \varsigma\tilde{E}_y) \quad (4.59a)$$

$$\tilde{H}_y = \frac{1}{\omega\mu_0}(\varsigma\tilde{E}_x - \xi\tilde{E}_z) \quad (4.59b)$$

$$\tilde{H}_z = \frac{1}{\omega\mu_0}(\xi\tilde{E}_y - \eta\tilde{E}_x) \quad (4.59c)$$

Direct substitution of Eq. (4.58) into (4.59a,b,c) yields

$$\begin{bmatrix} \tilde{H}_{xx}(\boldsymbol{K}) \\ \tilde{H}_{xy}(\boldsymbol{K}) \\ \tilde{H}_{xz}(\boldsymbol{K}) \end{bmatrix} = e^{-i(\xi x' + \eta y' + \varsigma z')} \begin{bmatrix} \eta(\varsigma\omega_{32} - \eta\omega_{33}) - \varsigma(\varsigma\omega_{22} - \eta\omega_{23}) \\ \varsigma(\varsigma\omega_{12} - \eta\omega_{13}) - \xi(\varsigma\omega_{32} - \eta\omega_{33}) \\ \xi(\varsigma\omega_{22} - \eta\omega_{23}) - \eta(\varsigma\omega_{12} - \eta\omega_{13}) \end{bmatrix} \quad (4.60)$$

Then the magnetic field components in the space-domain can be obtained from their spectral-domain counterparts in Eq. (4.60) by applying inverse Fourier transforms defined in Eq. (4.45)

$$H_{xx}(r) = \frac{1}{(2\pi)^3} \int_{-\infty}^{\infty} \int_{-\infty}^{\infty} \int_{-\infty}^{\infty} d\xi d\eta d\varsigma e^{i\xi(x-x')} e^{i\eta(y-y')} e^{i\varsigma(z-z')} (2\eta\varsigma\omega_{23} - \eta^2\omega_{33} - \varsigma^2\omega_{22})$$
(4.61)

$$H_{xy}(r) = \frac{1}{(2\pi)^3} \int_{-\infty}^{\infty} \int_{-\infty}^{\infty} \int_{-\infty}^{\infty} d\xi d\eta d\varsigma e^{i\xi(x-x')} e^{i\eta(y-y')} e^{i\varsigma(z-z')} [\varsigma(\varsigma\omega_{12} - \eta\omega_{13}) - \xi(\varsigma\omega_{32} - \eta\omega_{33})]$$
(4.62)

$$H_{xz}(r) = \frac{1}{(2\pi)^3} \int_{-\infty}^{\infty} \int_{-\infty}^{\infty} \int_{-\infty}^{\infty} d\xi d\eta d\varsigma e^{i\xi(x-x')} e^{i\eta(y-y')} e^{i\varsigma(z-z')} [\xi(\varsigma\omega_{22} - \eta\omega_{23}) - \eta(\varsigma\omega_{12} - \eta\omega_{13})]$$
(4.63)

As can be seen from Eqs. (4.61)–(4.63), there are triple infinite integrals of x, y, and z involved in the solution. In the numerical evaluation, a cylindrical transformation in the wave number space is invoked. Let ψ be the rotation angle in the $\xi - \eta$ plane, and we have

$$\xi = k \cos \psi \tag{4.64a}$$

$$\eta = k \sin \psi \tag{4.64b}$$

$$\int_{-\infty}^{\infty} \int_{-\infty}^{\infty} \int_{-\infty}^{\infty} d\xi d\eta d\varsigma = \int_0^{2\pi} d\psi \int_0^{\infty} k dk \int_{-\infty}^{\infty} d\varsigma \tag{4.64c}$$

Thus Eqs. (4.61)–(4.63) can be rewritten as

$$H_{xx}(r) = \frac{1}{(2\pi)^3} \int_0^{2\pi} d\psi \int_0^{\infty} k \, dk \int_{-\infty}^{\infty} d\varsigma e^{ik \cos \psi (x-x')} e^{ik \sin \psi (y-y')} e^{i\varsigma(z-z')}$$
$$(2k\varsigma \sin \psi \omega_{32} - k^2 \sin^2 \psi \omega_{33} - \varsigma^2 \omega_{22})$$
(4.65)

$$H_{xy}(r) = \frac{1}{(2\pi)^3} \int_0^{2\pi} d\psi \int_0^{\infty} k dk \int_{-\infty}^{\infty} d\varsigma e^{ik \cos \psi (x-x')} e^{ik \sin \psi (y-y')} e^{i\varsigma(z-z')}$$
$$(\varsigma^2 \omega_{12} - k\varsigma \sin \psi \omega_{13} - k\varsigma \cos \psi \omega_{32} + k^2 \cos \psi \sin \psi \omega_{33})$$
(4.66)

$$H_{xz}(r) = \frac{1}{(2\pi)^3} \int_0^{2\pi} d\psi \int_0^{\infty} dk k^2 \int_{-\infty}^{\infty} d\varsigma e^{ik \cos \psi (x-x')} e^{ik \sin \psi (y-y')} e^{i\varsigma(z-z')}$$
$$(k \sin^2 \psi \omega_{13} - k \sin \psi \cos \psi \omega_{23} - \varsigma \sin \psi \omega_{12} + \varsigma \cos \psi \omega_{22})$$
(4.67)

2. *The magnetic field components generated by a unit y-directed magnetic dipole* $M = (0, 1, 0)^T$

For a y-directed magnetic dipole $M = (0, 1, 0)^T$ at $r' = (x', y', z')$, following a similar derivation procedure, we can obtain the solution for the space-domain magnetic field components as follows

$$H_{yx}(r) = \frac{1}{(2\pi)^3} \int_{-\infty}^{\infty} \int_{-\infty}^{\infty} \int_{-\infty}^{\infty} d\xi d\eta d\varsigma e^{i\xi(x-x')} e^{i\eta(y-y')} e^{i\varsigma(z-z')} [\eta(\xi\omega_{33} - \varsigma\omega_{31}) - \varsigma(\xi\omega_{23} - \varsigma\omega_{21})]$$

(4.68)

$$H_{yy}(r) = \frac{1}{(2\pi)^3} \int_{-\infty}^{\infty} \int_{-\infty}^{\infty} \int_{-\infty}^{\infty} d\xi d\eta d\varsigma e^{i\xi(x-x')} e^{i\eta(y-y')} e^{i\varsigma(z-z')} (2\varsigma\xi\omega_{13} - \varsigma^2\omega_{11} - \xi^2\omega_{33})$$

(4.69)

$$H_{yz}(r) = \frac{1}{(2\pi)^3} \int_{-\infty}^{\infty} \int_{-\infty}^{\infty} \int_{-\infty}^{\infty} d\xi d\eta d\varsigma e^{i\xi(x-x')} e^{i\eta(y-y')} e^{i\varsigma(z-z')} [\xi(\xi\omega_{23} - \varsigma\omega_{21}) - \eta(\xi\omega_{13} - \varsigma\omega_{11})]$$

(4.70)

The following equations are actually used in the numerical evaluation by transforming the Cartesian coordinates into cylindrical coordinates

$$H_{yx}(r) = \frac{1}{(2\pi)^3} \int_0^{2\pi} d\psi \int_0^{\infty} dk k \int_{-\infty}^{\infty} d\varsigma e^{ik\cos\psi(x-x')} e^{ik\sin\psi(y-y')} e^{i\varsigma(z-z')}$$

$$(\varsigma^2 \omega_{21} - k\varsigma \sin\psi \omega_{31} - k\varsigma \cos\psi \omega_{23} + k^2 \cos\psi \sin\psi \omega_{33})$$

(4.71)

$$H_{yy}(r) = \frac{1}{(2\pi)^3} \int_0^{2\pi} d\psi \int_0^{\infty} dk k \int_{-\infty}^{\infty} d\varsigma e^{ik\cos\psi(x-x')} e^{ik\sin\psi(y-y')} e^{i\varsigma(z-z')}$$

$$(2k\varsigma \cos\psi \omega_{13} - k^2 \cos^2\psi \omega_{33} - \varsigma^2 \omega_{11})$$

(4.72)

$$H_{yz}(r) = \frac{1}{(2\pi)^3} \int_0^{2\pi} d\psi \int_0^{\infty} dk k^2 \int_{-\infty}^{\infty} d\varsigma e^{ik\cos\psi(x-x')} e^{ik\sin\psi(y-y')} e^{i\varsigma(z-z')}$$

$$(\varsigma \sin\psi \omega_{11} - \varsigma \cos\psi \omega_{21} - k \cos\psi \sin\psi \omega_{13} + k \cos^2\psi \omega_{23})$$

(4.73)

Note that H_{yx} has the same expression as H_{xy}. This is because of the reciprocity of the medium. We will see in the following equations that H_{xz} and H_{zx}, H_{yz}, and H_{zy} also have the same expression due to reciprocity.

3. *The magnetic field components generated by a unit z-directed magnetic dipole* $M = (0, 0, 1)^T$

Similarly, for a z-directed magnetic dipole $M = (0, 0, 1)^T$ at $r' = (x', y', z')$, the magnetic fields in the space-domain are

$$H_{zx}(r) = \frac{1}{(2\pi)^3} \int_{-\infty}^{\infty}\int_{-\infty}^{\infty}\int_{-\infty}^{\infty} d\xi d\eta d\varsigma e^{i\xi(x-x')} e^{i\eta(y-y')} e^{i\varsigma(z-z')} [\eta(\eta w_{31} - \xi w_{32}) - \varsigma(\eta w_{21} - \xi w_{22})] \tag{4.74}$$

$$H_{zy}(r) = \frac{1}{(2\pi)^3} \int_{-\infty}^{\infty}\int_{-\infty}^{\infty}\int_{-\infty}^{\infty} d\xi d\eta d\varsigma e^{i\xi(x-x')} e^{i\eta(y-y')} e^{i\varsigma(z-z')} [\varsigma(\eta w_{11} - \xi w_{12}) - \xi(\eta w_{31} - \xi w_{32})] \tag{4.75}$$

$$H_{zz}(r) = \frac{1}{(2\pi)^3} \int_{-\infty}^{\infty}\int_{-\infty}^{\infty}\int_{-\infty}^{\infty} d\xi d\eta d\varsigma e^{i\xi(x-x')} e^{i\eta(y-y')} e^{i\varsigma(z-z')} [\xi(\eta w_{21} - \xi w_{22}) - \eta(\eta w_{11} - \xi w_{12})] \tag{4.76}$$

Transforming the integral variables ξ and η into k and ψ, we have

$$H_{zx}(r) = \frac{1}{(2\pi)^3} \int_0^{2\pi} d\psi \int_0^{\infty} dk k^2 \int_{-\infty}^{\infty} d\varsigma e^{ik \cos \psi (x-x')} e^{ik \sin \psi (y-y')} e^{i\varsigma(z-z')} \tag{4.77}$$

$$(k \sin^2 \psi w_{31} - k \sin \psi \cos \psi w_{32} - \varsigma \sin \psi w_{21} + \varsigma \cos \psi w_{22})$$

$$H_{zy}(r) = \frac{1}{(2\pi)^3} \int_0^{2\pi} d\psi \int_0^{\infty} dk k^2 \int_{-\infty}^{\infty} d\varsigma e^{ik \cos \psi (x-x')} e^{ik \sin \psi (y-y')} e^{i\varsigma(z-z')} \tag{4.78}$$

$$(\varsigma \sin \psi w_{11} - \varsigma \cos \psi w_{12} - k \cos \psi \sin \psi w_{31} + k \cos^2 \psi w_{32})$$

$$H_{zz}(r) = \frac{1}{(2\pi)^3} \int_0^{2\pi} d\psi \int_0^{\infty} dk k^3 \int_{-\infty}^{\infty} d\varsigma e^{ik \cos \psi (x-x')} e^{ik \sin \psi (y-y')} e^{i\varsigma(z-z')} \tag{4.79}$$

$$(2 \sin \psi \cos \psi w_{12} - \cos^2 \psi w_{22} - \sin^2 \psi w_{11})$$

In fact, for the case where the instrument's transducer axes are aligned parallel to the principal axes of the conductivity tensor (i.e., all the dipping angle, azimuth angle, and tool angle are zero $\alpha = \beta = \gamma = 0$ degree), all the cross-coupling terms are zero. However, in general, the axes of the instrument are not parallel to the principal axes of the conductivity tensor, and the nondiagonal terms in the coupling matrix are not zero. Therefore it is necessary to find out the full coupling matrix in a more general case.

4.4.3 The full magnetic field response with arbitrary tool axis

In practice, the orientation of the transmitter and receiver coils is arbitrary with respect to the principal axes of the formation's conductivity tensor. In this section, we will consider the magnetic field response of a triaxial induction tool in a homogeneous biaxial anisotropic medium with arbitrarily oriented tool axis.

Fig. 4.17 [23] shows schematically the formation coordinate system described by (x, y, z) and a sonde coordinate system described by (x', y', z'). In Fig. 4.17, α, β, and γ denote the dipping, azimuthal, and orientation angle, respectively. Angle α is the relative deviation of the instrument axis z' with respect to the z axis of the conductivity tensor. Angle β is the angle between the projection of the instrument axis z' on the surface of the $x-y$ plane and x axis of the formation coordinate. Angle γ represents the rotation of the tool around the z' axis.

The formation bedding (unprimed) frame can be related to the sonde (primed) frame by a rotation matrix $\boldsymbol{R}$ given by:

$$\boldsymbol{R} = \begin{bmatrix} R_{11} & R_{12} & R_{13} \\ R_{21} & R_{22} & R_{23} \\ R_{31} & R_{32} & R_{33} \end{bmatrix}$$

$$= \begin{bmatrix} \cos\alpha\cos\beta\cos\gamma - \sin\beta\sin\gamma & -\cos\alpha\cos\beta\sin\gamma - \sin\beta\cos\gamma & \sin\alpha\cos\beta \\ \cos\alpha\sin\beta\cos\gamma + \cos\beta\sin\gamma & -\cos\alpha\sin\beta\sin\gamma + \cos\beta\cos\gamma & \sin\alpha\sin\beta \\ -\sin\alpha\cos\gamma & \sin\alpha\sin\gamma & \cos\alpha \end{bmatrix}$$

(4.80)

To find out the magnetic field response in the sonde system, the magnetic moments of the transmitter coils in the sonde coordinates are first transformed to

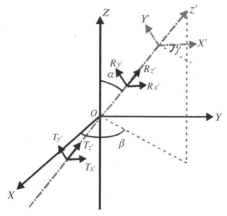

Figure 4.17 A schematic of the formation coordinate system (x, y, z) and sonde coordinate system (x', y', z').

effective magnetic moments in the formation coordinates by the rotation matrix. Then, the magnetic fields in the formation coordinates excited by the magnetic moments M can be readily obtained by

$$H = \hat{H}M \qquad (4.81)$$

where $\hat{H}$ is the dyadic corresponding to unit dipole source given by Eq. (4.56). Once the magnetic fields at the location of the receiver coils in the formation system are determined, the magnetic fields received at the receiver coils in the sonde system can be obtained by applying the inverse of the rotation matrix (the rotation matrix is orthogonal, therefore its inverse is equal to its transpose). Consequently, the coupling between the magnetic field components and the magnetic dipoles in the sonde system are given by Zhdanov et al. [24].

$$\hat{H}' = R^T \hat{H} R \qquad (4.82)$$

4.4.4 Computation of the triple integrals

In the previous section, we obtained the expressions for all nine components of the magnetic fields. As can be seen from Eqs. (4.65)–(4.67), (4.71)–(4.73), and (4.77)–(4.79), in order to compute the field quantities, we have to calculate integrals over k, ψ, and ς in the numerical evaluation. Since the integral over ψ is a definite integral, any numerical integral method is applicable. For the semi-infinite integral over k, a modified Gauss–Laguerre quadrature [25] is used. We found that the order of Gauss–Laguerre quadrature is mainly determined by the dipping angle. Fig. 4.18 shows the relative error of the magnetic field (imaginary part) as a function of the Gauss–Laguerre quadrature order for different dipping angles. We can see that as the dipping angle increases, larger order of Gauss–Laguerre quadrature is required to achieve a sufficient accuracy. When the dipping angle is 0, 60, 85, and 89 degrees, Gauss–Laguerre quadrature need 16, 16, 48, and 90 points to guarantee the relative error smaller than 0.4%.

For the infinite integral of ς, since the integrands become highly oscillatory as ς increases, special integration methods have to be considered. Here the integration over ς is performed using contour integration.

From Eq. (4.52), we can see that the integrands in Eqs. (4.65)–(4.67), (4.71)–(4.73), and (4.77)–(4.79) have four poles on the axial wave number plane: $\pm \varsigma_o$ and $\pm \varsigma_e$. The four poles correspond to the two eigenmodes for both forward and backward propagation, describing the two polarizations of the electromagnetic wave in the anisotropic medium. Assume $\varsigma_o^>$ ($\varsigma_e^>$) represents the one between ς_o (ς_e) and $-\varsigma_o$ ($-\varsigma_e$) whose imaginary part is greater than zero. For the case $z - z' > 0$, one obtains only contributions from two poles at $\varsigma_o^>$ and $\varsigma_e^>$, while for the case $z - z' < 0$, the contributions are from two poles at $-\varsigma_o^>$ and $-\varsigma_e^>$.

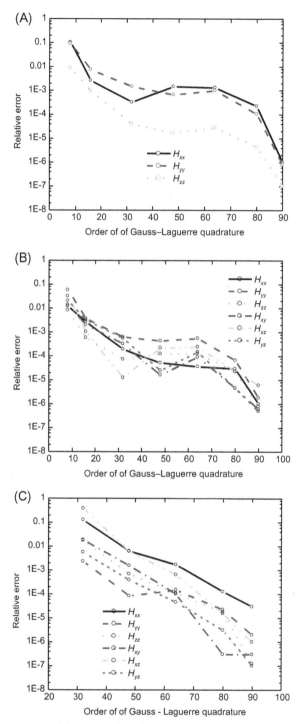

Figure 4.18 Relative errors as a function of Gauss–Laguerre quadrature order for different dipping angles ($\sigma_x = 500$ mS/m, $\sigma_y = 250$ mS/m, $\sigma_z = 125$ mS/m). (A) $\alpha = 0$ degree, (B) $\alpha = 60$ degrees, (C) $\alpha = 85$ degrees.

Using the contour integration [26] for ς, the result of the integration over ς in Eq. (4.65) is

$$\int_{-\infty}^{\infty} d\varsigma e^{i\varsigma(z-z')}(2k\varsigma \sin \psi \omega_{32} - k^2 \sin^2 \psi \omega_{33} - \varsigma^2 \omega_{22})$$
$$= 2\pi i \cdot \left\{ \frac{e^{i\varsigma(z-z')} \cdot [2k\varsigma \sin \psi \omega'_{32} - k^2 \sin^2 \psi \omega'_{33} - \varsigma^2 \omega'_{22}]}{k_z^2(\varsigma + \varsigma_o^>)(\varsigma + \varsigma_e^>)(\varsigma - \varsigma_e^>)} \bigg|_{\varsigma=\varsigma_o^>} \right.$$
$$\left. + \frac{e^{i\varsigma(z-z')} \cdot [2k\varsigma \sin \psi \omega'_{32} - k^2 \sin^2 \psi \omega'_{33} - \varsigma^2 \omega'_{22}]}{k_z^2(\varsigma + \varsigma_o^>)(\varsigma - \varsigma_o^>)(\varsigma + \varsigma_e^>)} \bigg|_{\varsigma=\varsigma_e^>} \right\}, \quad z - z' > 0 \quad (4.83)$$

$$\int_{-\infty}^{\infty} d\varsigma e^{i\varsigma(z-z')}(2k\varsigma \sin \psi \omega_{32} - k^2 \sin^2 \psi \omega_{33} - \varsigma^2 \omega_{22})$$
$$= 2\pi i \cdot \left\{ \frac{e^{i\varsigma(z-z')} \cdot [2k\varsigma \sin \psi \omega'_{32} - k^2 \sin^2 \psi \omega'_{33} - \varsigma^2 \omega'_{22}]}{k_z^2(\varsigma - \varsigma_o^>)(\varsigma + \varsigma_e^>)(\varsigma - \varsigma_e^>)} \bigg|_{\varsigma=-\varsigma_o^>} \right.$$
$$\left. + \frac{e^{i\varsigma(z-z')} \cdot [2k\varsigma \sin \psi \omega'_{32} - k^2 \sin^2 \psi \omega'_{33} - \varsigma^2 \omega'_{22}]}{k_z^2(\varsigma + \varsigma_o^>)(\varsigma - \varsigma_o^>)(\varsigma - \varsigma_e^>)} \bigg|_{\varsigma=-\varsigma_e^>} \right\}, \quad z - z' < 0 \quad (4.84)$$

where $\omega'_{32}, \omega'_{33}$, and ω'_{22} are the numerators of ω_{32}, ω_{33}, and ω_{22}, respectively.
Then, Eq. (4.65) can be rewritten as

$$H_{xx}(\mathbf{r}) = \frac{i}{(2\pi)^2} \int_0^{2\pi} d\psi \int_0^{\infty} kdk \cdot \left\{ \frac{e^{i\varsigma(z-z')} \cdot [2k\varsigma \sin \psi \omega'_{32} - k^2 \sin^2 \psi \omega'_{33} - \varsigma^2 \omega'_{22}]}{k_z^2(\varsigma + \varsigma_o^>)(\varsigma + \varsigma_e^>)(\varsigma - \varsigma_e^>)} \bigg|_{\varsigma=\varsigma_o^>} \right.$$
$$\left. + \frac{e^{i\varsigma(z-z')} \cdot [2k\varsigma \sin \psi \omega'_{32} - k^2 \sin^2 \psi \omega'_{33} - \varsigma^2 \omega'_{22}]}{k_z^2(\varsigma + \varsigma_o^>)(\varsigma - \varsigma_o^>)(\varsigma + \varsigma_e^>)} \bigg|_{\varsigma=\varsigma_e^>} \right\}, \quad z - z' < 0 \quad (4.85)$$

$$H_{xx}(\mathbf{r}) = \frac{i}{(2\pi)^2} \int_0^{2\pi} d\psi \int_0^{\infty} kdk \cdot \left\{ \frac{e^{i\varsigma(z-z')} \cdot [2k\varsigma \sin \psi \omega'_{32} - k^2 \sin^2 \psi \omega'_{33} - \varsigma^2 \omega'_{22}]}{k_z^2(\varsigma - \varsigma_o^>)(\varsigma + \varsigma_e^>)(\varsigma - \varsigma_e^>)} \bigg|_{\varsigma=-\varsigma_o^>} \right.$$
$$\left. + \frac{e^{i\varsigma(z-z')} \cdot [2k\varsigma \sin \psi \omega'_{32} - k^2 \sin^2 \psi \omega'_{33} - \varsigma^2 \omega'_{22}]}{k_z^2(\varsigma + \varsigma_o^>)(\varsigma - \varsigma_o^>)(\varsigma - \varsigma_e^>)} \bigg|_{\varsigma=-\varsigma_e^>} \right\}, \quad z - z' < 0 \quad (4.86)$$

The integration over ς in Eqs. (4.66)–(4.67), (4.71)–(4.73), and (4.77)–(4.79) can be performed by following the same procedure. It should be noted that the integrands are not separable functions of k, ψ, and ς, and therefore the integrals over k, ψ, and ς cannot be performed independently.

4.4.5 Numerical examples

In this section, we will present some numerical examples calculated using the theory described in Section 4.4.4.

Example 4.1

First, we consider isotropic and TI cases where exact solutions are available.

First, consider a homogeneous isotropic medium with conductivity of 500 mS/m. The relative dipping angle is 45 degrees and the frequency is 20 kHz. We change the spacing between the transmitter and receiver coils from 10 to 60 in. Fig. 4.19A shows the axial component H_{zz} obtained from the present code and the exact solution. It is observed that although totally different methods are used, the two results are almost the same, validating the present method.

Then, we consider a homogeneous TI medium with $\sigma_x = \sigma_y = 500$ mS/m and $\sigma_z = 125$ mS/m. The relative dipping angle is still 45 degrees and the frequency is 20 kHz. Fig. 4.19B shows the axial component H_{zz} obtained from the present code and the exact solution given by Moran and Gianzero [3]. Again, the results obtained from the two solutions are almost the same. The discrepancy begins from the sixth digit after the decimal point.

Example 4.2

Next, we will study the sensitivity of a real practical three-coil tool which is similar to the 3D Explorer tool jointly developed by Baker Atlas's and Shell [14,18] on various factors. The tool comprises one transmitter and two receivers respectively at 1 and 1.5 m from the transmitter. The tool response $\hat{\sigma}_a$ (apparent conductivity) is defined as

$$\hat{\sigma}_a = \begin{bmatrix} \sigma_a^{xx} & \sigma_a^{xy} & \sigma_a^{xz} \\ \sigma_a^{yx} & \sigma_a^{yy} & \sigma_a^{yz} \\ \sigma_a^{zx} & \sigma_a^{zy} & \sigma_a^{zz} \end{bmatrix} \quad (4.87)$$

where σ_a^{ij} is the apparent conductivity of the jth receiver when the ith transmitter is excited. The tool response $\hat{\sigma}_a$ is a function of the relative dip angle α, azimuth angle β, and the conductivity at each direction σ_x, σ_y, and σ_z. The sensitivity of $\hat{\sigma}_a$ to the dip angle, azimuth angle, and conductivities in each direction are the derivatives of the response with respect to these variables: $\frac{\partial \hat{\sigma}_a}{\partial \alpha}$ (mS/m/degree), $\frac{\partial \hat{\sigma}_a}{\partial \beta}$ (mS/m/degree), $\frac{\partial \hat{\sigma}_a}{\partial \sigma_x}, \frac{\partial \hat{\sigma}_a}{\partial \sigma_y}$, and $\frac{\partial \hat{\sigma}_a}{\partial \sigma_z}$. Consider a homogeneous biaxial formation with $\sigma_x = 500$ mS/m, $\sigma_y = 250$ mS/m, and $\sigma_z = 125$ mS/m, the sensitivity of the triaxial tool to α, β, σ_x, σ_y, and σ_z are shown in Figs. 4.20–4.24. In each figure, the horizontal axis is relative

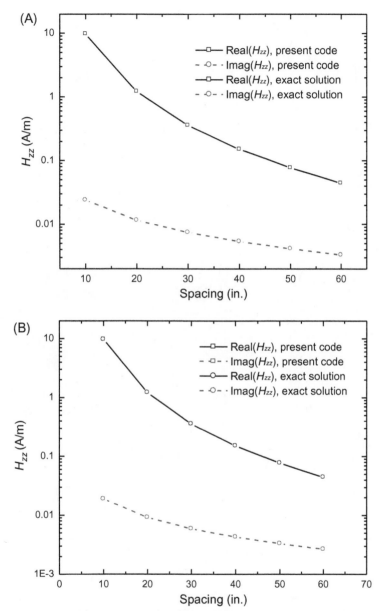

Figure 4.19 Comparison of the axial component H_{zz} obtained from the present method and the exact solution for (A) an isotropic medium; (B) a transverse isotropic medium.

dipping angle and the vertical axis is azimuth angle and the color represents the sensitivity. In Fig. 4.20, we show sensitivity functions for all nine components. We can see that all the cross pairs, xy and yx, xz and zx, yz and zy have the same sensitivity function in homogeneous formation. Therefore in Figs. 4.21–4.24, we only show five

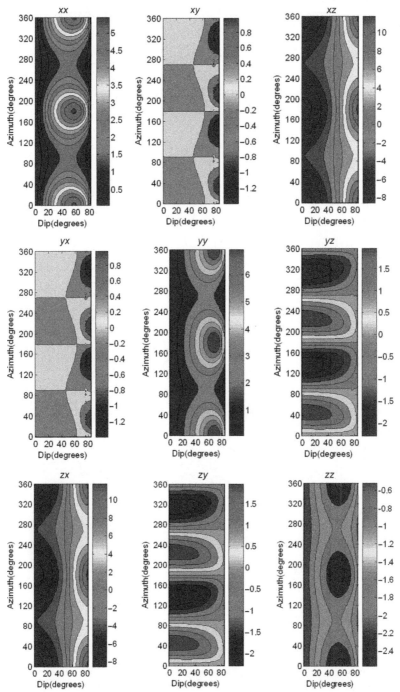

Figure 4.20 Sensitivity of a three-coil triaxial tool with respect to the dip angle $\frac{\partial \hat{\sigma}_a}{\partial \alpha}$ (mS/m/degree) in a homogeneous biaxial anisotropic formation.

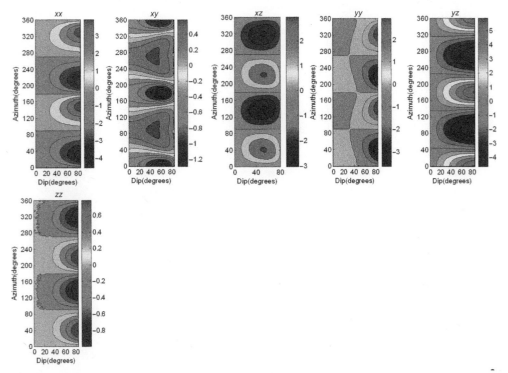

Figure 4.21 Sensitivity of a three-coil triaxial tool with respect to the azimuth angle $\frac{\partial \hat{\sigma}_a}{\partial \beta}$ (mS/m/degree) in a homogeneous biaxial anisotropic formation.

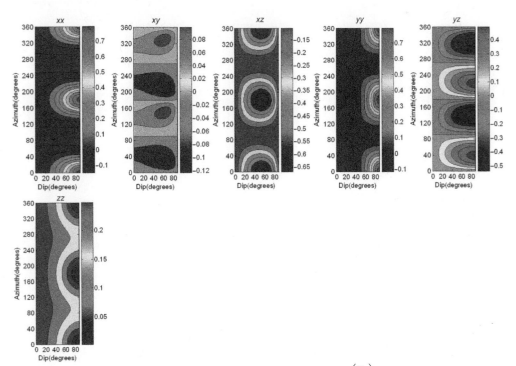

Figure 4.22 Sensitivity of a three-coil triaxial tool with respect to $\sigma_x \left(\frac{\partial \hat{\sigma}_a}{\partial \sigma_x} \right)$ in a homogeneous biaxial anisotropic formation.

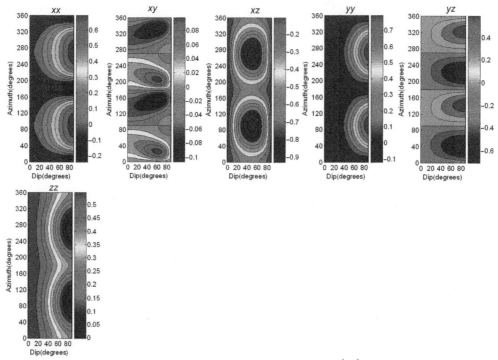

Figure 4.23 Sensitivity of a three-coil triaxial tool with respect to $\sigma_y \left(\frac{\partial \hat{\sigma}_a}{\partial \sigma_y} \right)$ in a homogeneous biaxial anisotropic formation.

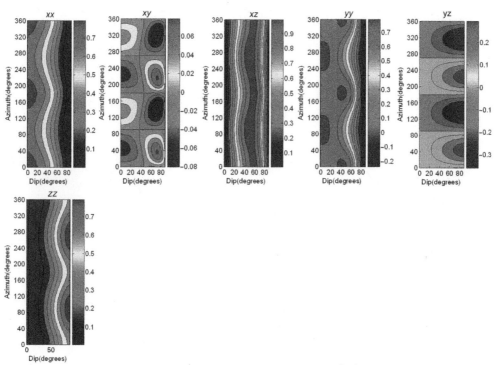

Figure 4.24 Sensitivity of a three-coil triaxial tool with respect to $\sigma_z \left(\frac{\partial \hat{\sigma}_a}{\partial \sigma_z} \right)$ in a homogeneous biaxial anisotropic formation.

components: xx, xy, xz, yy, yz, and zz for length limitation. From these figures, we can observe that:

1. Apparent conductivity is more sensitive to the dipping and azimuth angle than to formation conductivities.
2. The sensitivity of the cross couplings xz and zx, yz and zy are comparable to that of the diagonal coupling while the cross coupling xy and yx are less sensitive.

Example 4.3

Finally, we investigate the effects of frequencies on the responses of the same three-coil tool in Example 4.2. For clarity, we use resistivity instead of conductivity in this example. We consider two cases: (1) the formation is resistive and (2) the formation is conductive.

For resistive case, we assume the resistivities of the formation are $\rho_x = 200$ ohm-m, $\rho_y = 400$ ohm-m, and $\rho_z = 800$ ohm-m. Fig. 4.25 shows the apparent resistivity as frequency increases from 20 to 220 kHz when $\alpha = \beta = \gamma = 0$ degree. From the figure, we can see that at a low frequency (20 kHz), the transverse components ρ_a^{xx} and ρ_a^{yy} are directly proportional to ρ_z. Further, the transverse components ρ_a^{xx} and ρ_a^{yy} exhibit much stronger skin effect than the conventional coaxial component ρ_a^{zz}. To compensate for this effect, measurements at lower frequencies are preferred. For higher frequencies, data at multiple frequencies must be acquired and multifrequency skin-effect correction technique must be used. On the other hand, the coaxial component ρ_a^{zz} is less affected by the skin effect than ρ_a^{xx} and ρ_a^{yy}, and ρ_a^{zz} can reflect the geometric mean of the horizontal resistivities ($\sqrt{\rho_x \rho_y}$) within the whole frequency range 200–220 kHz.

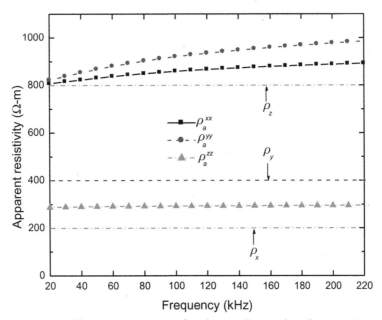

Figure 4.25 Frequency effect on responses of a three-coil triaxial tool in a resistive formation ($\alpha = \beta = \gamma = 0$ degrees).

Fig. 4.26 shows the apparent resistivity of the same tool for the same resistive formation at $\alpha = 75$ degrees, $\beta = 30$ degrees, and $\gamma = 0$ degree as frequency increases from 20 to 220 kHz. Comparison of Fig. 4.26 with Fig. 4.25 shows that in this case, both the diagonal and cross components of the apparent resistivity are less sensitive to frequency than in the case with zero dipping and zero azimuthal angles. It is also noticed that the cross

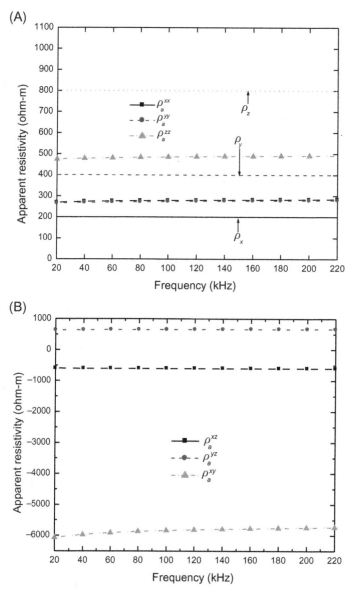

Figure 4.26 Frequency effect on responses of a three-coil triaxial tool in a resistive formation ($\alpha = 75$ degrees, $\beta = 30$ degrees, $\gamma = 0$ degree). (A) Diagonal terms of apparent resistivity, (B) Cross terms of apparent resistivity.

terms ρ_a^{xy} and ρ_a^{xz} are negative. Since the apparent resistivity are inversely proportional to the induced magnetic field, negative cross terms imply that the induced magnetic field is 180 degrees phase shifted with respect to the transmitter current.

For a relatively conductive case, the resistivities of the formation are supposed to be $\rho_x = 2$ ohm-m, $\rho_y = 4$ ohm-m, and $\rho_z = 8$ ohm-m. Fig. 4.27 shows the apparent

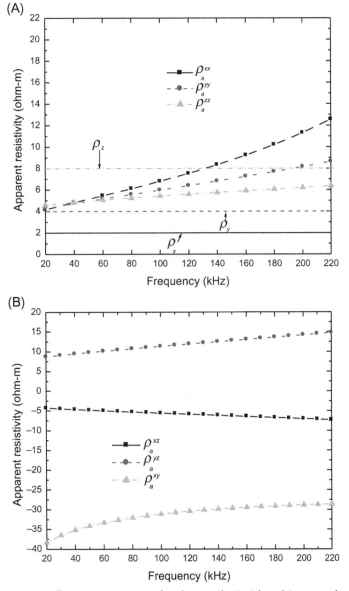

Figure 4.27 Frequency effect on responses of a three-coil triaxial tool in a conductive formation ($\alpha = 60$ degrees, $\beta = 30$ degrees, $\gamma = 0$ degree). (A) Diagonal terms of apparent resistivity, (B) Cross terms of apparent resistivity.

resistivity of the same tool for $\alpha = 60°, \beta = 30°$ and $\gamma = 0°$ as frequency increases from 20 to 220 kHz. It can be seen that all the diagonal components of the apparent resistivity ρ_a^{xx}, ρ_a^{yy}, and ρ_a^{zz} increase as the frequency increases. As for the cross components of the apparent resistivity, the amplitude (despite the phase shift with respect to the transmitter current) of ρ_a^{xz} and ρ_a^{yz} increase as the frequency increase while the amplitude of ρ_a^{xy} decreases as the frequency increases. This rule also applies to the resistive case, as we can see from Fig. 4.27.

REFERENCES

[1] N.M. Althausen, Electromagnetic wave propagation in layered-anisotropic media: Izvestia, Earth Phys. 8 (1969) 60−69.
[2] B.I. Anderson, T.D. Barber, M.G. Lüling, The response of induction tools to dipping, anisotropic formations, in: 36th Annual Logging Symposium Transactions, Society of Professional Well Log Analyst, Paper D, 1995.
[3] J.H. Moran, S. Gianzero, Effects of formation anisotropy on resistivity-logging measurements, Geophysics 44 (7) (1979) 1266−1286.
[4] J.H. Moran, S. Gianzero, Electrical anisotropy: its effect on well logs, in: A.A. Fitch (Ed.), Developments in Geophysical Exploration Methods-3, Applied Science Publishers, Ltd., Essex, 1982, pp. 195−238.
[5] W.C. Chew, S. Barone, B. Anderson, C. Hennessy, Diffraction of axisymmetric waves in a borehole by bed boundary discontinuities, Geophysics 49 (10) (1984) 1586−1595.
[6] R. Chemali, T. Helgesen, H. Meyer, C. Peveto, A. Poppitt, R. Randall, et al., Navigating and imaging in complex geology with azimuthal propagation resistivity while drilling, in: 2006 SPE ATCE, 24−27 September 2006, San Antonio, Texas.
[7] W.H. Meyer, E. Hart, K. Jensen, Geosteering with a combination of extra deep and azimuthal resistivity tools, in: 2008 SPE ATCE, Denver, Colorado.
[8] D.T. Macune, W.D. Flanagan, E. Choi, E. Marcellus, A compact compensated resistivity tool for logging while drilling, in: 2006 IADC/SPE, 21-23 February 2006, Miami, Florida, 2006.
[9] J. Lee, D. Macune, L. Qiu, C. Wei, Z. Yun, H. Li, Improving an established measurement while drilling system through a partnership for integrating a unique logging while drilling resistivity tool for applications in Chile, in: SPE 122420, 2009.
[10] M. Bittar, J. Klein, R. Beste, G. Hu, J. Pitcher, C. Golla, et al., A new azimuthal deep-reading resistivity tool for geosteering and advanced formation evaluation, in: SPE 109971, 2007.
[11] M. Bittar, F. Hveding, N. Clegg, J. Johnston, P. Solberg, G. Mangeroy, Maximizing reservoir contact in the Oseberg field using a new azimuthal deep-reading technology, in: SPE 116071, 2008.
[12] Q. Li, D. Omeragic, L. Chou, L. Yang, K. Duong, J. Smits, et al., New directional electromagnetic tool for proactive geosteering and accurate formation evaluation while drilling, in: The 46th SPWLA Annual Symposium, New Orleans, LA, 26−29 June, 2005.
[13] D. Omeragic, Q. Li, L. Chou, L. Yang, K. Duong, J. Smits, et al., Deep directional electromagnetic measurement for optimal well placement, in: SPE Annual Technical Conference and Exhibition, SPE 97045, 2005.
[14] B. Kriegshauser, O. Fanini, S. Forgang, G. Itskovich, M. Rabinovich, L. Tabarovsky, et al., A new multi-component induction logging tool to resolve anisotropic formation, in: SPWLA 41th Annual Symposium Transactions, Paper D, 2000.
[15] B. Anderson, T. Barber, T.M. Habashy, The interpretation and inversion of fully triaxial induction data: a sensitivity study, in: 36th Annual Logging Symposium Transactions, Society of Professional Well Log Analyst, Paper O, 2002.

[16] R. Rosthal, T. Barber, S. Bonner, K.-C. Chen, S. Davydycheva, G. Hazen, et al., Field test results of an experimental fully-triaxial induction tool, in: SPWLA 44th Annual Logging Symposium Transactions, Paper QQ, 2003.

[17] Z.Y. Zhang, L.M. Yu, B. Kriegshäuser, L. Tabarovsky, Determination of relative angles and anisotropic resistivity using multicomponent induction logging data, Geophysics 69 (4) (2003) 898−908.

[18] M. Rabinovich, L. Tabarovsky, B. Corley, J. van der Horst, M. Epov, Processing multi-component induction data for formation dips and anisotropy, Petrophysics 47 (6) (2006) 506−526.

[19] H. Wang, T. Barber, C. Morriss, R. Rosthal, R. Hayden, M. Markley, Determining anisotropic formation resistivity at any relative dip using a multiarray triaxial induction tool, in: 2006 SPE Annual Technical Conference and Exhibition, Paper SPE103113, 2006.

[20] M.M. Rabinovich, T. Gonfalini, B. Rocque, D. Corley, L. Georgi, et al., 2007, Multi-component induction logging: 10 years after, in: 48th Annual Logging Symposium, Society of Professional Well Log Analysts.

[21] S. Davydycheva, D. Homan, G. Minerbo, Triaxial induction tool with electrode sleeve, FD modeling in 3D geometries, J. Appl. Geophys. 67 (1) (2009) 98−108.

[22] S. Gianzero, D. Kennedy, L. Gao, L. SanMartin, The response of a triaxial induction sonde in a biaxial anisotropic medium, Petrophysics 43 (3) (2002) 172−184.

[23] L.L. Zhong, J. Li, L.C. Shen, R.C. Liu, Computation of triaxial induction logging tools in layered anisotropic dipping formations, IEEE Trans. Geosci. Remote Sens. 46 (4) (2008) 1148−1163.

[24] M. Zhdanov, D. Kennedy, E. Peksen, Foundation of tensor induction well-logging, Petrophysics 42 (6) (2001) 588−610.

[25] J. Burkardt, <http://people.scs.fsu.edu/~burkardt/datasets/quadrature_rules_laguerre/quadrature_rules_laguerre.html>, 2008.

[26] J.H. Zhang, W.Y. Qiu, Functions of a Complex Variable, CHEP and Springer, New York, NY, 2001, pp. 124−141 (Chapter 4).

APPENDIX A DERIVATION OF PARAMETERS a AND b IN EQS. (4.53)−(4.55)

The determinant of the coefficient matrix Ω can be written from Eq. (4.9) directly

$$\det \Omega = [k_x^2 - (\eta^2 + \varsigma^2)][k_y^2 - (\xi^2 + \varsigma^2)][k_z^2 - (\xi^2 + \eta^2)] + 2\xi^2 \eta^2 \varsigma^2 \\ - \xi^2 \varsigma^2 [k_y^2 - (\xi^2 + \varsigma^2)] - \eta^2 \varsigma^2 [k_x^2 - (\eta^2 + \varsigma^2)] - \xi^2 \eta^2 [k_z^2 - (\xi^2 + \eta^2)] \quad (A.1)$$

Arranging the right-hand side of Eq. (A.1) according to the mean of ς, Eq. (A.1) can be rewritten similarly, for a z-directed magnetic dipole as

$$\det \Omega = k_z^2 \varsigma^4 - \varsigma^2 [k_z^2 (k_x^2 + k_y^2) - \xi^2 (k_x^2 + k_z^2) - \eta^2 (k_y^2 + k_z^2)] \\ + (\xi^2 + \eta^2 - k_z^2)(\xi^2 k_x^2 + \eta^2 k_y^2 - k_x^2 k_y^2) \quad (A.2)$$

On the other hand, from Eq. (4.13), we obtain

$$\det \Omega = k_z^2 \varsigma^4 - 2 k_z^2 \varsigma^2 a + a^2 - b \quad (A.3)$$

Comparing the coefficients of ς in Eqs. (A.2) and (A.3) yields

$$-2 k_z^2 a = -[k_z^2 (k_x^2 + k_y^2) - \xi^2 (k_x^2 + k_z^2) - \eta^2 (k_y^2 + k_z^2)] \quad (A.4)$$

$$a^2 - b = (\xi^2 + \eta^2 - k_z^2)(\xi^2 k_x^2 + \eta^2 k_y^2 - k_x^2 k_y^2) \tag{A.5}$$

Therefore the parameters a and b are

$$a = \frac{k_z^2(k_x^2 + k_y^2) - \xi^2(k_x^2 + k_z^2) - \eta^2(k_y^2 + k_z^2)}{2k_z^2} \tag{A.6}$$

and

$$b = a^2 - \frac{(\xi^2 + \eta^2 - k_z^2)(\xi^2 k_x^2 + \eta^2 k_y^2 - k_x^2 k_y^2)}{k_z^2} \tag{A.7}$$

CHAPTER 5

Triaxial Induction Tool and Logging-While-Drilling Tool Response in a Transverse Isotropic-Layered Formation

Contents

5.1	Introduction	163
5.2	Summary of a Magnetic Dipole Source in a Transverse Isotropic Homogeneous Formation	165
5.3	Magnetic Dipole in a Layered Formation	166
	5.3.1 z-Directed magnetic dipole	167
	5.3.2 x-Directed magnetic dipole	168
	5.3.3 y-Directed magnetic dipole	169
	5.3.4 Magnitude of reflection and refraction magnetic fields	171
5.4	Convergence Algorithm	172
5.5	Simulation Results and Analysis	172
5.6	Analysis of Anisotropy Impact to the Resistivity LWD Tool	175
References		178
Appendix A Derivation of Hertz Vector Potential in Multiple Layer Formation		179

5.1 INTRODUCTION

In Chapter 4, Triaxial Induction and Logging-While-Drilling Resistivity Tool Response in Homogeneous Anisotropic Formations, we discussed the induction and logging-while-drilling (LWD) resistivity tool response in anisotropic homogenous formations. However, as pointed out in Chapter 1, Introduction to Well Logging, the earth formations mostly contain layers with different resistivities. The layers will greatly affect tool response, especially when layers are relatively thin, as in the case of sand-shale formations. Generally the axis of the logging tool or the borehole is deviated from the normal direction of the formation layers, resulting in a dipping angle of the tool with respect to the formation layers, which must be considered in the discussions. In this chapter, we will discuss the induction and LWD resistivity tool responses in a layered formation with dipping angles for the transverse isotropic (TI) formations.

For the simplicity, the discussions in this chapter assume that there is no borehole and mandrel in the formation. In other word, there is no change in the formation in the radial direction, so that one-dimensional (1D) analysis can be applied. For the cases where vertical layers exist, please refer to the later chapters of this book (see chapters: Induction and Logging-While-Drilling Tool Response in a Cylindrically Layered Isotropic Formation; and Induction and Logging-While-Drilling Resistivity Tool Response in a Two-Dimensional Isotropic Formation). For the formations that are layered and biaxially anisotropic, the tool response will be analyzed in Chapter 6, Triaxial Induction and Logging-While-Drilling Logging Tool Response in a Biaxial Anisotropic-Layered Formation.

It is challenging to interpret logs from triaxial induction and hence to characterize resistivity anisotropy in a formation. The primary critical component in log analysis is the ability to accurately predict the behavior of induced electromagnetic (EM) fields in anisotropic media. The other issue is how to achieve a reliable quantitative interpretation of the measured data. In this chapter, we will discuss an algorithm that can be used to model triaxial induction tool response in layered TI formation.

Theoretical study of triaxial induction logging was reported by Moran and Gianzero [1]. They concluded that anisotropic effect does affect induction logging devices. Later, Klein [2] developed a correction chart for anisotropy for an induction logging using similar analysis developed in Ref. [1]. However, the correction chart is difficult to apply in deviated borehole. In practical, deviated wells are commonly used to access and produce hydrocarbon reservoirs including the case of thinly layered formations; the tool seems to penetrate longer distances within the hydrocarbon-bearing layers due to the higher resistivity [2]. Multicomponent triaxial induction tools were designed to diagnose and measure horizontal and vertical conductivity of rock formations.

Graciet and Shen [3] discussed numerical solutions of traditional induction logging in layered anisotropic formation. Zhdanov et al. [4] has derived analytical solutions of triaxial induction logging in unbounded homogenous anisotropic formation. Wang et al. [5] introduces one efficient algorithm to simulate response of triaxial induction logging in layered cylindrical formation. In this approach, spectral Green functions are given from Maxwell equations. Sensitivity of triaxial logging in 1D-layered formation is discussed by Wang et al. [6]. Davydycheva et al. [7] used a finite difference scheme in analyzing the triaxial tool response in a three-dimensional (3D) anisotropic-layered formation. The simulation of triaxial induction response in deviated wells is still challenging since the associated 3D geometries is taxing on computational requirements whereby general-purpose 3D simulation algorithms often fail to calculate solutions in a limited amount of CPU time [8–10], which makes it difficult to be used in an inversion procedure.

5.2 SUMMARY OF A MAGNETIC DIPOLE SOURCE IN A TRANSVERSE ISOTROPIC HOMOGENEOUS FORMATION

In Chapter 4, Triaxial Induction and Logging-While-Drilling Resistivity Tool Response in Homogeneous Anisotropic Formations, we discussed the induction and LWD tool responses in an anisotropic homogeneous formation. The conductivity is represented by the second-rank tensor as,

$$\hat{\sigma} = \begin{bmatrix} \sigma_x & 0 & 0 \\ 0 & \sigma_y & 0 \\ 0 & 0 & \sigma_z \end{bmatrix} \quad (5.1)$$

For a TI formation, $\sigma_x = \sigma_y = \sigma_h$ is the horizontal conductivity and $\sigma_{zz} = \sigma_v$ is the vertical conductivity of the medium. In a biaxial isotropic formation, $\sigma_x \neq \sigma_y \neq \sigma_z$. In most cases, thin-bedded formations can be modeled as transversely isotropic formation where $\sigma_x = \sigma_y = \sigma_h \neq \sigma_z = \sigma_v$. The degree of anisotropy can be described by the coefficient of anisotropy given by,

$$\lambda = \sqrt{\frac{\sigma_h}{\sigma_v}} \quad (5.2)$$

Laboratory measurements have shown that λ may range from 1 to about 2.5 in different shale formations.

From Section 4.4.3, we can directly obtain the relationship between tool and formation coordinates. The magnetic moments $\overline{M}'$ of the sonde system is rotated into bedding coordinates:

$$\overline{M} = R\overline{M}' \quad (5.3)$$

where R is the rotation matrix converting the sonde coordinates to the bedding system given in Eq. (4.80). In Section 4.4.3, we assume $\hat{H}$ is magnetic field from unit magnetic dipole in bedding coordinates. Then the magnetic field response $\overline{H}$ in bedding system is obtained by

$$\overline{H} = \hat{H}\overline{M} \quad (4.81)$$

More explicitly,

$$\begin{pmatrix} H_x \\ H_y \\ H_z \end{pmatrix} = \begin{pmatrix} H^x_{xo} & H^y_{xo} & H^z_{xo} \\ H^x_{yo} & H^y_{yo} & H^z_{yo} \\ H^x_{zo} & H^y_{zo} & H^z_{zo} \end{pmatrix} \begin{pmatrix} M_x \\ M_y \\ M_z \end{pmatrix} \quad (5.4)$$

where $\hat{H}$ is defined by

$$\hat{H} = \begin{pmatrix} H^x_{xo} & H^y_{xo} & H^z_{xo} \\ H^x_{yo} & H^y_{yo} & H^z_{yo} \\ H^x_{zo} & H^y_{zo} & H^z_{zo} \end{pmatrix} \tag{5.5}$$

The final results should be in tool system. So we multiply transverse rotation matrix to transfer data under tool coordinates.

$$\overline{H}' = R^{-1}\overline{H} = R^{-1}\hat{H}R\overline{M}' \tag{5.6}$$

According to Eq. (5.6), we find that $\hat{H}$ is the key to derive full tool responses. If $\hat{H}$ is known, it is straight forward to employ Eq. (5.6) to obtain the expressions of nine magnetic field components in 1D multiple layered TI formation in formation coordinates. Next we will derive $\hat{H}$ in formation coordinate.

5.3 MAGNETIC DIPOLE IN A LAYERED FORMATION

Fig. 4.17 is a simplified schematic of an induction or LWD tool in a formation. Consider the formation layers are TI, and the logging tool may have three transmitter components and three receiver components in the direction of x, y, and z in tool-coordinate system. The approach of the solution is to use the homogenous solution in each layer given in Chapter 4, Triaxial Induction and Logging-While-Drilling Resistivity Tool Response in Homogeneous Anisotropic Formations, and match the boundary conditions at the layer boundaries.

One way to solve this problem is start from the source, describe the wave propagation in each layer from the source layer. This will yield the wave propagation algorithm. Since the wave propagation from the source is directly related to the location of the source, when the source moves to another logging point, the computation must be repeated for a new source location. On the other hand, once the formations are known, the layer conductivities are defined and the reflection and transmission coefficients will be determined and are independent of transmitter and receiver positions. Therefore, for each layer, when the source is moved from one logging point to another, we should be able to use the same set of transmission and reflection coefficients without going over to compute them again. This idea yields a second algorithm—general reflection and transmission method, which will be discussed in Chapter 6, Triaxial Induction and Logging-While-Drilling Logging Tool Response in a Biaxial Anisotropic-Layered Formation, when the formation is layered biaxial anisotropic.

5.3.1 z-Directed magnetic dipole

We can solve this EM problem using the Hertz potential as discussed in Chapter 4, Triaxial Induction and Logging-While-Drilling Resistivity Tool Response in Homogeneous Anisotropic Formations. From Section 4.1, we know that for the z-directed magnetic dipole, Hertz potential is given by Eq. (4.5),

$$\Pi = \Pi_z z \quad (5.7)$$

In a homogeneous media, the solutions to the vector potential is given in Chapter 4, Triaxial Induction and Logging-While-Drilling Resistivity Tool Response in Homogeneous Anisotropic Formations. In a layered formation, the vector potential and magnetic fields in the ith layer ($z_{i-1} < z < z_i$) are found to be

$$\pi_{zi} = \frac{1}{4\pi} \int_0^\infty \left(\frac{\beta_i}{\xi_{hi}} e^{-\xi_{hi}|z-z_0|} + F_i e^{-\xi_{hi}z} + G_i e^{\xi_{hi}z} \right) \alpha J_0(\alpha\rho) d\alpha \quad (5.8)$$

$$H_z^z = \frac{M_z}{4\pi} \int_0^\infty \left(\frac{\beta_i}{\xi_{hi}} e^{-\xi_{hi}|z-z_0|} + F_i e^{-\xi_{hi}z} + G_i e^{\xi_{hi}z} \right) \alpha^3 J_0(\alpha\rho) d\alpha \quad (5.9)$$

$$H_z^x = \frac{M_z}{4\pi} \int_0^\infty \xi_{hi} \left(\frac{\beta_i}{\xi_{hi}} \frac{|z-z_0|}{z-z_0} e^{-\xi_{hi}|z-z_0|} + F_i e^{-\xi_{hi}z} - G_i e^{\xi_{hi}z} \right) \alpha^2 \cos\varphi J_1(\alpha\rho) d\alpha \quad (5.10)$$

$$H_z^y = \frac{M_z}{4\pi} \int_0^\infty \xi_{hi} \left(\frac{\beta_i}{\xi_{hi}} \frac{|z-z_0|}{z-z_0} e^{-\xi_{hi}|z-z_0|} + F_i e^{-\xi_{hi}z} - G_i e^{\xi_{hi}z} \right) \alpha^2 \sin\varphi J_1(\alpha\rho) d\alpha \quad (5.11)$$

where

$$\xi_{hi} = (\alpha^2 - k_{hi}^2)^{1/2} \quad (5.12)$$

$$k_{hi} = (i\omega\mu\sigma_{hi})^{1/2} \quad (5.13)$$

$$\beta_i = \begin{cases} 0, & \text{if } M_z \text{ is not in the } i\text{th layer} \\ 1, & \text{if } M_z \text{ is in the } i\text{th layer} \end{cases} \quad (5.14)$$

In Eqs. (5.8)–(5.11), $J_0(\alpha\rho)$ is the zeroth-order Bessel function. ρ implies that equivalent transmitter and receiver in formation coordinate is not coaxial but should be parallel. F_i is the magnitude of reflection magnetic fields and G_i is the magnitude of refraction magnetic fields.

5.3.2 *x*-Directed magnetic dipole

For the *x*-directed magnetic dipole, Hertz potential is consisted by two parts, given by,

$$\Pi = \Pi_x x + \Pi_z z \tag{5.15}$$

In layered formation, the potential and magnetic fields in the *i*th layer $(z_{i-1} < z < z_i)$ are found to be:

$$\pi_{xi} = \frac{M_x}{4\pi \lambda_i} \int_0^\infty \left(\frac{\beta_i}{\xi_{vi}} e^{-\xi_{vi}|z-z_0|} + P_i e^{-\xi_{vi}\lambda_i z} + Q_i e^{\xi_{vi}\lambda_i z} \right) \alpha J_0(\alpha \rho) d\alpha \tag{5.16}$$

$$\pi_{zi} = \frac{M_x}{4\pi} \int_0^\infty (S_i e^{-\xi_{vi}\lambda_i z} + T_i e^{\xi_{vi}\lambda_i z} - \xi_{vi} P_i e^{-\xi_{vi}\lambda_i z} + \xi_{vi} Q_i e^{\xi_{vi}\lambda_i z}) \cos \phi J_1(\alpha \rho) d\alpha$$
$$+ \frac{M_x}{4\pi} \int_0^\infty \beta_i (e^{-\xi_{hii}|z-z_0|} - e^{-\xi_{vi}\lambda_i|z-z_0|}) \frac{|z-z_0|}{z-z_0} \cos \phi J_1(\alpha \rho) d\alpha \tag{5.17}$$

$$H_x^z = \frac{M_x}{4\pi} \cos\phi \int_0^\infty \left(\beta_i \frac{|z-z_0|}{z-z_0} e^{-\xi_{hi}|z-z_0|} + S_i e^{-\xi_{hi}z} + T_i e^{\xi_{hi}z} \right) \alpha^2 J_1(\alpha \rho) d\alpha \tag{5.18}$$

$$H_x^x = \frac{M_x}{4\pi} \int_0^\infty \left(\begin{array}{c} \frac{\beta_i \sin^2 \phi}{\lambda_i \xi_{vi}} k_h^2 e^{-\xi_{vi}\lambda_i|z-z_0|} - \beta_i \cos^2 \phi \xi_{hi} e^{-\xi_{hi}|z-z_0|} \\ + \frac{P_i}{\lambda_i} \sin^2 \phi k_{hi}^2 e^{-\xi_{vi}\lambda_i z} + \frac{Q_i}{\lambda_i} \sin^2 \phi k_{hi}^2 e^{\xi_{vi}\lambda_i z} \\ - S_i \cos^2 \phi \xi_{hi} e^{-\xi_{hi}z} + T_i \cos^2 \phi \xi_{hi} e^{\xi_{hi}z} + \end{array} \right) \alpha J_0(\alpha \rho) d\alpha$$

$$+ \frac{M_x}{4\pi \rho} \int_0^\infty \left(\begin{array}{c} \lambda_i \frac{\beta_i}{\xi_{vi}} k_v^2 e^{-\xi_{vi}\lambda_i|z-z_0|} + \beta_i \xi_{hi} e^{-\xi_{hi}|z-z_0|} \\ + P_i \lambda_i k_{vi}^2 e^{-\xi_{vi}\lambda_i z} + Q_i \lambda_i k_{vi}^2 e^{\xi_{vi}\lambda_i z} \\ + S_i \xi_{hi} e^{-\xi_{hi}z} - T_i \xi_{hi} e^{\xi_{hi}z} \end{array} \right) \cos 2\phi J_1(\alpha \rho) d\alpha \tag{5.19}$$

$$H_x^y = \frac{M_x}{4\pi\rho} \int_0^\infty \begin{pmatrix} \frac{\beta_i}{\lambda_i \xi_{vi}} k_h^2 e^{-\xi_{vi}\lambda_i|z-z_0|} - \beta_i \xi_{hi} e^{-\xi_{hi}|z-z_0|} \\ + \frac{P_i}{\lambda_i} k_{hi}^2 e^{-\xi_{vi}\lambda_i z} + \frac{Q_i}{\lambda_i} k_{hi}^2 e^{\xi_{vi}\lambda_i z} \\ - S_i \xi_{hi} e^{-\xi_{hi} z} + T_i \xi_{hi} e^{\xi_{hi} z} + \end{pmatrix} \alpha J_0(\alpha\rho) d\alpha$$

$$+ \frac{M_x}{4\pi\rho} \int_0^\infty \begin{pmatrix} \lambda_i \frac{\beta_i}{\xi_{vi}} k_v^2 e^{-\xi_{vi}\lambda_i|z-z_0|} + \beta_i \xi_{hi} e^{-\xi_{hi}|z-z_0|} \\ + P_i \lambda_i k_{vi}^2 e^{-\xi_{vi}\lambda_i z} + Q_i \lambda_i k_{vi}^2 e^{\xi_{vi}\lambda_i z} \\ + S_i \xi_{hi} e^{-\xi_{hi} z} - T_i \xi_{hi} e^{\xi_{hi} z} \end{pmatrix} \sin 2\phi J_1(\alpha\rho) d\alpha \quad (5.20)$$

where

$$\xi_{vi} = (\alpha^2 - k_{vi}^2)^{1/2} \quad (5.21)$$

$$k_{vi} = (i\omega\mu\sigma_{vi})^{1/2} \quad (5.22)$$

$$\lambda_i = \frac{k_{hi}}{k_{vi}} \quad (5.23)$$

$$\beta_i = \begin{cases} 0, & \text{if } M_x \text{ is not in the } i\text{th layer} \\ 1, & \text{if } M_x \text{ is in the } i\text{th layer} \end{cases} \quad (5.24)$$

In Eqs. (5.16)–(5.20), P_i and S_i are magnitude of reflection magnetic fields and Q_i and T_i are magnitude of refraction magnetic fields.

5.3.3 y-Directed magnetic dipole

For the y-directed magnetic dipole,

$$\Pi = \Pi_y y + \Pi_z z \quad (5.25)$$

In layered formation, within the ith layer of the formation, the potential and the magnetic fields are found to be:

$$\pi_{yi} = \frac{M_y}{4\pi\lambda_i} \int_0^\infty \left(\frac{\beta_i}{\xi_{vi}} e^{-\xi_{vi}|z-z_0|} + P_i e^{-\xi_{vi}\lambda_i z} + Q_i e^{\xi_{vi}\lambda_i z} \right) \alpha J_0(\alpha\rho) d\alpha \quad (5.26)$$

$$\pi_{zi} = \frac{M_y}{4\pi} \int_0^\infty (S_i e^{-\xi_{vi}\lambda_i z} + T_i e^{\xi_{vi}\lambda_i z} - \xi_{vi} P_i e^{-\xi_{vi}\lambda_i z} + \xi_{vi} Q_i e^{\xi_{vi}\lambda_i z}) \sin\phi J_1(\alpha\rho) d\alpha$$

$$+ \frac{M_y}{4\pi} \int_0^\infty \beta_i (e^{-\xi_{hi}|z-z_0|} - e^{-\xi_{vi}\lambda_i |z-z_0|}) \frac{|z-z_0|}{z-z_0} \sin\phi J_1(\alpha\rho) d\alpha \tag{5.27}$$

$$H_y^z = \frac{M_y}{4\pi} \sin\phi \int_0^\infty \left(\beta_i \frac{|z-z_0|}{z-z_0} e^{-\xi_{hi}|z-z_0|} + S_i e^{-\xi_{hi} z} + T_i e^{\xi_{hi} z} \right) \alpha^2 J_1(\alpha\rho) d\alpha \tag{5.28}$$

$$H_y^z = \frac{M_y}{4\pi} \sin\phi \int_0^\infty \left(\beta_i \frac{|z-z_0|}{z-z_0} e^{-\xi_{hi}|z-z_0|} + S_i e^{-\xi_{hi} z} + T_i e^{\xi_{hi} z} \right) \alpha^2 J_1(\alpha\rho) d\alpha \tag{5.29}$$

$$H_y^y = \frac{M_y}{4\pi} \int_0^\infty \left(\begin{array}{c} \frac{\beta_i \cos^2\phi}{\lambda_i \xi_{vi}} k_h^2 e^{-\xi_{vi}\lambda_i |z-z_0|} - \beta_i \sin^2\phi \xi_{hi} e^{-\xi_{hi}|z-z_0|} \\ + \frac{P_i}{\lambda_i} \cos^2\phi k_{hi}^2 e^{-\xi_{vi}\lambda_i z} + \frac{Q_i}{\lambda_i} \cos^2\phi k_{hi}^2 e^{\xi_{vi}\lambda_i z} \\ - S_i \sin^2\phi \xi_{hi} e^{-\xi_{hi} z} + T_i \sin^2\phi \xi_{hi} e^{\xi_{hi} z} + \end{array} \right) \alpha J_0(\alpha\rho) d\alpha$$

$$+ \frac{M_y}{4\pi\rho} \int_0^\infty \left(\begin{array}{c} \lambda_i \frac{\beta_i}{\xi_{vi}} k_v^2 e^{-\xi_{vi}\lambda_i |z-z_0|} + \beta_i \xi_{hi} e^{-\xi_{hi}|z-z_0|} \\ + P_i \lambda_i k_{vi}^2 e^{-\xi_{vi}\lambda_i z} + Q_i \lambda_i k_{vi}^2 e^{\xi_{vi}\lambda_i z} \\ + S_i \xi_{hi} e^{-\xi_{hi} z} - T_i \xi_{hi} e^{\xi_{hi} z} \end{array} \right) \cos 2\phi J_1(\alpha\rho) d\alpha \tag{5.30}$$

$$H_y^x = \frac{M_y}{4\pi} \int_0^\infty \left(\begin{array}{c} \frac{\beta_i}{\lambda_i \xi_{vi}} k_h^2 e^{-\xi_{vi}\lambda_i |z-z_0|} - \beta_i \xi_{hi} e^{-\xi_{hi}|z-z_0|} \\ + \frac{P_i}{\lambda_i} k_{hi}^2 e^{-\xi_{vi}\lambda_i z} + \frac{Q_i}{\lambda_i} k_{hi}^2 e^{\xi_{vi}\lambda_i z} \\ - S_i \xi_{hi} e^{-\xi_{hi} z} + T_i \xi_{hi} e^{\xi_{hi} z} + \end{array} \right) \sin\phi \cos\phi \alpha J_0(\alpha\rho) d\alpha$$

$$+ \frac{M_y}{4\pi\rho} \int_0^\infty \left(\begin{array}{c} \lambda_i \frac{\beta_i}{\xi_{vi}} k_v^2 e^{-\xi_{vi}\lambda_i |z-z_0|} + \beta_i \xi_{hi} e^{-\xi_{hi}|z-z_0|} \\ + P_i \lambda_i k_{vi}^2 e^{-\xi_{vi}\lambda_i z} + Q_i \lambda_i k_{vi}^2 e^{\xi_{vi}\lambda_i z} \\ + S_i \xi_{hi} e^{-\xi_{hi} z} - T_i \xi_{hi} e^{\xi_{hi} z} \end{array} \right) \sin 2\phi J_1(\alpha\rho) d\alpha \tag{5.31}$$

where

$$\beta_i = \begin{cases} 0, & \text{if } M_y \text{ is not in } i\text{th layer} \\ 1, & \text{if } M_y \text{ is in } i\text{th layer} \end{cases} \quad (5.32)$$

Similarly as x-directional dipole, in Eqs. (5.26)–(5.31), P_i and S_i are magnitude of reflection magnetic fields and Q_i and T_i are magnitude of refraction magnetic fields.

5.3.4 Magnitude of reflection and refraction magnetic fields

We use boundary conditions to determine previous unknown magnitude of reflection and refraction fields.

Let us examine P_i, Q_i, S_i, and T_i for x- and y-directional dipoles. From the continuity of electromagnetic fields at horizontal boundary $z = z_i$, the boundary conditions of the Hertz potential are as follows:

$$\mu_i \frac{\partial \pi_{yi}}{\partial z} = \mu_{i+1} \frac{\partial \pi_{y(i+1)}}{\partial z} \quad (5.33)$$

$$\mu_i \pi_{zi} = \mu_{i+1} \pi_{z(i+1)} \quad (5.34)$$

$$\lambda_i^2 \frac{\partial^2 \pi_{yi}}{\partial y^2} + \frac{\partial^2 \pi_{zi}}{\partial y \partial z} = \lambda_{i+1}^2 \frac{\partial^2 \pi_{y(i+1)}}{\partial y^2} + \frac{\partial^2 \pi_{z(i+1)}}{\partial y \partial z} \quad (5.35)$$

$$k_{hi}^2 \pi_{yi} = k_{h(i+1)}^2 \pi_{y(i+1)} \quad (5.36)$$

From Eqs. (5.33)–(5.36), P_i, Q_i, S_i, and T_i can be determined.

The next step is to calculate the coefficients F_i and G_i for z-directional dipole. From the continuity of electromagnetic fields at horizontal boundary $z = z_i$, the boundary conditions of the Hertz potential are

$$\frac{\partial \pi_{zi}}{\partial z} = \frac{\partial \pi_{z(i+1)}}{\partial z} \quad (5.37)$$

$$\mu_i \pi_{zi} = \mu_{i+1} \pi_{z(i+1)} \quad (5.38)$$

By applying boundary conditions, we can derive the final expression for coefficients F and G. Detailed derivation can be found in Ref. [11].

Table 5.1 Rule for choosing proper convergence algorithm

Dipping angle α (degrees)	Convergence algorithm
$\alpha \leq 79$ degrees	64 points Gauss-Quadrature
79 degrees $< \alpha \leq 82.5$ degrees	124 points Gauss-Quadrature
82.5 degrees $< \alpha \leq 89.9$ degrees	Fast Hankel transform

5.4 CONVERGENCE ALGORITHM

The operation speed of forward modeling plays key influence on inversion. The common fast convergence choice is fast Hankel transform. It is well known that the fast Hankel transform behaves like a digital filter. It is implemented by choosing sufficient effective points. However the disadvantage of the fast Hankel transform is slow convergence. To accelerate the convergence, we can employ Gauss-Quadrature algorithm, which incorporates less effective points for the integral. However the compromise may be the precision. Therefore using only one numerical algorithm to accelerate forward computation is not sufficient. It is noticed that at low dipping angle, Gauss-Quadrature algorithm is faster and simulation results is within 1% of fast Hankel transform. In highly deviated well, Gauss-Quadrature has no obvious help in improving convergence speed. Therefore it is beneficial to use fast Hankel transform and Gauss-Quadrature together. Table 5.1 provides the rules for choosing proper convergence algorithm.

5.5 SIMULATION RESULTS AND ANALYSIS

In this section, a few examples are presented to analyze the algorithm discussed in the previous sections. These examples are meant to verify and analyze the accuracy of the developed algorithm. Some published data are used as verification.

Example 5.1: Five-layer TI formation without dip

Fig. 5.1 shows one five-layer formation in Ref. [12]. In Ref. [12], finite difference numerical method was used to obtain the simulated results. The purpose of this example is to verify the developed algorithm.

In Fig. 5.1, the first layer, third layer, and fifth layer are isotropic layers with resistivities of 50 ohm-m. The second and fourth layers are anisotropic layers with horizontal resistivities of 3 ohm-m and vertical resistivities of 15 ohm-m. The thickness of the second layer and fourth layer are 0.73 and 3.66 m, respectively. In Ref. [12], the triaxial tool used has one transmitting and two receiving coils, with the second

Triaxial Induction Tool and Logging-While-Drilling Tool Response in a Transverse Isotropic-Layered Formation

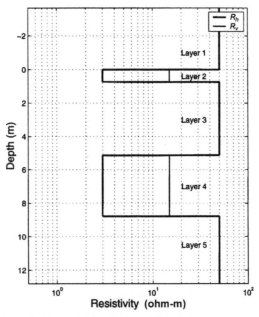

Figure 5.1 A 1D-layered model [12]. The second and fourth layers are anisotropic with horizontal resistivity (R_h) of 3 ohm-m and vertical resistivity (R_v) of 15 ohm-m. The thicknesses of the second layer and fourth layer are 0.73 m and 3.66 m, respectively. The uniform background resistivity is 50 ohm-m.

receiver as a bucking coil. The two receiver measurements are combined to give the final output,

$$H_p = H_{p1} - \frac{l_2^3}{l_1^3} \cdot H_{p2} \qquad (5.39)$$

where $p = xx$, yy, or zz and where $l_1 = 1.2$ m and $l_2 = 1.92$ m are the distances of the first and second receivers to the transmitter, respectively. Four different frequencies are considered: 14, 39, 77, 154 kHz, respectively. In the following, we will use this formation and tool setup to discuss the developed algorithm.

Consider the simple case with a dipping angle of 0 degree. Fig. 5.2 shows imaginary components of magnetic responses H_{xx} and H_{zz} in Ref. [12]. Left figure is H_{xx} and right figure represents H_{zz}.

Fig. 5.3 shows calculated imaginary parts of H_{xx} and H_{zz} at different frequencies using the forward model discussed in this chapter. Left figure is H_{xx} and right figure is H_{zz}. In Fig. 5.3, *dark blue line* represents response when frequency is 14 kHz. *Pink line* is responses at 39 kHz. *Yellow line* illustrates response when tool is working at 77 kHz. *Cyan blue line* is response at 154 kHz. Because the dipping angle is zero, all the cross components are zero.

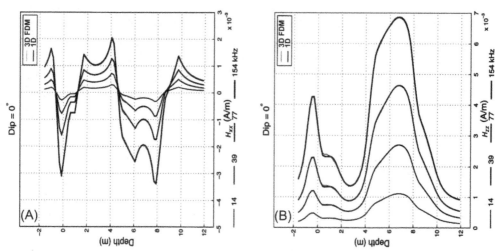

Figure 5.2 Imaginary components of (A) H_{xx}, (B) H_{zz} from Ref. [12] in a vertical well for the five-layer model in Fig. 5.1.

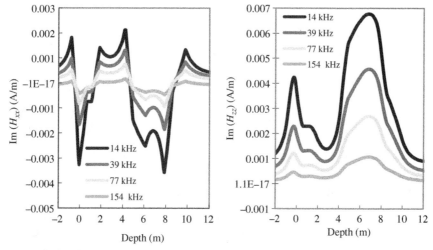

Figure 5.3 Calculated imaginary components of H_{xx}, H_{zz} in a vertical well from the 1D forward model discussed in this chapter for the five-layer formation in Fig. 5.1.

Now let's test the algorithm when the formation has a nonzero dipping angle. Fig. 5.4 shows the formation given in Ref. [13]. The formation is also a five-layer model. Layer one presents anisotropic shale with horizontal and vertical resistivity of 1.0 and 2.0 ohm-m, respectively. Layer two is a combination of 50% isotropic sand and 50% anisotropic shale. Shale resistivity is the same as layer one and the sand

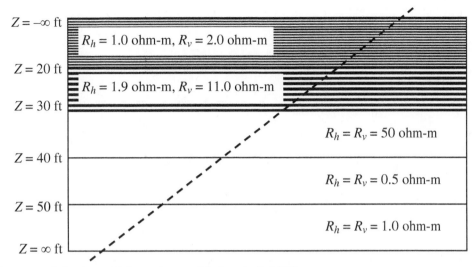

Figure 5.4 A five-layer 1D formation model given in Ref. [13].

resistivity is 20 ohm-m. Layer two has vertical resistivity of 11 ohm-m and horizontal resistivity of 1.9 ohm-m. Layer three presents isotropic oil-bearing sandstone that has 50 ohm-m resistivity. Layer four represents isotropic water saturated sand that has 0.5 ohm-m resistivity. Layer five is anisotropic shale of 1.0 ohm-m. On the other hand, the triaxial logging tool is consisted of one transmitter and one receiver. The distance from transmitter to receiver is 40 in. The operating frequency of the tool is 20 kHz.

Fig. 5.5 shows apparent conductivity from Ref. [13] when dipping angle and azimuthal angle are 75 and 30 degrees, respectively. Fig. 5.6 presents apparent conductivity from presented 1D forward algorithm. According to Figs. 5.5 and 5.6, the comparison proves that the developed analytical solution in this chapter provides accurate cross components as well.

5.6 ANALYSIS OF ANISOTROPY IMPACT TO THE RESISTIVITY LWD TOOL

The 1D analytic solution discussed in this chapter is a full wave solution without static or quasistatic approximation. Therefore this solution can be used in theory to solve higher frequency EM problems. One immediate application would be the LWD resistivity tool response. As we discussed in Chapter 4, Triaxial Induction and Logging-While-Drilling Resistivity Tool Response in Homogeneous Anisotropic Formations, LWD resistivity tool uses higher frequency in order to avoid induction effect due to the metal mandrel. The lower frequency end of a LWD tool is usually

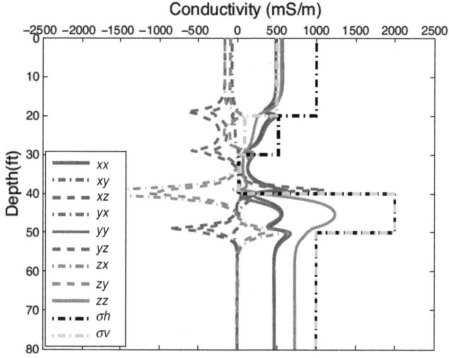

Figure 5.5 Apparent conductivity from Ref. [13] with both nonzero dipping and rotation angles in the five-layer formation model given in Fig. 5.4.

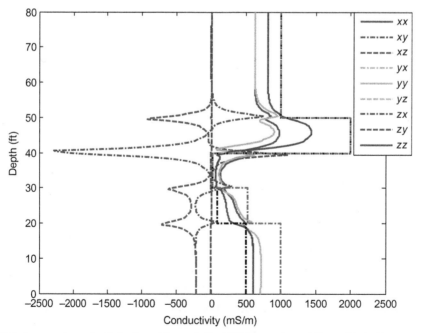

Figure 5.6 Apparent conductivity from the analytical 1D forward modeling with both nonzero dipping and rotation angles in the five-layer formation model given in Fig. 5.4.

400 kHz and higher end is 2 MHz. Chapter 4, Triaxial Induction and Logging-While-Drilling Resistivity Tool Response in Homogeneous Anisotropic Formations, discussed the conversion of phase difference and amplitude attenuation to the resistivity of the formation in an isotropic formation. In this section, let's use the analytic solution discussed in this chapter to observe the impact of the anisotropy to the LWD resistivity measurements. The operating frequency is 2 MHz in the following discussions. Distances from main transmitter to two receivers are 37 and 45 in., respectively. Figs. 5.7 and 5.8 show computed tool response in four different homogenous anisotropic formations. The four formations have the same horizontal resistivity but different vertical resistivity. The anisotropy ratio of the computation is defined by R_v/R_h.

From Figs. 5.7 and 5.8, we find that in a vertical well, no matter how the anisotropic ratio changes, phase shift and amplitude ratio are the same in all four cases. However, with the increased dipping angle, responses in anisotropic formations varies. Both amplitude and phase responses decrease in highly deviated well. This phenomenon can be understood since the LWD tool uses coil antennas and the signal received by the receiver antennas comes from the transmitter antenna and propagates through formations layers that are mainly perpendicular to the tool axis direction. When the

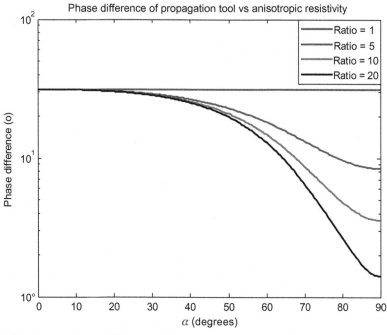

Figure 5.7 Calculated phase difference of basic propagation tool versus anisotropic ration in homogenous formation.

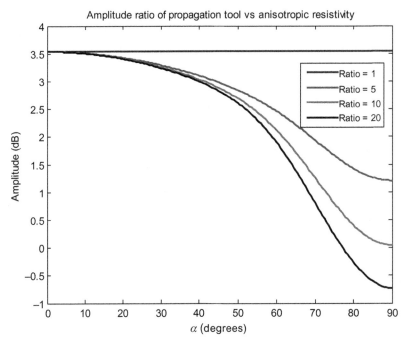

Figure 5.8 Calculated amplitude ratio of basic propagation tool versus anisotropic ration in homogenous formation.

dip is small, the wave path is dominated by the horizontal resistivity R_h. As the dip angle increases, the impact of vertical resistivity starts play a part. When the tool is in the horizontal position, or the dip is at 90 degrees, the propagation is dominated by the vertical resistivity R_v.

REFERENCES

[1] J.H. Moran, S. Gianzero, Effects of formation anisotropy on resistivity-logging measurements, Geophysics 44 (7) (1979) 1266—1286.
[2] J.D. Klein, Induction log anisotropy corrections, Log Anal. 34 (2) (1993) 18—27.
[3] S. Graciet, L.C. Shen, Theory and numerical simulation of induction and MWD resistivity tools in anisotropic dipping, Log Anal. 39 (no.1) (1998) 24—37.
[4] M. Zhdanov, D. Kennedy, E. Peksen, Foundations of tensor induction well-logging, Petrophysics 42 (2001) 588—610.
[5] G.L. Wang, C. Torres-Verdín, S. Gianzero, "Fast simulation of triaxial borehole induction measurements acquired in axially symmetrical and transversely isotropic media", Geophysics 74 (6) (2009) 233—249.
[6] H. Wang, T. Barber, K. Chen, et al., Triaxial induction logging: theory, modeling, inversion, and interpretation, in: International Oil & Gas Conference and Exhibition in China, Beijing, China, 2006.
[7] S. Davydycheva, D. Homan, G. Minerbo, "Triaxial induction tool with electrode sleeve: FD modeling in 3D geometries", J. Appl. Geophys. 67 (1) (2009) 98—108.

[8] A.B. Cheryauka, M.S. Zhdanov, Fast modeling of a tensor induction logging response in a horizontal well in inhomogeneous anisotropic formations, in: SPWLA 42nd Annual Logging Symposium, 2001.
[9] L. Yu, B. Kriegshauser, O. Fanini, J. Xiao, A fast inversion method for multicomponent induction log data, in: 71st Annual International Meeting, SEG, Expanded Abstracts, pp. 361–364, 2001.
[10] X. Lu, D. Alumbaugh, One-dimensional inversion of three component induction logging in anisotropic media, in: 71st Annual International Meeting, SEG, Expanded Abstracts, pp. 376–380, 2001.
[11] M. Huang, L.C. Shen, Computation of induction logs in multiple-layer dipping formation, IEEE Trans. Geosci. Remote Sens. 27 (3) (1989) 259.
[12] T. Wang, S. Fang, 3-D electromagnetic anisotropy modeling using finite differences, Geophysics 66 (5) (2001) 1386–1398.
[13] H. Wang, T. Barber, C. Morriss, R. Rosthal, R. Hayden, M. Markely, Determining anisotropic formation resistivity at any relative dip using a multiarray triaxial induction tool, in: SPE Annual Technical Conference and Exhibition, 2006.

APPENDIX A DERIVATION OF HERTZ VECTOR POTENTIAL IN MULTIPLE LAYER FORMATION

A.1 x-Direction magnetic dipole

We assume one horizontally oriented magnetic dipole $M = (M_x, 0, 0)$. According to Eq. (5.18):

$$\nabla_\lambda^2 \pi_x + k_v^2 \pi_x = -\frac{1}{\lambda^2} M_x \tag{A.1}$$

$$\nabla^2 \pi_z + k_h^2 \pi_z = (1 - \lambda^2) \frac{\partial^2 \pi_x}{\partial z \partial x} \tag{A.2}$$

where $\nabla_\lambda^2 = \left(\frac{\partial^2}{\partial x^2} + \frac{\partial^2}{\partial y^2} + \frac{1}{\lambda^2}\frac{\partial^2}{\partial z^2}\right)$, and $k_v^2 = i\omega\mu_0\sigma_v$.

To get the solution of Eq. (A.1), we first consider the homogeneous equations with zero on the right-hand sides:

$$\nabla_\lambda^2 \pi_x + k_v^2 \pi_x = 0 \tag{A.3}$$

We assume Hertz potential π_x in the form,

$$\pi_x = \int_0^\infty F(z)\alpha J_0(\alpha\rho) d\alpha \tag{A.4}$$

Since $\nabla_\lambda^2 \pi_x = \frac{\partial^2 \pi_x}{\partial x^2} + \frac{\partial^2 \pi_x}{\partial y^2} + \frac{\partial^2 \pi_x}{\lambda^2 \partial z^2}$, we first derive $\frac{\partial^2 \pi_x}{\partial x^2}$,

$$\frac{\partial^2 \pi_x}{\partial x^2} = \int_0^\infty F(z)\alpha \frac{\partial^2 J_0(\alpha\rho)}{\partial x^2} d\alpha \tag{A.5}$$

As we know,

$$\frac{\partial^2 J_0(\alpha\rho)}{\partial x^2} = -\frac{\alpha J_1(\alpha\rho)}{\rho} + \frac{2\alpha x^2 J_1(\alpha\rho)}{\rho^3} - \frac{\alpha^2 x^2 J_0(\alpha\rho)}{\rho^2} \qquad (A.6)$$

Then we get

$$\frac{\partial^2 \pi_x}{\partial x^2} = \int_0^\infty \left[-\frac{\alpha J_1(\alpha\rho)}{\rho} + \frac{2\alpha x^2 J_1(\alpha\rho)}{\rho^3} - \frac{\alpha^2 x^2}{\rho^2} J_0(\alpha\rho) \right] \alpha F(z) d\alpha \qquad (A.7)$$

Similarly, we obtain $\frac{\partial^2 \pi_x}{\partial y^2}$,

$$\frac{\partial^2 \pi_x}{\partial y^2} = \int_0^\infty \left[-\frac{\alpha J_1(\alpha\rho)}{\rho} + \frac{2\alpha y^2 J_1(\alpha\rho)}{\rho^3} - \frac{\alpha^2 y^2}{\rho^2} J_0(\alpha\rho) \right] \alpha F(z) d\alpha \qquad (A.8)$$

On the other hand,

$$\frac{\partial^2 \pi_x}{\lambda^2 \partial z^2} = \int_0^\infty \frac{\partial^2 F(z)}{\lambda^2 \partial z^2} \alpha J_0(\alpha\rho) d\alpha \qquad (A.9)$$

Therefore

$$\nabla_\lambda^2 \pi_x + k_\nu^2 \pi_x = \int_0^\infty \left[-\frac{\alpha J_1(\alpha\rho)}{\rho} + \frac{2\alpha x^2 J_1(\alpha\rho)}{\rho^3} - \frac{\alpha^2 x^2}{\rho^2} J_0(\alpha\rho) \right] \alpha F(z) d\alpha$$

$$+ \int_0^\infty \left[-\frac{\alpha J_1(\alpha\rho)}{\rho} + \frac{2\alpha y^2 J_1(\alpha\rho)}{\rho^3} - \frac{\alpha^2 y^2}{\rho^2} J_0(\alpha\rho) \right] \alpha F(z) d\alpha$$

$$+ \int_0^\infty \frac{\partial^2 F(z)}{\lambda^2 \partial z^2} \alpha J_0(\alpha\rho) d\alpha + k_\nu^2 \int_0^\infty F(z) \alpha J_0(\alpha\rho) d\alpha \qquad (A.10)$$

$$= \int_0^\infty (-\alpha^2 + k_\nu^2) J_0(\alpha\rho) \alpha F(z) d\alpha + \int_0^\infty \frac{\partial^2 F(z)}{\lambda^2 \partial z^2} \alpha J_0(\alpha\rho) d\alpha$$

$$= \int_0^\infty \left(-\xi_\nu^2 F(z) + \frac{\partial^2 F(z)}{\lambda^2 \partial z^2} \right) \alpha J_0(\alpha\rho) d\alpha$$

$$= 0$$

Then we get

$$F(z) = \frac{M_x}{4\pi\lambda} (P e^{-\xi_\nu \lambda z} + Q e^{\xi_\nu \lambda z}) \qquad (A.11)$$

We take account the constant number $\frac{M_x}{4\pi\lambda}$ for convenience expression. The solution of Eq. (A.3) is in the form:

$$\pi_{x,g} = \frac{M_x}{4\pi\lambda} \int_0^\infty (Pe^{-\xi_\nu \lambda z} + Qe^{\xi_\nu \lambda z})\alpha J_0(\alpha\rho)d\alpha \qquad (A.12)$$

where $\pi_{x,g}$ represents general solution.

The particular solution of Eq. (A.1) is well known:

$$\pi_{x,p} = \frac{M_x}{4\pi\lambda}\frac{e^{ik_\nu s}}{s} = \frac{M_x}{4\pi\lambda}\int_0^\infty \frac{1}{\xi_\nu} e^{-\xi_\nu \lambda |z-z_0|}\alpha J_0(\alpha\rho)d\alpha \qquad (A.13)$$

The final solution of Eq. (A.1) is found to be as follows:

$$\pi_x = \pi_{x,g} + \pi_{x,p} = \frac{M_x}{4\pi\lambda}\int_0^\infty \left(\frac{1}{\xi_\nu} e^{-\xi_\nu \lambda |z-z_0|} + Pe^{-\xi_\nu \lambda z} + Qe^{\xi_\nu \lambda z}\right)\alpha J_0(\alpha\rho)d\alpha \qquad (A.14)$$

Next let us see how to solve Eq. (A.2).

$$\nabla^2 \pi_z + k_h^2 \pi_z = (1-\lambda^2)\frac{\partial^2(\pi_{x,g}+\pi_{x,p})}{\partial z \partial x} = (1-\lambda^2)\frac{\partial^2 \pi_{x,g}}{\partial z \partial x} + (1-\lambda^2)\frac{\partial^2 \pi_{x,p}}{\partial z \partial x} \qquad (A.15)$$

Now we rewrite Eq. (A.15) into two independent equations:

$$\nabla^2 \pi_{z,g} + k_h^2 \pi_{z,g} = (1-\lambda^2)\frac{\partial^2 \pi_{x,g}}{\partial z \partial x} \qquad (A.16)$$

$$\nabla^2 \pi_{z,p} + k_h^2 \pi_{z,p} = (1-\lambda^2)\frac{\partial^2 \pi_{x,p}}{\partial z \partial x} \qquad (A.17)$$

Note: The solution of Eq. (A.15) should be the sum of $\pi_{z,g}, \pi_{z,p}$.

First we try to get $\pi_{z,g}$. Let

$$\pi_{z,g} = \frac{x}{\rho}\frac{M_x}{4\pi}\int_0^\infty E(z)J_1(\alpha\rho)d\alpha = -\frac{\partial}{\partial x}\frac{M_x}{4\pi}\int_0^\infty \frac{E(z)}{\alpha}J_0(\alpha\rho)d\alpha \qquad (A.18)$$

Then we know

$$\frac{\partial^2 \pi_{z,g}}{\partial x^2} = -\frac{\partial}{\partial x}\frac{M_x}{4\pi}\int_0^\infty \frac{E(z)}{\alpha}\frac{\partial^2 J_0(\alpha\rho)}{\partial x^2}d\alpha$$

$$= \frac{\partial}{\partial x}\frac{M_x}{4\pi}\int_0^\infty E(z)\left[\frac{1}{\rho}J_1(\alpha\rho) + \frac{\alpha x^2}{\rho^2}J_0(\alpha\rho) - \frac{2x^3}{\rho^3}J_1(\alpha\rho)\right]d\alpha \qquad (A.19)$$

$$\frac{\partial^2 \pi_{z,g}}{\partial y^2} = -\frac{\partial}{\partial x} \frac{M_x}{4\pi} \int_0^\infty \frac{E(z)}{\alpha} \frac{\partial^2 J_0(\alpha\rho)}{\partial y^2} d\alpha$$

$$= \frac{\partial}{\partial x} \frac{M_x}{4\pi} \int_0^\infty E(z) \left[\frac{1}{\rho} J_1(\alpha\rho) + \frac{\alpha y^2}{\rho^2} J_0(\alpha\rho) - \frac{2y^3}{\rho^3} J_1(\alpha\rho) \right] d\alpha \quad \text{(A.20)}$$

$$\frac{\partial^2 \pi_{z,g}}{\partial z^2} = -\frac{\partial}{\partial x} \frac{M_x}{4\pi} \int_0^\infty \frac{d^2 E(z)}{\alpha dz^2} J_0(\alpha\rho) d\alpha \quad \text{(A.21)}$$

Therefore

$$\nabla^2 \pi_{z,g} + k_h^2 \pi_{z,g} = \frac{\partial}{\partial x} \frac{M_x}{4\pi} \int_0^\infty \left[\frac{E(z)}{\alpha} \xi_h^2 J_0(\alpha\rho) - \frac{d^2 E(z)}{\alpha dz^2} J_0(\alpha\rho) \right] d\alpha \quad \text{(A.22)}$$

As we know

$$(1-\lambda^2) \frac{\partial^2 \pi_{x,p}}{\partial z \partial x} = (1-\lambda^2) \frac{\partial}{\partial x} \frac{M_x}{4\pi\lambda} \int_0^\infty (\xi_v \lambda P e^{-\xi_v \lambda z} + \xi_v \lambda Q e^{\xi_v \lambda z}) \alpha J_0(\alpha\rho) d\alpha \quad \text{(A.23)}$$

Then we get this equation

$$\frac{\partial}{\partial x} \frac{M_x}{4\pi} \int_0^\infty \left[\frac{E(z)}{\alpha} \xi_h^2 - \frac{d^2 E(z)}{\alpha dz^2} \right] J_0(\alpha\rho) d\alpha = (1-\lambda^2) \frac{\partial}{\partial x} \frac{M_x}{4\pi\lambda} \int_0^\infty (P e^{-\xi_v \lambda z} + Q e^{\xi_v \lambda z}) \xi_v \lambda \alpha J_0(\alpha\rho) d\alpha$$
$$\text{(A.24)}$$

Therefore

$$\frac{d^2 E(z)}{dz^2} - E(z) \xi_h^2 = (1-\lambda^2)(P e^{-\xi_v \lambda z} - Q e^{\xi_v \lambda z}) \xi_v \alpha^2 \quad \text{(A.25)}$$

The solution of above equation is in the form:

$$E(z) = S e^{-\xi_h z} + T e^{\xi_h z} + E^*(z) \quad \text{(A.26)}$$

We apply constant variation method to get $E^*(z)$. Let

$$E^*(z) = S(z) e^{-\xi_h z} + T(z) e^{\xi_h z} \quad \text{(A.27)}$$

$S(z), T(z)$ should satisfy these linear equations:

$$S'(z) e^{-\xi_h z} + T'(z) e^{\xi_h z} = 0 \quad \text{(A.28)}$$

$$S'(z)(-\xi_h) e^{-\xi_h z} + T'(z) \xi_h e^{\xi_h z} = (1-\lambda^2)(P e^{-\xi_v \lambda z} - Q e^{\xi_v \lambda z}) \xi_v \alpha^2 \quad \text{(A.29)}$$

Then we get

$$S(z) = -\frac{(1-\lambda^2)\alpha^2\xi_v}{2\xi_h}\left[\frac{Pe^{-\xi_v\lambda z+\xi_h z}}{-\xi_v\lambda+\xi_h} - \frac{Qe^{\xi_v\lambda z+\xi_h z}}{\xi_v\lambda+\xi_h}\right]$$
$$= -\frac{(1-\lambda^2)\alpha^2\xi_v}{2\xi_h(\xi_h^2-\xi_v^2\lambda^2)}\left[Pe^{-\xi_v\lambda z+\xi_h z}(\xi_v\lambda+\xi_h) - Qe^{\xi_v\lambda z+\xi_h z}(-\xi_v\lambda+\xi_h)\right] \quad (A.30)$$

$$T(z) = \frac{(1-\lambda^2)\alpha^2\xi_v}{2\xi_h}\left[\frac{Pe^{-\xi_v\lambda z-\xi_h z}}{-\xi_v\lambda-\xi_h} - \frac{Qe^{\xi_v\lambda z-\xi_h z}}{\xi_v\lambda-\xi_h}\right]$$
$$= \frac{(1-\lambda^2)\alpha^2\xi_v}{2\xi_h(\xi_h^2-\xi_v^2\lambda^2)}\left[Pe^{-\xi_v\lambda z+\xi_h z}(\xi_v\lambda-\xi_h) - Qe^{\xi_v\lambda z+\xi_h z}(-\xi_v\lambda-\xi_h)\right] \quad (A.31)$$

Since

$$\frac{(1-\lambda^2)\alpha^2}{\xi_h^2-\xi_v^2\lambda^2} = \frac{(1-\lambda^2)\alpha^2}{(\alpha^2-k_h^2)-(\alpha^2-k_v^2)\lambda^2}$$
$$= \frac{(1-\lambda^2)\alpha^2}{(\alpha^2-\alpha^2\lambda^2)-(k_h^2-k_v^2\lambda^2)} \quad (A.32)$$
$$= \frac{(1-\lambda^2)\alpha^2}{(\alpha^2-\alpha^2\lambda^2)-(k_h^2-k_h^2)}$$
$$= 1$$

So

$$S(z) = -\frac{\xi_v}{2\xi_h}\left[Pe^{-\xi_v\lambda z+\xi_h z}(\xi_v\lambda+\xi_h) - Qe^{\xi_v\lambda z+\xi_h z}(-\xi_v\lambda+\xi_h)\right] \quad (A.33)$$

$$T(z) = \frac{\xi_v}{2\xi_h}\left[Pe^{-\xi_v\lambda z+\xi_h z}(\xi_v\lambda-\xi_h) - Qe^{\xi_v\lambda z+\xi_h z}(-\xi_v\lambda-\xi_h)\right] \quad (A.34)$$

Then

$$E^*(z) = S(z)e^{-\xi_h z} + T(z)e^{\xi_h z} = \xi_v[-Pe^{-\xi_v\lambda z} + Qe^{\xi_v\lambda z}] \quad (A.35)$$

The solution of equation is in the form:

$$E(z) = Se^{-\xi_h z} + Te^{\xi_h z} + \xi_v[-Pe^{-\xi_v\lambda z} + Qe^{\xi_v\lambda z}] \quad (A.36)$$

So we know

$$\pi_{z,g} = \frac{x}{\rho}\frac{M_x}{4\pi}\int_0^\infty [Se^{-\xi_h z} + Te^{\xi_h z} + \xi_\nu(-Pe^{-\xi_\nu \lambda z} + Qe^{\xi_\nu \lambda z})]J_1(\alpha\rho)d\alpha \qquad (A.37)$$

Next we need to derive $\pi_{z,p}$. Let

$$\pi_{z,p} = \frac{x}{\rho}\frac{M_x}{4\pi}\int_0^\infty C(z)J_1(\alpha\rho)d\alpha = -\frac{\partial}{\partial x}\frac{M_x}{4\pi}\int_0^\infty \frac{C(z)}{\alpha}J_0(\alpha\rho)d\alpha \qquad (A.38)$$

Similarly we get

$$\nabla^2 \pi_{z,p} + k_h^2 \pi_{z,p} = \frac{\partial}{\partial x}\frac{M_x}{4\pi}\int_0^\infty \left[C(z)\xi_h^2 - \frac{d^2C(z)}{dz^2}\right]\frac{J_0(\alpha\rho)}{\alpha}d\alpha \qquad (A.39)$$

And

$$\frac{\partial \pi_{x,p}}{\partial z} = -\frac{M_x}{4\pi}\int_0^\infty \frac{|z-z_0|}{z-z_0}e^{-\xi_\nu \lambda |z-z_0|}\alpha J_0(\alpha\rho)d\alpha \qquad (A.40)$$

Equation is rewritten in the form:

$$\frac{\partial}{\partial x}\frac{M_x}{4\pi}\int_0^\infty \left[C(z)\xi_h^2 - \frac{d^2C(z)}{dz^2}\right]\frac{J_0(\alpha\rho)}{\alpha}d\alpha = -\frac{\partial}{\partial x}\frac{M_x}{4\pi}\int_0^\infty \frac{|z-z_0|}{z-z_0}e^{-\xi_\nu \lambda |z-z_0|}\alpha J_0(\alpha\rho)d\alpha$$

$$(A.41)$$

So we get the new equation

$$\frac{d^2C(z)}{dz^2} - C(z)\xi_h^2 = (\lambda^2 - 1)\frac{|z-z_0|}{z-z_0}e^{-\xi_\nu \lambda |z-z_0|}\alpha^2 \qquad (A.42)$$

Similarly we apply constant variance method to get $C(z)$.

$$C(z) = Ae^{-\xi_h |z-z_0|} + Be^{\xi_h |z-z_0|} - \frac{|z-z_0|}{z-z_0}e^{-\xi_\nu \lambda |z-z_0|} \qquad (A.43)$$

Then we choose $A = \frac{|z-z_0|}{z-z_0}$, $B = 0$ according to boundary condition of source in homogenous formation.

The particular solution is in the form:

$$\pi_{z,p} = \frac{x}{\rho}\frac{M_x}{4\pi}\int_0^\infty \frac{|z-z_0|}{z-z_0}(e^{-\xi_h |z-z_0|} - e^{-\xi_\nu \lambda |z-z_0|})J_1(\alpha\rho)d\alpha \qquad (A.44)$$

The final solution is

$$\pi_z = \frac{x}{\rho}\frac{M_x}{4\pi}\int_0^\infty \left[Se^{-\xi_h z} + Te^{\xi_h z} + \xi_\nu(-Pe^{-\xi_\nu \lambda z} + Qe^{\xi_\nu \lambda z})\right] J_1(\alpha\rho)d\alpha \quad (A.45)$$

$$+ \frac{x}{\rho}\frac{M_x}{4\pi}\int_0^\infty \frac{|z-z_0|}{z-z_0}(e^{-\xi_h|z-z_0|} - e^{-\xi_\nu \lambda|z-z_0|}) J_1(\alpha\rho)d\alpha$$

A.2 x-Direction magnetic dipole

According to x dipole, in the similar way, we can get Hertz vector potential.

$$\pi_y = \frac{M_y}{4\pi\lambda}\int_0^\infty \left(\frac{1}{\xi_\nu}e^{-\xi_\nu \lambda|z-z_0|} + Pe^{-\xi_\nu \lambda z} + Qe^{\xi_\nu \lambda z}\right)\alpha J_0(\alpha\rho)d\alpha \quad (A.46)$$

$$\pi_z = \frac{y}{\rho}\frac{M_y}{4\pi}\int_0^\infty \left[Se^{-\xi_h z} + Te^{\xi_h z} + \xi_\nu(-Pe^{-\xi_\nu \lambda z} + Qe^{\xi_\nu \lambda z})\right] J_1(\alpha\rho)d\alpha \quad (A.47)$$

$$+ \frac{y}{\rho}\frac{M_y}{4\pi}\int_0^\infty \frac{|z-z_0|}{z-z_0}(e^{-\xi_h|z-z_0|} - e^{-\xi_\nu \lambda|z-z_0|}) J_1(\alpha\rho)d\alpha$$

A.3 z-Direction magnetic dipole

We assume one horizontally oriented magnetic dipole $M = (0, 0, M_z)$. According to Eq. (5.18):

$$\nabla^2 \pi_z + k_h^2 \pi_z = -M_z \quad (A.48)$$

We first consider the first equation with zero on the right-hand sides:

$$\nabla^2 \pi_{z,g} + k_h^2 \pi_{z,g} = 0 \quad (A.49)$$

If we assume $\pi_{z,g}$ in the form:

$$\pi_{z,g} = \int_0^\infty D(z)\alpha J_0(\alpha\rho)d\alpha \quad (A.50)$$

It is easy to get,

$$D(z) = Fe^{-\xi_h z} + Ge^{\xi_h z} \quad (A.51)$$

The general solution is shown as,

$$\pi_{z,g} = \int_0^\infty (Fe^{-\xi_h z} + Ge^{\xi_h z})\alpha J_0(\alpha\rho)d\alpha \qquad (A.52)$$

Since we have known that

$$\pi_{z,p} = \frac{M_z}{4\pi\lambda}\frac{e^{ik_v s}}{s} = \frac{M_x}{4\pi\lambda}\int_0^\infty \frac{1}{\xi_v} e^{-\xi_v \lambda|z-z_0|}\alpha J_0(\alpha\rho)d\alpha \qquad (A.53)$$

It is easy to get the final solution of Eq. (A.48)

$$\pi_z = \frac{M_x}{4\pi\lambda}\int_0^\infty \left(\frac{1}{\xi_v}e^{-\xi_v \lambda|z-z_0|} + Fe^{-\xi_h z} + Ge^{\xi_h z}\right)\alpha J_0(\alpha\rho)d\alpha \qquad (A.54)$$

CHAPTER 6

Triaxial Induction and Logging-While-Drilling Logging Tool Response in a Biaxial Anisotropic-Layered Formation

Contents

6.1 Spectral-Domain Solution to Maxwell's Equations in a Homogeneous Biaxial Anisotropic Medium	188
6.2 Propagation in Unbounded Medium	193
6.3 Propagation in Layered Medium	193
6.4 Computation of the Double Integrals	196
6.5 Numerical Examples	197
References	202
Appendix A Derivation of Matrix A	202

In Chapter 4, Triaxial Induction and Logging-While-Drilling Resistivity Tool Response in Homogeneous Anisotropic Formations, the triaxial induction tool performance was discussed in a homogeneous biaxial formation. In Chapter 5, Triaxial Induction Tool and Logging-While-Drilling Tool Response in a Transverse Isotropic Layered Formation, we studied an analytical method to solve the problem of triaxial logging tools in a layered transverse isotropic (TI) formation. In many cases, the formation gets more complicated. For example, in a layered TI formation, if cracks present as discussed in Chapter 4, Triaxial Induction and Logging-While-Drilling Resistivity Tool Response in Homogeneous Anisotropic Formations, the formation can be considered to be a biaxial anisotropy. It is important to study the tool response in a biaxial-layered formation. Due to the complexity of the environment, the mathematic formulations become more involved. Fortunately, if we ignore borehole effect and eccentricity, analytic solutions can be found to the problem. Even though numerical solutions can be applied, the analytical solutions are always advantageous in terms of computation speed and ease of use. The analytical solution may be a good approach for inversion. In this chapter, we will derive the full magnetic field response of a triaxial induction sonde in a layered biaxial anisotropic medium. The derivation of the triaxial induction tool response in a layered biaxial anisotropic formation is divided into four steps. The source is assumed to be a magnetic dipole with three components. The first step is to find the solutions to the Maxwell's equations in a homogeneous biaxial anisotropic media based on the electric field. Using this solution, the second step observes the wave-propagation characteristics in an unbounded biaxial anisotropic formation. The third step brings the boundary into the consideration and transmission and reflection coefficients are found using the concept of generalized transmission and reflection matrix, with which the fields can be obtained in any layer. Final step is to derive the expressions of the magnetic fields by substituting the electric field obtained in the previous steps into the Maxwell's equations with tensor conductivities. In the following sections, we will describe these four steps in detail.

6.1 SPECTRAL-DOMAIN SOLUTION TO MAXWELL'S EQUATIONS IN A HOMOGENEOUS BIAXIAL ANISOTROPIC MEDIUM

Similar to the approach described in previous chapters, assuming the harmonic-time dependence to be $e^{-j\omega t}$, Maxwell's equations for the electric and magnetic fields in the space domain are:

$$\nabla \times \vec{H}(\vec{r}) = -j\omega\tilde{\varepsilon}\vec{E}(\vec{r}) + \vec{J}_s(\vec{r}) \tag{6.1a}$$

$$\nabla \times \vec{E}(\vec{r}) = j\omega\mu_0 \vec{H}(\vec{r}) + j\omega\mu_0 \vec{M}_s(\vec{r}) \tag{6.1b}$$

where μ_0 is the magnetic permeability of the vacuum, $\vec{r} = (x, y, z)$ is the position vector, $\vec{M}_s(\vec{r})$ is the magnetic-source flux density, and $\vec{J}_s(\vec{r})$ is the electric-source current density, $\tilde{\varepsilon}$ is the complex dielectric constant tensor in the principal axis defined as

$$\tilde{\varepsilon} = \begin{bmatrix} \tilde{\varepsilon}_x & 0 & 0 \\ 0 & \tilde{\varepsilon}_y & 0 \\ 0 & 0 & \tilde{\varepsilon}_z \end{bmatrix}, \quad \tilde{\varepsilon}_x = \bar{\varepsilon}_x + j\frac{\sigma_x}{\omega}, \quad \tilde{\varepsilon}_y = \bar{\varepsilon}_y + j\frac{\sigma_y}{\omega}, \quad \tilde{\varepsilon}_z = \bar{\varepsilon}_z + j\frac{\sigma_z}{\omega} \tag{6.2}$$

We only consider conductivity and dielectric permittivity anisotropy, no anisotropy considered in magnetic permeability. For the induction logging problems discussed in this chapter we assume that $\vec{J}_s(\vec{r}) = 0$, which means only magnetic dipoles are used to represent induction coils. $\vec{E}(\vec{r})$, $\vec{H}(\vec{r})$, and $\vec{M}(\vec{r})$ can be expressed in terms of

Fourier transforms in two horizontal directions of their spectral-domain counterparts $\tilde{E}(k_x, k_y, z)$, $\tilde{H}(k_x, k_y, z)$, and $\tilde{M}_s(k_x, k_y, z)$:

$$\vec{E}(\vec{r}) = \frac{1}{(2\pi)^2} \int_{-\infty}^{+\infty} \int_{-\infty}^{+\infty} e^{j(k_x x + k_y y)} dk_x dk_y \cdot \tilde{E}(k_x, k_y) \quad (6.3a)$$

$$\vec{H}(\vec{r}) = \frac{1}{(2\pi)^2} \int_{-\infty}^{+\infty} \int_{-\infty}^{+\infty} e^{j(k_x x + k_y y)} dk_x dk_y \cdot \tilde{H}(k_x, k_y) \quad (6.3b)$$

$$\vec{M}_s(\vec{r}) = \frac{1}{(2\pi)^2} \int_{-\infty}^{+\infty} \int_{-\infty}^{+\infty} e^{j(k_x x + k_y y)} dk_x dk_y \cdot \tilde{M}_s(k_x, k_y) \quad (6.3c)$$

For mathematical convenience, we first solve Eq. (6.1a,b) in the spectral domain and then use Eq. (6.3a–c) to obtain the space-domain solutions from their spectral-domain counterparts. Using the relations $\frac{\partial}{\partial x} = jk_x$ and $\frac{\partial}{\partial y} = jk_y$, we can obtain:

$$\nabla \times \vec{E} = \begin{pmatrix} \frac{\partial E_z}{\partial y} - \frac{\partial E_y}{\partial z} \\ \frac{\partial E_x}{\partial z} - \frac{\partial E_z}{\partial x} \\ \frac{\partial E_y}{\partial x} - \frac{\partial E_x}{\partial y} \end{pmatrix} = \begin{pmatrix} jk_y E_z - \frac{\partial E_y}{\partial z} \\ \frac{\partial E_x}{\partial z} - jk_x E_z \\ jk_x E_y - jk_y E_x \end{pmatrix}, \quad \nabla \times \vec{H} = \begin{pmatrix} \frac{\partial H_z}{\partial y} - \frac{\partial H_y}{\partial z} \\ \frac{\partial H_x}{\partial z} - \frac{\partial H_z}{\partial x} \\ \frac{\partial H_y}{\partial x} - \frac{\partial H_x}{\partial y} \end{pmatrix} = \begin{pmatrix} jk_y H_z - \frac{\partial H_y}{\partial z} \\ \frac{\partial H_x}{\partial z} - jk_x H_z \\ jk_x H_y - jk_y H_x \end{pmatrix}$$

(6.4)

Substituting Eqs. (6.3a–c) and (6.4) into (6.1a,b), four equations can be obtained through several steps:

$$\frac{\partial}{\partial z} \tilde{E}_x = \frac{1}{-j\omega \tilde{\varepsilon}_z} \left[k_y(k_x \tilde{H}_x + k_y \tilde{H}_y) - (k_x^2 + k_y^2)\tilde{H}_y \right] + j\omega \mu_0 \tilde{H}_y + j\omega \mu_0 \tilde{M}_y \quad (6.5a)$$

$$\frac{\partial}{\partial z} \tilde{E}_y = \frac{1}{-j\omega \tilde{\varepsilon}_z} \left[-k_x(k_x \tilde{H}_x + k_y \tilde{H}_y) + (k_x^2 + k_y^2)\tilde{H}_x \right] - j\omega \mu_0 \tilde{H}_x - j\omega \mu_0 \tilde{M}_x \quad (6.5b)$$

$$\frac{\partial}{\partial z}(-\tilde{H}_y) = \frac{1}{j\omega \mu_0} \left[k_x(k_x \tilde{E}_x + k_y \tilde{E}_y) - (k_x^2 + k_y^2)\tilde{E}_x \right] - j\omega \tilde{\varepsilon}_x \tilde{E}_x + jk_y \tilde{M}_z \quad (6.5c)$$

$$\frac{\partial}{\partial z} \tilde{H}_x = \frac{1}{j\omega \mu_0} \left[k_y(k_x \tilde{E}_x + k_y \tilde{E}_y) - (k_x^2 + k_y^2)\tilde{E}_y \right] - j\omega \tilde{\varepsilon}_y \tilde{E}_y - jk_x \tilde{M}_z \quad (6.5d)$$

Then, equations of transverse electromagnetic (EM) fields in matrix form are found to be:

$$\frac{d}{dz}\begin{bmatrix}\tilde{E}_x \\ \tilde{E}_y \\ -\tilde{H}_y \\ \tilde{H}_x\end{bmatrix} + j\omega \begin{bmatrix} 0 & 0 & \mu_0 - \frac{k_x^2}{\omega^2 \tilde{\varepsilon}_z} & -\frac{k_x k_y}{\omega^2 \tilde{\varepsilon}_z} \\ 0 & 0 & -\frac{k_x k_y}{\omega^2 \tilde{\varepsilon}_z} & \mu_0 - \frac{k_y^2}{\omega^2 \tilde{\varepsilon}_z} \\ \tilde{\varepsilon}_x - \frac{k_y^2}{\omega^2 \mu_0} & \frac{k_x k_y}{\omega^2 \mu_0} & 0 & 0 \\ \frac{k_x k_y}{\omega^2 \mu_0} & \tilde{\varepsilon}_y - \frac{k_x^2}{\omega^2 \mu_0} & 0 & 0 \end{bmatrix}\begin{bmatrix}\tilde{E}_x \\ \tilde{E}_y \\ -\tilde{H}_y \\ \tilde{H}_x\end{bmatrix} = \begin{bmatrix}j\omega\mu_0 \tilde{M}_y \\ -j\omega\mu_0 \tilde{M}_x \\ jk_y \tilde{M}_z \\ -jk_x \tilde{M}_z\end{bmatrix} \quad (6.6)$$

which can be simplified as:

$$\frac{d}{dz}\begin{bmatrix}\mathbf{f_E} \\ \mathbf{f_H}\end{bmatrix} + j\omega\begin{bmatrix}\mathbf{0} & \mathbf{C_1} \\ \mathbf{C_2} & \mathbf{0}\end{bmatrix}\begin{bmatrix}\mathbf{f_E} \\ \mathbf{f_H}\end{bmatrix} = \begin{bmatrix}\mathbf{s_H} \\ \mathbf{s_V}\end{bmatrix} \quad (6.7)$$

where $\mathbf{0}$ is a 2×2 zero matrix, and

$$\mathbf{f_E} = \begin{bmatrix}\tilde{E}_x \\ \tilde{E}_y\end{bmatrix}, \quad \mathbf{f_H} = \begin{bmatrix}-\tilde{H}_y \\ \tilde{H}_x\end{bmatrix}, \quad \mathbf{s_H} = \begin{bmatrix}j\omega\mu_0\tilde{M}_y \\ -j\omega\mu_0\tilde{M}_x\end{bmatrix}, \quad \mathbf{s_V} = \begin{bmatrix}j\omega q_y \tilde{M}_z \\ -j\omega q_x \tilde{M}_z\end{bmatrix}, \quad (6.8)$$

$$\mathbf{C_1} = \begin{bmatrix}\mu_0 - \frac{q_x^2}{\tilde{\varepsilon}_z} & -\frac{q_x q_y}{\tilde{\varepsilon}_z} \\ -\frac{q_x q_y}{\tilde{\varepsilon}_z} & \mu_0 - \frac{q_y^2}{\tilde{\varepsilon}_z}\end{bmatrix}, \quad \mathbf{C_2} = \begin{bmatrix}\tilde{\varepsilon}_x - \frac{q_y^2}{\mu_0} & \frac{q_x q_y}{\mu_0} \\ \frac{q_x q_y}{\mu_0} & \tilde{\varepsilon}_y - \frac{q_x^2}{\mu_0}\end{bmatrix} \quad (6.9)$$

in which $q_x = k_x/\omega$ and $q_y = k_y/\omega$. $\tilde{H}_z$ can be obtained by

$$\tilde{H}_z = \frac{1}{\omega\mu_0}(k_x \tilde{E}_y - k_y \tilde{E}_x) - \tilde{M}_z \quad (6.10)$$

By denoting

$$\mathbf{f} = \begin{bmatrix}\mathbf{f_E} \\ \mathbf{f_H}\end{bmatrix}, \quad \mathbf{s} = \begin{bmatrix}\mathbf{s_H} \\ \mathbf{s_V}\end{bmatrix}, \quad \text{and} \quad \mathbf{C} = \begin{bmatrix}\mathbf{0} & \mathbf{C_1} \\ \mathbf{C_2} & \mathbf{0}\end{bmatrix} \quad (6.11)$$

Eq. (6.7) can be written as:

$$\left(\frac{d}{dz}\mathbf{I} + j\omega\mathbf{C}\right)\mathbf{f} = \mathbf{s} \tag{6.12}$$

where $\mathbf{C}$ can be diagonalized as $\mathbf{C} = \mathbf{A}\Lambda\mathbf{A}^{-1}$, Λ is a diagonal matrix composed of four eigenvalues of $\mathbf{C}$, $\mathbf{A}$ is composed of corresponding four-column eigenvectors, and $\mathbf{A}^{-1}$ is the inverse matrix of $\mathbf{A}$ (Appendix A). The Eigen equation of $\mathbf{C}$ is:

$$q_z^4 + \left(-\mu_0(\tilde{\varepsilon}_x + \tilde{\varepsilon}_y) + (q_x^2 + q_y^2) + \frac{\tilde{\varepsilon}_x q_x^2 + \tilde{\varepsilon}_y q_y^2}{\tilde{\varepsilon}_z}\right) q_z^2$$
$$+ \mu_0^2 \tilde{\varepsilon}_x \tilde{\varepsilon}_y - \mu_0 \left(\tilde{\varepsilon}_x q_x^2 + \tilde{\varepsilon}_y q_y^2 + \frac{\tilde{\varepsilon}_x \tilde{\varepsilon}_y}{\tilde{\varepsilon}_z}(q_y^2 + q_x^2)\right) + (\tilde{\varepsilon}_x q_x^2 + \tilde{\varepsilon}_y q_y^2)\frac{q_y^2 + q_x^2}{\tilde{\varepsilon}_z} = 0 \tag{6.13}$$

The solutions can be obtained easily:

$$q_{z1} = +\sqrt{\frac{b + \sqrt{c}}{2\tilde{\varepsilon}_z}}, \quad q_{z2} = +\sqrt{\frac{b - \sqrt{c}}{2\tilde{\varepsilon}_z}}, \quad q_{z3} = -\sqrt{\frac{b + \sqrt{c}}{2\tilde{\varepsilon}_z}}, \quad q_{z4} = -\sqrt{\frac{b - \sqrt{c}}{2\tilde{\varepsilon}_z}} \tag{6.14}$$

in which

$$b = \mu_0 \tilde{\varepsilon}_z(\tilde{\varepsilon}_x + \tilde{\varepsilon}_y) - \tilde{\varepsilon}_z(q_x^2 + q_y^2) - (\tilde{\varepsilon}_x q_x^2 + \tilde{\varepsilon}_y q_y^2) \tag{6.15}$$

$$c = \left[-\mu_0 \tilde{\varepsilon}_z(\tilde{\varepsilon}_x - \tilde{\varepsilon}_y) + (\tilde{\varepsilon}_x - \tilde{\varepsilon}_z)q_x^2 - (\tilde{\varepsilon}_y - \tilde{\varepsilon}_z)q_y^2\right]^2 + 4(\tilde{\varepsilon}_x - \tilde{\varepsilon}_z)(\tilde{\varepsilon}_y - \tilde{\varepsilon}_z)q_x^2 q_y^2 \tag{6.16}$$

Denote $\mathbf{q}_z = \begin{bmatrix} q_{z1} & 0 \\ 0 & q_{z2} \end{bmatrix}$, then Λ can be written as:

$$\Lambda = \begin{bmatrix} \mathbf{q}_z & 0 \\ 0 & -\mathbf{q}_z \end{bmatrix} \tag{6.17}$$

Assuming $\mathbf{f} = \mathbf{A}\mathbf{w}$ [1,2], which means

$$\mathbf{w} = \mathbf{A}^{-1}\mathbf{f} = \begin{bmatrix} \mathbf{u} \\ \mathbf{d} \end{bmatrix} \tag{6.18}$$

where $\mathbf{u}$ represents the up-going wave and $\mathbf{d}$ is the down-going wave. Substituting Eq. (6.18) into (6.12) and using expression $\mathbf{C} = \mathbf{A}\Lambda\mathbf{A}^{-1}$, results a differential equation:

$$\frac{d\mathbf{w}}{dz} = -j\omega\Lambda\mathbf{w} + \mathbf{A}^{-1}\mathbf{s} - \mathbf{A}^{-1}\frac{d\mathbf{A}}{dz}\mathbf{w} \tag{6.19}$$

For the induction logging problems consisting of one group of transmitter coils considered in this chapter, when the coils are located at (x_s, y_s, z_s), $\tilde{M}_s(k_x, k_y, z)$ can be obtained by Eq. (6.3c):

$$\tilde{M}_s(k_x, k_y, z) = \int_{-\infty}^{+\infty} \int_{-\infty}^{+\infty} \vec{M}_s(x, y, z) e^{-j(k_x x + k_y y)} dx dy = \vec{M}_s(x_s, y_s, z) \delta(z - z_s) e^{-j(k_x x_s + k_y y_s)} \quad (6.20)$$

To obtain the full coupling matrix connecting source excitations to magnetic field response, we need to consider three-directional magnetic dipoles along the x axis (XMD), y axis (YMD), and z axis (ZMD) separately. For XMD located at (x_s, y_s, z_s), the magnetic-source flux density in the spectral domain and the source terms, respectively are:

$$\tilde{M}_s(k_x, k_y, z) = \hat{x} M_x(x_s, y_s, z) \delta(z - z_s) e^{-j(k_x x_s + k_y y_s)} \quad (6.21)$$

$$\mathbf{s_H} = \begin{bmatrix} 0 \\ -j\omega\mu_0 \tilde{M}_x \end{bmatrix}, \quad \mathbf{s_V} = \begin{bmatrix} 0 \\ 0 \end{bmatrix} \quad (6.22)$$

For YMD, expressions for $\tilde{M}_s$, $\mathbf{s_H}$, and $\mathbf{s_V}$ are:

$$\tilde{M}_s(k_x, k_y, z) = \hat{y} M_y(x_s, y_s, z) \delta(z - z_s) e^{-j(k_x x_s + k_y y_s)} \quad (6.23)$$

$$\mathbf{s_H} = \begin{bmatrix} j\omega\mu_0 \tilde{M}_y \\ 0 \end{bmatrix}, \quad \mathbf{s_V} = \begin{bmatrix} 0 \\ 0 \end{bmatrix} \quad (6.24)$$

Similarly, for ZMD, expressions for $\tilde{M}_s$, $\mathbf{s_H}$, and $\mathbf{s_V}$ are:

$$\tilde{M}_s(k_x, k_y, z) = \hat{z} M_z(x_s, y_s, z) \delta(z - z_s) e^{-j(k_x x_s + k_y y_s)} \quad (6.25)$$

$$\mathbf{s_H} = \begin{bmatrix} 0 \\ 0 \end{bmatrix}, \quad \mathbf{s_V} = \begin{bmatrix} j\omega q_y \tilde{M}_z \\ -j\omega q_x \tilde{M}_z \end{bmatrix} \quad (6.26)$$

Here $\mathbf{s_H}$ and $\mathbf{s_V}$ are known. $\mathbf{A}$ is determined by the properties of the medium. In homogenous medium, we know that $\frac{d\mathbf{A}}{dz} = 0$. Thus Eq. (6.19) becomes

$$\frac{d\mathbf{w}}{dz} = -j\omega \Lambda \mathbf{w} + \mathbf{A}^{-1} \mathbf{s} \delta(z - z_s) \quad (6.27)$$

The solution to Eq. (6.27) in different cases will be discussed in Section 6.2.

6.2 PROPAGATION IN UNBOUNDED MEDIUM

The simple case is a source in a homogenous unbounded medium. Take the integration of both sides of Eq. (6.27) from z_s^- to z_s^+:

$$\int_{z_s^-}^{z_s^+} \frac{d\mathbf{w}}{dz} dz = \int_{z_s^-}^{z_s^+} (-j\omega\Lambda\mathbf{w})dz + \int_{z_s^-}^{z_s^+} \mathbf{A}^{-1}\mathbf{s}\delta(z-z_s)dz \quad (6.28)$$

we obtain the following equation:

$$\mathbf{w}(z_s^+) - \mathbf{w}(z_s^-) = \mathbf{A}^{-1}\mathbf{s} \quad (6.29)$$

A discontinuity of $\mathbf{w}(z)$ is found through z_s. In source-free homogenous region, Eq. (6.27) becomes:

$$\frac{d\mathbf{w}}{dz} = -j\omega\Lambda\mathbf{w} \quad (6.30)$$

for which the solution is:

$$\mathbf{w}(z) = e^{-j\omega\Lambda(z-z_0)}\mathbf{w}(z_0) \quad (6.31)$$

Let $\mathbf{P}(z,z_0) = e^{-j\omega\Lambda(z-z_0)}$ denote the propagation matrix from z_0 to z, its expression is:

$$\mathbf{P}(z,z_0) = \begin{pmatrix} e^{-j\omega\mathbf{q_z}(z-z_0)} & 0 \\ 0 & e^{j\omega\mathbf{q_z}(z-z_0)} \end{pmatrix} \quad (6.32)$$

And Eq. (6.31) becomes:

$$\mathbf{w}(z) = \mathbf{P}(z,z_0)\mathbf{w}(z_0) \quad (6.33)$$

For convenience, $\widehat{\mathbf{P}}(z,z_0)$ represents the up-going propagation matrix when $z_0 < z$ and $\widetilde{\mathbf{P}}(z,z_0)$ is the down-going propagation matrix when $z_0 > z$.

6.3 PROPAGATION IN LAYERED MEDIUM

Formulations in an unbounded medium are given in Section 6.2. However the source is usually located in multilayered medium. Without loss of generalities, in an N-layer medium, the transmitters are located in region m, as shown in Fig. 6.1.

Considering a single boundary at horizontal interface between region j and region $j+1$, or $z = z_j$, we have the EM fields' continuity conditions:

$$\mathbf{f}(z_j^-) = \mathbf{f}(z_j^+) \quad (6.34)$$

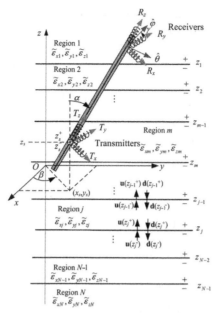

Figure 6.1 A triaxial induction tool is embedded in the layered medium.

which means,

$$\mathbf{A}_{j+1}\mathbf{w}\left(z_j^-\right) = \mathbf{A}_j\mathbf{w}\left(z_j^+\right) \tag{6.35}$$

It can also be written as:

$$\mathbf{A}_{j+1}\left[\breve{\mathbf{P}}\left(z_j^-, z_j^+\right)\mathbf{w}\left(z_j^+\right)\right] = \mathbf{A}_j\mathbf{w}\left(z_j^+\right) \tag{6.36}$$

or

$$\mathbf{A}_{j+1}\mathbf{w}\left(z_j^-\right) = \mathbf{A}_j\left[\widehat{\mathbf{P}}\left(z_j^+, z_j^-\right)\mathbf{w}\left(z_j^-\right)\right] \tag{6.37}$$

where $\breve{\mathbf{P}}(z_j^-, z_j^+)$ is the down-going propagation matrix from z_j^+ to z_j^- and $\widehat{\mathbf{P}}(z_j^+, z_j^-)$ is the up-going propagation matrix from z_j^- to z_j^+. Solving for $\widehat{\mathbf{P}}(z_j^+, z_j^-)$ and $\breve{\mathbf{P}}(z_j^-, z_j^+)$, yields:

$$\widehat{\mathbf{P}}\left(z_j^+, z_j^-\right) = \mathbf{A}_j^{-1}\mathbf{A}_{j+1}, \quad \breve{\mathbf{P}}\left(z_j^-, z_j^+\right) = \mathbf{A}_{j+1}^{-1}\mathbf{A}_j \tag{6.38}$$

Obviously, they have the relationship:

$$\widehat{\mathbf{P}}(z_j^+, z_j^-) = \breve{\mathbf{P}}(z_j^-, z_j^+)^{-1} \tag{6.39}$$

Using the propagation matrix in homogenous medium and that at the interface of layered medium, we have the relations as follows:

$$\mathbf{w}(z_{j-1}^-) = \widehat{\mathbf{P}}(z_{j-1}^-, z_j^+) \cdot \widehat{\mathbf{P}}(z_j^+, z_j^-) \cdot \mathbf{w}(z_j^-), \tag{6.40}$$

$$\mathbf{w}(z_j^+) = \widecheck{\mathbf{P}}(z_j^+, z_{j-1}^+) \cdot \widecheck{\mathbf{P}}(z_{j-1}^-, z_{j-1}^+) \cdot \mathbf{w}(z_{j-1}^+), \tag{6.41}$$

Let $\widehat{\mathbf{R}}_n$ denote the generalized reflection matrix for up-going wave at z_n^- ($n = 1, 2, \ldots N-1$). The down-going wave is a consequence of the reflection of the up-going wave, so $\mathbf{d}(z_n^-) = \widehat{\mathbf{R}}_n \mathbf{u}(z_n^-)$. Similarly, let $\widecheck{\mathbf{R}}_n$ denote the generalized reflection matrix for down-going wave at z_n^+ ($n = 1, 2, \ldots N-1$). The up-going wave is a consequence of the reflection of the down-going wave, i.e., $\mathbf{u}(z_n^+) = \widecheck{\mathbf{R}}_n \mathbf{d}(z_n^+)$.

Therefore, Eq. (6.40) can be expanded as:

$$\begin{pmatrix} \mathbf{u}(z_{j-1}^-) \\ \widehat{\mathbf{R}}_{j-1}\mathbf{u}(z_{j-1}^-) \end{pmatrix} = \begin{pmatrix} e^{-j\omega \mathbf{q}_{z,j} h_j} & 0 \\ 0 & e^{j\omega \mathbf{q}_{z,j} h_j} \end{pmatrix} \cdot \begin{pmatrix} \widehat{P}_{11} & \widehat{P}_{12} \\ \widehat{P}_{21} & \widehat{P}_{22} \end{pmatrix}_j \cdot \begin{pmatrix} \mathbf{u}(z_j^-) \\ \widehat{\mathbf{R}}_j \mathbf{u}(z_j^-) \end{pmatrix} \tag{6.42}$$

in which h_j is the thickness of layer j, $h_j = z_{j-1} - z_j$ and $\widehat{\mathbf{P}}(z_j^+, z_j^-) = \begin{pmatrix} \widehat{P}_{11} \widehat{P}_{12} \\ \widehat{P}_{21} \widehat{P}_{22} \end{pmatrix}_j$ can be solved by Eq. (6.38). Then, two recursive equations can be obtained:

$$\mathbf{u}(z_{j-1}^-) = e^{-j\omega \mathbf{q}_{z,j} h_j}[\widehat{P}_{11,j} + \widehat{P}_{12,j}\widehat{\mathbf{R}}_j]\mathbf{u}(z_j^-) \tag{6.43}$$

$$\left[e^{j\omega \mathbf{q}_{z,j} h_j}\widehat{P}_{22,j} - \widehat{\mathbf{R}}_{j-1}e^{-j\omega \mathbf{q}_{z,j} h_j}\widehat{P}_{12,j}\right]^{-1}\left[\widehat{\mathbf{R}}_{j-1}e^{-j\omega \mathbf{q}_{z,j} h_j}\widehat{P}_{11,j} - e^{j\omega \mathbf{q}_{z,j} h_j}\widehat{P}_{21,j}\right] = \widehat{\mathbf{R}}_j \tag{6.44}$$

As shown in Fig. 6.1, region 1 extends to infinity, so $\widehat{\mathbf{R}}_0 = 0$. Up-going reflection matrix at each interface z_n, $\widehat{\mathbf{R}}_n$ ($n = 1, 2, \ldots N-1$) can be obtained by recursive relation (Eq. 6.44).

Likewise, Eq. (6.41) can be expanded as:

$$\begin{pmatrix} \widecheck{\mathbf{R}}_j \mathbf{d}(z_j^+) \\ \mathbf{d}(z_j^+) \end{pmatrix} = \begin{pmatrix} e^{j\omega \mathbf{q}_{z,j} h_j} & 0 \\ 0 & e^{-j\omega \mathbf{q}_{z,j} h_j} \end{pmatrix} \cdot \begin{pmatrix} \widecheck{P}_{11} & \widecheck{P}_{12} \\ \widecheck{P}_{21} & \widecheck{P}_{22} \end{pmatrix}_{j-1} \cdot \begin{pmatrix} \widecheck{\mathbf{R}}_{j-1}\mathbf{d}(z_{j-1}^+) \\ \mathbf{d}(z_{j-1}^+) \end{pmatrix} \tag{6.45}$$

where $\widecheck{\mathbf{P}}(z_{j-1}^-, z_{j-1}^+) = \begin{pmatrix} \widecheck{P}_{11} & \widecheck{P}_{12} \\ \widecheck{P}_{21} & \widecheck{P}_{22} \end{pmatrix}_{j-1}$ can be obtained by Eq. (6.38) by replacing j with $j-1$. Two recursive equations can be derived:

$$\mathbf{d}(z_j^+) = e^{-j\omega \mathbf{q}_{z,j} h_j}(\widecheck{P}_{21,j-1}\widecheck{\mathbf{R}}_{j-1} + \widecheck{P}_{22,j-1})\mathbf{d}(z_{j-1}^+) \tag{6.46}$$

$$\left[e^{j\omega \mathbf{q}_{z,j} h_j}\widecheck{P}_{11} - \widecheck{\mathbf{R}}_j e^{-j\omega \mathbf{q}_{z,j} h_j}\widecheck{P}_{21}\right]^{-1}\left[\widecheck{\mathbf{R}}_j e^{-j\omega \mathbf{q}_{z,j} h_j}\widecheck{P}_{22} - e^{j\omega \mathbf{q}_{z,j} h_j}\widecheck{P}_{12}\right] = \widecheck{\mathbf{R}}_{j-1} \tag{6.47}$$

As shown in Fig. 6.1, region N extends to infinity, so $\breve{\mathbf{R}}_N = 0$. Then, according to Eq. (6.47), down-going reflection matrix at each interface z_n, $\breve{\mathbf{R}}_n$ ($n = 1, 2, \ldots N-1$) can be obtained.

The solutions in region m have the relations:

$$\mathbf{w}(z_s^+) = \breve{\mathbf{P}}(z_s^+, z_{m-1}^-)\mathbf{w}(z_{m-1}^-), \quad \mathbf{w}(z_{m-1}^-) = \begin{pmatrix} \mathbf{u}(z_{m-1}^-) \\ \widehat{R}_{m-1}\mathbf{u}(z_{m-1}^-) \end{pmatrix} \quad (6.48)$$

and

$$\mathbf{w}(z_s^-) = \widehat{\mathbf{P}}(z_s^-, z_m^+)\mathbf{w}(z_m^+), \quad \mathbf{w}(z_m^-) = \begin{pmatrix} \breve{R}_m \mathbf{d}(z_m^+) \\ \mathbf{d}(z_m^+) \end{pmatrix} \quad (6.49)$$

After substituting Eqs. (6.48) and (6.49) back into Eq. (6.29), we can find

$$\breve{\mathbf{P}}(z_s^+, z_{m-1}^-)\begin{pmatrix} \mathbf{u}(z_{m-1}^-) \\ \widehat{R}_{m-1}\mathbf{u}(z_{m-1}^-) \end{pmatrix} - \widehat{\mathbf{P}}(z_s^-, z_m^+)\begin{pmatrix} \breve{R}_m \mathbf{d}(z_m^+) \\ \mathbf{d}(z_m^+) \end{pmatrix} = \mathbf{A}_m^{-1}\begin{pmatrix} s_H \\ s_V \end{pmatrix} \quad (6.50)$$

where

$$\breve{\mathbf{P}}(z_s^+, z_{m-1}^-) = \begin{pmatrix} e^{-j\omega \mathbf{q}_{z,m}(z_s^+ - z_{m-1}^-)} & 0 \\ 0 & e^{j\omega \mathbf{q}_{z,m}(z_s^+ - z_{m-1}^-)} \end{pmatrix}, \quad \widehat{\mathbf{P}}(z_s^-, z_m^+) = \begin{pmatrix} e^{-j\omega \mathbf{q}_{z,m}(z_s^- - z_m^+)} & 0 \\ 0 & e^{j\omega \mathbf{q}_{z,m}(z_s^- - z_m^+)} \end{pmatrix} \quad (6.51)$$

Eq. (6.50) is a system of linear equations about $\mathbf{u}(z_{m-1}^-)$ and $\mathbf{d}(z_m^+)$. $\breve{R}_m$ and $\widehat{R}_{m-1}$ can be derived according to Eqs. (6.44) and (6.47), respectively. Then, $\mathbf{u}(z_{m-1}^-)$ and $\mathbf{d}(z_m^+)$ can be solved.

Once the expressions of $\mathbf{u}(z_{m-1}^-)$ and $\mathbf{d}(z_m^+)$ are known, $\mathbf{w}(z)$ in each region n can be derived through propagation matrix and reflection matrix. By using expression $\mathbf{f}(z) = A_n \mathbf{w}(z)$, $\tilde{E}_x(k_x, k_y, z)$, $\tilde{E}_y(k_x, k_y, z)$, $\tilde{H}_x(k_x, k_y, z)$, and $\tilde{H}_y(k_x, k_y, z)$ are known, $\tilde{H}_z(k_x, k_y, z)$ can be obtained by Eq. (6.10). Components of magnetic fields in the space domain can be obtained by Eq. (6.3a–c). As can be seen from Eq. (6.3a–c), to compute the field quantities, we have to calculate integrals over k_x and k_y using the numerical evaluation.

6.4 COMPUTATION OF THE DOUBLE INTEGRALS

A cylindrical transformation in the wavenumber space is invoked. Let φ be the rotation angel in the $kx - ky$ plane, and we have:

$$k_x = k_\rho \cos \varphi, \quad k_y = k_\rho \sin \varphi \quad (6.52)$$

$$\int_{-\infty}^{+\infty} \int_{-\infty}^{+\infty} e^{j(k_x x + k_y y)} dk_x dk_y = \int_0^{2\pi} \int_0^\infty e^{jk_\rho(\cos \varphi x + \sin \varphi y)} dk_\rho d\varphi \quad (6.53)$$

The integral over φ is a definite integral, and it can be calculated using Gauss–Legendre quadrature method. For the semi-infinite integral over k_ρ, we use a modified Gauss–Laguerre quadrature method [3]. Generalized Gauss–Laguerre quadrature rule is an extension of the Gaussian quadrature method for approximating the value of integrals of the following kind:

$$\int_0^\infty x^\alpha e^{-x} f(x) dx \approx \sum_{j=1}^n w_j f(x_j) \quad (6.54)$$

where $\alpha > -1$, abscissas x_j is the jth root of Laguerre polynomial $L_n(x)$ and weights w_j is given by

$$w_j = \frac{x_j}{(n+1)^2 [L_{n+1}(x_j)]^2} \quad (6.55)$$

A modified Gauss–Laguerre quadrature method can be used to evaluate such integral:

$$\int_0^\infty f(x) dx \approx \sum_{j=1}^n w'_j f(x_j) \quad (6.56)$$

for smooth $f(x)$, and

$$w'_j = \frac{w_j}{x^\alpha e^{-x_j}} \quad (6.57)$$

In practice, the orientations of the transmitter and receiver coils are arbitrary with respect to the principal axes of the formation's conductivity tensor. The magnetic field response of a triaxial induction tool with arbitrarily oriented tool axis can be obtained as Section 4.4.3.

6.5 NUMERICAL EXAMPLES

In this section, we will present some numerical examples calculated using the theory described in the Section 6.4.

Example 6.1

The proposed algorithm is applied to the homogeneous biaxial anisotropic formation. In this case, we consider a homogenous biaxial formation with $R_x = 0.25$ ohm-m, $R_y = 1$ ohm-m, and $R_z = 2$ ohm-m, assuming a 2C-40 triaxial induction tool operating at 20 kHz. The relative dipping angle is still 60 degrees. Table 6.1 shows comparison with the results by Yuan et al. and Davydycheva et al. The medium does not change in the y-direction, the components H_{xy}, H_{yx}, H_{yz}, and H_{zy} are equal to

Table 6.1 Comparison of the present 1D modeling method with that of Yuan et al. [4] and Davydycheva et al. [5]: magnetic field of 2 coil-40-in. (1 m) sonde at 20 kHz in homogeneous 60-degree dipping biaxially anisotropic medium with $R_x = 0.25$, $R_y = 1$, and $R_z = 2$ ohm-m

Method	H_{xx} (A/m)	H_{yy} (A/m)	H_{zz} (A/m)	H_{xz} (A/m)	H_{zx} (A/m)
Yuan et al.	−0.080065 + 0.010509i	−0.079415 + 0.005938i	0.149391 + 0.011546i	0.0018858 − 0.0061081i	0.0018858 − 0.0061081i
Davydycheva et al.	−0.080065 + 0.010523i	−0.079416 + 0.005902i	0.149391 + 0.011554i	0.0018858 − 0.0061190i	0.0018858 − 0.0061190i
Present method	−0.0800647 + 0.0105088i	−0.0794169 + 0.00593138i	0.1493904 + 0.0115457i	0.00188551 − 0.00610858i	0.00188551 − 0.00610858i

zero; they are not given in the table. The agreement between the results obtained by using the two methods is satisfactory. Fifty points are used for Gauss–Laguerre quadrature method, and the maximum difference is less than 10^{-5}. The proposed method works well in homogenous biaxial medium.

Example 6.2

Then, a fully layered biaxial anisotropic case is studied. We use the triaxial induction tool presented in Ref. [6] to do the simulations. The operation frequency for this tool is 26.8 kHz. For convenience sake without losing generality, only two spacings are tested. The long spacing is 54 in. with bucking coil at 39 in. and the short spacing is 21 in. with bucking coil at 15.8 in.

We model a three-layer anisotropic formation. The upper and lower layers are both isotropic homogeneous medium. The parameter of these three layers are $R_1 = R_3 = 10$ ohm-m for the upper and lower layers, $R_x = 10$, $R_y = 1$, and $R_z = 4$ ohm-m for the middle layer, with fracturing across the horizontal x axis, $\varepsilon_r = \mu_r = 1$ for all the layers. During the process of simulation, we neglect the borehole geometry and only consider the depth variation.

Fig. 6.2A–H shows the comparisons of our method with the code developed by Well Logging Lab at the University of Houston [6]. Dashed and triangle-marked lines correspond to the real part of the conductivity for 54-in. array, with a bucking receiver spacing of 39 in., and solid and circle-marked lines show 21-in. measurements, with a bucking receiver spacing of 15.8 in. Fig. 6.2A–C shows the nonzero components when the tool dip angle is 0 degree and the middle layer is 2 ft thick, whereas Fig. 6.2D–H illustrates a similar case when the tool dip angle is 60 degrees and the middle layer is 16 ft thick (in xz-plane). As we can see, all the results show good agreements. We show all nonzero couplings xx, yy, zz, and $(xz \pm zx)/2$ (note that $xy = yx = yz = zy = 0$ due to the symmetry with respect to the plane $y = 0$). Fifty points are used for Gauss–Laguerre quadrature method in all of the calculations, and the maximum relative error is 2.3% at 60-degree dip and 0.2% at 0-degree dip. As the dipping angle increases, a higher order of Gauss–Laguerre quadrature method is required to achieve sufficient accuracy, as

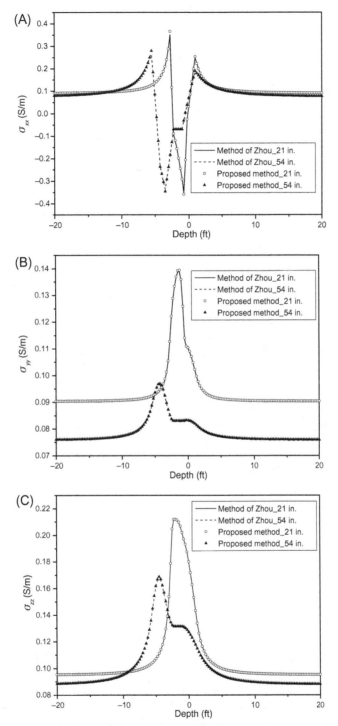

Figure 6.2 Apparent conductivity of three-layer model. (A) *xx* component at 0-degree dip (2 ft), (B) *yy* component at 0-degree dip (2 ft), (C) *zz* component at 0-degree dip (2 ft), (D) *xx* component at 60-degree dip (16 ft), (E) *yy* component at 60-degree dip (16 ft), (F) *zz* component at 60-degree dip (16 ft), (G) (*xz* ± *zx*)/2 component at 60-degree dip (16 ft, 21 in.), and (H) (*xz* ± *zx*)/2 component at 60-degree dip (16 ft, 54 in.).

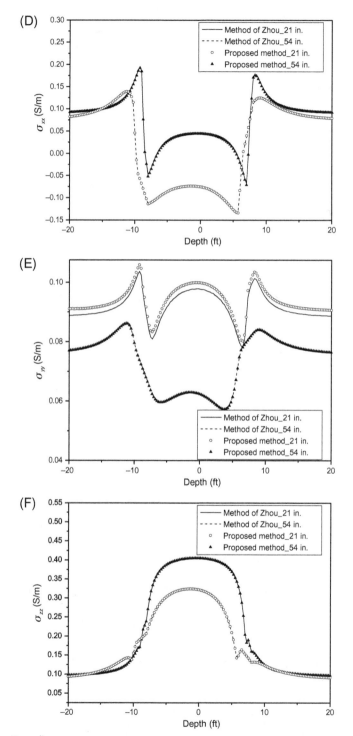

Figure 6.2 (Continued)

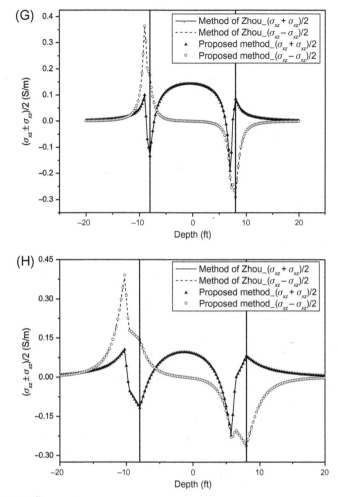

Figure 6.2 (Continued)

presented by Yuan et al. [4]. The computation time of the new method is about 0.03 s per logging point on a 2.5-GHz laptop computer.

As in TI case, half-difference of the cross components $(xz - zx)/2$ in the biaxial case has clear spikes at the bed boundaries, being close to zero in a distance of about 1–1.5 spacings from the bed boundaries. This behavior was observed by Davydycheva et al. for the case of TI-anisotropic formation, and this example demonstrates that this feature is not significantly affected by the fracturing. However the half-sum $(xz + zx)/2$, known to be responsible for the formation anisotropy detection, is significantly affected. If it was negative in the middle of the TI-anisotropic bed, it appears positive in the fractured bed. This feature can serve as a possible indicator of the biaxial anisotropic formation.

REFERENCES

[1] W.C. Chew, Waves and Fields in Inhomogenous Media, Wiley-IEEE Press, New York, NY, 1999.
[2] L.O. Løseth, B. Ursin, Electromagnetic fields in planarly layered anisotropic media, Geophys. J. Int. 170 (2007) 44–80.
[3] J. Burkardt, Quadrature rules of Gauss-Laguerre type. <http://people.sc.fsu.edu/~jburkardt/datasets/quadrature_rules_laguerre/quadrature_rules_laguerre.html>.
[4] N. Yuan, X.C. Nie, R. Liu, C.W. Qiu, Simulation of full responses of a triaxial induction tool in a homogeneous biaxial anisotropic formation, Geophysics 75 (2) (2010) E101–E114, 10.1190/1.3336959.
[5] S. Davydycheva, T. Wang, A fast modeling method to solve Maxwell's equations in 1D layered biaxial anisotropic medium, Geophysics 76 (5) (2011) F293–F302, 10.1190/GEO2010-0280.1.
[6] M. Zhou, Simulation of triaxial induction tool response in biaxial anisotropic formation (Master thesis), Well Logging Lab., Department of Electrical and Computer Engineering, University of Houston, Houston, TX, 2016.

APPENDIX A DERIVATION OF MATRIX A

Denote

$$\mathbf{A} = \begin{bmatrix} a_{11} & a_{12} & a_{13} & a_{14} \\ a_{21} & a_{22} & a_{23} & a_{24} \\ a_{31} & a_{32} & a_{33} & a_{34} \\ a_{41} & a_{42} & a_{43} & a_{44} \end{bmatrix} \tag{A.1}$$

According to Eq. (6.14), $q_{z3} = -q_{z1}$ and $q_{z4} = -q_{z2}$. Relations can be found:

$$\text{If } a_{11} = a_{13}, a_{12} = a_{14}, \quad \text{then } a_{21} = a_{23}, a_{22} = a_{24} \tag{A.2}$$

$$\text{If } a_{33} = -a_{31}, a_{34} = -a_{32}, \quad \text{then } a_{43} = -a_{41}, a_{44} = -a_{42} \tag{A.3}$$

Then we denote $\mathbf{A} = \frac{1}{\sqrt{2}}\begin{bmatrix} A_E & A_E \\ A_H & -A_H \end{bmatrix}$. Using expression $\mathbf{C} = \mathbf{A}\Lambda\mathbf{A}^{-1}$, we can obtain:

$$\begin{bmatrix} 0 & C_1 \\ C_2 & 0 \end{bmatrix} \frac{1}{\sqrt{2}} \begin{bmatrix} A_E & A_E \\ A_H & -A_H \end{bmatrix} = \frac{1}{\sqrt{2}} \begin{bmatrix} A_E & A_E \\ A_H & -A_H \end{bmatrix} \begin{bmatrix} \mathbf{q}_z & 0 \\ 0 & -\mathbf{q}_z \end{bmatrix} \tag{A.4}$$

or,

$$\begin{bmatrix} C_1 A_H & -C_1 A_H \\ C_2 A_E & C_2 A_E \end{bmatrix} = \begin{bmatrix} A_E \mathbf{q}_z & -A_E \mathbf{q}_z \\ A_H \mathbf{q}_z & A_H \mathbf{q}_z \end{bmatrix} \tag{A.5}$$

That means,

$$C_1 A_H = A_E \mathbf{q}_z, \quad \text{or} \quad C_1 = A_E \mathbf{q}_z A_H^{-1}$$
$$C_2 A_E = A_H \mathbf{q}_z, \qquad\qquad C_2 = A_H \mathbf{q}_z A_E^{-1} \qquad (A.6)$$

Then relations can be established:

$$C_1 C_2 = A_E \mathbf{q}_z A_H^{-1} A_H \mathbf{q}_z A_E^{-1} = A_E \mathbf{q}_z^2 A_E^{-1}$$
$$C_2 C_1 = A_H \mathbf{q}_z A_E^{-1} A_E \mathbf{q}_z A_H^{-1} = A_H \mathbf{q}_z^2 A_H^{-1} \qquad (A.7)$$

And

$$C_1 C_2 A_E = A_E \mathbf{q}_z^2$$
$$C_2 C_1 A_H = A_H \mathbf{q}_z^2 \qquad (A.8)$$

According to Eq. (6.9),

$$C_1 C_2 = \begin{bmatrix} \left(\mu_0 - \dfrac{q_x^2}{\tilde{\varepsilon}_z}\right)\left(\tilde{\varepsilon}_x - \dfrac{q_y^2}{\mu_0}\right) - \dfrac{q_x q_y}{\tilde{\varepsilon}_z}\dfrac{q_x q_y}{\mu_0} & \left(\mu_0 - \dfrac{q_x^2}{\tilde{\varepsilon}_z}\right)\dfrac{q_x q_y}{\mu_0} - \dfrac{q_x q_y}{\tilde{\varepsilon}_z}\left(\tilde{\varepsilon}_y - \dfrac{q_x^2}{\mu_0}\right) \\ -\dfrac{q_x q_y}{\tilde{\varepsilon}_z}\left(\tilde{\varepsilon}_x - \dfrac{q_y^2}{\mu_0}\right) + \dfrac{q_x q_y}{\mu_0}\left(\mu_0 - \dfrac{q_y^2}{\tilde{\varepsilon}_z}\right) & -\dfrac{q_x q_y}{\tilde{\varepsilon}_z}\dfrac{q_x q_y}{\mu_0} + \left(\mu_0 - \dfrac{q_y^2}{\tilde{\varepsilon}_z}\right)\left(\tilde{\varepsilon}_y - \dfrac{q_x^2}{\mu_0}\right) \end{bmatrix},$$
$$(A.9)$$

$$C_2 C_1 = \begin{bmatrix} \left(\tilde{\varepsilon}_x - \dfrac{q_y^2}{\mu_0}\right)\left(\mu_0 - \dfrac{q_x^2}{\tilde{\varepsilon}_z}\right) - \dfrac{q_x q_y}{\tilde{\varepsilon}_z}\dfrac{q_x q_y}{\mu_0} & -\dfrac{q_x q_y}{\tilde{\varepsilon}_z}\left(\tilde{\varepsilon}_x - \dfrac{q_y^2}{\mu_0}\right) + \dfrac{q_x q_y}{\mu_0}\left(\mu_0 - \dfrac{q_y^2}{\tilde{\varepsilon}_z}\right) \\ \dfrac{q_x q_y}{\mu_0}\left(\mu_0 - \dfrac{q_x^2}{\tilde{\varepsilon}_z}\right) - \dfrac{q_x q_y}{\tilde{\varepsilon}_z}\left(\tilde{\varepsilon}_y - \dfrac{q_x^2}{\mu_0}\right) & -\dfrac{q_x q_y}{\tilde{\varepsilon}_z}\dfrac{q_x q_y}{\mu_0} + \left(\tilde{\varepsilon}_y - \dfrac{q_x^2}{\mu_0}\right)\left(\mu_0 - \dfrac{q_y^2}{\tilde{\varepsilon}_z}\right) \end{bmatrix},$$
$$(A.10)$$

Let $C_1 C_2 = \begin{bmatrix} \gamma_{11} & \gamma_{12} \\ \gamma_{21} & \gamma_{22} \end{bmatrix}$ and $C_2 C_1 = \begin{bmatrix} \gamma_{11} & \gamma_{21} \\ \gamma_{12} & \gamma_{22} \end{bmatrix} = (C_1 C_2)^T$, then Eq. (A.8) becomes:

$$C_1 C_2 A_E = A_E \mathbf{q}_z^2$$
$$(C_1 C_2)^T A_H = A_H \mathbf{q}_z^2 \qquad (A.11)$$

Easily, we know:

$$A_E^{-1} = (A_H)^T \qquad (A.12)$$

Thus $\mathbf{A}$ and $\mathbf{A}^{-1}$ can be written as:

$$\mathbf{A} = \frac{1}{\sqrt{2}}\begin{bmatrix} A_E & A_E \\ A_H & -A_H \end{bmatrix} = \frac{1}{\sqrt{2}}\begin{bmatrix} A_E & A_E \\ (A_E^{-1})^T & -(A_E^{-1})^T \end{bmatrix} \quad (A.13)$$

$$\mathbf{A}^{-1} = \frac{1}{\sqrt{2}}\begin{bmatrix} A_E^{-1} & A_H^{-1} \\ A_E^{-1} & -A_H^{-1} \end{bmatrix} = \frac{1}{\sqrt{2}}\begin{bmatrix} (A_H)^T & (A_E)^T \\ (A_H)^T & -(A_E)^T \end{bmatrix} = \frac{1}{\sqrt{2}}\begin{bmatrix} A_E^{-1} & (A_E)^T \\ A_E^{-1} & -(A_E)^T \end{bmatrix} \quad (A.14)$$

According to $C_1 C_2 A_E = A_E \mathbf{q}_z^2$, a linear system of equations about the elements of A_E can be obtained:

$$\begin{bmatrix} \gamma_{11} & \gamma_{12} \\ \gamma_{21} & \gamma_{22} \end{bmatrix}\begin{bmatrix} a_{11} & a_{12} \\ a_{21} & a_{22} \end{bmatrix} = \begin{bmatrix} a_{11} & a_{12} \\ a_{21} & a_{22} \end{bmatrix}\begin{bmatrix} q_{z1}^2 & 0 \\ 0 & q_{z2}^2 \end{bmatrix} \quad (A.15)$$

It can be expanded as:

$$\begin{bmatrix} -q_{z1}^2+\gamma_{11} & 0 & \gamma_{12} & 0 \\ 0 & -q_{z2}^2+\gamma_{11} & 0 & \gamma_{12} \\ \gamma_{21} & 0 & -q_{z1}^2+\gamma_{22} & 0 \\ 0 & \gamma_{21} & 0 & -q_{z2}^2+\gamma_{22} \end{bmatrix}\begin{bmatrix} a_{11} \\ a_{12} \\ a_{21} \\ a_{22} \end{bmatrix} = \begin{bmatrix} 0 \\ 0 \\ 0 \\ 0 \end{bmatrix} \quad (A.16)$$

Thus four relations can be obtained:

$$\frac{a_{21}}{a_{11}} = \frac{-q_{z1}^2+\gamma_{11}}{-\gamma_{12}} = \frac{-\mu_0\tilde{\varepsilon}_x\tilde{\varepsilon}_z + \tilde{\varepsilon}_x q_x^2 + \tilde{\varepsilon}_z(q_y^2+q_{z1}^2)}{q_x q_y(\tilde{\varepsilon}_z - \tilde{\varepsilon}_y)} \quad (A.17)$$

$$\frac{a_{21}}{a_{11}} = \frac{-\gamma_{21}}{-q_{z1}^2+\gamma_{22}} = \frac{q_x q_y(\tilde{\varepsilon}_z - \tilde{\varepsilon}_x)}{-\mu_0\tilde{\varepsilon}_z\tilde{\varepsilon}_y + \tilde{\varepsilon}_y q_y^2 + \tilde{\varepsilon}_z(q_x^2+q_{z1}^2)} \quad (A.18)$$

$$\frac{a_{22}}{a_{12}} = \frac{-q_{z2}^2+\gamma_{11}}{-\gamma_{12}} = \frac{-\mu_0\tilde{\varepsilon}_x\tilde{\varepsilon}_z + \tilde{\varepsilon}_x q_x^2 + \tilde{\varepsilon}_z(q_y^2+q_{z2}^2)}{q_x q_y(\tilde{\varepsilon}_z - \tilde{\varepsilon}_y)} \quad (A.19)$$

$$\frac{a_{22}}{a_{12}} = \frac{-\gamma_{21}}{-q_{z2}^2+\gamma_{22}} = \frac{q_x q_y(\tilde{\varepsilon}_z - \tilde{\varepsilon}_x)}{-\mu_0\tilde{\varepsilon}_z\tilde{\varepsilon}_y + \tilde{\varepsilon}_y q_y^2 + \tilde{\varepsilon}_z(q_x^2+q_{z2}^2)} \quad (A.20)$$

Combined with the relation $(A_E)^{-1} = (A_H)^T$, A_E and A_H can be determined. Then the expressions of $\mathbf{A}$ and $\mathbf{A}^{-1}$ can be obtained by Eqs. (A.13) and (A.14).

CHAPTER 7

Induction and LWD Tool Response in a Cylindrically Layered Isotropic Formation

Contents

7.1 Introduction	205
7.2 Induction and LWD Tool Response in a Four-Layer Cylindrical Medium	206
7.2.1 Geometrical configuration of the four-layer model	207
7.2.2 Solution method of the induction and LWD tool response in a four-layer cylindrical formation	208
7.2.3 Borehole and mandrel effects to LWD and induction tool responses	214
7.2.4 Influence of the mandrel conductivity to the LWD tool performance	217
7.3 Response of Induction and LWD Tools in Arbitrary Cylindrically Layered Media	217
7.3.1 Geometrical configuration	224
7.3.2 Methodology	225
7.3.3 Discussions of convergence, accuracy, and numerical computation	230
7.3.4 Simulation results	233
7.4 Conclusions	241
References	242
Appendix A Derivation for the Magnetic Fields in Spectral Domain	243
Appendix B Derivation for the Expression of Electrical Field for the Homogeneous Formation in Spectral Domain	245
Appendix C Derivation for the Expression of Electrical Field for Arbitrary Cylindrical Layered Formations in Spectral Domain	247

7.1 INTRODUCTION

The logging problems in a vertical borehole often involve both horizontal layers (formation layers) and cylindrical layers (due to borehole mud and invasion zones). Therefore we can separate the complicated logging problems into two categories: cylindrically layered formation around the logging tools and the horizontal layers without borehole and invasion zones. Both cases can be considered as one-dimensional problems and can be solved using analytical methods. In Chapter 5, Triaxial Induction Tool and Logging-While-Drilling Tool Response in a Transverse

Isotropic-Layered Formation and Chapter 6, Triaxial Induction and Logging-While-Drilling Logging Tool Response in a Biaxial Anisotropic-Layered Formation, we thoroughly discussed the analytical solutions to the response of induction and logging-while-drilling (LWD) tools in a horizontally layered formation. In this chapter, we will consider the analytical solutions in a cylindrically layered formation.

An accurate analytical solution to the response of induction and LWD tools in cylindrically layered media is essential in well logging [1,2]. The cylindrically layered formation can be used in many different situations to evaluate depth of investigation of a tool, or it can be used directly in an inversion process as forward modeling. It can also be used to compute the borehole correction and conversion chart of a logging tool since the analytic solution can be very accurate. Other qualitative analysis of logging problems, such as magnetic permeability effect, invasion effect, and borehole effect, can also be obtained using this method. Due to the fast computation speed, it can be used as the forward modeling algorithm in inversion of the radial profile of the formation [3,4].

In this chapter, we will discuss an analytic algorithm, which can calculate the response of induction or LWD tools in cylindrically layered media. The challenge is to establish the analytical model and discuss the convergence conditions. Using spectral expression of the Helmholtz equations, it can be derived that the solution to the tool response in a cylindrically layered formation is an inverse Fourier transform of Bessel functions [5–7]. The scaled Bessel function routines are used to avoid overflow/underflow for large/small arguments. The inverse Fourier transform is done using numerical integration. To overcome the oscillating nature of the integrand in the inverse Fourier transform and to reach the best possible accuracy, a separation of the integrand into a directly coupled wave and reflected wave is necessary and effective. The directly coupled part can be evaluated by direct integration in the space domain while the reflection part requires choosing an appropriate cutoff value and number of sample points for the integrand to achieve a convergent solution.

In Section 7.2, the solution to the response of induction and LWD tools with a mandrel, borehole, and homogeneous formation is solved. There are four cylindrical layers considered: mandrel as the first and second layer, mud as the third layer, and homogeneous formation as the fourth layer. In Section 7.3, the arbitrary number of cylindrically layered media is discussed to expand the algorithm in Section 7.2 to include multiinvasion cases [8].

7.2 INDUCTION AND LWD TOOL RESPONSE IN A FOUR-LAYER CYLINDRICAL MEDIUM

The discussion in Section 7.1 solves the problem of induction and LWD tool response in a homogeneous formation with a metal mandrel. This geometry can be used to compute the effect of mandrel. In many cases of LWD tool applications, the antennas are installed in grooves on the mandrel. The effects of grooves must be considered. If the effect of grooves is to be considered, the problem becomes a two-dimensional problem, which will be discussed in Chapter 8, Induction and Logging-While-Drilling Resistivity Tool Response in a Two-Dimensional Isotropic Formation. Taking into account the mandrel, borehole, invasion zone, and the formation, we will consider a four-layer medium. We will develop an analytic solution to calculate the electromagnetic fields and apparent conductivity of induction and LWD sondes in a four-layer cylindrical medium. Exact expressions in the form of modified Bessel functions for the field will be derived. Each region is characterized by its conductivity and magnetic permeability. Note that the model does not include eccentricity. When the tool becomes eccentric in the borehole, the problem becomes a three-dimensional problem and analytic solutions are rather complicated.

7.2.1 Geometrical configuration of the four-layer model

This algorithm solves a cylindrically layered formation problem. There are four layers in the formation, with each layer representing a different medium. A cylindrical coordinate system (r, φ, z) is used in the analysis for convenience. Fig. 7.1 shows the structure for the LWD case. A transmitter and two receivers are mounted coaxially on a conducting mandrel. Mud exists both inside and outside the mandrel. The outmost region denotes the homogeneous earth formation extending to infinity. A time-harmonic current is applied to the transmitter antenna. Electromagnetic fields are

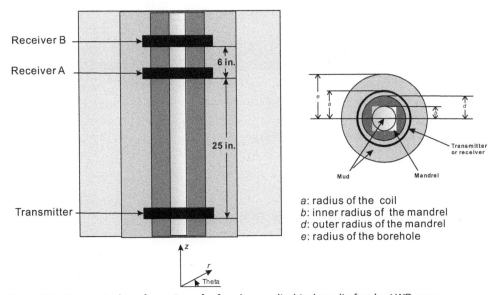

Figure 7.1 Geometrical configuration of a four-layer cylindrical media for the LWD case.

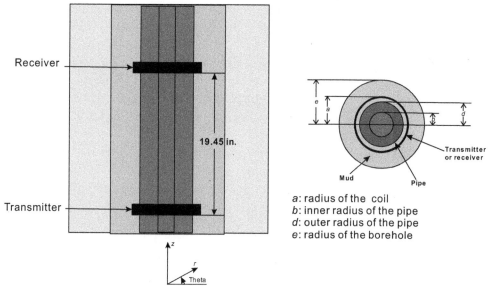

Figure 7.2 Geometrical configuration of a four-layer cylindrical media for induction logging.

generated by the current loop and propagate through the earth formation. The amplitude ratio and phase difference of the voltages are measured by the Receiver A and Receiver B and used to obtain the formation information.

For the induction case shown in Fig. 7.2, a transmitter and a receiver are mounted coaxially on the pipe. The pipe is solid, so the two innermost formation regions are given the same electrical properties. The mud surrounds the pipe, and the homogeneous medium in *purple* (light gray in print versions) is the outmost formation.

7.2.2 Solution method of the induction and LWD tool response in a four-layer cylindrical formation

Rewriting the Maxwell's equation in cylindrical coordinates, and noting that the radius of the antenna coil is a, we begin with the differential equation for the axial component of the electric field when Fourier transform is applied to z coordinate [1,9]:

$$\left(\frac{\partial^2}{\partial r^2} + \frac{1}{r}\frac{\partial}{\partial r} - \frac{1}{r^2} - j\omega\mu\sigma + \omega^2\mu\varepsilon - k_z^2\right)\tilde{E}_\varphi(r, k_z) = j\omega\mu\delta(r - a) \qquad (7.1)$$

where k_z is the wave number in the z direction, and $\tilde{E}_\varphi(r, k_z)$ is the Fourier transform of the electric field applied to z in the φ direction. First, we consider the source-free form of Eq. (7.1)

$$\left(\frac{\partial^2}{\partial r^2} + \frac{1}{r}\frac{\partial}{\partial r} - \frac{1}{r^2} - j\omega\mu\sigma + \omega^2\mu\varepsilon - k_z^2\right)\tilde{E}_\varphi(r, k_z) = 0 \qquad (7.2)$$

Define the propagation constant in the ith region as

$$k_i^2 = -j\omega\mu\sigma_i + \omega^2\mu_i\varepsilon_i \tag{7.3}$$

where $(\sigma_i, \mu_i, \varepsilon_i)$ are the conductivity, magnetic permeability, and permittivity of the ith region. Hence Eq. (7.2) becomes

$$\left(\frac{\partial^2}{\partial r^2} + \frac{1}{r}\frac{\partial}{\partial r} - \frac{1}{r^2} + k_i^2 - k_z^2\right)\tilde{E}_\varphi(r, k_z) = 0 \tag{7.4}$$

substitution of $\gamma_i^2 = k_z^2 - k_i^2$ gives

$$\left(\frac{\partial^2}{\partial r^2} + \frac{1}{r}\frac{\partial}{\partial r} - \frac{1}{r^2} - \gamma_i^2\right)\tilde{E}_\varphi(r, k_z) = 0 \tag{7.5}$$

The general solution for Eq. (7.5) is

$$\tilde{E}_\varphi(r, k_z) = B(\gamma_i r) = [f_i^- I_1(\gamma_i r) + f_i^+ K_1(\gamma_i r)] \tag{7.6}$$

Then the general solution in the space domain is

$$E_\varphi(r, z) = \int_{-\infty}^{+\infty} [f_i^- I_1(\gamma_i r) + f_i^+ K_1(\gamma_i r)]e^{jk_z z} dk_z \tag{7.7}$$

where f_i^-, f_i^+ are unknown coefficients for region i, and $I_1(\gamma_i r), K_1(\gamma_i r)$ are modified Bessel functions of the first kind. For the innermost region, there is only an incoming wave, therefore in region I we have

$$\tilde{E}_{\varphi 1}(r, k_z) = (-j\omega\mu_1 a)f_1(k_z)I_1(\gamma_1 r) \text{ in region I} \tag{7.8}$$

For the outermost region (region IV) there is only an outgoing wave, hence

$$\tilde{E}_{\varphi 5}(r, k_z) = (-j\omega\mu_4 a)f_4(k_z)K_1(\gamma_4 r) \text{ in region IV} \tag{7.9}$$

For the region in which the coils are located, we have to consider two separate solutions. The solution is

$$\tilde{E}_{\varphi 3}(r, k_z) = (-j\omega\mu_3 a)[f_3^-(k_z)I_1(\gamma_3 r) + f_3^+(k_z)K_1(\gamma_3 r) + K_1(\gamma_3 a)I_1(\gamma_3 r)] \tag{7.10}$$

inside the current loop and

$$\tilde{E}_{\varphi 4}(r, k_z) = (-j\omega\mu_3 a)[f_3^-(k_z)I_1(\gamma_3 r) + f_3^+(k_z)K_1(\gamma_3 r) + I_1(\gamma_3 a)K_1(\gamma_3 r)] \tag{7.11}$$

outside the current loop.

The field induced by the source is given by

$$\tilde{E}_\varphi(r, k_z) = -j\omega\mu a \begin{cases} I_1(\gamma_3 r)K_1(\gamma_3 a), & r < a, \\ I_1(\gamma_3 a)K_1(\gamma_3 r), & r > a. \end{cases} \tag{7.12}$$

Then the general solution for each region is as follows:

$$\tilde{E}_{\varphi 1}(r, k_z) = (-j\omega\mu_1 a)f_1(k_z)I_1(\gamma_1 r),$$

$$\tilde{E}_{\varphi 2}(r, k_z) = (-j\omega\mu_2 a)[f_2^-(k_z)I_1(\gamma_2 r) + f_2^+(k_z)K_1(\gamma_2 r)],$$

$$\tilde{E}_{\varphi 3}(r, k_z) = (-j\omega\mu_3 a)[f_3^-(k_z)I_1(\gamma_3 r) + f_3^+(k_z)K_1(\gamma_3 r) + K_1(\gamma_3 a)I_1(\gamma_3 r)], \quad (7.13)$$

$$\tilde{E}_{\varphi 4}(r, k_z) = (-j\omega\mu_3 a)[f_3^-(k_z)I_1(\gamma_3 r) + f_3^+(k_z)K_1(\gamma_3 r) + I_1(\gamma_3 a)K_1(\gamma_3 r)],$$

and $\tilde{E}_{\varphi 5}(r, k_z) = (-j\omega\mu_4 a)f_4(k_z)K_1(\gamma_4 r)$

According to Maxwell's equations, the magnetic field is related to the electrical field by

$$\underline{H} = \frac{1}{-j\omega\mu_i}\nabla \times \underline{E} \quad (7.14)$$

$$= (a\gamma_i)\left[f_i^-(k_z)I_0(\gamma_i r) - f_i^+(k_z)K_0(\gamma_i r)\right]$$

Please refer to Appendix A for detailed derivations.

Therefore the general solution to the H field in each region is:

$$\tilde{H}_{z1}(r, k_z) = a\gamma_1 f_1(k_z)I_0(\gamma_1 r),$$

$$\tilde{H}_{z2}(r, k_z) = a\gamma_2[f_2^-(k_z)I_0(\gamma_2 r) - f_2^+(k_z)K_0(\gamma_2 r)],$$

$$\tilde{H}_{z3}(r, k_z) = a\gamma_3[f_3^-(k_z)I_0(\gamma_3 r) - f_3^+(k_z)K_0(\gamma_3 r) + K_1(\gamma_3 a)I_0(\gamma_3 r)], \quad (7.15)$$

$$\tilde{H}_{z4}(r, k_z) = a\gamma_3[f_3^-(k_z)I_0(\gamma_3 r) - f_3^+(k_z)K_0(\gamma_3 r) - I_1(\gamma_3 a)K_0(\gamma_3 r)],$$

and $\tilde{H}_{z5}(r, k_z) = a\gamma_4[-f_4(k_z)K_0(\gamma_4 r)]$.

The tangential electric and magnetic fields are continuous on each boundary, i.e.,

$$\tilde{E}_\varphi(r, k_z)\Big|_{r=b^-}^{b^+} = 0,$$

$$\tilde{H}_z(r, k_z)\Big|_{r=b^-}^{b^+} = 0,$$

$$\tilde{E}_\varphi(r, k_z)\Big|_{r=d^-}^{d^+} = 0,$$

$$\tilde{H}_z(r, k_z)\Big|_{r=d^-}^{d^+} = 0, \quad (7.16)$$

$$\tilde{E}_\varphi(r, k_z)\Big|_{r=e^-}^{e^+} = 0,$$

and $\tilde{H}_z(r, k_z)\Big|_{r=e^-}^{e^+} = 0.$

Apply the boundary conditions (7.16) to the general solutions in Eq. (7.15) yields

$$(-j\omega\mu_1 a)f_1(k_z)I_1(\gamma_1 b) = (-j\omega\mu_2 a)[f_2^-(k_z)I_1(\gamma_2 b) + f_2^+(k_z)K_1(\gamma_2 b)]$$

$$\gamma_1 f_1(k_z)I_0(\gamma_1 b) = \gamma_2[f_2^-(k_z)I_0(\gamma_2 b) - f_2^+(k_z)K_0(\gamma_2 b)],$$

$$(-j\omega\mu_2 a)[f_2^-(k_z)I_1(\gamma_2 d) + f_2^+(k_z)K_1(\gamma_2 d)]$$
$$= (-j\omega\mu_3 a)[f_3^-(k_z)I_1(\gamma_3 d) + f_3^+(k_z)K_1(\gamma_3 d) + K_1(\gamma_3 a)I_1(\gamma_3 d)],$$

$$\gamma_2[f_2^-(k_z)I_0(\gamma_2 d) - f_2^+(k_z)K_0(\gamma_2 d)]$$
$$= \gamma_3[f_3^-(k_z)I_0(\gamma_3 d) - f_3^+(k_z)K_0(\gamma_3 d) + K_1(\gamma_3 a)I_0(\gamma_3 d)],$$

$$(-j\omega\mu_3 a)[f_3^-(k_z)I_1(\gamma_3 e) + f_3^+(k_z)K_1(\gamma_3 e) + I_1(\gamma_3 a)K_1(\gamma_3 e)]$$
$$= (-j\omega\mu_4 a)f_4(k_z)K_1(\gamma_4 e),$$

and $\gamma_3[f_3^-(k_z)I_0(\gamma_3 e) - f_3^+(k_z)K_0(\gamma_3 e) - I_1(\gamma_3 a)K_0(\gamma_3 e)] = \gamma_4[-f_4(k_z)K_0(\gamma_4 e)].$

(7.17)

The above equations can be written in a matrix given by

$$\begin{bmatrix} \mu_1 I_1(\gamma_1 b) & -\mu_2 I_1(\gamma_2 b) & -\mu_2 K_1(\gamma_2 b) & 0 & 0 & 0 \\ \gamma_1 I_0(\gamma_1 b) & -\gamma_2 I_0(\gamma_2 b) & \gamma_2 K_0(\gamma_2 b) & 0 & 0 & 0 \\ 0 & \mu_2 I_1(\gamma_2 d) & \mu_2 K_1(\gamma_2 d) & -\mu_3 I_1(\gamma_3 d) & -\mu_3 K_1(\gamma_3 d) & 0 \\ 0 & \gamma_2 I_0(\gamma_2 d) & -\gamma_2 K_0(\gamma_2 d) & -\gamma_3 I_0(\gamma_3 d) & \gamma_3 K_0(\gamma_3 d) & 0 \\ 0 & 0 & 0 & \mu_3 I_1(\gamma_3 e) & \mu_3 K_1(\gamma_3 e) & -\mu_4 K_1(\gamma_4 e) \\ 0 & 0 & 0 & \gamma_3 I_0(\gamma_3 e) & -\gamma_3 K_0(\gamma_3 e) & \gamma_4 K_0(\gamma_4 e) \end{bmatrix}$$

$\uparrow \qquad \uparrow \qquad \uparrow \qquad \uparrow \qquad \uparrow \qquad \uparrow$

$e^{-\text{abs}(\text{real}(\gamma_1 b))} \quad e^{-\text{abs}(\text{real}(\gamma_2 d))} \quad e^{\gamma_2 b} \quad e^{-\text{abs}(\text{real}(\gamma_3 e))} \quad e^{\gamma_3 d} \quad e^{-\gamma_4 e}$

$$\cdot \begin{bmatrix} f_1(k_z) \\ f_2^-(k_z) \\ f_2^+(k_z) \\ f_3^-(k_z) \\ f_3^+(k_z) \\ f_4(k_z) \end{bmatrix} = \begin{bmatrix} 0 \\ 0 \\ \mu_3 I_1(\gamma_3 d)K_1(\gamma_3 a) \\ \gamma_3 I_0(\gamma_3 d)K_1(\gamma_3 a) \\ -\mu_3 I_1(\gamma_3 a)K_1(\gamma_3 e) \\ \gamma_3 I_1(\gamma_3 a)K_0(\gamma_3 e) \end{bmatrix}$$

$\qquad\qquad\qquad\qquad\qquad\qquad \uparrow \qquad\qquad \uparrow$

$e^{-\text{abs}(\text{real}(\gamma_3 d))} \quad \text{or} \quad e^{-\text{abs}(\text{real}(\gamma_3 a))} \quad e^{\gamma_3 a} \text{ or } e^{\gamma_3 e}$

(7.18)

To avoid computational difficulties, we scale the modified Bessel functions. The last row in Eq. (7.18) shows the scaling factor for each column. Expressed in terms of the scaled Bessel functions $\bar{I}$ and $\bar{K}$ Eq. (7.18) becomes

$$\begin{bmatrix} \mu_1\bar{I}_1(\gamma_1 b) & -\mu_2\bar{I}_1(\gamma_2 b) & -\mu_2\bar{K}_1(\gamma_2 b) & 0 & 0 & 0 \\ \gamma_1\bar{I}_0(\gamma_1 b) & -\gamma_2\bar{I}_0(\gamma_2 b) & \gamma_2\bar{K}_0(\gamma_2 b) & 0 & 0 & 0 \\ 0 & \mu_2\bar{I}_1(\gamma_2 d) & \mu_2\bar{K}_1(\gamma_2 d) & -\mu_3\bar{I}_1(\gamma_3 d) & -\mu_3\bar{K}_1(\gamma_3 d) & 0 \\ 0 & \gamma_2\bar{I}_0(\gamma_2 d) & -\gamma_2\bar{K}_0(\gamma_2 d) & -\gamma_3\bar{I}_0(\gamma_3 d) & \gamma_3\bar{K}_0(\gamma_3 d) & 0 \\ 0 & 0 & 0 & \mu_3\bar{I}_1(\gamma_3 e) & \mu_3\bar{K}_1(\gamma_3 e) & -\mu_4\bar{K}_1(\gamma_4 e) \\ 0 & 0 & 0 & \gamma_3\bar{I}_0(\gamma_3 e) & -\gamma_3\bar{K}_0(\gamma_3 e) & \gamma_4\bar{K}_0(\gamma_4 e) \end{bmatrix} \begin{bmatrix} \bar{f}_1(k_z) \\ \bar{f}_2^-(k_z) \\ \bar{f}_2^+(k_z) \\ \bar{f}_3^-(k_z) \\ \bar{f}_3^+(k_z) \\ \bar{f}_4(k_z) \end{bmatrix}$$

$$\uparrow \quad \uparrow \quad \uparrow \quad \uparrow \quad \uparrow \quad \uparrow$$
$$e^{-\alpha_1 b} \quad e^{-\alpha_2 d} \quad e^{\gamma_2 b} \quad e^{-\alpha_3 e} \quad e^{\gamma_3 d} \quad e^{-\gamma_4 e}$$

$$= \begin{bmatrix} 0 \\ 0 \\ \mu_3\bar{I}_1(\gamma_3 d)\bar{K}_1(\gamma_3 a)e^{\alpha_3 d - \gamma_3 a} \\ \gamma_3\bar{I}_0(\gamma_3 d)\bar{K}_1(\gamma_3 a)e^{\alpha_3 d - \gamma_3 a} \\ -\mu_3\bar{I}_1(\gamma_3 a)\bar{K}_1(\gamma_3 e)e^{\alpha_3 a - \gamma_3 e} \\ \gamma_3\bar{I}_1(\gamma_3 a)\bar{K}_0(\gamma_3 e)e^{\alpha_3 a - \gamma_3 e} \end{bmatrix}$$

$$\uparrow \quad \uparrow$$
$$e^{-\alpha_3 d} \text{ or } e^{-\alpha_3 a} \quad e^{\gamma_3 a} \text{ or } e^{\gamma_3 e}$$

(7.19)

The relations between the scaled and unscaled unknowns are:

$$f_1(k_z) = \bar{f}_1(k_z)e^{-\text{abs}(\text{real}(\gamma_1 b))} = \bar{f}_1(k_z)e^{-\alpha_3 b},$$
$$f_2^-(k_z) = \bar{f}_2^-(k_z)e^{-\text{abs}(\text{real}(\gamma_2 d))} = \bar{f}_2^-(k_z)e^{-\alpha_2 d},$$
$$f_2^+(k_z) = \bar{f}_2^+(k_z)e^{\gamma_2 b},$$
$$f_3^-(k_z) = \bar{f}_3^-(k_z)e^{-\text{abs}(\text{real}(\gamma_3 e))} = \bar{f}_3^-(k_z)e^{-\alpha_3 e},$$
$$f_3^+(k_z) = \bar{f}_3^+(k_z)e^{\gamma_3 d},$$
$$\text{and } f_4^+(k_z) = \bar{f}_4(k_z)e^{-\gamma_4 e}.$$

(7.20)

The scaling factors with abs(real(γ_i)) in the index of the exponent are replaced with α_i, where α_i is the real part of γ_i, $\gamma_i = \alpha_i + i\beta_i$. The scaled Bessel functions can be computed by Amos's subroutine.

Substituting Eq. (7.20) into the general expressions for the electromagnetic fields (7.13), we obtain

$$\tilde{E}_{\varphi 1}(r,k_z) = (-j\omega\mu_1 a)\bar{f}_1(k_z)\bar{I}_1(\gamma_1 r)e^{[\text{abs}(\text{real}(\gamma_1 r))-\text{abs}(\text{real}(\gamma_1 b))]},$$
$$\tilde{E}_{\varphi 2}(r,k_z) = (-j\omega\mu_2 a)[\bar{f}_2^-(k_z)\bar{I}_1(\gamma_2 r)e^{[\text{abs}(\text{real}(\gamma_2 r))-\text{abs}(\text{real}(\gamma_2 d))]} + \bar{f}_2^+(k_z)\bar{K}_1(\gamma_2 r)e^{\gamma_2(b-r)}],$$
$$\tilde{E}_{\varphi 3}(r,k_z) = (-j\omega\mu_3 a)[\bar{f}_3^-(k_z)\bar{I}_1(\gamma_3 r)e^{\text{abs}(\text{real}(\gamma_3 r))-\text{abs}(\text{real}(\gamma_3 e))}$$
$$+ \bar{f}_3^+(k_z)\bar{K}_1(\gamma_3 r)e^{\gamma_3(d-r)} + \bar{K}_1(\gamma_3 a)\bar{I}_1(\gamma_3 r)e^{[\text{abs}(\text{real}(\gamma_3 r))-\gamma_3 a]}],$$
$$\tilde{E}_{\varphi 4}(r,k_z) = (-j\omega\mu_3 a)[\bar{f}_3^-(k_z)\bar{I}_1(\gamma_3 r)e^{\gamma_3 r-\text{abs}(\text{real}(\gamma_3 e))}$$
$$+ \bar{f}_3^+(k_z)\bar{K}_1(\gamma_3 r)e^{\gamma_3(d-r)} + \bar{I}_1(\gamma_3 a)\bar{K}_1(\gamma_3 r)e^{[\text{abs}(\text{real}(\gamma_3 a))-\gamma_3 r]}],$$
$$\tilde{E}_{\varphi 5}(r,k_z) = (-j\omega\mu_4 a)\bar{f}_4(k_z)\bar{K}_1(\gamma_4 r)e^{-\gamma_4 r-\gamma_4 e},$$

(7.21)

$$\tilde{H}_{z1}(r,k_z) = a\gamma_1 \bar{f}_1(k_z)\bar{I}_0(\gamma_1 r)e^{[\text{abs}(\text{real}(\gamma_1 r))-\text{abs}(\text{real}(\gamma_1 b))]},$$
$$\tilde{H}_{z2}(r,k_z) = a\gamma_2[\bar{f}_2^-(k_z)\bar{I}_0(\gamma_2 r)e^{[\text{abs}(\text{real}(\gamma_2 r))-\text{abs}(\text{real}(\gamma_2 d))]} - \bar{f}_2^+(k_z)\bar{K}_0(\gamma_2 r)e^{\gamma_2(b-r)}],$$
$$\tilde{H}_{z3}(r,k_z) = a\gamma_3[\bar{f}_3^-(k_z)\bar{I}_0(\gamma_3 r)e^{\text{abs}(\text{real}(\gamma_3 r))-\text{abs}(\text{real}(\gamma_3 e))}$$
$$- \bar{f}_3^+(k_z)\bar{K}_0(\gamma_3 r)e^{\gamma_3(d-r)} + \bar{K}_1(\gamma_3 a)\bar{I}_0(\gamma_3 r)e^{\text{abs}(\text{real}(\gamma_3 r))-\gamma_3 a}],$$
$$\tilde{H}_{z4}(r,k_z) = a\gamma_3[\bar{f}_3^-(k_z)\bar{I}_0(\gamma_3 r)e^{\text{abs}(\text{real}(\gamma_3 r))-\text{abs}(\text{real}(\gamma_3 e))}$$
$$- \bar{f}_3(k_z)\bar{K}_0(\gamma_3 r)e^{\gamma_3(d-r)} - \bar{I}_1(\gamma_3 a)\bar{K}_0(\gamma_3 r)e^{\text{abs}(\text{real}(\gamma_3 a))-\gamma_3 r}],$$
$$\text{and } \tilde{H}_{z5}(r,k_z) = a\gamma_4[-\bar{f}_4(k_z)\bar{K}_0(\gamma_4 r)e^{-\gamma_4 r-\gamma_4 e}].$$

(7.22)

Please refer to Appendix B for a detailed derivation.

Now the problem is reduced to solving the linear equations for the unknowns. By solving the linear equations, we can obtain the expressions for $\tilde{E}_{\varphi i}(r,k_z)$ and $\tilde{H}_{zi}(r,k_z)$. Then we use the following expressions:

$$E_\varphi(r,z) = \int_{-\infty}^{\infty} \tilde{E}_\varphi(r,k_z)e^{jk_z z}dk_z \qquad (7.23)$$

and

$$H_z(r,z) = \int_{-\infty}^{\infty} \tilde{H}_z(r,k_z)e^{jk_z z}dk_z \qquad (7.24)$$

to do the inverse Fourier transform and solve for $E_\varphi(r,z)$ and $H_z(r,z)$ in each region.

To save computation time we employ

$$E_\varphi(r,z) = 2\int_0^\infty \tilde{E}_\varphi(r,k_z)\cos(k_z z)dk_z \qquad (7.25)$$

and

$$H_z(r,z) = 2\int_0^\infty \tilde{H}_z(r,k_z)\cos(k_z z)dk_z \qquad (7.26)$$

instead. Eqs. (7.25) and (7.26) can be solved by using inverse Fourier transform.

7.2.3 Borehole and mandrel effects to LWD and induction tool responses

7.2.3.1 The LWD resistivity tool response with borehole mud and mandrel

In Chapter 4, Triaxial Induction and Logging-While-Drilling Resistivity Tool Response in Homogeneous Anisotropic Formations, we discussed the LWD resistivity tool response when the tool is in a homogenous formation. To simplify the problem, we ignored the effect of the mandrel in the computation. We also noticed that the conversion charts which convert the measured attenuation and phase difference by the LWD tool to the formation resistivity is based on the homogeneous assumption. However, as we can see, the borehole mud and mandrel will have impact to the LWD response. With the model we established in this chapter, we are able to take into consideration of the mandrel and borehole effects. In this section, we calculate the amplitude ratio and phase difference of two receivers versus the conductivity of the formation when borehole and mandrel are present. Apparently, as the conductivity of the borehole mud increases, the impact to the tool response will increase. However the collar of the LWD tool is a fixed variable and the impact to the tool response should be constant.

To study the tool response as a function of the LWD mandrel and the borehole mud [10], we consider a basic LWD tool. The frequency used in these examples is 2 MHz for the LWD resistivity tool. The tool diameter is 6.75 in., the antenna diameter is 6.5 in., and the borehole diameter is 8.5 in., respectively. We first consider the mandrel impact to the tool response. Usually, the mandrel is made of nonmagnetic stainless steel. The conductivity of the material is about 10^6 S/m. Fig. 7.6A and B shows the comparison between the tool responses in terms of phase and attenuation resistivity with and without mandrel when the homogeneous formation resistivity varies from 0.1 to 1000 ohm-m. Here the phase resistivity and attenuation resistivity mean the resistivities derived from the signal phase and signal attenuation, respectively.

Fig. 7.3A and B is the apparent phase and attenuation resistivity as functions of formation resistivity at different borehole resistivities when mandrel exists. These figures are used to study the tool response when both mandrel and borehole exist. The conductivity of the mud changes from 0.1 to 2 ohm-m while the formation resistivity changes from 0.1 to 1000 ohm-m. The apparent resistivity measured by the tool in a homogeneous formation without mandrel is also plotted in the figures. It is seen that the influence of the borehole and mandrel becomes significant when the formation resistivity increases. To overcome this problem, borehole correction is necessary to interpret the measured resistivity. In logging and drilling, the borehole diameter is usually known and the resistivity of the mud at downhole temperature can be obtained by its resistivity measured on the surface and an empirical formula. If the tool responses are computed and plotted against different borehole sizes, formation resistivity, and borehole mud, the formation resistivity can be obtained from the plots. These plots are usually called borehole correction charts. Fig. 7.4 shows a few examples of the borehole correction charts for a commercial LWD resistivity tool.

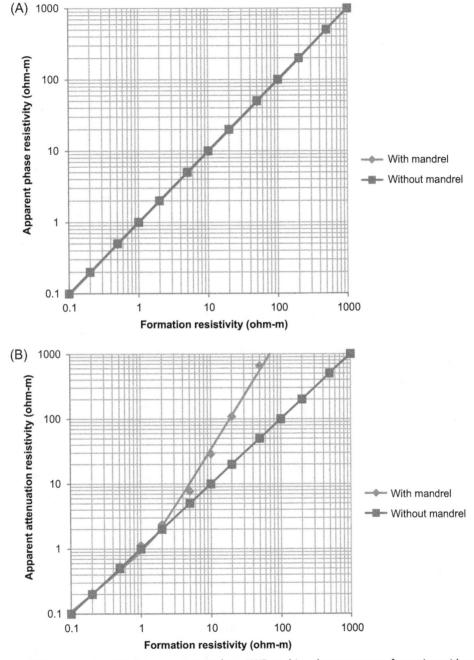

Figure 7.3 Apparent resistivity measurement by a LWD tool in a homogeneous formation with and without mandrel. The frequency is 2 MHz, mandrel diameter is 6.5 in., the antenna diameter is 6.75 in. (A) Apparent phase resistivity and (B) apparent attenuation resistivity.

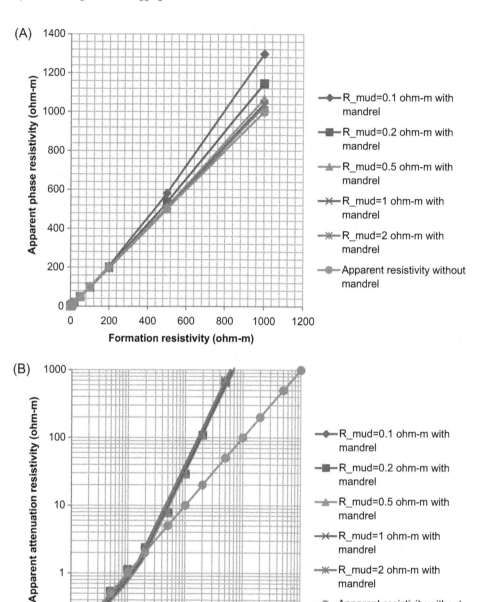

Figure 7.4 Apparent resistivity measurement by a LWD tool in a homogeneous formation with mandrel and borehole mud. The frequency is 2 MHz, tool diameter is 6.5 in., the antenna diameter is 6.75 in., and the borehole diameter is 8.5 in. (A) Apparent phase resistivity and (B) apparent attenuation resistivity.

7.2.3.2 The borehole effects to an induction logging tool

From Section 7.2.3.1, we can see that the borehole mud and borehole size have noticeable effects to the LWD measurements. Both amplitude ratio based and phase difference based apparent resistivity change with the borehole mud resistivity. This phenomenon can also be seen in the induction logging tool. Fig. 7.5 shows the real part of the apparent conductivity as a function of the formation resistivity. The frequency is also 20 kHz. The radius for the coil, inner mandrel, outer mandrel, and the borehole are 2.25, 0.887, 2, and 3 in., respectively. Consider an array induction tool with different arrays. Array 1, 2, 3 has a TR distance of 10, 18, and 30 in., respectively. The corresponding bucking coils are located at 7.143, 12.185, and 20.503 in., respectively. Fig. 7.6 shows the apparent resistivity measured by these three arrays when the borehole diameter is 10 in. while mud resistivity changes from 0.1 to 1000 ohm-m. It is clearly seen that when the TR distance is short, the borehole mud resistivity has greater impact to the measured resistivity. Lower mud resistivity will greatly offset the readings of the formation resistivity, as shown in Fig. 7.6A. As the distance between the transmitter and receiver increases, the borehole effect is less significant as shown in Fig. 7.6B and C.

Fig. 7.7 shows the apparent resistivity of the same array induction tool when the mud resistivity is fixed at 1 ohm-m while changing the diameter of the borehole. Similar conclusion can be drawn from these figures that the short array has greater influence from the borehole while the longer arrays have less impact from the borehole size.

7.2.4 Influence of the mandrel conductivity to the LWD tool performance

In tool design, the LWD tool has a metal mandrel as the tool collar. LWD resistivity antennas are imbedded inside the slotted mandrel for mechanical considerations (see chapter: Triaxial Induction and Logging-While-Drilling Resistivity Tool Response in Homogeneous Anisotropic Formations). The mandrel is usually made of stainless steel. The conductivity of the stainless steel may have impact to the tool performance due to the vicinity to the antennas. To investigate the impact of the collar conductivity to the tool response in different mud, the tool performance in terms of amplitude ratio and phase differences are computed as a function of formation conductivity at various collar conductivity values. Figs. 7.8—7.11 show the computed results. From these figures, we can clearly see that for mandrel conductivity changing from 10^4 to 10^6 S/m, the curves overlap to each other in both water-based mud and oil-based mud, which implies that the mandrel conductivity has little influence to the tool performance for most stainless steel collar.

7.3 RESPONSE OF INDUCTION AND LWD TOOLS IN ARBITRARY CYLINDRICALLY LAYERED MEDIA

The four-layer model in Section 7.2 can solve most practical cases where invasions are not considered. However, mud cakes and invasions are inevitable in some cases, especially in the situation of highly permeable formations. To solve the invasion

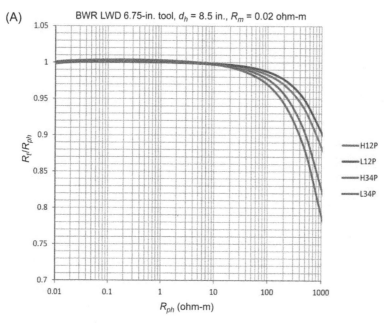

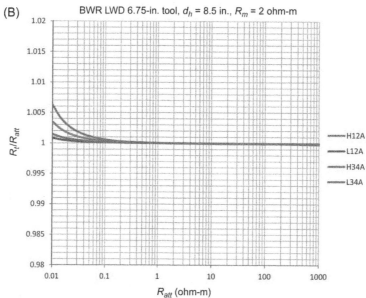

Figure 7.5 Borehole correction charts for MPR 6.7-in. LWD resistivity tool. The tool dimensions and measurement information are given in Fig. 4.3 and Table 4.2. H12A is the attenuation resistivity at 2 MHz for long-spacing antenna pairs, L12A is the attenuation resistivity at 400 kHz for long-spacing antenna pairs, H34A is the attenuation resistivity at 2 MHz for short-spacing antenna pairs, and L34A is the attenuation resistivity at 400 kHz for long-spacing antenna pairs. H12P is the phase difference resistivity at 2 MHz for long-spacing antenna pairs, L12P is the phase difference resistivity at 400 kHz for long-spacing antenna pairs, H34P is the phase difference resistivity at 2 MHz for short-spacing antenna pairs, and L34P is the phase difference resistivity at 400 kHz for long-spacing antenna pairs. The borehole diameter is 8.5 in. (A) Attenuation borehole correction charts for mud resistivity is 0.02 ohm-m. (B) Phase difference resistivity borehole correction charts for mud resistivity is 0.02 ohm-m. (C) Attenuation borehole correction charts for mud resistivity is 2 ohm-m. (D) Phase difference resistivity borehole correction charts for mud resistivity is 2 ohm-m.

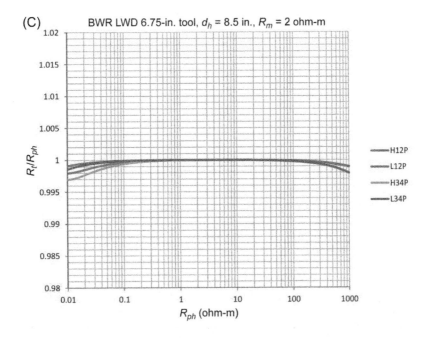

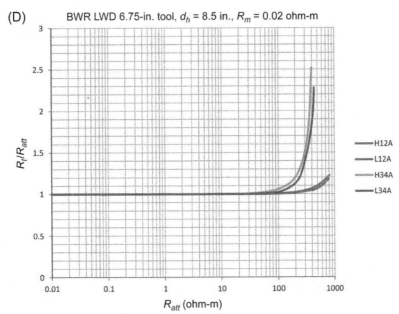

Figure 7.5 (Continued)

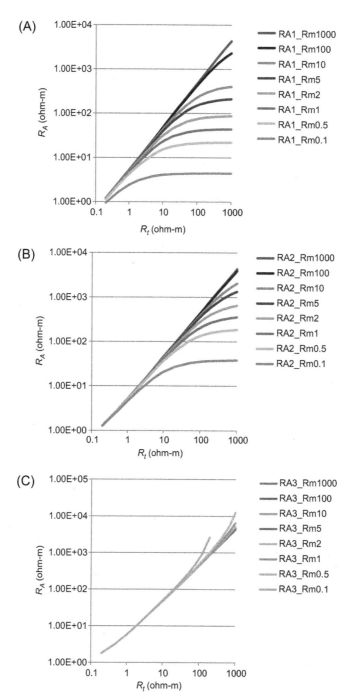

Figure 7.6 Apparent resistivity versus the formation resistivity of an array induction tool when the borehole size is fixed at 10 in. and mud resistivity changes from 0.1 to 1000 ohm-m. (A) TR spacing is 10 in.; (B) TR spacing is 18 in.; and (C) TR spacing is 30 in.

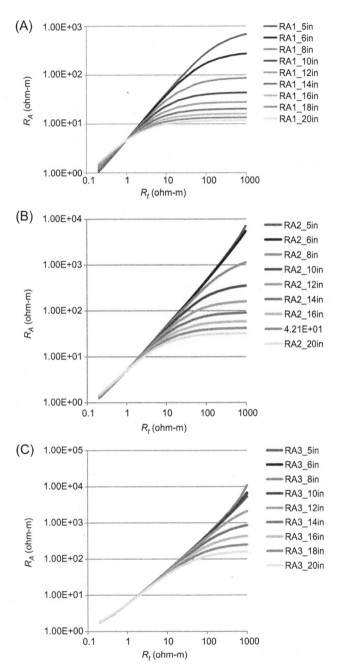

Figure 7.7 Apparent resistivity versus the formation resistivity of an array induction tool when the mud resistivity is fixed at 10 ohm-m and borehole diameter changes from 5 to 20 in. (A) TR spacing is 10 in.; (B) TR spacing is 18 in.; and (C) TR spacing is 30 in.

222 Theory of Electromagnetic Well Logging

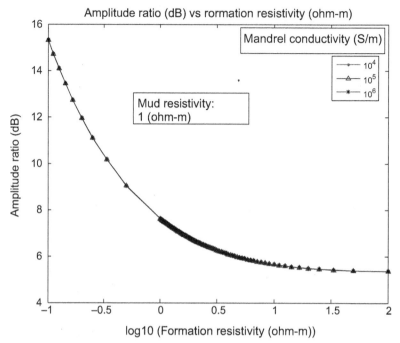

Figure 7.8 Amplitude ratio versus formation resistivity in water-based mud.

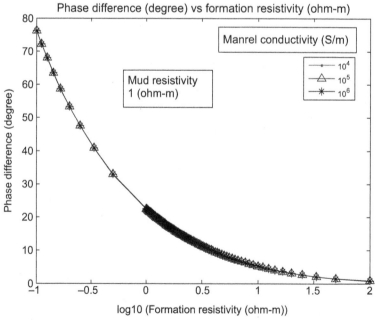

Figure 7.9 Phase difference versus formation resistivity in water-based mud.

Induction and LWD Tool Response in a Cylindrically Layered Isotropic Formation   223

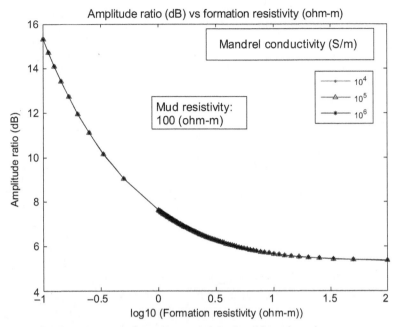

Figure 7.10 Amplitude ratio versus formation resistivity in oil-based mud.

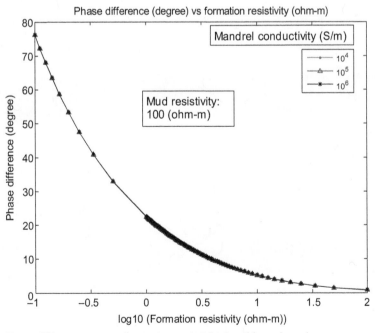

Figure 7.11 Phase difference versus formation resistivity in oil-based mud.

problem, the four-layer case must be extended to arbitrary multiple cylindrical layers. In this section, we will discuss the extension of the four-layer formation to multilayer cases. The four-layer equations are reformulated to fit the multiple layer formation. Computations based on the new equations are carried out. A more thorough convergence investigation is conducted making sure the algorithm is convergent in different circumstances.

7.3.1 Geometrical configuration

Consider arbitrary multiple layers in the formation, with each layer representing a different medium. Again, cylindrical coordinate system (r, φ, z) is used in the analysis. Fig. 7.12 shows the geometric structure of the model for the LWD case. A transmitter and two receivers are mounted coaxially on a conducting mandrel. Mud exists both inside and outside the mandrel. There are arbitrary multiple layers outside the borehole denoted with *gray shade* in the figure. A time-harmonic (2 MHz) current is induced on the transmitter. Electromagnetic fields are generated by the current and propagate through the earth formation. The amplitude ratio and phase difference of the voltages induced in Receiver A and Receiver B are measured and used to obtain the formation information.

Similarly, the case of induction tool is shown in Fig. 7.13, a transmitter and a receiver are mounted coaxially on the tool body. The inner mandrel is assumed to be

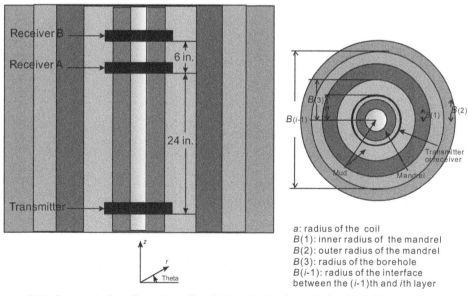

a: radius of the coil
$B(1)$: inner radius of the mandrel
$B(2)$: outer radius of the mandrel
$B(3)$: radius of the borehole
$B(i\text{-}1)$: radius of the interface between the $(i\text{-}1)$th and ith layer

Figure 7.12 Geometrical configuration of multiple cylindrical media for the LWD case.

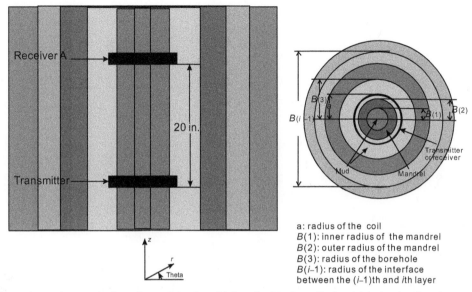

Figure 7.13 Geometrical configuration of multiple cylindrical media for the induction logging case.

solid, so the two innermost regions are assigned with the same electrical properties. The mud surrounds the pipe, and arbitrary multiple cylindrical layers encircle the borehole.

7.3.2 Methodology

With the practice in the four-layer case, the arbitrary layer can be easily handled. By comparing the geometry between these two situations, Eq. (7.13) can be expanded into arbitrary layers:

$$\tilde{E}_{\varphi 1}(r, k_z) = (-j\omega\mu_1 a)f_1(k_z)I_1(\gamma_1 r),$$
$$\tilde{E}_{\varphi 2}(r, k_z) = (-j\omega\mu_2 a)[f_2^-(k_z)I_1(\gamma_2 r) + f_2^+(k_z)K_1(\gamma_2 r)],$$
$$\tilde{E}_{\varphi 3}(r, k_z) = (-j\omega\mu_3 a)[f_3^-(k_z)I_1(\gamma_3 r) + f_3^+(k_z)K_1(\gamma_3 r) + K_1(\gamma_3 a)I_1(\gamma_3 r)],$$
$$\tilde{E}_{\varphi 4}(r, k_z) = (-j\omega\mu_3 a)[f_3^-(k_z)I_1(\gamma_3 r) + f_3^+(k_z)K_1(\gamma_3 r) + I_1(\gamma_3 a)K_1(\gamma_3 r)],$$

(7.27)

$$\vdots$$

$$\text{and } \tilde{E}_{\varphi n+1}(r, k_z) = (-j\omega\mu_n a)f_n(k_z)K_1(\gamma_n r).$$

Similarly, the general solutions for H in each region are:

$$\tilde{H}_{z1}(r, k_z) = a\gamma_1 f_1(k_z) I_0(\gamma_1 r),$$

$$\tilde{H}_{z2}(r, k_z) = a\gamma_2 [f_2^-(k_z) I_0(\gamma_2 r) - f_2^+(k_z) K_0(\gamma_2 r)],$$

$$\tilde{H}_{z3}(r, k_z) = a\gamma_3 [f_3^-(k_z) I_0(\gamma_3 r) - f_3^+(k_z) K_0(\gamma_3 r) + K_1(\gamma_3 a) I_0(\gamma_3 r)],$$

$$\tilde{H}_{z4}(r, k_z) = a\gamma_3 [f_3^-(k_z) I_0(\gamma_3 r) - f_3^+(k_z) K_0(\gamma_3 r) - I_1(\gamma_3 a) K_0(\gamma_3 r)], \quad (7.28)$$

$$\vdots$$

and $\tilde{H}_{zn+1}(r, k_z) = a\gamma_n [-f_n(k_z) K_0(\gamma_n r)].$

The electric and magnetic fields are continuous on each boundary, i.e.,

$$\tilde{E}_{\varphi 1}(r, k_z) \Big|_{r=\text{boundary radius}(1)^-}^{\text{boundary radius}(1)^+} = 0,$$

$$\tilde{H}_{z1}(r, k_z) \Big|_{r=\text{boundary radius}(1)^-}^{\text{boundary radius}(1)^+} = 0,$$

$$\tilde{E}_{\varphi 1}(r, k_z) \Big|_{r=\text{boundary radius}(2)^-}^{\text{boundary radius}(2)^+} = 0,$$

$$\tilde{H}_{z1}(r, k_z) \Big|_{r=\text{boundary radius}(2)^-}^{\text{boundary radius}(2)^+} = 0, \quad (7.29)$$

$$\vdots$$

$$\tilde{E}_{\varphi(n+1)}(r, k_z) \Big|_{r=\text{boundary radius}(n-1)^-}^{\text{boundary radius}(n-1)^+} = 0,$$

and $\tilde{H}_{z(n+1)}(r, k_z) \Big|_{r=\text{boundary radius}(n-1)^-}^{\text{boundary radius}(n-1)^+} = 0.$

Applying the boundary conditions (7.29) to the general solutions (7.27) yields

$$(-j\omega\mu_1 a)f_1(k_z)I_1(\gamma_1 b) = (-j\omega\mu_2 a)[f_2^-(k_z)I_1(\gamma_2 b) + f_2^+(k_z)K_1(\gamma_2 b)],$$

$$\gamma_1 f_1(k_z)I_0(\gamma_1 b) = \gamma_2[f_2^-(k_z)I_0(\gamma_2 b) - f_2^+(k_z)K_0(\gamma_2 b)],$$

$$(-j\omega\mu_2 a)[f_2^-(k_z)I_1(\gamma_2 d) + f_2^+(k_z)K_1(\gamma_2 d)],$$

$$= (-j\omega\mu_3 a)[f_3^-(k_z)I_1(\gamma_3 d) + f_3^+(k_z)K_1(\gamma_3 d) + K_1(\gamma_3 a)I_1(\gamma_3 d)],$$

$$\gamma_2[f_2^-(k_z)I_0(\gamma_2 d) - f_2^+(k_z)K_0(\gamma_2 d)],$$

$$= \gamma_3[f_3^-(k_z)I_0(\gamma_3 d) - f_3^+(k_z)K_0(\gamma_3 d) + K_1(\gamma_3 a)I_0(\gamma_3 d)],$$

$$\vdots \qquad (7.30)$$

$$(-j\omega\mu_{n-1} a)[f_{n-1}^-(k_z)I_1(\gamma_{n-1} e) + f_{n-1}^+(k_z)K_1(\gamma_{n-1} e) + I_1(\gamma_{n-1} a)K_1(\gamma_{n-1} e)]$$

$$= (-j\omega\mu_n a)f_n(k_z)K_1(\gamma_n e),$$

and $\gamma_{n-1}[f_{n-1}^-(k_z)I_0(\gamma_{n-1} e) - f_{n-1}^+(k_z)K_0(\gamma_{n-1} e) - I_1(\gamma_{n-1} a)K_0(\gamma_{n-1} e)]$

$$= \gamma_n[-f_n(k_z)K_0(\gamma_n e)].$$

Eq. (7.30) can be written in matrix form as

$$\begin{bmatrix} f_1 & f_2^- & f_2^+ & f_3^- & f_3^+ & f_4^- & f_4^+ & f_5^- & f_5^+ & \cdots & f_{n-2}^- & f_{n-2}^+ & f_{n-1}^- & f_{n-1}^+ & f_n \end{bmatrix}$$

$$\begin{pmatrix} \text{boundary 1} \\ \text{boundary 2} \\ \text{boundary 3} \\ \text{boundary 4} \\ \vdots \\ \text{boundary m} \\ \vdots \\ \text{boundary (n-1)} \end{pmatrix} \begin{bmatrix} \Box & \Box & \Box & & & & & & & & & & & & \\ & \Box & \Box & \Box & & & & & & & & & & & \\ & & \Box & \Box & \Box & \Box & & & & & & & & & \\ & & & \Box & \Box & \Box & \Box & & & & & & & & \\ & & & & \Box & \Box & \Box & \Box & & & & & & & \\ & & & & & \Box & \Box & \Box & \Box & & & & & & \\ & & & & & & & & \ddots & & & & & & \\ & & & & & & & & & & \Box & \Box & \Box & \Box & \\ & & & & & & & & & & & \Box & \Box & \Box & \Box \\ & & & & & & & & & & & & \Box & \Box & \Box \end{bmatrix} \begin{bmatrix} f_1 \\ f_2^- \\ f_2^+ \\ f_3^- \\ f_3^+ \\ f_4^- \\ f_4^+ \\ f_5^- \\ f_5^+ \\ \vdots \\ f_{n-2}^- \\ f_{n-2}^+ \\ f_{n-1}^- \\ f_{n-1}^+ \\ f_n \end{bmatrix} = \begin{bmatrix} 0 \\ 0 \\ \mu_3 I_1(\gamma_3 B(2))K_1(\gamma_3 R) \\ \gamma_3 I_0(\gamma_3 B(2))K_1(\gamma_3 R) \\ -\mu_3 I_1(\gamma_3 R)K_1(\gamma_3 B(3)) \\ \gamma_3 I_1(\gamma_3 R)K_0(\gamma_3 B(3)) \\ 0 \\ 0 \\ \vdots \\ \\ \\ \\ 0 \\ 0 \end{bmatrix} \quad (7.31)$$

Since accurate numerical results cannot be obtained directly due to overflow/underflow, we scale the modified Bessel functions and the scaling factors for each column are shown at the bottom of

$$\begin{matrix} & f_1 & f_2^- & f_2^+ & f_3^- & f_3^+ & \cdots & f_{n-1}^- & f_{n-1}^+ & f_n \\ \text{boundary 1} & \mu_1 \bar{I}_1(\gamma_1 B(1)) & -\mu_2 \bar{I}_1(\gamma_2 B(1)) & -\mu_2 \bar{K}_1(\gamma_2 B(1)) & 0 & 0 & & & & \\ \text{boundary 2} & \gamma_1 \bar{I}_0(\gamma_1 B(1)) & -\gamma_2 \bar{I}_0(\gamma_2 B(1)) & \gamma_2 \bar{K}_0(\gamma_2 B(1)) & 0 & 0 & & & & \\ \text{boundary m} & 0 & \mu_2 \bar{I}_1(\gamma_2 B(2)) & \mu_2 \bar{K}_1(\gamma_2 B(2)) & -\mu_3 \bar{I}_1(\gamma_3 B(2)) & -\mu_3 \bar{K}_1(\gamma_3 B(2)) & & & & \\ \vdots & 0 & \gamma_2 \bar{I}_0(\gamma_2 B(2)) & -\gamma_2 \bar{K}_0(\gamma_2 B(2)) & -\gamma_3 \bar{I}_0(\gamma_3 B(2)) & \gamma_3 \bar{K}_0(\gamma_3 B(2)) & & & & \\ \text{boundary (n-1)} & & & & & & & \mu_{n-1} \bar{I}_1(\gamma_{n-1} B(n-1)) & \mu_{n-1} \bar{K}_1(\gamma_{n-1} B(n-1)) & -\mu_n \bar{K}_1(\gamma_n B(n-1)) \\ & & & & & & & \gamma_{n-1} \bar{I}_0(\gamma_{n-1} B(n-1)) & -\gamma_{n-1} \bar{K}_0(\gamma_{n-1} B(n-1)) & \gamma_n \bar{K}_0(\gamma_n B(n-1)) \end{matrix}$$

$$\begin{matrix} \uparrow & \uparrow & \uparrow & \uparrow & \uparrow & & \uparrow & \uparrow & \uparrow \\ e^{-\text{abs}(\text{real}(\gamma_1 B(1)))} & e^{-\text{abs}(\text{real}(\gamma_2 B(2)))} & e^{\gamma_2 B(1)} & e^{-\text{abs}(\text{real}(\gamma_3 B(3)))} & e^{\gamma_3 B(2)} & & e^{-\text{abs}(\text{real}(\gamma_{n-1} B(n-1)))} & e^{\gamma_{n-1} B(n-2)} & e^{\gamma_n B(n-1)} \end{matrix}$$

$$\begin{bmatrix} \bar{f}_1(k_z) \\ \bar{f}_2^-(k_z) \\ \bar{f}_2^+(k_z) \\ \bar{f}_3^-(k_z) \\ \bar{f}_3^+(k_z) \\ \cdots \\ \bar{f}_{n-1}^-(k_z) \\ \bar{f}_{n-1}^+(k_z) \\ \bar{f}_n(k_z) \end{bmatrix} = \begin{bmatrix} 0 \\ 0 \\ \mu_3 \bar{I}_1(\gamma_3 B(2)) \bar{K}_1(\gamma_3 R) e^{\text{abs}(\text{real}(\gamma_3 B(2))) - \gamma_3 R} \\ \gamma_3 \bar{I}_0(\gamma_3 B(2)) \bar{K}_1(\gamma_3 R) e^{\text{abs}(\text{real}(\gamma_3 B(2))) - \gamma_3 R} \\ -\mu_3 \bar{I}_1(\gamma_3 R) \bar{K}_1(\gamma_3 B(3)) e^{\text{abs}(\text{real}(\gamma_3 R)) - \gamma_3 B(3)} \\ \gamma_3 \bar{I}_1(\gamma_3 R) \bar{K}_0(\gamma_3 B(3)) e^{\text{abs}(\text{real}(\gamma_3 R)) - \gamma_3 B(3)} \\ \cdots \\ 0 \\ 0 \\ 0 \end{bmatrix}$$

$$\begin{matrix} \uparrow & & \uparrow \\ e^{-\text{abs}(\text{real}(\gamma_3 B(2)))} \text{ or } e^{-\text{abs}(\text{real}(\gamma_3 R))} & & e^{\gamma_3 R} \text{ or } e^{\gamma_3 B(3)} \end{matrix}$$

(7.32)

The relations between the scaled and unscaled unknowns are:

$$f_1(k_z) = \overline{f}_1(k_z) e^{-\text{abs}(\text{real}(\gamma_1 B(1)))} = \overline{f}_1(k_z) e^{-\alpha_1 B(1)},$$

$$f_2^-(k_z) = \overline{f}_2^-(k_z) e^{-\text{abs}(\text{real}(\gamma_2 B(2)))} = \overline{f}_2^-(k_z) e^{-\alpha_2 B(2)},$$

$$f_2^+(k_z) = \overline{f}_2^+(k_z) e^{\gamma_2 B(1)},$$

$$f_3^-(k_z) = \overline{f}_3^-(k_z) e^{-\text{abs}(\text{real}(\gamma_3 B(3)))} = \overline{f}_3^-(k_z) e^{-\alpha_3 B(3)},$$

$$f_3^+(k_z) = \overline{f}_3^+(k_z) e^{\gamma_3 B(2)},$$

$$\vdots$$

$$f_{n-1}^-(k_z) = \overline{f}_{n-1}^-(k_z) e^{-\text{abs}(\text{real}(\gamma_{n-1} B(n-1)))} = \overline{f}_{n-1}^-(k_z) e^{-\alpha_{n-1} B(n-1)},$$

$$f_{n-1}^+(k_z) = \overline{f}_{n-1}^+(k_z) e^{\gamma_{n-1} B(n-2)},$$

$$\text{and } f_n(k_z) = \overline{f}_n(k_z) e^{-\gamma_n B(n-1)}.$$

(7.33)

The scaling factors with $\text{abs}(\text{real}(\gamma_i))$ in the index of the exponent are replaced with α_i, where α_i is the real part of γ_i, $\gamma_i = \alpha_i + i\beta_i$. The scaled Bessel functions are computed by Amos's subroutine when a flag is set.

Substituting Eq. (7.33) into the general expressions for the electromagnetic fields, we obtain

$$\tilde{E}_{\varphi 1}(r, k_z) = (-j\omega\mu_1 R)\overline{f}_1(k_z)\overline{I}_1(\gamma_1 r) e^{[\alpha_1(r-B(1))]},$$

$$\tilde{E}_{\varphi 2}(r, k_z) = (-j\omega\mu_2 R)[\overline{f}_2^-(k_z)\overline{I}_1(\gamma_2 r) e^{[\alpha_2(r-B(2))]} + \overline{f}_2^+(k_z)\overline{K}_1(\gamma_2 r) e^{\gamma_2(B(1)-r)}],$$

$$\tilde{E}_{\varphi 3}(r, k_z) = (-j\omega\mu_3 R)[\overline{f}_3^-(k_z)\overline{I}_1(\gamma_3 r) e^{[\alpha_3(r-B(3))]}$$

$$+ \overline{f}_3^+(k_z)\overline{K}_1(\gamma_3 r) e^{\gamma_3(B(2)-r)} + \overline{K}_1(\gamma_3 R)\overline{I}_1(\gamma_3 r) e^{[\alpha_3 r - \gamma_3 R]}],$$

$$\tilde{E}_{\varphi 4}(r, k_z) = (-j\omega\mu_3 R)[\overline{f}_3^-(k_z)\overline{I}_1(\gamma_3 r) e^{(\gamma_3 r - \alpha_3 B(3))}$$

$$+ \overline{f}_3^+(k_z)\overline{K}_1(\gamma_3 r) e^{\gamma_3(B(2)-r)} + \overline{I}_1(\gamma_3 R)\overline{K}_1(\gamma_3 r) e^{[\alpha_3 R - \gamma_3 r]}],$$

$$\vdots$$

$$\tilde{E}_{\varphi n}(r, k_z) = (-j\omega\mu_{n-1} R)[\overline{f}_{n-1}^-(k_z)\overline{I}_1(\gamma_{n-1} r) e^{[\gamma_{n-1} r - \alpha_{n-1} B(n-1)]} + \overline{f}_{n-1}^+(k_z)\overline{K}_1(\gamma_{n-1} r) e^{\gamma_{n-1}(B(n-2)-r)}],$$

$$\tilde{E}_{\varphi(n+1)}(r, k_z) = (-j\omega\mu_n R)\overline{f}_n(k_z)\overline{K}_1(\gamma_n r) e^{-\gamma_n r - \gamma_n B(n-1)},$$

(7.34)

$$\tilde{H}_{z1}(r,k_z) = a\gamma_1 \bar{f}_1(k_z)\bar{I}_0(\gamma_1 r)e^{[\alpha_1(r-B(1))]},$$

$$\tilde{H}_{z2}(r,k_z) = a\gamma_2 \bar{f}_2^-(k_z)\bar{I}_0(\gamma_2 r)e^{[\text{abs}(\text{real}(\gamma_2 r))-\text{abs}(\text{real}(\gamma_2 B(2)))]} - \bar{f}_2^+(k_z)\overline{K}_0(\gamma_2 r)e^{\gamma_2(B(1)-r)})$$

$$\tilde{H}_{z3}(r,k_z) = a\gamma_3[\bar{f}_3^-(k_z)\bar{I}_0(\gamma_3 r)e^{\alpha_3(r-B(3))}$$
$$- \bar{f}_3^+(k_z)\overline{K}_0(\gamma_3 r)e^{\gamma_3(B(2)-r)} + \overline{K}_1(\gamma_3 R)\bar{I}_0(\gamma_3 r)e^{\alpha_3 r - \gamma_3 R}],$$

$$\tilde{H}_{z4}(r,k_z) = a\gamma_3[\bar{f}_3^-(k_z)\bar{I}_0(\gamma_3 r)e^{\alpha_3(r-B(3))}$$
$$- \bar{f}_3(k_z)\overline{K}_0(\gamma_3 r)e^{\gamma_3(B(2)-r)} - \bar{I}_1(\gamma_3 R)\overline{K}_0(\gamma_3 r)e^{(\alpha_3 R - \gamma_3 r)}],$$

$$\vdots$$

$$\tilde{H}_{zn}(r,k_z) = a\gamma_{n-1}[\bar{f}_{n-1}^-(k_z)\bar{I}_0(\gamma_{n-1}r)e^{\alpha_{n-1}(r-B(n-1))} - \bar{f}_{n-1}(k_z)\overline{K}_0(\gamma_{n-1}r)e^{\gamma_{n-1}(B(n-2)-r)}],$$

$$\text{and } \tilde{H}_{z(n+1)}(r,k_z) = a\gamma_n[-\bar{f}_n(k_z)\overline{K}_0(\gamma_n r)e^{-\gamma_n r - \gamma_n B(n-1)}].$$

(7.35)

Please refer to Appendix C for detailed derivation.

Now the problem is reduced to solving the linear equations for the unknowns. By solving the linear equations, we can get the expressions for $\tilde{E}_{\varphi i}(r,k_z)$ and $\tilde{H}_{zi}(r,k_z)$. Then we use the following expressions to do the inverse Fourier transform to get $E_\varphi(r,z)$ and $H_z(r,z)$ in each region from Eqs. (7.33)–(7.35).

7.3.3 Discussions of convergence, accuracy, and numerical computation

From the derivations in Section 7.3.2, we may notice that the analytic method seems to be accurate and no approximations. However, since this method is based on the spectrum solution and inverse Fourier transform must be used in the conversion of the spectral quantities to the spatial quantities as given in Eqs. (7.33)–(7.35). As we have seen, Eqs. (7.33)–(7.35) are integrations in spectral domain. Since numerical integrations are involved, the number of sampling points and the spectral cutoff will greatly impact to the accuracy and convergence of the algorithm [11–18]. As the properties of the spectral functions of the field quantities are directly related to the formations and geometric structure of the tool and layers, different cases should be considered to investigate the performance of the algorithm.

To make this algorithm applicable to various cases, different tests must be performed, including extreme cases such as high conductivity contrast between mandrel and mud, or mud and formation. Permutations and combinations of possible values of mandrel/pipe, mud, and formation are employed to assure accuracy. Tables 7.1 and 7.2 show the possible values of the conductivity of each layer for testing. This is a quantitative test concerning the spectral cutoff of the integrand in the IFFT (Inverse Fast Fourier Transform).

To investigate the accuracy of the algorithm, the following stratagem is used: each set of the testing data is compared with the one that has the same parameters but different cutoff and sample rate in the spatial domain. If the difference in the results are within the tolerance (10^{-6}), the cutoff and sample rate are considered adequate.

Figs. 7.14 and 7.15 show the relative errors of the amplitude ratio and phase difference between a set of data with cutoff = 6000 and 2^{14} samples, and another set of data with cutoff = 3000 and 2^{13} samples in the LWD case. Correspondingly, the comparison for the induction tool is shown in Fig. 7.16. From Figs. 7.14–7.16, we can see

Table 7.1 Permutations and combinations of layer conductivity for an LWD tool

Layer	Parameter
	Conductivity (S/m)
Mandrel	10^3, 10^4, 10^5, 10^6, 10^7
Mud	5×10^{-4}, 10^{-3}, 10^{-2}, 10^{-1}, 1, 10
Formation	5×10^{-4}, 10^{-3}, 10^{-2}, 10^{-1}, 1, 10

Table 7.2 Permutations and combinations of layer conductivity values for an induction tool

Layer	Parameter
	Conductivity (S/m)
Pipe	10^3, 10^4, 10^5, 10^6, 10^7
Mud	5×10^{-4}, 10^{-3}, 10^{-2}, 10^{-1}, 1, 10
Formation	5×10^{-4}, 10^{-3}, 10^{-2}, 10^{-1}, 1, 10

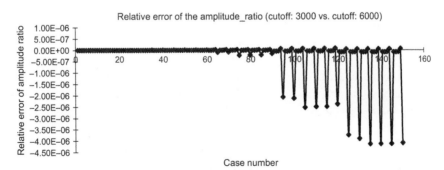

Figure 7.14 Relative error of the amplitude ratio between two sets of data: (1) cutoff = 3000, number of sample points = 2^{13}; (2) cutoff = 6000, number of sample points = 2^{14}.

that the algorithm is fairly stable and is not very sensitive to the number of sampling points and cutoff values when they are sufficiently large.

Furthermore, to investigate the convergence of the algorithm in details, for each case in Tables 7.3 and 7.4, a series of tests varying the spectral cutoff and number of sample points are carried out. We fix the cutoff and keep doubling the number of sample points until the result converges (a fractional error of 10^{-6} is employed in this test). Then we double the cutoff and repeat the process until the results converge

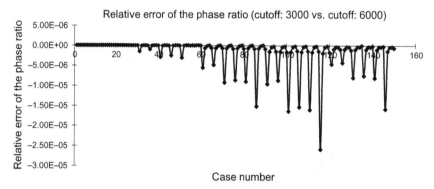

Figure 7.15 Relative error of the phase difference between two sets of data: (1) cutoff = 3000, number of sample points = 2^{13}; (2) cutoff = 6000, number of sample points = 2^{14}.

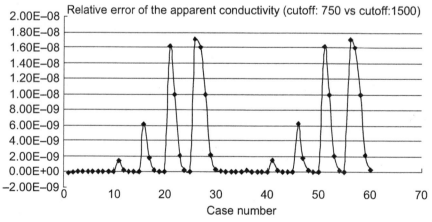

Figure 7.16 Relative error of the apparent conductivity between two sets of data: (1) cutoff = 750, number of sample points = 2^{11}; (2) cutoff = 1500, number of sample points = 2^{12}.

Table 7.3 An example of the testing procedure

Layer	Parameters			
	Conductivity (S/m)	Relative dielectric constant (unit)	Relative magnetic permeability (unit)	Boundary radius (in.)
Mandrel	1.0d6	1.0	1.0	2.0
Mud	5.0d-4	1.0	1.0	3.0
Formation	5.0d-4	1.0	1.0	Infinity

overall. A value of 12,000 was used for the cutoff limit and the results turned out to be sufficiently accurate.

Table 7.4 shows the results obtained from the test and the data in Table 7.3 are put into Figs. 7.17 and 7.18 to give a better understanding. We first fix the cutoff and make the number of sample points double each time until the results obtained from two consecutive tests are sufficiently close. Each line in Fig. 7.17 shows the amplitude ratio obtained for a given number of samples while Fig. 7.18 shows the phase difference. We can see that the lines converge to the same value when the sample points are enough. The longer the cutoff value for the IFFT is, the more sample points are needed for the result to converge. On the other hand, if the cutoff is not long enough, the convergence value will be different than that obtained from a higher cutoff. Hence the cutoff point should be carefully selected. For the induction case, the results converge faster than the LWD case, so cutoff value of 3000 was used. A cutoff of 6000 was used for the LWD tool.

7.3.4 Simulation results
7.3.4.1 Induction and LWD tool response with complex invasion profile
This example investigates the effect of invasion for an induction tool and LWD tool. A seven-cylindrical layer case is investigated. The conductivities of the layers are illustrated in Fig. 7.19 and Table 7.5. The mud conductivity is 5 S/m and from Fig. 7.19 we can see that the conductivities of the layers decrease gradually, which models the invasion. Table 7.5 gives the actual data for each layered formation. Fig. 7.20 is the simulated apparent conductivity obtained by using an induction tool (Fig. 7.13) as a function of the formation conductivity (Layer 7 in Table 7.5) with invasion profile given in Fig. 7.19 and specified in Table 7.5. Fig. 7.21A and B is responses of LWD tool (Fig. 7.12).

Table 7.4 Results for different combinations of sampling and cutoff for one set of formation parameters

Number of samples	Cutoff	Amplitude ratio (dB)	Phase difference (degrees)
750	750	5.503957734	179.4247123
1500	750	5.553337652	179.7868177
3000	750	5.560213448	179.8205103
6000	750	5.561144068	179.8198486
12,000	750	5.561129172	179.8195332
24,000	750	5.561129556	179.8195368
750	1500	4.944647966	175.742042
1500	1500	5.50397677	179.4245829
3000	1500	5.553356587	179.7866894
6000	1500	5.560232351	179.8203823
12,000	1500	5.561162962	179.8197206
24,000	1500	5.561148063	179.8194052
48,000	1500	5.561148446	179.8194089
750	3000	−1.869599316	168.0032419
1500	3000	4.944647966	175.742042
3000	3000	5.50397677	179.4245829
6000	3000	5.553356586	179.7866894
12,000	3000	5.560232351	179.8203823
24,000	3000	5.561162962	179.8197206
48,000	3000	5.561148063	179.8194053
96,000	3000	5.561148446	179.8194089
750	6000	−26.0126458	169.0196675
1500	6000	−1.869599316	168.0032419
3000	6000	4.944647966	175.742042
6000	6000	5.50397677	179.4245829
12,000	6000	5.553356586	179.7866894
24,000	6000	5.560232351	179.8203823
48,000	6000	5.561162962	179.8197206
96,000	6000	5.561148063	179.8194053
192,000	6000	5.561148446	179.8194089
750	12,000	−34.28829668	169.6834079
1500	12,000	−26.0126458	169.0196675
3000	12,000	−1.869599316	168.0032419
6000	12,000	4.944647966	175.742042
12,000	12,000	5.50397677	179.4245829
24,000	12,000	5.553356586	179.7866894
48,000	12,000	5.560232351	179.8203823
96,000	12,000	5.561162962	179.8197206
192,000	12,000	5.561148063	179.8194053
384,000	12,000	5.561148446	179.8194089

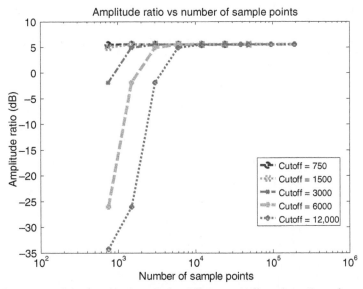

Figure 7.17 Illustration of the amplitude ratio for different cutoffs and number of sample points.

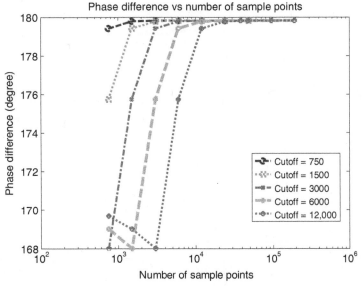

Figure 7.18 Illustration of the phase difference for different cutoffs and number of sample points.

7.3.4.2 Magnetic mud influence to an induction and LWD tool response

Magnetic materials are often used in a drilling process, which may impact to the response of an induction and LWD resistivity logging tool. Using the developed algorithm, the effect of magnetic mud can be evaluated. Table 7.7 gives the layer parameters, and Fig. 7.22 shows apparent conductivity versus the formation conductivity for an induction case.

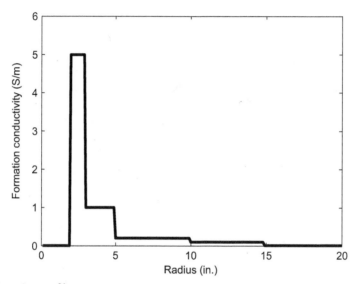

Figure 7.19 Invasion profile.

Table 7.5 Layered formation for an induction invasion case

Layer	Parameters			
	Conductivity (S/m)	Relative dielectric constant (unit)	Relative magnetic permeability (unit)	Boundary radius (in.)
Layer 1 and 2: Pipe	10^{-7}	1.0	1.0	2.0
Layer 3: Mud	5.0	1.0	1.0	3.0
Layer 4: Invasion 1	1.0	1.0	1.0	5.0
Layer 5: Invasion 2	0.2	1.0	1.0	10.0
Layer 6: Invasion 3	0.1	1.0	1.0	15.0
Layer 7: Formation	0.01	1.0	1.0	Infinity

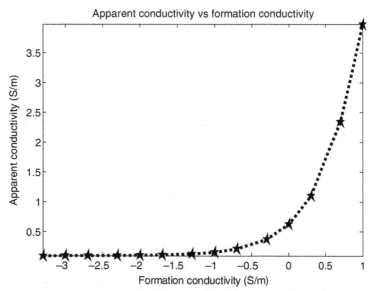

Figure 7.20 The apparent conductivity of the induction tool shown in Fig. 7.13 in a multilayer invasion profile shown in Fig. 7.19 and Table 7.5 as a function of formation conductivity.

The magnetic mud will also impact the response of an LWD tool. Table 7.8 gives the layer parameters and Fig. 7.23 shows the amplitude ratio and phase difference versus the formation conductivity for an LWD tool defined in Fig. 7.12. From Figs. 7.22 and 7.23, we can clearly see that the magnetic material in the mud will greatly impact the tool response of an induction and LWD tools. The impact of the magnetic mud is very similar to the change in the mud conductivity. We can see this conclusion from the skin depth as given in Eq. (2.51):

$$d_s = \sqrt{\frac{2}{\omega \mu \sigma}} \qquad (2.51)$$

The skin depth, which determines the real and imaginary part of the complex propagating constant k, makes the change in both induction and LWD tool response due to the mud magnetic permeability the same way as the conductivity of the mud does.

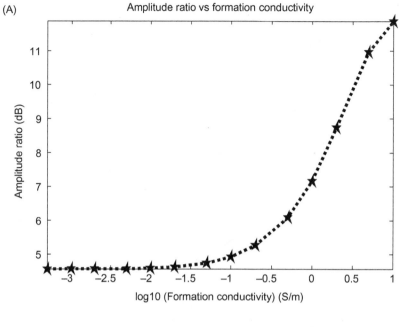

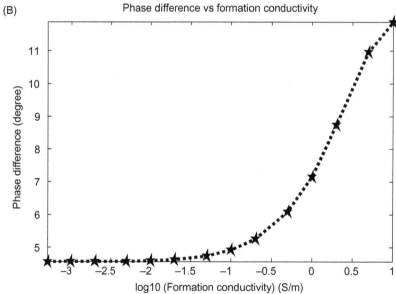

Figure 7.21 The amplitude ratio and phase difference of the LWD tool described in Fig. 7.12 in a multilayer invasion profile shown in Fig. 7.19 and Table 7.6. (A) Amplitude ratio as a function of formation conductivity and (B) phase difference as a function of formation conductivity.

Table 7.6 Layer formation for an LWD invasion case

Layer	Parameters			
	Conductivity (S/m)	Relative dielectric constant (unit)	Relative magnetic permeability (unit)	Boundary radius (in.)
Layer 1 and 2: Pipe	10^7	1.0	1.0	2.0
Layer 3: Mud	5.0	1.0	1.0	3.0
Layer 4: Invasion 1	1.0	1.0	1.0	5.0
Layer 5: Invasion 2	0.2	1.0	1.0	10.0
Layer 6: Invasion 3	0.1	1.0	1.0	15.0
Layer 7: Formation	0.01	1.0	1.0	Infinity

Table 7.7 Layered formation parameters to investigate magnetic mud effect for an induction logging tool

Layer	Parameters		
	Conductivity (S/m)	Relative magnetic permeability (unit)	Boundary radius (in.)
Layer 1: Mandrel	10^{-6}	1.0	1.18
Layer 2: Insulator layer	10^{-6}	1.0	2.0
Layer 3: Mud	5.0	1–500	3.0
Layer 4: Formation	0.2	1.0	Infinity

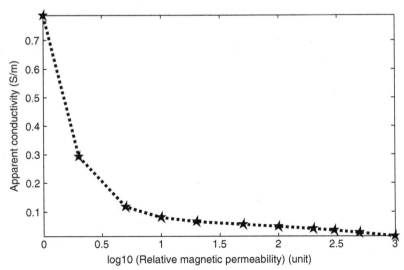

Figure 7.22 Induction tool response with magnetic mud. The tool information is given in Fig. 7.13 and the formation layers are described in Table 7.7.

Table 7.8 Layered formation for investigating the response of an LWD tool with a magnetic mud

Layer	Parameters		
	Conductivity (S/m)	Relative magnetic permeability (unit)	Boundary radius (in.)
Layer 1 and 2: Pipe	10^6	1.0	2.0
Layer 3: Mud	5.0	1–500	3.0
Layer 4: Formation	0.2	1.0	Infinity

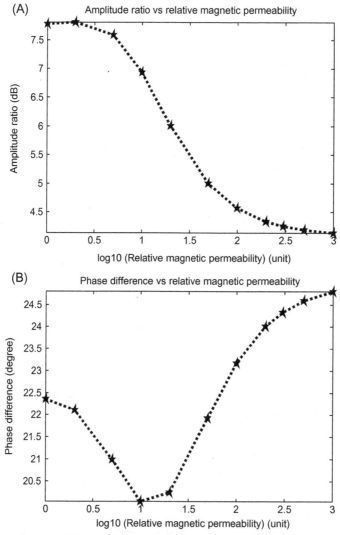

Figure 7.23 The influence of the magnetic permeability of the mud to the LWD tool response (A) amplitude ratio and (B) phase difference The tool information is given in Fig. 7.12 and the formation layers are described in Table 7.8.

7.4 CONCLUSIONS

In this chapter, we discussed induction and LWD resistivity tool response in a cylindrically layered media. To make the analysis easier, two cases are discussed. The first case considers a mandrel, insulation layer in the tool antenna, mud conductivity, and the homogenous formation. The second model considers an arbitrarily layered media and can be used to study invasion effects.

Note that in the practical solution, calculations of Hankel functions and inverse Fourier transforms are applied to obtain the results. Windowing effects in applying the inverse FFT are thoroughly discussed and three windows are compared. Electromagnetic fields and voltages on the receiving coils can be calculated using the proposed algorithm.

There is always a doubt about the perfect electric conductor assumption of the mandrel conductivity. We investigated the response of induction and LWD tools in a homogeneous formation with varying mandrel conductivity and mud conductivity. There are four layers in the model, the mandrel is the first and second, the mud is the third, and the homogenous formation is the fourth. The innermost layer is reserved for possible content of mud but is considered as mandrel in the test cases. It is found that the mandrel conductivity has little impact to the tool response when the mandrel conductivity changes from 10^3 to 10^6 S/m.

The algorithm is developed by solving Maxwell's equations in a cylindrically layered media. A second-order partial differential equation is obtained by substituting the magnetic field into the expression for the electric field. The solution to the partial differential equation must satisfy the boundary conditions. The boundary conditions are based on the fact that the tangential components of the electromagnetic fields are continuous on the boundaries. Linear equations are then formed. The solutions to the linear equations are the Fourier components of the electric field. By implementing the FFT (Fast Fourier Transform), the space domain solution is achieved.

Some examples for both induction and LWD tools are illustrated in the result part of this chapter. A preliminary convergence investigation is carried out by changing the sample rate. The cutoff issue is brought forward in this chapter and investigated by doubling it to test the convergence.

An in-depth convergence investigation is carried out. The cutoff for the integrand and the number of sample points are considered and tested in an iterative method to ensure the convergence of the result.

A customized accuracy can be achieved by different techniques: scaling of the Bessel functions, separating the direct coupling and reflection part in the integrand of the inverse Fourier transform, careful selection of cutoff and number of sample points for the inverse Fourier transform.

REFERENCES

[1] L. Shen, Theory of Induction and MWD Tools in Cylindrically Layered Medium, Well Logging Technical Report, 1993.
[2] J. Moran, K. Kunz, Basic theory of induction logging and application to study of two-coil sondes, Geophysics 27 (1962) 829–858.
[3] A. Bhardwaj, 1-D inversion of tri-axial induction logs in anisotropic medium (Master's thesis), Department of Electrical Engineering, University of Houston, 2007.
[4] B. Anderson, Modeling and inversion methods for the interpretation of resistivity logging tool response (Ph.D.), Delft University of Technology, 2001.
[5] V. Nikitina, The general solution of an axially symmetrical problem in induction logging theory, Izvest. Akad. Nauk SSSR. Ser. Geofiz. 4 (1960) 607–616.
[6] A. Kaufmann, Theory of Induction Logging, Siberian Dept. of Nauka Press, Novosibersk, 1965.
[7] Q. Liu, W. Chew, Diffraction of non-axisymmetric waves in cylindrically layered media by horizontal discontinuities, Radio Sci. 27 (1992) 569–581.
[8] M. Paszynski, L. Demkowicz, D. Pardo, Verification of Goal-Oriented HP-Adaptivity, The University of Texas at Austin, 2005.
[9] S. Gianzero, B. Anderson, An integral transform solution to the fundamental problem in resistivity logging, Geophysics 47 (1982) 946–956.
[10] X. Wu, T. Habashy, Influence of steel casings on electromagnetic signals, Geophysics 59 (3) (1994) 378–390.
[11] K. Michalski, J. Mosig, Multilayered media Green's functions in integral equation formulations, IEEE Trans. Antennas Propagat. 45 (March 1997) 508–519.
[12] K. Michalski, Extrapolation methods for sommerfeld integral tails, IEEE Trans. Antennas Propagat. 46 (10) (1998).
[13] I. Longman, Note on a method for computing infinite integrals of oscillatory functions, Proc. Cambridge Phil. Soc. 52 (1956) 764–768.
[14] M. Blakemore, G. Evans, J. Hyslop, Comparison of some methods for evaluating infinite range oscillatory integrals, J. Comput. Phys. 22 (1976) 352–376.
[15] S. Lucas, H. Stone, Evaluating infinite integrals involving Bessel functions of arbitrary order, J. Comput. Appl. Math. 64 (1995) 217–231.
[16] J. Mosig, Integral equation technique, in: T. Itoh (Ed.), Numerical Techniques for Microwave and Millimeter-Wave Passive Structures, Wiley, New York, NY, 1989, pp. 133–213.
[17] T. Espelid, K. Overholt, DQAINF: an algorithm for automatic integration of infinite oscillating tails, Num. Algorithms 8 (1994) 83–101.
[18] J. Lyness, Integrating some infinite oscillating tails, J. Comput. Appl. Math. 12/13 (1985) 109–117.

APPENDIX A DERIVATION FOR THE MAGNETIC FIELDS IN SPECTRAL DOMAIN

The magnetic field is given in terms of the electric field by

$$\underline{H} = \frac{1}{-j\omega\mu_i} \nabla \times \underline{E}$$

$$= \frac{1}{-j\omega\mu_i} \nabla \times (E_\varphi \hat{\varphi})$$

$$= \frac{1}{-j\omega\mu_i} \frac{1}{r} \frac{\partial(rE_\varphi)}{\partial r}$$

$$= \frac{1}{-j\omega\mu_i} \frac{1}{r} \left(E_\varphi + r\frac{\partial E_\varphi}{\partial r} \right)$$

$$= \frac{1}{-j\omega\mu_i} \frac{1}{r} \left\{ \begin{array}{l} (-j\omega\mu_i a)[f_i^-(k_z)I_1(\gamma_i r) + f_i^+(k_z)K_1(\gamma_i r)] \\ + r\frac{\partial}{\partial r}(-j\omega\mu_i a)\left[f_i^-(k_z)I_1(\gamma_i r) + f_i^+(k_z)K_1(\gamma_i r)\right] \end{array} \right\}$$

$$= \frac{1}{-j\omega\mu_i} \frac{1}{r} (-j\omega\mu_i a) \left\{ \begin{array}{l} [f_i^-(k_z)I_1(\gamma_i r) + f_i^+(k_z)K_1(\gamma_i r)] \\ + r\frac{\partial}{\partial r}\left[f_i^-(k_z)I_1(\gamma_i r) + f_i^+(k_z)K_1(\gamma_i r)\right] \end{array} \right\}$$

$$= \frac{a}{r} \left\{ \begin{array}{l} [f_i^-(k_z)I_1(\gamma_i r) + f_i^+(k_z)K_1(\gamma_i r)] \\ + r\gamma_i \frac{\partial}{\partial(\gamma_i r)}\left[f_i^-(k_z)I_1(\gamma_i r) + f_i^+(k_z)K_1(\gamma_i r)\right] \end{array} \right\}$$

using recurrence formula:

$$I'_n(x) = I_{n-1}(x) - \frac{n}{x}I_n(x)$$

$$K'_n(x) = -K_{n-1}(x) - \frac{n}{x}K_n(x)$$

$$I'_1(x) = I_0(x) - \frac{1}{x}I_1(x)$$

and $K'_1(x) = -K_0(x) - \frac{1}{x}K_1(x)$

$$= \frac{a}{r}\left\{\begin{array}{l}[f_i^-(k_z)I_1(\gamma_i r) + f_i^+(k_z)K_1(\gamma_i r)] \\ +r\gamma_i[f_i^-(k_z)\left\{I_0(\gamma_i r) - \frac{1}{\gamma_i r}I_1(\gamma_i r)\right\} \\ +f_i^+(k_z)\left\{-K_0(\gamma_i r) - \frac{1}{\gamma_i r}K_1(\gamma_i r)\right\}]\end{array}\right\}$$

$$= \frac{a}{r}\left\{\begin{array}{l}[f_i^-(k_z)I_1(\gamma_i r) + f_i^+(k_z)K_1(\gamma_i r)] \\ +[(r\gamma_i)f_i^-(k_z)\left\{I_0(\gamma_i r) - \frac{1}{\gamma_i r}I_1(\gamma_i r)\right\} \\ +(r\gamma_i)f_i^+(k_z)\left\{-K_0(\gamma_i r) - \frac{1}{\gamma_i r}K_1(\gamma_i r)\right\}]\end{array}\right\}$$

$$= \frac{a}{r}\left\{\begin{array}{l}[f_i^-(k_z)I_1(\gamma_i r) + f_i^+(k_z)K_1(\gamma_i r)] \\ +[(r\gamma_i)f_i^-(k_z)I_0(\gamma_i r) - (r\gamma_i)f_i^-(k_z)\frac{1}{\gamma_i r}I_1(\gamma_i r)] \\ [-(r\gamma_i)f_i^+(k_z)K_0(\gamma_i r) - (r\gamma_i)f_i^+(k_z)\frac{1}{\gamma_i r}K_1(\gamma_i r)]\end{array}\right\}$$

$$= \frac{a}{r}\left\{\begin{array}{l}[\cancel{f_i^-(k_z)I_1(\gamma_i r)} + \cancel{f_i^+(k_z)K_1(\gamma_i r)}] \\ +[(r\gamma_i)f_i^-(k_z)I_0(\gamma_i r) - \cancel{f_i^-(k_z)I_1(\gamma_i r)}] \\ [-(r\gamma_i)f_i^+(k_z)K_0(\gamma_i r) - \cancel{f_i^+(k_z)K_1(\gamma_i r)}]\end{array}\right\}$$

$$= \frac{a}{r}[(r\gamma_i)f_i^-(k_z)I_0(\gamma_i r) - (r\gamma_i)f_i^+(k_z)K_0(\gamma_i r)]$$

$$= (a\gamma_i)[f_i^-(k_z)I_0(\gamma_i r) - f_i^+(k_z)K_0(\gamma_i r)],$$

APPENDIX B DERIVATION FOR THE EXPRESSION OF ELECTRICAL FIELD FOR THE HOMOGENEOUS FORMATION IN SPECTRAL DOMAIN

The electrical fields for each region are given by

$$\tilde{E}_{\varphi 1}(r, k_z) = (-j\omega\mu_1 a)f_1(k_z)I_1(\gamma_1 r)$$
$$= (-j\omega\mu_1 a)\overline{f}_1(k_z)e^{-\text{abs}(\text{real}(\gamma_1 b))}\overline{I}_1(\gamma_1 r)e^{\text{abs}(\text{real}(\gamma_1 r))}$$
$$= (-j\omega\mu_1 a)\overline{f}_1(k_z)\overline{I}_1(\gamma_1 r)e^{[\text{abs}(\text{real}(\gamma_1 r))-\text{abs}(\text{real}(\gamma_1 b))]}$$
$$= (-j\omega\mu_1 a)\overline{f}_1(k_z)\overline{I}_1(\gamma_1 r)e^{\alpha_1(r-b)}$$

$$\tilde{E}_{\varphi 2}(r, k_z) = (-j\omega\mu_2 a)[f_2^-(k_z)I_1(\gamma_2 r) + f_2^+(k_z)K_1(\gamma_2 r)]$$
$$= (-j\omega\mu_2 a)[\overline{f}_2^-(k_z)e^{-\text{abs}(\text{real}(\gamma_2 d))}\overline{I}_1(\gamma_2 r)e^{\text{abs}(\text{real}(\gamma_2 r))} + \overline{f}_2^+(k_z)e^{\gamma_2 b}\overline{K}_1(\gamma_2 r)e^{-\gamma_2 r}]$$
$$= (-j\omega\mu_2 a)[\overline{f}_2^-(k_z)\overline{I}_1(\gamma_2 r)e^{[\text{abs}(\text{real}(\gamma_2 r))-\text{abs}(\text{real}(\gamma_2 d))]} + \overline{f}_2^+(k_z)\overline{K}_1(\gamma_2 r)e^{\gamma_2(b-r)}]$$
$$= (-j\omega\mu_2 a)[\overline{f}_2^-(k_z)\overline{I}_1(\gamma_2 r)e^{\alpha_2(r-d)} + \overline{f}_2^+(k_z)\overline{K}_1(\gamma_2 r)e^{\gamma_2(b-r)}]$$

$$\tilde{E}_{\varphi 3}(r, k_z) = (-j\omega\mu_3 a)[f_3^-(k_z)I_1(\gamma_3 r) + f_3^+(k_z)K_1(\gamma_3 r) + K_1(\gamma_3 a)I_1(\gamma_3 r)]$$
$$= (-j\omega\mu_3 a)[\overline{f}_3^-(k_z)e^{-\text{abs}(\text{real}(\gamma_3 e))}\overline{I}_1(\gamma_3 r)e^{\text{abs}(\text{real}(\gamma_3 r))}$$
$$+ \overline{f}_3^+(k_z)e^{\gamma_3 d}\overline{K}_1(\gamma_3 r)e^{-\gamma_3 r} + \overline{K}_1(\gamma_3 a)e^{-\gamma_3 a}\overline{I}_1(\gamma_3 r)e^{\text{abs}(\text{real}(\gamma_3 r))}]$$
$$= (-j\omega\mu_3 a)[\overline{f}_3^-(k_z)\overline{I}_1(\gamma_3 r)e^{\text{abs}(\text{real}(\gamma_3 r))-\text{abs}(\text{real}(\gamma_3 e))}$$
$$+ \overline{f}_3^+(k_z)\overline{K}_1(\gamma_3 r)e^{\gamma_3(d-r)} + \overline{K}_1(\gamma_3 a)\overline{I}_1(\gamma_3 r)e^{[\text{abs}(\text{real}(\gamma_3 r))-\gamma_3 a]}]$$
$$= (-j\omega\mu_3 a)[\overline{f}_3^-(k_z)\overline{I}_1(\gamma_3 r)e^{\alpha_3(r-e)}$$
$$+ \overline{f}_3^+(k_z)\overline{K}_1(\gamma_3 r)e^{\gamma_3(d-r)} + \overline{K}_1(\gamma_3 a)\overline{I}_1(\gamma_3 r)e^{(\alpha_3 r-\gamma_3 a)}]$$

$$\tilde{E}_{\varphi 4}(r, k_z) = (-j\omega\mu_3 a)[f_3^-(k_z)I_1(\gamma_3 r) + f_3^+(k_z)K_1(\gamma_3 r) + I_1(\gamma_3 a)K_1(\gamma_3 r)]$$
$$= (-j\omega\mu_3 a)[\overline{f}_3^-(k_z)e^{-\text{abs}(\text{real}(\gamma_3 e))}\overline{I}_1(\gamma_3 r)e^{\gamma_3 r}$$
$$+ \overline{f}_3^+(k_z)e^{\gamma_3 d}\overline{K}_1(\gamma_3 r)e^{-\gamma_3 r} + \overline{I}_1(\gamma_3 a)e^{(\text{abs}(\text{real}(\gamma_3 a)))}\overline{K}_1(\gamma_3 r)e^{-\gamma_3 r}]$$
$$= (-j\omega\mu_3 a)[\overline{f}_3^-(k_z)\overline{I}_1(\gamma_3 r)e^{\gamma_3 r-\text{abs}(\text{real}(\gamma_3 e))}$$
$$+ \overline{f}_3^+(k_z)\overline{K}_1(\gamma_3 r)e^{\gamma_3(d-r)} + \overline{I}_1(\gamma_3 a)\overline{K}_1(\gamma_3 r)e^{[\text{abs}(\text{real}(\gamma_3 a))-\gamma_3 r]}]$$
$$= (-j\omega\mu_3 a)[\overline{f}_3^-(k_z)\overline{I}_1(\gamma_3 r)e^{(\gamma_3 r-\alpha_3 e)}$$
$$+ \overline{f}_3^+(k_z)\overline{K}_1(\gamma_3 r)e^{\gamma_3(d-r)} + \overline{I}_1(\gamma_3 a)\overline{K}_1(\gamma_3 r)e^{(\alpha_3 a-\gamma_3 r)}]$$

$$\tilde{E}_{\varphi 5}(r, k_z) = (-j\omega\mu_4 a)f_4(k_z)K_1(\gamma_4 r)$$
$$= (-j\omega\mu_4 a)\overline{f}_4(k_z)e^{-\gamma_4 e}\overline{K}_1(\gamma_4 r)e^{-\gamma_4 r}$$
$$= (-j\omega\mu_4 a)\overline{f}_4(k_z)\overline{K}_1(\gamma_4 r)e^{-\gamma_4 r-\gamma_4 e}$$

and the magnetic fields are given by

$$\tilde{H}_{z1}(r, k_z) = a\gamma_1 f_1(k_z) I_0(\gamma_1 r)$$
$$= a\gamma_1 \bar{f}_1(k_z) e^{-\text{abs}(\text{real}(\gamma_1 b))} \bar{I}_0(\gamma_1 r) e^{\text{abs}(\text{real}(\gamma_1 r))}$$
$$= a\gamma_1 \bar{f}_1(k_z) \bar{I}_0(\gamma_1 r) e^{[\text{abs}(\text{real}(\gamma_1 r)) - \text{abs}(\text{real}(\gamma_1 b))]}$$
$$= a\gamma_1 \bar{f}_1(k_z) \bar{I}_0(\gamma_1 r) e^{[\alpha_1(r-b)]}$$

$$\tilde{H}_{z2}(r, k_z) = a\gamma_2 [f_2^-(k_z) I_0(\gamma_2 r) - f_2^+(k_z) K_0(\gamma_2 r)]$$
$$= a\gamma_2 [\bar{f}_2^-(k_z) e^{-\text{abs}(\text{real}(\gamma_2 d))} \bar{I}_0(\gamma_2 r) e^{\text{abs}(\text{real}(\gamma_2 r))} - \bar{f}_2^+(k_z) e^{\gamma_2 b} \bar{K}_0(\gamma_2 r) e^{-\gamma_2 r}]$$
$$= a\gamma_2 [\bar{f}_2^-(k_z) \bar{I}_0(\gamma_2 r) e^{[\text{abs}(\text{real}(\gamma_2 r)) - \text{abs}(\text{real}(\gamma_2 d))]} - \bar{f}_2^+(k_z) \bar{K}_0(\gamma_2 r) e^{\gamma_2 (b-r)}]$$
$$= a\gamma_2 [\bar{f}_2^-(k_z) \bar{I}_0(\gamma_2 r) e^{[\alpha_2(r-d)]} - \bar{f}_2^+(k_z) \bar{K}_0(\gamma_2 r) e^{\gamma_2 (b-r)}]$$

$$\tilde{H}_{z3}(r, k_z) = a\gamma_3 [f_3^-(k_z) I_0(\gamma_3 r) - f_3^+(k_z) K_0(\gamma_3 r) + K_1(\gamma_3 a) I_0(\gamma_3 r)]$$
$$= a\gamma_3 [\bar{f}_3^-(k_z) e^{-\text{abs}(\text{real}(\gamma_3 e))} \bar{I}_0(\gamma_3 r) e^{\text{abs}(\text{real}(\gamma_3 r))}$$
$$- \bar{f}_3^+(k_z) e^{\gamma_3 d} \bar{K}_0(\gamma_3 r) e^{-\gamma_3 r} + \bar{K}_1(\gamma_3 a) e^{-\gamma_3 a} \bar{I}_0(\gamma_3 r) e^{\text{abs}(\text{real}(\gamma_3 r))}]$$
$$= a\gamma_3 [\bar{f}_3^-(k_z) \bar{I}_0(\gamma_3 r) e^{\text{abs}(\text{real}(\gamma_3 r)) - \text{abs}(\text{real}(\gamma_3 e))}$$
$$- \bar{f}_3^+(k_z) \bar{K}_0(\gamma_3 r) e^{\gamma_3 (d-r)} + \bar{K}_1(\gamma_3 a) \bar{I}_0(\gamma_3 r) e^{\text{abs}(\text{real}(\gamma_3 r)) - \gamma_3 a}]$$
$$= a\gamma_3 [\bar{f}_3^-(k_z) \bar{I}_0(\gamma_3 r) e^{[\alpha_3(r-e)]}$$
$$- \bar{f}_3^+(k_z) \bar{K}_0(\gamma_3 r) e^{\gamma_3 (d-r)} + \bar{K}_1(\gamma_3 a) \bar{I}_0(\gamma_3 r) e^{\alpha_3 r - \gamma_3 a}]$$

$$\tilde{H}_{z4}(r, k_z) = a\gamma_3 [f_3^-(k_z) I_0(\gamma_3 r) - f_3^+(k_z) K_0(\gamma_3 r) - I_1(\gamma_3 a) K_0(\gamma_3 r)]$$
$$= a\gamma_3 [\bar{f}_3^-(k_z) e^{-\text{abs}(\text{real}(\gamma_3 e))} \bar{I}_0(\gamma_3 r) e^{\text{abs}(\text{real}(\gamma_3 r))}$$
$$- \bar{f}_3^+(k_z) e^{\gamma_3 d} \bar{K}_0(\gamma_3 r) e^{-\gamma_3 r} - \bar{I}_1(\gamma_3 a) e^{\text{abs}(\text{real}(\gamma_3 a))} \bar{K}_0(\gamma_3 r) e^{-\gamma_3 r}]$$
$$= a\gamma_3 [\bar{f}_3^-(k_z) \bar{I}_0(\gamma_3 r) e^{\text{abs}(\text{real}(\gamma_3 r)) - \text{abs}(\text{real}(\gamma_3 e))}$$
$$- \bar{f}_3^+(k_z) \bar{K}_0(\gamma_3 r) e^{\gamma_3 d - \gamma_3 r} - \bar{I}_1(\gamma_3 a) \bar{K}_0(\gamma_3 r) e^{\text{abs}(\text{real}(\gamma_3 a)) - \gamma_3 r}]$$
$$= a\gamma_3 [\bar{f}_3^-(k_z) \bar{I}_0(\gamma_3 r) e^{[\alpha_3(r-e)]}$$
$$- \bar{f}_3(k_z) \bar{K}_0(\gamma_3 r) e^{\gamma_3 (d-r)} - \bar{I}_1(\gamma_3 a) \bar{K}_0(\gamma_3 r) e^{(\alpha_3 a - \gamma_3 r)}]$$

$$\tilde{H}_{z5}(r, k_z) = a\gamma_4 [-f_4(k_z) K_0(\gamma_4 r)]$$
$$= a\gamma_4 [-\bar{f}_4(k_z) e^{-\gamma_4 e} \bar{K}_0(\gamma_4 r) e^{-\gamma_4 r}]$$
$$= a\gamma_4 [-\bar{f}_4(k_z) \bar{K}_0(\gamma_4 r) e^{-\gamma_4 r - \gamma_4 e}]$$

APPENDIX C DERIVATION FOR THE EXPRESSION OF ELECTRICAL FIELD FOR ARBITRARY CYLINDRICAL LAYERED FORMATIONS IN SPECTRAL DOMAIN

The electrical fields in each layer for arbitrary-layered formations are given by

$$\tilde{E}_{\varphi 1}(r,k_z) = (-j\omega\mu_1 R)f_1(k_z)I_1(\gamma_1 r)$$
$$= (-j\omega\mu_1 R)\bar{f}_1(k_z)e^{-\text{abs}(\text{real}(\gamma_1 B(1)))}\bar{I}_1(\gamma_1 r)e^{\text{abs}(\text{real}(\gamma_1 r))}$$
$$= (-j\omega\mu_1 R)\bar{f}_1(k_z)\bar{I}_1(\gamma_1 r)e^{[\text{abs}(\text{real}(\gamma_1 r))-\text{abs}(\text{real}(\gamma_1 B(1)))]}$$
$$= (-j\omega\mu_1 R)\bar{f}_1(k_z)\bar{I}_1(\gamma_1 r)e^{[\alpha_1(r-B(1))]}$$

$$\tilde{E}_{\varphi 2}(r,k_z) = (-j\omega\mu_2 R)[f_2^-(k_z)I_1(\gamma_2 r) + f_2^+(k_z)K_1(\gamma_2 r)]$$
$$= (-j\omega\mu_2 R)[\bar{f}_2^-(k_z)e^{-\text{abs}(\text{real}(\gamma_2 B(2)))}\bar{I}_1(\gamma_2 r)e^{\text{abs}(\text{real}(\gamma_2 r))} + \bar{f}_2^+(k_z)e^{\gamma_2 B(1)}\bar{K}_1(\gamma_2 r)e^{-\gamma_2 r}]$$
$$= (-j\omega\mu_2 R)[\bar{f}_2^-(k_z)\bar{I}_1(\gamma_2 r)e^{[\text{abs}(\text{real}(\gamma_2 r))-\text{abs}(\text{real}(\gamma_2 B(2)))]} + \bar{f}_2^+(k_z)\bar{K}_1(\gamma_2 r)e^{\gamma_2(B(1)-r)}]$$
$$= (-j\omega\mu_2 R)[\bar{f}_2^-(k_z)\bar{I}_1(\gamma_2 r)e^{[\alpha_2(r-B(2))]} + \bar{f}_2^+(k_z)\bar{K}_1(\gamma_2 r)e^{\gamma_2(B(1)-r)}]$$

$$\tilde{E}_{\varphi 3}(r,k_z) = (-j\omega\mu_3 R)[f_3^-(k_z)I_1(\gamma_3 r) + f_3^+(k_z)K_1(\gamma_3 r) + K_1(\gamma_3 R)I_1(\gamma_3 r)]$$
$$= (-j\omega\mu_3 R)[\bar{f}_3^-(k_z)e^{-\text{abs}(\text{real}(\gamma_3 B(3)))}\bar{I}_1(\gamma_3 r)e^{\text{abs}(\text{real}(\gamma_3 r))}$$
$$+ \bar{f}_3^+(k_z)e^{\gamma_3 B(2)}\bar{K}_1(\gamma_3 r)e^{-\gamma_3 r} + \bar{K}_1(\gamma_3 R)e^{-\gamma_3 R}\bar{I}_1(\gamma_3 r)e^{\text{abs}(\text{real}(\gamma_3 r))}]$$
$$= (-j\omega\mu_3 R)[\bar{f}_3^-(k_z)\bar{I}_1(\gamma_3 r)e^{\text{abs}(\text{real}(\gamma_3 r))-\text{abs}(\text{real}(\gamma_3 B(3)))}$$
$$+ \bar{f}_3^+(k_z)\bar{K}_1(\gamma_3 r)e^{\gamma_3(B(2)-r)} + \bar{K}_1(\gamma_3 R)\bar{I}_1(\gamma_3 r)e^{[\text{abs}(\text{real}(\gamma_3 r))-\gamma_3 R]}]$$
$$= (-j\omega\mu_3 R)[\bar{f}_3^-(k_z)\bar{I}_1(\gamma_3 r)e^{[\alpha_3(r-B(3))]}$$
$$+ \bar{f}_3^+(k_z)\bar{K}_1(\gamma_3 r)e^{\gamma_3(B(2)-r)} + \bar{K}_1(\gamma_3 R)\bar{I}_1(\gamma_3 r)e^{[\alpha_3 r-\gamma_3 R]}]$$

$$\tilde{E}_{\varphi 4}(r,k_z) = (-j\omega\mu_3 R)[f_3^-(k_z)I_1(\gamma_3 r) + f_3^+(k_z)K_1(\gamma_3 r) + I_1(\gamma_3 R)K_1(\gamma_3 r)]$$
$$= (-j\omega\mu_3 R)[\bar{f}_3^-(k_z)e^{-\text{abs}(\text{real}(\gamma_3 B(3)))}\bar{I}_1(\gamma_3 r)e^{\gamma_3 r}$$
$$+ \bar{f}_3^+(k_z)e^{\gamma_3 B(2)}\bar{K}_1(\gamma_3 r)e^{-\gamma_3 r} + \bar{I}_1(\gamma_3 R)e^{(\text{abs}(\text{real}(\gamma_3 R)))}\bar{K}_1(\gamma_3 r)e^{-\gamma_3 r}]$$
$$= (-j\omega\mu_3 R)[\bar{f}_3^-(k_z)\bar{I}_1(\gamma_3 r)e^{\gamma_3 r-\text{abs}(\text{real}(\gamma_3 B(3)))}$$
$$+ \bar{f}_3^+(k_z)\bar{K}_1(\gamma_3 r)e^{\gamma_3(B(2)-r)} + \bar{I}_1(\gamma_3 R)\bar{K}_1(\gamma_3 r)e^{[\text{abs}(\text{real}(\gamma_3 R))-\gamma_3 r]}]$$
$$= (-j\omega\mu_3 R)[\bar{f}_3^-(k_z)\bar{I}_1(\gamma_3 r)e^{[\gamma_3 r-\alpha_3 B(3)]}$$
$$+ \bar{f}_3^+(k_z)\bar{K}_1(\gamma_3 r)e^{\gamma_3(B(2)-r)} + \bar{I}_1(\gamma_3 R)\bar{K}_1(\gamma_3 r)e^{[\alpha_3 R-\gamma_3 r]}]$$

⋮

$$\tilde{E}_{\varphi n}(r,k_z) = (-j\omega\mu_{n-1}R)[f_{n-1}^-(k_z)I_1(\gamma_{n-1}r) + f_{n-1}^+(k_z)K_1(\gamma_{n-1}r)]$$

$$= (-j\omega\mu_{n-1}R)[\overline{f}_{n-1}^-(k_z)e^{-\mathrm{abs}(\mathrm{real}(\gamma_{n-1}B(n-1)))}\overline{I}_1(\gamma_{n-1}r)e^{\gamma_{n-1}r}$$

$$+ \overline{f}_{n-1}^+(k_z)e^{\gamma_{n-1}B(n-2)}\overline{K}_1(\gamma_{n-1}r)e^{-\gamma_{n-1}r}]$$

$$= (-j\omega\mu_{n-1}R)[\overline{f}_{n-1}^-(k_z)\overline{I}_1(\gamma_{n-1}r)e^{\gamma_{n-1}r-\mathrm{abs}(\mathrm{real}(\gamma_{n-1}B(n-1)))}$$

$$+ \overline{f}_{n-1}^+(k_z)\overline{K}_1(\gamma_{n-1}r)e^{\gamma_{n-1}(B(n-2)-r)}]$$

$$= (-j\omega\mu_{n-1}R)[\overline{f}_{n-1}^-(k_z)\overline{I}_1(\gamma_{n-1}r)e^{[\gamma_{n-1}r-\alpha_{n-1}B(n-1)]}$$

$$+ \overline{f}_{n-1}^+(k_z)\overline{K}_1(\gamma_{n-1}r)e^{\gamma_{n-1}(B(n-2)-r)}]$$

$$\tilde{E}_{\varphi(n+1)}(r,k_z) = (-j\omega\mu_n R)f_n(k_z)K_1(\gamma_n r)$$

$$= (-j\omega\mu_n R)\overline{f}_n(k_z)e^{-\gamma_n B(n-1)}\overline{K}_1(\gamma_n r)e^{-\gamma_n r}$$

$$= (-j\omega\mu_n R)\overline{f}_n(k_z)\overline{K}_1(\gamma_n r)e^{-\gamma_n r-\gamma_n B(n-1)}$$

The magnetic fields in each layer for arbitrary-layered formations are given by

$$\tilde{H}_{z1}(r,k_z) = a\gamma_1 f_1(k_z)I_0(\gamma_1 r)$$

$$= a\gamma_1 \overline{f}_1(k_z)e^{-\mathrm{abs}(\mathrm{real}(\gamma_1 B(1)))}\overline{I}_0(\gamma_1 r)e^{\mathrm{abs}(\mathrm{real}(\gamma_1 r))}$$

$$= a\gamma_1 \overline{f}_1(k_z)\overline{I}_0(\gamma_1 r)e^{[\mathrm{abs}(\mathrm{real}(\gamma_1 r))-\mathrm{abs}(\mathrm{real}(\gamma_1 B(1)))]}$$

$$= a\gamma_1 \overline{f}_1(k_z)\overline{I}_0(\gamma_1 r)e^{[\alpha_1(r-B(1))]}$$

$$\tilde{H}_{z2}(r,k_z) = a\gamma_2[f_2^-(k_z)I_0(\gamma_2 r) - f_2^+(k_z)K_0(\gamma_2 r)]$$

$$= a\gamma_2[\overline{f}_2^-(k_z)e^{-\mathrm{abs}(\mathrm{real}(\gamma_2 B(2)))}\overline{I}_0(\gamma_2 r)e^{\mathrm{abs}(\mathrm{real}(\gamma_2 r))} - \overline{f}_2^+(k_z)e^{\gamma_2 B(1)}\overline{K}_0(\gamma_2 r)e^{-\gamma_2 r}]$$

$$= a\gamma_2[\overline{f}_2^-(k_z)\overline{I}_0(\gamma_2 r)e^{[\mathrm{abs}(\mathrm{real}(\gamma_2 r))-\mathrm{abs}(\mathrm{real}(\gamma_2 B(2)))]} - \overline{f}_2^+(k_z)\overline{K}_0(\gamma_2 r)e^{\gamma_2(B(1)-r)}]$$

$$= a\gamma_2[\overline{f}_2^-(k_z)\overline{I}_0(\gamma_2 r)e^{[\alpha_2(r-B(2))]} - \overline{f}_2^+(k_z)\overline{K}_0(\gamma_2 r)e^{\gamma_2(B(1)-r)}]$$

$$\tilde{H}_{z3}(r,k_z) = a\gamma_3[f_3^-(k_z)I_0(\gamma_3 r) - f_3^+(k_z)K_0(\gamma_3 r) + K_1(\gamma_3 R)I_0(\gamma_3 r)]$$

$$= a\gamma_3[\overline{f}_3^-(k_z)e^{-\mathrm{abs}(\mathrm{real}(\gamma_3 B(3)))}\overline{I}_0(\gamma_3 r)e^{\mathrm{abs}(\mathrm{real}(\gamma_3 r))}$$

$$- \overline{f}_3^+(k_z)e^{\gamma_3 B(2)}\overline{K}_0(\gamma_3 r)e^{-\gamma_3 r} + \overline{K}_1(\gamma_3 R)e^{-\gamma_3 a}\overline{I}_0(\gamma_3 r)e^{\mathrm{abs}(\mathrm{real}(\gamma_3 r))}]$$

$$= a\gamma_3[\overline{f}_3^-(k_z)\overline{I}_0(\gamma_3 r)e^{\mathrm{abs}(\mathrm{real}(\gamma_3 r))-\mathrm{abs}(\mathrm{real}(\gamma_3 B(3)))}$$

$$- \overline{f}_3^+(k_z)\overline{K}_0(\gamma_3 r)e^{\gamma_3(B(2)-r)} + \overline{K}_1(\gamma_3 R)\overline{I}_0(\gamma_3 r)e^{\mathrm{abs}(\mathrm{real}(\gamma_3 r))-\gamma_3 R}]$$

$$= a\gamma_3[\overline{f}_3^-(k_z)\overline{I}_0(\gamma_3 r)e^{[\alpha_3(r-B(3))]}$$

$$- \overline{f}_3^+(k_z)\overline{K}_0(\gamma_3 r)e^{\gamma_3(B(2)-r)} + \overline{K}_1(\gamma_3 R)\overline{I}_0(\gamma_3 r)e^{(\alpha_3 r-\gamma_3 R)}]$$

$$\tilde{H}_{z4}(r, k_z) = a\gamma_3[f_3^-(k_z)I_0(\gamma_3 r) - f_3^+(k_z)K_0(\gamma_3 r) - I_1(\gamma_3 R)K_0(\gamma_3 r)]$$
$$= a\gamma_3[\bar{f}_3^-(k_z)e^{-\text{abs}(\text{real}(\gamma_3 B(3)))}\bar{I}_0(\gamma_3 r)e^{\text{abs}(\text{real}(\gamma_3 r))}$$
$$- \bar{f}_3^+(k_z)e^{\gamma_3 B(2)}\bar{K}_0(\gamma_3 r)e^{-\gamma_3 r} - \bar{I}_1(\gamma_3 R)e^{\text{abs}(\text{real}(\gamma_3 R))}\bar{K}_0(\gamma_3 r)e^{-\gamma_3 r}]$$
$$= a\gamma_3[\bar{f}_3^-(k_z)\bar{I}_0(\gamma_3 r)e^{\text{abs}(\text{real}(\gamma_3 r)) - \text{abs}(\text{real}(\gamma_3 B(3)))}$$
$$- \bar{f}_3(k_z)\bar{K}_0(\gamma_3 r)e^{\gamma_3(B(2)-r)} - \bar{I}_1(\gamma_3 R)\bar{K}_0(\gamma_3 r)e^{\text{abs}(\text{real}(\gamma_3 R)) - \gamma_3 r}]$$
$$= a\gamma_3[\bar{f}_3^-(k_z)\bar{I}_0(\gamma_3 r)e^{[\alpha_3(r-B(3))]}$$
$$- \bar{f}_3(k_z)\bar{K}_0(\gamma_3 r)e^{\gamma_3(B(2)-r)} - \bar{I}_1(\gamma_3 R)\bar{K}_0(\gamma_3 r)e^{\alpha_3 R - \gamma_3 r}]$$
$$\vdots$$
$$\tilde{H}_{zn}(r, k_z) = a\gamma_{n-1}[f_{n-1}^-(k_z)I_0(\gamma_{n-1} r) - f_{n-1}^+(k_z)K_0(\gamma_{n-1} r)]$$
$$= a\gamma_{n-1}[\bar{f}_{n-1}^-(k_z)e^{-\text{abs}(\text{real}(\gamma_{n-1} B(n-1)))}\bar{I}_0(\gamma_{n-1} r)e^{\text{abs}(\text{real}(\gamma_{n-1} r))}$$
$$- \bar{f}_{n-1}^+(k_z)e^{\gamma_{n-1} B(n-2)}\bar{K}_0(\gamma_{n-1} r)e^{-\gamma_{n-1} r}]$$
$$= a\gamma_{n-1}[\bar{f}_{n-1}^-(k_z)\bar{I}_0(\gamma_{n-1} r)e^{\text{abs}(\text{real}(\gamma_{n-1} r)) - \text{abs}(\text{real}(\gamma_{n-1} B(n-1)))}$$
$$- \bar{f}_{n-1}(k_z)\bar{K}_0(\gamma_{n-1} r)e^{\gamma_{n-1}(B(n-2)-r)}]$$
$$= a\gamma_{n-1}[\bar{f}_{n-1}^-(k_z)\bar{I}_0(\gamma_{n-1} r)e^{[\alpha_{n-1}(r-B(n-1))]} - \bar{f}_{n-1}(k_z)\bar{K}_0(\gamma_{n-1} r)e^{\gamma_{n-1}(B(n-2)-r)}]$$
$$\tilde{H}_{z(n+1)}(r, k_z) = a\gamma_n[-f_n(k_z)K_0(\gamma_n r)]$$
$$= a\gamma_n[-\bar{f}_n(k_z)e^{-\gamma_n B(n-1)}\bar{K}_0(\gamma_n r)e^{-\gamma_n r}]$$
$$= a\gamma_n[-\bar{f}_n(k_z)\bar{K}_0(\gamma_n r)e^{-\gamma_n r - \gamma_n B(n-1)}]$$

CHAPTER 8

Induction and Logging-While-Drilling Resistivity Tool Response in a Two-Dimensional Isotropic Formation

Contents

8.1 Introduction	251
8.2 Formulations	252
8.3 Numerical Consideration	263
8.4 Verifications	265
8.5 Array Induction Logs	270
8.6 Measurement-While-Drilling Logs	273
8.7 Simulation of Effects of Mandrel Grooves on MWD Conductivity Logs	281
8.7.1 Theoretical MWD models	281
8.7.2 The groove effects in a homogeneous medium	283
8.7.3 The groove effects in a formation with a borehole	284
8.7.4 The effects of the conversion table	287
8.8 Summary	292
References	293

8.1 INTRODUCTION

In the previous chapters, we discussed the induction and LWD tool responses in a homogenous and a one-dimensional formations including horizontally layered and cylindrically layered formations. In well logging, formations with both cylindrical and planar boundaries are often encountered. In general, this case is a three-dimensional (3D) problem and analytical method is difficult to apply. However, if the tool is centered and the borehole is vertical, the system will be axially symmetrical, in which case an analytical solution can be found. In this chapter, we will discuss the solutions of the induction and logging-while-drilling (LWD) resistivity tool response in a vertically and radially layered formation without eccentricity. Many simulation methods to model the induction and LWD logs in such formations have been developed. If we assume the vertically layered formation has axial symmetry, the 3D problem is simplified to a two-dimensional (2D) problem. Due to the absorption of electromagnetic (EM) waves in the formation, the waves propagating in the formation decay exponentially. A forward computation method may be unstable if the wave propagation is not treated properly. In this chapter, we present a stable algorithm to solve this problem by using a horizontal eigenmode expansion method. This method was first discussed by Chew [1–4] and Pai [5,6].

The eigenmode expansion method is an extension of the method of separation of variables. The method can be used in two different ways: to expand the field into horizontal eigenmodes (in ρ direction) or into vertical eigenmodes (in z direction). Chew et al. [1] first proposed to expand the ρ dependence of the field into a series of eigenstates to solve a single vertical boundary problem numerically. Later, this method was developed to analyze complicated borehole environments with multilayer formations [2–4]. This method proved to be much more efficient than the finite element method [7]. Pai [5,6] and Li and Shen [8] solved the same problem using a vertical eigenmode expansion method. In Pai's work, the field is expressed in terms of vertical eigenmodes which are obtained analytically. Horizontal modes propagate in the vertical direction while vertical modes propagate in the horizontal direction. Modes are reflected at the layer boundaries. Mathematical description of the wave propagation and the reflection from layer to layer determines the stability of the algorithm.

In this chapter, we use horizontal eigenstates to solve the 2D well logging problem in a formation with both horizontal and vertical layers. The wave propagation from layer to layer is modeled by a three-layer module, which guarantees the stability of the computation. Each horizontal layer is divided into subregions in the ρ direction. The eigenmodes are found numerically in each layer, and the fields are expressed as a sum of the eigenmodes.

8.2 FORMULATIONS

Fig. 8.1 shows the configuration of an induction tool in a formation with multilayer, multiinvasion zones. The source is a coil at the center of the borehole carrying a uniform current. Comparing this structure with that in Fig. 5.1, we notice that there are both cylindrical and horizontal layers in Fig. 8.1. Chapter 5, Triaxial Induction Tool and Logging-While-Drilling Tool Response in a Transverse Isotropic-Layered Formation and Chapter 6 assume infinite layer boundaries in the z and ρ directions. Chapter 7, Induction and Logging-While-Drilling Tool Response in a Cylindrically Layered Isotropic Formation, assumes boundaries in the radial direction only but no horizontal layers. In this chapter, we explore the analytical method that can handle both radial and horizontal layers. Note that the cylindrical symmetry of the system makes the E field have only one component, which is E_ϕ. Due to the azimuthal symmetry of

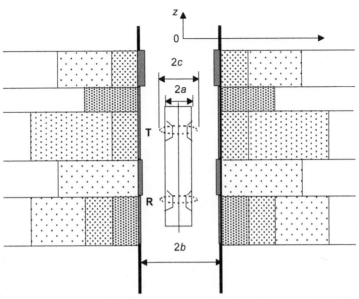

Figure 8.1 A formation with a borehole (in the middle), vertical layers, and multiinvasion zones. The radii of transmitting and receiving coils are c. The radius of the mandrel is a. The radius of borehole is b.

the formation, and assume that the magnetic permeability is constant everywhere, the electrical field in the formation satisfies Eq. (7.4) and can be written as:

$$\nabla \times \nabla \times \overline{\mathbf{E}} - k^2 \overline{\mathbf{E}} = -j\omega\mu\overline{\mathbf{J}} \qquad (8.1)$$

where μ is the magnetic permeability, and $\overline{\mathbf{J}}$ is the current density in the transmitting coil. It is assumed that $\overline{\mathbf{J}}$ is the harmonic current and has only a ϕ component. For a coil carrying a current I having a radius of ρ_t and located at z_t, which is the same case with Eq. (7.6). Therefore by rearranging the terms in Eq. (7.6), Eq. (8.1) can be written as

$$\left[\frac{\partial^2}{\partial \rho^2} + \frac{1}{\rho}\frac{\partial}{\partial \rho} - \frac{1}{\rho^2} + \frac{\partial^2}{\partial z^2} + k^2\right] E_\phi = j\omega\mu I \delta(\rho - \rho_t)\delta(z - z_t) \qquad (8.2)$$

where δ is the Dirac delta function. The wave number k is defined as

$$k^2 = \omega^2 \mu \varepsilon^* \qquad (8.3)$$

where

$$\varepsilon^* = \varepsilon - j\frac{\sigma}{\omega} \qquad (8.4)$$

The dielectric permittivity and conductivity of the medium are ε and σ, respectively. Generally, both ε and σ are functions of ρ and z.

In the horizontal layer i, the solution to Eq. (8.2) can be expressed as a series of horizontal eigenstates,

$$E_{\phi i} = \sum_{\alpha=1}^{N} \left[a_{\alpha i} e^{-k_{\alpha i} z} + b_{\alpha i} e^{jk_{\alpha i} z} \right] \frac{1}{\rho} f_{\alpha i}(\rho) \qquad (8.5)$$

where $a_{\alpha i}$ and $b_{\alpha i}$ are unknown constants independent of ρ and z, $k_{\alpha i}$ is the eigenvalue and $f_{\alpha i}(\rho)$ is the eigenfunction of mode i. The $k_{\alpha i}$ has a negative imaginary part. Therefore the terms $a_{\alpha i} e^{-jk_{\alpha i} z}$ and $b_{\alpha i} e^{jk_{\alpha i} z}$ represent decaying waves propagating in the positive z and negative z directions, respectively. It follows that

$$\left[\frac{\partial^2}{\partial \rho^2} + \frac{1}{\rho} \frac{\partial}{\partial \rho} - \frac{1}{\rho^2} + \frac{\partial^2}{\partial z^2} + k^2 \right] f_{\alpha i}(\rho) = 0 \qquad (8.6)$$

The function $f_{\alpha i}(\rho)$ is defined over a finite region $[\rho_{\min}, \rho_{\max}]$, and satisfies the following boundary conditions:

$$f_{\alpha i}(\rho_{\min} = 0) = 0, \quad \text{for induction tool} \qquad (8.7a)$$

$$f_{\alpha i}(\rho_{\min} = \rho_m) = 0, \quad \text{for MWD tool} \qquad (8.7b)$$

and

$$f_{\alpha i}(\rho_{\max}) = 0 \qquad (8.7c)$$

where ρ_m is the radius of the mandrel of the measurement-while-drilling (MWD) tool.

To obtain the solution of Eq. (8.6), a numerical scheme is used so that solutions can be sought for μ and ε, which are independent of ρ. First, a set of basis functions, which are complete over the interval, is used to expand the function as

$$f_{\alpha i}(\rho) = \sum_{n=1}^{\infty} b_{\alpha n} g_n(\rho) \qquad (8.8)$$

The boundary conditions on $g_n(\rho)$ are

$$g_n(\rho_{\min} = 0) = 0, \quad \text{for induction tool} \qquad (8.9a)$$

$$g_n(\rho_{\min} = \rho_m) = 0, \quad \text{for MWD tool} \qquad (8.9b)$$

$$g_n(\rho_{\max}) = 0 \qquad (8.9c)$$

In lossy media, due to the absorption of formation to EM waves, the waves propagating in the formation decay exponentially. Beyond a certain distance, the waves become very weak. Therefore the summation in Eq. (8.8) can be truncated to its first N terms. Substituting Eq. (8.8) into (8.6), it can be rewritten as

$$\sum_{n=1}^{N} b_{\alpha n}\left[\frac{\partial^2}{\partial \rho^2} + \frac{1}{\rho}\frac{\partial}{\partial \rho} - \frac{1}{\rho^2} + \frac{\partial^2}{\partial z^2} + k^2\right] g_n(\rho) = 0 \qquad (8.10)$$

The inner product is defined as

$$\langle g_m, g_n \rangle = \int_0^\infty d\rho \frac{1}{\rho\mu} g_m(\rho) g_n(\rho) \qquad (8.11)$$

Eq. (8.10) is multiplied by $(1/\rho\mu) g_m(\rho)$ and integrated from 0 to ∞, one obtains a matrix equation as follows:

$$\sum_{n=1}^{N} b_{\alpha n}[B_{nm} - k_{\alpha i}^2 G_{nm}] = 0 \qquad (8.12)$$

where

$$B_{nm} = \left\langle g_m, \rho\mu \frac{\partial}{\partial \rho} \frac{1}{\rho\mu} \frac{\partial}{\partial \rho} g_n \right\rangle + \omega^2 \langle g_m, \mu\varepsilon g_n \rangle \qquad (8.13)$$

and

$$G_{nm} = \langle g_m, g_n \rangle \qquad (8.14)$$

Using integration by parts, it is shown that

$$B_{nm} = \int_0^\infty d\rho \frac{1}{\rho\mu} g'_m(\rho) g'_n(\rho) + \omega^2 \int_0^\infty d\rho \frac{\varepsilon}{\rho} g_m(\rho) g_n(\rho) \qquad (8.15)$$

where the primes indicate derivatives with respect to the argument of the function. Hence, with the definition of the inner product in Eq. (8.11), B_{nm} and G_{nm} are symmetric matrices. They are also the matrix representations of the differential operator in Eq. (8.6). The symmetry of B_{nm} and G_{nm} is a consequence of the self-adjointness or reciprocal nature of the problem. Therefore g_m and g_n are not necessarily orthogonal to each other. The eigenvalues $k_{\alpha i}$ in Eq. (8.12) are obtained by solving

$$[\overline{G}^{-1} \cdot \overline{B} - k_{\alpha i}^2 \overline{I}] \overline{b}_\alpha = 0 \qquad (8.16)$$

where
- $\overline{B}$ is a matrix of $N \times N$ with $B_{nm} = B_{mn}$,
- $\overline{G}$ is a matrix of $N \times N$ with $G_{nm} = G_{mn}$, and
- $\overline{b}_\alpha$ is an eigenvector.

Eigenvectors obtained from Eq. (8.16) are orthogonal with respect to $\overline{G}$:

$$\overline{b}_\alpha^t \overline{G} \overline{b}_\beta = \delta_{\alpha\beta} D_\alpha \tag{8.17}$$

Because of this, it also can be proved that the eigenfunctions satisfy the orthogonality relation:

$$\int_{\rho_{\min}}^{\rho_{\max}} \frac{1}{\rho\mu} f_{\alpha i}(\rho) f_{\beta i}(\rho) d\rho = \sum_{n=1}^{N} \sum_{m=1}^{N} \overline{b}_{\alpha n} \overline{b}_{m\beta}^t \int_0^\infty d\rho \frac{1}{\rho\mu} g_n(\rho) g_m(\rho)$$
$$= \sum_{n=1}^{N} \sum_{m=1}^{N} \overline{b}_{\alpha n} G_{nm} b_{m\beta}^t = \delta_{\alpha\beta} D_\alpha \tag{8.18}$$

Usually, the finite region $[\rho_{\min}, \rho_{\max}]$ is divided into subregions. Triangular functions may be used as basis functions in the subregions. A series of triangular functions is shown in Fig. 8.2.

To satisfy the condition at ρ_0, a quadratic function is chosen in the first subregion as

$$g_1(\rho) = \begin{cases} \dfrac{\rho^2}{\rho_1^2} & 0 < \rho \leq \rho_1 \\ \dfrac{\rho - \rho_2}{\rho_1 - \rho_2} & \rho_1 < \rho \leq \rho_2 \end{cases} \tag{8.19}$$

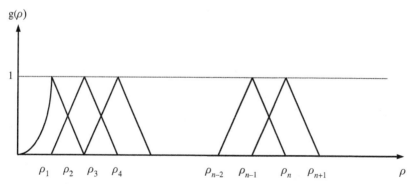

Figure 8.2 A series of triangular functions as a basis function set.

In other subregions, the basis functions are chosen as

$$g_k(\rho) = \begin{cases} \dfrac{\rho - \rho_{k-1}}{\rho_k - \rho_{k-1}} & \rho_{k-1} < \rho \le \rho_k \\ \dfrac{\rho - \rho_{k+1}}{\rho_k - \rho_{k+1}} & \rho_k < \rho \le \rho_{k+1} \end{cases} \quad (8.20)$$

With the availability of the basis functions, the elements of the matrix $\overline{G}$ and $\overline{B}$ are found by

$$G_{11} = \frac{1}{\mu}\left\{\frac{1}{4} + \frac{1}{(\rho_1 - \rho_2)^2}\left[\frac{1}{2}(\rho_2^2 - \rho_1^2) - 2\rho_2(\rho_2 - \rho_1) + \rho_2^2 \ln\left(\frac{\rho_2}{\rho_1}\right)\right]\right\} \quad (8.21a)$$

$$G_{ii} = \frac{1}{\mu(\rho_i - \rho_{i-1})^2}\left[\frac{1}{2}(\rho_i^2 - \rho_{i-1}^2) - 2\rho_{i-1}(\rho_i - \rho_{i-1}) + \rho_{i-1}^2 \ln\left(\frac{\rho_i}{\rho_{i-1}}\right)\right]$$
$$+ \frac{1}{\mu(\rho_i - \rho_{i+1})^2}\left[\frac{1}{2}(\rho_{i+1}^2 - \rho_i^2) - 2\rho_{i+1}(\rho_{i+1} - \rho_i) + \rho_{i+1}^2 \ln\left(\frac{\rho_{i+1}}{\rho_i}\right)\right] \quad (8.21b)$$

$$G_{ii+1} = \frac{1}{\mu(\rho_i - \rho_{i+1})^2}\left[\frac{1}{2}(\rho_{i+1}^2 - \rho_i^2) - \rho_i\rho_{i+1} \ln\left(\frac{\rho_{i+1}}{\rho_i}\right)\right] \quad (8.21c)$$

$$G_{i+1i} = G_{ii+1} \quad (8.21d)$$

and

$$B_{11} = -\frac{2}{\mu\rho_1^2} + \frac{1}{\mu(\rho_1 - \rho_2)^2}\ln\left(\frac{\rho_2}{\rho_1}\right)$$
$$+ \omega^2\left\{\frac{\varepsilon_1}{4} + \frac{\varepsilon_2}{(\rho_1 - \rho_2)^2}\left[\frac{1}{2}(\rho_2^2 - \rho_1^2) - 2\rho_2(\rho_2 - \rho_1) + \rho_2^2 \ln\left(\frac{\rho_2}{\rho_1}\right)\right]\right\} \quad (8.22a)$$

$$B_{ii} = -\frac{1}{\mu(\rho_i - \rho_{i-1})^2}\ln\left(\frac{\rho_i}{\rho_{i-1}}\right) - \frac{1}{\mu(\rho_{i+1} - \rho_i)^2}\ln\left(\frac{\rho_{i+1}}{\rho_i}\right)$$
$$+ \frac{\omega^2\varepsilon_i}{(\rho_i - \rho_{i-1})^2}\left[\frac{1}{2}(\rho_i^2 - \rho_{i-1}^2) - 2\rho_{i-1}(\rho_i - \rho_{i-1}) + \rho_{i-1}^2 \ln\left(\frac{\rho_i}{\rho_{i-1}}\right)\right] \quad (8.22b)$$
$$+ \frac{\omega^2\varepsilon_{i+1}}{(\rho_i - \rho_{i+1})^2}\left[\frac{1}{2}(\rho_{i+1}^2 - \rho_i^2) - 2\rho_{i+1}(\rho_{i+1} - \rho_i) + \rho_{i+1}^2 \ln\left(\frac{\rho_{i+1}}{\rho_i}\right)\right]$$

$$B_{ii+1} = \frac{1}{\mu(\rho_i-\rho_{i+1})^2}\ln\left(\frac{\rho_{i+1}}{\rho_i}\right) + \frac{\omega^2\varepsilon_{i+1}}{(\rho_i-\rho_{i+1})^2}\left[\frac{1}{2}(\rho_{i+1}^2-\rho_i^2) - \rho_i\rho_{i+1}\ln\left(\frac{\rho_{i+1}}{\rho_i}\right)\right] \quad (8.22c)$$

$$B_{i+1i} = B_{ii+1} \quad (8.22d)$$

With the availability of the eigenfunctions and eigenvalues, an approximate solution to Eq. (8.2) can be found. First, a homogeneous formation is considered. In this case, the electric field is expanded in terms of eigenfunctions:

$$\rho E_\phi = \sum_{\alpha=1}^{N} a_\alpha f_\alpha(\rho) \quad (8.23)$$

where a_α is a function of z. Substituting Eq. (8.23) into (8.2), gives

$$\sum_{\alpha=1}^{N}\left[\frac{\partial^2}{\partial\rho^2} + \frac{1}{\rho}\frac{\partial}{\partial\rho} - \frac{1}{\rho^2} + \frac{\partial^2}{\partial z^2} + k^2\right] a_\alpha f_\alpha(\rho) = j\omega\mu I\rho\delta(\rho-\rho_t)\delta(z-z_t). \quad (8.24)$$

The two sides of Eq. (8.24) are multiplied by $(1/\rho\mu)f_\beta(\rho)$ and integrated from $\rho_{\min}$ to ∞, then

$$\sum_{\alpha=1}^{N}\left\{\frac{\partial^2}{\partial z^2}\left[a_\alpha \int_{\rho_{\min}}^{\infty}\frac{d\rho}{\rho\mu}f_\beta(\rho)f_\alpha(\rho)\right] a_\alpha \left[\int_{\rho_{\min}}^{\infty}\frac{d\rho}{\rho\mu}f_\beta(\rho)\left(\frac{\partial^2}{\partial\rho^2} + \frac{1}{\rho}\frac{\partial}{\partial\rho} - \frac{1}{\rho^2} + k^2\right)f_\alpha(\rho)\right]\right\}$$

$$= j\omega I\delta(z-z_t)f_\beta(\rho_t) \quad (8.25)$$

Since

$$\int_{\rho_{\min}}^{\infty}\frac{d\rho}{\rho\mu}f_\beta(\rho)\left(\frac{\partial^2}{\partial\rho^2} + \frac{1}{\rho}\frac{\partial}{\partial\rho} - \frac{1}{\rho^2} + k^2\right)f_\alpha(\rho) = k_\beta^2\int_{\rho_{\min}}^{\infty}\frac{d\rho}{\rho\mu}f_\beta(\rho)f_\alpha(\rho) = k_\beta^2\delta_{\alpha\beta}D_\alpha \quad (8.26)$$

Eq. (8.25) becomes

$$\left[\frac{\partial^2}{\partial z^2} + k_\beta^2\right]a_\beta = j\omega I\delta(z-z_t)f_\beta(\rho_t)D_\beta^{-1} \quad (8.27)$$

Solving Eq. (8.27) yields

$$a_\beta = \frac{-\omega I f_\beta(\rho_t)}{2k_\beta D_\beta} e^{jk_\beta|z-z_t|} \tag{8.28}$$

The approximate solution to Eq. (8.2) is

$$\rho E_\phi = -\frac{\omega I}{2} \sum_{\alpha=1}^{N} \frac{f_\alpha(\rho_t) f_\alpha(\rho)}{k_\alpha D_\alpha} e^{jk_\beta|z-z_t|} \tag{8.29}$$

The accuracy of the solution can be increased if more terms in the expansion of Eq. (8.23) are used.

In the presence of a horizontal layer boundary at z_0 between layer i and layer $i+1$ shown in Fig. 8.3, the tangential electric field E_ϕ and magnetic field H_ρ satisfy the following boundary conditions:

$$E_{\phi i}\big|_{z=z_0} = E_{\phi i+1}\big|_{z=z_0} \tag{8.30}$$

$$-\frac{1}{j\omega\mu_i}\frac{\partial E_{\phi i}}{\partial z}\bigg|_{z=z_0} + \frac{1}{j\omega\mu_{i+1}}\frac{\partial E_{\phi i+1}}{\partial z}\bigg|_{z=z_0} = -J_\phi\big|_{z=z_0} \tag{8.31}$$

where subscripts i and $i+1$ denote variables in layer i and layer $i+1$, respectively.

Substituting Eq. (8.5) into (8.30) and (8.31) and taking inner products, the coefficients $a_{\alpha i}$ and $b_{\alpha i}$ in layer i can be expressed as functions of $a_{\alpha i+1}$ and $b_{\alpha i+1}$:

$$a_{\beta i} e^{-jk_{\beta i} z_i} = \sum_{\alpha=1}^{N} \frac{1}{2D_{\beta i} k_{\beta i}} \left\{ a_{\alpha i+1} e^{-jk_{\alpha i+1} z_i} \left[k_{\beta i} \left(\overline{C}_{\beta i+1}^1\right)^t \overline{G}_i^t \overline{C}_{\alpha i} + k_{\beta i+1} \left(\overline{C}_{\beta i+1}\right)^t \overline{G}_{i+1}^t \overline{C}_{\alpha i} \right] \right.$$
$$\left. \times b_{\alpha i+1} e^{jk_{\alpha i+1} z_i} \left[k_{\beta i} \left(\overline{C}_{\beta i+1}\right)^t \overline{G}_i^t \overline{C}_{\alpha i} - k_{\beta i+1} \left(\overline{C}_{\beta i+1}\right)^t \overline{G}_{i+1}^t \overline{C}_{\alpha i} \right] \right\} + \overline{F}_i$$
$$\tag{8.32}$$

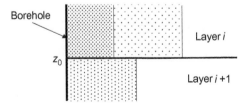

Figure 8.3 Boundary condition.

$$b_{\beta i}e^{jk_{\beta i}z_i} = \sum_{\alpha=1}^{N} \frac{1}{2D_{\beta i}k_{\beta i}} \left\{ a_{\alpha i+1}e^{-jk_{\alpha i+1}z_i} \left[k_{\beta i}\left(\overline{C}_{\beta i+1}\right)^t \overline{G}_i^t \overline{C}_{\alpha i} - k_{\beta i+1}\left(\overline{C}_{\beta i+1}\right)^t \overline{G}_{i+1}^t \overline{C}_{\alpha i} \right] \right.$$

$$\left. + b_{\alpha i+1}e^{jk_{\alpha i+1}z_i} \left[k_{\beta i}(\overline{C}_{\beta i+1})^t \overline{G}_i^t \overline{C}_{\alpha i} + k_{\beta i+1}(\overline{C}_{\beta i+1})^t \overline{G}_{i+1}^t \overline{C}_{\alpha i} \right] \right\} - \overline{F}_i$$

(8.33)

where

$$D_{\beta i} = \frac{1}{\mu} \int_{\rho_{min}}^{\rho_{max}} J_\phi \frac{1}{\rho} f_{\beta i}^2(\rho) d\rho = \overline{C}_{\beta i}^t \overline{G}_i^t \overline{C}_{\beta i} \qquad (8.34)$$

$$\overline{F}_i = \int_{\rho_{min}}^{\rho_{max}} J_\phi(z_0) \frac{\omega}{2k_{\beta i}D_{\beta i}} f_{\beta i}(\rho) d\rho \qquad (8.35)$$

The vectors $\overline{C}_{\alpha i}$ and $\overline{C}_{\alpha i+1}$ are eigenvectors corresponding to eigenvalues $k_{\alpha i}$ and $k_{\alpha i+1}$ in layer i and $i+1$, respectively. The $\overline{C}_\alpha^t$ is the transpose of $\overline{C}_\alpha$. The matrix $\overline{G}$ is called base matrix whose elements are

$$G_{\alpha\beta} = \int_{\rho_{min}}^{\rho_{max}} \frac{1}{\mu\rho} g_\alpha(\rho) g_\beta(\rho) d\rho \qquad (8.36)$$

Define the wave vectors as

$$\overline{V}(z_i) = [a_1 e^{-jk_{1i}z_i} \ldots \ldots a_N e^{-jk_{Ni}z_i}]^t \qquad (8.37)$$

$$\overline{U}(z_i) = [b_1 e^{jk_{1i}z_i} \ldots \ldots b_N e^{jk_{Ni}z_i}]^t \qquad (8.38)$$

where $\overline{V}(z_i)$ and $\overline{U}(z_i)$ represent the fields propagating in positive z and negative z directions, respectively.

Eqs. (8.10) and (8.11) can be expressed in a matrix form:

$$\begin{bmatrix} \overline{V}^i(z_i) \\ \overline{U}^i(z_i) \end{bmatrix} = \begin{bmatrix} \overline{C}_{dd} & \overline{C}_{du} \\ \overline{C}_{ud} & \overline{C}_{uu} \end{bmatrix} \begin{bmatrix} \overline{V}^{i+1}(z_i) \\ \overline{U}^{i+1}(z_i) \end{bmatrix} + \begin{bmatrix} \overline{F}_i \\ -\overline{F}_i \end{bmatrix} \qquad (8.39)$$

where

$$(\overline{C}_{dd})_{\alpha\beta} = \frac{1}{2D_{\beta i}k_{\beta i}} \left[k_{\beta i}(\overline{C}_{\beta i+1})^t \overline{G}_i^t \overline{C}_{\alpha i} + k_{\beta i+1}(\overline{C}_{\beta i+1})^t \overline{G}_{i+1}^t \overline{C}_{\alpha i} \right] \qquad (8.40)$$

$$\left(\overline{C}_{du}\right)_{\alpha\beta} = \frac{1}{2D_{\beta i}k_{\beta i}} \left[k_{\beta i}\left(\overline{C}_{\beta i+1}\right)^t \overline{G}_i^t \overline{C}_{\alpha i} - k_{\beta i+1}\left(\overline{C}_{\beta i+1}\right)^t \overline{G}_{i+1}^t \overline{C}_{\alpha i} \right] \qquad (8.41)$$

$$\left(\overline{C}_{ud}\right)_{\alpha\beta} = \left(\overline{C}_{du}\right)_{\alpha\beta} \qquad (8.42)$$

$$\left(\overline{C}_{uu}\right)_{\alpha\beta} = \left(\overline{C}_{dd}\right)_{\alpha\beta} \qquad (8.43)$$

At each vertical boundary, the fields are reflected and transmitted. If $\overline{V}^i(z)$ represents the incident field and there is no source at the boundary, $\overline{U}^i(z)$ is zero. The matrix equation, (8.39), is reduced to

$$\begin{bmatrix} \overline{V}^i(z_i) \\ \overline{U}^i(z_i) \end{bmatrix} = \begin{bmatrix} \overline{C}_{dd} & \overline{C}_{du} \\ \overline{C}_{ud} & \overline{C}_{uu} \end{bmatrix} \begin{bmatrix} \overline{V}^{i+1}(z_i) \\ 0 \end{bmatrix} \qquad (8.44)$$

Then, we have

$$\overline{U}^i(z) = \overline{C}_{du}\overline{C}_{uu}^{-1}\overline{V}^i(z) = \overline{R}_{i,i+1}\overline{V}^i(z) \qquad (8.45)$$

$$\overline{V}^{i+1}(z) = \overline{C}_{uu}^{-1}\overline{V}^i(z) = \overline{T}_{i,i+1}\overline{V}^i(z) \qquad (8.46)$$

where $\overline{C}_{uu}^{-1}$ is the inverse of matrix $\overline{C}_{uu}$. The $\overline{R}_{i,i+1}$ and $\overline{T}_{i,i+1}$ are defined as reflection and transmission matrices at the vertical boundary i. In general, $\overline{R}_{i,i+1}$ and $\overline{T}_{i,i+1}$ are not diagonal except in the case the formations are only horizontally layered. The matrix nature of Eqs. (8.45) and (8.46) implies that each incident eigenmode will generate a complete set of modes in the transmitted and reflected waves.

Once the eigenmodes in each layer are found, the wave propagation in layers is converted to mode propagation. All modes attenuate in the direction in which they propagate. Higher modes decay faster than lower modes. Therefore after a certain propagation distance, the lower modes are the dominant components. Note that transmitted waves keep the same direction of propagation when passing through a layer boundary while reflected waves change directions. One way to describe the mode propagating through layers is to categorize the waves into "up-going" (positive z direction) and "down-going" (negative z direction) waves, and then match the boundary conditions at layer boundaries to find unknown coefficients [1–4,7]. The above method, though mathematically elegant, may be numerically unstable. The instability is caused by the way the wave propagation is treated. As the up-going or down-going waves include exponentially increasing and decreasing modes simultaneously, the coefficients may be well beyond the maximum and minimum of a computer. Any kind of truncation will result in severe instability of the

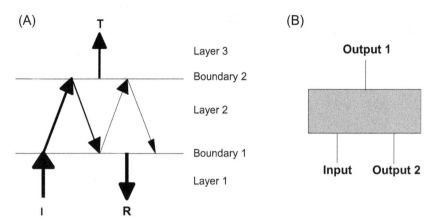

Figure 8.4 (A) The three-layer model. (B) The three-port circuit.

algorithm. For example, $e^{50} \times e^{-50} = 1.0$, if the real e^{-50} is truncated to e^{-30}, then result will be changed to e^{20} much greater than 1.0, the expected value. Therefore, after a series of calculations, the result may be beyond the maximum of the computer and result in a floating point error.

To avoid the instability problem, the wave must be traced so that only exponentially decaying waves are involved in the computation. Consider a three-layer formation shown in Fig. 8.4A. The vector I represents the incident field, the vector R is defined as reflected field from boundary 1, which is a summation of reflected field of incident I at the boundary 1 and all transmitted fields in layer 2. The vector T is defined as transmitted field from boundary 2, which is a summation of transmitted field of all modes in layer 2. This three-layer model can be described as a three-port network as shown in Fig. 8.4B. Input, Output 1, and Output 2 represent I, T, and R, respectively. In the lossy medium, the field decays in the propagating direction. Therefore, after a certain number of reflections between boundaries 1 and 2, the field becomes negligible. The accuracy of the T and R can be increased by taking more reflections between boundaries. In each of the three-layer modules, the computation of T and R traces the wave until the wave is negligible. Using the wave-tracing algorithm, the computation is guaranteed to be convergent.

In the multilayer case, layers are partitioned into a series of three-layer components. The three-layer model is computed for each module. Fig. 8.5 shows the strategy of simulating a multilayer case using the three-layer modules. The vector I represents the incident field from the transmitter. In comparing the output $X1$ of the module 1 with the incident wave I, if the difference between them is less than a given tolerance ε_d, then the calculation is stopped. If not, module 2 is implemented for calculation. If the difference between $Z1$ and I is greater than the ε_d, a third stage is automatically

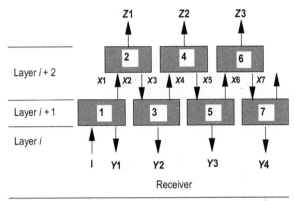

Figure 8.5 The dynamic network module to compute multilayer formation using a three-layer model.

added. This process continues until the criterion is satisfied. Similarly, the reflected field **Y2** can be calculated by using module 3. If the field **X3** is too large, module 4 and module 5 may be necessary. The same is true for modules 6 and 7. Combining **Y**1, **Y**2, **Y**3, and **Y**4, the field at the position of the receiver can be obtained. For the layers below the receiver, the procedures are the same. Since all the modes decay in the process, this method is very stable.

8.3 NUMERICAL CONSIDERATION

The eigenvectors can be obtained from Eq. (8.16), which can be used to derive the eigenfunctions in Eq. (8.8). If N basis functions are used in Eq. (8.8), Eq. (8.16) will produce N eigenvalues and eigenvectors. Hence, there are N eigenfunctions available. With the availability of the eigenfunctions, eigenvectors, and eigenvalues, the approximate solution to Eq. (8.2) can be obtained. Therefore the selection of the basis function is very important. Usually, the triangular functions are used as basis functions which are equivalent to the piecewise linear approximation of the field. When it is assumed that the inhomogeneous formation around the borehole are piecewise constant functions, the integrations in Eqs. (8.11) and (8.15) can be performed analytically. This enables the matrix elements to be computed efficiently. The number of basis functions is one of the keys determining the computation accuracy. Table 8.1 shows the comparison of apparent conductivities computed from different number of basis functions. The formation profile is shown in Fig. 8.6 and the tool is the 2C40, which has one transmitter and one receiver separated by 40 in. Table 8.1 shows that the accuracy of the computation depends on the number of eigenmodes. Generally speaking, the more eigenmodes, the more accurate the result

Table 8.1 Comparison of the apparent conductivities computed from different numbers of eigenmodes

Depth (ft)	Apparent conductivity (mho/m)			
	No. of mode = 20	No. of mode = 30	No. of mode = 40	No. of mode = 50
−10	0.794890332	0.802456596	0.815997171	0.816167546
−9.5	0.793447136	0.801011542	0.814527555	0.814699495
−9	0.791494125	0.799057786	0.812540576	0.812713911
−8.5	0.788851441	0.796415939	0.809853747	0.810028182
−8	0.785262805	0.792830021	0.806206629	0.806381641
−7.5	0.780353031	0.78792446	0.801217055	0.801391747
−7	0.7735541	0.781128976	0.794304685	0.794477623
−6.5	0.763969136	0.77153996	0.784550021	0.784719135
−6	0.750105282	0.757648175	0.770417732	0.770580496
−5.5	0.729313176	0.736771855	0.749181457	0.749335522
−5	0.696618453	0.703940797	0.715790753	0.715934771
−4.5	0.645219897	0.652964203	0.663896229	0.664030231
−4	0.584512182	0.591994373	0.601898445	0.602022908
−3.5	0.522664553	0.529470138	0.538330826	0.538445754
−3	0.460953757	0.467010332	0.474823554	0.474928886
−2.5	0.399907138	0.405213221	0.411990307	0.412086164
−2	0.340131312	0.344756612	0.350525722	0.350612507
−1.5	0.283160314	0.28751852	0.292323723	0.292402086
−1	0.237688765	0.242467652	0.246456774	0.246527421
−0.5	0.214051228	0.21869786	0.22228398	0.222348721
0	0.206947335	0.211537565	0.215001235	0.215063734
0.5	0.214051228	0.21869786	0.22228398	0.222348721
1	0.237688765	0.242467652	0.246456774	0.246527421
1.5	0.283160314	0.28751852	0.292323723	0.292402086
2	0.340131312	0.344756612	0.350525722	0.350612507
2.5	0.399907138	0.405213221	0.411990307	0.412086164
3	0.460953757	0.467010332	0.474823554	0.474928886
3.5	0.522664553	0.529470138	0.538330826	0.538445754
4	0.584512182	0.591994373	0.601898445	0.602022908
4.5	0.645219897	0.652964203	0.663896229	0.664030231
5	0.696618453	0.703940797	0.715790753	0.715934771
5.5	0.729313176	0.736771855	0.749181457	0.749335522
6	0.750105282	0.757648175	0.770417732	0.770580496
6.5	0.763969136	0.77153996	0.784550021	0.784719135
7	0.7735541	0.781128976	0.794304685	0.794477623
7.5	0.780353031	0.78792446	0.801217055	0.801391747
8	0.785262805	0.792830021	0.806206629	0.806381641
8.5	0.788851441	0.796415939	0.809853747	0.810028182
9	0.791494125	0.799057786	0.812540576	0.812713911
9.5	0.793447136	0.801011542	0.814527555	0.814699495
10	0.794890332	0.802456596	0.815997171	0.816167546

The formation is shown in Fig. 8.6 and the tool is the 2C40 which has one transmitter and one receiver separated by 40 in.

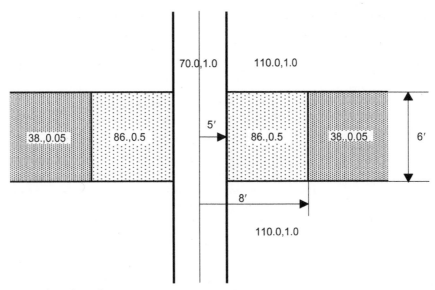

Figure 8.6 A three-layer formation.

is. Moreover, the result is consistent if 40 or more eigenmodes are used. Therefore a minimum of 40 eigenmodes is usually needed.

In theory, it is necessary to approximate the field with triangular functions over an infinite region in the ρ direction, i.e., from the center of the borehole (induction tool) or the metal surface of MWD tool $\rho = \rho_{min}$ to $\rho = \infty$. But, in the lossy media, due to the absorption of formation to EM waves, the waves propagating in the formation decay exponentially. Therefore it is only needed to approximate the field over a finite region. Table 8.2 shows the comparison of the apparent conductivities computed from some different regions. The formation is shown in Fig. 8.6 and tool is the 2C40. It is seen that the accuracy increases when large region is taken, and the variation is very small when the region is bigger than 10 m. In the proposed algorithm, the region is set to 15 m. Also, since the measurement of the field is done in the borehole, it is required to approximate the field more accurately close to the borehole. This can be achieved by choosing a small step size for the triangular function inside and close to the borehole while a larger step size is used far from the borehole.

8.4 VERIFICATIONS

To verify the proposed algorithm, induction tool responses to a thin layer sandwiched between two shoulder beds shown in Fig. 8.7 are simulated. The formation shown in Fig. 8.7A has a 5-in. radius borehole, an invasion zone of 8 in., and a bed thickness of 1 ft. The dielectric permittivities and conductivities are 70.0 and 1.0 mho/m for

Table 8.2 Comparison of the apparent conductivities computed from different radial distances

Depth (ft)	Apparent conductivity (mho/m)				
	$\rho_{max} = 5.0$ m	$\rho_{max} = 7.0$ m	$\rho_{max} = 10.0$ m	$\rho_{max} = 15.0$ m	$\rho_{max} = 20.0$ m
−10	0.790956433	0.807204376	0.816066248	0.815997171	0.814141163
−9.5	0.789340982	0.805554415	0.814565895	0.814527555	0.812666984
−9	0.787214616	0.803381992	0.812545442	0.812540576	0.810673778
−8.5	0.784396805	0.8005051	0.809822089	0.809853747	0.807978354
−8	0.78062909	0.796663776	0.806135007	0.806206629	0.80431946
−7.5	0.775531313	0.791476394	0.801101638	0.801217055	0.799314263
−7	0.768525268	0.784363491	0.794141534	0.794304685	0.792382599
−6.5	0.758694036	0.77440755	0.784336141	0.784550021	0.782607683
−6	0.744506045	0.76007806	0.770153864	0.770417732	0.768463074
−5.5	0.723241019	0.738659454	0.748877999	0.749181457	0.747243911
−5	0.689857517	0.705120648	0.715476142	0.715790753	0.713938271
−4.5	0.638012521	0.65312753	0.663611554	0.663896229	0.662218139
−4	0.576094196	0.591064175	0.601662171	0.601898445	0.600427384
−3.5	0.512624671	0.527453188	0.538148532	0.538330826	0.537071282
−3	0.449229082	0.463922958	0.474699873	0.474823554	0.473778115
−2.5	0.386516824	0.401085967	0.411929912	0.411990307	0.41116146
−2	0.325178173	0.339635649	0.350533402	0.350525722	0.349915958
−1.5	0.267101376	0.281463893	0.292403477	0.292323723	0.291933778
−1	0.221342411	0.23563	0.246599728	0.246456774	0.246258687
−0.5	0.197235391	0.211470375	0.222457789	0.22228398	0.222190162
0	0.189973996	0.204189337	0.215182531	0.215001235	0.214936508
0.5	0.197235391	0.211470375	0.222457789	0.22228398	0.222190162
1	0.221342411	0.23563	0.246599728	0.246456774	0.246258687
1.5	0.267101376	0.281463893	0.292403477	0.292323723	0.291933778
2	0.325178173	0.339635649	0.350533402	0.350525722	0.349915959
2.5	0.386516824	0.401085967	0.411929912	0.411990307	0.41116146
3	0.449229082	0.463922958	0.474699873	0.474823554	0.473778116
3.5	0.512624671	0.527453188	0.538148532	0.538330826	0.537071282
4	0.576094196	0.591064175	0.601662171	0.601898445	0.600427384
4.5	0.638012521	0.65312753	0.663611554	0.663896229	0.662218139
5	0.689857517	0.705120648	0.715476142	0.715790753	0.713938271
5.5	0.723241019	0.738659454	0.748877999	0.749181457	0.747243911
6	0.744506045	0.76007806	0.770153864	0.770417732	0.768463074
6.5	0.758694036	0.77440755	0.784336141	0.784550021	0.782607683
7	0.768525268	0.784363491	0.794141534	0.794304685	0.792382599
7.5	0.775531313	0.791476394	0.801101638	0.801217055	0.799314263
8	0.78062909	0.796663776	0.806135007	0.806206629	0.80431946
8.5	0.784396805	0.8005051	0.809822089	0.809853747	0.807978354
9	0.787214616	0.803381992	0.812545442	0.812540576	0.810673778
9.5	0.789340982	0.805554415	0.814565895	0.814527555	0.812666984
10	0.790956433	0.807204376	0.816066248	0.815997171	0.814141163

The formation is shown in Fig. 8.6 and the tool is the 2C40 which has on transmitter and one receiver separated by 40 in.

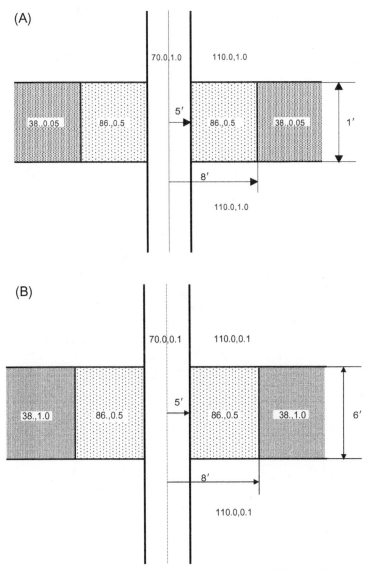

Figure 8.7 Three-layer formations (A) layer thickness is 1 foot and (B) layer thickness is 6 feet.

borehole, 110.0 and 1.0 mho/m for two shoulder beds, 86.0 and 0.5 for the invasion zone, and 38.0 and 0.05 for the thin bed, respectively. Shown in Fig. 8.8A is the computed apparent conductivity (denoted by σ_a) using a 2C40 tool. The 2C40 tool is a two-coil tool, which has one transmitter and one receiver separated by 40 in. The operating frequency of the tool is 20 kHz. It is assumed that the transmitting and receiving coils are all coaxial with the z axis. The computed log is also compared

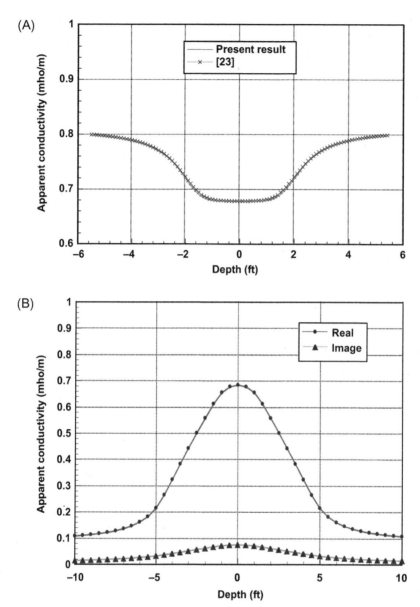

Figure 8.8 (A) Comparison of the computed logs obtained by using the proposed method and Ref. [8]. The tool is the 2C40 tool. The profile of the formation is shown in Fig. 8.7A. (B) The computed logs obtained by using the proposed method. The profile of the formation is shown in Fig. 8.7B.

with the one obtained by using a vertical eigenstate method [8] and the agreement is excellent. In this example, the number of modes used is 40, computation time is about 20 seconds on an Alpha DEC workstation for 120 logging points. Another case shown in Fig. 8.7B consists of 0.1 S/m mud. The computed logs are shown in Fig. 8.8B.

The second example is the comparison of the results from the proposed algorithm and the method described in Chapter 7, Induction and Logging-While-Drilling Tool Response in a Cylindrically Layered Isotropic Formation. The formation only consists of cylindrical layers without horizontal beds. The radius of the borehole is 5 in., and the thickness of invasion zone varies from 8 to 50 in. Here, the numbers of modes are chosen as 20 and 50, respectively. The tool is the 2C40. The results show that the relative error is less than 1% when 50 modes are used, but for 20 modes, the relative error is a little larger. However the latter is much faster than the former. Consequently, different modes can be selected according to the accuracy and speed requirements (Table 8.3).

Table 8.3 Comparison of the results computed from the HEM and the CIND method

Invasion zone (in.)	20 Modes (S/m)	50 Modes (S/m)	CIND (S/m)	Error for 20 modes (%)	Error for 50 modes (%)
8"	0.06596	0.06994	0.0707	6.7	1.07
10"	0.07617	0.08068	0.0812	6.19	0.64
12"	0.08839	0.09307	0.0935	5.14	0.46
14"	0.10166	0.10677	0.107	4.99	0.21
16"	0.11582	0.12133	0.1213	4.51	0.22
18"	0.13036	0.13628	0.136	4.14	0.2
20"	0.145	0.15125	0.1508	3.84	0.29
22"	0.1592	0.16598	0.1653	3.69	0.41
24"	0.17282	0.18027	0.1793	3.61	0.54
26"	0.18704	0.194	0.1928	2.99	0.62
28"	0.19953	0.20706	0.2057	3	0.66
30"	0.21132	0.21948	0.2179	3.02	0.73
32"	0.22241	0.23115	0.2294	3.05	0.76
34"	0.23283	0.24221	0.2402	3.07	0.83
36"	0.24262	0.25257	0.2505	3.15	0.83
38"	0.25331	0.26233	0.26	2.57	0.89
40"	0.26225	0.27147	0.269	2.51	0.92
42"	0.27066	0.28007	0.2775	2.46	0.93
44"	0.27858	0.28813	0.2854	2.4	0.96
46"	0.28603	0.29572	0.2929	2.35	0.96
48"	0.29306	0.30283	0.2999	2.28	0.98
50"	0.29973	0.30953	0.3065	2.21	0.99
Average error (%)				3.55	0.69

The CIND method can calculate the responses of the induction tool in the cylindrical medium.

8.5 ARRAY INDUCTION LOGS

The third example is the simulation of an array induction log in a formation with multilayer, multiinvasion zones shown in Fig. 8.9. The formation has a borehole, two shoulder beds, and multiinvasion zones. The radius of the borehole is 10 in. Some layers have a thickness of 1 ft while the others are 2-ft thick. The conductivities of each bed and shoulders are shown in Fig. 8.9 with the units of mhos per meter. The relative dielectric permittivities of all layers are set as 1.0. The array induction tool has four pairs of receiving coils and one transmitting coil as shown in Fig. 8.10. Shown in Fig. 8.11 are curves of real and imaginary parts of computed apparent conductivities corresponding to different pairs of receiving coils. The results show that a different investigation depth can be obtained by changing the distances between the transmitter and the receivers. From these curves, different interpretations for the formation can be reached. The pairs with shorter distances carry more information about the invasion zones, and pairs with longer distances bear more information about virgin zones. Therefore the array induction tool provides more detailed information about the formation and can be used to image the formation.

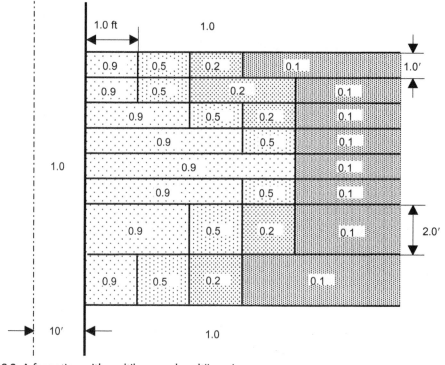

Figure 8.9 A formation with multilayer and multiinvasion zones.

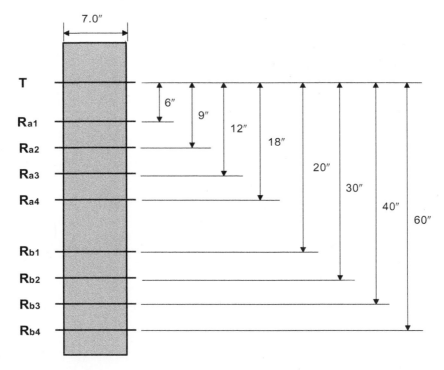

Number of turns:
Transmitter coil: 1
Receiver coils:

Coil	Ra	Rb	Rc	Rd
1	−18.31	−24.13	−28.19	−33.12
2	1	1	1	1

Figure 8.10 The configuration of an array induction tool.

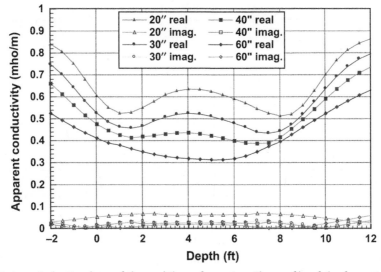

Figure 8.11 Array induction logs of the multilayer formation. The profile of the formation is shown in Fig. 8.9 and the tool is shown in Fig. 8.10.

272 Theory of Electromagnetic Well Logging

The fourth example is also the simulation of another kind of array induction log in the formation with multilayer, multiinvasion zones which is provided by the Shell Research, KSEPL. The profiles of borehole, invasion zones, and true formation are shown in Fig. 8.12, Fig. 8.13, and Fig. 8.14, respectively. The radius of the borehole is 4 in. Some layers in the formations are very thin, less than 0.5 ft. The contrast

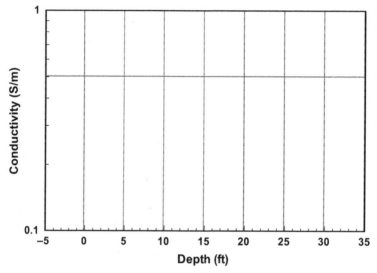

Figure 8.12 The profile of borehole in fourth example.

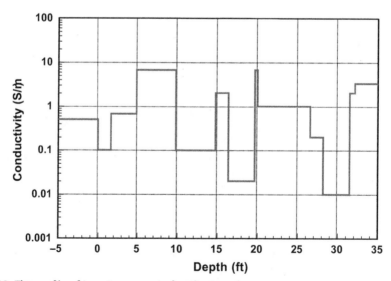

Figure 8.13 The profile of invasion zones in fourth example.

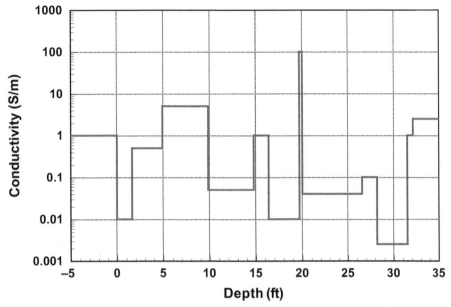

Figure 8.14 The profile of true formation in fourth example.

of conductivity to the neighbor formation is up to 1000 times. The relative dielectric permittivities of all layers are set as 1.0. The array induction tool has one pair of receiving coils and one transmitting coil as shown in Fig. 8.15. The turns of the receivers are 11.66351 and 10.00000, respectively. The turn of the transmitter is 1. The radii of the transmitter and receiver are 0.1 in. The operating frequency of the tool is 50 kHz. Shown in Fig. 8.16 is the curve of the computed apparent conductivity. From this example, it can be seen that the algorithm works very well and is stable even though the layer of the formation is very thin and the contrast is very large.

8.6 MEASUREMENT-WHILE-DRILLING LOGS

Several types of resistivity sensors are available on MWD tools. The coil-type tools used for MWD application are much different from induction tools in several ways, although both kinds of tools use similar coils to transmit and receive electromagnetic waves. First, the MWD tool is operated at 2 MHz instead of 20 kHz used by induction tools. Second, the MWD tool has been designed to account for the presence of a very conductive mandrel, whereas the ordinary induction tool has an insulating mandrel. Physically speaking, the electromagnetic fields satisfy the different boundary conditions which have been introduced in Section 8.2. A typical

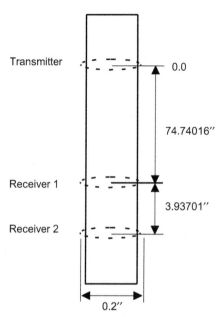

Figure 8.15 The parameters of an array induction tool used in fourth example (provided by the Shell Research, KSEPT).

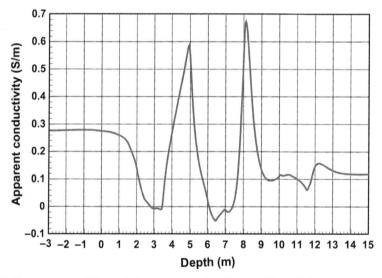

Figure 8.16 The computed log of fourth example. The profiles of formation are shown in Figs. 8.12–Fig. 8.14. The tool used is shown in Fig. 8.15.

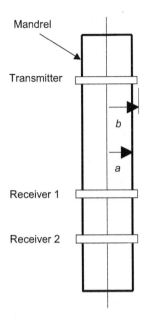

Figure 8.17 A coil-type MWD tool.

coil-type MWD tool is shown in Fig. 8.17. It consists of a transmitting coil of radius b placed over a conductive drill mandrel of radius a. The two receiving coils have identical radius and they are mounted concentrically around the mandrel. In some models, pairs of receiving and transmitting coils are placed symmetrically in a so-called borehole-compensated arrangement [9–20].

The MWD technique developed in this study is used to calculate the responses of MWD tool in multiinvasion beds. The MWD tool measures the phase difference and the amplitude ratio of the voltages at different pairs of receiving coils. Usually, present MWD logs do not display these raw data. Instead, some conversion algorithm is used to display the measured data in terms of apparent resistivities. In this study, the following method is used to create a conversion table, and the raw data are converted to the apparent conductivities according to this table. The amplitude ratio and the phase difference between the voltages at the two receiving coils are first obtained when the tool is in a homogeneous medium with various conductivity values. These data are obtained by computer simulation. Figs. 8.18 and 8.19 show the curves of amplitude ratio and phase difference as the functions of the conductivity, and are obtained from one kind of MWD tool shown in Fig. 8.20. The conductivity of the homogeneous medium varies from 0.001 to 5.0 S/m. When the tool is in actual operation, the phase difference and the amplitude ratio are measured, and then converted into apparent

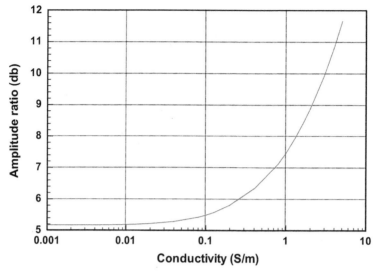

Figure 8.18 The amplitude ratio of the two receivers when the MWD tool is in a homogeneous medium. The specifications of MWD tool are shown in Fig. 8.20.

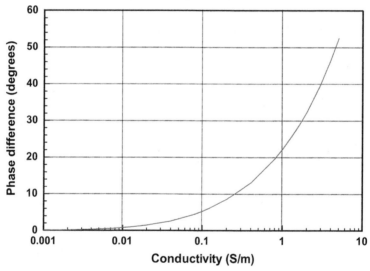

Figure 8.19 The phase difference between the two receivers when the MWD tool is in a homogeneous medium. The specifications of MWD tool are shown in Fig. 8.20.

conductivities by the conversion table. The terms σ_a and σ_p are used to represent apparent conductivities obtained this way using the amplitude ratio and the phase difference, respectively. Details of the principle and the conversion scheme are described in Chapter 4, Triaxial Induction and Logging-While-Drilling Resistivity Tool Response in Homogeneous Anisotropic Formations.

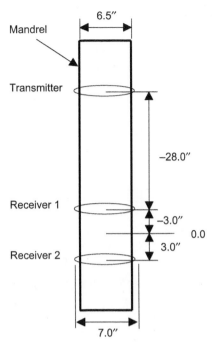

Figure 8.20 The specifications of an MWD tool.

Fig. 8.21 shows the computed MWD logs in a three-layer formation shown in Fig. 8.22. The tool has one transmitter and two receivers, also shown in Fig. 8.22. The curve symbolized by dots is the apparent conductivity converted from the amplitude ratio of the voltages in the two receivers, σ_a. The other is the apparent conductivity computed from the phase difference, σ_p. The logging range is from -5.0 to 5.0 ft.

Fig. 8.23 shows the computed MWD logs for an MWD tool having two transmitting coils and two receiving coils operating in the borehole-compensated mode. The specifications of the MWD tool are shown in Fig. 8.24. The formation is still a three-layer formation shown in Fig. 8.22. The logging range is from -60 to 60 in.

Fig. 8.25 shows the MWD logs for multilayer, multiinvasion zone formation shown in Fig. 8.26. The tool is the borehole-compensated MWD tool shown in Fig. 8.24. The formation has a borehole with a radius of 5 in. and a thickness of 2.0 feet in the center layers. The mud conductivity and dielectric permittivity are 10.0 S/m and 70.0 ε_0, respectively. The dielectric permittivities of all layers are 10.0 ε_0. The conductivities of every part in the formation are shown in Fig. 8.26. The logging range is from -10.0 to 10.0 ft.

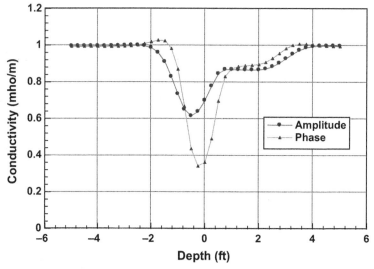

Figure 8.21 The MWD logs of the three-layer formation shown in Fig. 8.22.

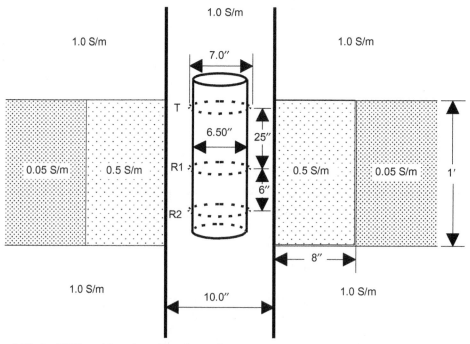

Figure 8.22 An MWD tool in a three-layer formation.

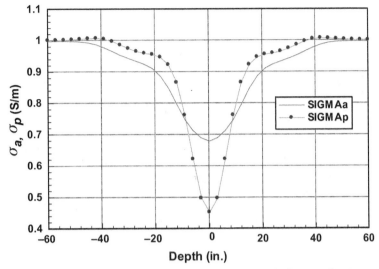

Figure 8.23 The MWD logs for a borehole-compensated MWD tool. The specifications of MWD tool are shown in Fig. 8.24.

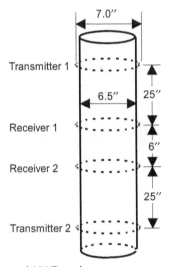

Figure 8.24 A borehole-compensated MWD tool.

Note that the simulation method presented above is based on a horizontal eigenmode expansion method. In this method, waves are treated as decaying waves. The three-layer module is used to solve a multilayer problem. Using this method, corresponding computer codes for the mixed-boundary problem are stable and versatile.

280 Theory of Electromagnetic Well Logging

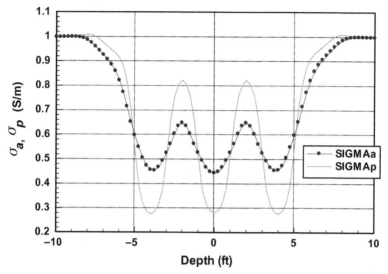

Figure 8.25 The MWD logs for a multilayer, multiinvasion zone formation shown in Fig. 8.26. The MWD tool is a borehole-compensated tool shown in Fig. 8.24.

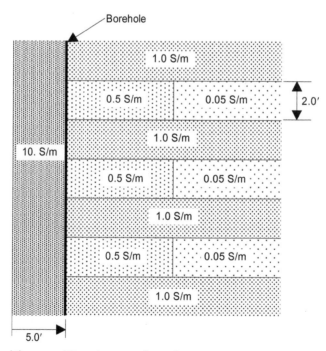

Figure 8.26 A multilayer, multiinvasion zone formation.

Theoretically, an infinite number of modes are needed. In practice, since the formation is lossy, the field decays exponentially in a few skin depths from the center of the borehole. Therefore only a finite number of modes are necessary. In the examples shown in this thesis, only 20 eigenmodes are used. Smaller grids are chosen near the borehole and greater grids in areas far from the borehole. From the simulated results, it is found that the developed algorithm is stable and has a satisfactory accuracy.

8.7 SIMULATION OF EFFECTS OF MANDREL GROOVES ON MWD CONDUCTIVITY LOGS

Many MWD tools carry coil-type conductivity sensors consisting of transmitting and receiving coils located in shallow grooves on the drill mandrel. The traditional analysis of this type of conductivity sensor neglects the effect of the groove. It only considers the constant diameter of the mandrel. In this chapter, the horizontal eigenstate method is used to perform the analysis of the effects of mandrel grooves.

8.7.1 Theoretical MWD models

The two MWD models are used in this chapter. These two models are shown in Fig. 8.27 and Fig. 8.28, respectively. They are all composed of one transmitting coil and two receiving coils wound around the steel mandrel. The first model shown in Fig. 8.27 has a uniform mandrel, i.e., the diameter of the mandrel is constant. This model is used in most MWD investigations. The second model shown in Fig. 8.28 has coils recessed in shallow grooves. This model accounts for the effects of the grooves. The drill mandrel has varying diameter and finite conductivity. It produces small but measurable effects in the phase and attenuation measured between the receivers. In the case of the uniform-mandrel MWD model, the horizontal eigenstate method can be directly used with satisfying the boundary conditions on the surface of mandrel, i.e., the tangential electric fields on the surface are equal to zero. In the case of the MWD tool with grooves, the boundary conditions are a little complex. In this situation, the horizontal eigenstate method cannot be used directly. To overcome this problem, a modified model shown in Fig. 8.29 may be used. In this modified model, it is assumed that the homogeneous media is around the tool and divided into seven layers. There are two regions in layers 1, 3, 5, and 7. The region 1 is part of the steel mandrel and treated as an invasion zone. The mandrel is characterized by $\sigma_m = 5 \times 10^6$ S/m, $\mu_m = 100\mu_0$, and $\varepsilon_m = 80\varepsilon_0$, where μ_0 and ε_0 are free-space magnetic permeability and dielectric permittivity, respectively. The region 2 is the formation. The layers 2, 4,

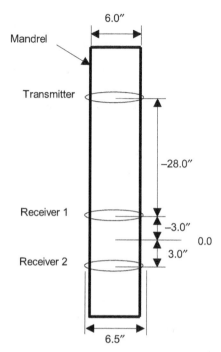

Figure 8.27 An MWD tool with constant-diameter mandrel.

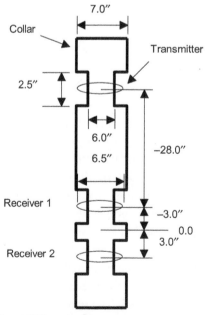

Figure 8.28 The geometry of an MWD tool with grooves.

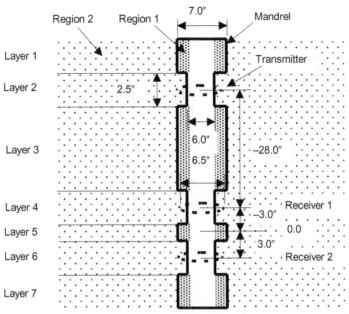

Figure 8.29 The modified model of an MWD tool. The mandrel parameter: $\sigma_m = 5 \times 10^6$ S/m, $\mu_m = 100\mu_0$, $\varepsilon_m = 80\varepsilon_0$.

and 6 only include the formation. In this chapter, the effects of the mandrel grooves are studied using this modified MWD model.

8.7.2 The groove effects in a homogeneous medium

To study the effects of the mandrel grooves, the tool is first examined in a homogeneous medium and the conductivity of the medium varies from 0.01 to 10 mho/m. The operating frequency of the MWD tool is 2 MHz. The amplitudes and phases of the electric fields are detected by receivers 1 and 2. Fig. 8.30A and B shows the amplitude ratios (db) and the phase differences (degrees), respectively. The amplitude curves shown in Fig. 8.30A are in terms of $20 \log(V_1/V_2)$, where V_1 and V_2 represent the induced voltages in the coils of receiver 1 and receiver 2, respectively. To demonstrate the details in the small amount of conductivity, the log scale is used on horizontal axis. Fig. 8.31 and Fig. 8.32 show the amplitudes and phases detected by receiver 1 and receiver 2, respectively. These figures show a great change effected by the MWD grooves on the amplitudes at receivers 1 and 2, individually, but the amplitude ratio compensates for this variation and has just a small difference. For the phases in both receivers and the phase difference, the changes are very small, and sometimes it cannot be distinguished. Therefore, in this case, the effects of the grooves are a little larger on the amplitude parts than on the phase parts.

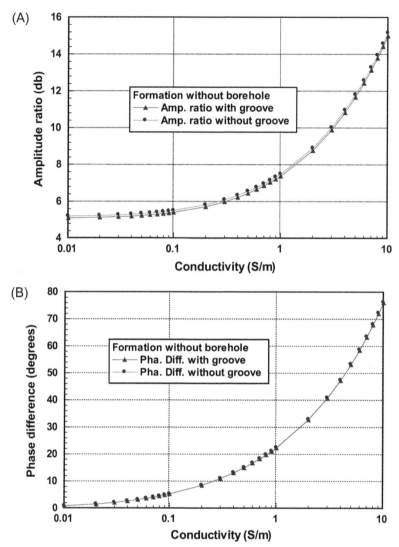

Figure 8.30 (A) Comparison of the amplitude ratios of the MWD tool with or without grooves in a homogeneous medium. (B) Comparison of the phase differences of the MWD tool with or without grooves in a homogeneous formation.

8.7.3 The groove effects in a formation with a borehole

Boreholes can significantly affect the quality of electromagnetic propagation measurements. Fig. 8.33 shows comparison of amplitudes ratios (dB) and phase differences of MWD tools with or without grooves. The formation is a homogeneous formation with a 5-in. radius borehole. The conductivity of the mud is 10.0 S/m.

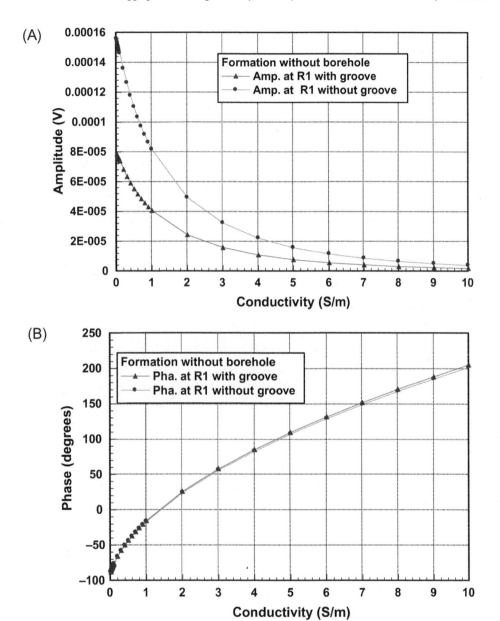

Figure 8.31 (A) Comparison of amplitudes of the MWD tool with or without grooves at receiver 1. (B) Comparison of the phases of the MWD tool with or without grooves at receiver 1.

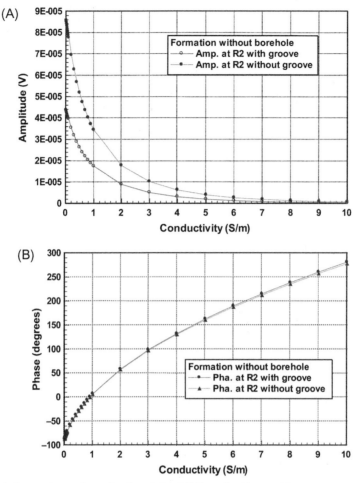

Figure 8.32 (A) Comparison of amplitudes of the MWD tool with or without grooves at receiver 2. (B) Comparison of the phases of the MWD tool with or without grooves at receiver 2.

The conductivity of the homogeneous formation varies from 0.01 to 10.0 S/m. The MWD tool is shown in Fig. 8.28. From Fig. 8.33, it can be seen that the amplitude ratios and the phase differences change very little.

Fig. 8.34 shows comparison of amplitudes ratios (dB) and phase differences of MWD tools with or without grooves. The formation is a three-layer formation with a 5-in. radius borehole shown in Fig. 8.35. The mud conductivity is 10.0 S/m. The shoulder bed conductivity is 1.0 S/m. The conductivity of the center bed varies from 0.01 to 10 S/m, and the thickness of the center bed is 4.0 ft. It also can be seen that the results are similar to the previous cases.

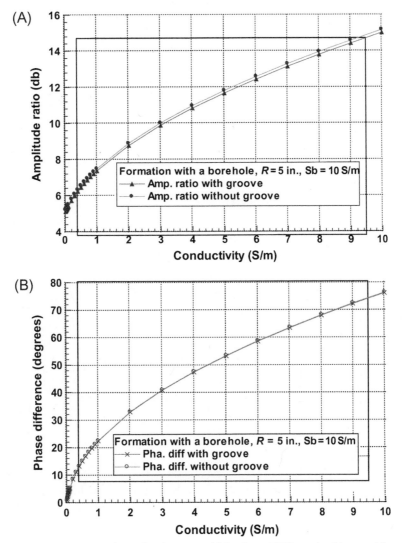

Figure 8.33 (A) Comparison of amplitude ratios (db) of an MWD tool with or without grooves when a borehole exists. (B) Comparison of the phase differences (degrees) of the MWD tool with or without grooves when a borehole exists.

8.7.4 The effects of the conversion table

The remarks concerning the coil-type MWD sensor are made in the context of phase and attenuation. Although these quantities are measured in the actual tools, the present MWD logs do not display these raw data. Instead, some conversion algorithms are used to convert the measured data in terms of apparent conductivities.

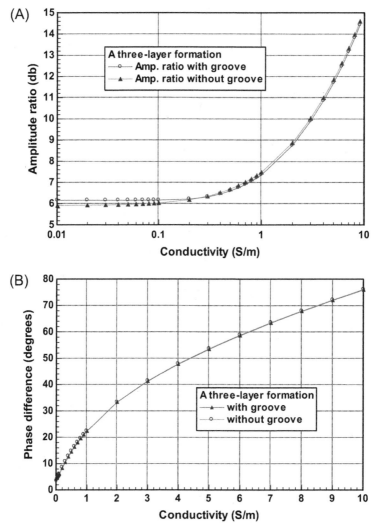

Figure 8.34 (A) Comparison of amplitude ratios (db) of the MWD tool with or without grooves in a three-layer formation. (B) Comparison of the phase differences (degrees) of the MWD tool with or without grooves in a three-layer formation.

The following algorithm is used in this study. The phase difference and amplitude ratio between the voltages at the two receiving coils are first obtained when the tool is in the homogeneous medium with various conductivity values. These data are obtained by running the simulation program. Usually, different MWD tools have different conversion tables. Table 8.4 shows a conversion table where the amplitude ratio and phase difference are the functions of the conductivity of the homogeneous

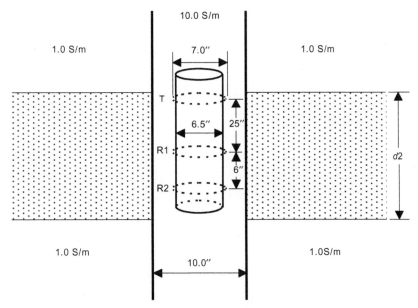

Thickness of the bed $d2 = 4$ ft.

Figure 8.35 A three-layer formation used to study the effects of grooves.

Table 8.4 A conductivity conversion table

Conductivity (S/m)	Amplitude ratio (db)	Phase difference (degrees)
1.00E-03	5.157740024	9.21E-02
2.00E-03	5.158831459	0.182743313
4.00E-03	5.16257533	0.356534202
8.00E-03	5.172843652	0.673835336
1.00E-02	5.178488877	0.821669343
1.25E-02	5.185819848	0.999368732
2.00E-02	5.209208606	1.496203628
4.00E-02	5.276172463	2.639912581
8.00E-02	5.411543219	4.485574102
0.1	5.477330654	5.275433264
0.125	5.557320459	6.178416718
0.2	5.782813268	8.500483404
0.4	6.300351405	13.17444843
0.8	7.116846783	19.78785793
1	7.461172961	22.45407378
1.3	7.928705722	25.99822132
1.6	8.353590572	29.14929572
2	8.870924498	32.90625344
3	9.99025488	40.8352475
4	10.94212938	47.48455322
5	11.78570781	53.34400281

medium. When the tool is in actual operation, the amplitude ratio and phase difference are measured, and then converted into conductivity values according to the conversion table, similar to Table 8.4. The apparent conductivity obtained by using the amplitude ratio is denoted as σ_a, and by using phase difference, is denoted as σ_p. Also, the σ_a is called as amplitude-based apparent conductivity, and σ_p is called as phase-based apparent conductivity.

Table 8.5 and Table 8.6 show the conversion tables generated by MWD tools without or with grooves, respectively. The σ_a and σ_p shown in Fig. 8.37 are converted from Table 8.5 and Table 8.6, respectively. The formation is shown in Fig. 8.36. The mud conductivity is 10.0 S/m. The shoulder bed conductivity is 1.0 S/m. The conductivity of center bed is 0.05 S/m, and the thickness of center bed is 6.0 ft. The conductivity of invasion zone is 0.5 S/m, and is 15-in. thick. From Fig. 8.37, it can be seen that the different conversion tables can cause a big variations in the value of σ_a, but not much in the value of σ_p. These results can be proved from Tables 8.5 and 8.6.

Table 8.5 A conversion table generated by an MWD tool without grooves

Conductivity (S/m)	Amplitude ratio (dB)	Phase difference (degrees)
1.00E-03	5.385023319	8.30E-02
2.00E-03	5.385558953	0.165743959
4.00E-03	5.387643952	0.329366483
8.00E-03	5.395194544	0.643872542
1.00E-02	5.400192773	0.793107774
1.25E-02	5.407151732	0.971931389
2.00E-02	5.430214068	1.464897082
4.00E-02	5.494753805	2.591727828
8.00E-02	5.626740837	4.431591433
0.1	5.692070389	5.220330625
0.125	5.772161093	6.120289503
0.2	5.999761356	8.415445606
0.4	6.514162713	12.95101609
0.8	7.305493676	19.47526977
1	7.647286845	22.13914433
1.3	8.117278542	25.62845854
1.6	8.541489011	28.67106024
2	9.047429991	32.27002708
3	10.11483816	39.97174709
4	11.01578629	46.56393076
5	11.81123728	52.44622103

Table 8.6 A conversion table generated by an MWD tool with grooves

Conductivity (S/m)	Amplitude ratio (dB)	Phase difference (degrees)
0.001	5.086	0.093
0.002	5.087	0.183
0.004	5.091	0.356
0.008	5.101	0.671
0.01	5.107	0.818
0.0125	5.114	0.995
0.02	5.137	1.489
0.04	5.204	2.625
0.08	5.339	4.457
0.1	5.404	5.241
0.125	5.484	6.138
0.2	5.707	8.443
0.4	6.221	13.084
0.8	7.031	19.655
1	7.372	22.306
1.3	7.835	25.833
1.6	8.256	28.972
2	8.769	32.718
3	9.88	40.629
4	10.825	47.266
5	11.663	53.117

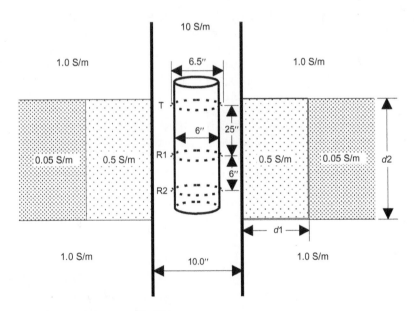

1. Invasion zone $d1$: 15 in.
2. Thickness of the bed $d2$: 6 (ft).
3. Frequency: 2 MHz.

Figure 8.36 A three-layer formation with a borehole and invasion zones used to study the effects of different conversion tables.

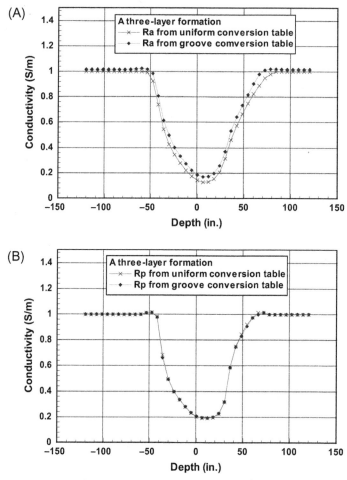

Figure 8.37 (A) Comparison of amplitude conductivities (S/m) of the MWD tool with or without grooves using different conversion tables. The formation is shown in Fig. 8.36. (B) Comparison of phase conductivities (S/m) of the MWD tool with or without grooves using different conversion tables. The formation is shown in Fig. 8.36.

8.8 SUMMARY

A stable numerical method to simulate the induction and MWD logs is discussed in this chapter. The algorithm is based on a horizontal eigenmode expansion method. The formation can have both cylindrical and planar boundaries. To reduce the complexity, the three-dimensional problem is simplified to a two-dimensional problem by assuming that the formation has an axial symmetry. Note that this method only applies in the vertical well without any dipping angles. When dipping angle involves, the problem becomes three-dimensional problems and numerical methods are preferred.

In the lossy medium, the propagating waves decay exponentially. Therefore a simulation method may be unstable if the wave propagation is not treated properly. The algorithm discussed in this chapter is stable because the waves are treated as decaying waves along the propagating direction. Based on this idea, a three-layer module is used to solve the multilayer problem. From the results, it can be seen that the algorithm is accurate, stable, and versatile.

Theoretically, infinite number of eigenmodes and the infinite distance in the radial direction are needed. In practice, since the formation is lossy, the wave decays exponentially from the center of the borehole. Therefore only finite number of eigenmodes and a finite distance are necessary. Usually, more eigenmodes can obtain more accurate simulation, but the computation time is relatively longer. From the simulation results, it is found that the results are very consistent when the number of eigenmodes is greater than 40, and the error of the results is less than 2% when the number of eigenmodes is equal to 20. The latter consumes much less computation time, so the accuracy and the computation time for practical use may be compromised.

The MWD tool simulation technique developed in this chapter is used to calculate the responses of the MWD tool in a multilayer, multiinvasion zone formation. The MWD tool measures the phase difference and the amplitude ratio of the voltages at a pair of receiving coils. These data are then converted to the phase-based apparent conductivity and the amplitude-based apparent conductivity by using the homogeneous medium as a reference. To reduce the skin depth and the field penetration into the mandrel, the operating frequency of the MWD tool is relatively higher, usually 2 MHz. The accuracy of the MWD measurement is discussed and compared with the simulated data obtained from the other algorithms. The results agree very well. A borehole-compensated MWD tool is used to study the effect of the borehole. The result is satisfactory.

The MWD tools discussed above are all uniform-mandrel models, e.g., the diameter of the mandrel is constant. In a practical model, the coils of the transmitters and the receivers are recessed into shallow grooves. In this study, a stable numerical algorithm based on a horizontal eigenmode method is used to investigate the effects of the grooves in a homogeneous formation and a three-layer formation. From the simulation results, it can be seen that the effects of the grooves are large on the individual voltages induced in the receiving coils, but very small on the amplitude ratio and the phase difference.

REFERENCES

[1] W.C. Chew, S. Barone, B. Anderson, C. Hennessy, Diffraction of axisymmetric waves in a borehole by bed boundary discontinuities, Geophysics 49 (10) (1984) 1586–1595.
[2] W.C. Chew, B. Anderson, Propagation of electromagnetic waves through geological beds in a geophysical probing environment, Radio Sci. 20 (1985) 611–621.

[3] Q.H. Liu, W.C. Chew, M.R. Taherian, K.A. Safinya, A modeling study of electromagnetic propagation tool in complicated borehole environments, Log Anal. 30 (6) (1989) 424–436.
[4] W.C. Chew, Z. Nie, Q.H. Liu, B. Anderson, An efficient solution for the response of electrical well logging tools in a complex environment, IEEE Trans. Geosci. Remote Sens. 29 (1991) 308–313.
[5] D.M. Pai, Induction log modeling using vertical eigenstates, IEEE Trans. Geosci. Remote Sens. 29 (1991) 209–213.
[6] D.M. Pai, J. Ahmad, W. David Kennedy, Two-dimensional induction log modeling using a coupled-mode, multiple-reflection series method, Geophysics 58 (4) (1993) 466–474.
[7] B. Anderson, S.K. Chang, Synthetic induction logs by the finite element method, Log Anal. 23 (6) (1982) 17–26.
[8] J. Li, L.C. Shen, Vertical eigenstate method for simulation of induction and MWD resistivity sensors, IEEE Trans. Geosci. Remote Sens. 31 (2) (1993) 399–406.
[9] Y.M. Jan, R.L. Campbell Jr., Borehole correction of MWD gamma ray and resistivity logs, paper PP, in: 25th Annual Logging Symposium Transactions: Society of Professional Well Log Analyst, 1984, pp. PP1–20.
[10] M.E. Cobern, E.B. Nuckols, Application of MWD resistivity logs to evaluation of formation invasion, paper OO, in: 26th Annual Logging Symposium Transactions: Society of Professional Well Log Analyst, 1985, pp. OO1–16.
[11] D. Coop, L.C. Shen, F.S.C. Huang, The theory of 2MHz resistivity tool and its application to measurement-while-drilling, Log Anal. 25 (3) (1984) 35–46.
[12] S. Gianzero, R. Chemali, Y. Lin, S.M. Su, A new resistivity tool for measurement-while-drilling, paper W, in: 26th Annual Logging Symposium Transactions: Society of Professional Well Log Analyst, 1985.
[13] S. Gianzero, R. Chemali, S.M. Su, Determining the invasion near the borehole with the M.W.D. toroid sonde, paper A, in: 27th Annual Logging Symposium Transactions: Society of Professional Well Log Analyst, June 1986.
[14] S. Bonner, A. Bagersh, B. Clark, G. Dajee, M. Dennison, J.S. Hall, J. Jundt, J. Lovell, R. Rosthal, D. Allen, A new generation of electrode resistivity measurements for formation evaluation while drilling, paper OO, in: 35th Annual logging Symposium Transactions: Society of Professional Well Log Analyst, June 1994.
[15] P. Rodney, R. Bartel, Design of a propagating wave resistivity sensor in order to minimize the influence of borehole fluids on the sensor response, SPE-18114, in: SPE 63rd Annual Technical Conference Proceedings: Society of Petroleum Engineers, 1988.
[16] B. Clark, D.F. Allen, D. Best, S.D. Bonner, J. Jundt, M.G. Luling, M.O. Ross, Electromagnetic propagation logging while drilling—theory and experiment, SPE-18117, in: SPE 63rd Annual Technical Conference Proceedings: Society of Petroleum Engineers, 1988, pp. 103–117.
[17] L.C. Shen, Theory of a coil-type resistivity sensor for MWD application, Log Anal. 32 (5) (1991) 603–611.
[18] L.C. Shen, Investigation depth of coil-type MWD resistivity sensor, in: Trans. SPWLA 32nd Logging Symposium, Paper C, 1991.
[19] J. Li, L.C. Shen, MWD resistivity logs in invade beds, Log Anal. 34 (2) (1993) 15–17.
[20] B. Clark, SPE, D.F. Allen, D.L. Best, SPE, S.D. Bonner, SPE, J. Jundt, M.G. Luling, M.O. Ross, Schlumberger-LWD, Electromagnetic propagation logging while drilling: theory and experiment, SPE Form. Eval. 5 (3), 1990, pp. 263–271.

CHAPTER 9

Theory of Inversion for Triaxial Induction and Logging-While-Drilling Logging Data in One- and Two-Dimensional Formations

Contents

9.1	Introduction	296
9.2	Gauss—Newton Algorithm	297
9.3	Cholesky Factorization	299
9.4	Line Search	301
9.5	Jacobian Matrix	302
9.6	Constraints	303
9.7	Initial Values	304
	9.7.1 Inverting for initial values	304
	9.7.2 Initial boundary locations	306
	9.7.3 Boundary merge	308
	9.7.4 Noise analysis	309
9.8	Inversion Results and Analysis	310
	9.8.1 Synthetic log inversion using all nine components of the magnetic field	310
	9.8.2 Inversion of synthetic logs using diagonal components of the magnetic fields	315
	9.8.3 Inversion of a five-layer synthetic formation	319
	9.8.4 Inversion in a 15-layer synthetic formation	324
	9.8.5 Inversion of real, isotropic formation with synthetic data	325
	9.8.6 Inversion of real log data	330
9.9	Inversion of Induction Logs in a Two-Dimensional Formation	330
	9.9.1 Theory of 2D induction log inversion	334
	9.9.2 Results from the least squares inversion	338
9.10	Summary	347
	References	348

9.1 INTRODUCTION

As discussed in Chapter 2, Fundamentals of Electromagnetic Fields Induction Logging Tools, and previous other chapters, the data measured by logging tools can reflect the formation resistivity distribution near the borehole. The resistivity distribution measured by the tools is affected by borehole mud, borehole geometry, tool design, and formation near the measurement points, and away from the measurement points. The measured apparent resistivity, by definition, describes formation resistivity accurately only when the formation is homogeneous. However, in logging data interpretation, it is desired to have true formation resistivity distribution when the formations are not homogeneous. Many efforts have been devoted to make the measured logs as close as real formation. One of the most rigorous methods is to use mathematical inversion method. The inversion process can be simply considered as an optimization process. The common procedure is to keep changing the tool response to an artificial formation until the calculated tool response matches the real tool response obtained from measurements. The last formation structure which has a response that replicates the measured logs is the results of the inversion process.

In practice, the inversion results may not be as ideal as described earlier. Errors and mismatch between the computer model and the real logs will always exist and can be significant. The comparison between simulated and real logs should decrease in a way that considers the global effects with conditions. One of the most popular method of the comparison is the least square method, which calls for minimizing the differences between the simulated data from a test formation structure and the measured logs in a least square fashion. The inversion process using least square method minimizes the least square differences between the computer logs and measured logs by varying parameters of the formation parameters such as layers, resistivity, etc. to a satisfactory criterion. Note that as most optimization method, the least square solution may not be unique. Therefore it is usually a common practice to enforce additional regularization conditions to the least squares equation.

Consider the least squares method as an example of the inversion process. At the first step, an initial guessed model of the formation is established, based on which, tool response (or computer logs) is calculated using a forward modeling method in a given range of depth. The second step is to find the differences between the computer logs and the measured logs in terms of least squares fashion at the data points interested. If the difference is greater than a given criterion, the initial model is modified and a second forward computation of the artificial logs is conducted. This iterative computation continues until the given criterion is satisfied and the iteration stops. The artificial formation resistivity distribution is the result of the inversion process.

Please note that there are several possibilities for an inversion method:
1. The process may not be convergent, which means the iteration continues but no final results satisfies the given criterion. Therefore the inversion process is failed. In this case, a new inversion process should be initiated with a new initial guess. In most cases, the initial guess is the apparent resistivity measured by the tool.
2. Mathematically, the inversion algorithm is a nonlinear optimization. Due to the nonuniqueness of the mathematic optimization, the final results of the inversion may not be the true formation. In other word, the true solution is not found but another solution is found by the inversion algorithm which satisfies the given criterion. In this case, additional regularization conditions should be applied and a new inversion process should be initiated.
3. Due to the large number of the forward modeling computation, the inversion speed can be very slow. Fast forward modeling method is critical to the inversion algorithm. In most cases, one-dimensional (1D) forward modeling is used due to its speed.
4. Additional regularization methods should be included in the inversion process. These regularization methods may be empirical. The purpose of the regularization method is to obtain a solution that is close to the reality such as maximum flatness, maximum oil, or minimum oil [1].

In this chapter, we will go through the inversion method in detail assuming a forward modeling algorithm is known.

9.2 GAUSS–NEWTON ALGORITHM

As described earlier, the inversion process is a nonlinear optimization in terms of mathematics. In this section, we will frame the inversion problem into a mathematic algorithm. One of the nonlinear programming method is Gauss–Newton algorithm.

Consider a series of measured logs m, which is assembled in a vector defined by

$$m = \left[\text{Im}(H_{xx,1}), \text{Im}(H_{xy,1}), \text{Im}(H_{xz,1}), \ldots, \text{Im}(H_{zz,9 \times NR}) \right]^T \qquad (9.1)$$

where the superscript T indicates transposition and NR is the number of logging points. Hence, in total we should have $9 \times NR$ corresponding to the nine components of the magnetic field measured by the triaxial induction logging tool. In the framework of the inversion, these measurement data is assumed to be borehole corrected and the invasion effect is ignored, which makes the problem a 1D formation.

Consider a transverse isotropic formation, each layer is characterized by its horizontal conductivity, vertical conductivity, and bed boundary positions. This leads to a total of $3 \times L - 1$ parameters for an L-layer formation. Besides, dipping angle and rotation angle play an important role in the responses of triaxial induction tool.

Hence we should have $N = 3 \times L + 1$ parameters to be inverted in the 1D inversion. The parameter vector **x** is the vector of unknown parameters defined as

$$\mathbf{x} = [x_1, x_2, \ldots, x_n]^T = \begin{bmatrix} \log(\alpha), \log(\gamma), \log(Z_1), \ldots, \log(Z_{(N-1)/3}), \log(R_{hl}), \\ \log(R_{vl}), \ldots, \log(R_{h(N-1)/3}), \log(R_{v(N-1)/3}) \end{bmatrix}^T \quad (9.2)$$

Logarithm is used to rescale all parameters within the proper magnitude range. The set of measured data points is denoted by the vector M, such that

$$M = [m_1, m_2, \ldots, m_N]^T$$

$$= \begin{bmatrix} \mathrm{Im}(H_{XX1}), \mathrm{Im}(H_{XY1}), \mathrm{Im}(H_{XZ1}), \mathrm{Im}(H_{YX1}), \mathrm{Im}(H_{YY1}), \mathrm{Im}(H_{YZ1}), \\ \mathrm{Im}(H_{ZX1}), \mathrm{Im}(H_{ZY1}), \mathrm{Im}(H_{ZZ1}), \ldots, \mathrm{Im}(H_{XXNR}), \ldots, \mathrm{Im}(H_{ZZNR}) \end{bmatrix} \quad (9.3)$$

From Chapter 1, Introduction to Well Logging, we know that the measured voltage from the induced signals are proportional to the imaginary of the magnetic field. The real part of the magnetic field reflects the directly coupled signals from the transmitter. Therefore, in the inversion process, we only need imaginary part of the measured magnetic field.

We approach this nonlinear problem iteratively to minimize this objective function (cost function) in a least squares fashion:

$$C(\mathbf{x}) = \frac{1}{2} R(\mathbf{x})^T R(\mathbf{x}) \quad (9.4)$$

in which the residual function R is defined as $R(\mathbf{x}) = S(\mathbf{x}) - M$. Here $S(\mathbf{x})$ is the simulated tool response corresponding to a particular value of the unknown parameter **x**.

Cost function is one important conception in inversion. It is used to determine total error between calculated log and the measured log. The smaller the cost function is, the better inversion results we may obtain. Hence it is necessary for us to find one appropriate inversion algorithm to minimize the cost function. One of the effective optimization algorithm is the Gauss—Newton minimization approach. According to Taylor expansion, we approximate the cost function with a local quadratic model as [2]:

$$C(x) \approx \frac{1}{2} R^T(x_c) R(x_c) + g^T(x_c)(x - x_c) + \frac{1}{2}(x - x_c)^T \overline{\overline{H}}(x_c)(x - x_c) \quad (9.5)$$

in which $g(x) = \nabla C(x) = \overline{\overline{J}}^T(x) R(x)$ is the gradient of the cost function, **C(x)** and $\overline{\overline{H}}(x) = \nabla \nabla C(x)$ is the Hessian of cost function **C(x)** which is given by:

$$\overline{\overline{H}}(x) = \overline{\overline{J}}^T(x) \overline{\overline{J}}(x) + \overline{\overline{S}}(x) \approx \overline{\overline{J}}^T(x) \overline{\overline{J}}(x) + \mu I \quad (9.6)$$

where $\bar{\bar{S}}(x) = \sum_{i=1}^{9 \times NR} r_i(x) \nabla^2 r_i(x)$ denotes the second-order information in $\bar{\bar{H}}(x)$. In Eq. (9.6), we introduce positive constant μ. By determining $\mu > 0$, $\bar{\bar{H}}(x) \approx \bar{\bar{J}}^T(x) \bar{\bar{J}}(x) + \mu I$ is positive definite. We apply Cholesky factorization algorithm to update μ. Then Eq. (9.5) becomes:

$$C(x) \approx \frac{1}{2} R^T(x_c) R(x_c) + R^T(x_c) \bar{\bar{J}}(x_c)(x - x_c) + \frac{1}{2}(x - x_c)^T \left(\bar{\bar{J}}^T(x_c) \bar{\bar{J}}(x_c) + \mu I \right)(x - x_c) \tag{9.7}$$

Thus the solution of Eq. (9.7) is

$$x_+ \approx x_c - \left(\bar{\bar{J}}^T(x_c) \bar{\bar{J}}(x_c) + \mu I \right)^{-1} \bar{\bar{J}}^T(x_c) R(x_c) \tag{9.8}$$

9.3 CHOLESKY FACTORIZATION

As mentioned in Section 9.2, the modified Gauss–Newton step is solved efficiently by using the Cholesky decomposition of the modified $\bar{\bar{H}}(x)$. In this algorithm, we introduce the Gill and Murray Cholesky decomposition based on Gerschgorin bounds [3] to $\bar{\bar{H}}(x)$.

The classic Cholesky decomposition algorithm assumes a positive definite matrix and symmetric variance matrix (C). It then proceeds via the matrix decomposition

$$\underset{(k \times k)}{C} = \underset{(k \times k)}{L} \underset{(k \times k)}{D} \underset{(k \times k)}{L'} \tag{9.9}$$

The basic Cholesky procedure is a one-pass algorithm that generates two output matrices, which can then be combined for the desired "square root" matrix. The algorithm moves down the main diagonal of the input matrix determining diagonal values of D and triangular values of L from the current column of the C and previously calculated components of L and C. Thus the procedure is necessarily sensitive to values in the original matrix and previously calculated values in the D and L matrices. There are k stages in the algorithm corresponding to the k dimensionality of the input matrix. The jth step $(1 \leq j \leq k)$ is characterized by two operations:

$$D_{j,j} = C_{j,j} - \sum_{l=1}^{j-1} L_{j,l}^2 D_{l,l} \tag{9.10}$$

and

$$L_{i,j} = \frac{\left[C_{i,j} - \sum_{l=1}^{j-1} L_{j,l} L_{i,l} D_{l,l} \right]}{D_{j,j}}, \quad i = j+1, \ldots, k \tag{9.11}$$

where D is a positive diagonal matrix so that on completion of the algorithm, its square root is multiplies by L to give the Cholesky decomposition. From this algorithm it is easy to see why the Cholesky algorithm cannot tolerate singular or nonpositive definite input matrices. Singular matrices cause a divide-by-zero problem in Eq. (9.11), and nonpositive definite matrices cause the sum in Eq. (9.10) to be greater than $C_{j,j}$, causing negative diagonal values.

Hence Gill and Murray [4,5] introduced an algorithm to find a nonnegative diagonal matrix, E, such that $C + E$ is positive definite and the diagonal values of E are as small as possible. In order to make $C + E$ positive definite, the Gill and Murray Cholesky decomposition takes the greatest negative eigenvalue of C, λ_1, and assigning $E = -(\lambda_1 + \varepsilon)I$, where ε is a small positive increment. However, this approach (implemented in various computer programs, such as the Gauss "max-like" model) produces E values that are much larger than required, and therefore the $C + E$ matrix is much less like C than it could be.

To see Gill et al.'s [4,5] approach, we can rewrite the Cholesky algorithm provided as Eqs. (9.10) and (9.11) in matrix notation. The jth submatrix of its application at the jth step is

$$C_j = \begin{bmatrix} c_{j,j} & c'_j \\ c_j & c_{j+1} \end{bmatrix} \qquad (9.12)$$

where $c_{j,j}$ is the jth pivot diagonal, c'_j is the row vector to the right of $c_{j,j}$, which is the transpose of the c_j column vector beneath $c_{j,j}$, and c_{j+1} is the $(j + 1)$th submatrix. The jth row of the L matrix is calculated by: $L_{j,j} = \sqrt{c_{j,j}}$ and $L_{(j+1):k,j} = c_{(j+1):k,j}/L_{j,j}$. The $(j + 1)$th submatrix is then updated by

$$c^*_{j+1} = c_{j+1} - \frac{c_j c'_j}{L^2_{j,j}} \qquad (9.13)$$

Suppose that at each iteration we defined $L_{j,j} = \sqrt{c_{j,j} + \delta_j}$, where δ_j is a small positive integer sufficiently large so that $C_{j+1} > c_j c'_j / L^2_{j,j}$. This would obviously ensure that each of the j iterations does not produce a negative diagonal value or divide-by-zero operation. However the size of δ_j is difficult to determine and involves trade-off between satisfaction with the current iteration and satisfaction bigger than zero, subsequent diagonal values are greatly increased through the operation of Eq. (9.13). Conversely, we do not want to be adding large δ_j values on any given iteration.

Gill et al. [4,5] note the effect of the e_j vector on subsequent iterations and suggest that minimizing the summed effect of δ_j is equivalent to minimizing the effect of the vector maximum norm of c_j, $\|c_j\|_\infty$, at each iteration. This is done at the jth step by making δ_j the smallest nonnegative value satisfying

$$\|c_j\|_\infty \beta^{-2} - c_{j,j} \leq \delta_j \qquad (9.14)$$

where

$$\beta = \max \begin{cases} \max(\mathrm{diag}(C)) \\ \max(\mathrm{notdiag}(C))\sqrt{k^2-1} \\ \varepsilon_m \end{cases} \quad (9.15)$$

where ε_m is the smallest positive number that can be represented on the computer used to implement the algorithm (normally called the machine epsilon). This algorithm always produced a factorization and has the advantage of not modifying already positive definite C matrices. However the bounds in Eq. (9.15) have been shown to be nonoptimal and thus provide $C + E$ that is again farther from C than necessary.

So we apply the Gerschgorin circle theorem to determine upper bounds in Eq. (9.15). In Gerschgorin circle theorem, δ_j is determined by

$$\delta_j = \max\left(\varepsilon_m - C_{j,j} + \max\left(\|c_j\|, (\varepsilon_m)^{1/3}\max(\mathrm{diag}(C)), E_{j-1,i-1}\right)\right) \quad (9.16)$$

9.4 LINE SEARCH

Eq. (9.18) gives us Newton direction $P \approx x_+ - x_c$. Usually this step cannot promise minimum value of the cost function because of poor match between exact cost function and quadratic approximation. To speed up the minimization process, we incorporate a line search along the direction of Gauss–Newton step to guarantee a reduced cost function after each iteration until cost function satisfies:

$$C(x_k + \lambda P_k) \leq C(x_k) + \alpha \lambda_k \delta C_{k+1} \quad (9.17)$$

where $\alpha \in \{0, 1\}$, λ_k is the kth line search step. In practice, α is always set very small, e.g., $\alpha = 10^{-4}$. Starting at $x_{k+1} = x_k + \lambda_k P_k$, cost function $C(x)$ can be expressed as quadratic form of step length λ, as in Eq. (9.18)

$$C(\lambda) = C(X_k + \lambda P_k) \approx a + b\lambda + c\lambda^2 \quad (9.18)$$

in which a, b, c are constant determined from the current information on cost function $C(\lambda)$,

$$a = C(\lambda = 0) = C(x_k) \quad (9.19)$$

$$b = \left.\frac{dC(\lambda)}{d\lambda}\right|_{\lambda=0} = g^T(x_k)p_k \quad (9.20)$$

$$c = \frac{1}{\left\{\lambda_k^{(m)}\right\}^2}\left[C\left(x_k + \lambda_k^{(m)}P_k\right) - C(x_k) - \lambda_k^{(m)}\delta C_{k+1}\right] \quad (9.21)$$

Thus, $\lambda_k^{(m+1)}$, which is the minimum of $C(\lambda)$, for $m = 0, 1, 2,\ldots$ is given by

$$\lambda_k^{(m+1)} = -\frac{b}{2c} = \frac{\left\{\lambda_k^{(m)}\right\}^2 \delta C_{k+1}}{2\left[C(x_k + \lambda_k^{(m)} P_k) - C(x_k) - \lambda_k^{(m)} \delta C_{k+1}\right]} \quad (9.22)$$

Thus we start with $\lambda_k^{(0)} = 1$ and proceed with the backtracking procedure of Eq. (9.18) until Eq. (9.17) is satisfied. To take advantage of the newly acquired information on the cost function beyond the first backtrack and improve accuracy, we replace the quadratic approximation of Eq. (9.18) with the cubical form Eq. (9.23).

$$C(\lambda) = C(x_k + \lambda P_k) \approx a + b\lambda + c_2\lambda^2 + d\lambda^3 \quad (9.23)$$

where

$$\begin{matrix} c \\ d \end{matrix} = \frac{1}{\lambda_2 - \lambda_1} \begin{bmatrix} -\lambda_1/\lambda_2^2 & \lambda_2/\lambda_1^2 \\ 1/\lambda_2^2 & -1/\lambda_1^2 \end{bmatrix} \cdot \begin{bmatrix} C(\lambda_2) - \lambda_2 b - C(X_k) \\ C(\lambda_1) - \lambda_1 b - C(X_k) \end{bmatrix} \quad (9.24)$$

λ_1, λ_2 are two previous subsequent search steps.

The final solution for $\lambda_k^{(m+1)}$ is

$$\lambda_k^{(m+1)} = \frac{-c + \sqrt{c^2 - 3db}}{3d} \quad (9.25)$$

9.5 JACOBIAN MATRIX

In Eq. (9.17), Jacobian matrix is given by

$$J(x) = \begin{bmatrix} \partial s_1/\partial x_1 & \cdots & \partial s_1/\partial x_i & \cdots & \partial s_1/\partial x_N \\ \vdots & \ddots & \vdots & \ddots & \vdots \\ \partial s_j/\partial x_1 & \cdots & \partial s_j/\partial x_i & \cdots & \partial s_j/\partial x_N \\ \vdots & \ddots & \vdots & \ddots & \vdots \\ \partial s_{9 \times NR}/\partial x_1 & \cdots & \partial s_{9 \times NR}/\partial x_i & \cdots & \partial s_{9 \times NR}/\partial x_N \end{bmatrix} \quad (9.26)$$

where every entry of the Jacobian matrix is estimated through a finite difference computation,

$$\frac{\partial s_j(x)}{\partial x_i} \approx \frac{s_j[(1 + \Delta)x_i] - s_j(x_i)}{\Delta x_i} \quad (9.27)$$

In practical implementation Δ is very small, say, 10^{-4}. As is clear from Eq. (9.26), the computation of this Jacobian matrix will dominate the total computation time of the inversion procedure. In each Gauss–Newton step we need to solve a $9 \times NR \times N$ forward problem to construct the Jacobian matrix.

9.6 CONSTRAINTS

In the minimization process, the variables can be any value. However, in physical problems, the variables are physical parameters and they are bounded. To limit the range of the variable in the minimization process, an easier way is to apply a nonlinear transformation to impose a priori information of maximum and minimum bounds on the unknown parameters. Consider the following transformation

$$x_i = \frac{x_i^{\max} + x_i^{\min}}{2} + \frac{x_i^{\max} - x_i^{\min}}{2} \sin(c_i) \quad -\infty < c_i < +\infty \quad (9.28)$$

where $x_i^{\max}, x_i^{\min}$ are the upper and lower bounds of the physical parameter x_i. It is clear that

$$x_i \to x_i^{\min}, \quad \text{as } \sin(c_i) \to -1 \quad (9.29)$$

$$x_i \to x_i^{\max}, \quad \text{as } \sin(c_i) \to +1 \quad (9.30)$$

In the inversion algorithm, the inversion computation updates the artificial unknown parameters c_i instead of the physical parameters x_i. For each step, c_i is transformed to x_i which is used to compute forward modeling. To reduce the complexity of the computation, consider

$$\frac{\partial s_j}{\partial c_j} = \frac{dx_i}{xc_j} \frac{\partial s_j}{\partial x_i} = \sqrt{(x_i^{\max} - x_i)(x_i - x_i^{\min})} \frac{\partial s_j}{\partial x_i} \quad (9.31)$$

The two successive iterates $x_{i,k+1}$ and $x_{i,k}$ of x_i are related by

$$\begin{aligned} x_{i,k+1} &= \frac{x_i^{\max} + x_i^{\min}}{2} + \frac{x_i^{\max} - x_i^{\min}}{2} \sin(c_{i,k+1}) \\ &= \frac{x_i^{\max} + x_i^{\min}}{2} + \frac{x_i^{\max} - x_i^{\min}}{2} \sin(c_{i,k} + q_{i,k}) \end{aligned} \quad (9.32)$$

where

$$c_i = \arcsin\left(\frac{2x_{i,k} - x_i^{\max} - x_i^{\min}}{x_i^{\max} - x_i^{\min}}\right) \quad (9.33)$$

and where $q_{i,k} = c_{i,k+1} - c_{i,k}$ is the Gauss–Newton search step in c_i toward the minimum of the cost functional in Eq. (9.18). This Gauss–Newton direction in x_i is related to the Gauss–Newton direction in c_i through the following relation

$$p_i = q_i \frac{dx_i}{dc_i} \tag{9.34}$$

Hence, by using the relationship of Eq. (9.34) into Eq. (9.32), we obtain the following relationship between the two successive iterates $x_{i,k+1}$ and $x_{i,k}$ of x_i (assuming an adjustable step length γ_k along the search direction x_i):

$$x_{i,k+1} = \frac{x_i^{\max} + x_i^{\min}}{2} + \left(x_{i,k} - \frac{x_i^{\max} + x_i^{\min}}{2}\right)\cos\left(\frac{\nu_k p_{i,k}}{\gamma_k}\right) + \gamma_k \sin\left(\frac{\nu_k p_{i,k}}{\gamma_k}\right) \tag{9.35}$$

where

$$\gamma_k = \sqrt{\left(x_i^{\max} - x_{i,k}\right)\left(x_{i,k} - x_i^{\min}\right)} \tag{9.36}$$

Thus in the inversion process it is not necessary to compute either c_i or q_i explicitly. This will reduce the round-off errors caused by introduction of the nonlinear function.

9.7 INITIAL VALUES

9.7.1 Inverting for initial values

Initial values of the variables must be given before the inversion starts. As described in the previous sections, the inversion process is a nonunique mathematic problem by nature. The minimization may converge to a local minima, which will not be the desired solution, which is a global minima. Therefore selecting correct initial values of the variables helps in arriving correct answer. For the inversion of induction logging data, since we cannot determine the exact number of layers of a practical case, a natural selection of the initial value would be a homogeneous formation, which can be considered as an average resistivity of the formation to be considered. This process is also called zero-dimensional inversion. The apparent difference between the zero-dimensional inversion and 1D inversion is that the zero-dimensional inversion inverts parameters based on each logging point under the assumption that the formation is homogenous. If boundary is eliminated in the zero-dimensional inversion, at each logging point, four parameters (dipping angle, rotation angle, horizontal conductivity, vertical conductivity) are obtained. The algorithm and flowchart of zero-dimensional inversion are similar to 1D inversion, as shown in Fig. 9.1.

Once the zero-dimensional inversion method is determined, the results of the inverted data can be used as initial values of the 1D inversion in each 1D layer. On the other hand, since the zero-dimensional inversion is relatively simple, and the

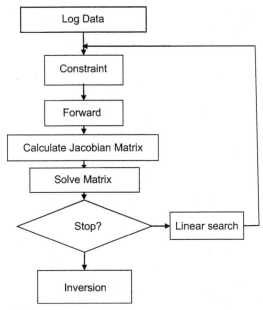

Figure 9.1 Flowchart of zero-dimensional inversion.

measured magnetic field H may be directly converted to the desired initial values using approximate analytic expressions [4,5] α, γ, σ_h, and σ_v:

$$\alpha = a\tan\left[\frac{2H^t_{xz_i}}{H^t_{xx_i} - H^c_{yy_i}}\right] \tag{9.37}$$

$$\gamma = a\tan\left(\frac{2H^c_{xy_i}}{H^c_{xx_i} - H^c_{yy_i}}\right) \tag{9.38}$$

$$\sigma_h = \frac{4\pi l}{\omega\mu_0}\left[\text{Im}(H^{x'}_{x'}) + \frac{1}{2}\text{Im}(H^{z'}_{z'}) + \sqrt{\left(\text{Im}(H^{x'}_{x'}) - \frac{1}{2}\text{Im}(H^{z'}_{z'})\right)^2 + 2\text{Im}(H^{x'}_{z'})^2}\right] \tag{9.39}$$

$$\lambda^2 = 256\pi^2 l^2 \sigma^2_{ha}/\text{Im}(H^{z'}_{z'})\left(\text{Im}(H^{x'}_{x'}) + \text{Im}(H^{y'}_{y'}) + \text{Im}(H^{z'}_{z'}) - \frac{\omega\mu_0}{4\pi l}\sigma_h\right) \tag{9.40}$$

$$\sigma_v = \frac{1}{\lambda^2}\sigma_h \tag{9.41}$$

where superscripts t and c represent the borehole and tool coordinates.

By using either method discussed above, the approximate average values of α, γ, σ_h, and σ_v are used as initial values in each layer. To obtain the real formation structure, we also need to define the initial boundaries before the 1D inversion starts.

9.7.2 Initial boundary locations

In Chapter 5, Triaxial Induction Tool and Logging-While-Drilling Tool Response in a Transverse Isotropic-Layered Formation andChapter 6, Triaxial Induction and Logging-While-Drilling Logging Tool Response in a Biaxial Anisotropic-Layered Formation, we noticed that the induction tool is very sensitive to the formation boundaries. Therefore the apparent conductivities, which are derived from the measured magnetic fields, also carry the boundary information. The initial boundary location can be obtained by using the measured logs. Measured logs can provide the boundary information but requires process to retrieve the boundary locations. Two methods are usually employed to determine the initial boundaries.

One way to retrieve the boundary information is to observe the weighted vertical and horizontal apparent resistivity $2\sigma_v - \sigma_h$ [6–10]. It was found that the weighted difference $2\sigma_v - \sigma_h$ is very sensitive to the boundary change. According to Eq. (9.42), we add a window with n points on logging curve $2\sigma_v - \sigma_h$ around the ith log point. Then apply Eq. (9.43) to calculate variance distribution in this window. By repeating the previous process, the variance curve based on totally logging points is computed. Finally, peak points are picked up and initial boundaries are placed on peaks in the variance curves.

$$a_i = \frac{1}{n} \sum_{j=i-n/2}^{i+n/2} \log\left(\left|2\sigma_{v,j} - \sigma_{h,j}\right|\right) \tag{9.42}$$

$$v_i^2 = \frac{1}{n} \sum_{j=i-n/2}^{i+n/2} \left[\log\left(\left|2\sigma_{v,j} - \sigma_{h,j}\right|\right) - a_i\right]^2 \tag{9.43}$$

For instance, if we have two-layered anisotropic formation as shown in Fig. 9.2, true boundary is placed at 10 ft. Horizontal conductivities in each layer are 1 and 0.55 S/m, respectively. Vertical conductivities in each layer are 0.55 and 0.1818 S/m, respectively.

Consider a dipping angle of 60 degrees. Tool spacing is selected as 40 in. Operating frequency is 20 kHz. Variance curve is given in Fig. 9.3. Two local maxima are found at the true boundary (10 ft).

The disadvantage of this boundary finding method is the instability. As we know, inversion results from zero-dimensional always have errors and the errors are not

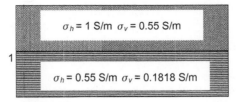

Figure 9.2 Two-layer anisotropic formation.

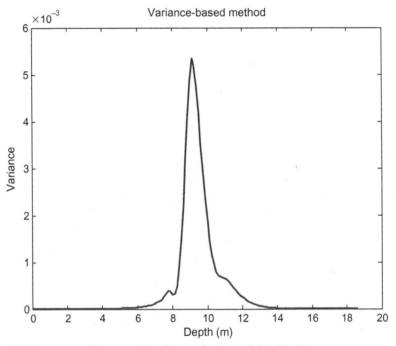

Figure 9.3 Variance curve of $2\sigma_h - \sigma_v$ for the two-layer model in Fig. 9.2.

insignificant. In this case, we may not completely rely on the variance-based method. Since accurate boundary locations are essential to the accurate inversion, it is necessary to apply cross components as the second solution.

By careful study of the cross components, it is found that H_{xz} and H_{zx} have higher sensitivity to boundary locations at any dipping angle. As shown in Fig. 9.3, we should notice that imaginary part of cross components H_{xz} and H_{zx} have symmetric sharp pulses near the boundary. This is known as horn effect, which is determined by the internal property of triaxial induction tool. Therefore a combined cross components (H_{xz}, H_{zx}) with variance-based method to detect initial boundary locations are preferable (Fig. 9.4).

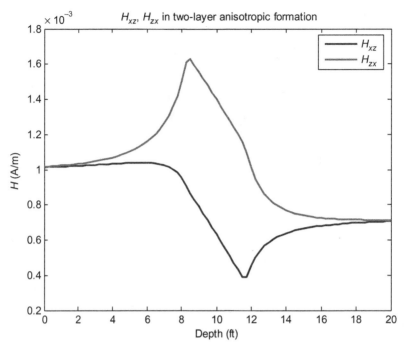

Figure 9.4 Imaginary part of cross components (H_{xz}, H_{zx}) for the two-layer model in Fig. 9.2.

9.7.3 Boundary merge

In the boundary location determinate, it is inevitable to have redundant boundaries, which means more variables in the optimization process, and thus, the reduction in the inversion efficiency. Usually redundant initial boundaries can be detected in zero-dimensional inversion if same conductivities are found between adjacent layers. Hence it is necessary for us to use proper procedure to merge the same formation layers. One way to do so is to use golden section search to reduce the number of initial boundaries.

In this method, we first set up a threshold (e.g., 1 ft) to trigger boundary merge procedure. If a layer boundary is found to be located at z_i by using the initial boundary location procedure discussed above using cross components, the next boundary is located at z_{i+1}. When $z_{i+1} < z_i + 1'$, the boundary position z_{i+1} must merge with z_i. A new location of boundary z'_i will be generated, which is in between z_i and z_{i+1} with the conductivity values of the layer z_i. The boundary merge process will find the best location between z_i and z_{i+1}, so that the cost function reaches the minimum. Therefore this process is a single-variable optimization. For simplicity, the golden section search is usually used. The basic idea of the golden search can be shown in Fig. 9.5. The diagram in Fig. 9.5 illustrates a single step in the technique for finding a minimum.

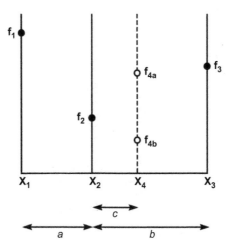

Figure 9.5 Diagram of a golden section search.

For instance, the functional values of **f(x)** are on the vertical axis, and the horizontal axis is the **x** parameter. The value of **f(x)** has already been evaluated at the three points: x_1, x_2, and x_3. Since f_2 is smaller than either f_1 or f_3, it is clear that a minimum lies inside the interval from x_1 to x_3. The next step in the minimization process is to "probe" the function by evaluating it at a new value of **x**, namely x_4. It is most efficient to choose x_4 somewhere inside the largest interval, i.e., between x_2 and x_3. From the diagram, it is clear that if the function yields f_{4a}, then a minimum lies between x_1 and x_4 and the new triplet of points will be x_1, x_2, and x_4. However if the function yields the value f_{4b}, then a minimum lies between x_2 and x_3 and the new triplet of points will be x_2, x_4, and x_3. Thus, in either case, we can construct a new narrower search interval that is guaranteed to contain the function's minimum.

9.7.4 Noise analysis

The behavior of the inversion of a triaxial array in the presence of noise and error must be evaluated. The noise will play an important part when the formation resistivity is high. The noise level of the electronics due to temperature and interference between signal channels in the tool will cause the inversion to generate errors. For induction-type measurements (especially triaxial arrays), there may be other sources which will cause the inversion errors, such as borehole correction, eccentricity, and borehole rugosity. A rugose borehole will serve as a source of noise, and errors in a smooth-hole borehole correction algorithm will produce onset errors.

For a triaxial array with colocated x, y, and z coils, the borehole noise will be correlated in all the measurements. To simulate this type of noise, an array of random numbers between −1 and +1 will be generated using a white-noise distribution. Triaxial array will be scaled to 0−100% of the mean value of each triaxial measure channel (including cross components) and added to the computed response data. This modified data set will be then used as input to the inversion algorithm. This type of noise is called as coherent noise.

If the x, y, and z coils are not colocated, or if the tool is moving with an irregular speed, the noise will be incoherent. To simulate incoherent noise, an array of different random numbers will be generated for each measurement channel and then scaled and added as above [9,10]. We call it as incoherent noise. In the inversion algorithm, these two types of noises should be considered to simulate noise from the tool.

9.8 INVERSION RESULTS AND ANALYSIS

In this section, we will validate the inversion method using synthetic data and field log. In the following examples, initial values are obtained by the zero-dimensional inversion.

9.8.1 Synthetic log inversion using all nine components of the magnetic field

In this example, a triaxial 2C40 tool is used for simplicity, which means the distance between the transmitters and receivers is 40 in. and all three components are assumed to be colocated. The operating frequency is 20 kHz. The raw data is obtained by using the 1D forward modeling. The formation model is a three-layer anisotropic model, as shown in Fig. 9.6. In the inversion process, all nine components of the measured magnetic field are used at each logging point.

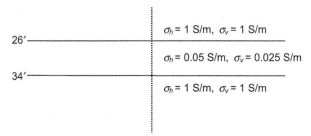

Figure 9.6 A three-layer anisotropic model.

9.8.1.1 Case 1—Dipping angle: 30 degrees, rotation angle: 60 degrees

In Case 1, the well is deviated at 30 degrees and the tool rotation angle is 60 degrees. Fig. 9.7 shows inversion results.

In Fig. 9.7, the *blue line* (black in print versions) represents the true formation resistivity distribution. The inverted resistivity is denoted by the *red dashed lines* (gray dashed line in print versions) while initial guess is shown by the *red dotted lines* (gray dotted line in print versions). Without special indication, the same drawing mechanism is used to indicate the true resistivity, initial model and inverted resistivity in all the examples. From Fig. 9.7, we can see that the discrepancy between the inverted values for R_h, R_v, the bed boundary locations and the corresponding true formation parameters are within 1%. The initial guess for the dipping angle, rotation angle and the inverted dipping angle, rotation angle are 32.28, 34.62, 30, and 60 degrees, respectively. The inversion converged in 5 iterations.

To investigate noise tolerance of the inversion, 5% coherent noise is added to the simulated log data and the inversion is repeated. The inversion results are shown in Fig. 9.8.

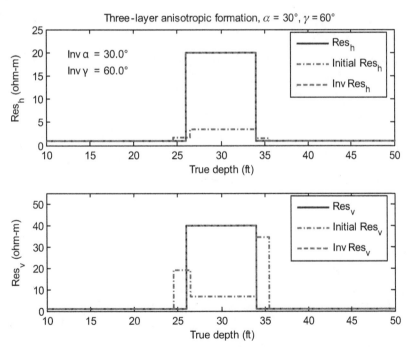

Figure 9.7 Inverted resistivity for Case 1 with raw data for the model in Fig. 9.6. The dipping angle and rotation angle are 30 and 60 degrees, respectively. The raw data is obtained from the 1D forward modeling.

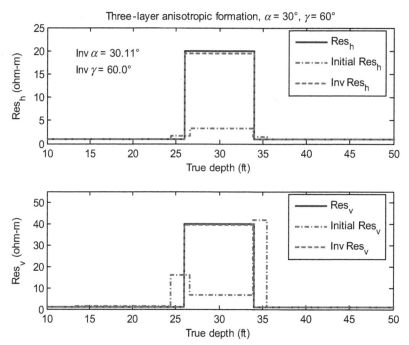

Figure 9.8 Inverted resistivity for Case 1 for the model in Fig. 9.6. Data is contaminated by 5% coherent noise.

In this case, the inversion also converged in 5 iterations. The initial dipping angle and rotation angle are 32.29 and 34.45 degrees, respectively while the inverted dipping angle and rotation angle are 30.11 and 60 degrees, respectively. We can see that the present inversion method can still obtain satisfactory results in the presence of coherent noise.

In the next example, a 5% incoherent noise to raw data is added and the inversion is conducted using the noise-added logs. Inversion results are shown in Fig. 9.9.

From Fig. 9.9, we can see that the agreement between the inverted resistivity and the true formation resistivity are very good. Initial guesses of the dipping angle and rotation angle are 30.83 and 29.21 degrees, respectively. The inverted dipping angle and rotation angle are 30.12 and 60 degrees, respectively. Converged results are obtained in 8 iterations. Comparing Figs. 9.8 and 9.9, we find that the inversion results in Fig. 9.8 are closer to the true formation model than those in Fig. 9.9, implying that the influence from the incoherent noise is stronger than that from the coherent noise.

9.8.1.2 Case 2—Dipping angle: 80 degrees, rotation angle: 30 degrees

In general, high dipping angle will have greater simulation and inversion errors. In Case 2, let us consider the same formation as in Case 1 except that the well is

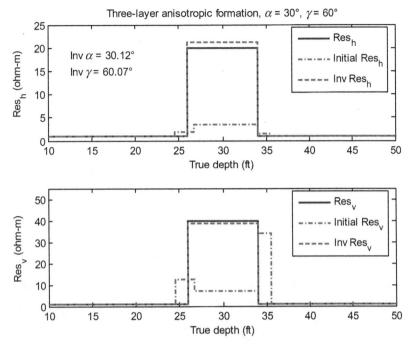

Figure 9.9 Inverted resistivity for Case 1 for the model in Fig. 9.6. Data is contaminated by 5% incoherent noise.

deviated at a greater dipping angle of 80 degrees and the tool rotation is kept at 30 degrees. Fig. 9.10 shows the inverted results using simulated log from the forward modeling.

From Fig. 9.10, it is observed that although the initial guess provides redundant initial boundaries, the algorithm determines the boundary effectively and still yield very good inversion result. In this example, the initial guess of dipping angle and rotation angle are 38.54 and 29.15 degrees, respectively. The inverted dipping angle and rotation angle are 80 and 30 degrees, respectively. The errors for all the inverted parameters (R_h, R_v, dip, rotation, and bed boundaries) are within 1%. The results converged in 5 iterations.

When the noise level is increased to 5%, the inversion results are given in Fig. 9.11.

From Fig. 9.11, it is seen that the initial values are further away from the solution since multiple initial layers are generated from the zero-dimensional inversion. However, stable and reasonable inversion results are obtained. The inverted dipping and rotation angle are 79.91 and 30 degrees, respectively while the initial dipping and rotation angle are 38.40 and 29.33 degrees, respectively. The inversion results converged in 7 iterations.

314 Theory of Electromagnetic Well Logging

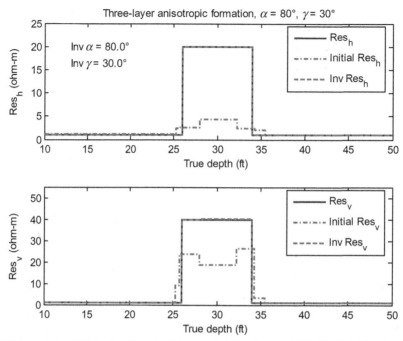

Figure 9.10 Inverted resistivity for Case 2 with raw data for the model in Fig. 9.6. The dipping angle and rotation angle are 80 and 30 degrees, respectively. The raw data is from 1D forward modeling.

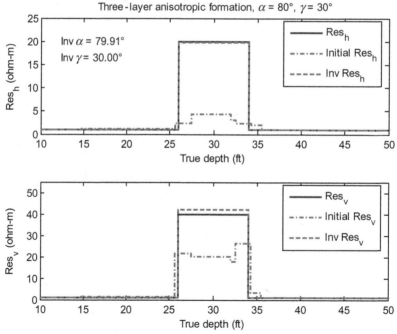

Figure 9.11 Inverted resistivity for Case 2 for the model in Fig. 9.6. Data is contaminated by 5% coherent noise.

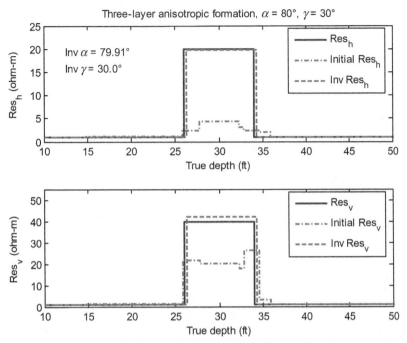

Figure 9.12 Inverted resistivity for Case 2 for the model in Fig. 9.6. Data is contaminated by 5% incoherent noise.

If 5% incoherent noise is added to the raw data, the inversion becomes more challenging. Fig. 9.12 shows the inversion results. It can be seen that the inversion resistivities match well with the formation resistivities. The initial guess of dipping angle, rotation angle, inverted dipping angle, and rotation angle are 37.87, 68.80, 80, and 30 degrees, respectively. The inversion results are converged in 7 iterations.

9.8.2 Inversion of synthetic logs using diagonal components of the magnetic fields

We can use only diagonal terms of magnetic fields instead of full nine components in the inversion process [5] (see Chapter 5: Triaxial Induction Tool and LWD Tool Response in a Transverse Isotropic-Layered Formation). The diagonal terms have sufficient sensitivity to the rotation angle to provide a reasonably good solution as long as the initial guess of the rotation angle error is within 20 degrees from the actual solution. The cross components are only used as a boundary indicator as discussed in Section 9.7.2. In the following discussion, the diagonal terms of the raw data is used to simplify the inversion process. The true formation is the same as that discussed in Section 9.8.1.

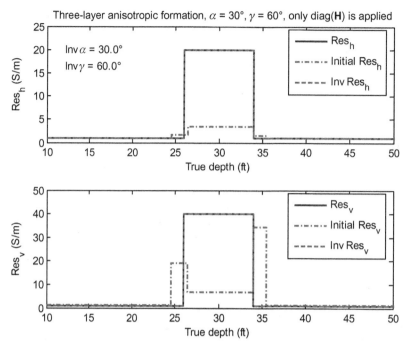

Figure 9.13 Inverted resistivity using only diagonal terms of **H** for Case 1 for the model in Fig. 5.1. Raw data is from 1D forward modeling.

9.8.2.1 Case 1. Dipping angle: 30 degrees, rotation angle: 60 degrees—formation as shown in Fig. 9.6

Similar to Section 9.8.1, the first example is a low-dip case, in which the dipping angle and rotation angle are 30 and 60 degrees, respectively. Only the diagonal terms of the tensor **H** are used in the inversion.

Fig. 9.13 shows the inversion results. From Fig. 9.13, it is seen that the inversion results match very well with the true parameters. Initial guess of dipping angle, rotation angle, inverted dipping angle, and rotation angle are 32.28, 34.62, 30.00, and 60.00 degrees, respectively. The algorithm converged in 6 iterations. From this example, it is found that although the cross components of the magnetic field are not used in the inversion process, the inversion still yields a reasonably accurate solution, which confirms Anderson's conclusion when no noise is added to the log data.

To investigate noise performance of the method, coherent and noncoherent noises are added to the simulated log data. Fig. 9.14 shows the inversion results for Case 1 also using only diagonal terms of **H** but with 5% coherent noise.

From Fig. 9.14, we can see that the inversion results match well with true parameters. Initial guess of dipping angle, rotation angle, inverted dipping angle, and rotation angle are 32.30, 34.43, 29.76, and 59.99 degrees, respectively. The inversion

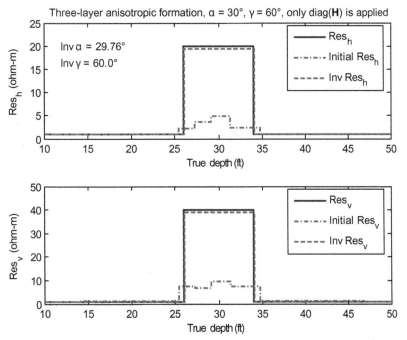

Figure 9.14 Inverted resistivity using only diagonal terms of **H** for Case 1 for the model in Fig. 9.6. Data is contaminated by 5% coherent noise.

is convergent in 6 iterations. This example implies that the inversion code still works stably and yields satisfactory results using only diagonal terms of magnetic field in the presence of coherent noise.

Next, we add 5% incoherent noise to diagonal terms of **H**. The inversion results are shown in Fig. 9.15. The inversion results have good agreement with the true formation. Initial guess of dipping angle, rotation angle, inverted dipping angle, and rotation angle are 30.74, 26.81, 30.36, and 59.87 degrees, respectively. The inversion converges in 9 iterations.

9.8.2.2 Dipping angle: 80 degrees, rotation angle: 30 degrees—formation as shown in Fig. 9.6

In greater dipping angle, the inversion process is conducted using only diagonal components of the H fields. Using the same data with Section 9.8.1.2, the true dipping angle and rotation angle are 80 and 30 degrees, respectively.

Fig. 9.16 shows the inversion results using only diagonal terms of **H** in the 1D inversion without noise. In Fig. 9.16, initial guess of dipping angle, rotation angle, inverted dipping angle, and rotation angle are 38.54, 29.15, 80, and 30 degrees, respectively.

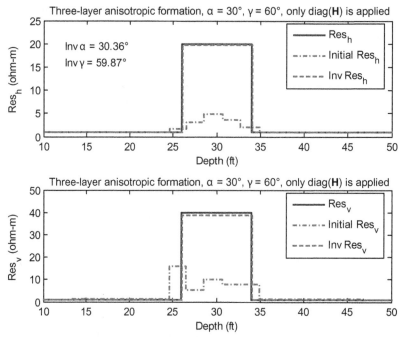

Figure 9.15 Inverted resistivity using only diagonal terms of **H** for the Case 1 for the model in Fig. 9.6. Data is contaminated by 5% incoherent noise.

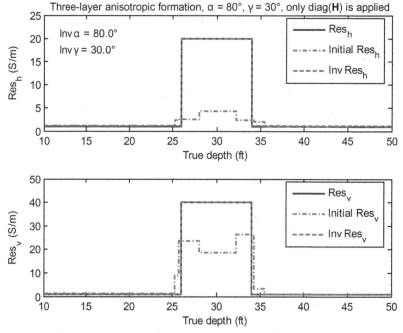

Figure 9.16 Inverted resistivity using only diagonal terms of **H** for the model in Fig. 9.16. Data is from 1D forward model TRITI10.

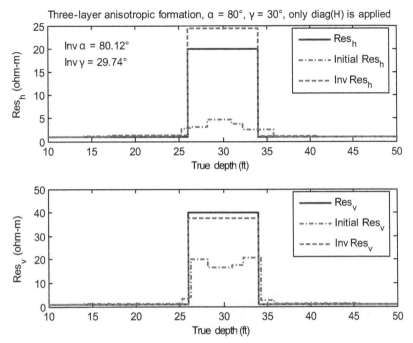

Figure 9.17 Inverted resistivity using only diagonal terms of **H** for the model in Fig. 9.6. Data is contaminated by 5% coherent noise.

Fig. 9.17 shows the inversion results with 5% coherent noise. In Fig. 9.17, initial guess of dipping angle, rotation angle, inverted dipping angle, and rotation angle are 35.99, 60.50, 80.12, and 29.74 degrees, respectively.

Fig. 9.18 shows the inversion results with 5% incoherent noise. In Fig. 9.18, initial guess of dipping angle, rotation angle, inverted dipping angle, and rotation angle are 38.40, 36.18, 80.12, and 30.03 degrees, respectively.

From Figs. 9.16–9.18, we can see that satisfactory inversion results are obtained in all the cases. The iteration numbers for Figs. 9.16, 9.17, and 9.18 are 10, 12, and 23, respectively.

9.8.3 Inversion of a five-layer synthetic formation

In this section, we consider a five-layer synthetic formation called Oklahoma model as shown in Fig. 9.19. The conductivities of the model is indicated in the figure. All the layers are anisotropic with conductivity contrast ratio of 2. The raw data is obtained from 1D forward modeling The spacing between the transmitter and receiver is 40 in. and the frequency is 20 kHz.

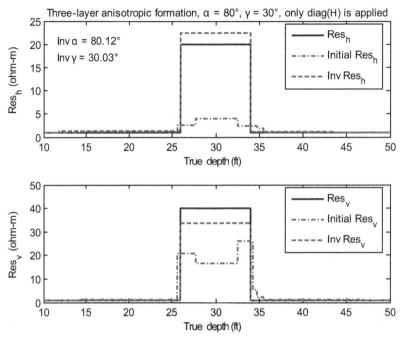

Figure 9.18 Inverted resistivity using only diagonal terms of **H** for the model in Fig. 9.6. Data is contaminated by 5% incoherent noise.

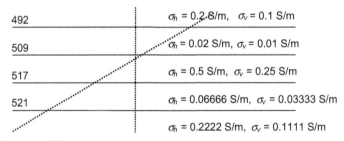

Figure 9.19 A five-layer Oklahoma model.

9.8.3.1 Dipping angle: 30 degrees, rotation angle: 0 degrees

In the first case, we assume that the well deviation is 30 degrees. Fig. 9.20 shows the inversion results.

In Fig. 9.20, initial guess of dipping angle, rotation angle, inverted dipping angle, and rotation angle are 41.49, 0.0057, 30, and 0 degrees, respectively. The inversion is converged in seven times.

Fig. 9.21 plotted the inversion results with 5% coherent noise added to raw data. In Fig. 9.21, the initial guess of dipping angle, rotation angle, inverted dipping angle,

Theory of Inversion for Triaxial Induction and LWD Logging Data in 1D and 2D Formations 321

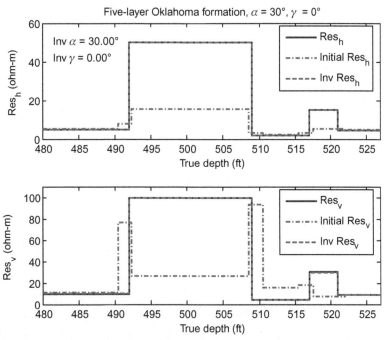

Figure 9.20 Inverted resistivity for the model in Fig. 9.19. The dipping angle and rotation angle are 30 and 0 degrees, respectively, no noise is added to the logs. The synthetic log data is obtained by using forward modeling.

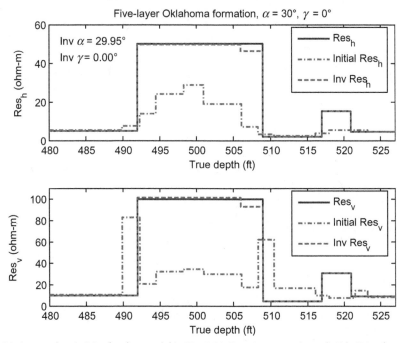

Figure 9.21 Inverted resistivity for the model in Fig. 9.19. Data is contaminated with 5% coherent noise.

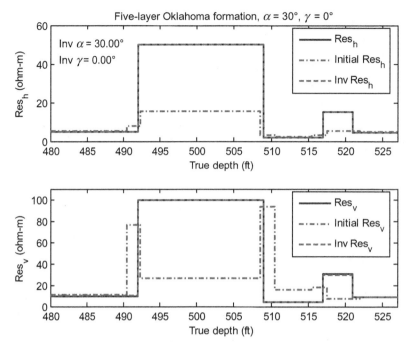

Figure 9.22 Inverted resistivity for the model in Fig. 9.19. Data is contaminated with 5% incoherent noise.

and rotation angle are 41.76, 0.005, 29.95, and 0 degrees, respectively. The inversion is converged in 10 iterations.

Fig. 9.22 shows the inversion results with 5% incoherent noise added to raw data. Initial guess of dipping angle, rotation angle, inverted dipping angle, and rotation angle are 41.72, 0.005, 29.92, and 0 degrees, respectively. The inversion is converged in 9 iterations.

9.8.3.2 Dipping angle: 85 degrees, rotation angle: 0 degrees

In the second case the dipping angle is 85 degrees. Fig. 9.23 shows inversion results with raw data. In Fig. 9.23, the initial guess of dipping angle, rotation angle, inverted dipping angle, and rotation angle are 82.12, 0.005, 85, and 0 degrees, respectively. The inversion is converged in 7 iterations.

Fig. 9.24 shows inversion results with 5% coherent noise. In Fig. 9.24, the initial guess of dipping angle, rotation angle, inverted dipping angle, and rotation angle are 82.10, 0.005, 84.98, and 0 degrees, respectively. The inversion is converged in seven iterations.

Fig. 9.25 shows inversion results with 5% incoherent noise added to raw data. In Fig. 9.25, the initial guess of dipping angle, rotation angle, inverted dipping angle,

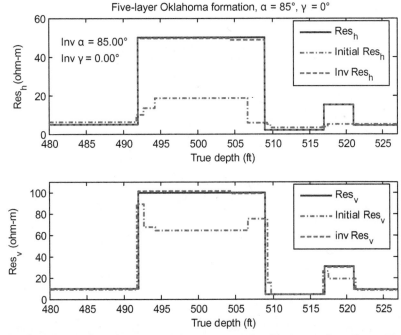

Figure 9.23 Inverted resistivity for the model in Fig. 9.19. The dipping angle and rotation angle are 85 and 0 degrees, respectively. No noise is added. The synthetic logs are obtained from the forward modeling.

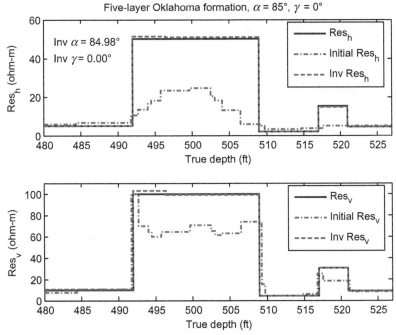

Figure 9.24 Inverted resistivity for the model in Fig. 9.19. Data is contaminated with 5% coherent noise.

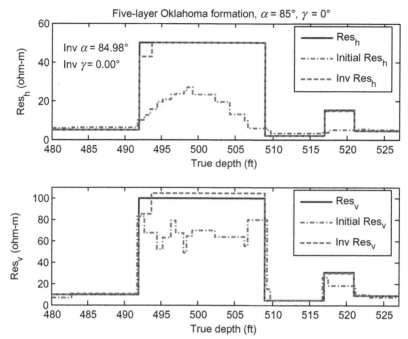

Figure 9.25 Inverted resistivity for the model in Fig. 9.19. Data is contaminated with 5% incoherent noise.

and rotation angle are 82.01, 0.005, 84.98, and 0 degrees, respectively. In all the three figures, the inversion results agree very well with the true parameters.

9.8.4 Inversion in a 15-layer synthetic formation

Oklahoma model is a benchmark model often used to validate inversion algorithm [9,10]. In this example, a 15-layer Oklahoma formation is discussed and the inversion algorithm to the synthetic logs in the formation as shown in Fig. 9.26. Again the distance between the transmitter and receiver is 40 in. and the frequency is 20 kHz.

To test the algorithm in different cases, the synthetic data used in this section is obtained from the method based on Chapter 6, Triaxial Induction and Logging-While-Drilling Logging Tool Response in a Biaxial Anisotropic-Layered Formation, of this book. Fig. 9.27 shows the input log. Since we assume the borehole is vertical, all cross components are zero in the isotropic formation.

Fig. 9.28 shows the inversion results. The inversion process converges in 28 iterations. From Fig. 9.28, it is found that the inverted resistivity is very close to the true resistivity in low-resistivity zones while far from the true resistivity in high-resistivity zones. This is reasonable since induction logging tool has better sensitivity to conductive layer than resistive layer. When the formation resistivity is greater than

492	$\sigma_h = 0.2$ S/m, $\sigma_v = 0.2$ S/m
509	$\sigma_h = 0.02$ S/m, $\sigma_v = 0.02$ S/m
517	$\sigma_h = 0.5$ S/m, $\sigma_v = 0.5$ S/m
521	$\sigma_h = 0.06666$ S/m, $\sigma_v = 0.06666$ S/m
524	$\sigma_h = 0.2222$ S/m, $\sigma_v = 0.2222$ S/m
531	$\sigma_h = 0.01$ S/m, $\sigma_v = 0.01$ S/m
535	$\sigma_h = 0.2858$ S/m, $\sigma_v = 0.2858$ S/m
541	$\sigma_h = 0.00222$ S/m, $\sigma_v = 0.00222$ S/m
544	$\sigma_h = 0.0332$ S/m, $\sigma_v = 0.0332$ S/m
549	$\sigma_h = 0.00166$ S/m, $\sigma_v = 0.00166$ S/m
556	$\sigma_h = 0.05$ S/m, $\sigma_v = 0.05$ S/m
574	$\sigma_h = 0.00134$ S/m, $\sigma_v = 0.00134$ S/m
582	$\sigma_h = 0.005$ S/m, $\sigma_v = 0.005$ S/m
589	$\sigma_h = 0.13334$ S/m, $\sigma_v = 0.13334$ S/m
	$\sigma_h = 0.002$ S/m, $\sigma_v = 0.002$ S/m

Figure 9.26 A 15-layer Oklahoma model.

100 ohm-m, the resolution of induction logging tool is significantly decreased. To investigate inversion results, we convert the inverted resistivity into conductivity, as shown in Fig. 9.29. To investigate the inversion result, we use the formation parameters obtained from the inversion as the input of the forward modeling and calculate the apparent conductivity. Fig. 9.30 compares the input log with the calculated log from the inverted formation parameters.

9.8.5 Inversion of real, isotropic formation with synthetic data

In this example, the geometric formation is constructed according to Devine test site of BP America [10] as shown in Fig. 9.31. The simulated log is obtained by using the two-dimensional (2D) forward modeling method using finite element method discussed [11]. The borehole is assumed to be vertical. The frequency is 512 Hz and the distance between the transmitter and receiver is 40 in. The simulated log is shown in Fig. 9.32.

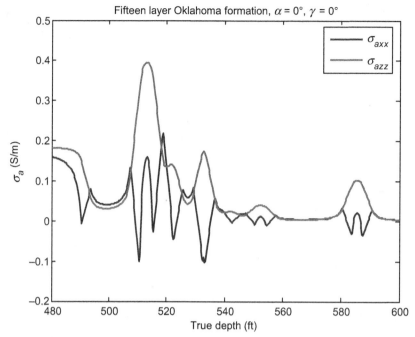

Figure 9.27 Simulated log obtained by using the method discussed in Chapter 6, Triaxial Induction and Logging-While-Drilling Logging Tool Response in a Biaxial Anisotropic-Layered Formation, for the 15-layer Oklahoma model in Fig. 9.26.

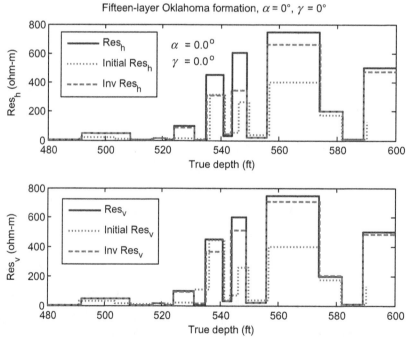

Figure 9.28 Inverted resistivity for the 15-layer Oklahoma model in Fig. 9.26. Raw data is provided by INDTRI.

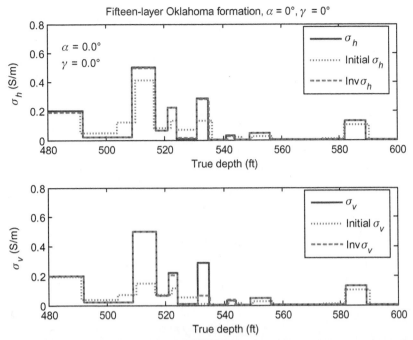

Figure 9.29 Inverted conductivity for the model in Fig. 9.26. Raw data is obtained by using the method discussed in Chapter 6, Triaxial Induction and Logging-While-Drilling Logging Tool Response in a Biaxial Anisotropic-Layered Formation.

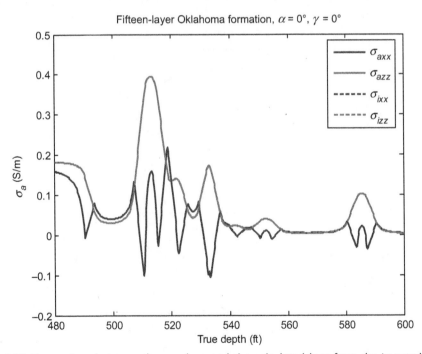

Figure 9.30 Comparison between the raw logs and the calculated logs from the inverted formation parameters.

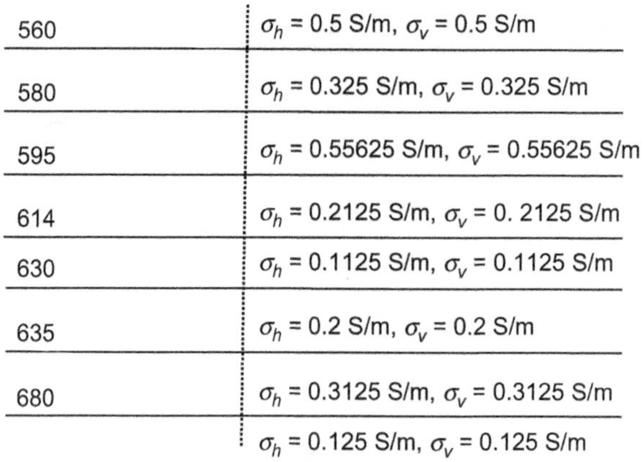

560
σ_h = 0.5 S/m, σ_v = 0.5 S/m
580
σ_h = 0.325 S/m, σ_v = 0.325 S/m
595
σ_h = 0.55625 S/m, σ_v = 0.55625 S/m
614
σ_h = 0.2125 S/m, σ_v = 0. 2125 S/m
630
σ_h = 0.1125 S/m, σ_v = 0.1125 S/m
635
σ_h = 0.2 S/m, σ_v = 0.2 S/m
680
σ_h = 0.3125 S/m, σ_v = 0.3125 S/m
σ_h = 0.125 S/m, σ_v = 0.125 S/m

Figure 9.31 Eight-layer Devine test site formation.

Eight-layer Devine test site formation, α = 0°, γ= 0°
σ_a (S/m)
0.6
0.55
0.5
0.45
0.4
0.35
0.3
0.25
0.2
0.15
0.1
500
520
540
560
580
600
620
640
660
680
700
True depth (m)

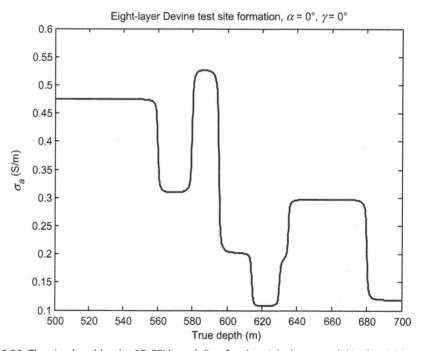

Figure 9.32 The simulated log by 2D-FEM modeling for the eight-layer model in Fig. 9.31.

Fig. 9.33 shows the inversion results with 5% coherent noise. Fig. 9.34 compares the contaminated input data (5% coherent noise) with the calculated log. The initial guess of Fig. 9.33 is modified from zero-dimensional inversion by manually merging boundaries.

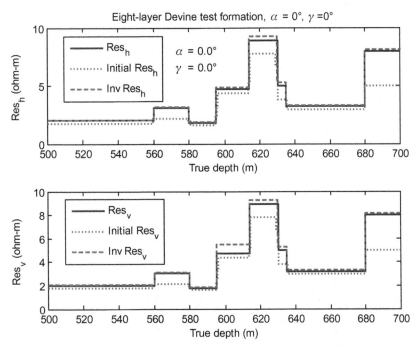

Figure 9.33 Inverted resistivity for the eight-layer model in Fig. 9.31. Raw data is contaminated by 5% coherent noise.

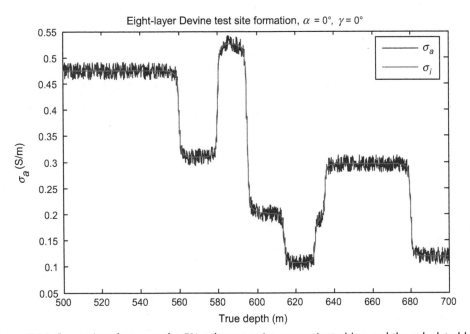

Figure 9.34 Comparison between the 5% coherent-noise contaminated log and the calculated log for the 8-layer Devine test site model in Fig. 9.31.

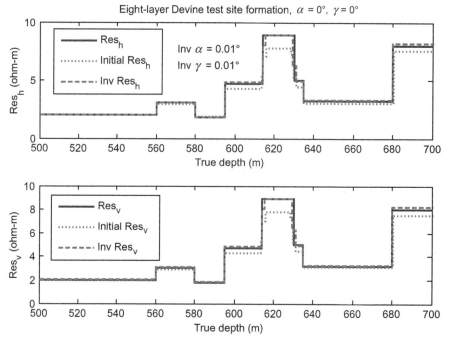

Figure 9.35 Inverted resistivity for the eight-layer model in Fig. 9.31. Raw data is contaminated by 5% incoherent noise.

Fig. 9.35 shows the inversion results with 5% incoherent noise. Fig. 9.36 compares the contaminated input data (5% incoherent noise) with the calculated log. Fig. 9.35 uses the same initial guess as for Fig. 9.33.

9.8.6 Inversion of real log data

Finally we apply the inversion method to invert an induction log taken from Well No. 36-6, East Newkirk, Oklahoma [1], as shown in Fig. 9.37.

In Fig. 9.37, the *solid line* represents the field log from 6FF40. The *dashed line* represents the inverted log from Zhang et al. [1]. We use polynomial expansion to fit the field log and plot it in Fig. 9.38. Fig. 9.39 shows the inversion results using the proposed algorithm. Fig. 9.40 compares the field log with the calculated log from the inverted parameters. By comparing Figs. 9.37 and 9.40, we can see that proposed inversion method has better agreement between the field log and the calculated log than the published paper.

9.9 INVERSION OF INDUCTION LOGS IN A TWO-DIMENSIONAL FORMATION

In the previous sections, we discussed inversion method of triaxial induction measurements in a 1D formation without borehole and invasion zones. Most of the cases, the

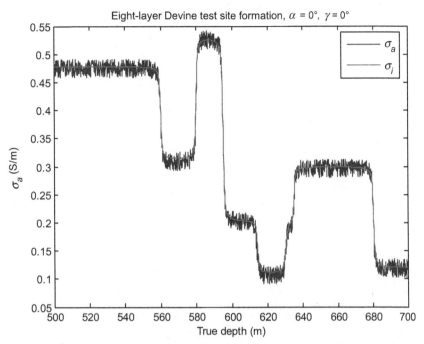

Figure 9.36 Comparison between the 5% incoherent-noise contaminated log and the calculated log for the eight-layer Devine test site model in Fig. 9.31.

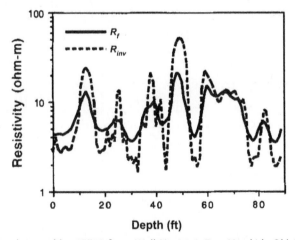

Figure 9.37 Field log detected by 6FF40 from Well No 36-6, East Newkirk, Oklahoma [1].

1D inversion is used to solve a practical inversion problem. However, sometimes, the influence of borehole and invasion zones must be considered, which involves a 2D formation: borehole, invasion zones, and vertical layers. In this section, we will discuss the 2D inversion method using array induction logging data. Due to the limit of

332 Theory of Electromagnetic Well Logging

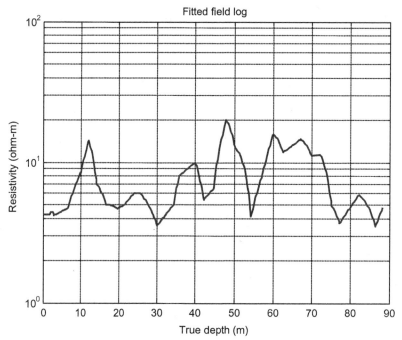

Figure 9.38 Fitted field log in Fig. 9.37.

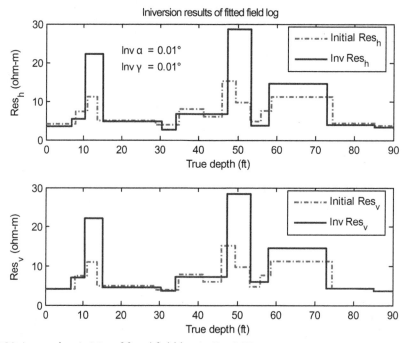

Figure 9.39 Inverted resistivity of fitted field log in Fig. 9.37.

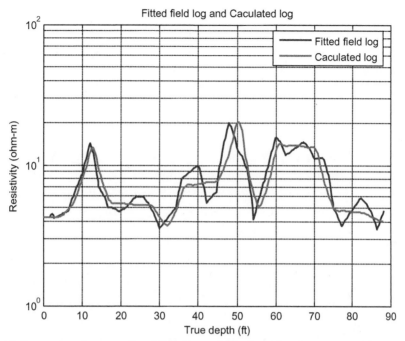

Figure 9.40 Comparison between fitted field log and calculated log from inverted formation.

analytical solutions, the inversion discussed in this section is only limited to vertical borehole in an isotropic 2D formation.

As discussed in the previous sections, inversion of induction logs is a nonlinear problem. The purpose of the inversion is to determine the unknown electric properties of the formation from the measured logs. Starting from an estimated initial formation profile, a simulation program (forward modeling) is called to compute the logs based on the initial formation profile. The computed logs are compared with the measured logs. A misfit, which measures the difference between the measured and computed logs, is defined as the least squares norm of these two data sets. If the misfit does not satisfy a given error tolerance, a new formation profile is estimated and the computation is repeated until the tolerance criterion is met. The last estimated profile is considered to be the profile which is closest to the practical formation. This process may diverge if the inversion problem is not properly setup, or the formed model is not accurate. The divergence of the inversion usually results in the ill conditioning of the linear system, which is an approximation of the nonlinear problem.

In Sections 9.1–9.8, we can see that in 1D inversion algorithm, the process is based on an optimization, trying to minimize the cost function in order to obtain a stable solution. The forward modeling is embedded into the inversion process. In this section, we will discuss another inversion method, which is based on a scattering field in the targeted area. In 2D inversion, the interested space is divided into small elements, each has an unknown conductivity value. The idea of inversion is very easy to understand: each element of the space contributes to the received signal as a scatterer in a uniform background conductivity. The scattered field by this element can be expressed as its first-order approximation and it is the difference between the total field and the field when the source is in the homogeneous background. Since this process is called Born Approximation, therefore, the inversion based on this method is called Born inversion.

As we have seen, several inversion algorithms [12—18] have been developed. Caorsi et al. [13,14] proposed an inversion algorithm, which subdivides the nonlinear problem into two linear steps and uses the moment method with pseudoinverse transformation to solve the ill conditioned problems which occur when the kernel properties of the integral equation of the electric field are transferred into the matrix equation. Some researchers [15—18] have investigated algorithms based on successively linearizing the nonlinear integral equation. To solve this problem, Chew et al. proposed a Born iterative method (BIM) [15] and a distorted Born iterative method (DBIM) [16]. They showed that the former is more noise-resistant than the latter, but the latter converges much faster. To greatly enhance the convergence speed of the DBIM, they applied the fast recursive aggregate T-matrix algorithm together with conjugate gradient method into the solution of the direct scattering part in the iterative inverse scattering algorithm [17]. Liu et al. [18] developed an efficient noise-tolerant, 2D conductivity-imaging technique to convert the data measured by an electromagnetic (EM) cross-hole tool. This method is especially useful when the formations are mostly layered. Zhang et al. [1] proposed three algorithms, which are maximum flatness, maximum oil, and minimum oil algorithms, to convert induction logs. These algorithms are efficient and stable because noise in the field logs have little effect on the result.

As discussed in the previous sections, the inversion process is treated as an optimization problem. The horizontal eigenstate method discussed in Chapter 8, Induction and Logging-While-Drilling Resistivity Tool Response in a Two-Dimensional Isotropic Formation, is used as forward modeling in the 2D inversion. A linearization method is used to obtain a set of linear equations. These inversion equations are solved with a least squares method.

9.9.1 Theory of 2D induction log inversion

Fig. 9.41 shows the flowchart of the 2D inversion algorithm. Compared with Fig. 9.1 of 1D inversion, it is very similar. Generally, the method needs a certain number of iterations before the final inverted formation can be obtained. In each iteration, the background formation profile is updated automatically, and the computed logs will be calculated based on the new background formation.

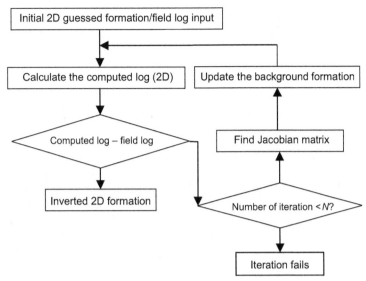

Figure 9.41 The flowchart of 2D induction log inversion.

From Chapter 1, Introduction to Well Logging, for the coil-type induction tools, such as 2C40, 6FF40, and Array Induction tools, the tools directly measure the induced voltages in the receiving coils. The apparent conductivity σ_a is defined as

$$\sigma_a = \frac{1}{L}\sum_{i=1}^{N_t}\sum_{j=1}^{N_r} NT_i NR_j V_{ij} \qquad (9.44)$$

where N_t = number of transmitting coils;
N_r = number of receiving coils;
NT_i = number of turns of the ith transmitting coil;
NR_j = number of turns of the jth receiving coil;
V_{ij} = voltage in the jth receiving coil induced by the ith transmitting coil;
and L = normalization constant determined by the tool parameters.

From Eq. (9.44), it is seen that the apparent conductivity is a linear function of voltage V_{ij}. Usually, voltage V_{ij} is a nonlinear function of the formation conductivities. The axisymmetric inhomogeneous medium considered here is a three-dimensional (3D) medium with rational symmetry. For convenience in the following discussion, the cylindrical coordinate (ρ, ϕ, z) is chosen such that z is the rotation axis. In this work, the magnetic permeability μ and the dielectric constant ε are assumed to be constants μ_0 and ε_0, respectively. In a 2D formation, the conductivity $\sigma(\rho, z)$ of the medium is the function of ρ and z. In the inversion process, the formation is divided into many small grids along the ρ and z directions. It is assumed that the conductivity in each grid is constant. These grids are used not only in the inversion, but also in the forward problem. Therefore the apparent conductivity σ_a can be expressed as

$$\sigma_a = \sigma_a(\sigma_1, \sigma_2, \ldots, \sigma_N) \qquad (9.45)$$

where N is the number of grids.

As we know, the nonlinear problem is very difficult to be solved. To obtain the inversion result, e.g., the formation profile, a linear relationship between σ and ν is desired. As discussed in Section 9.1, to keep inverted conductivity positive, an exponential transformation is used as follows:

$$\sigma_i = \pi \exp(q_i) \qquad (9.46)$$

The apparent conductivity σ_a is expanded in a Taylor's series,

$$\sigma_a(\overline{\sigma}^{(i+1)}) = \sigma_a(\overline{\sigma}^{(i)}) + \sum_{n=1}^{N} \frac{\partial \sigma_a(\overline{\sigma}^{(i)})}{\partial \sigma_n} \Delta q_n^{(i)} + \cdots \qquad (9.47)$$

where $\overline{\sigma}$ is a vector $(\sigma_1, \sigma_2, \ldots, \sigma_N)$, N is the number of the grids, i is the iteration index and

$$\Delta q_n^{(i)} = q_n^{(i+1)} - q_n^{(i)} \qquad (9.48)$$

If the high-order differential terms in Eq. (9.48) are truncated, the first two terms are only considered. Eq. (9.48) becomes a linear equation,

$$\sigma_a(\overline{\sigma}^{(i+1)}) \approx \sigma_a(\overline{\sigma}^{(i)}) + \sum_{n=1}^{N} \frac{\partial \sigma_a(\overline{\sigma}^{(i)})}{\partial \sigma_n} \Delta q_n^{(i)} \qquad (9.49)$$

To calculate the differential term, an approximation is used,

$$A_n^{(i)} = \frac{\partial \sigma_a(\overline{\sigma}^{(i)})}{\partial \sigma_n} = \frac{\sigma_a^{(i)}\big|_{\sigma_n + \Delta \sigma_n} - \sigma_a^{(i)}\big|_{\sigma_n}}{\Delta \sigma_n} \qquad (9.50)$$

where $\Delta \sigma_n$ is a very small value, usually equal to a 10th of σ_n. The terms $\sigma_a^{(i)}\big|_{\sigma_n + \Delta \sigma_n}$ and $\sigma_a^{(i)}\big|_{\sigma_n}$ are the apparent conductivities computed by the forward modeling at the ith iteration when the conductivities of the nth grid are $\sigma_n + \Delta \sigma_n$ and σ_n, respectively.

In Eq. (9.49), the term $\sigma_a(\sigma^{(i)})$ is the apparent conductivity computed from the formation profile at the ith iteration by a forward modeling. This term is represented as $\sigma_{ac}^{(i)}$. The final purpose of the inversion is to let the computed log $\sigma_{ac}^{(i)}$ approach the field log σ_{af} within some error tolerance. Here, the σ_{af} is equal to $\sigma_a(\overline{\sigma}^{(i+1)})$. Therefore Eq. (9.49) is modified as

$$\sigma_{af} - \sigma_{ac}^{(i)} = \sum_{n=1}^{N} A_n^{(i)} \Delta q_n^{(i)} \qquad (9.51)$$

For M logging points, the equations can be written as

$$\begin{aligned}
\sigma_{af,1} - \sigma_{ac,1}^{(i)} &= \sum_{n=1}^{N} A_{1n}^{(i)} \Delta q_n^{(i)}, \\
\sigma_{af,2} - \sigma_{ac,2}^{(i)} &= \sum_{n=1}^{N} A_{2n}^{(i)} \Delta q_n^{(i)}, \\
&\vdots \\
\sigma_{af,M} - \sigma_{ac,M}^{(i)} &= \sum_{n=1}^{N} A_{Mn}^{(i)} \Delta q_n^{(i)}
\end{aligned} \qquad (9.52)$$

where

$$A_{mn}^{(i)} = \frac{\partial \sigma_{ac}(\overline{\sigma}^{(i)})}{\partial \sigma_n} = \frac{\sigma_{ac,m}^{(i)}\big|_{\sigma_n + \Delta\sigma_n} - \sigma_{ac,m}^{(i)}\big|_{\sigma_n}}{\Delta \sigma_n} \qquad (9.53)$$

with $m = 1, 2, \ldots, M$.

The Jacobian matrix is defined as

$$[\mathbf{A}^{(i)}] = \begin{bmatrix} A_{11}^{(i)} & A_{12}^{(i)} & \cdots & A_{1N}^{(i)} \\ A_{21}^{(i)} & A_{22}^{(i)} & \cdots & A_{2N}^{(i)} \\ \cdots & \cdots & \cdots & \cdots \\ A_{M1}^{(i)} & A_{M2}^{(i)} & \cdots & A_{MN}^{(i)} \end{bmatrix} \qquad (9.54)$$

From Eq. (9.52), it can be seen that the unknown $\Delta q_n^{(i)}$ ($n = 1, 2,\ldots, N$) has a linear relationship with the difference between the computed logs and the field logs. Therefore Eq. (9.52) can be represented in the following linear matrix equation:

$$[\mathbf{A}^{(i)}][\Delta \mathbf{q}^{(i)}] = [\mathbf{B}^{(i)}] \qquad (9.55)$$

where

$$B_m^{(i)} = \sigma_{af,m} - \sigma_{ac,m}^{(i)}, \quad m = 1, 2, \ldots, M \qquad (9.56)$$

To solve Eq. (9.55), the least squares method can be used when there are more logging points than the number of unknowns. The solution yields

$$[\Delta \mathbf{q}^{(i)}] = \left\{ [\mathbf{A}^{(i)}]^T [\mathbf{A}^{(i)}] \right\}^{-1} [\mathbf{A}^{(i)}]^T [\mathbf{B}^{(i)}] \quad (9.57)$$

After the $[\Delta \mathbf{q}^{(i)}]$ is found, the conductivities of the formation at the *i*th iteration can be obtained by

$$q_n^{(i+1)} = q_n^{(i)} + \Delta q_n^{(i)}, \quad n = 1, 2, \ldots, N \quad (9.58)$$

This formation profile is also used as the background formation in the next iteration.

In the inversion process, it is impossible to require the computed logs to be exactly equal to the field logs. Usually, the standard deviation between the field logs and the computed logs is defined. In this study, the root-mean-square difference is used to describe how close the computed logs are to the field logs. The criteria for rms error is given by

$$S = \sqrt{\frac{1}{M} \sum_{m=1}^{M} \left(\sigma_{af,m} - \sigma_{ac,m}^{(i)} \right)^2} \quad (9.59)$$

The relative rms error is also defined as

$$R_s = \frac{1}{\sigma_{aver}} \sqrt{\frac{1}{M} \sum_{m=1}^{M} \left(\sigma_{af,m} - \sigma_{ac,m}^{(i)} \right)^2} \quad (9.60)$$

where σ_{aver} is the average of the conductivities of the formation. In practice, S and R_s should be minimized.

9.9.2 Results from the least squares inversion

As an application, some numerical results are shown in this section. The tool used in the inversion algorithm is shown in Fig. 9.42. It consists of one transmitting coil and four receiving coils separated by different spaces. The operating frequency is 20 kHz. At this frequency, the conduction current is dominant over the displacement current. Therefore it is assumed that the relative dielectric constant and magnetic permeability of the formation are all equal to 1, and that the conductivity σ is a function of the space. The background conductivity for all the examples shown below is the average of the field logs. The logging range is from −4.0 to

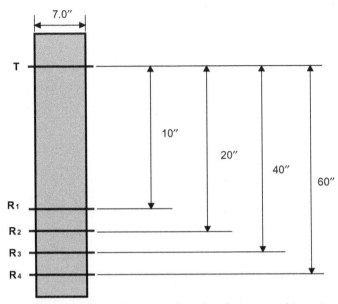

Figure 9.42 The configuration of an induction tool used in the proposed inversion algorithm.

4.0 ft. The vertical grids covering this area is −4.0, −3.0, −2.0, −1.0, 0.0, 1.0, 2.0, 3.0, 4.0, and the radial grids vary according to the variation of the formation. The density of logging point is two points per foot. Therefore from −4.0 to 4.0, there are 17 logging points. Moreover, the four receiving coils of the tool are treated individually. This makes the total number of the logs equal to 68. In each iteration of the inversion, the horizontal eigenstate method with 20 eigenmodes is used to solve the forward problem. The horizontal eigenstate method has already been introduced in Chapter 8, Induction and Logging-While-Drilling Resistivity Tool Response in a Two-Dimensional Isotropic Formation. In all the following examples, the synthetic data obtained by the horizontal eigenstate method are used as the field logs.

Fig. 9.43 shows the profile of an original formation. It consists of eight horizontal layers and a borehole. The conductivity of the mud is 1.0 S/m. The radius of the borehole is R_b. The tool used is shown in Fig. 9.42. In this case, the number of unknowns is 8, and the number of the logging data is 68, which is much greater than that of unknowns. Table 9.1 shows the comparison of the inverted formation with the original formation at the different radii of the boreholes, which are 3, 6, and 10 in., respectively. It can be seen that the inverted formation agrees very well with the original formation, and the effect of the borehole is very small. The number of iterations is 4.

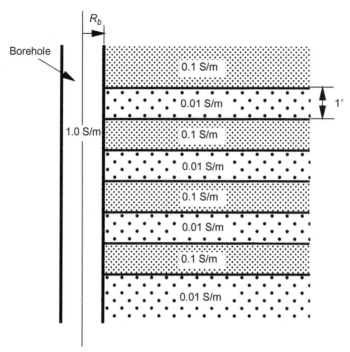

Figure 9.43 A multilayered formation with a borehole. The radius of the borehole is R_b. The conductivity of the mud is 1.0 S/m.

Table 9.1 Comparison of the inverted formation and the original formation with different radii of boreholes

σ (S/m) (Original formation)	Inverted σ (S/m) $R_b = 3$ in.	Inverted σ (S/m) $R_b = 6$ in.	Inverted σ (S/m) $R_b = 10$ in.
0.1	0.1000097307	0.1000119845	0.1000200722
0.01	0.0099945024	0.0099911419	0.0099741320
0.1	0.100006756	0.1000091909	0.1000222446
0.01	0.0100033515	0.0100031811	0.0099990906
0.1	0.1000038676	0.1000041877	0.1000057122
0.01	0.0100097072	0.0100124188	0.0100229887
0.1	0.0999961061	0.0999941198	0.0999859646
0.01	0.0100117233	0.0100132176	0.0100179193
Error S (rms)	7.9691×10^{-6}	6.626×10^{-6}	8.215×10^{-6}
The relative error (rms)	1.4489×10^{-4}	1.2047×10^{-4}	1.4936×10^{-4}

The conductivity of the mud is 1.0 S/m. The formation profile is shown in Fig. 9.43. The configuration of the tool is shown in Fig. 9.42. The number of iterations is 4.

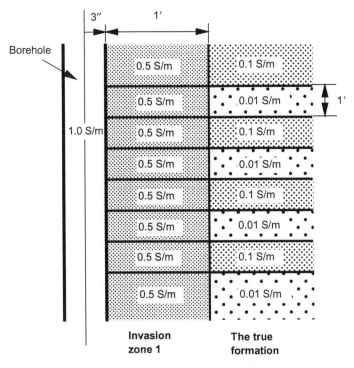

Figure 9.44 A multilayered formation with a borehole and one invasion zone. The borehole conductivity and the radius is 1.0 S/m and 3 in., respectively.

Fig. 9.44 shows a multilayered formation with a borehole and an invasion zone. The radius of the borehole is 3 in. The conductivity of the mud is 1.0 S/m. The area is divided into vertical grids at -4, -3, -2, -1, 0, 1, 2, 3, 4 ft and radial grids at 0.25, 1.25, 50.0 ft. The total number of unknowns is $2 \times 8 = 16$. The tool shown in Fig. 9.42 is used. Table 9.2 shows the comparison of the inverted formation and the original formation. Note that the inversion result is very satisfactory. Both the rms error and the relative rms error are very small. The number of iterations is 4.

Fig. 9.45 shows a multilayered formation with a borehole and two invasion zones. The conductivity of the borehole mud is 1.0 S/m. The radius of the borehole is 3 in. The area is divided into vertical grids at -4, -3, -2, -1, 0, 1, 2, 3, 4 ft and radial grids at 0.25, 1.25, 2.25, 50.0 ft. Therefore the total number of unknowns is 24. The inversion results listed in Table 9.3 shows that this inversion is a good reconstruction of the original formation. A similar case shown in Fig. 9.46 has the same grids as shown in Fig. 9.47. Table 9.4 shows the inverted formation. To help clarifying the

Table 9.2 Comparison of the inverted formation and the original formation with a borehole and one invasion zone

σ (S/m) (Original formation)		Inverted σ (S/m)	
Invasion zone 1	The true formation	Invasion zone 1	The true formation
0.5	0.1	0.49999929307	0.1000035328
0.5	0.01	0.5000045263	0.0099930353
0.5	0.1	0.5000022436	0.0999894943
0.5	0.01	0.4999958411	0.0099930698
0.5	0.1	0.4999908484	0.1000560934
0.5	0.01	0.5000040401	0.0099351052
0.5	0.1	0.4999902367	0.1000541831
0.5	0.01	0.4999920622	0.0100022864
Error S (rms)	2.78×10^{-6}	The relative error (rms)	1.018×10^{-5}

The profile of the formation is shown in Fig. 9.44. The number of iterations is 4.

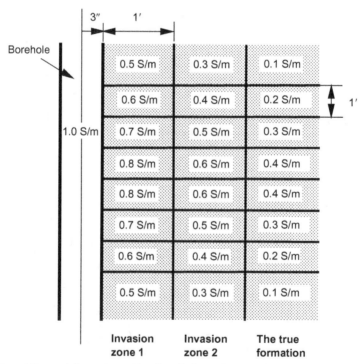

Figure 9.45 A multilayered formation with a borehole and two invasion zones.

geometry of this problem, the inverted results are plotted in a 3D fashion shown in Fig. 9.47. The iteration number of these two cases is 4.

Fig. 9.48 shows a more complicated formation. The properties of the borehole is the same as in the previous cases. The area has vertical grids at −4, −3, −2, −1, 0, 1,

Table 9.3 The inverted formation

	Inverted σ (S/m)	
Invasion zone 1	Invasion zone 2	The true formation
0.5000319847	0.29997086937	0.1000819438
0.5999746681	0.4001300832	0.2005230156
0.6999975645	0.5000828020	0.2985360239
0.7999956481	0.5998741369	0.4019490050
0.7999858562	0.6002771091	0.3978510109
0.7000088223	0.4997962259	0.3015698478
0.6000152169	0.3999027804	0.1995702623
0.4999997639	0.3000206938	0.0999608143
	Error S (rms)	3.64×10^{-6}
	The relative error (rms)	8.089×10^{-6}

The profile of the original formation is shown in Fig. 9.45. The radius of the borehole is 6 in. The number of iterations is 4.

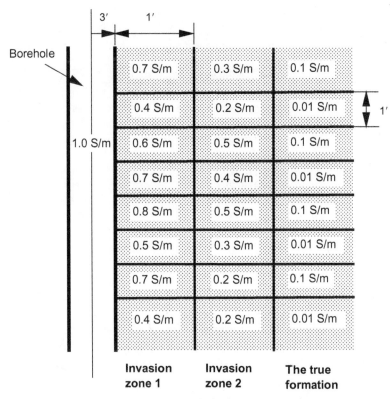

Figure 9.46 A multilayered formation with a borehole and two invasion zones.

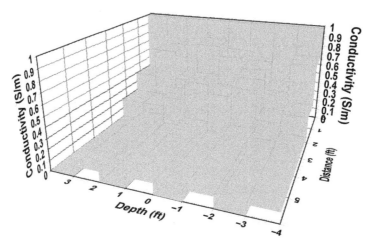

Figure 9.47 The inverted formation. The original formation is shown in Fig. 9.46. The radius of the borehole is 3 in. The rms error and the relative rms error are 9.802×10^{-7} and 3.001×10^{-6}, respectively. The number of iterations is 5.

Table 9.4 The inverted formation

	Inverted σ (S/m)	
Invasion zone 1	**Invasion zone 2**	**The true formation**
0.6999957282	0.3001143124	0.1001481682
0.3999895350	0.2002863838	0.0070674481
0.6000451007	0.4991864919	0.1061830994
0.6999536176	0.4009394205	0.0024918031
0.8000416203	0.4991869939	0.1066045346
0.4999817806	0.3004138133	0.0059920343
0.6999998859	0.1999751263	0.1012736338
0.4000008718	0.1999479711	0.0100017998
	Error S (rms)	9.802×10^{-7}
	The relative error (rms)	3.001×10^{-6}

The profile of the original formation is shown in Fig. 9.46. The configuration of the tool is shown in Fig. 9.42. The number of iterations is 5.

2, 3, 4 ft and radial grids at 0.25, 1.25, 2.25, 3.25, 50.0 ft. The number of unknowns is $4 \times 8 = 32$. The tool used is the same as in Fig. 9.42. Table 9.5 shows the inverted data and the rms errors. Fig. 9.49 is the plotted 3D inverted formation. Fig. 9.50 shows the comparisons between the original formation and inverted formation in different invasion zones. From these figures, it can be seen that the inverted data are very close to that of the original formation. The number of iterations is 4.

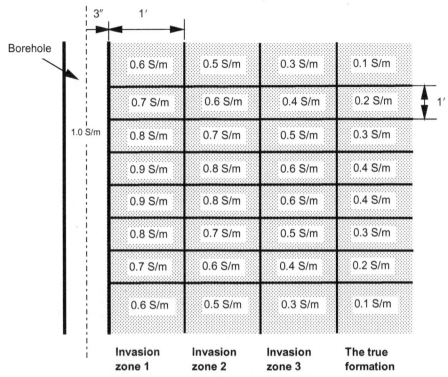

Figure 9.48 A multilayered formation with a borehole and multiinvasion zones.

Table 9.5 The inverted formation

	Inverted σ (S/m)		
Invasion zone 1	**Invasion zone 2**	**Invasion zone 3**	**The true formation**
0.5999720039	0.5003470469	0.2985737802	0.1005337628
0.7000019493	0.5999490684	0.4002519658	0.2016742466
0.8000035047	0.6999252275	0.5008416415	0.2953497958
0.8999940723	0.8001203766	0.5988146034	0.4054811442
0.9000031797	0.7999320215	0.6008443959	0.3954168174
0.7999970169	0.7000442705	0.4995616508	0.3027096789
0.6999970646	0.6000569681	0.3996362446	0.1995456151
0.5999989712	0.5000206553	0.2999356002	0.1000209362
		Error S (rms)	1.41×10^{-5}
		The relative error (rms)	2.686×10^{-5}

The profile of the original formation is shown in Fig. 9.48. The configuration of the tool is shown in Fig. 9.42. The number of iterations is 4.

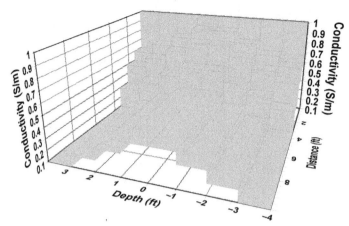

Figure 9.49 The inverted formation. The original formation is shown in Fig. 9.48. The rms error and the relative rms error are 1.41×10^{-5} and 2.686×10^{-5}, respectively.

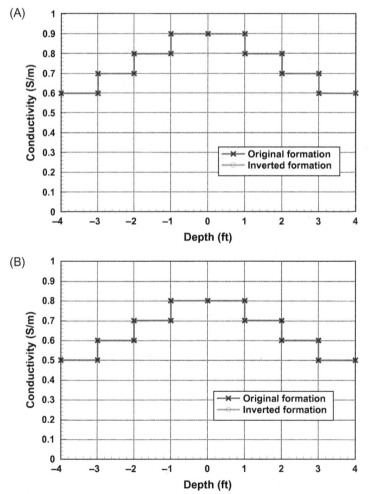

Figure 9.50 Comparison of the original formation shown in Fig. 9.48 and the inverted formation: (A) in invasion zone 1, (B) in invasion zone 2, (C) in invasion zone 3, (D) in invasion zone 4.

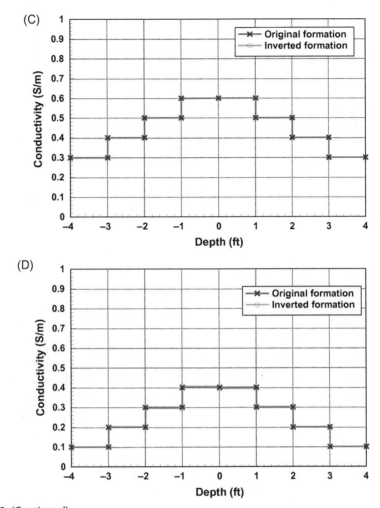

Figure 9.50 (Continued)

9.10 SUMMARY

An automatic inversion algorithm is applied to solve a 2D inverse problem. The field logs at the rotation axis of the axisymmetric inhomogeneous medium are used to reconstruct the conductivity of the formation. In each iteration, the nonlinear inverse problem is linearized, and the Jacobian matrix is calculated. The numerical horizontal eigenstate method is used to solve the forward problem. Several results are shown to demonstrate the inversion of conductivity profiles using induction measurements. It is found that the inverted formation agrees very well with the original formation.

An inversion algorithm for processing induction logs is also discussed in this chapter. The field logs at the axisymmetric inhomogeneous medium are used to reconstruct the conductivity of the medium. This nonlinear inversion problem is linearized by an exponential transformation and a Taylor series expansion. After this processing, all equations involved in the inversion algorithm become linear with the exception of calculating the computed logs which are obtained from an efficient horizontal eigenstate method. In each iteration, the Jacobian matrix is calculated by the forward modeling, and then a small variation in conductivity is computed through the least squares method and used to update the background medium. The purpose of the inversion is to minimize the difference between the computed logs and the field logs. This step takes a lot of computation time. To speed up the inversion process, the average of the field logs is used as the initial inversion background which is very efficient and only a few iterations are needed for the inversion convergence. The results shown in this study are completed within 6 iterations. The accuracy of the inversion result is relatively high. The induction tool used in the inversion consists of one transmitting coil and four receiving coils separated by different spaces. To increase the number of the equations and ensure the accuracy of the results, the measured data of each receiving coil are used individually.

For a formation without deviation and anisotropy, 2D inversion of the formation resistivity distribution can be applied. The 2D inversion is based on a Born approximation and the use of iterative inversion. The forward model is also an approximate model using horizontal eigenmode expansion method. The horizontal eigenmode expansion method expands the EM field in the radial direction and uses mode propagation in the vertical direction in an axially symmetrical formation. In the horizontal direction, numerical method is used and analytical method is used in the vertical direction. The inversion of the 2D formation resistivity distribution is based on the Born approximation to linearize the 2D nonlinear inversion problem. Both numerical and field results are discussed in this chapter.

REFERENCES

[1] Y.C. Zhang, L.C. Shen, C. Liu, Inversion of induction logs based on maximum flatness, maximum oil, and minimum oil algorithms, Geophysics 59 (9) (1994) 1320–1326.
[2] J.E. Dennis Jr., R.B. Schnabel, Numerical Methods for Unconstrained Optimization and Nonlinear Equations, Prentice-Hall, Engle-Wood Cliffs, NJ, 1983.
[3] M. Altman, J. Gill, M. McDonald, Numerical Issues in Statistical Computing for the Social Scientist, John Wiley & Sons, Inc., New Jersey, 2004.
[4] P.E. Gill, W. Murray, Newton-type methods for unconstrained and linearly constrained optimization, Math. Prog. 28 (1974) 311–350.
[5] P.E. Gill, W. Murray, M.H. Wright, Practical Optimization, Academic Press, London, 1981.
[6] M. Zhdanov, D. Kennedy, E. Peksen, Foundations of tensor induction well-logging, Petrophysics 42 (2001) 588–610.

[7] Z. Zhang, L. Yu, B. Kriegshauser, L. Tabarovsky, Determination of relative angles and anisotropic resistivity using multicomponent induction logging data, Geophysics 69 (2004) 898–908.
[8] R. Rosthal, T. Barber, S. Bonner, Field test results of an experimental fully-triaxial induction tool, in: SPWLA 44th Annual Logging Symposium June 2003.
[9] B.I. Anderson, T.D. Barber, T.M. Habashy, The interpretation and inversion of fully triaxial induction data: a sensitivity study, in: SPWLA 43rd Annual Logging Symposium, 2002.
[10] R. Freedman, G.N. Minerbo, Maximum entropy inversion of induction-log data, SPE Form. Eval. 6 (2) (1991) 259–268.
[11] N. Yuan, X.C. Nie, R. Liu, Improvement of 1D simulation codes for induction, MWD and triaxial tools in multi-layered dipping beds, Well Logging Laboratory Technical Report, pp. 32–71, 2010.
[12] M.J. Wilt, D.L. Alumbaugh, H.F. Morrison, A. Becker, K.H. Lee, M. Deszcz-Pan, Crosswell electromagnetic tomography: system design considerations and field results, Geophysics 60 (1995) 871–885.
[13] S. Caorsi, G.L. Gragnani, M. Pastorino, A multiview microwave imaging system for two-dimensional penetrable objects, IEEE Trans. Microw. Theory Technol. 39 (May 1991) 845–851.
[14] S. Caorsi, G.L. Gragnam, and M. Pastorino, Numerical solution to three-dimensional inverse scattering for dielectric reconstruction purposes, in: IEEE Proceedings, vol. 139, February 1992, pp. 45–52.
[15] W.C. Chew, Y.M. Wang, Reconstruction of two-dimensional permittivity distribution using the distorted Born iterative method, IEEE Trans. Med. Imaging 9 (June 1990) 218–225.
[16] Y.M. Wang, W.C. Chew, An iterative solution of two-dimensional electromagnetic inverse scattering problem, Int. J. Imaging Syst. Technol. 1 (1989) 100–108.
[17] Y.M. Wang, W.C. Chew, Accelerating the iterative inverse scattering algorithms by using the fast recursive aggregate T-matrix algorithm, Radio Sci. 27 (March–April 1992) 109–116.
[18] C. Liu, L.C. Shen, Y.C. Zhang, Monitoring saltwater injection using conductivity images obtained by electromagnetic cross-hole measurements, Radio Sci. 30 (5) (1995) 1405–1415.

CHAPTER 10

The Application of Image Theory in Geosteering

Contents

10.1 Introduction	351
10.2 Theory of Forward Modeling Using Image Theory	353
10.2.1 Review of traditional image theory	353
10.2.2 Complex image theory in nonperfect medium	354
10.3 Simulation Results and Discussions	361
10.3.1 One-dimensional formation model	361
10.3.2 Tool configuration	362
10.3.3 Simulation results	363
10.3.4 Discussion	377
10.4 Boundary Distance Inversion	391
10.4.1 Theory of inversion	391
10.4.2 Workflow of inversion problem	392
10.4.3 Processing flow of boundary detection in geosteering	392
10.4.4 Bolzano bisection method	393
10.4.5 Simulation results	393
10.4.6 Simulation results with noise added	400
10.5 Conclusion	404
References	405

10.1 INTRODUCTION

From Chapter 9, Theory of Inversion for Triaxial Induction and Logging-While-Drilling Logging Data in One- and Two-Dimensional Formations, we can see that the inversion in either one-dimensional (1D) or two-dimensional formations are rather mathematically complicated and time consuming due to the fact that the forward modeling is called numerously in each minimization step. Jacobian matrix is usually computed using numerical differentiation, which also uses forward modeling. Unfortunately, the forward modeling of the induction or logging-while-drilling (LWD) tools are relatively slow. In the forward modeling the most time-consuming part is the numerical integration. For real-time inversion, fast forward modeling and inversion method must be used. In this chapter, we will discuss a fast forward modeling and inversion algorithm that uses image theory to avoid numerical integration. This method is an approximate solution but with an accuracy acceptable for most geosteering applications.

We know that the geosteering is a real-time process used to control and adjust the direction of the drilling bit in a horizontal or deviated well in order to keep the drilling in the target layers as shown in Fig. 10.1. One of the most challenging issues in geosteering is the boundary detection, which calculates the tool distance away from the upper or lower boundary from the measured resistivity data. To implement the real-time control of the drilling process, the forward modeling must be fast enough with compromise on the accuracy and complexity of the formation. In this chapter, the complex image theory in lossy media is introduced to simplify the forward model, which reduces the simulation time and improves the real-time controllability of the geosteering system. This method is implemented in both two-layer and three-layer cases. The accuracy is tested at different frequencies and conductivity combinations. Compared with the results from the full solution, the complex image method has satisfactory accuracy and much less computation time. The error only appeared in the small area when the tool is too close to the boundaries.

Directional resistivity has been applied in geosteering in past years to interpret the measurements used to obtain the distance to bed, dipping angles and anisotropy, among others [1]. Li et al. presented a differential measurement approach, based on the standard propagation resistivity tool, placing two tilted antennas on a drill string, to obtain the bed information by the ratio of two signals at different tool azimuth angles [2]. The measurements contain both the direction-sensitive information and

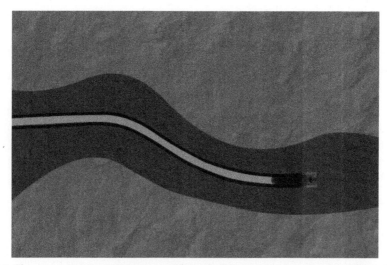

Figure 10.1 The geosteering system is a guiding device to provide accurate information of the downhole drilling so that the drill bit is kept inside the production zone.

direction-insensitive information, by using postprocessing. In 2006 Wang proposed a new approach that employs an orthogonal transmitter and receiver antennas [3]. The voltage signals from a main receiver antenna and a bucking antenna directly represent directional sensitive information.

Due to the requirement of real time, a fast forward modeling method is desirable. Currently, most of the forward modeling is based on the full solution. In 2005 Omeragic proposed a model-based (parametric) inversion method to detect distances to both upper and lower shoulder beds [2]. In 2006 Wang showed the inversion of distance to a bed based on a full 1D forward model [3]. In 2007 Wang first adopted the image theory to interpret the directional resistivity measurement and showed that the image theory could be used as a quantitative computation method.

The conventional image theory is used to simplify the inhomogeneous problem to the homogenous problem when the source is over a perfect electric conductor (PEC) or perfect magnetic conductor (PMC) interface. In 1969 Wait extended the approximate discrete image theory to a finite conducting interface [4]. Then Bannister further developed this extension to arbitrary sources [5]. In 1984 Lindell and Alanen studied the continuous exact image source over a dissipate plane [6].

The application of the image source could extremely simplify the forward modeling and speed up the calculation. Wang published more cases verifying the feasibility of the complex image in well logging [7]. This method is powerful for both qualitative and quantitative analysis of a logging response in a stratified formation.

Generally, image theory is transferring the inhomogeneous problem to a homogeneous problem by setting up an image source. Then the homogeneous space Green's function can be used to solve the field distribution, which is much easier and faster than the full solution.

10.2 THEORY OF FORWARD MODELING USING IMAGE THEORY
10.2.1 Review of traditional image theory

Generally, image theory is to convert an inhomogeneous problem to a homogeneous one by introducing an image source. Then the homogeneous space Green's function could be used to solve the field distribution, which is much easier and faster than the full solution.

The conventional image theory is referring to one electrical dipole over the PEC interface, shown in Fig. 10.2. There is no field in the lower half-space. The field of upper half-space can be calculated by replacing the interface by introducing an image source at lower space and applying the homogeneous Green's function. The field in upper space will be the summation of the fields generated by both sources. The two-layer inhomogeneous problem is then converted into a homogeneous problem.

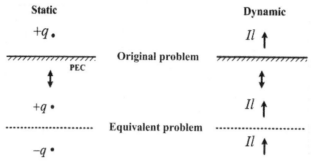

Figure 10.2 Image theory of PEC interface.

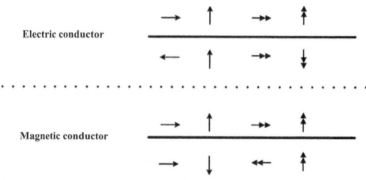

Figure 10.3 Summary of image theory.

More generally, the image sources of electrical dipoles (represented by *single arrows*) and magnetic dipoles (represented by *double arrows*) over PEC and PMC, respectively are shown in Fig. 10.3. Over the electric conductor, the image sources of the horizontal electrical dipole and vertical magnetic dipole have the opposite direction from the original sources. This agrees with the fact that the tangential current does not radiate along the PEC plane. Similarly, when the vertical electrical dipole and horizontal magnetic dipole (HMD) are over the magnetic conductor, there is no field radiation either.

10.2.2 Complex image theory in nonperfect medium

For deriving the complex image theory used into the application of geosteering, the transmitter of the directional resistivity logging tool is exacted to a horizontally placed magnetic dipole source. This assumption is consistent with the real implement of the transmitter antenna, which is a coil antenna around the tool body.

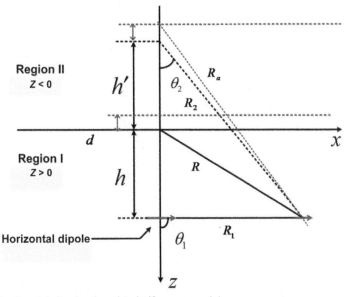

Figure 10.4 A horizontal dipole placed in half-space model.

10.2.2.1 Horizontal dipole in half-space

Let a horizontal electrical dipole of moment p be placed at $(0, 0, h)$ pointing at positive x axis as shown in Fig. 10.4. It has been derived that the Hertz potential functions in this two regions satisfying the following two equations, respectively,

$$\nabla^2 \mathbf{\Pi}_1 + \gamma_1^2 \mathbf{\Pi}_1 = \mathbf{p}\delta(x)\delta(y)\delta(z-h), \quad z \geq 0 \text{ and} \quad (10.1a)$$

$$\nabla^2 \mathbf{\Pi}_2 + \gamma_2^2 \mathbf{\Pi}_2 = 0, \quad z \leq 0 \quad (10.1b)$$

where $\gamma_1^2 = \omega^2 \mu_0(\varepsilon_1 - j\sigma_1/\omega)$ and $\gamma_2^2 = \omega^2 \mu_0(\varepsilon_2 - j\sigma_2/\omega)$.

The Hertz potential in region I could be decomposed into two directions

$$\mathbf{\Pi}_1 = x' \Pi_{1x} + z' \Pi_{1z} \quad (10.2)$$

10.2.2.2 Dipole in lossless half-space

If the horizontal dipole is within the air and above a conductive media, as shown in Fig. 10.4, $\gamma_1^2 = \gamma_0^2 = \omega^2 \mu_0 \varepsilon_0$ and $\gamma_2^2 = \gamma^2 = \omega^2 \mu_0 (\varepsilon - j\sigma/\omega)$. In spectral domain, the Hertz potential expressions for HMD are [8],

$$\Pi_{1x} = \frac{p}{4\pi} \left[\frac{e^{-\gamma_0 R_1}}{R_1} + \int_0^\infty \frac{u_0 - u}{u_0 + u} J_0(\lambda\rho) e^{-u_0(z+h)} \frac{\lambda}{u_0} d\lambda \right] \text{ and} \quad (10.3a)$$

$$\Pi_{1z} = \frac{p}{2\pi} \left[\int_0^\infty \frac{(u - u_0)e^{-u_0(z+h)}}{(\gamma_0^2 u + \gamma^2 u_0)} J_1(\lambda\rho) \lambda^2 d\lambda \right] \cos\phi \quad (10.3b)$$

where

$$R_1 = [\rho^2 + (z-h)^2]^{1/2},$$
$$u_0 = (\lambda^2 - \gamma_0^2)^{1/2}, \text{ and}$$
$$u = (\lambda^2 - \gamma^2)^{1/2}.$$

For the application in well logging, most cases are low frequency and satisfy the quasistatic condition, where we can assume $u_0 \approx \lambda$, then

$$P_m = \int_0^\infty \frac{u_0 - u}{u_0 + u} J_0(\lambda\rho) e^{-u_0(z+h)} \frac{\lambda}{u_0} d\lambda \approx -\int_0^\infty \frac{u - \lambda}{u + \lambda} J_0(\lambda\rho) e^{-\lambda(z+h)} d\lambda \quad (10.4)$$

The Taylor series expansion of the function $f(\lambda)$ can be written in the form

$$f(\lambda) = e^{\lambda d_{\text{shift}}} \frac{u - \lambda}{u + \lambda} = \sum_{n=0}^\infty a_n \lambda^n \quad (10.5)$$

where $d_{\text{shift}} = (1 - j)\delta$ and $a_n = (1/n!)f^{(n)}(0)$.

Approximate using only the first term and consequently,

$$\Pi_{1x} \approx \frac{p}{4\pi}\left(\frac{1}{R_1} - \frac{1}{R_a}\right) \text{ and} \quad (10.6)$$

$$\Pi_{1z} \approx \frac{p\cos\phi}{4\pi\rho}\left[\frac{(d_{\text{shift}} + z + h)}{R_a} - \frac{(z+h)}{R_2}\right] \quad (10.7)$$

where $R_a = [\rho^2 + (z + h + d_{\text{shift}})^2]^{1/2}$ and $R_2 = [\rho^2 + (z + h)^2]^{1/2}$. Because the boundary shift d_{shift} is very small, the difference between R_a and R_2 is almost zero, which means

$$\frac{(d_{\text{shift}}t + z + h)}{R_a} - \frac{(z+h)}{R_2} \approx 0 \quad (10.8)$$

Then, with the assumption of quasistatic, the Hertz potential of the horizontal dipole placed in a two-layer half-space media can be simplified to one component

$$\Pi_{1x} \approx \frac{p}{4\pi}\left(\frac{1}{R_1} - \frac{1}{R_a}\right) \text{ and} \quad (10.9)$$

$$\Pi_{1z} \approx 0 \quad (10.10)$$

Therefore the Sommerfeld integral is simplified to a summation of two terms. Both are in x direction and located at $z = h$ and $z = -(h + d_{\text{shift}})$, respectively. The total field is the superposition of the fields radiated by the two discrete sources in the homogeneous medium.

We can extend this case when the boundary is not perfect conductive and the source region is within the relative low-conductive media. The nonperfect conductive boundary can be approximated as a perfect conductive boundary by introducing a complex depth shift d_{shift}. By shifting the boundary, the conventional image theory could be applied.

The equivalent two-layer model is shown in Fig. 10.5, in which the remote bed (upper layer, where the image source is located) is much more conductive than the near bed (lower layer, where the original source is located). Bannister gave the shifts for horizontal and vertical magnetic dipoles, respectively. They are

$$d_{\text{VMD}_{\text{shift}}} = \frac{1}{\sqrt{k_b^2 - k_n^2}} \quad \text{and} \quad d_{\text{HMD}_{\text{shift}}} = \frac{\sqrt{k_b^2 - k_n^2}}{-k_n^2} \tag{10.11}$$

where $k_b^2 = \omega^2 \mu \varepsilon_b - j\omega\mu\sigma_b$ and $k_n^2 = \omega^2 \mu \varepsilon_n - j\omega\mu\sigma_n$ are the wave numbers of near bed and the remote bed. If we further assume that the remote bed is sufficiently more conductive than the near bed. Then the shift distance can be simplified to

$$d_{\text{VMD}_{\text{shift}}} = d_{\text{HMD}_{\text{shift}}} \approx \frac{1}{jk_n} \tag{10.12}$$

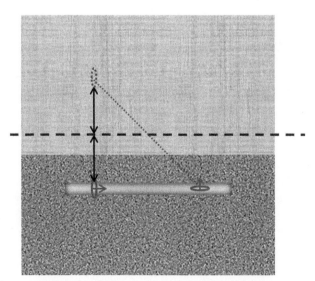

Figure 10.5 Two-layer equivalent model by applying the image theory.

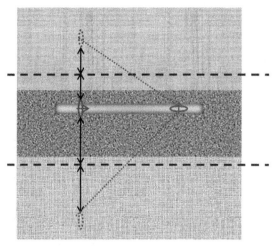

Figure 10.6 Three-layer equivalent model by applying the image theory.

For the logging tool with spacing L, the H field received by the receiver is

$$H_{xz} = \frac{P}{4\pi} \frac{e^{-jk_b r}}{r^3} (k_b^2 r^2 - 3jk_b r - 3) \frac{(2d_{\text{shift}} + 2h)L}{r} \quad \text{and} \tag{10.13a}$$

$$H_{xx} = \frac{P}{4\pi} \left[\frac{e^{-jk_b r}}{r^3} (k_b^2 r^2 - 3jk_b r - 3) \left(\frac{2d_{\text{shift}} + 2h}{r} \right)^2 + \frac{2e^{-jk_b r}}{r^3} (jk_b r + 1) \right]$$
$$+ \frac{P}{4\pi} \frac{e^{-jk_b L}}{L^3} (jk_b L + 1) \tag{10.13b}$$

Consider a three-layer model as shown in Fig. 10.6. In this model, for each boundary, only the first image is considered. According to the application condition of the approximated image theory, the middle layer, where the drilling bit stays, has higher resistivity compared with the other two adjacent layers. Then the three-layer model is simplified into a homogeneous model with three sources.

10.2.2.3 Dipole in the dissipative media

Consider another case when region I is dissipative while that region II is nonconductive. The parameters of those two regions are $\gamma_1^2 = \gamma^2 = \omega^2 \mu_0 (\varepsilon - j\sigma/\omega)$ and $\gamma_2^2 = \gamma_0^2 = \omega^2 \mu_0 \varepsilon_0$. Then, the Hertz potential in the two regions become

$$\Pi_{1x} = \frac{p}{4\pi} \left[\frac{e^{-\gamma R_1}}{R_1} + \int_0^\infty \frac{u - u_0}{u + u_0} J_0(\lambda \rho) e^{-u(z+h)} \frac{\lambda}{u} d\lambda \right] \quad \text{and} \tag{10.14a}$$

$$\Pi_{1z} = \frac{p}{2\pi} \left[\int_0^\infty \frac{(u_0 - u) e^{-u(z+h)}}{(\gamma_0^2 u + \gamma^2 u_0)} J_1(\lambda \rho) \lambda^2 d\lambda \right] \cos \phi \tag{10.14b}$$

where

$$R_0 = [\rho^2 + (z-h)^2]^{1/2},$$
$$u_0 = (\lambda^2 - \gamma_0^2)^{1/2}, \text{ and}$$
$$u = (\lambda^2 - \gamma^2)^{1/2}.$$

Define $n = (\varepsilon_r - j\sigma/\omega\varepsilon_0)^{1/2}$, $\gamma = n\gamma_0$ and apply Sommerfeld identity (10.15) in Eq. (10.14),

$$\frac{e^{-\gamma_1 R_2}}{R_2} = \int_0^\infty \frac{J_0(\lambda\rho)}{u} e^{-u|z+h|} \lambda d\lambda \tag{10.15}$$

The Hertz potential can be rewritten as

$$\Pi_{1x} = \frac{p}{4\pi}\left[\frac{e^{-\gamma R_1}}{R_1} - \frac{e^{-\gamma R_2}}{R_2} + 2\int_0^\infty \frac{J_0(\lambda\rho)e^{[-u(z+h)]}}{u+u_0} \lambda d\lambda\right] \text{ and} \tag{10.16a}$$

$$\Pi_{1z} = -\frac{p}{2\pi}(1-n^2)\left[\int_0^\infty \frac{J_1(\lambda\rho)e^{-u(z+h)}}{(u+u_0)(n^2 u_0 + u)} \lambda^2 d\lambda\right]\cos\phi \tag{10.16b}$$

where $R_1 = [\rho^2 + (z-h)^2]^{1/2}$ and $R_2 = [\rho^2 + (z+h)^2]^{1/2}$. To simplify these expressions, by using Lien's method, define the abbreviations,

$$G_1 = \frac{e^{-\gamma R_1}}{R_1}, \quad G_2 = \frac{e^{-\gamma R_2}}{R_2},$$

$$U = 2\int_0^\infty \frac{J_1(\lambda\rho)e^{[-u(z+h)]}}{u+u_0} \lambda d\lambda \text{ and}$$

$$W = -2(1-n^2)\left[\int_0^\infty \frac{J_1(\lambda\rho)e^{-u(z+h)}}{(u+u_0)(n^2 u_0 + u)} \lambda^2 d\lambda\right]$$
$$= 2(1-n^2)\frac{\partial}{\partial x}\int_0^\infty \frac{J_0(\lambda\rho)e^{-u(z+h)}}{(u+u_0)(n^2 u_0 + u)} \lambda d\lambda$$

The Hertz potential function $\mathbf{\Pi}_1$ is then given by

$$\mathbf{\Pi}_1 = \frac{p}{4\pi}(G_1 - G_2 + U)x' + Wz' \tag{10.17}$$

The field components in region I can be found by following Norton method in cylindrical coordinates

$$\frac{\partial W}{\partial z} = -2\frac{\partial}{\partial x}\int_0^\infty \left(\frac{1}{u_0 + u} - \frac{n^2}{n^2 u_0 + u}\right) J_0(\lambda\rho) e^{-u(z+h)} \lambda d\lambda \qquad (10.18)$$

Define $V = 2n^2 \int_0^\infty \frac{J_0(\lambda\rho) e^{-u(z+h)}}{n^2 u_0 + u} \lambda d\lambda$, then

$$\frac{\partial W}{\partial z} = -\frac{\partial U}{\partial x} + \frac{\partial V}{\partial x} \qquad (10.19)$$

The divergence of $\mathbf{\Pi}_1$, therefore, is given by

$$\nabla \cdot \mathbf{\Pi}_1 = \frac{\partial}{\partial x}(G_1 - G_2 + V) \qquad (10.20)$$

Because $H_1 = \nabla(\nabla \cdot \mathbf{\Pi}_1) + \gamma^2 \mathbf{\Pi}_1$, the z component is

$$H_{1z} = \gamma^2 W + \frac{\partial^2}{\partial x \partial z}(G_1 - G_2 + V) \qquad (10.21)$$

Transfer the solution into cylindrical coordinates and apply the Sommerfeld integral representation of spherical wave function, Eq. (10.15), the integral in Eq. (10.18) becomes

$$-2\frac{\partial}{\partial x}\int_0^\infty \left(1 - \frac{u}{n^2 u_0 + u}\right) J_0(\lambda\rho) e^{-u(z+h)} \lambda d\lambda = 2\frac{\partial^2 G_2}{\partial z \partial \rho} - \frac{1}{n^2}\frac{\partial^2 V}{\partial z \partial \rho} \qquad (10.22)$$

Furthermore, in the low-frequency assumption, using the leading term of the asymptotic expression of the Bessel function, the approximate expression of V is given

$$H_{1z} = \frac{\partial^2}{\partial z \partial \rho}\left(G_1 + G_2 - \frac{V}{n^2}\right)\cos\phi \qquad (10.23)$$

From Eq. (10.23), the cross z component is generated by original source G_1, image source G_2, and a correction term related to V. Roy Harold Lien gave the low-frequency approximation of the integral V under the assumption that $|n^2| \gg 1$ and $|jnkR_2/2| \gg 1$, the leading term approximation is given as

$$V = 2k_0 \rho^{-1} e^{-j\gamma(z+h)} \text{ and} \qquad (10.24)$$

$$H_{1z} = \frac{\partial^2}{\partial z \partial \rho}(G_1 + G_2)\cos\phi - \frac{2k_0}{n^2 \rho^2}\gamma e^{-j\gamma(z+h)}\cos\phi \qquad (10.25)$$

Considering the tool is always located in xz plane and parallel with x axis, $\phi = 0$ degree, we will have

$$H_{1z} = \frac{\partial^2}{\partial z \partial \rho}(G_1 + G_2) + H_c \text{ and} \tag{10.26}$$

$$H_c = -\frac{2k_0}{n^2 \rho^2} \gamma e^{-jr(z+h)} \tag{10.27}$$

The final expressions for the magnetic field in region I, for the case $\sigma/\omega\varepsilon \ll 1$, are generated by the original source placed at $z = h$, an image place at $z = -h$ and a correction term expressed in Eq. (10.27).

Then, the H field received by the receiver in cylindrical coordinator is

$$H_{xz} = \frac{P}{4\pi} \left\{ \frac{1}{2}\left[-n^2k^2 + j3nkr_1^{-1} + 3r_1^{-2}\right]\sin 2\theta_1 G_1 \right. $$
$$\left. + \frac{1}{2}\left[-n^2k^2 + j3nkr_2^{-1} + 3r_2^{-2}\right]\sin 2\theta_2 G_2 - \frac{1}{n^2}\frac{\partial^2 V}{\partial z \partial \rho} \right\} \text{ and} \tag{10.28a}$$

$$H_{xx} = \frac{P}{4\pi}\left\{\left[n^2k^2\cos^2\theta_1 + jnk(2-3\cos^2\theta_1)r_1^{-1} + (2-3\cos^2\theta_1)r_1^{-2}\right]G_1 \right.$$
$$\left. + \left[n^2k^2 - j3nkr_2^{-1} - 3r_2^{-2}\right]\cos^2\theta_2 G_2 - \frac{1}{\rho}\frac{\partial V}{\partial \rho}\right\} \tag{10.28b}$$

where θ_1 and θ_2 are the angles shown in Fig. 10.4.

10.3 SIMULATION RESULTS AND DISCUSSIONS

10.3.1 One-dimensional formation model

If the borehole is neglected and only the depth variation is considered, an isotropic formation could be modeled as a layered medium, as shown in Fig. 10.7. In this model, the z direction is the depth direction. In the application of geosteering, the tool is always kept in the production layer, which means in most cases, the tool is placed horizontally. This assumption applies to all following testing examples. The testing points will be along the z direction. Then the received signal is a function of distance from the tool position to the boundary. According to the electrical properties of the near bed and remote bed, two cases are possible. One is when the tool is within the high-resistive bed.

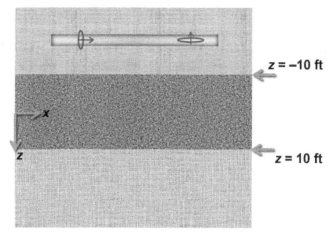

Figure 10.7 Three-layer model.

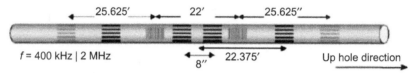

Figure 10.8 Azitrack tool configuration.

The other is when the tool is in the high-conductive bed. For these two cases, different approximations must be used to apply the complex image method. For each case, the simulation results generated by the complex image method will be presented together with the results obtained by a full solution code, which is named INDTRI developed by the Well Logging Lab at the University of Houston. The Hankel integral is solved by 283 points fast Hankel transform. The relative permittivity and permeability of each layer are set to be 1. That is because, firstly, the tool is working at relatively low frequency, the effect of permittivity is not significant. Secondly, in the most cases, the earth is nonmagnetic. We can always neglect the permeability of the earth.

10.3.2 Tool configuration

Consider the Azitrack directional resistivity tool discussed in Chapter 4, Triaxial Induction and Logging-While-Drilling Resistivity Tool Response in Homogeneous Anisotropic Formations, by Baker Hughes as an example. Fig. 10.8 shows tool structure. As described in Chapter 4, Triaxial Induction and Logging-While-Drilling Resistivity Tool Response in Homogeneous Anisotropic Formations, this tool works at two frequencies, 2 MHz and 400 kHz. The configuration is symmetric, which is called a compensated LWD configuration. As shown in Fig. 10.8, there are several different spacings. For convenience sake without losing generality, only two spacings are considered in the following discussions. The long spacing is 33.375 in. and the short spacing is 22.265 in.

In the following simulation, actually, we only consider the radiation of dipole source and neglect the effect of mandrel, the reason why we can use this assumption is, firstly, compared with the geological size of the formation, the size of the mandrel can be neglected. Secondly, the effect of mandrel is to enhance or reduce the magnitude of the field, but not change the distribution. The third one is the effect of the mandrel can be compensated by the symmetrical configuration. So, in the following simulation, we can neglect the effect of the mandrel.

10.3.3 Simulation results

10.3.3.1 $R_1 = R_3 = 1$ ohm-m, $R_3 = 100$ ohm-m

Consider the three-layer model, as shown in Fig. 10.7. The high-resistivity layer is in the middle. The parameters of these three layers are $\varepsilon_{r1} = \varepsilon_{r3} = 1$, $\mu_{r1} = \mu_{r3} = 1$, and $\sigma_1 = \sigma_3 = 1$ for the upper and lower layer, $\varepsilon_{r2} = 1$, $\mu_{r2} = 1$, and $\sigma_2 = 0.01$ for the middle layer. The boundaries are at $z = 10$ ft and $z = -10$ ft. This is the general case when the drilling bit is in the high-resistivity layer.

1. Frequency = 2 MHz, spacing = 33.375 in.

   Fig. 10.9 shows simulation results of the cross component H_{zx}. The *pink circle* (light gray in print versions) indicates result calculated by approximated method and the *blue dash line* (dark gray in print versions) is the result of a full solution. From the results, we can see that the image method works pretty well. Even when the tool crosses the boundary, there is only small error between the approximation and full solution. Here, the results show that when the tool is working at 2 MHz, the long-spacing channel works well. The cross component could represent the boundary information. Compared with the full solution results, there is no much error.

   Fig. 10.10 shows the phase shift and attenuation of the compensated propagation tool. These two parameters will be used in inversing the apparent resistivity of the formation. Compared with the full solution results, there is noticeable error when the logging tool is close to the boundary. Because our three-layer model is symmetrical, the simulation results are also symmetrical.

2. Frequency = 2 MHz, spacing = 22.265 in.

   Figs. 10.11 and 10.12 give the simulation results when the logging tool is working at 2 MHz and the spacing is short. Compared with the full solution results, the approximation method also works well. Only small error appears around boundary. Compared with the long-spacing case, cross component H_{zx} has a little more error at the boundary. Phase shift and attenuation are a little bit better. Although in this case, the error of cross component is a little bit larger, it does not affect the boundary information.

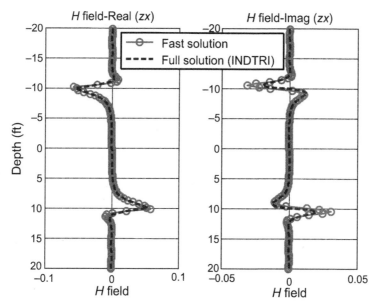

Figure 10.9 Tool response H_{zx} component ($\sigma_1 = \sigma_3 = 1$, $\sigma_2 = 0.01$, 2 MHz, long).

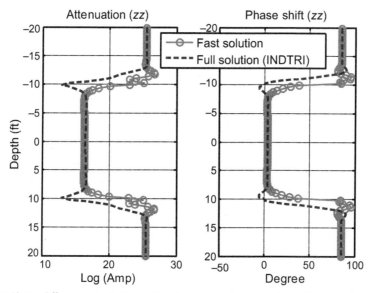

Figure 10.10 Phase difference and attenuation ($\sigma_1 = \sigma_3 = 1$, $\sigma_2 = 0.01$, 2 MHz, long).

The Application of Image Theory in Geosteering 365

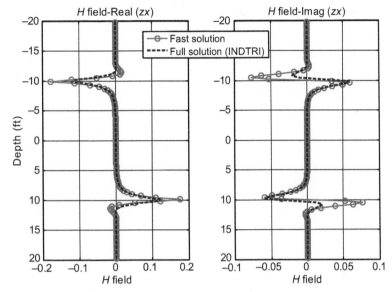

Figure 10.11 Tool response H_{zx} component ($\sigma_1 = \sigma_3 = 1$, $\sigma_2 = 0.01$, 2 MHz, short).

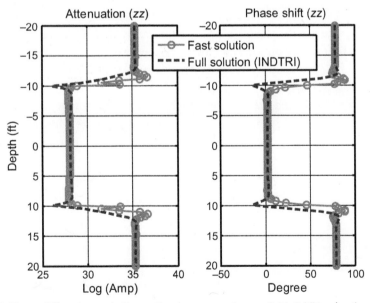

Figure 10.12 Phase difference and attenuation ($\sigma_1 = \sigma_3 = 1$, $\sigma_2 = 0.01$, 2 MHz, short).

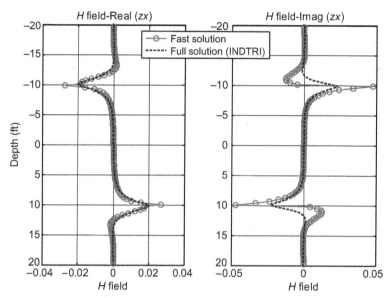

Figure 10.13 Tool response H_{zx} component ($\sigma_1 = \sigma_3 = 1$, $\sigma_2 = 0.01$, 400 kHz, long).

3. Frequency = 400 kHz, spacing = 33.375 in.

   When the tool is working at 400 kHz and the spacing is long, from Fig. 10.13, the real part of the cross component H_{zx} still matches well with full solution results, even when the logging point is at the boundary. Image part of H_{zx} has obvious error. Based on this property, we consider that the boundary inversion could be developed only in terms of the real part of H_{zx}.

   Fig. 10.14 shows the phase shift and attenuation of the logging tool, when it is working at 400 kHz and the spacing is long. The results show that, in most range, the simulation results of the approximation method agree with the full solution results. Only exception is around the boundaries, where noticeable error is witnessed.

4. Frequency = 400 kHz, spacing = 22.265 in.

   When the tool is working at 400 kHz with short spacing, the cross component H_{zx} is not as good as before, as shown in Fig. 10.15. Not only imaginary part, but also real part of H_{zx} deviates from full solution around the boundaries. However, this error only exists within the area 2 in. away from the boundaries. For the application of geosteering, this distance is relatively small. So, this error is acceptable.

   Fig. 10.16 shows the phase shift and attenuation when the tool is working at 400 kHz and with short spacing. Compared with other channels, the simulation results of the approximation method show enough agreement with the full solution results. Error only occurs near the boundaries. Based on the phase shift and attenuation, the apparent conductivities of the three layers could be inversed.

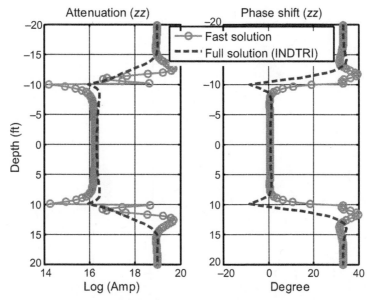

Figure 10.14 Phase difference and attenuation ($\sigma_1 = \sigma_3 = 1$, $\sigma_2 = 0.01$, 400 kHz, long).

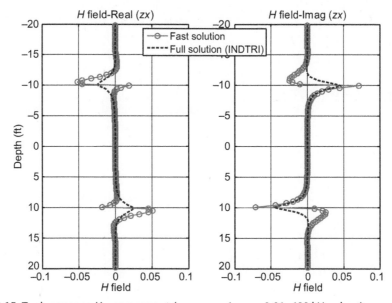

Figure 10.15 Tool response H_{zx} component ($\sigma_1 = \sigma_3 = 1$, $\sigma_2 = 0.01$, 400 kHz, short).

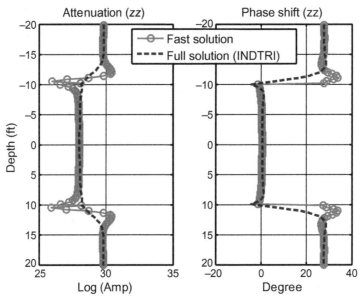

Figure 10.16 Phase difference and attenuation ($\sigma_1 = \sigma_3 = 1$, $\sigma_2 = 0.01$, 400 kHz, short).

10.3.3.2 $R_1 = R_3 = 1$ ohm-m, $R_3 = 10$ ohm-m

Consider the three-layer model, as shown in Fig. 10.7. The resistivity of the middle layer is less than case in Section 10.3.3.1. The parameters of these three layers are $\varepsilon_{r1} = \varepsilon_{r3} = 1$, $\mu_{r1} = \mu_{r3} = 1$, and $\sigma_1 = \sigma_3 = 1$ for the upper and lower layer, $\varepsilon_{r2} = 1$, $\mu_{r2} = 1$, and $\sigma_2 = 0.1$ for the middle layer. The boundaries are at $z = 10$ ft and $z = -10$ ft. This is also the general case when the drilling bit is in the high-resistivity layer.

1. Frequency = 2 MHz, spacing = 33.375 in.

    Figs. 10.17 and 10.18 give the simulation results when the tool is working at relative high frequency and long spacing, where the middle layer of the formation is relatively less resistive. The figure shows that the cross component H_{zx} simulated by approximation method has a good agreement with the data given by full solution. This means this approximation method can be applied into the forward modeling of geosteering tool and the simulation results are good enough to be used to extract boundary information.
2. Frequency = 2 MHz, spacing = 22.265 in.
3. Frequency = 400 kHz, spacing = 33.375 in.
4. Frequency = 400 kHz, spacing = 22.265 in.

    Figs. 10.19–10.24 show the simulation results when the tool is working in other three channels. As in the first case, the approximation method works well in all other three channels, 2 MHz with short spacing, 400 kHz with long spacing, and 400 kHz with short spacing. Although when the frequency is lower, the error

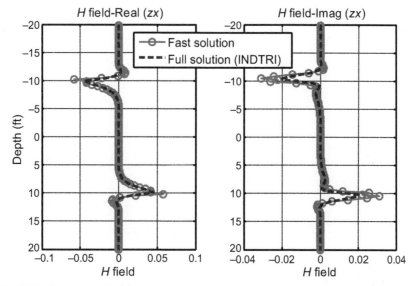

Figure 10.17 Tool response H_{zx} component ($\sigma_1 = \sigma_3 = 1$, $\sigma_2 = 0.1$, 2 MHz, long).

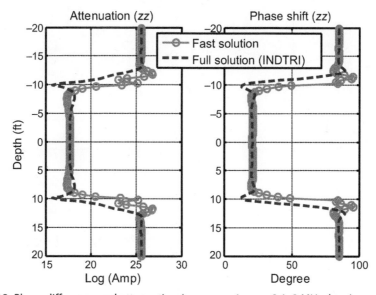

Figure 10.18 Phase difference and attenuation ($\sigma_1 = \sigma_3 = 1$, $\sigma_2 = 0.1$, 2 MHz, long).

around boundaries becomes larger, however, the accuracy is still within an acceptable range. From the results of phase shift and attenuation, the apparent resistivity of each layer can be inverted. Based on the inverted resistivity and the cross component H_{zx} data, the boundary information can be extracted.

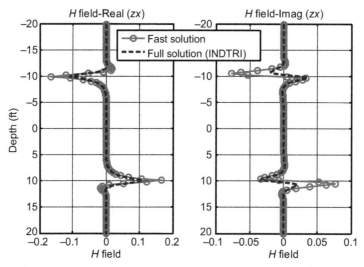

Figure 10.19 Tool response H_{zx} component ($\sigma_1 = \sigma_3 = 1$, $\sigma_2 = 0.1$, 2 MHz, short).

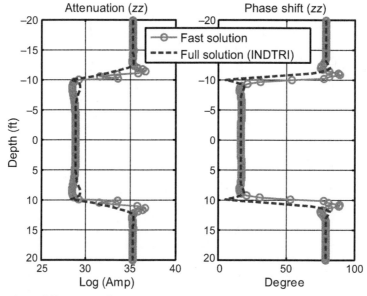

Figure 10.20 Phase difference and attenuation ($\sigma_1 = \sigma_3 = 1$, $\sigma_2 = 0.1$, 2 MHz, short).

10.3.3.3 $R_1 = R_3 = 100$ ohm-m, $R_2 = 1$ ohm-m

In this case, the formation model is also three layer. The difference between this and previous two cases is that, in this case, the middle layer is of high conductive and two remote layers are of relatively high resistivity. In this case, the parameters of these three layers are $\varepsilon_{r1} = \varepsilon_{r3} = 1$, $\mu_{r1} = \mu_{r3} = 1$, and $\sigma_1 = \sigma_3 = 0.01$ for the upper and lower

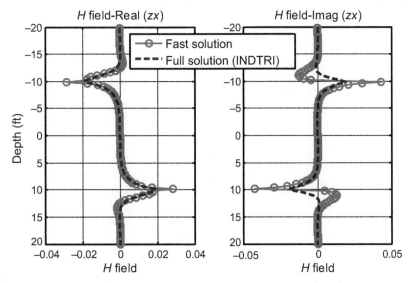

Figure 10.21 Tool response H_{zx} component ($\sigma_1 = \sigma_3 = 1$, $\sigma_2 = 0.1$, 400 kHz, long).

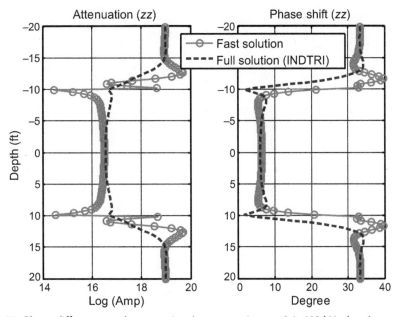

Figure 10.22 Phase difference and attenuation ($\sigma_1 = \sigma_3 = 1$, $\sigma_2 = 0.1$, 400 kHz, long).

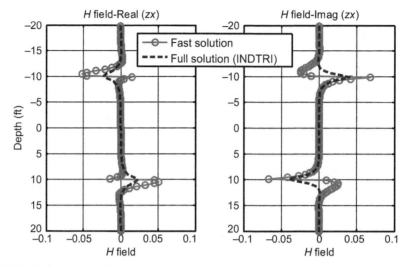

Figure 10.23 Tool response H_{zx} component ($\sigma_1 = \sigma_3 = 1$, $\sigma_2 = 0.1$, 400 kHz, short).

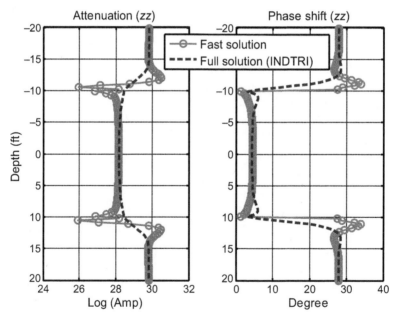

Figure 10.24 Phase difference and attenuation ($\sigma_1 = \sigma_3 = 1$, $\sigma_2 = 0.1$, 400 kHz, short).

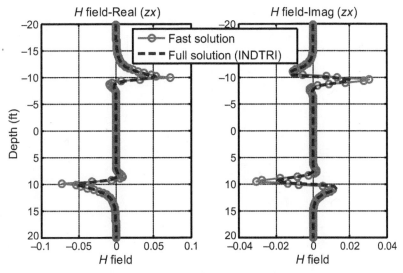

Figure 10.25 Tool response H_{zx} component ($\sigma_1 = \sigma_3 = 0.01$, $\sigma_2 = 1$, 2 MHz, long).

layer, $\varepsilon_{r2} = 1$, $\mu_{r2} = 1$, and $\sigma_2 = 1$ for the middle layer. The boundaries are at $z = 10$ ft and $z = -10$ ft. This is also the general case when the drilling bit is in the high-resistivity layer.

1. Frequency = 2 MHz, spacing = 33.375 in.
2. Frequency = 2 MHz, spacing = 22.265 in.
3. Frequency = 400 kHz, spacing = 33.375 in.
4. Frequency = 400 kHz, spacing = 22.265 in.

Figs. 10.25–10.32 show the simulation results of full channels, when the middle layer of the formation is high conductive. Because the three-layer model is treated as the combination of two independent boundaries, the simulation results are similar to the ones where the middle layer is of high resistive. We find that the approximation method works well in this case. The cross component H_{zx}, phase shift and attenuation are all good enough to be used into boundary detection.

10.3.3.4 $R_1 = 1$ ohm-m, $R_1 = 20$ ohm-m, $R_1 = 0.5$ ohm-m

In all the previous three cases, the upper layer and lower layer have the same conductivity, which means the models are all symmetrical. The unsymmetrical case is also tested. The parameters of these three layers are $\varepsilon_{r1} = \varepsilon_{r2} = \varepsilon_{r3} = 1$, $\mu_{r1} = \mu_{r2} = \mu_{r3} = 1$, and $\sigma_1 = 1, \sigma_2 = 0.05, \sigma_3 = 2$. The boundaries are at $z = 10$ ft and $z = -10$ ft. This case is more general as in real application.

Similarly, four channels are all tested in this case. As is expected, the approximation method also works well in this unsymmetrical formation. The approximation method

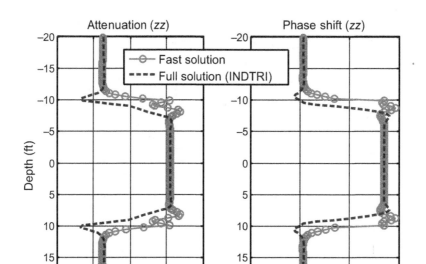

Figure 10.26 Phase difference and attenuation ($\sigma_1 = \sigma_3 = 0.01$, $\sigma_2 = 1$, 2 MHz, long).

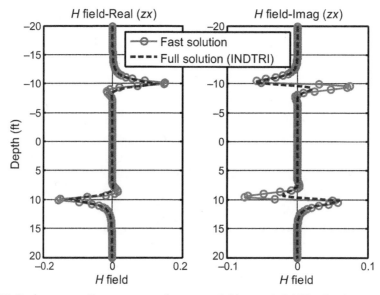

Figure 10.27 Tool response H_{zx} component ($\sigma_1 = \sigma_3 = 0.01$, $\sigma_2 = 1$, 2 MHz, short).

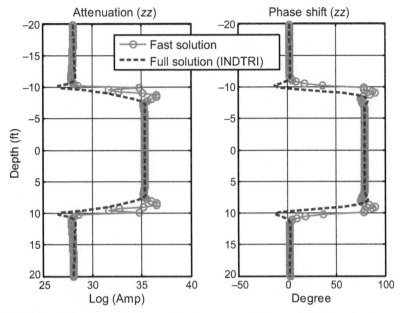

Figure 10.28 Phase difference and attenuation ($\sigma_1 = \sigma_3 = 0.01$, $\sigma_2 = 1$, 2 MHz, short).

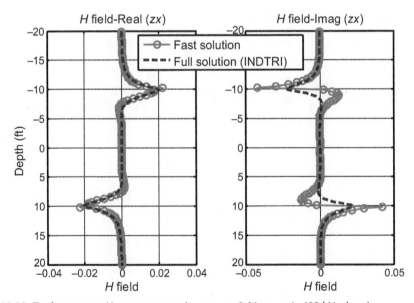

Figure 10.29 Tool response H_{zx} component ($\sigma_1 = \sigma_3 = 0.01$, $\sigma_2 = 1$, 400 kHz, long).

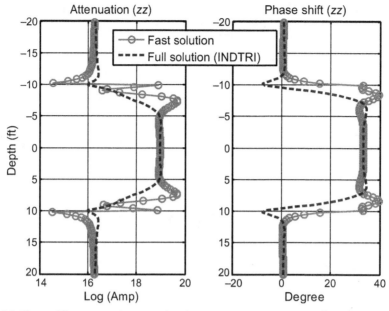

Figure 10.30 Phase difference and attenuation ($\sigma_1 = \sigma_3 = 0.01$, $\sigma_2 = 1$, 400 kHz, long).

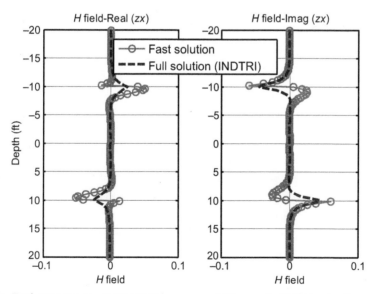

Figure 10.31 Tool response H_{zx} component ($\sigma_1 = \sigma_3 = 0.01$, $\sigma_2 = 1$, 400 kHz, short).

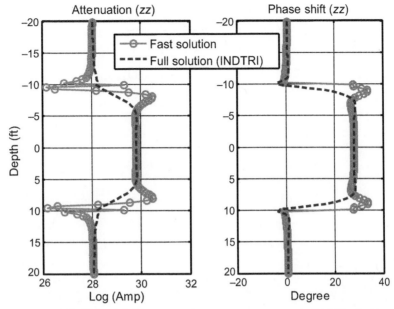

Figure 10.32 Phase difference and attenuation ($\sigma_1 = \sigma_3 = 0.01$, $\sigma_2 = 1$, 400 kHz, short).

can be also used into the layer with relatively high conductivity. One additional term was introduced to correct the image results. The real part of cross components H_{zx} has more accuracy than the imaginary part, which indicates that it is better to extract the boundary information only from the real part of the signal. Compared with the cross component, although there is a little bit more error of phase shift and attenuation, the logging values away from boundaries are good enough to invert the apparent resistivity of the formation. So, the approximation method can be used into the application of geosteering. The boundary information can be extracted from the cross component.

1. Frequency = 2 MHz, spacing = 33.375 in. (Figs. 10.33 and 10.34)
2. Frequency = 2 MHz, spacing = 22.265 in. (Figs. 10.35 and 10.36)
3. Frequency = 400 kHz, spacing = 33.375 in. (Figs. 10.37 and 10.38)
4. Frequency = 400 kHz, spacing = 22.265 in. (Figs. 10.39 and 10.40)

10.3.4 Discussion

According to the simulation results, the real part of cross component H_{zx} shows that nonzero values only exist near boundary. In the area far away from boundary, the values of H_{zx} are zero. This is the advantage of the orthogonal configuration tool. The cross component is only sensitive to the boundary. When the tool is approaching the boundary, the cross component will increase. Then when the distance from the

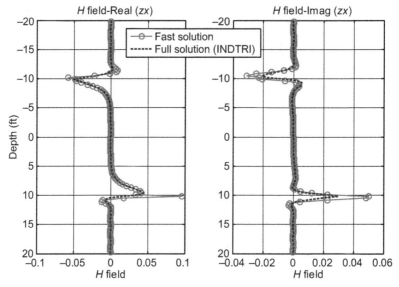

Figure 10.33 Tool response H_{zx} component ($\sigma_1 = 1$, $\sigma_2 = 0.05$, $\sigma_3 = 2$, 2 MHz, long).

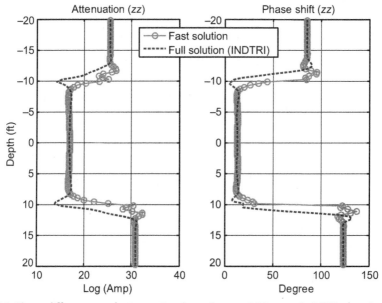

Figure 10.34 Phase difference and attenuation ($\sigma_1 = 1$, $\sigma_2 = 0.05$, $\sigma_3 = 2$, 2 MHz, long).

The Application of Image Theory in Geosteering 379

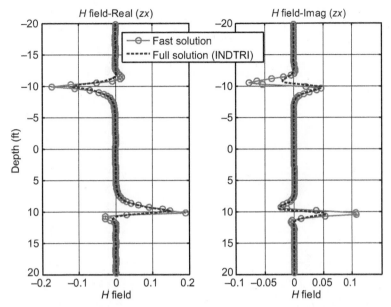

Figure 10.35 Tool response H_{zx} component ($\sigma_1 = 1$, $\sigma_2 = 0.05$, $\sigma_3 = 2$, 2 MHz, short).

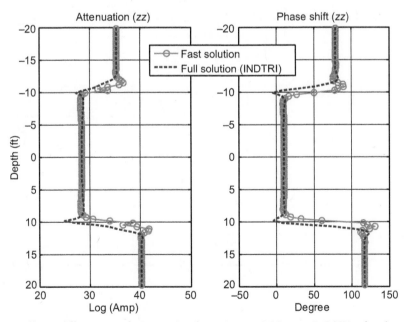

Figure 10.36 Phase difference and attenuation ($\sigma_1 = 1$, $\sigma_2 = 0.05$, $\sigma_3 = 2$, 2 MHz, short).

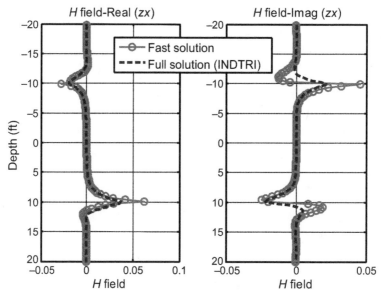

Figure 10.37 Tool response H_{zx} component ($\sigma_1 = 1$, $\sigma_2 = 0.05$, $\sigma_3 = 2$, 400 kHz, long).

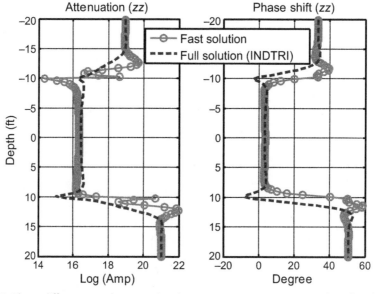

Figure 10.38 Phase difference and attenuation ($\sigma_1 = 1$, $\sigma_2 = 0.05$, $\sigma_3 = 2$, 400 kHz, long).

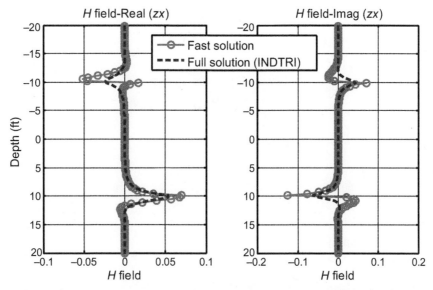

Figure 10.39 Tool response H_{zx} component ($\sigma_1 = 1$, $\sigma_2 = 0.05$, $\sigma_3 = 2$, 400 kHz, short).

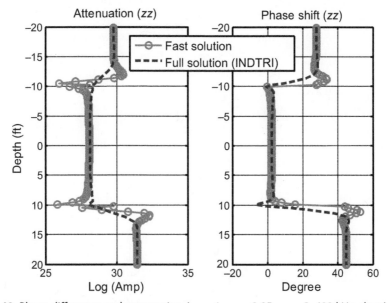

Figure 10.40 Phase difference and attenuation ($\sigma_1 = 1$, $\sigma_2 = 0.05$, $\sigma_3 = 2$, 400 kHz, short).

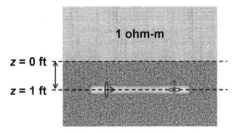

Figure 10.41 Observation point 1 ft away from boundary.

drilling bit to the boundary is larger than a specific value, the tool cannot detect the boundary any more. Based on the simulation results we have, tool's sensitivity of boundary is affected by the combination of frequency and spacing. Besides, the conductivity of formation also affects the tool response.

10.3.4.1 Effects of conductivity contrast

In this section, the tolerance of the image method at different conductivity contrast will be investigated for the two-layer model, as shown in Fig. 10.41. The upper layer of the model is a low-conductivity layer. The resistivity of the lower layer varies from 10 ohm-m to 1 kohm-m. The tolerance of the image method is tested at 2 MHz and antenna spacing is 34 in.

Fig. 10.42 shows the absolute error and relative error between the approximation results and full solution results at the observation point 1 ft and 2 ft away from the boundary, respectively. The results show that when the resistivity ratio between the upper layer and the lower layer is increased, the error between the approximation method and the full solution converges. The relative error of the xz component is smaller when the resistivity ratio of the upper layer and the lower layer is larger. At the observation point 1 ft away from the boundary, when the resistivity ratio between the upper layer and the lower layer is more than 100, the absolute error is less than 0.0123; the relative error between the approximation method and the full solution is about 15%. When the observation point is at the area 2 ft away from the boundary, the error will be less.

10.3.4.2 Frequency

In terms of practical application, assume that the current excited into the transmitter is 200 mA. The area of antenna is 2.5 in.^{2}. Then, the moment of single-turn antenna is about 3.2e-4 A m^2. Then, the H_{zx} data in Figs. 10.8 and 10.12 can be converted to the received voltage signal, shown in Fig. 10.43. For evaluating the sensitivity of the boundary detection, the detectable minimum signal power should be considered. Currently, the minimum detectable voltage is about 100 nV.

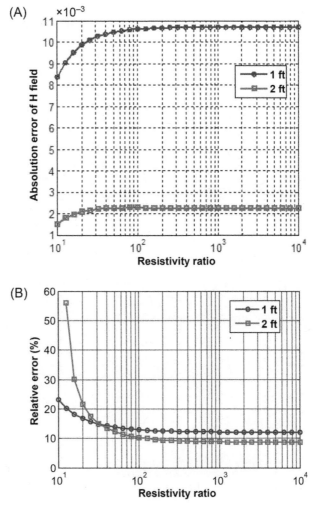

Figure 10.42 (A) Absolute error of 2 MHz tool at 1 ft away from boundary, (B) relative error of 2 MHz tool at 1 ft away from boundary.

In Fig. 10.43, the H_{zx} is converted into voltage by considering that the transmitter has only single turn and its moment is 3.2e-4. The parameters of the formation is $\sigma_1 = \sigma_3 = 1$ and $\sigma_2 = 0.01$. The *yellow line* (gray in print versions) shows the minimum voltage value that can be detected by the sensor.

As shown in Fig. 10.43, in high-resistive area, tool responses at two working frequencies have similar sensitivities. The signal of cross component fades to zero at the position about 5 ft away from the boundary. In the high-conductive range, the signal decays even further. The detectable distances are around 2–4 ft. In this range, the high frequency signal decays faster, so the relatively low-frequency working channel has better sensitivity. Tool can detect further at the relatively low frequency.

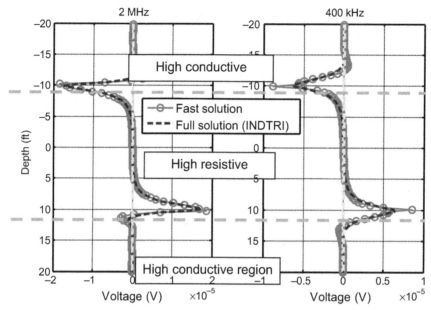

Figure 10.43 Voltage signal generated by cross component H_{zx} of single-turn transmitter ($\sigma_1 = \sigma_3 = 1$, $\sigma_2 = 0.01$).

Figure 10.44 Two-layer model with boundary at $z = 0$ ft.

Similarly, by comparing the simulation results in other formations, we can always get at least 5 ft detectable distance in high-resistive region. The detectable distances in high-conductive region are different caused by the different working frequencies. Low-frequency channel has larger detectable distance.

10.3.4.3 Spacing

To investigate the effect of spacing, two-layer model with only one boundary is considered, as shown in Fig. 10.44. The parameters of the two layers are $\sigma_1 = 1$ and $\sigma_2 = 0.01$. The boundary is located at $z = 0$. For testing the effect of spacing, frequency should be fixed. The fixed frequency is chosen to be 400 kHz, simply because the detectable distance is larger at low frequency. The spacing range is from 33 to 55 in.

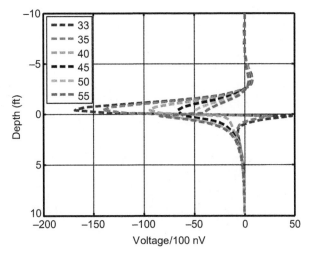

Figure 10.45 Cross-component response versus 100 nV in different spacing (400 kHz).

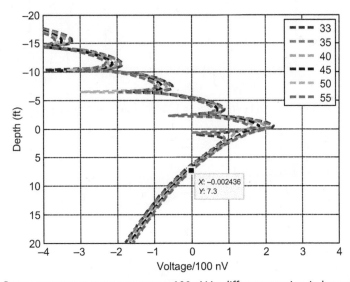

Figure 10.46 Cross-component response versus 100 nV in different spacing in log scale (400 kHz).

Fig. 10.45 shows the cross-component simulation results with different spacing. The results are all divided by 100 nV, which is the minimum detectable voltage in application. When the spacing is larger, the peak at the boundary is lower. In the high-resistive region, the detectable distance is larger. On the contrary, in the high-conductive region, the detectable distance becomes smaller. Replot the results in log scale in Fig. 10.46. It is easy to find that when spacing is 55 in., the detectable distance is about 7 ft. It is also noticed that the detectable distance is not sensitive to the spacing. That probably because the wavelength effect, which is compared with the wavelength, the spacing is relatively small. The property is good for tool design, which means the tool does not need to be too long.

Figs. 10.47 and 10.48 show the same results when the tool is working at 2 MHz. Compared with the results at 400 kHz, when both transmitter and receiver have one turn, with spacing 55 in., the detectable distance of both frequencies are around 7 ft.

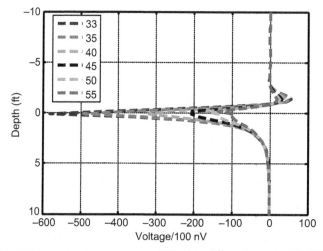

Figure 10.47 Cross-component response versus 100 nV in different spacing (2 MHz).

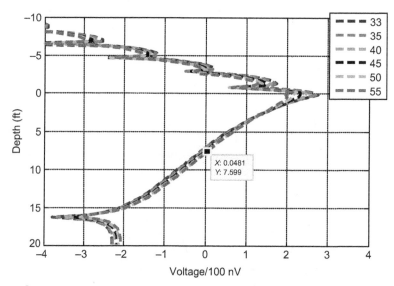

Figure 10.48 Cross-component response versus 100 nV in different spacing in log scale (2 MHz).

Table 10.1 Computation speed testing

Logging points	*6000*	*60,000*	*600,000*
Image method(s)	0.12	0.59	5.37
Full solution(s)	8.67	86.17	859.88
Speed ratio	72	150	160

The values in italics indicates comparison for the logging points of 6000, 60,000, and 600,000.

However, when transmitter has 10 turns, the receiving signal will be enlarged 10 times. In this situation, the tool working at 400 kHz has larger detectable distance than the tool working at 2 MHz.

10.3.4.4 Calculation speed

Table 10.1 shows the CPU time comparison between the image method and the full solution for different numbers of iterations. The results show that the image method is much faster than the full solution. When iteration time is 1000, the image method is 160 times faster than the full solution. In addition, when iterative times increase, the image method will have a greater advantage in computation speed.

Testing model: Three-layered model
Testing tool: Frequency = 2 MHz   Antenna Spacing = 19 in.

10.3.4.5 Logging with high deviated angle

Until now, all simulation cases assume a horizontal well. However, in the real application, most cases are not in exactly horizontal situation. To further understand the effectiveness of the image theory method, the well with high deviated angle is investigated. The schematic of the well with high deviated angle is shown in Fig. 10.49,

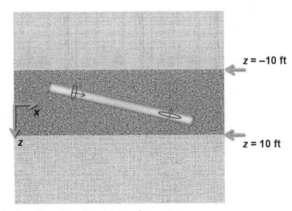

Figure 10.49 High deviated well in three-layer formation.

where the tool is not exactly horizontally placed. The dipping angle of the simulated tool is from 60 to 85 degrees. For convenience, phase shift and amplitude ratio is not shown. Only the cross-component response is shown below.

The three-layer 1D model is shown in Fig. 10.49. The parameters of this formation are $\varepsilon_{r1} = \varepsilon_{r3} = 1$, $\mu_{r1} = \mu_{r3} = 1$, and $\sigma_1 = \sigma_3 = 1$ for the upper and lower layer, $\varepsilon_{r2} = 1$, $\mu_{r2} = 1$, and $\sigma_2 = 0.01$ for the middle layer. The boundaries are at $z = 10$ ft and $z = -10$ ft. The logging is working at 2 MHz and the spacing is 34 in.

1. Dipping angle = 85 degrees

   Fig. 10.50 shows the cross component of the simulation results when the dipping angle is 85 degrees. Because in this case, the tool is almost horizontally placed, the simulation results look similar as the case when the logging tool is placed exactly horizontally. Comparing the image theory method with the full solution, there is not much difference. The complex image theory works pretty well when the logging tool is placed almost horizontally.

2. Dipping = 75 degrees

   Fig. 10.51 shows the simulation results when the dipping angle is reduced to 75 degrees. In this case, the complex image method works well too. The fast solution shows enough agreement with the full solution. In addition, because the tool is not horizontal with the bed boundaries, although the formation is symmetry, the simulation results become asymmetry. The cross component shows stronger response at the lower boundary.

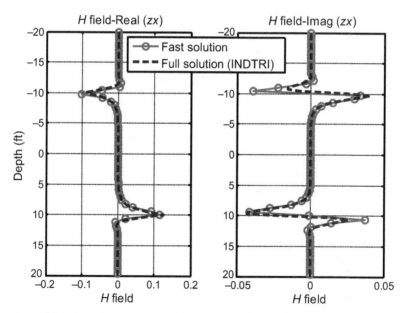

Figure 10.50 Tool response H_{zx} component (dipping = 85 degrees).

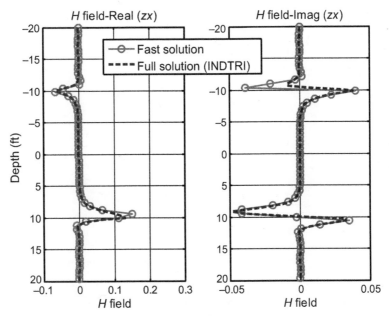

Figure 10.51 Tool response H_{zx} component (dipping = 75 degrees).

3. Dipping angle = 65 degrees
   When the dipping angle goes to 65 degrees, the asymmetry of the cross component is more obvious, as shown in Fig. 10.52. In this case, the complex image method gives the large peak when the tool is across the lower boundary. But, in the area near the upper boundary, the simulation results from complex image method can follow the full solution results closely. The reason is that, when the dipping angle is less than 90 degrees, for the upper boundary, the transmitter is closer to the boundary and the receiver is relatively further away from the boundary. In opposite, for the lower boundary, the receiver is closer to the boundary than the transmitter. As the instruction shown before, the complex image method has larger error in the area near the boundary. So, in Fig. 10.52, there is lager error appearing near the lower boundary. But, as shown in Fig. 10.42, the error is still within 2–3 ft from the boundary. This is acceptable in the application of geosteering system.
4. Dipping angle = 60 degrees
   Similarly, Fig. 10.53 shows the cross component of the simulation results when the dipping angle is 60 degrees. Because the dipping angle is much less than horizontal case, the simulation results show more obvious asymmetric and the complex image method gives larger error near the lower boundary. It is already shown

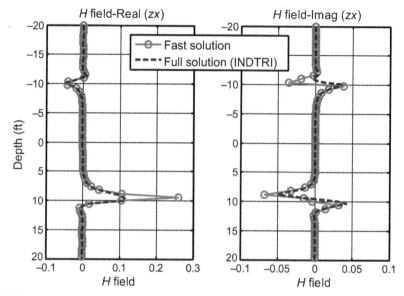

Figure 10.52 Tool response H_{zx} component (dipping = 65 degrees).

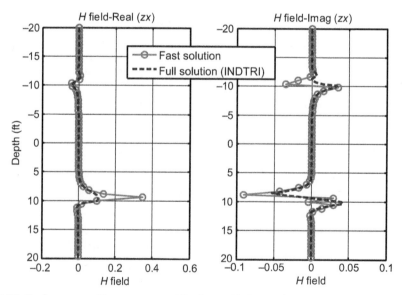

Figure 10.53 Tool response H_{zx} component (dipping = 60 degrees).

in Fig. 10.49 that the middle layer is relative high-resistive layer. Zoom in the Fig. 10.53 and show the middle layer only in Fig. 10.54. The simulation results show that the complex image method works well in the relative high-resistive layer, even when the dipping angle is 60 degrees. The simulation results of

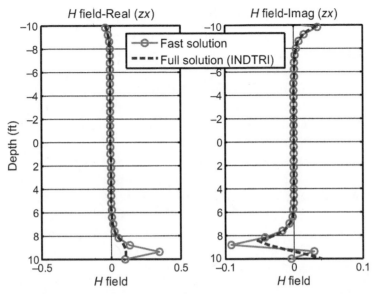

Figure 10.54 Tool response H_{zx} component in the middle layer (dipping = 60 degrees).

complex image method can follow the variation of the results from full solution in the most area. Error only occurs near the lower boundary and the error area is within 2 ft away from the boundary. Based on the discussion above, it can be concluded that the complex image method can work in the highly deviated well with dipping angle varying from 60 to 90 degrees.

10.4 BOUNDARY DISTANCE INVERSION

10.4.1 Theory of inversion

The inversion methods have been extensively discussed in Chapter 9, Theory of Inversion for Triaxial Induction and Logging-While-Drilling Logging Data in One- and Two-Dimensional Formations. In Chapter 9, Theory of Inversion for Triaxial Induction and Logging-While-Drilling Logging Data in One- and Two-Dimensional Formations, our interests are the formation conductivities. In geosteering, the most important parameters are the distance of the boundaries from the tool. Due to the real-time control requirements, the inversion must be fast enough and can be done in downhole. The downhole processors are usually not as powerful as the one on the surface, the inversion process must be simple and fast. In this chapter, we discuss a special inversion method, which is similar to the methods discussed in Chapter 9, Theory of Inversion for Triaxial Induction and Logging-While-Drilling Logging Data in One- and Two-Dimensional Formations, but with significant smaller number of variables.

10.4.2 Workflow of inversion problem

For real-time drilling direction adjustment, geosteering system is a negative feedback control system, which adjusts the direction of the drilling bit based on the real-time data collected from downhole. Such data includes the real-time position of the drilling bit and its distance away from the boundary. Boundary Detection is thus the key part of this system. A fast and accurate method is essential for real-time control.

Boundary detection is usually modeled as an inversion problem. In an iterative manner, we are to minimize the difference between the data collected from the receiving antenna and the simulation results from the forward modeling in certain tolerance. The value of parameters, e.g., the distance to boundary, is calculated as a by-product in the minimization process. Fig. 10.55 shows the flowchart of the inversion process used in this chapter, which generally includes forward modeling and model correction. Because of such an iterative procedure, real-time system requires that the forward modeling, which calculates the field distribution of dipole in multilayered media, to be fast and accurate.

10.4.3 Processing flow of boundary detection in geosteering

In the geosteering system, there are three steps to process the measurement before going to the boundary distance inversion. Those steps help the system to get the basic information of the formation and initialize the simulation model used into the boundary distance inversion. Fig. 10.56 gives the general flow of such process. Firstly, logging data is collected by the receiver. Secondly, a brief geological model of the

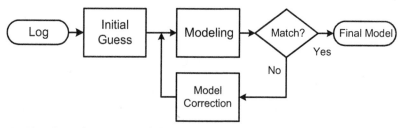

Figure 10.55 Flowchart of inversion problem.

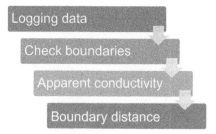

Figure 10.56 Boundary distance inversion flow.

formation is generated from the logging data. This step mainly focuses on finding the positions of all boundaries. Thirdly, based on the phase shift and attenuation of each layer, the apparent resistivity of each layer can be inverted out.

After those three steps, the depth of the boundary, the apparent conductivities of both layers, and the logging curves are ready. The only unknown is the distance from the drilling bit to the boundary. Then follow the flowchart shown in Fig. 10.55, by iterating the forward modeling, the optimized boundary distance can be inverted out. Because the boundary inversion in last step is supposed to be finished downhole, the fast forward modeling is required. The complex image method discussed in this chapter is used to speed up the last step, which is the boundary distance inversion.

10.4.4 Bolzano bisection method

The bisection method is a root-finding method that repeatedly bisects an interval and then selects a subinterval in which a root must lie for further processing.

For a real variable x, where f is a continuous function $f(x) = 0$ defined on an interval $[a, b]$ and $f(a)f(b) < 0$. Then, $f(x)$ has at least one root in $[a, b]$. The procedure of the bisection is shown below.

Firstly, let $[a, b] = [a_1, b_1]$, denote the middle point of $[a, b]$ as p_1,

$$p_1 = \frac{a_1 + b_1}{2} \tag{10.29}$$

Give a threshold of length (TOL) (small enough). Plug p_1 back into the equation. If $|f(p_1)| < TOL$, then p_1 is the approximate root of the equation $f(x) = 0$. If $|f(p_1)| > TOL$, we will search the root in the interval $[a_1, p_1]$ or $[p_1, b_1]$.

Secondly, if $f(p_1)f(b_1) > 0$, the root will be in the interval $[a_1, p_1]$. Else, the root will be in the interval $[p_1, b_1]$. Then the searching region is reduced by half. Repeat the previous steps, the approximate root with acceptable error will be found.

10.4.5 Simulation results

The parameters of logging tool used in following cases:

Frequency = 2 MHz, spacing = 36.375 in., and the dipping angle is 90 degrees.

10.4.5.1 Sensitivity of depth

Sensitivity of depth is a parameter defined by $\Delta H/\Delta d$, which represents the variation speed of H field along with varying of depth. Higher sensitivity of depth contributes higher convergence speed of boundary distance inversion. It is an important parameter to choose the component used into boundary distance inversion.

One two-layer model was used to test the inversion process. As shown in Fig. 10.57, the two-layer model has one boundary at $z = 0$ ft. The resistivities of the

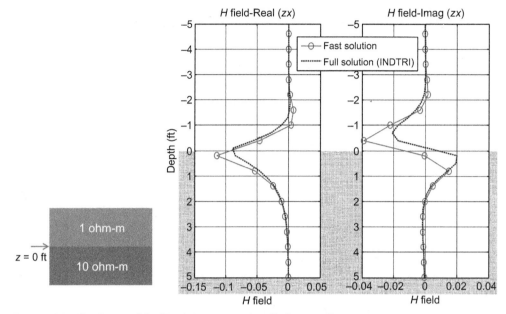

Figure 10.57 Testing model of inversion processing ($R_1:R_2 = 1:10$).

two layers are 1 and 10 ohm-m, respectively. The right-hand side of Fig. 10.57 shows the simulation results of H_{zx}.

Fig. 10.58 shows the depth sensitivity of the real part and amplitude of cross component, respectively. Because the real part and imaginary part of H_{zx} almost follow the same trend, there is not much difference shown in Fig. 10.58. From 0 to 4 ft, the absolute value of sensitivity is all larger than 0. This indicates that, in this range of depth, both the real part and the amplitude of the cross component can be used to inverse the boundary distance. Replot the depth sensitivity in Fig. 10.59, which only shows the depth from 4 to 10 ft. From this figure, it is easy to see that, the depth sensitivity of the amplitude of cross component decreases as the depth increases. But the absolute value of amplitude sensitivity is always larger than zero. However the sensitivity of the real part of cross component moves closer to zero when the depth is larger than 6 ft. That means, with a depth larger than 6 ft, the real part of the cross component is nonsensitive to the boundary. With a depth from 6 to 10 ft, the amplitude of cross component gives better performance.

10.4.5.2 $R_1 = 1$ ohm-m, $R_2 = 10$ ohm-m

Based on the processing flow shown in Fig. 10.56, the last step is to calculate the distance from the drilling bit to the boundary. In this processing, the apparent resistivity of the formation and the depth of the boundary are already known. The distance is the only unknown. For each distance, there will be a received H_{zx} corresponding.

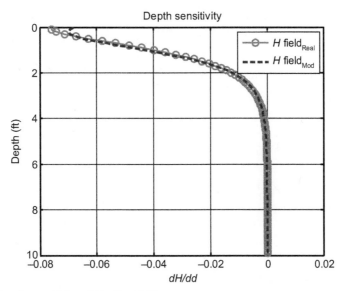

Figure 10.58 Depth sensitivity of H_{zx} (0–10 ft).

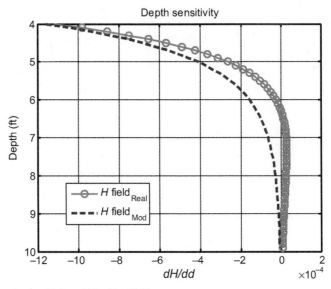

Figure 10.59 Depth sensitivity of H_{zx} (4–10 ft).

Then based on this model, by combining the received signal, the distance away from the boundary can be calculated. Generally, it can be calculated by solving one unknown equation. The problem is that this equation is of high order. The unknown cannot be solved explicitly. In this part, bisection method is used to calculate the distance.

Assume the unknown distance is d. We thus have a high-order equation,

$$f(d) = V_r \qquad (10.30)$$

where V_r is the measurement of received H_{zx}. Rewrite the equation as

$$f(d) - V_r = 0 \qquad (10.31)$$

Then the problem is to find the root of Eq. (10.31). Here, the H_{zx} measurement is obtained from the analytical full solution. The inversion process is running the forward modeling with complex image theory. Because we already know the position of the boundary, in this case, the boundary is at $z = 0$; and in the most case, the logging tool is working in the high-resistivity side, in this case, the high-resistivity side is located in the area $z > 0$. In this part, only lower half-space was tested. In the most cases, the tool is working in the high-resistive layer and approaching to boundary.

1. Distance inversion from the real part of the H_{zx}

   Table 10.2 shows the inversion results, when the logging points are located at 5.0, 4.0, 3.0, and 2.0 ft away from the boundary. The real parts of the H_{zx} are all within the detectable range. The results show that when the distance from the drilling bit to the boundary is within 4 ft, the inversion method can find the distance accurately. However, when the distance between the drilling bit and the boundary is close to 5 ft or higher, the inversion process gives higher error. Fig. 10.60 shows the real and imaginary parts of the cross component H_{zx} in the range from 4 ft to 10 ft. It's clear that, starting from 5 ft, the real part of H_{zx} is not monotonic. There exist two roots of Eq. (47). This is why the error becomes larger around this depth.

2. Distance inversion from the amplitude of the H_{zx}

   To solve the problems appearing in case (1), the amplitude of H_{zx} is used in the distance inversion. Fig. 10.61 gives the amplitude of the cross component H_{zx}. It shows that the amplitude of H_{zx} has a single value in each side of the boundary. The curve is monotonic. Fig. 10.62 is the zoom in figure of the amplitude in the range from 4 to 10 ft. It is obvious that the curve is monotonic and is always larger than zero. Then, the boundary distance is inversed based on the amplitude of the cross component H_{zx}.

Table 10.2 Distance inversion table (H_{zx}_real, $R_1:R_2 = 1:10$)

Distance (ft)	H_{zx}_real (abs)	Voltage (nV)	Inversion results (ft)	Error (%)
5.0	0.0006	192	4.1414	17.17
4.0	0.0007	224	4.0518	1.30
3.0	0.0031	992	3.0444	1.48
2.0	0.0111	3552	2.0489	2.45

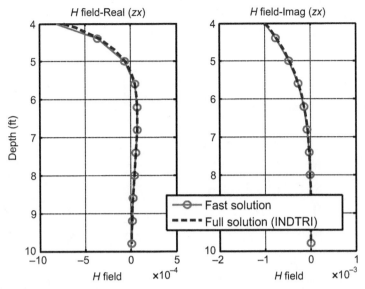

Figure 10.60 Zoom in cross component H_{zx} of tool response.

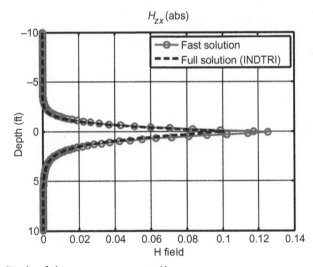

Figure 10.61 Amplitude of the cross component H_{zx}.

Table 10.3 shows the inversion results based on the amplitude of H_{zx}. Compared with the results in Table 10.2, the inversion method based on amplitude of H_{zx} is faster and more accurate. The algorithm can even handle the case when the logging point is 10 ft away from the boundary and keeps the relative error within 1%.

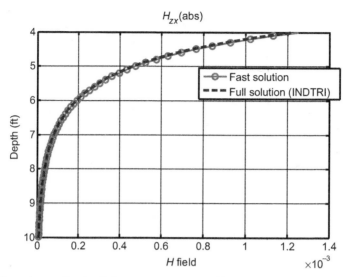

Figure 10.62 Zoom in amplitude of the cross component H_{zx}.

Table 10.3 Distance inversion table (H_{zx}_abs, R_1:R_2 = 1:10)

Distance (ft)	H_{zx} (abs)	Voltage (nV)	Inversion results (ft)	Error (%)
10.0	7.7888e-06	2.4924	10.0418	0.42
9.0	1.6558e-05	5.2985	9.0403	0.45
8.0	3.6071e-05	11.5427	8.0394	0.49
7.0	8.1004e-05	25.9213	7.0392	0.56
6.0	0.0002	64.00	6.0384	0.64
5.0	0.0005	160.00	5.0390	0.78
4.0	0.0012	384.00	4.0381	0.95
3.0	0.0035	1120.00	3.0401	1.34
2.0	0.0111	3552.00	2.0502	2.51
1.0	0.0379	12,128.00	1.0983	9.83

Fig. 10.63 shows the relative error of the inversion results in Table 10.3, which shows that, when the drilling bit is located in the range 4 to 10 ft away from the boundary, the relative error of the boundary distance inversion is within 1%. When the drilling bit is close to the boundary, within the area 2–4 ft away from the boundary, the relative error is larger, but still within 3%. When the drilling bit moves to the area 1 ft away from the boundary, the boundary distance inversion is not as accurate. The relative error goes up to 10%. That is because the complex image theory does not work well around boundary. For the two-layer formation, shown in Fig. 10.57, considering the minimum detectable voltage 100 nV, for the transmitter antenna with 3.2×10^5 A-m, the detectable distance is about 6 ft. Neglect the limitation of

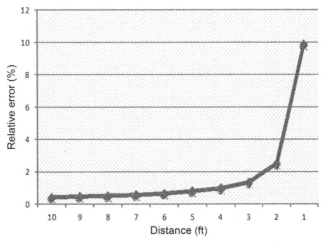

Figure 10.63 Relative error of the inversion results (H_{zx}_abs, $R_1:R_2$ = 1:10).

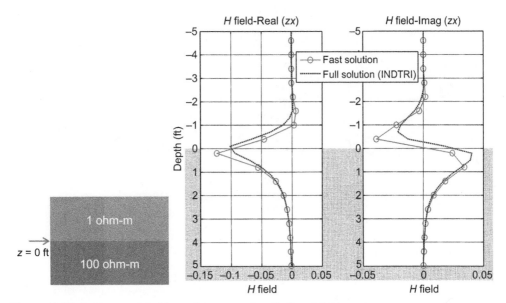

Figure 10.64 Testing model of inversion processing ($R_1:R_2$ = 1:100).

minimum detectable voltage, the inversion process can work even when the logging point is 10 ft far away to the boundary.

10.4.5.3 R_1 = 1 ohm-m, R_2 = 100 ohm-m

Similar as Section 10.4.5.1, retest the inversion method in the case when the lower layer is 100 ohm-m and the upper layer is 1 ohm-m. The boundary is still at $z = 0$. Fig. 10.64 shows the formation model and the simulation results.

Table 10.4 Distance inversion table (H_{zx}_abs, $R_1:R_2 = 1:100$)

Distance (ft)	H_{zx} (abs)	Voltage (nV)	Inversion results (ft)	Error (%)
10.0	8.9276e-05	28.57	10.0135	0.14
9.0	1.3812e-04	44.20	9.0141	0.17
8.0	2.2105e-04	70.74	8.0147	0.18
7.0	3.6903e-04	118.09	7.0160	0.23
6.0	6.4986e-04	207.96	6.0166	0.28
5.0	12.2592e-04	392.29	5.0179	0.36
4.0	25.3043e-04	809.74	4.0196	0.49
3.0	58.7665e-04	1880.53	3.0227	0.76
2.0	0.0158	5056.00	2.0320	1.60
1.0	0.0476	15,232.00	1.0740	7.40

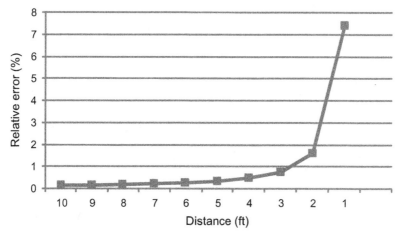

Figure 10.65 Relative error of the inversion results (H_{zx}_real, $R_1:R_2 = 1:100$).

Table 10.4 shows the comparison between the inverted distance and actual distance. The results show that the inversion results are pretty close to the actual distance. Except the logging point at 1 ft away from the boundary, which is very close to the boundary, in the distance range from 2 ft to 10 ft, the relative error stays within 2% as is shown in Fig. 10.65.

Replot the relative error curves of the two cases in Fig. 10.66. It shows that when the conductivity contrast of the two layers is larger, the relative error of the inversion results is smaller.

10.4.6 Simulation results with noise added

The behavior of the inversion method in the presence of noise and error is also evaluated. To simulate the noise, an array of random number between -1 and $+1$ was

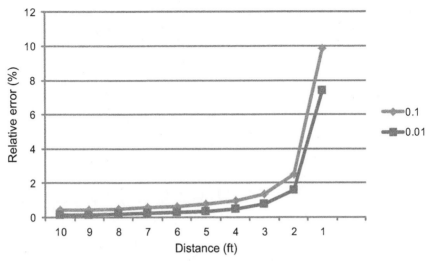

Figure 10.66 Comparison of the relative error in different formation.

Table 10.5 Distance inversion table with 1% noise added (H_{zx}_abs, R_1:R_2 = 1:10)

Distance (ft)	H_{zx}_ideal (abs)	Voltage (nV)	H_{zx}_noise (1%)	Inversion results (ft)	Error (%)
10.0	7.7888e-06	2.4924	6.7448e-06	9.8850	1.15
9.0	1.6558e-05	5.2985	1.9574e-05	9.0272	0.30
8.0	3.6071e-05	11.5427	3.5953e-05	8.0038	0.05
7.0	8.1004e-05	25.9213	7.8162e-05	7.0756	1.08
6.0	1.8906e-04	64.00	1.8968e-04	6.0496	0.83
5.0	4.6384e-04	160.00	4.6693e-04	5.0448	0.90
4.0	1.2157e-03	384.00	1.2137e-03	4.0403	1.01
3.0	3.4783e-03	1120.00	3.4781e-03	3.0401	1.34
2.0	1.1089e-02	3552.00	1.1092e-02	2.0502	2.51

generated using a white-noise generator. This array was scaled to 1−10% of the minimum detectable voltage 100 nV. Convert the voltage to the H field. The array was scaled to 1−10% of 3.125e-04. Use the scaled array as a noise. Add this noise to the data simulated from the analytical full solution as a measured data. Then plug this data into the inversion processing. By iterating the forward modeling developed based on the complex image theory, calculate the distance away from the boundary.

10.4.6.1 R_1 = 1 ohm-m, R_2 = 10 ohm-m

Considering the effect of noise, reprocess the two-layer model in Fig. 10.57. Table 10.5 shows the cross component with 1% noise and the inversion results

Table 10.6 Distance inversion table with 5% noise added (H_{zx}_abs, $R_1:R_2 = 1:10$)

Distance (ft)	H_{zx}_ideal (abs)	Voltage (nV)	H_{zx}_noise (5%)	Inversion results (ft)	Error (%)
10.0	7.7888e-06	2.4924	7.1562e-06	12.0000	20
9.0	1.6558e-05	5.2985	2.4941e-05	8.3353	7.38
8.0	3.6071e-05	11.5427	3.8697e-05	7.9036	1.21
7.0	8.1004e-05	25.9213	6.8256e-05	7.2593	3.70
6.0	1.8906e-04	64.00	1.7833e-04	6.1237	2.06
5.0	4.6384e-04	160.00	4.7825e-04	5.0273	0.55
4.0	1.2157e-03	384.00	1.2240e-03	4.0483	1.21
3.0	3.4783e-03	1120.00	3.4874e-03	3.0401	1.34
2.0	1.1089e-02	3552.00	1.1088e-02	2.0505	2.53

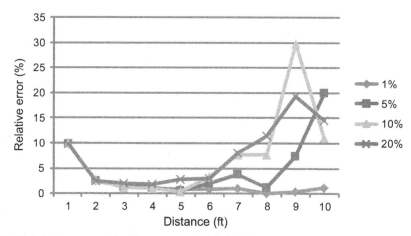

Figure 10.67 Relative error with different noise added ($R_1:R_2 = 1:10$).

generated using the noised data. The inversion results show that, less than 1% noise, the inversion method still gives reliable result. The relative error between the inversed distance and the accurate distance stays within 3%.

Table 10.6 shows processing results when the noise is increased to 5%. In this case, because the added noise is at the same order of the ideal data far from the boundary, the noise causes higher error to the inversion results. The relative errors of the logging points within 6 ft away from boundary still remain within 3%. For logging points away from boundary for more than 6 ft, the relative error can be as high as 20%.

Fig. 10.67 shows the curves of relative error when different percentages of noise are added to the ideal data. It is easy to see that when the added noise is increasing, the relative error is larger. The noise effects more in the area relatively further away from boundary than the area close to the boundary. That is because, in noise study, the noise level is fixed, but for a fixed level of transmitter power, the received signal reduces a lot as the distance from the boundary is enlarged. The noise has more effect in the area further away from the boundary.

10.4.6.2 $R_1 = 1$ ohm-m, $R_2 = 100$ ohm-m

When the two-layer model shown in Fig. 10.64, where the conductivity of low medium is 100 ohm-m, compared with the two-layer model in Fig. 10.57, where the conductivity of low medium is 10 ohm-m, the received cross component has larger amplitude at the relative far area. For example, at the observation point 10 ft away from the boundary, in the case with 100 ohm-m lower medium, the amplitude of the cross component is 8.9276e-05. In the case with 10 ohm-m lower medium, the amplitude of the cross component is only 7.7888e-06 which is one magnitude lower. That means, in the same noisy environment, the logging tool has better performance in the case with larger conductivity contrast. This conclusion agrees with the results shown in the Section 10.3.4.

Since in the high-conductivity contrast formation, the cross component is stronger. The amount of noise in this formation is started to be added from 5%.

Table 10.7 shows the ideal data of the cross component, data with 5% noise of the cross component, inversion distance generated from the noised data as measurements and the relative errors between the inversed distance and real positions of the logging points.

Fig. 10.68 shows the curves of relative error when the added noise is increased to 10%, 20%, and 50%. It shows that, when the noise is increased to 20%, in the most area, the relative error of the inversion distance still remains within 5%. When the noise is increased to 50%, there is huge error for the testing logging points 6 ft or further away from boundary. Within 6 ft, the relative errors are always less than 5%.

Table 10.7 Distance inversion table with 5% noise added (H_{zx}_abs, R_1:R_2 = 1:100)

Distance (ft)	H_{zx} (abs)	Voltage (nV)	H_{zx}_noise (5%)	Inversion results (ft)	Error (%)
10.0	8.9276e-05	28.57	8.5212e-05	9.8007	1.99
9.0	1.3812e-04	44.20	1.4027e-04	9.1521	1.69
8.0	2.2105e-04	70.74	2.2331e-04	8.0140	0.18
7.0	3.6903e-04	118.09	3.6820e-04	6.9739	0.37
6.0	6.4986e-04	207.96	6.5107e-04	5.9904	0.16
5.0	1.2259e-03	392.29	1.2238e-03	5.0085	0.17
4.0	2.5304e-03	809.74	2.5257e-03	4.0163	0.41
3.0	5.8767e-03	1880.53	5.8691e-03	3.0256	0.85
2.0	1.5786e-02	5056.00	1.5799e-02	2.0320	1.60

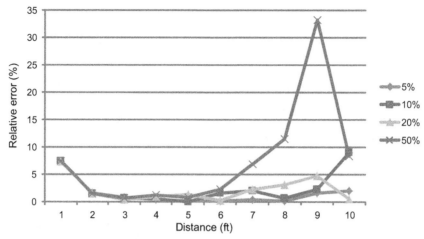

Figure 10.68 Relative error with different noise added ($R_1:R_2 = 1:100$).

10.5 CONCLUSION

Image theory, as a method used to simplify the inhomogeneous media, can be applied in geosteering to speed up the simulation. The advantage of this theory is the simplicity in formulation and fast in computation.

The complex image approximation method was tested at 2 MHz and 400 kHz, respectively. Compared with the full solution results, the complex image method has very good agreement at both frequencies. Error only occurs near boundary. However, in the application of geosteering, the error is acceptable. It works better at higher frequencies than lower frequencies. This is because when the frequency is higher, the skin depth of the formation is shorter, which is closer to a perfect conductor, and therefore the image theory is more accurate.

The accuracy of the complex image theory also depends on the conductivities of both layers. When the conductivity difference between the upper layer and the lower layer increases, the error decreases. The absolute error and relative error are collected at different observation points. The error is larger when the drilling bit is closer to the boundaries. For the 2 MHz tool, when the logging point is more than 2 ft away from the boundary, the relative error is less than 10%. For the 400 kHz tool, this distance is increased to 3 ft. Compared with the full solution method, the complex image approximation method can significantly speed up the simulation. In the testing with 1000 iteration and 600,000 logging points in total, the complex image method is more than 100 times faster than the full solution. This difference in efficiency is also enlarged along with the increase of the logging point. This method can be used in real-time data inversion of the distance to boundary computation in a geosteering system.

Effects of frequency and spacing are investigated. For one-turn antenna, with area 2.5 in.² and excited by 200 mA, the general detectable distance in high-resistive layer is about 5 ft. When tool is working at 400 kHz, longer spacing gives larger detectable distance. The simulation shows that when the spacing is 55 in., the detectable distance is about 7 ft.

Inversion process is given in last part. Two-layer model with boundary at $z = 0$ is tested. Boundary distance is inversed based on the amplitude curve of the cross component H_{zx}. The inversion results show that the inversion code works well in the distance range from 2 ft to 10 ft. The relative error is kept in 2%. By comparing the relative error from different formation combination, it can be concluded that larger conductivity contrast of the formation contributes more accurate inversion results.

The effect of noise was discussed in inversion processing. A random white noise with amplitude 100 nV, scaled from 1% to 50%, was added into the analytical full solution data to test the antinoise capacity. The relative errors of inversion results generated using ideal data with different amount noise added in two different formations are calculated and plotted. The results show that the proposed method is more robust in formations with higher conductivity contrast. Compared with the area relatively further away from the boundary, the relative error can be kept in a lower range in the area within 6 ft away from the boundary.

REFERENCES

[1] M.S. Efnik, M. Hamawi, A. Shamri, A. Madjidi, C. Shade, Using new advances in LWD technology for geosteering and geologic modeling, in: SPE/IADC 57537, 1999.
[2] Q. Li, D. Omeragic, L. Chou, L. Yang, K. Duong, J. Smits, et al., New directional electromagnetic tool for proactive geo-steering and accurate formation evaluation while drilling, in: 46th SPWLA Annual Logging Symposium, Paper UU, 2005.
[3] T. Wang, H. Meyer, L. Yu, Dipping bed response and inversion for distance to bed for a new while-drilling resistivity measurement, in: SEG New Orleans 2006 Annual Meeting, 2006, pp. 416–420.
[4] J.R. Wait, Image theory of a quasi-static magnetic dipole over a dissipative half-space, Electron. Lett. 5 (13) (1969) 281–282.
[5] P.R. Bannister, Summary of image field expressions for the quasi-static fields of antennas at or above the earth's surface, Proc. IEEE 67 (7) (1979) 1001–1008.
[6] I.V. Lindell, E. Alanen, Exact image theory for the Sommerfeld half-space problem, Part I: Vertical magnetic dipole, IEEE Trans. Antennas Propag. AP-32 (2) (1984) 126–133.
[7] Q.Z. Dong, T. Wang, A fast forward model for simulating a layered medium using the complex image theory, in: SEG San Antonio 2011 Annual Meeting, 2011, pp. 573–577.
[8] R.H. Lien, Radiation from a horizontal dipole in a semi-infinite dissipative medium, J. Appl. Phys. 24 (1) (1953) 1–4.

CHAPTER 11

Ahead-of-the-Bit Tools and Far Detection Electromagnetic Tools

Contents

11.1 Introduction	407
11.2 Ahead-of-the-Bit Field Distribution of LWD Tools	410
11.3 Toroidal Transmitter	415
11.4 Boundary Detection Using Orthogonal Antennas	420
11.5 Deep-Looking Directional Resistivity Tool	424
11.5.1 Physics of the directional resistivity tool	425
11.5.2 Forward modeling of a deep-looking tool with tilted antennas	429
11.6 Distance Inversion Based on the Gauss—Newton Algorithm	435
11.6.1 Summary of Gauss—Newton Method	435
11.6.2 Inversion in two-layer formations	438
11.6.3 Inversion in three-layer formations	441
11.7 Conclusions	444
References	445

11.1 INTRODUCTION

As described in Chapter 10, The Application of Image Theory in Geosteering, logging-while-drilling (LWD) technology has been used as a geosteering aid in directional drilling. Based on real-time measurements provided by LWD tools, operators are able to make better-informed drilling decisions to improve drilling efficiency as well as to reduce safety risks. The drilling engineers would like to "see" as far as possible the formation boundaries around the drilling bit. To implement the far detections, one of the useful methods is directional electromagnetic (EM) method.

In this chapter, a comprehensive investigation is conducted on the use of directional LWD resistivity tools in geosteering, especially the application in detecting remote bed boundaries. By looking into the electromagnetic field of various tool configurations, an independent evaluation is provided on the downhole boundary detection capability of multiple types of resistivity logging tools, as well as their applicability in different drilling environments.

To explore the potential of predicting formation properties in front of the drill bit, tool responses are first modeled with different downhole electromagnetic transmitters in homogeneous formation, where the ahead-of-the-bit field distribution is investigated. Field attenuation rates are compared among different tools, and the influence of borehole conductivity is studied. Next, tool responses are modeled in two-layer formation models to evaluate their boundary detection capabilities. The look-ahead capabilities are compared between tools with axially symmetrical antennas, with boundaries perpendicularly approached by the tool. Also, cross-component measurements are studied for tools using orthogonal antennas with boundaries parallel with the tool axis. After that, the deep-looking capability of a new directional resistivity tool using ultralong spacings and low frequencies is explored. Tool responses for different configuration parameters and drilling environments are calculated and discussed. At last, an inversion algorithm based on the Gauss—Newton method is developed to invert the boundary distance from the tool response, which can be either applied in two-layer or three-layer formations.

This chapter addresses the challenge of using LWD resistivity tools to predict formation anomalies ahead of or around the drill bit. Through the simulation results, one can gain an organized knowledge on the characteristics of LWD tools in terms of boundary detection capability. The detailed comparison results between tools of different types establish a missing link in the research of deep resistivity tools, and provide an objective reference for future designs of downhole boundary detection methods. The investigation of the deep-looking directional resistivity tool has demonstrated that an ultralong detection range can be achieved with azimuthal sensitivity using frequency-domain excitation sources.

The rapid development of LWD technology made the geosteering possible. Based on the real-time data gathered with LWD tools, better-informed drilling decisions can be made to improve drilling efficiency as well as to reduce safety risks. Due to its electromagnetic nature, LWD resistivity tools typically have a longer detection range than that of other LWD tools (e.g., acoustic, gamma ray, nuclear magnetic resonance (NMR)), and hence play an important role in geosteering applications. With early detection of approaching bed boundaries, the operator can accurately control the drilling direction, steering the bit onto the optimal well path, or away from unwanted formation structures.

Many examples have shown that it is advantageous to detect a formation anomaly ahead of or around a drill bit, such as a bypassed reservoir, an overpressured zone, a fault, or a salt dome. However, for conventional LWD resistivity tools, the response is mainly contributed by the formation volume around the tool, and cannot directly "look ahead." Payton et al. [1] proposed to use a transient electromagnetic method to detect boundaries. Unlike traditional resistivity logging tools which use frequency-domain excitation to generate electromagnetic field, the transient electromagnetic method adopts a time-domain excitation, which employs pulse signals or other periodical waveforms as a source and measures the returned broadband response. This technique is able to detect formation anomalies up to a hundred meters away. Banning et al. [2] further explored the potential of applying this method in detecting formation anomalies ahead of and around the drill bit, measuring both of direction and distance information. By monitoring the temporal change of received voltage, one can separate the responses of different spatial areas, and the data in later time stages contain information of remote bed boundaries. Theoretically, the transient method is capable of providing information about formation anomalies ahead of the bit, but it would require complicated downhole sensors and advanced LWD telemetry method to transmit the large volume of data to the surface if used to assist geosteering operations. At the time of writing, transient electromagnetic tools have not yet been commercialized.

In frequency domain, the earliest possible measurement of the formation being drilled is provided by the Resistivity-at-the-Bit tool developed by Bonner et al. [3]. The concept of the tool is based on the earlier work by Gianzero et al. [4], replacing the traditional coil antennas of resistivity tools with toroidal transmitters. A low-frequency axial current is driven through the drill bit, into the formation, and then flows back to the collar. When the tool is mounted closely to the bit, this "at-the-bit" measurement can be used as a reference for geosteering, or rather, geostopping. Field tests show that this type of resistivity measurement is earlier than any other measurements, but the response still lags behind the actually drilled spot by several inches. Bittar et al. [5] proposed that ahead-of-the-bit boundaries can be indicated by the relative difference between measurements by multispacing toroidal transmitters. For coil tools, Zhou et al. [6] briefly discussed the electromagnetic field ahead of the drill bit in 2000, and investigated the deep-looking limits of frequency- and time-domain methods.

Although it is not common to directly measure the ahead-of-the-bit formation volume with resistivity logging tools, the look-ahead capability can be acquired in an indirect way. In highly deviated wells with nearby bed boundaries, if the tool can provide deep measurements to detect lateral boundaries, the apparent distance from the bit to the boundary can be calculated with a given dip angle, i.e., the "look-around" capability can be converted to a "look-ahead" distance. Therefore the deep-looking capability would benefit from the increase of the radial depth of investigation (DOI) for LWD tools used as a geosteering aid. This can be done by increasing the transmitter−receiver spacing and using lower frequencies, as is applied on the deep resistivity tool in Seydoux et al. [7]. This tool responds rather early to approaching boundaries, claiming a detection range of 30 m, but the measurements lack directionality due to the axial symmetry of antennas. To provide directional information, an LWD tool with azimuthal sensitivity was developed to assist geosteering practice [8].

This tool uses cross-component measurements to distinguish between the boundaries approached from above and below the tool, and claims to detect boundaries that are 10–15 ft away from the borehole. Until recently, a new ultradeep directional tool is developed achieving a detection range of up to 30 m [9], but the tool physics has not yet been disclosed.

The objective of this chapter is to conduct a comprehensive investigation on the use of LWD resistivity logging tools in geosteering, especially the application in detecting remote bed boundaries. By looking into the electromagnetic field of various tool configurations, an independent evaluation is performed on the downhole boundary detection capability of different resistivity logging tools and their applicability in various drilling environments.

From the EM theory, to look ahead is rather difficult. First of all, we have to generate a field that can reach the front of the drill bit, or, at least, have part of the EM energy reaching the formation ahead of the drill bit. In other word, the antennas installed on the drill collar must have some kind of directivity. As we know, the directivity of antennas is based on the superposition of the waves with different phases. For low-frequency EM field, making directional antennas are very difficult in a limited space since the wavelength is too long. The other way to "focus" the field is to use multiple electrodes on the drill collar and control the potentials at each electrode so that the current flow is pushed forward to the designed direction. This idea has been successfully implemented in the laterolog tools in wireline (and LWD) to focus the field into the radial direction. Unfortunately, there is no space to place any electrode in the drill string to focus the field in the direction of drilling. In the first part of this chapter, we will discuss the sources available for possible look-ahead tools. Then we will investigate how these sources and detectors perform for the look-ahead detection. Finally, we will study the directional tool configuration for geosteering applications.

11.2 AHEAD-OF-THE-BIT FIELD DISTRIBUTION OF LWD TOOLS

From the previous chapters, we noticed that four different sources are used in EM logging: electrodes (laterolog, microresistivity imager), coils (induction tools, LWD resistivity), RF antennas (dielectric tools), and toroidal coils (near-bit resistivity). As we know, skin depth is the key to the detection range. Lower frequency will "see" further. Therefore only low-frequency tools are possible to penetrate through conductive formation and reach further. Although DC sources can be used as a source, the implementation will be difficult due to noise and DC bias shift in the receiver circuits. Therefore most DC logging tools use low-frequency AC instead.

Mostly used sources in the EM logging tools are coils. From previous chapters, we know that the coils are used in many different tools including induction, and LWD resistivity. The main reason we use coils in the logging tools is that coils fit the geometry naturally. Although the electrical dipole may be used for transmitting sources, the geometry of electric dipoles are not convenient in fitting into the logging tools. From a circuit point of view, coils are inductors; when frequency is low, the impedance will be low and therefore, the coil antennas are used only at kilohertz to low megahertz range. At higher frequencies, the coils will have a self-resonant frequency, which makes the coil no longer stable as an inductor. In LWD tools, ferrites are added to the antennas to increase signal transmitting and receiving efficiency with a compromise in temperature stability since the magnetic permeability of the ferrites is a nonlinear function of temperature. Therefore, in order to overcome the temperature issue, most LWD tools have to use compensation method to remove the temperature instability caused by antennas and circuits.

Another choice of antenna for logging tools is toroidal antenna. Toroidal antenna has the geometry of a coil but equivalent to an electric dipole source as shown in Fig. 11.1. A resistivity logging tool with toroidal antennas was introduced by Gianzero et al. [4]. A toroidal antenna is a winding of loops of conductive wire around a ring of material with a high value of magnetic permeability. The concept of using toroidal transmitters and receivers for induction logging was first proposed by Arps [10]. An alternating current flows through the wire to generate an alternating magnetic field inside the torus, which in turn induce radial and axial currents in the surrounding formations. The induced current can flow along the conductive drill collar, and then form a return path in the formations. Using toroidal sensors, ring electrodes, or button electrodes as receivers, one can derive the formation conductivity from the voltage. A practical implementation of a toroidal antenna is shown in Fig. 11.1, where the counter windings can effectively minimize the z-direction magnetic dipole component. The core is usually made of either ferrite materials or mu-metal to increase radiation efficiency.

The toroidal tool added a new direction to resistivity logging and formation evaluation, and has shown its advantages when using conductive mud in highly resistive

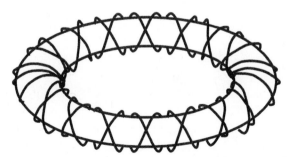

Figure 11.1 A practical implementation of a toroidal antenna with contra-wound wires.

formations. Further adaptations of toroidal tools are made to improve the reliability and accuracy of the resistivity measurements in Bonner et al. [3] and Bittar and Hu [11].

A toroidal antenna can be modeled as a magnetic current loop. Although magnetic current does not exist in the physical world, it can serve as an intermediate variable in analytical calculations.

Similar to the formulations of Chapter 2, Fundamentals of Electromagnetic Fields Induction Logging Tools, the solution to the field generated by a magnetic ring source can be obtained by using the duality theorem in electromagnetism (Section 3.2, Harrington, RF), Maxwell's equations can be solved to find the field of a toroidal transmitter. For a magnetic current loop, the field generated is dual to the electrical current loop. In the spherical coordinates, the magnetic field generated by the electrical current loop in a cylindrical coordinates is given in Eq. (2.36):

$$\mathbf{E}_\phi = \frac{-j\omega\mu I_T A_T N_T}{4\pi} \frac{\rho e^{-jk\sqrt{\rho^2+z^2}}(1+jk\sqrt{\rho^2+z^2})}{(\rho^2+z^2)^{3/2}} \qquad (2.36)$$

Converting the expression to the spherical coordinates, we have

$$E_\phi = \frac{\omega\mu I A_T N_T k}{4\pi r} e^{-jkr}\left(1 - \frac{j}{kr}\right)\sin(\theta) \qquad (11.1a)$$

The magnetic components of the current loop can also be found directly from the E field expression:

$$H_r = \frac{I A_T N_T k}{2\pi r^2} e^{-jkr}\left(j + \frac{1}{rk}\right)\cos(\theta) \qquad (11.1b)$$

$$H_\theta = \frac{I A_T N_T k}{4\pi r} e^{-jkr}\left(-k + \frac{j}{r} + \frac{1}{r^2}\right)\sin(\theta) \qquad (11.1c)$$

Using the duality theorem for the magnetic current loop Eq. (11.1a,b,c) becomes

$$H_\phi(r) = \frac{\omega\mu k \gamma_s}{4\pi r} e^{-jkr}\left(1 - \frac{j}{kr}\right)\sin(\theta) \qquad (11.2a)$$

$$E_r(r) = \frac{I\gamma_s k}{2\pi r^2} e^{-jkr}\left(j + \frac{1}{rk}\right)\cos(\theta) \qquad (11.2b)$$

$$E_\theta = \frac{I A_T N_T k}{4\pi r} e^{-jkr}\left(-k + \frac{j}{r} + \frac{1}{r^2}\right)\sin(\theta) \qquad (11.2c)$$

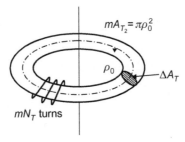

Figure 11.2 Toroidal transmitter.

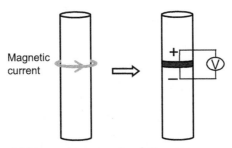

Figure 11.3 Modeling a toroidal transmitter as an insulating gap.

after the duality substitutions are applied, where $\gamma_s = N_T I_T A_T \cdot \Delta A_T$ is the moment of the toroidal antenna. Here ΔA_T is the cross-sectional area, and A_T is the area which is limited by the central line of the toroidal antenna, as illustrated in Fig. 11.2. Using a similar toroidal receiver along tool axis, the induced voltage can be expressed by

$$V = j\omega\mu N_R H_\phi \Delta A_R \tag{11.3}$$

where ΔA_R is the cross-section area of the toroidal receiver.

In numerical modeling, a toroidal transmitter can also be modeled as an insulating gap, as illustrated in Fig. 11.3. The tool is separated into two parts, with an alternating voltage source connecting to both sides of an imbedded insulator. Such a gap structure has been seen in LWD telemetry, used to transmit data from downhole equipment to the surface.

Fig. 11.4 shows the field distribution near the transmitter antenna when toroids are used in both water-based mud (WBM) and oil-based mud (OBM). We can clearly see that *E field* actually goes into the formation in the front of the drill bit. An 8.5-in. borehole is included in the model, filled with two types of mud: WBM of conductivity 10 S/m (resistivity 0.1 ohm-m) and OBM of conductivity 0.001 S/m (1000 ohm-m). From Fig. 11.4, we can see that in the OBM case, the most field does not go into the formation in the radial direction; instead the field is pushed along the drill string and to

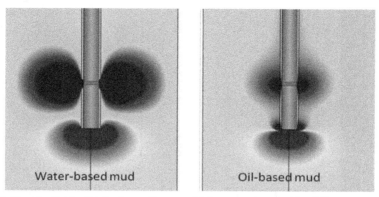

Figure 11.4 Electric field distribution of a toroidal transmitter with different types of mud. The formation conductivity is 20 ohm-m and the transmitter is 0.8 m away from the drill bit.

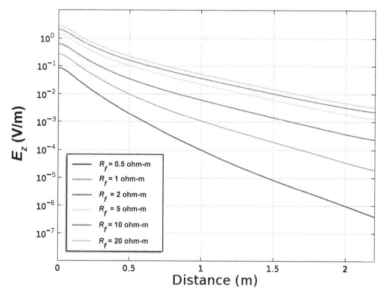

Figure 11.5 Ahead-of-the-bit field of a toroidal transmitter (insulating gap antenna) with water-based mud.

the front of the drill bit. However, in the WBM case, the field has more energy going into the formation radially.

As shown in Fig. 11.3, theoretically, the toroid antenna is equivalent to the gap around the drill collar. Figs. 11.5 and 11.6 show the E_z field ahead of the drill bit with a gap antenna in different formation conductivities and when borehole mud is WBM and OBM. We can see that the field attenuation reduces when the resistivity of the formation increases.

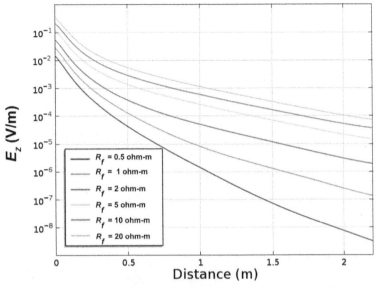

Figure 11.6 Ahead-of-the-bit field of a toroidal transmitter (insulating gap antenna) with oil-based mud.

11.3 TOROIDAL TRANSMITTER

For a resistivity tool using a toroidal transmitter, several types of receiver can be used: toroidal receiver, ring electrode, and button electrode. Fig. 11.7 shows the current distribution of a toroidal transmitter in a homogeneous medium. Both radial and axial currents can be measured and used to calculate the formation resistivity. The axial currents can be measured by a single toroidal receiver similar to the transmitter as discussed in Section 11.1, or by using a pair of toroidal receivers and taking a voltage difference [4]. If the transmitter is close to the bit face, the bit can be seen as an electrode, conducting the currents into the formation ahead. The effective bit electric length depends on the formation conductivity and the collar resistivity, and can be extended by a conductive borehole [3].

The axial currents on the drill collar can be measured by a ring electrode or a button electrode. The electrodes are insulated from the drill collar but held at the same potential with the collar so that the original current distribution is not disturbed. The button electrode has an azimuthal sensitivity, while the ring electrode provides an azimuthally average measurement. The button electrodes are widely used for borehole resistivity imagings. Due to the fact that the drill collar can be rotating, LWD resistivity imagers provide continuous borehole coverage that captures near-wellbore information. The detailed discussions of the imaging tools will be given in Chapter 15, Laterolog Tools and Array Laterolog Tools, of this book.

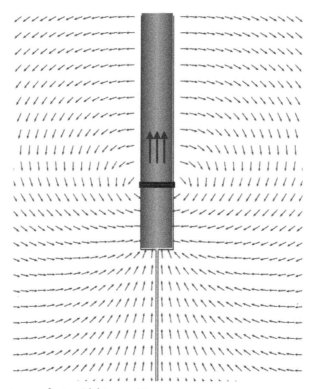

Figure 11.7 Current map of a toroidal transmitter.

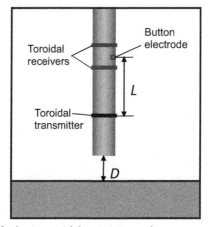

Figure 11.8 Configuration of a basic toroidal resistivity tool.

To investigate the sensitivity of a toroidal system to the formation change in front of a drill bit, a numerical model is established. Fig. 11.8 shows the configuration of the toroidal tool used in the modeling. A toroidal transmitter is modeled as a magnetic current around the drill string, and two toroidal receivers are situated above the

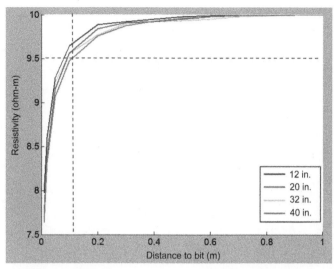

Figure 11.9 Response of a 10-kHz toroidal tool in a 10-ohm-m bed with a 1-ohm-m boundary ahead of the bit, measured by toroidal receivers.

transmitter, measuring the axial currents flowing along the collar. For comparison, a button electrode is also used to measure the radial currents flowing out of the drill string. The button electrode is in the middle between the two toroidal receivers. The transmitter−receiver spacing L is defined as the distance between the button electrode and the transmitter.

A dynamic meshing approach is used similar to the modeling of coil tools. As the conductive boundary moves toward the tool, the azimuthal magnetic field H_φ at the toroidal receivers and the radial current density J_r at the button electrode are measured. The apparent resistivity can be transformed either from the magnetic field difference of the two receivers, or from the radial current.

Figs. 11.9 and 11.10 show the responses of the tool with four different spacings: 12, 20, 32, and 40 in. The distance between the two toroidal receivers is 8 in. Measurements are acquired at 10 kHz. A two-layer formation model is used, with a local bed resistivity 10 ohm-m, and a 1-ohm-m bed boundary perpendicular to the tool axis, i.e., the formation conductivity contrast $\lambda = 10$. The results show that measurements provided by the toroidal receiver pair and the button electrode are very similar. This equivalence can be explained by the relationship of the axial and radial currents. If the axial currents at the lower and upper toroidal receivers are I_{z1} and I_{z2}, respectively, and the radial current flowing off the drill string from the area between the two receivers is I_r, the following equation holds [5]:

$$I_r = I_{z1} - I_{z2} \quad (11.4)$$

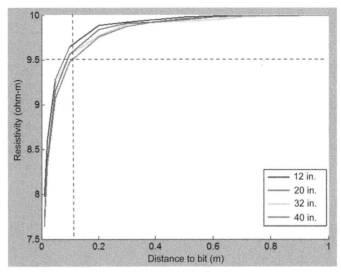

Figure 11.10 Response of a 10-kHz toroidal tool in a 10-ohm-m bed with a 1-ohm-m boundary ahead of the bit, measured by button electrode.

The results also demonstrate that the tool response is not very sensitive to the spacing change. For the 10-kHz tool with 40-in. spacing, the look-ahead distance is about 0.11 m when the resistivity drops from 10 to 9.5 ohm-m. Note that for 10-kHz coil tools, the response is much lazier. The 5% resistivity threshold cannot be reached even when the bit hits the boundary. From this perspective, the toroidal tool gives an earlier response to ahead-of-the-bit boundaries than traditional coil tools.

Fig. 11.11 shows the tool response at three different operating frequencies: 10, 20, and 50 kHz. The transmitter−receiver spacing is 32 in. Due to the equivalence of the toroidal receiver and button electrode measurements, here only the button electrode response, i.e., the radial current resistivity, is plotted. The data show that the sensitivity to the boundaries ahead improves as the frequency increases, but does not benefit as much as coil tools. At 50 kHz, the detection range is about 0.18 m.

Fig. 11.12 shows the influence of the formation conductivity contrast on the tool response. The local-layer resistivity remains at 10 ohm-m, while the ahead-of-the-bit boundary resistivity is set to 1, 0.5, and 0.2 ohm-m. Compared with coil tools, formation conductivity contrast barely affects the tool response. For a 20-kHz toroidal tool with 32-in. spacing, when λ changes from 10 to 50, the detection range only increases from 0.13 to 0.17 m.

As a summary, the look-ahead response of toroidal tools is not very sensitive to the adjustment of tool configuration. Once the operating frequency of the transmitter is determined, the tool response is relatively steady. The improvement of detection range by increasing the transmitter−receiver spacing and formation conductivity contrast is

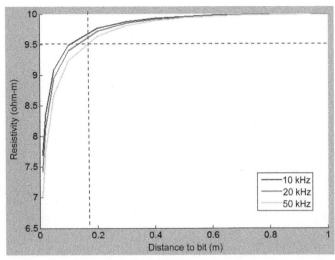

Figure 11.11 Response of a 32-in. toroidal tool in a 10-ohm-m bed with a 1-ohm-m boundary ahead of the bit.

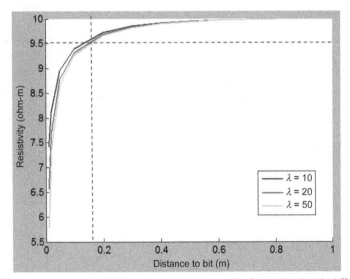

Figure 11.12 Response of a 32-in., 20-kHz toroidal tool in a 10-ohm-m bed with different boundary conductivities.

not very obvious as what we observe on coil tools. For lower frequencies and lower formation conductivity contrasts, the sensitivity of the coil tools is not sufficient to identify approaching boundaries, while toroidal tools show an earlier response on resistivity logs. For coil tools, the sensitivity to formation conductivity contrast makes it possible to predict the bed conductivity before the bit penetrates the layer ahead,

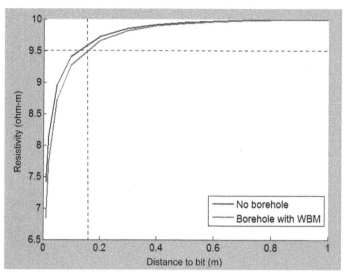

Figure 11.13 Response of a 20-kHz, 32-in. toroidal tool in a 10-ohm-m bed with a 1-ohm-m boundary ahead of the bit.

given that the boundary positions are preacquired and adequately accurate. On the other hand, if the boundary approached by the bit is unexpected, the distance inverted from toroidal tool responses should be more reliable.

Fig. 11.13 shows the response of a 20-kHz, 32-in. toroidal tool in the same two-layer formation model but with a borehole included. The borehole is filled with WBM ($R_m = 0.1$ ohm-m). As discussed previously, the WBM-filled borehole is beneficial for the look-ahead capabilities of toroidal tools in homogeneous formations, since the ahead-of-the-bit field attenuation is weaker with the existence of conductive mud. However, Fig. 11.13 demonstrates that the tool response is not very sensitive to borehole conductivity change in terms of boundary detection. Compared with the case where no borehole is included, the detection range defined by a 5% resistivity drop increases from 0.13 to 0.17 m with a conductive borehole.

11.4 BOUNDARY DETECTION USING ORTHOGONAL ANTENNAS

With the advancement of directional drilling technology, many wells are now designed with high angles or horizontally. In this way, production can be maximized in thin pay zones, making the drilling procedure more economic. For induction tools using coaxial antennas, the detection sensitivity of horizontal boundaries is closely related to the radial DOI of the tool, and has been well discussed [12,13]. For normal LWD tools, the radial detection range is generally a few feet. In this section, the tool

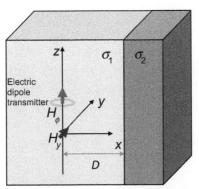

Figure 11.14 Three-dimensional model with a z-direction toroidal transmitter and a y-direction coil receiver.

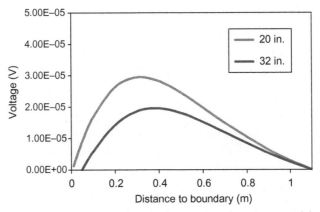

Figure 11.15 Voltage measured by a 20-kHz tool using a z-direction toroidal transmitter and a y-direction coil receiver in a 10/1 formation model with boundary parallel with the tool axis.

response with orthogonal antennas is investigated to explore its horizontal boundary detection capabilities.

First, the response of a tool using an axial toroidal transmitter and a transverse coil receiver is studied. A three-dimensional (3D) model is built in COMSOL, illustrated in Fig. 11.14. The transmitter is modeled as a unit-magnitude electric dipole in z direction. In homogeneous media, the generated magnetic field is in φ direction, so the magnetic field measured by the y-direction coil H_y is zero. With the existence of a boundary which is parallel with XZ plane, H_y is no longer zero, and hence can be used as an indicator to the boundary distance.

Fig. 11.15 shows the tool response as the boundary moves toward the tool. The local layer is 10 ohm-m, and the target layer is 1 ohm-m. Two transmitter—receiver spacings are used: 20 and 32 in. Measurements are acquired at 20 kHz. The results show that the received voltage responds to the boundary at a relatively early position. With a detection threshold of 2 μV, the detection range for 20 in. is about 1.0 m. However the response is not a monotonic function of the distance to the boundary. As the conductive layer approaches, the voltage first increases, and then drops down after a peak value is reached. This phenomenon can be observed from the H_y distribution plotted in Fig. 11.16, where the *white arrow* represents the direction of magnetic field on the $x = 0$ plane. When the boundary is relatively far from the tool, the H_y field has an elliptical shape, as shown in the left figure. However, if the tool is too close to the boundary, the middle part of the field becomes narrower, resembling a spindle-torus shape. From Fig. 11.15, one can also see that the short-spacing signal is stronger and more sensitive to the distance, since the propagation attenuation is lower.

A second model is illustrated in Fig. 11.17. The tool uses an axial coil transmitter, which is modeled as a unit-magnitude magnetic dipole, and a transverse coil receiver in x direction. In homogeneous formations, the cross components H_x and H_y should be zero. However, with a boundary that is parallel with the YZ plane existing

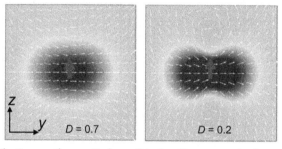

Figure 11.16 H_y distribution on the $x = 0$ plane, D is the distance to the boundary.

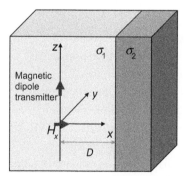

Figure 11.17 Three-dimensional model with a z-direction coil transmitter and an x-direction coil receiver.

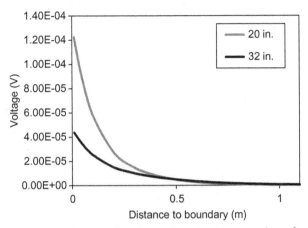

Figure 11.18 Voltage measured by an *x*-direction coil receiver in a two-layer formation model.

in proximity, the magnetic field H_x is no longer zero and can be used as an indicator of the boundary-to-tool distance.

Fig. 11.18 shows the voltage measurement of the *x*-directed coil as the boundary position changes. The same two-layer model is used, and the tool also operates at 20 kHz. The results show that the response monotonically increases as the boundary moves toward the tool. With the same 2-μV voltage threshold, the detection range is about 0.65 m, which appears to be shorter than the previous model. However the monotonicity of the response reduces the uncertainty of distance inversion, which is desired in geosteering operations. To take advantage of this monotonic feature, one can increase the signal strength by increasing the number of coil turns to reach a certain detection threshold.

As a summary, we can see that the look-ahead capability of coil and toroidal tools investigated in a formation model where the boundary is perpendicular to the tool axis have some capability in detecting boundaries in the front of the drill bit. The results have shown that with axially symmetrical antennas, the ability of detecting boundaries ahead of the bit is very limited. For a 20-kHz coil tool, a detection range of 0.33 m can be reached under favorable conditions, while a toroidal tool typically detects up to 0.17 m. It is also observed that the coil tool response is easily affected by formation conductivity contrast λ. When λ is not high enough, the sensitivity is rather low and cannot indicate boundaries ahead. On the other hand, the response of toroidal tools is relatively independent of λ. If the conductivity of the approaching boundary is uncertain, the toroidal tool response can be relied on to reduce the ambiguity brought by different boundary conductivities.

From the 3D model built for exploring the feasibility of detection horizontal boundaries with orthogonal antennas shows that the orthogonal antennas are promising.

The results show that for both coil and toroidal transmitters, the voltage received by a transverse receivers can be used to indicate an approaching horizontal boundary. However, for toroidal transmitters, the response is not monotonic, which may cause problems in distance inversion. Therefore it is more beneficial to detect horizontal boundaries with coil tools using orthogonal antennas.

11.5 DEEP-LOOKING DIRECTIONAL RESISTIVITY TOOL

As analyzed in the previous sections, the LWD resistivity tools have some look-ahead capabilities. However the direct look-ahead capability is greatly limited by the geometry of the cylindrical tool structure. It is almost impossible to place antenna arrays over the cross section of the drill bit. However, as we discussed in the previous sections, the LWD resistivity tools are sensitive to the boundaries on the side of the tool. We know that the DOI is largely dependent on the operating frequency (skin depth) and transmitter—receiver spacing. The cross component of the tool is sensitive to the boundaries on the side. If we build a tool, that has (1) long T-R spacing; (2) cross-component measurement; (3) low frequency, this tool may be used to detect side formation boundaries far from the tool. In this section, we will discuss the performance of such long detection LWD tool.

Traditional LWD resistivity tools use transmitters and receivers that have the same polarization. As discussed in the previous chapters, a generic LWD tool consists of at least one transmitter coil and two receiver coils, and calculates amplitude ratio and phase difference of the induced voltage at the two receivers. Commercial tools are usually equipped with an array of antennas in order to take measurements with different DOIs at the same time, and operate at multiple frequencies, too. The operating frequency of LWD tools is usually higher (400 kHz—2 MHz) than that of wireline induction tools (10—120 kHz), which overcomes the effects brought by the metal tool body, making the DOI of LWD tools comparable with induction tools (2—5 ft).

Due to its electromagnetic nature, LWD resistivity tools can reach much further distance than other LWD tools (e.g., gamma ray, NMR, etc.), and hence become critical in making real-time decisions for such applications. In highly deviated wells where the tool is not strictly parallel to the formation boundaries, the capability of detecting a boundary around the tool can be leveraged to predict the ahead-of-the-bit distance to the boundary. As shown in Fig. 11.19, if the look-around distance D_{ar} and the relative formation dip α can be obtained, the look-ahead distance D_{ah} can be expressed by

$$D_{ah} = \frac{D_{ar}}{\tan \alpha} - D_{tb} \tag{11.5}$$

where D_{tb} represents the distance from transmitter coil to the drill bit. Since D_{ar} is a function of the radial DOI, such pseudo look-ahead capability can be enhanced by expanding the DOI of LWD tools.

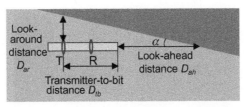

Figure 11.19 Look-ahead distance converted from look-around distance.

In 2003 an ultradeep LWD tool was proposed to facilitate reservoir navigation applications [7]. Operating at lower frequencies (2, 10, and 100 kHz), the radial response of the tool is much larger than traditional LWD tools. The transmitters and receivers are manufactured on individual subs, so that in theory they could be placed anywhere behind the drill bit. In a case where two transmitter and one receiver subs are used, and the distances between the transmitters and the receiver are around 21 and 11 m, respectively, a detection range of tens of meters is reported in field tests.

However a major disadvantage of this type of tools is that it lacks azimuthal directionality. Due to the axial symmetry of the tool configuration, the resistivity measurement taken by coaxial antennas is an average value reflecting the properties of the bulk formation volume around the borehole. Any anomalies that are in the way of the eddy currents induced in the formation will affect the tool response, in spite of the azimuthal direction in which they are located.

As shown in Figs. 11.20 and 11.21, a 36-in., 400-kHz propagation tool approaches the boundary with the same relative dip angle $\alpha = 60$ degrees, while the azimuthal positions of the boundary are 180 degrees different. The tool responses are exactly the same. This axisymmetric feature might be acceptable in formation evaluation, but can be problematic in geosteering applications. To make steering decisions (to drill upward or downward) with this type of tools, some related geological knowledge must be obtained beforehand (e.g., the existence of an oil–water contact).

11.5.1 Physics of the directional resistivity tool

Inspired by wireline triaxial induction tools, multicomponent measurements are also integrated by LWD tools. With transverse or tilted coil antennas as receivers, an azimuthal sensitivity can be obtained, which is beneficial for geosteering applications [8].

The azimuthal sensitivity of a 36-in., 400-kHz tool with an axial coil transmitter and a transverse coil receiver is illustrated in Fig. 11.22. The tool response is denoted as the cross component H_{zx}, with the subscript z representing the z-direction transmitter, and x the transverse receiver. In a homogeneous formation, there is no x-direction magnetic field. With the existence of a boundary nearby, H_{zx} measures a nonzero value, which can be used as an indicator of boundary detection.

To simulate the tool response in Fig. 11.22, the tool is placed in a resistive formation bed of 10 ohm-m, parallel with a conductive bed of 1 ohm-m. The distance

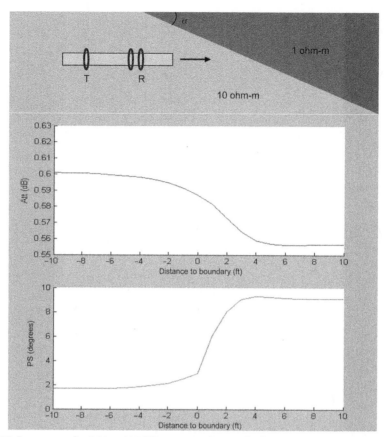

Figure 11.20 Response of a 36-in., 400-kHz propagation tool when crossing a 10/1 formation with boundary approaching from above ($\alpha = 60$ degrees).

between the tool and the boundary remains constant. While the tool rotates with the transverse receiver pointing to different angles, both the real part (*blue line* (dark gray in print versions)) and the imaginary part (*red line* (light gray in print versions)) of H_{zx} varies as a function of $\cos\varphi$, where φ represents the azimuth angle. This means in the two scenarios illustrated by Fig. 11.23, where the boundary locates at a certain distance above and below the horizontal tool, the tool responses will show different signs. This nice feature is obviously beneficial for steering purposes. With adequate detection range, real-time drilling decisions can be made based on these measurements.

LWD tools with tilted coil antennas have similar benefits as transverse antennas do. A tilted coil points neither along the tool axis nor sideways, but is mounted with a certain angle (typically 45 degrees). The implementation of transverse and tilted coils can both be realized with slots on the tool, and covered by specialized shields, as

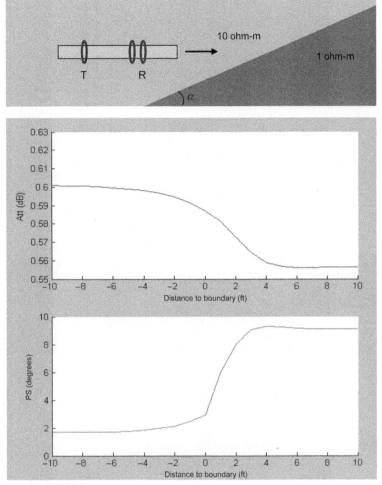

Figure 11.21 Response of a 36-in., 400-kHz propagation tool when crossing a 10/1 formation with boundary approaching from below ($\alpha = 60$ degrees).

shown in Fig. 11.24 [14]. While the drilling string rotates, azimuthal measurements can be acquired by the receivers.

The directional measurements taken by tilted coil receivers actually result from the sign change of the cross component H_{zx}, as discussed above. The only difference is that H_{zz} is also involved here. As shown in Fig. 11.25, the magnetic field received at the tilted antenna can be seen as a synthesis of both H_{zx} and H_{zz}, and can be expressed by

$$H_{\text{up}} = H_{zx} \sin \varphi + H_{zz} \cos \varphi \tag{11.6}$$

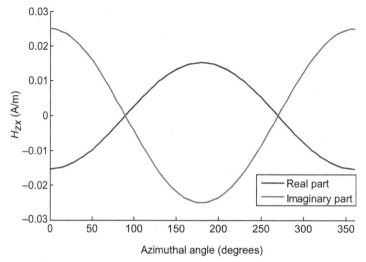

Figure 11.22 Azimuthal sensitivity of H_{zx} for a 36-in., 400-kHz tool with a horizontal boundary. $\lambda = 10$.

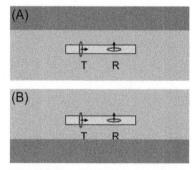

Figure 11.23 Boundary above (A) and below (B) the transverse coil receiver.

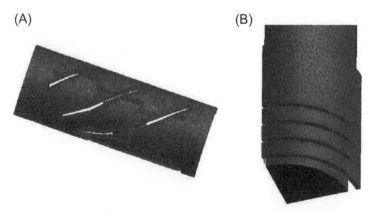

Figure 11.24 (A) Tilted coil and (B) transverse coil shielded with slots [14].

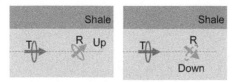

Figure 11.25 Directional measurements with tilted antennas.

and

$$H_{\text{down}} = -H_{zx} \sin \varphi + H_{zz} \cos \varphi \quad (11.7)$$

where φ is the angle between the receiver coil axis and the tool axis.

Therefore the directional measurements (amplitude ratio and phase shift) can be calculated from

$$\frac{H_{\text{up}}}{H_{\text{down}}} = \frac{H_{zz} \cos \varphi + H_{zx} \sin \varphi}{H_{zz} \cos \varphi - H_{zx} \sin \varphi} = 1 + \frac{2H_{zx} \sin \varphi}{H_{zz} \cos \varphi - H_{zx} \sin \varphi} \quad (11.8)$$

The directional resistivity tools provided by service companies proved to be useful in many field tests. However, with the normal LWD frequency and tool spacing, these tools can only detect up to 21 ft. Deeper detection range cannot be reached with such tool configuration.

Since 2010, a new deep-looking tool was developed and reported to have the capability of detecting up to 30-m boundaries [9]. It has the azimuthal sensitivity as the directional resistivity tools do, but the detailed physics has not been disclosed yet.

11.5.2 Forward modeling of a deep-looking tool with tilted antennas

For induction tools, it is well known that DOI is a function of spacing and the skin depth. This rule applies to LWD directional resistivity tools as well. To reach a further boundary, a lower frequency and a longer distance between transmitters and receivers must be adopted. Based on the principles of the directional resistivity tool, the responses are simulated for lower frequencies and with longer spacings, and it proved to be effective in detecting further boundaries.

The forward modeling of the tool is based on a computer code named **TRITI2011_series** [15]. It analytically calculates the response of triaxial induction tools in one-dimensional multilayered formations. Using Eqs. (11.6)–(11.8), we can simulate the response of a tilted coil receiver and obtain the directional measurements.

First, the influence of spacing and frequency is investigated using a horizontal tool with a tilted coil antenna. As illustrated in Fig. 11.26, a three-layer formation model was used, with the middle layer having a resistivity of 10 ohm-m, and the conductive shoulder beds are 1 and 2 ohm-m, respectively.

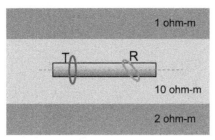

Figure 11.26 A three-layer formation model used to simulate the tool response.

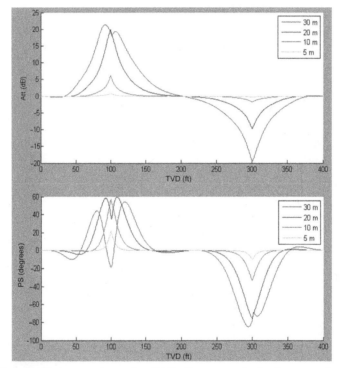

Figure 11.27 Deep-looking responses of a 5-kHz frequency tool when crossing a 200-ft bed. Resistivity from left to right: 1, 10, 2 ohm-m.

The results are shown in Figs. 11.27–11.29. The resistive middle layer is 200 ft thick. The tool starts from 100 ft above the 1-ohm-m boundary, and ends at 100 ft below the 2-ohm-m boundary, remaining parallel with the boundaries at all times. Four different spacings are used: 5, 10, 20, and 30 m. The operating frequencies are 5, 2, and 1 kHz.

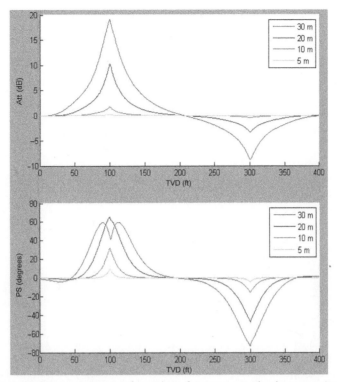

Figure 11.28 Deep-looking responses of a 2-kHz frequency tool when crossing a 200-ft bed. Resistivity from left to right: 1, 10, 2 ohm-m.

Both attenuation ratio and phase shift measurements can be used to detect approaching boundaries. Once the tool reaches the resistive zone, the signs of the responses can generally indicate whether the boundary is approaching from above or below. With a certain threshold (e.g., 0.05 dB for attenuation ratio, 0.15 degrees for phase shift), one can define a maximum detection range for the tool.

The responses are functions of spacing, frequency, and formation conductivity contrast. Long-spacing measurements usually show a larger detection range. For 2-kHz frequency and 20-m spacing, the tool can detect boundaries that are around 100 ft away (Fig. 11.28). However, in thinner layers, such responses may be affected by both shoulder beds and hence pose challenges for interpretation. In that case, shorter spacing measurements would be more reliable. In practice, multispacing measurements are recommended to adapt to different formation thicknesses.

Although lower frequencies seemingly expand the detection range, the signal amplitude of the responses decreases accordingly, especially for attenuation ratio. This feature calls for a trade-off in practical tool configuration.

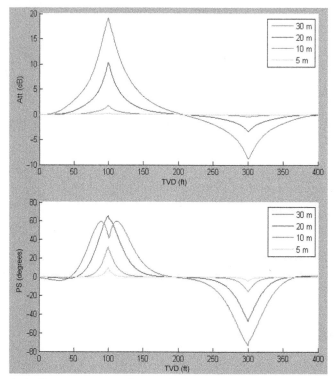

Figure 11.29 Deep-looking responses of a 1-kHz frequency tool when crossing a 200-ft bed. Resistivity from left to right: 1, 10, 2 ohm-m.

Figs. 11.30 and 11.31 plot the responses of the same spacing, but with different operation frequencies. While the tool is within the resistive layer, the monotonicity of the signal depends on the product of wave number k and spacing L, where k can be written as

$$k = (1-j)\frac{1}{\delta} \qquad (11.9)$$

Here δ is the skin depth that can be expressed by

$$\delta = \sqrt{\frac{2}{\omega\mu\sigma}} \qquad (11.10)$$

If L is too long, or either of frequency and conductivity is too high, the tool response loses its monotonicity and becomes complicated to interpret. Therefore frequency and spacing should be carefully selected in practical tool design.

Ahead-of-the-Bit Tools and Far Detection Electromagnetic Tools 433

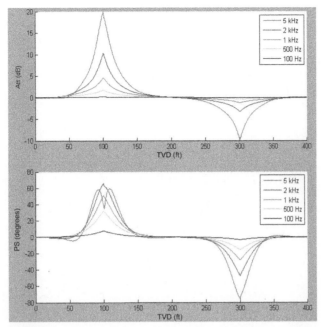

Figure 11.30 Deep-looking responses of a 20-m tool when crossing a 200-ft bed. Resistivity from left to right: 1, 10, 2 ohm-m.

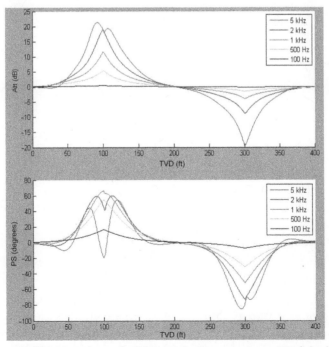

Figure 11.31 Deep-looking responses of a 30-m tool when crossing a 200-ft bed. Resistivity from left to right: 1, 10, 2 ohm-m.

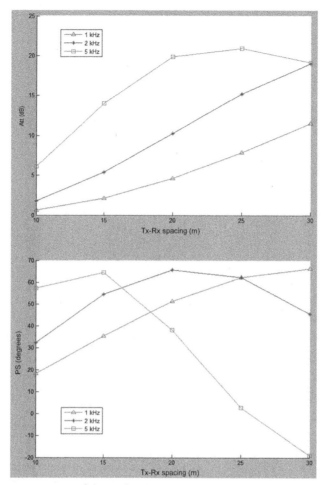

Figure 11.32 The monotonicity of the tool responses at the 1-ohm-m boundary.

Fig. 11.32 plots the tool responses at the 1-ohm-m boundary. As the spacing goes up, the amplitude of the high-frequency responses are more easily affected by the decreasing skin depth (especially for the phase shift measurements), while the lower frequency responses remain monotonically increasing.

Figs. 11.33–11.35 further illustrate the influence of bed thickness. The middle layer starts from 100 ft, and ends at 200, 150, and 120 ft. The results show that if the bed is much shorter than the detection range of a particular tool configuration, the tool response is affected by both shoulder beds, and the sign of the signal may not be a reliable indicator of boundary location. An accurate interpretation will rely on a full inversion for boundary distance.

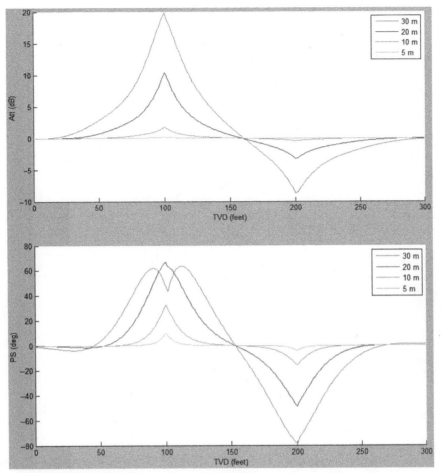

Figure 11.33 Deep-looking responses for 2-kHz frequency when crossing a 100-ft bed. Resistivity from left to right: 1, 10, 2 ohm-m.

11.6 DISTANCE INVERSION BASED ON THE GAUSS–NEWTON ALGORITHM

11.6.1 Summary of Gauss–Newton Method

To determine the boundary distances from the responses of the deep-looking tool, the inversion algorithm discussed in Chapter 9, Theory of Inversion for Triaxial Induction and Logging-While-Drilling Logging Data in One- and Two-Dimensional Formations and Chapter 10, The Application of Image Theory in Geosteering, based on the Gauss–Newton method is used. As described in Chapter 9, Theory of Inversion for Triaxial Induction and Logging-While-Drilling Logging Data in One- and Two-Dimensional Formations and Chapter 10, The Application of Image Theory

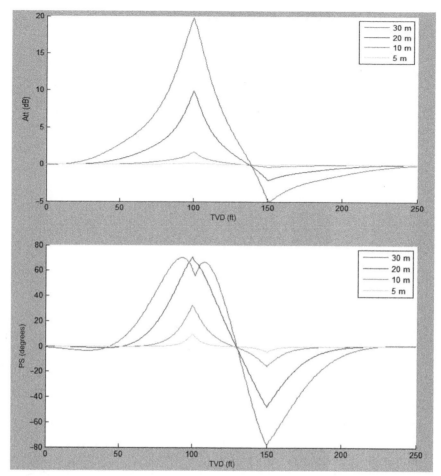

Figure 11.34 Deep-looking responses for 2-kHz frequency when crossing a 50-ft bed. Resistivity from left to right: 1, 10, 2 ohm-m.

in Geosteering, Gauss−Newton algorithm is a method to find a minimum of a function, which is usually the difference function between real measurements and analytical results calculated from models, by computing first-order derivatives. Detailed formulations are given in Chapter 9, Theory of Inversion for Triaxial Induction and Logging-While-Drilling Logging Data in One- and Two-Dimensional Formations and Chapter 10, The Application of Image Theory in Geosteering, a summary of the equations used in for this purpose can be summarized as follows.

The Gauss−Newton method iteratively searches for a minimum of the sum of the squares

$$S(\beta) = \sum_i^m r_i^2(\beta) \tag{11.11}$$

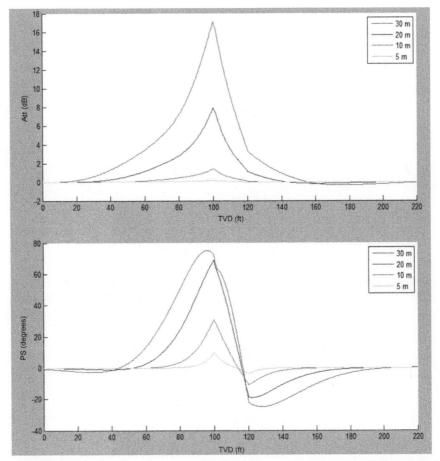

Figure 11.35 Deep-looking responses for 2-kHz frequency when crossing a 20-ft bed. Resistivity from left to right: 1, 10, 2 ohm-m.

where

$$r_i(\boldsymbol{\beta}) = y_i - f_i(\boldsymbol{\beta}) \qquad (11.12)$$

is the residue function, and

$$\boldsymbol{\beta} = (\beta_1, \beta_2, \ldots, \beta_n) \qquad (11.13)$$

is the desired variable vector. To obtain a converged solution, m must be larger or equal to n.

With an initial guess $\boldsymbol{\beta}^{(0)}$, the iteration process can be expressed by

$$\boldsymbol{\beta}^{(k+1)} = \boldsymbol{\beta}^{(k)} + (\mathbf{J_f}^T \mathbf{J_f})^{-1} \mathbf{J_f} \mathbf{r} \qquad (11.14)$$

where

$$\mathbf{J_f} = \frac{\partial f_i(\beta_i^{(k)})}{\partial \beta_i} \qquad (11.15)$$

is the Jacobian matrix of the function **f**.

11.6.2 Inversion in two-layer formations

First, a two-layer formation model is used, where the conductivities of both layers are assumed to be known, and only the distance to the boundary is inverted. The tool remains parallel to the boundary.

The residual function here can be expressed by

$$\mathbf{r} = \begin{pmatrix} r_1 \\ r_2 \end{pmatrix} = \begin{pmatrix} y_{Att} - f_{Att}(d) \\ y_{PS} - f_{PS}(d) \end{pmatrix} \qquad (11.16)$$

where $f_{Att}(d)$ and $f_{PS}(d)$ are the attenuation ratio and the phase shift responses calculated from the forward model, and y_{Att} and y_{PS} are the actual measurements. The Jacobian matrix here is represented by

$$\mathbf{J_f} = \begin{pmatrix} f'_{Att}(d^{(k)}) \\ f'_{PS}(d^{(k)}) \end{pmatrix} \qquad (11.17)$$

where

$$f'_{Att}(d^{(k)}) = \frac{f_{Att}((1+\Delta)d^{(k)}) - f_{Att}(d^{(k)})}{\Delta d^{(k)}} \qquad (11.18)$$

and

$$f'_{PS}(d^{(k)}) = \frac{f_{PS}((1+\Delta)d^{(k)}) - f_{PS}(d^{(k)})}{\Delta d^{(k)}} \qquad (11.19)$$

The inversion algorithm was tested for the four scenarios depicted in Fig. 11.36. The same conductivity difference $\lambda = 10$ was applied in all four cases, with the resistive layer representing a hydrocarbon reservoir, the conductive layer a shale bed above or the OWC below. Fig. 11.36A and B captures two typical cases for reservoir navigation in a thick bed, while Fig. 11.36C and D could happen if the original resistive target was missed, for which the tool response is also worth investigating.

The inversion results for a horizontal tool with 5-kHz frequency and 20-m spacing are listed below. The initial guess for boundary distance was 10 ft in most cases except for those with a different $d^{(0)}$ mentioned under "remarks." The increment is set as $\Delta = 10^{-4}$, and the iteration ceases when $d^{(k+1)} - d^{(k)} < 0.5$ ft.

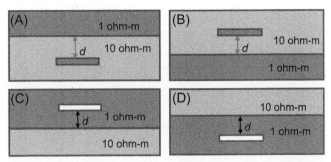

Figure 11.36 Two-layer models used for inversion. (A) Tool is in high resistivity layer and the conductive boundary is above; (B) Tool is in high resistivity layer and the conductive boundary is below; (C) Tool is in low resistivity layer and the resistive boundary is below; (D) Tool is in low resistivity layer and the resistive boundary is above.

Table 11.1 Two-layer inversion results for 5-kHz frequency, 20-m spacing tool in resistive layer (conductive boundary above)

Distance (ft)	Number of iterations	Inverted distance (ft)	Remarks
5	3	4.9998	$d^{(0)} = 2$
15	3	15.0015	
30	5	30.0000	
50	6	49.9993	
75	5	75.0008	
85	6	85.0026	
100	7	99.9989	

Table 11.2 Two-layer inversion results for 5-kHz frequency, 20-m spacing tool in resistive layer (conductive boundary below)

Distance (ft)	Number of iterations	Inverted distance (ft)	Remarks
5	6	4.9791	$d^{(0)} = 10$
15	3	14.9994	
30	5	30.0013	
50	4	49.9985	
75	6	74.9993	
85	7	85.0090	
100	9	100.0005	

Tables 11.1 and 11.2 show the inversion results when the tool is in the 10-ohm-m layer. In most of its detection range, the inversion algorithm works very well, with the relative error below 1%. However, with the tool very close from the boundary, a change of initial guess is needed to obtain the desired results. This is because the Gauss–Newton method stops searching whenever a stationary point of **S** is met, which could be the result from a local extremum.

When the tool is close to the boundaries, inversion results can be affected by the multivalued attribute of the tool responses as discussed in previous sections, and converge to a different distance, which may lead to similar responses. On the other hand, if the tool is too far away from the boundary in which the signal is rather weak, the iteration becomes slow and may lead to unreliable results as well.

Tables 11.3 and 11.4 show the inversion results when the tool is in the 1-ohm-m layer. Compared with the two cases above, the multivalued problem appeared at a closer distance, where the initial guess of distance had to be changed to find the expected solution. Due to the high conductivity, the skin depth is relatively short in the 1-ohm-m formation. When the distance is beyond 75 ft from the boundary, the responses are rather weak, which makes the single-valued solution more difficult to obtain.

In practice, the initial guess of the distance could be adjusted with reference to other available information, such as geological maps, depth measurements, and other types of logs. Another solution is to integrate multispacing, multifrequency responses into the inversion process. In that way, the ambiguity caused by possible multivalued problems could be removed.

Table 11.3 Two-layer inversion results for 5-kHz frequency, 20-m spacing tool in conductive layer (resistive boundary below)

Distance (ft)	Number of iterations	Inverted distance (ft)	Remarks
5	4	4.9780	
15	3	14.9999	
30	5	29.9996	
50	3	50.0115	
75	4	74.9842	$d^{(0)} = 50$
85	8	84.9995	$d^{(0)} = 50$
100	10	90.4007	$d^{(0)} = 50$

Table 11.4 Two-layer inversion results for 5-kHz frequency, 20-m spacing tool in conductive layer (resistive boundary above)

Distance (ft)	Number of iterations	Inverted distance (ft)	Remarks
5	4	4.9894	
15	3	15.0002	
30	3	29.9999	$d^{(0)} = 20$
50	3	50.0103	
75	5	74.9871	
85	8	84.9849	
100	3	100.0013	$d^{(0)} = 90$

To validate this statement, the inversion algorithm is modified with four log inputs, so the residual vector can be expressed by

$$\mathbf{r} = \begin{pmatrix} r_1 \\ r_2 \\ r_3 \\ r_4 \end{pmatrix} = \begin{pmatrix} y_{Att1} - f_{Att1}(d) \\ y_{PS1} - f_{PS1}(d) \\ y_{Att2} - f_{Att2}(d) \\ y_{PS2} - f_{PS2}(d) \end{pmatrix} \quad (11.20)$$

where $f_{Att1}(d)$ and $f_{PS1}(d)$ represent the tool responses of the first frequency or spacing, and $f_{Att2}(d)$ and $f_{PS2}(d)$ the second. Combining responses of 20- and 30-m spacings, we repeated the experiments in Fig. 11.36A and D, where the multivalued problems appeared for near-boundary inversions. Results are shown in Tables 11.5 and 11.6, in which the initial guess was all set as 10 ft. Compared with Tables 11.1 and 11.4, one can see that the multivalued problem has been solved with little sacrifice on accuracy.

11.6.3 Inversion in three-layer formations

In this section, an inversion algorithm is developed for a three-layer formation model shown in Fig. 11.37. Assuming the tool is parallel to the boundaries, and all three bed resistivities are known, the Gauss–Newton method can be used to invert the distances to the two shoulder bed boundaries at the same time.

Table 11.5 Two-layer inversion results for 5-kHz frequency, 20- and 30-m spacings combined tool in resistive layer (conductive boundary above)

Distance (ft)	Number of iterations	Inverted distance (ft)
5	3	5.0001
15	3	14.9991
30	3	30.0009

Table 11.6 Two-layer inversion results for 5-kHz frequency, 20- and 30-m spacings combined tool in conductive layer (resistive boundary above)

Distance (ft)	Number of iterations	Inverted distance (ft)
5	3	4.9986
15	2	14.9989
30	5	29.9987

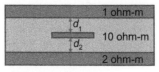

Figure 11.37 Three-layer model used for inversion.

Since there are two variables, d_1 and d_2, we have
$$\beta = (d_1, d_2) \tag{11.21}$$
and
$$\mathbf{r} = \begin{pmatrix} r_1 \\ r_2 \end{pmatrix} = \begin{pmatrix} y_{Att} - f_{Att}(d_1, d_2) \\ y_{PS} - f_{PS}(d_1, d_2) \end{pmatrix} \tag{11.22}$$

The Jacobian matrix becomes
$$\mathbf{J_f} = \begin{pmatrix} \dfrac{\partial f_{Att}}{\partial d_1} & \dfrac{\partial f_{Att}}{\partial d_2} \\ \dfrac{\partial f_{PS}}{\partial d_1} & \dfrac{\partial f_{PS}}{\partial d_2} \end{pmatrix} \tag{11.23}$$

where
$$\frac{\partial f_{Att}(d_1^{(k)}, d_2^{(k)})}{\partial d_1} = \frac{f_{Att}((1+\Delta)d_1^{(k)}, d_2^{(k)}) - f_{Att}(d_1^{(k)}, d_2^{(k)})}{\Delta d_1^{(k)}} \tag{11.24}$$

$$\frac{\partial f_{Att}(d_1^{(k)}, d_2^{(k)})}{\partial d_2} = \frac{f_{Att}(d_1^{(k)}, (1+\Delta)d_2^{(k)}) - f_{Att}(d_1^{(k)}, d_2^{(k)})}{\Delta d_2^{(k)}} \tag{11.25}$$

$$\frac{\partial f_{PS}(d_1^{(k)}, d_2^{(k)})}{\partial d_1} = \frac{f_{PS}((1+\Delta)d_1^{(k)}, d_2^{(k)}) - f_{PS}(d_1^{(k)}, d_2^{(k)})}{\Delta d_1^{(k)}} \tag{11.26}$$

$$\frac{\partial f_{PS}(d_1^{(k)}, d_2^{(k)})}{\partial d_2} = \frac{f_{PS}(d_1^{(k)}, (1+\Delta)d_2^{(k)}) - f_{PS}(d_1^{(k)}, d_2^{(k)})}{\Delta d_2^{(k)}} \tag{11.27}$$

The inversion results for a horizontal tool with 2-kHz frequency and 20-m spacing are listed in Table 11.7. The initial guess for both d_1 and d_2 was 10 ft except for those mentioned under "remarks." The increment is set as $\Delta = 10^{-4}$, and the iteration ceases when $d_1^{(k+1)} - d_1^{(k)} < 0.5$ ft, and $d_2^{(k+1)} - d_2^{(k)} < 0.5$ ft.

The results show that the inversion algorithm is greatly affected by the multivalued attribute of the tool responses, and are sensitive to the initial guess of d_1 and d_2. To solve this problem, one can again combine multifrequency and multispacing responses as input variables, as discussed in the last section. Alternatively, it can be done by using the bed thickness of the middle layer, which could usually be acquired from geological maps, as a constraint, to help eliminate the nonuniqueness of inversion.

With known bed thickness of the middle layer, similar experiments to those in Table 11.7 are performed, and the results are listed in Table 11.8. Inversion at all points used the same initial guess for d_1 and d_2 (10 ft). Compared with Table 11.7, we can see that not only the nonuniqueness problem is solved, but the iteration process converges faster in most cases.

Table 11.7 Three-layer inversion results for 2-kHz frequency, 20-m spacing tool in middle resistive layer

Distance to upper boundary (ft)	Distance to lower boundary (ft)	Number of iterations	Inverted distance to upper boundary (ft)	Inverted distance to lower boundary (ft)	Remarks
8	12	3	8.0001	11.9967	$d_1^{(0)} = 5$, $d_2^{(0)} = 10$
10	40	5	10.0028	40.0159	
20	30	4	19.9863	29.9481	$d_1^{(0)} = 10$, $d_2^{(0)} = 50$
20	80	5	20.0003	79.9837	
40	60	5	40.0110	60.0713	$d_1^{(0)} = 30$, $d_2^{(0)} = 70$
80	20	3	79.9664	19.9964	

Table 11.8 Three-layer inversion results for 2-kHz frequency, 20-m spacing tool in middle resistive layer, with known bed thickness

Distance to upper boundary (ft)	Distance to lower boundary (ft)	Number of iterations	Inverted distance to upper boundary (ft)	Inverted distance to lower boundary (ft)
8	12	2	8.0009	11.9991
10	40	1	10.0007	39.9993
20	30	3	20.0001	29.9999
20	80	2	20.0001	79.9999
40	60	4	39.9996	60.0004
80	20	4	80.0046	19.9954

Table 11.9 Three-layer inversion results for 2-kHz frequency, 20-m spacing, 60-degree dip angle tool in middle resistive layer, with known bed thickness

Distance to upper boundary (ft)	Distance to lower boundary (ft)	Number of iteration	Inverted distance to upper boundary (ft)	Inverted distance to lower boundary (ft)
10	40	1	9.9993	40.0007
20	30	3	20.0025	29.9975
20	80	2	19.9999	80.0001
40	60	4	40.0015	59.9985
80	20	4	79.9989	20.0011

Note that the inversion algorithm developed here not only applies to horizontal wells, but can also be generalized to other dip angles. Table 11.9 shows an example where the tool is not parallel to the boundaries, but has a 60-degree angle. The inversion results are equivalently accurate and reliable.

11.7 CONCLUSIONS

LWD resistivity tools play an important role in geosteering practice. With real-time information about the formation anomalies in front of or around the bit, the operator is able to make better-informed decisions, making the drilling process more efficient and economic. In this chapter, a study is conducted on the boundary detection capabilities of LWD tools with different configurations. By modeling the tool responses with analytical and numerical simulations, a better insight is developed on the applicability of LWD tools in various drilling environments.

To evaluate the potential of predicting formation properties in front of the drill bit, tools with different downhole electromagnetic transmitters are first modeled in homogeneous formation, and the ahead-of-the-bit field distribution is investigated. For both coil and toroidal antennas, the field attenuation follows a similar pattern that is determined by formation resistivity and operating frequency, while the comparison results have shown that coil antennas have a lower attenuation rate with respect to distance from the bit. The borehole conductivity barely affects the performance of coil tool in terms of look-ahead capability, but the attenuation of toroidal transmitters can be significantly improved by using more conductive mud.

Simulation results in two-layer formation models have further shown different behaviors of coil and toroidal tools. First, the look-ahead ability is compared between tools using axially symmetrical antennas. In favorable conditions, a 20-kHz coil tool can detect boundaries 0.33 m away from the bit. The detection range of toroidal tools is shorter on average, typically about 0.17 m, but is rather independent of the drilling environment. If the formation conductivity contrast is too low, the response of a coil tool may not be sensitive enough to indicate an approaching boundary, while toroidal tools can still provide look-ahead responses. Next, horizontal boundary detection capability is tested with tools using orthogonal antennas. The results have demonstrated that such cross-component measurements can be used for boundary detection with both coil and toroidal antennas, but the response of coil tools is seen as a better indicator of boundary distance due to its monotonicity. Based on this observation, further investigations are conducted on the boundary detection capability of coil tools using multidirectional measurements.

It has been demonstrated that the deep-looking capability can be achieved by applying ultralong transmitter—receiver spacings and low frequencies to the directional resistivity tool using transverse or tilted receivers. For a 20-m, 2-kHz tool, boundaries that are approximately 100 ft away can be detected using the developed inversion algorithm. This work provides a new perspective for the research of resistivity logging tools. With the rapid advancement of modern drilling technology, LWD tools are expected to not only perform effective measurements around the wellbore, but also facilitate the drilling operations by predicting formation anomalies in advance. Based on the simulations, one can conclude that horizontal bed boundaries can be best detected and located by ultralong, low-frequency tools using multidirectional measurements, while the look-ahead capability of resistivity tools is rather limited with current frequency-domain excitation methods. To further expand the detection range ahead of the bit, the transient electromagnetic method using time-domain excitation sources may provide a solution for the next-generation LWD resistivity tools.

REFERENCES

[1] C.C. Payton, K.M. Strack, L.A. Tabarovsky, Method and Apparatus for Measuring Transient Electromagnetic and Electrical Energy Components Propagated in an Earth Formation. Patent US5955884 A, 1999.

[2] E. Banning, T. Hagiwara, R. Ostermeier, System and Method for Locating an Anomaly Ahead of a Drill Bit. Patent US20050092487 A1, 2005.

[3] S. Bonner, A. Bagersh, B. Clark, G. Dajee, M. Dennison, J.S. Hall, J. Jundt, J. Lovell, R. Rosthal, D. Allen, A New Generation of Electrode Resistivity Measurements for Formation Evaluation While Drilling, in: SPWLA 35th Annual Logging Symposium, Tulsa, Oklahoma, 1994.

[4] S. Gianzero, R. Chemali, Y. Lin, S.M. Su, M. Foster, A new resistivity tool for measurement-while-drilling, in: SPWLA 26th Annual Logging Symposium, Dallas, Texas, 1985.

[5] M.S. Bittar, G. Hu, W.E. Hendricks, Look-Ahead Boundary Detection and Distance Measurement. Patent US20100176812 A1, 2010.

[6] Q. Zhou, D. Gregory, S. Chen, W.C. Chew, Investigation on electromagnetic measurement ahead of drill-bit, in: Geoscience and Remote Sensing Symposium, 2000.

[7] J. Seydoux, J. Tabanou, L. Ortenzi, J.M. Denichou, Y. De Laet, D. Omeragic, M. Iversen, M. Fejerskov, A deep-resistivity logging-while-drilling device for proactive geosteering, in: Offshore Technology Conference, 5/5/2003, Houston, Texas, 2003.

[8] Q. Li, D. Omeragic, L. Chou, L. Yang, K. Duong, New directional electromagnetic tool for proactive geosteering and accurate formation evaluation while drilling, in: SPWLA 46th Annual Logging Symposium, 26–29 June, New Orleans, Louisiana, 2005.

[9] R. Beer, L.C.T. da Cunha, A.M.V. Coutinho, M.R. Schmitt, J. Seydoux, E. Legendre, J. Yang, Q. Li, Cas, A.C. da Silva, P. Ferraris, E. Barbosa, A.B.F. Guedes, Geosteering and/or reservoir characterization the prowess of new-generation LWD tools, in SPWLA 51st Annual Logging Symposium, 19–23 June, Perth, Australia, 2010.

[10] J.J. Arps, Inductive Resistivity Guard Logging Apparatus Including Toroidal Coils Mounted on a Conductive Stem. Patent US3305771 A, 1967.

[11] M. Bittar, G. Hu, The effects of rock anisotropy on LWD toroidal resistivity sensors, in: SPWLA 45th Annual Logging Symposium, Noordwijk, Netherlands, 2004.

[12] W.H. Meyer, Interpretation of propagation resistivity logs in high angle wells, in: SPWLA 39th Annual Logging Symposium, Keystone, Colorado, 1998.

[13] M. Rabinovich, D. Beard, I. Geldmacher, L. Tabarovsky, M. Fidan, Interpretation of induction logging data in horizontal wells, in: SPWLA 41st Annual Logging Symposium, Dallas, Texas, 2000.

[14] D. Omeragic, Q. Li, L. Chou, L. Yang, K. Duong, J.W. Smits, T. Lau, C. Liu, R. Dworak, V. Dreuillault, J. Yang, H. Ye, Deep directional electromagnetic measurements for optimal well placement, in: SPE Annual Technical Conference and Exhibition, 9–12 October, Dallas, Texas, 2005.

[15] N. Yuan, X. Nie, R.C. Liu, Response of a Triaxial Induction Logging Tool in a Homogeneous Biaxial Anisotropic Formation, in: SEG Annual Meeting, Houston, Texas, 2009.

[16] B.I. Anderson, Modeling and Inversion Methods for the Interpretation of Resistivity Logging Tool Response, Delft University Press, Delft, 2001.

[17] C. Bell, J. Hampson, P. Eadsforth, R.E. Chemali, T.B. Helgesen, W.H. Meyer, R. Randall, C. Peveto, A. Poppitt, J. Signorelli, T. Wang, Navigating and imaging in complex geology with azimuthal propagation resistivity while drilling, in: SPE Annual Technical Conference and Exhibition, 24–27 September, San Antonio, Texas, 2006.

[18] M. Bittar, Electromagnetic Wave Resistivity Tool Having a Tilted Antenna for Determining the Horizontal and Vertical Resistivities and Relative Dip Angle in Anisotropic Earth Formations. Patent US6163155 A, 2000.
[19] H.G. Doll, Electrical Resistivity Well Logging Method and Apparatus. Patent US2712627 A, 1955.
[20] H.G. Doll, Introduction to Induction Logging and Application to Logging of Wells Drilled With Oil Base Mud, Am. Inst. Min. Metall. Pet. Eng. 1 (06) (1949) 148–162.
[21] M. Gorek, C. Fulda, Method and Apparatus for Determining Formation Resistivity Ahead of the Bit and Azimuthal at the Bit. Patent US7554329 B2, 2009.
[22] T. Habashy, B.I. Anderson, Reconciling differences in depth of investigation between 2-Mhz phase shift and attenuation resistivity measurements, in: SPWLA 32nd Annual Logging Symposium, Midland, Texas, 1991.
[23] R.H. Hardman, L.C. Shen, Theory of induction sonde in dipping beds, Geophysics 51 (03) (1986) 800–809.
[24] T.B. Helgesen, C. Fulda, W.H. Meyer, A.K. Thorsen, M. Iversen, Reservoir navigation with an extra deep resistivity LWD service, in: SPWLA 46th Annual Logging Symposium, New Orleans, Louisiana, 2005.
[25] A. Karinski, A. Mousatov, Feasibility of vertical-resistivity determination by the LWD sonde with toroidal antennas for oil-base drilling fluid, in: SPWLA 43rd Annual Logging Symposium, Oiso, Japan, 2002.
[26] A. Karinski, A. Mousatov, Vertical resistivity estimation with toroidal antennas in transversely isotropic media, in: SPWLA 42nd Annual Logging Symposium, Houston, Texas, 2001.
[27] W.D. Kennedy, B. Corley, S. Painchaud, G. Nardi, E. Hart, Geosteering using deep resistivity images from azimuthal and multiple propagation resistivity, in: SPWLA 50th Annual Logging Symposium, The Woodlands, Texas, 2009.
[28] B. Kriegshauser, O. Fanini, S. Forgang, G. Itskovich, M. Rabinovich, L. Tabarovsky, L. Yu, M. Epov, P. Gupta, J. v d Horst, A new multicomponent induction logging tool to resolve anisotropic formations, in: SPWLA 41st Annual Logging Symposium, Dallas, Texas, 2000.
[29] K.T. McDonald, Electromagnetic fields of a small helical toroidal antenna. [Online]. Available: <http://www.hep.princeton.edu/~mcdonald/examples/cwhta.pdf>, 2008.
[30] J.H. Moran, K.S. Kunz, Basic theory of induction logging and application to study of two-coil sondes, Geophysics 27 (6) (1962) 829–858.
[31] J.H. Moran, Induction logging—geometrical factors with skin effect, Log Anal. 23 (06) (1982) 4–10.
[32] D. Omeragic, T. Habashy, Y. Chen, V. Polyakov, C. Kuo, R. Altman, D. Hupp, C. Maeso, 3D reservoir characterization and well placement in complex scenarios using azimuthal measurements while drilling, in: SPWLA 50th Annual Logging Symposium, The Woodlands, Texas, 2009.
[33] D. Omeragic, T.M. Habashy, C. Esmersoy, Method for Calculating a Distance Between a Well Logging Instrument and a Formation Boundary by Inversion Processing Measurements From the Logging Instrument. Patent US6594584 B1, 2003.
[34] M. Rabinovich, L. Fei, J. Lofts, S. Martakov, Deep? How Deep and What? The Vagaries and Myths of "Look Around" Deep-Resistivity Measurements While Drilling, in: SPWLA 52nd Annual Logging Symposium, 14–18 May, Colorado Springs, Colorado, 2011.
[35] P.F. Rodney, M.M. Wisler, Electromagnetic Wave Resistivity MWD Tool, SPE Drill. Eng. 1 (05) (1986) 337–346.
[36] J. Seydoux, D. Omeragic, D.M. Homan, Directional Resistivity Measurement for Well Placement and Formation Evaluation. Patent US20110238312 A1, 2011.
[37] B.R. Spies, T.M. Habashy, Sensitivity analysis of cross-well electromagnetics, in: SEG Annual Meeting, New Orleans, Louisiana, 1992.
[38] H. Wang, P. Wu, R. Rosthal, G. Minerbo, T. Barber, Modeling and understanding the triaxial induction logging in borehole environment with dip anisotropic formation, in: SEG Annual Meeting, Las Vegas, Nevada, 2008.
[39] M. Zhdanov, D. Kennedy, E. Peksen, Foundations of tensor induction well-logging, Soc. Petrophys. Well-Log Anal. 42 (06) (2001) 588–610.

CHAPTER 12

Principle of Dielectric Logging Tools

Contents

12.1	Introduction	447
12.2	History of Dielectric Tool Study	449
12.3	Frequency Selection of a Dielectric Tool	451
12.4	Antenna Spacing	451
12.5	Sensitivity Analysis	454
	12.5.1 Isotropic formation	454
12.6	Sensitivity Analysis in Anisotropic Formation	459
12.7	Dielectric Logging Tool Design and Modeling Using Three-Dimensional Numerical Modeling Software Package	463
12.8	Cavity-Backed Slot Antenna	467
12.9	Effects of the Pad	470
12.10	Borehole Mud Influence	475
12.11	Vertical Resolution	479
12.12	Mud Cake and Invasion	483
12.13	Depth of Investigation	491
12.14	Applications of Dielectric Tools	495
12.15	Summary	496
References		497
Appendix		498

12.1 INTRODUCTION

In the previous chapters, we discussed electromagnet (EM)-based logging tools working at relatively lower frequencies. In Chapter 2, Fundamentals of Electromagnetic Fields Induction Logging Tools and Chapter 3, Electrical Properties of Sediment Rocks: Mixing Laws and Measurement Methods, we noticed that in EM wave propagation, the conductivity is dominant (conduction current) at lower frequencies. However, when the frequency is getting higher, e.g., in megahertz like in logging-while-drilling (LWD) resistivity frequencies, displacement current will have to be considered, which is determined by the dielectric characteristics of the formation. When the frequency further increases, the displacement current's role becomes more important than the conduction current in the EM propagation process. Therefore, from measurement point of view, higher frequency is preferred when we want to measure the dielectric constant of the formation. As discussed in Chapter 3, Electrical Properties of Sediment Rocks: Mixing Laws and Measurement Methods, the dielectric constant of oil and gas are relatively small (1−3) whereas the water has very high dielectric constant (70−81). It is very natural to use dielectric measurement to identify water versus oil and gas. On the other hand, the conductivities of the fresh water and oil or gas are very similar. Unfortunately, the conventional logging tools such as induction, LWD, and laterolog tools are not sensitive to the dielectric constant change due to the low frequency operation. It is seen from Chapter 3, Electrical Properties of Sediment Rocks: Mixing Laws and Measurement Methods, that the dielectric constant is not really a constant, it changes with the frequency. The dispersion characteristics of the dielectric constant may also be used to identify the formations. Due to the high-frequency nature of the dielectric tools, the investigation depth is rather limited (a few inches, see Table 1.2).

As we know, the borehole mud has high content of water for water-based mud (WBM) and is very lossy to the EM signals, and has high dielectric constant. To measure the dielectric properties of the formation, the EM signals generated by the tool must be able to penetrate through the mud layer before reaching the formation, which makes the tool to be designed as a pad-type tool so that it can be mechanically pushed against the borehole wall and reduce the influence from the mud to the tool performance. Therefore the dielectric tools are made as wireline tools. Actually the dielectric tool can not only measure the dielectric constant of the formation, but also measure formation resistivity simultaneously, and further derive out the water-saturated formation porosity based on the measured dielectric constant and resistivity. When used, the tool pad is pushed against the borehole wall by the pusher mechanism.

Due to the high frequency, dielectric tools usually use cavity-backed slot antennas. The cavity is filled with ceramic materials with high dielectric constant (as high as 100) to reduce the antenna dimensions since the dimensions of the antenna is determined by the operating frequency and is inversely proportional to the square root of the relative dielectric constant of the material filled inside the cavity. The radiation of EM field is from the slot opening of the cavity (Fig. 12.1). It is easy to see that the E field in the slot has vertical component and the H field is in the direction of the slot. Due to the small size of the slot, we can approximate the H field in the slot by a uniform magnetic dipole and the effect of the pad can be replaced by an image dipole. Therefore the analysis of the dielectric tool becomes very simple. All the analysis method discussed in the previous chapters using magnetic dipoles can be applied. If detailed and accurate analysis or design is necessary, three-dimensional (3D) analysis must be used.

New-generation dielectric tools are also used in detecting shale reservoir, heavy oil, and residue oil in invasion zones. This chapter will investigate both design and simulation of novel array dielectric tools and dielectric constant of formation cores by lab measurements.

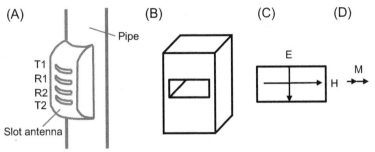

Figure 12.1 The dielectric antenna pad and the antenna equivalent dipole. (A) The generic antenna pad; (B) the slot antenna; (C) the field inside the slot; and (D) the equivalent dipole antenna of the slot.

Applications of dielectric tools can date back to late 1970s when their outstanding advantage of distinguishing hydrocarbon and fresh water was demonstrated. Later during the past decade, development of new-generation dielectric tools hits a peak because of their possible applications in detecting shale oil and gas. Shale reservoir has a tight formation structure with shallow invasion. Hence, shallow measurement from dielectric tools provides useful formation information to the log analysists. It is highly appreciated for supplemental information it provided, such as fluid-saturated formation porosity, other than resistivity and dielectric constant at various high frequencies. Also, relevance of dielectric to Cation Exchange Capacity (CEC) adds its value to rock texture interpretation. Nowadays, edging techniques in tool design finally matches data acquisition accuracy and overcomes rugosity. Logging data from the tools are combined with shallow Nuclear Magnetic Resonant (NMR) tools and Microresistivity tools, as well as Array Induction Logging tools to offer thorough petrophysical formation interpretation. In Chapter 3, Electrical Properties of Sediment Rocks: Mixing Laws and Measurement Methods, we discussed the dielectric properties of mixtures in details. It can be seen that the dielectric constant of a material is directly related to its physical characteristics. In this chapter, we will use examples to explain the tool performance instead of analytical analysis since the dielectric tool can be described by the equations discussed in the previous chapters using magnetic dipole approximation.

12.2 HISTORY OF DIELECTRIC TOOL STUDY

The first generation of dielectric tools dates back to 1970s. At that time, limitation of electronic techniques could not make it widely used. Inversion and interpretation of the data are based on traveling time and phase difference from analog traces. A main application target of dielectric tools was to distinguish fresh water-saturated formation and hydrocarbon bearing one. Nowadays, with the hunger of human to the fossil energy, new development of exploration technologies has made significant achievements. Therefore dielectric tools have found their applications.

Schlumberger has their deep propagation tool (DPT) in 1984, which operated at 25 MHz and measured propagation time as tool responses. The DPT tool operates at 1.1 GHz using a slot antenna. Baker Atlas filed a mandrel-type dielectric tool pattern since 1984, which operates at two frequencies: 47 and 200 MHz. Shen proposed multiple frequency tools [1] as research direction in the future. Many papers were published to support the field logging tools by lab measurements of various rocks for their dielectric characteristics [2–4].

Recently, new-generation dielectric tools are equipped with antenna arrays, with multiple frequency channels, multiple spacings, and both longitudinal and transverse antennas. This leads to better vertical resolution and capability in exploring anisotropy and more complex formation structures. Schlumberger's commercialized Dielectric Scanner in 2008 and found many applications since it can measure the dielectric dispersion curve [5–16]. The data interpretation of such tools are specifically effective if NMR data is combined [17–21]. Halliburton released its high frequency dielectric tool (HFDT) [22] in 2011, working at a single frequency, 1 GHz, provides quite accurate dielectric and conductivity measurements. Fig. 12.2 shows Shen's generic tool, HFDT tool, and Dielectric Scanner.

Many recent publications presented successful cases of dielectric tool applications in identifying thin shale beds [6,8,17], detecting heavy oil [11–13,15,19–21], and providing heavy oil quick look methodology. Answer products for dielectric tools are

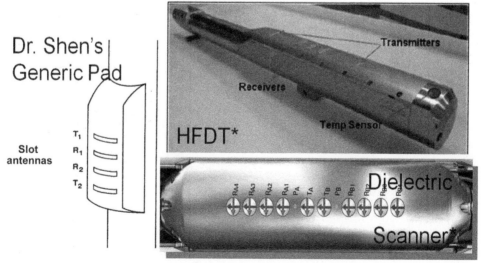

Figure 12.2 Dielectric tools.
* HFDT = High frequency dielectric tool

more focused on extracting dielectric dispersion within given frequency range and convert it to texture parameters such as tortuosity and CEC as well as water saturation. Meanwhile, cross interpretation, which combines dielectric tools with Microresistivity, Density, or NMR tools, is adopted by log analysts now to gain more information of the reservoir [17−21].

12.3 FREQUENCY SELECTION OF A DIELECTRIC TOOL

Frequencies above LWD frequency are required so that sufficient influence from dielectric constant can be dominant in the tool response. From Chapter 2, Fundamentals of Electromagnetic Fields Induction Logging Tools and Chapter 3, Electrical Properties of Sediment Rocks: Mixing Laws and Measurement Methods, we can see that the complex dielectric constant is a function of dielectric constant and the conductivity as shown in Eq. (2.16). Higher frequencies will make the tool more sensitive to dielectric change with the compromise in the investigation depth. Therefore the selection of frequency in a dielectric tool has to consider the compromise between investigation depth and sensitivity to the formation dielectric constant. The other factor to select a multifrequency dielectric tool is to be able to measure the dielectric dispersion curve. Most dielectric tools choose tens of MHz to 1 GHz as the operating frequency range.

12.4 ANTENNA SPACING

Obviously, pad size limits the antenna spacing. A pad-type tool in the industry is normally around 20 in. for mechanical reasons. Thus, farthest antenna can be placed at around 8 in. away from center point.

Spacing should be correlated to frequency as well. It is not a wise choice to use the same spacing for all the frequencies. Multifrequency tools used several frequencies with multiple antenna spacing. Fig. 12.3 is an example of a two-spacing dielectric tool response when formation parameter changes at 20 MHz and 1 GHz operating frequencies. Note that the tool responses are described by the phase shift and amplitude ratio of the two received signals, and both are functions of formation dielectric constant and conductivity. The tool responses are flat as to the change in formation conductivity at low-conductivity region and change rapidly when formation conductivity reaches to a threshold, which is a function of frequency and dielectric constant.

As we have seen from previous chapters, spacing between the receiver pairs also determines vertical resolution of the tool. The differential electromagnetic field between two receivers is concentrated in the middle area of the two receivers.

Similarly, spacing between the transmitter and the receiver in the same group impacts Depth of Investigation (DOI). DOI increases with transmitter−receiver

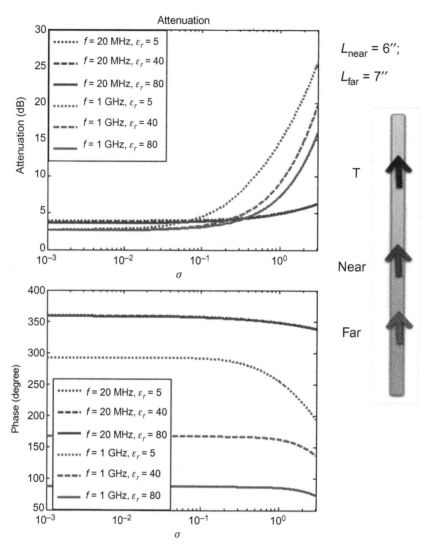

Figure 12.3 Tool responses at 20 MHz and 1 GHz with the same antenna spacing.

spacing. Contradictorily, from the aspect of signal level, receiver signal could drop significantly if receiver is too far away from the transmitter, causing intolerance of noise. As a result, tool design is to make a compromise between DOI and signal level.

Another issue to keep in mind in designing a propagation type of tools is to avoid phase wrap at relative higher frequencies. For example, at 1 GHz, under the condition that distance between two receivers in the center is 1.5 in., phase difference would exceed more than one cycle, i.e., 2π, for conductive formation or formation with greater dielectric constant, which will lead to the increased wavenumber. The phase-wrap cut-off spacing for 200 MHz is 4 in.

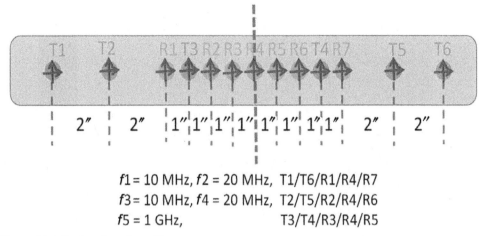

Figure 12.4 Tool pad configuration.

Detailed commercial dielectric tool configuration is not available in public literature. To investigate the performance of such tools, we use a generic model with multifrequency, multispacing construction. Symmetric structure and relative measurements can reduce the instability of tool electronics. Therefore a generic dielectric tool which will be discussed in this chapter uses five frequencies: 10, 20, 100, 200 MHz, and 1 GHz. The spacing and antenna polarization is shown in Fig. 12.4.

In this chapter, we will discuss the dielectric logging tool response in different formations using the array dielectric tool model in Fig. 12.4. We will explore complex resistivity in a wide range of complex formation environment, including anisotropy, dip, laminated layers, borehole, and invasion zones.

Consider a simple dielectric tool model, which is shown in Fig. 12.3. For simple analysis, we can consider the field is a plane wave and the formation is homogenous. The field at the receivers can be expressed as follows:

$$E_{r1} = E_t e^{j(k_r + jk_i)(z_t - z_{r1})} \quad (12.1a)$$

$$E_{r2} = E_t e^{j(k_r + jk_i)(z_t - z_{r1})} \quad (12.1b)$$

where E_{r1} and E_{r2} are the received field strength at the receiver 1 and receiver 2, respectively. k_r is the propagation constant $k_r = \omega\sqrt{\mu\varepsilon}$, where ε is the complex dielectric constant, μ is the magnetic permeability, and ω is the angular frequency. E_t is the field strength at the transmitter location. If we use the amplitude ratio and phase difference between the two receivers, we have,

$$DB = 20\log_{10} k_i(z_{r2}/z_{r1}) \quad (12.2a)$$

$$\varphi = k_r(z_{r2} - z_{r1}) \quad (12.2b)$$

Since both k_r and k_i are functions of dielectric constant and conductivity of the formation, the values of the dielectric constant and conductivity can be obtained by solving the simultaneous equations given in (12.2a,b). For the simple case, when the formation is lossless, we can directly solve Eq. (12.2b) and obtain the dielectric constant value of the formation from measured phase difference ϕ:

$$\varepsilon_r = \left(\frac{\phi}{\omega(z_{r2}-z_{r1})}\right)^2 \tag{12.3}$$

The Eqs. (12.2a) and (12.2b) can be solved numerically to obtain both dielectric constant and conductivity from the measurements.

12.5 SENSITIVITY ANALYSIS

Similar to Chapter 4, Triaxial Induction and Logging-While-Drilling Resistivity Tool Response in Homogeneous Anisotropic Formations, since the dielectric tool can be approximated by using magnetic dipole in a homogeneous formation, the analytical solution used in Chapter 4, Triaxial Induction and Logging-While-Drilling Resistivity Tool Response in Homogeneous Anisotropic Formations, can be used to study the tool performance except that the frequencies are different. This approximation may not be accurate since there are issues of tool body, borehole mud, and tool eccentricity, which are not considered in the dipole model. However, it can be used to analyze general tool characterization in terms of relative parameters. In this section, we use the analytical solutions described in Chapter 4, Triaxial Induction and Logging-While-Drilling Resistivity Tool Response in Homogeneous Anisotropic Formations, and apply to the dielectric tool analysis.

Sensitivity is an important building block for the tool design and data inversion. In tool design, we want to design a tool which is sensitive to the formation parameters that we are interested in such as dielectric constant and conductivity of the formation. However, we also want the tool is not so sensitive to the parameters that are not of interests such as borehole mud and surface roughness. Based on sensitivity of tool responses to conductivity and permittivity in homogeneous formation, we can obtain the information of the measurement quantity (amplitude and phase, or real and imaginary part of the measured signal) change rate with respect to the formation parameter change. Combined with error matrix and singular decomposition analysis, it can be used for tool stability study and data inversion stability analysis. The sensitivity analysis is an important step in optimization of the tool design.

12.5.1 Isotropic formation

Let us consider a simple tool design example: a dielectric tool operating at 1 GHz with its two transmitters located 3 in. from the pad center symmetrically, and three

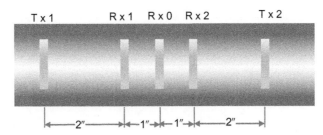

Figure 12.5 A schematic of a dielectric tool pad. The antennas are cavity-backed slot antennas and they are modeled by magnetic dipoles in the sensitivity analysis.

receivers at the tool center with a 1 in. spacing to each other as shown in Fig. 12.5. In homogeneous isotropic formation, symmetry results in exactly the same signal from top and bottom receivers. Consequently, attenuation and phase difference are measured from top to middle receivers. Results are shown as following: the top two pictures in Fig. 12.6 are raw responses attenuation and phase difference; remaining four are plots of sensitivity Jacobian matrix.

First, symmetric characteristic of tool response sensitivity can be observed. Attenuation sensitivity to conductivity is the same as phase sensitivity to permittivity; similarly, attenuation sensitivity to permittivity is similar to phase sensitivity to conductivity. Reciprocal sensitivities share exactly the same sensitive zones. For example, the most sensitive region of $\partial A/\partial \sigma$ and $\partial P/\partial(\omega\varepsilon)$ are both concentrated near the left bottom corner of the plot, where both conductivity and dielectric are small. This is the most sensitive region of $\partial A/\partial(\omega\varepsilon)$ and $\partial P/\partial\sigma$ as well. In addition, this is also where tool obtains best signal strength if we look at signal plots.

Another interesting observation is that from the contour plot (Fig. 12.7), contour of attenuation sensitivity to dielectric follows the trend of tangential delta $\frac{\sigma}{\omega\varepsilon}$. The same applies to phase sensitivity to conductivity. In Fig. 12.8, contour of tool response differentiation with respect to both conductivity and permittivity are plotted together. *Dotted curves* are response sensitivity to permittivity and *solid curves* are to conductivity. Again, it validates the reciprocity in a more quantitative way. Growth of relative permittivity causes decline of signal level as well as tool sensitivity. In other word, dominant factor to $\partial P/\partial\sigma$ and $\partial A/\partial(\omega\varepsilon)$ is conductivity at this frequency.

Other channels are similar to 1 GHz, sensitive region moves to less conductive area. As frequency decreases, influence from dielectric reduces. Attenuation can only sense variation in conductivity; phase still shows dielectric information. Due to reciprocity, only phase sensitivity is presented in the next few figures. Pattern of $\partial P/\partial\sigma$ keeps moving to the left and pattern of $\partial P/\partial(\omega\varepsilon)$ extends to lower boundary with decreasing of frequency. This causes a contradiction: the value of partial differentiation at lower frequencies is much smaller than the high-frequency

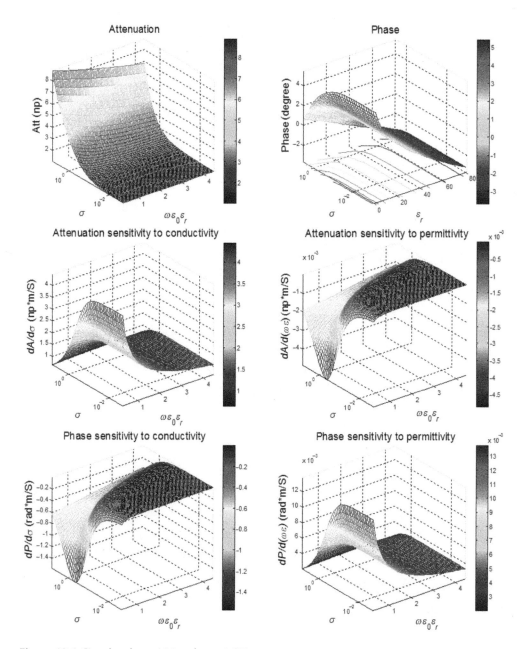

Figure 12.6 Signal and sensitivity plot at 1 GHz.

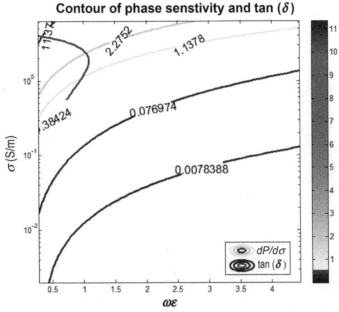

Figure 12.7 Comparison of tangential delta with attenuation sensitivity to dielectric constant and conductivity.

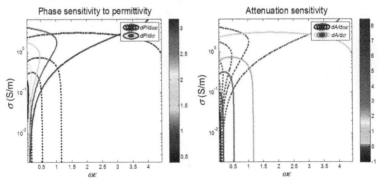

Figure 12.8 Sensitivity contour plot.

situation; however, regions with higher dielectric constant turn to relatively more sensitive area.

To summarize capability of the tool in detecting formation conductivity and permittivity, all the five channels are investigated and main specifications are listed in Table 12.1. Conclusion can be drawn that high-frequency phase difference is the most sensitive channel to dielectric property, which is as expected.

Table 12.1 Summary of sensitivity of all channels

Freq (D_t/D_r)	10 MHz (8″/4″)	20 MHz (8″/4″)	100 MHz (6″/2″)	200 MHz (6″/2″)	1 GHz (3″/2″)
A (dB)	1.5E+01 – 2.7E+01	1.30E+01 – 3.3E+01	7.50E+00 – 2.9E+01	7.20E+00 – 3.8E+01	7.10E+00 – 4.2E+01
P (degree)	1.40E−01 – 9.0E+01	2.80E−01 – 1.4E+02	9.00E−01 – 1.5E+02	3.40E+00 – 2.2E+02	2.00E+01 – 3.4E+02
$\|dA/d\sigma\|$ max	8.40E−02	1.70E−01	2.00E−01	2.80E−01	3.20E−01
Sensitive region	Resistive, relative low dielectric	Resistive, relative low dielectric	Resistive, relative low dielectric	Resistive, low dielectric	Resistive, low dielectric
$\|dA/d(\omega\varepsilon)\|$ max	3.90E−08	1.30E−07	2.60E−06	1.70E−06	1.60E−06
Sensitive region	Resistive, relative low dielectric	Resistive, relative low dielectric	Resistive, low dielectric	Resistive, low dielectric	Relative resistive, low dielectric
$\|dP/d\sigma\|$ max	1.30E−01	1.10E−01	8.40E−02	1.40E−02	5.20E−04
$\|dP/d(\omega\varepsilon)\|$ max	2.60E−08	2.10E−07	6.10E−06	3.50E−05	9.80E−04

12.6 SENSITIVITY ANALYSIS IN ANISOTROPIC FORMATION

Due to the complexity of response equations, it is easier to use numerical differentiation to find Jacobian matrix for anisotropic formations. Let J_{ij} represents partial differentiation of ith response in $\bar{S}$ vector to jth parameter in vector $\bar{x}$, the sensitivity of attenuation and phase of XX, XZ, and ZZ components of the tool response with respect to dipping angle and conductivity can be expressed as J_{i1} and J_{i2}. Fig. 12.9 shows a 1 GHz response of a multicomponent dielectric tool. In this case, only dipping angle and conductivity are variables, isotropic relative dielectric constant is fixed at 5 and anisotropy ratio assumed to be a constant 5 (Figs. 12.10–12.12).

Moreover, applying Hessian singular value decomposition analysis provides standard deviation of each parameter. The smaller standard deviation is, the more stable inversion could be. From the plot, dielectric inversion is most stable, followed by horizontal conductivity, dip, and anisotropy ratio. In this case, standard deviations of dip and anisotropy ratio monotonously decline with conductivity variations and drops to relative lower value at higher conductivity. In contrast, dielectric constant

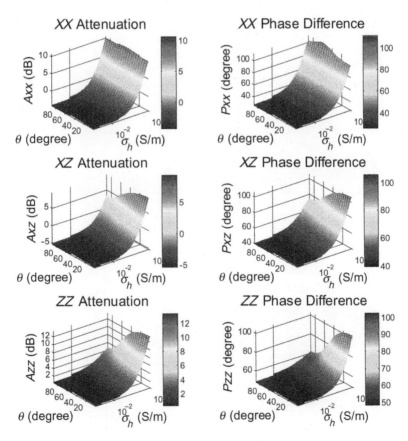

Figure 12.9 Multicomponent dielectric tool responses at 1 GHz.

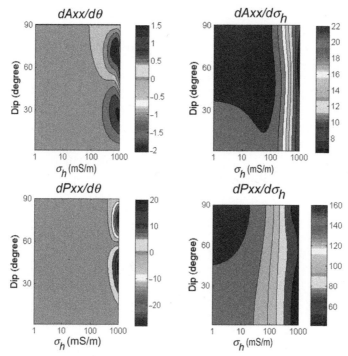

Figure 12.10 Sensitivity of *XX* responses at 1 GHz.

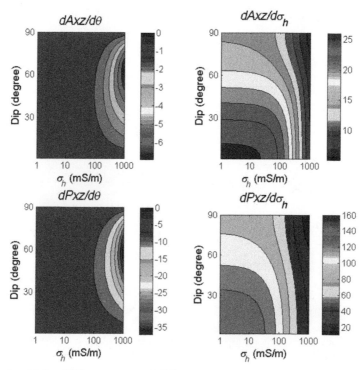

Figure 12.11 Sensitivity of *XZ* responses at 1 GHz.

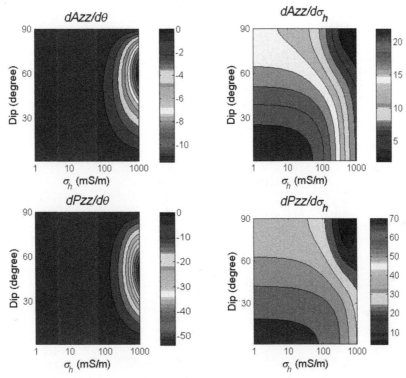

Figure 12.12 Sensitivity of ZZ responses at 1 GHz.

increases monotonously. As formation conductivity increases, standard deviations of all the four parameters would fall below certain criteria, making the inversion feasible.

To further study dielectric and anisotropy effects, more cases with changing relative dielectric constant or anisotropy ratio are computed and results are plotted in Figs. 12.13 and 12.14. Figs. 12.13 and 12.14 are similar to the cases in Fig. 12.15 but targeted to high dielectric formation. In these figures, we can clearly see that the tool's responses to the formation parameters are not smooth. Ripples appear in all standard deviation curves. The ripples occur at more conductive region when the dipping angle is relatively small. Previous observations that the conductive formation leads to more stable inversion still hold. In the meantime, it can be noticed from standard deviation plots that greater dipping angles enhance standard deviations of the tool response. This indicates that it would be difficult in finding convergent inversion solutions in highly deviated wells. Comparing these two graphs with Fig. 12.15, it is clear that high dielectric constant in the formation pushes the response ripples in inversion to more resistive area. In another word, the "appearing" monotonous phenomenon is caused by low dielectric constant of the formations; if very salty formation is encountered, we might reach the bumpy area as well for the previous situation.

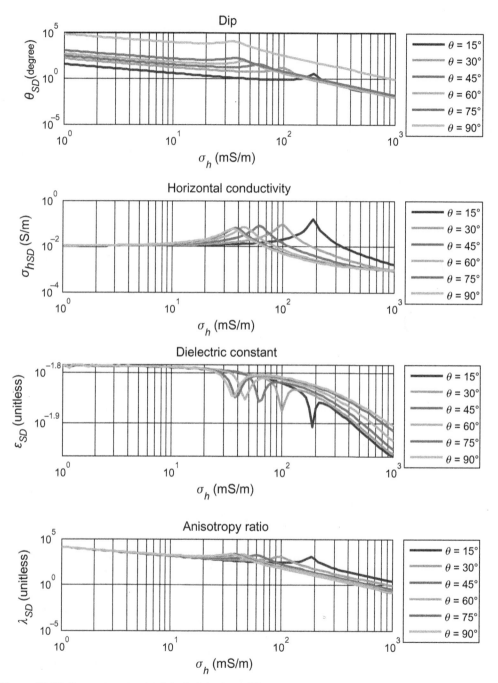

Figure 12.13 Parameter standard deviations ($\varepsilon_r = 45$).

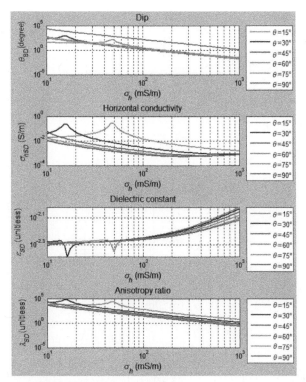

Figure 12.14 Parameter standard deviations ($\varepsilon_r = 75$).

Figs. 12.16 and 12.17 show the tool response to anisotropy parameters of the formation. The most dominant impact from the anisotropy of the formation is the decrease of the standard deviations versus dipping and anisotropy ratio. This is quite intuitive since θ and λ are intimately correlated with anisotropy. On the other hand, it does not affect standard deviation as much as the dielectric constant and dipping angle.

From the analysis given above, when considering the data inversion of the dielectric tool, especially inversion of both dielectric constant and conductivity, it is possible and stable using the multicomponent antenna design. Best results can be obtained if relative dip angle between tool and formation is not too large, formation is relatively conductive, dielectric constant is relatively low, and anisotropy contrast is large enough. It should be noticed that this is a single-antenna array analysis; better solution can be obtained when combining all measurements channels.

12.7 DIELECTRIC LOGGING TOOL DESIGN AND MODELING USING THREE-DIMENSIONAL NUMERICAL MODELING SOFTWARE PACKAGE

In previous sections, we studied the dielectric logging tool performance using a one-dimensional dipole model. As discussed, the purpose of the analysis is to understand the tool performance in different formations. However, to further study a practical

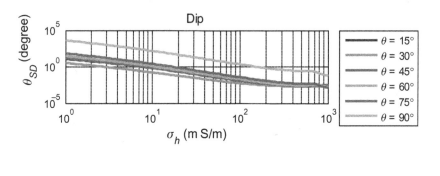

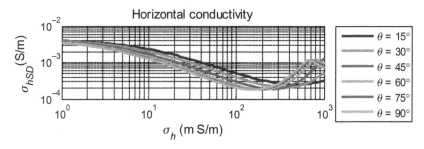

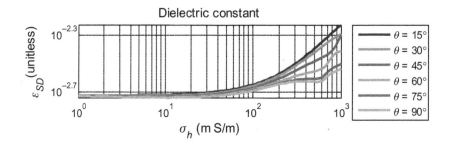

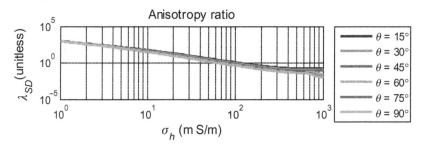

Figure 12.15 Formation parameter standard deviation from 1 GHz measurements.

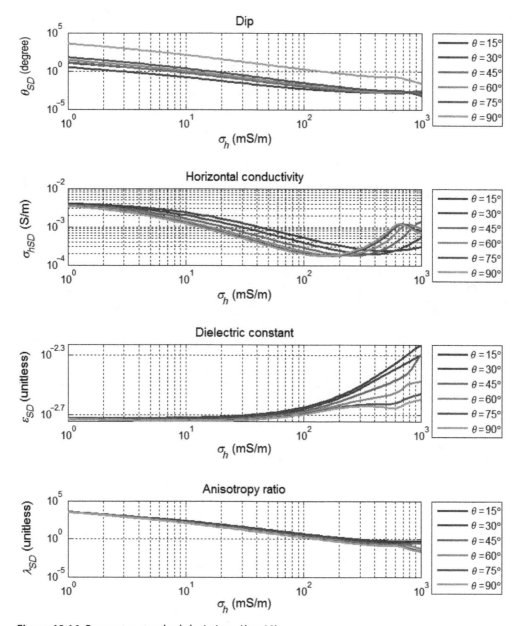

Figure 12.16 Parameter standard deviations ($\lambda = 10$).

dielectric logging tool, a 3D modeling software must be used. In this section, we will use commercial 3D EM simulation software to approach the tool design and simulation. The COMSOL Multiphysics is a popular numeric simulation software providing solutions for multiphysics modeling. The RF Module solves RF and

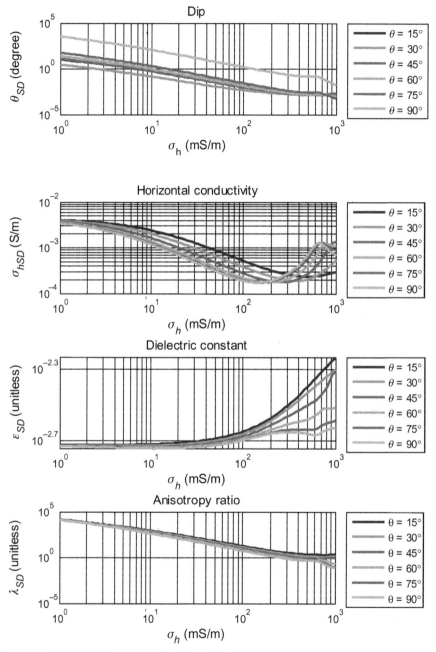

Figure 12.17 Parameter standard deviations ($\lambda = 20$).

microwave applications' problems by solving Maxwell equations numerically. LiveLink in MATLAB communicates with COMSOL Multiphysics and can enhance the functions of both software packages. Users can adjust model settings, analyze data, and create customized interface in MATLAB for the model. In this section, we will discuss the detailed dielectric tool design method using COMSOL RF module and LiveLink and study the tool response of the Array Dielectric Tool in various earth formations.

With the freedom of 3D modeling, it is possible to simulate the dielectric pad as it is in real situation and the dipole assumption is not necessary. This is more realistic than using analytical models as described in the previous chapter. However the compromise is the time including modeling complexity and CPU time. Compared with analytical solutions, the numerical solutions are more flexible, but analytical models have more clear physical meanings. Therefore, for the sake of understanding the tool concept, analytical model is preferred. For actual tool design, numerical methods are more practical. Based on the numerical simulations, influence from formation structures can be studied, such as pad impact, borehole mud, mud cake, invasion, and layered beds.

12.8 CAVITY-BACKED SLOT ANTENNA

For dielectric pads working at relatively high frequencies, cavity-backed slot antennas are commonly used. Fig. 12.18 shows the cavity-backed antenna model. Material filling the cavity is high dielectric ceramic with relative dielectric constant around 100 to reduce the physical dimensions. The slot is excited by TE_{10} waveguide mode as shown. In this case, the antenna can be modeled as a magnetic dipole orientated along the slot.

Fig. 12.19 is the field distribution near the antenna in air at 1 GHz; in Fig. 12.20, far field polar plot. In Fig. 12.20, the E field amplitude is displayed in comparison with magnetic dipole antenna. Radiation patterns of the cavity-backed antenna and the magnetic dipole antenna have similar performance. However, EM field amplitude

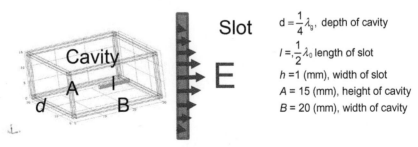

Figure 12.18 Cavity-backed slot antenna used in dielectric tool design.

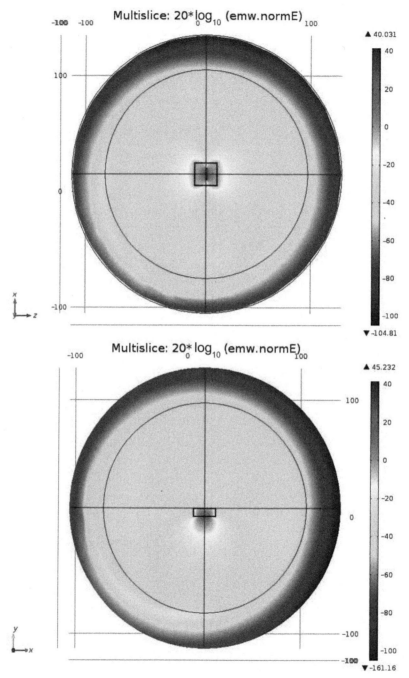

Figure 12.19 Electrical field |E| distribution of the cavity-backed slot antenna in air.

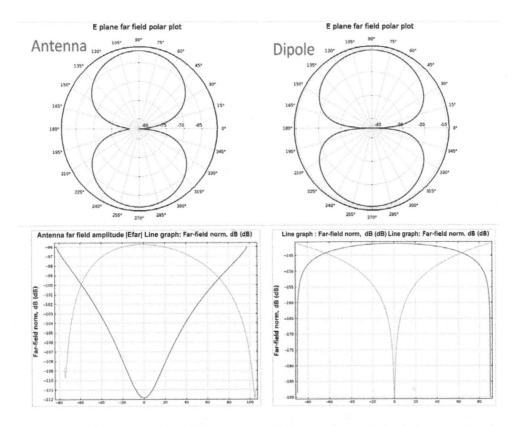

Figure 12.20 Comparison of radiation patterns between the cavity-backed antenna and a magnetic dipole antenna.

from the cavity-backed antenna is about 60 dB higher. This is because the resonance cavity with high dielectric constant materials enhanced the field radiation and has much higher radiation efficiency than the magnetic dipole.

To better understand the antenna performance in different formations, let us first evaluate the received signals at a fixed transmitter—receiver distance in a homogeneous formation. Assume the operating frequency is 1 GHz, T-R distance is 2 in., and ignore the impact of the pad. Fig. 12.21 shows the field distribution along the receiving slot antenna aperture. It can be seen that the field is a TE_{10} mode. The formation relative dielectric constant changes from 5, which is usually a parameter representing a slightly wet sand layer or shale, to 25, which is a parameter of a water-saturated sand layer. Conductivity changes from 1 to 1000 ohm-m. Received signal grows with respect to dielectric constant and decreases with formation conductivity.

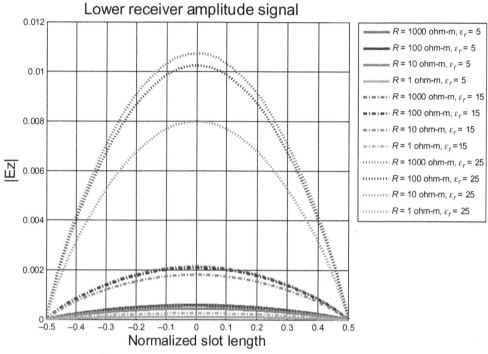

Figure 12.21 Electrical potential long the receiver antenna slot.

To further demonstrate that the discussions we used in the previous sections using dipole models are correct, we use the 3D numerical modeling to verify the dipole simulation. Consider the tool structure showing in Fig. 12.5. We name the receiver antennas as upper, middle, and lower antennas. The formation dielectric constant is selected at 5, 15, and 25, respectively while conductivities changes from 0.001 to 10, representing resistivity of 0.1–1000 ohm-m. Amplitude ratio and phase difference between receivers are computed using 3D numerical simulation and analytical dipole model. In Fig. 12.22, we can see the comparison between both attenuation and phase from the cavity-backed slot antenna array and dipole-based analytical solutions. Antenna responses agree very well with dipole results. Based on the above discussions, it can be seen that the cavity-backed slot antennas can be used for practical tool design since it has greater sensitivity compared with dipole antennas.

12.9 EFFECTS OF THE PAD

The dielectric constant logging tool is a pad-type tool. It is natural to ask ourselves that how much impact the tool pad can generate to the tool performance. For pad-type tools, a mechanical arm from tool body pushes the pad against borehole wall tightly to reduce the impact from the borehole mud. The pad surface is machined to

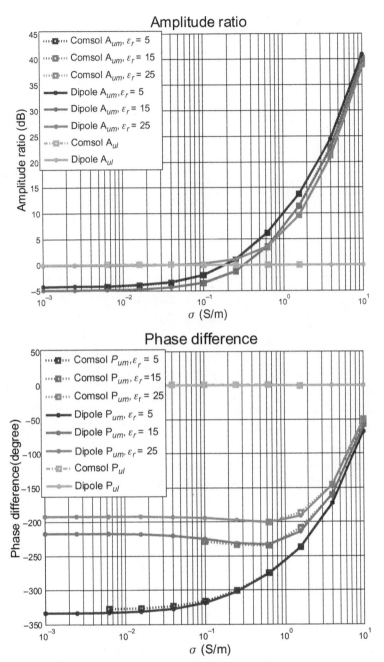

Figure 12.22 Attenuation and phase difference of the tool model showing in Fig. 12.5 computed by using both 3D numerical simulation and the analytical dipole model. The operating frequency of the dielectric tool is 1 GHz.

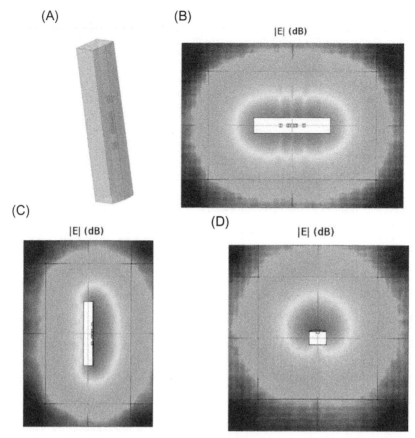

Figure 12.23 The *E* field near the pad in a homogeneous formation with dielectric constant of 25 and conductivity of 1 S/m. (A) The COMSOL model; (B) E distribution viewing from the back of the pad; (C) side view of the *E* field; and (D) the *E* field viewing from the tool axis.

fit the curve of the borehole. To model the pad body, it is assumed that the shape of the pad is a rectangular for simplicity. The pad surface can be treated as perfect electric conductor. Therefore most electromagnetic field penetrates into the formation from the front of the pad. Nevertheless, scattering around the edge and behind the pad is inevitable due to the finite area of pad surface.

Assume the pad is 20 in. long, 4 in. thick, and 3 in. wide. In practice, a piece of nonconductive and abrasion-resistive materials is installed in the front of the antennas to resist abrasion during logging and match the surface curve of a standard 6-in. borehole. The formation is a homogeneous formation with its conductivity of 1 S/m and the dielectric constant of 25. With the model described above, 3D simulation is done using COMSOL software package for the tool and the amplitude of the *E* field at the receiver antennas is plotted in Fig. 12.23. Fig. 12.23A is the geometry constructed in COMSOL.

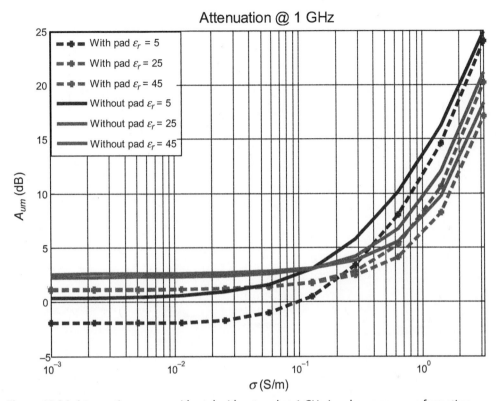

Figure 12.24 Attenuation curves with and without pad at 1 GHz in a homogeneous formation.

Fig 12.23B–D is the amplitudes of the E field viewing from different angles. From the field plots, it is seen that the field is concentrated around the antenna but radiates into the formation. There are fields leaking from the antenna to the back of the pad. However the field is rather weak. The main lobe of the radiation is in the front side of the antenna pad.

To better understand the impact of the pad to the tool performance, similar simulations are conducted as that showing in Fig. 12.22. In this scenario, the tool pad shown above is placed in the center of homogeneous formation. Results are plotted together with the curves without the pad as described in Figs. 12.24 and 12.25. In Figs. 12.24 and 12.25, *dashed lines* with markers are from the models with the pad body while *solid curves* are the results without the pad. Tool responses with the pad body and without the pad body are plotted against conductivity and at three different dielectric constant: $\varepsilon_r = 5$, $\varepsilon_r = 25$, and $\varepsilon_r = 45$, respectively. From these plots, we can see that the phase difference between receivers does not change much due to the existence of the pad. On the other hand, attenuation curves have greater differences from the ones without the pad. Separation between *dashed and solid curves* can be observed when 1 GHz antenna array is modeled, especially when material is resistive.

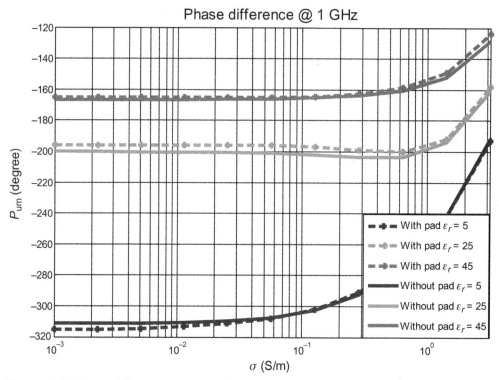

Figure 12.25 Phase difference curves with and without pad at 1 GHz in a homogeneous formation.

For more conductive or higher dielectric formation, the difference becomes greater. This is due to the fact that the amplitude of the received signals is affected by an image source generated by the pad surface whereas the phase is not.

The discrepancy increases at lower frequencies. Figs. 12.26 and 12.27 are the received signals at 10 MHz, attenuation from two models are further away from each other. This is because at lower frequencies the cancelation effect of the image source due to the pad surface and the antenna enhances since the wavelength is longer. By careful observation of the phase plots in Figs. 12.25 and 12.27, we can clearly see that in both 10 MHz and 1 GHz, the phase difference of the dielectric tool with and without the pad has equivalent performance.

To use the dipole approximation for inversion, the image source can be added to replace the pad surface to reduce the discrepancy in amplitude ratio. If the dipole model is used without the image, attenuation correction must be conducted when applying dipole models in tool analysis or inversion. Corresponding correction chart from tool responses to dipole results can be obtained from calibration in homogeneous formation before the tool is used.

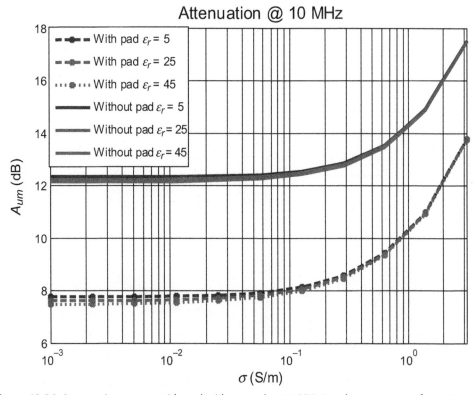

Figure 12.26 Attenuation curves with and without pad at 10 MHz in a homogeneous formation.

12.10 BOREHOLE MUD INFLUENCE

As discussed in the previous sections, the dielectric logging tools are pad-based wireline tools. A pusher pushes the antenna pad against the borehole wall. However, due to the roughness of the borehole wall, mud may exist between the antenna surface and the borehole. Even though, the pad-type tools are designed to be able to overcome borehole rugosity, the small gap between the pad and the borehole will have impact to the measurement results. In this section, a 3D geometry of pad, borehole, and formation is built with COMSOL. Material filled in the borehole areas could be WBM, oil-based mud (OBM), or the same with formation property to stands for homogeneous circumstance. Electrical properties of WBM and OBM are listed in Table 12.2 for studies in this or the following sections.

Using the mud given in Table 12.2, the received E field can be calculated. Fig. 12.28 shows the E field distribution in these three scenarios with identical formation property. From top to bottom is OBM, without mud, and WBM cases, respectively; while from left to right is top, side, and front view of field distribution. For WBM, radiation pattern

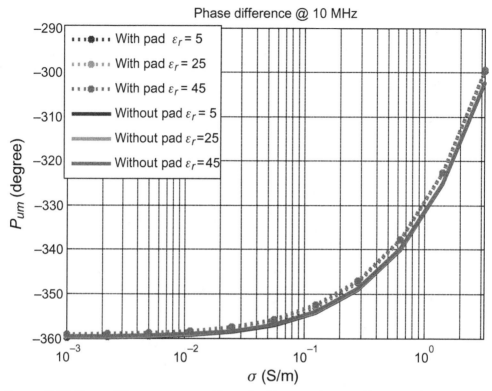

Figure 12.27 Phase difference curves with and without pad at 10 MHz in a homogeneous formation.

Table 12.2 Electrical properties of different mud

	WBM		OBM	
	Mud	Mud cake	Mud	Mud cake
ε_r	40	20	2.5	5
σ	5	1	0.001	0.005

from transmitters is similar to the homogeneous case. Moreover, field is more concentrated in front of the pad in WBM. On the contrary, scattering electromagnetic wave inside the borehole is much stronger in OBM. Nevertheless, strength of the signal ted into formation at all three cases are close to each other, as seen from the E field amplitude plot curves. In Fig. 12.28, it is assumed the pad is closely in contact with the borehole wall and there is no gap between the two surfaces.

Figs. 12.29 and 12.30 show the attenuation and phase differences of the two receivers when borehole mud presents compared with the case without mud at 1 GHz. The curves overlap to each other, which means the mud in the borehole has no

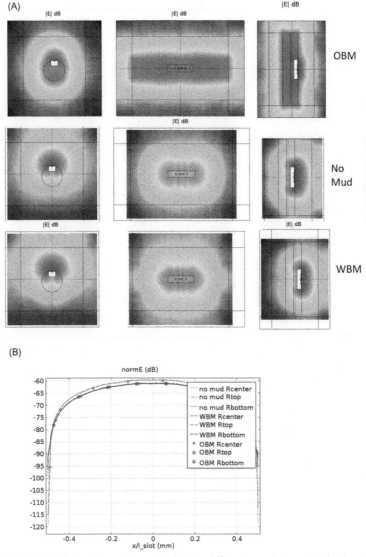

Figure 12.28 (A) Field plot and receiver signal with different mud types in the borehole. There is no gap between the pad surface and the borehole wall. The operating frequency is 1 GHz and the dielectric constant and conductivity of the formation is 25 and 1 S/m, respectively. (B) Field distribution when the pad is in oil-based mud, air, and water-based mud given in Table 12.2.

impact to the tool performance if there is no gap between the tool surface and the borehole wall. However, this is not always true due to borehole roughness and tool surface fitting to the borehole wall. In practice, there is always a small gap between the tool surface and the borehole wall. Apparently, WBM gap will have greater impact to the tool measurements than the OBM.

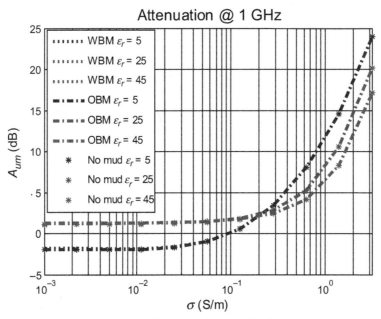

Figure 12.29 The amplitude ratio in dB of the received signals. There is no gap assumed between the pad surface and the borehole wall. The operating frequency is 1 GHz and the dielectric constant and conductivity of the formation is 25 and 1 S/m, respectively.

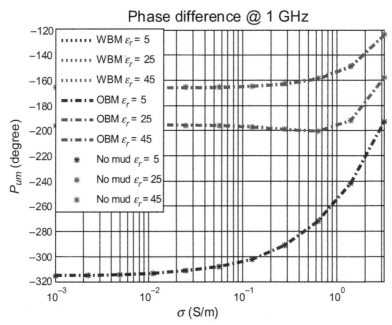

Figure 12.30 The phase difference of the received signals. There is no gap assumed between the pad surface and the borehole wall. The operating frequency is 1 GHz and the dielectric constant and conductivity of the formation is 25 and 1 S/m, respectively.

12.11 VERTICAL RESOLUTION

Another critical tool specification is the vertical resolution. As seen from previous chapters, the vertical resolution of a logging tool is largely dependent on the transmitter and receiver spacing and frequency. It is not reasonable to give a fixed quantity of vertical resolution of all measurement channels since they are different in spacing and frequencies. To better understand the vertical resolution of the dielectric logging tool, numerical simulations can be used. Since the vertical resolution is only dealing with vertically layered formations, it is possible to use the analytical solution. However, field distribution plot in the formation will give us insight of the vertical resolution problems. Therefore we select to use 3D numerical simulation package instead of analytical solution. To simplify the procedure, we simulate the tool responses when operating in stratified formation with thin conductive or dielectric layers for each individual antenna array.

Consider the formation model showing in Fig. 12.31. Let us discuss the case when a thin layer sandwiched by two semi-infinitively thick shoulder beds. If we run the tool through this formation, we can record the tool response. By changing the thickness of the middle layer, the vertical resolution of the antenna array can be obtained.

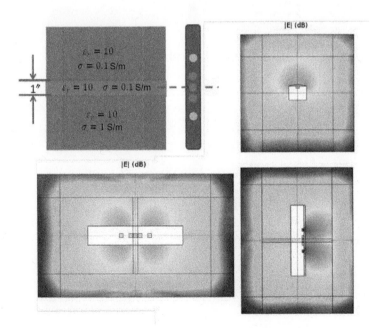

Figure 12.31 Field distribution when the tool is in a three-layer formation. The center layer is 1″ thick and the tool center coincides with the formation center. The center formation is a conductive layer with a conductivity of 0.1 S/m while the two shoulder beds are 1 S/m.

Fig. 12.31 also shows the E field distribution of the tool when in the layered formation. The purpose of these figures is to understand better of the tool performance in a layered formation. We can see that the field plots. In the model, a thin 1-in. horizontal layer with 0.1 S/m is sandwiched by two shoulder formations with 1 S/m conductivity. The dielectric constants of all three layers are the same. Field distribution at two different positions on the logging trajectory is computed. Fig. 12.32 is when the tool is in the center of the conductive bed; and Fig. 12.33 is the case where the bed is almost over tool pad.

From the EM field plots, we can see that the field is symmetric around the tool when at the center of the 1″ layer; on the other hand, distorted field is observed if the top transmitter is beneath the thin layer, causing a bit more scattering as well. Reflection from boundaries contributes to the distortion and scattering of electromagnetic waves.

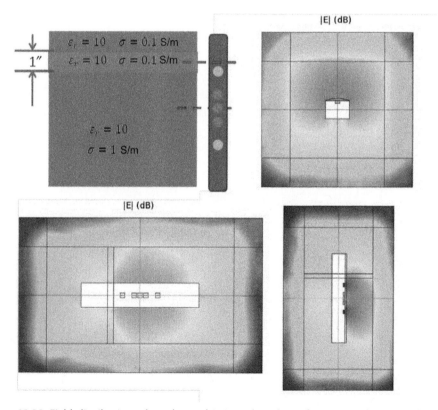

Figure 12.32 Field distribution when the tool is in a three-layer formation. The center layer is 1″ thick and the tool center is above the formation center. The center formation is a conductive layer with a conductivity of 0.1 S/m while the two shoulder beds are 1 S/m.

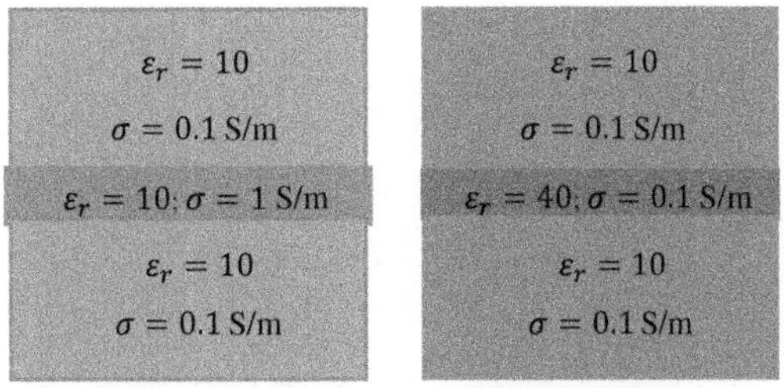

Figure 12.33 Thin layer and shoulder bed parameters of the formation used to study vertical resolution of the dielectric tool. Note that for conductivity contrast study, dielectric constant of each layer is kept constant and vice versa for dielectric constant contrast.

To investigate tool response for different layer thickness, synthetic logging tracks obtained from the designed dielectric tool should be studied. Assuming the tool is in a seven-layer formation structure with three target layers of thickness 8″, 4″, and 1″, respectively, the computed tool responses are as shown in Figs. 12.34 and 12.35. Shoulder beds are thick enough so that they will not have any impact to the thin layer responses. Conductivity contrast layers and dielectric contrast layers are studied separately. To study the tool response to conductivity contrast, the dielectric constant of each formation layer is set to constant and vice versa for dielectric constant contrast. The formation dipping is not considered to reduce complexity and only XX and ZZ are of interest. Responses from all five frequency channels are plotted together in each logging track.

From the simulated log, it can be seen that for a fixed T-R spacing, 1 GHz channel has the highest vertical resolution while 10 MHz channel could not function as well as others in detecting thin layers. 1 GHz attenuation is very sensitive detecting the 1″ conductive bed. It reaches a peak at the boundary locations. Similar behavior can be seen from the phase difference curves. Meanwhile, lower frequency channels are less responsive, especially when beds are relatively thin. For the given formation configuration, 100- and 200-MHz responses are able to detect 4″ layer; however, 10- and 20-MHz attenuations suffer difficulty in exploring layers thinner than 1 ft.

In general, dielectric contrast can be reflected from 1 GHz phase difference. Difference in its value is obvious between adjacent layers. Attenuation of the same channels is helpful as well. But the lowest two frequencies almost lost sensitivity in this situation. Further study can go on to testify signal contribution in inversion; Hessian sensitivity analysis mentioned in Chapter 11, Ahead-of-the-Bit Tools and Far Detection Electromagnetic Tools, can be applied to conduct the research.

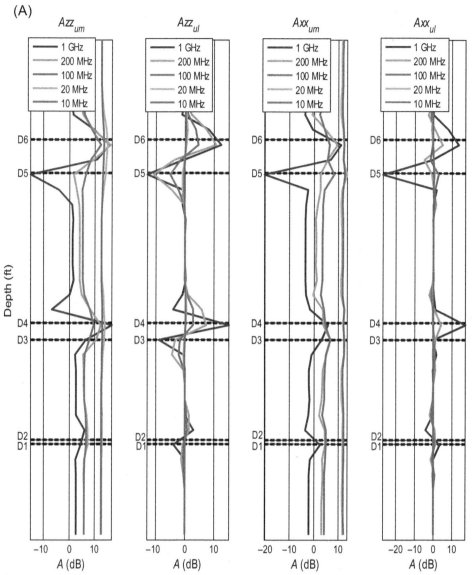

Figure 12.34 Synthetic log with conductivity thin layers showing the tool's capability in detecting vertical thin layers. Three thin layers are located in a homogeneous background formation with the conductivity of 0.1 S/m and dielectric constant of 15. The thickness of the three layers are 1″, 4″, and 8″, respectively. The thin layers have a dielectric constant of 15 and conductivity of 1 S/m for all three thin layers. (A) Attenuation and (B) phase shift of the received signals.

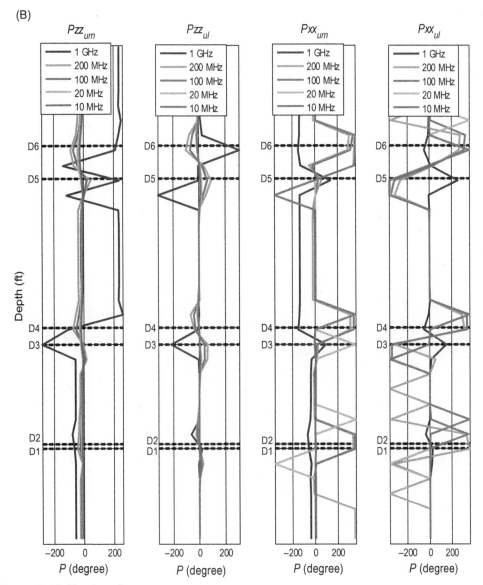

Figure 12.34 (Continued)

12.12 MUD CAKE AND INVASION

In the previous sections, we can find that the pad body and the borehole mud will have impact on the dielectric logging tool response. Due to the fact that the dielectric logging tool is a wireline tool, mud cakes and invasion should be considered since it will build up during and after drilling. Since the dielectric logging tool is very

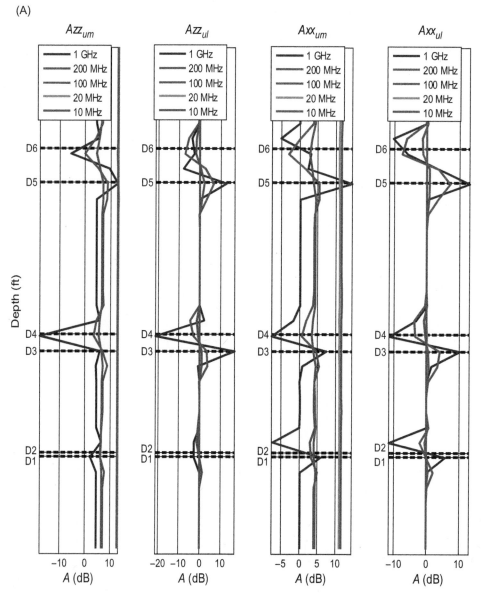

Figure 12.35 Synthetic log with dielectric thin layers showing the tool's capability in detecting vertical thin layers. Three thin layers are located in a homogeneous background formation with the conductivity of 0.1 S/m and dielectric constant of 15. The thickness of the three layers are 1″, 4″, and 8″, respectively. The thin layers have a conductivity of 0.1 S/m and dielectric constant of 45 and for all three thin layers. (A) Attenuation and (B) phase shift of the received signals.

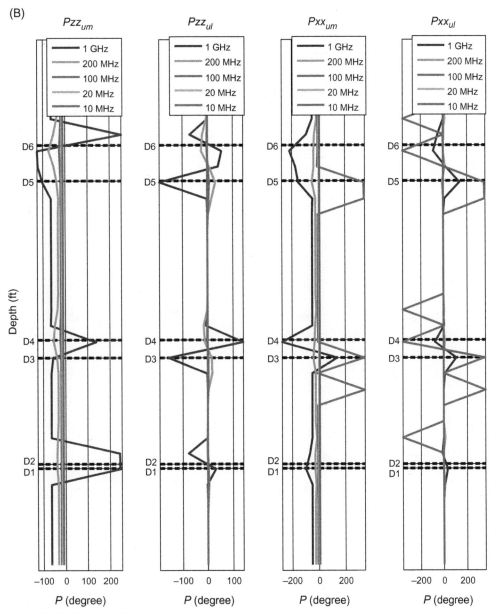

Figure 12.35 (Continued)

sensitive to the materials that are nearest to the tool surface, the effects of mud cake and invasion profiles must be investigated. To simplify the analysis, it is possible to use a two-dimensional (2D) axial symmetrical system to do the analysis without losing generosity. Therefore we adopt 2D axis-symmetric cylindrical model to estimate tool

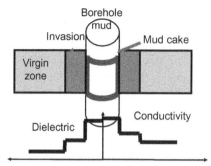

Figure 12.36 A schematic of mud cake and invasion profile.

performance with existence of mud cake and annulus invasion. Based on the model, we treat the tool pad as a solid copper cylinder coated with a thin layer of insulating materials. Antennas are simplified as magnetic dipoles floating on top of the pad surface. This is not as realistic as 3D models, but it is a compromise between accuracy and computation complexity. A general description of the model is shown in Fig. 12.36.

An example of electrical properties of the mud cake is summarized in Table 12.2. To investigate the effects of mud cake at different thickness, four different mud cake thicknesses are considered, which are 0.25″, 0.5″, 0.75″, and 1″.

Mud property will influence the tool performance since it presents a shield to the formation to be investigated. Figs. 12.37 and 12.38 show the phase difference and amplitude ratio of the two receivers at 1 GHz and 200 MHz with mud cake inserted in between the dielectric tool pad and the formation. For different formation dielectric constants (5, 25, and 45), representing different formation water contents, Figs. 12.37 and 12.38 show the tool response as a function of formation conductivity for different mud properties. The tool response without mud cake is also plotted in the same figures for comparison. If we carefully examine the structure of the mud cake and invasion profile from the point of impedance, an equivalent circuit can be obtained for the transmitter and receiver antenna system. Fig. 12.39 shows an equivalent circuit diagram of the transmitter antenna. Where Z_a is the impedance of the antenna, Z_m is the impedance of the mud cake, and Z_f is the impedance of the formation. If the formation impedance is considered to be the load of the antenna system, the Z_m presents matching impedance between Z_a and Z_f.

From the plots, we can clearly see that if water content in the formation is not high, in other words, relative dielectric constant of the formation is low, attenuation faces declination with existence of mud; value from OBM is closer to homogeneous conditions. Meanwhile, phase difference increases in WBM due to skin effect. As dielectric enhances, wavelength decreases; as a result, wave becomes oscillating more

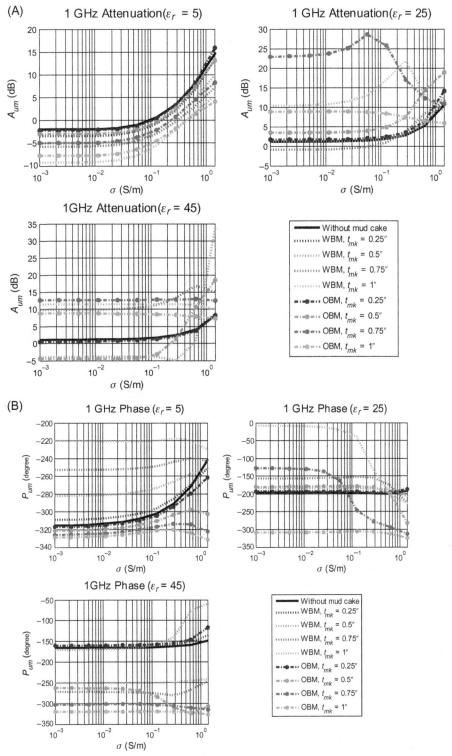

Figure 12.37 Impact of mud cake to the tool performance with different mud cake thickness at 1 GHz as a function of formation conductivity for given formation dielectric constant of 5, 25, and 45, respectively. (A) Amplitude ratio and (B) phase shift.

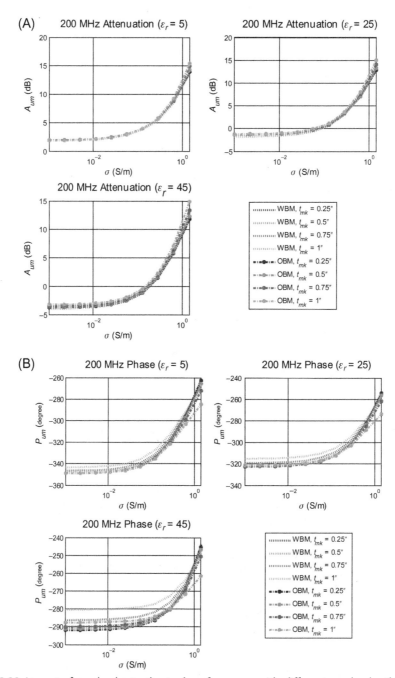

Figure 12.38 Impact of mud cake to the tool performance with different mud cake thickness at 200 MHz as a function of formation conductivity for given formation dielectric constant of 5, 25, and 45, respectively. (A) Amplitude ratio and (B) phase shift.

As we discussed earlier, Eq. (12.4) assumes the field is a plane wave. $Z_m(d)$ is the mud impedance at the mud—formation interface and d is the thickness of the mud layer. Z_m is the mud impedance at the interface between the tool and the mud. The impedance of the formation can be defined as

$$Z_F = \sqrt{\frac{\mu}{\varepsilon}} = \frac{\mu}{\sqrt{\mu\varepsilon}} = \frac{\omega\mu}{k_r + jk_i} \qquad (12.5)$$

Looking from the antenna, the impedance matching condition is

$$Z_a = Z_F + Z_m$$

Note that Z_m is a function of mud cake thickness and dielectric constant and conductivity of mud cake. Therefore, when these parameters change, the matching conditions to the antenna will also change.

However, for other frequencies, influence from mud cake is not as severe as that in the 1 GHz channels. 200 MHz example is displayed as following. As frequency goes lower, field can penetrate further into the formation, making the tool better capable at detecting virgin zone properties.

In order to study tool performance if mud filtrate invades into formation, we present results from invasion model as well. Annulus invasion is assumed here, so conductivity and permittivity of the invasion zone is regarded as half the sum of mud and virgin zone, i.e., $\varepsilon_{\text{invasion}} = \frac{\varepsilon_{\text{frm}} + \varepsilon_{\text{mud}}}{2}$ and $\sigma_{\text{invasion}} = \frac{\sigma_{\text{frm}} + \sigma_{\text{mud}}}{2}$. To be practical, 0.5-in. mud cake is placed in front of borehole wall and invasion zone radius is less than 4 in. for all channels. At 1 GHz, largest invasion radius considered is 2 in. since the measurement is shallow.

Still, salty WBM and OBM act differently. Interesting thing is that all the OBM curves overlap with each other. For WBM, thickness of invasion zone makes a big difference. Both attenuation and phase at this frequency become less sensitive to formation properties. Other measurements are organized in the appendix. Based on this, shale oil or gas where formation is too tight to invade is better working environment for dielectric tools. Also, combination of all measurement could offer possibility to inverse for invasion zone property (Figs. 12.41 and 12.42).

12.13 DEPTH OF INVESTIGATION

As discussed in the previous chapters, DOI quantitatively measures how far in radial direction that electromagnetic field generated by tool can penetrate. Industry benchmark is to use a simplified two radial layer structure. The inner layer has resistivity of 1 ohm-m while outer layer has resistivity 10 ohm-m. With boundary getting farther away from the borehole, measured apparent resistivity would decrease from

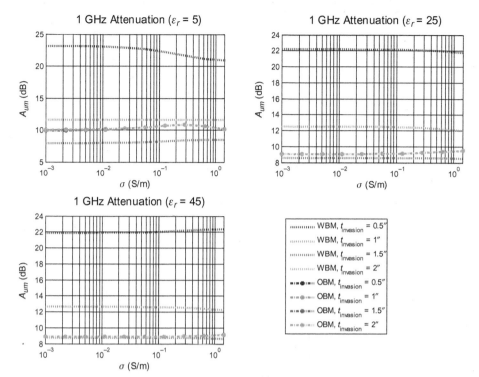

Figure 12.41 1 GHz attenuation with invasion zone.

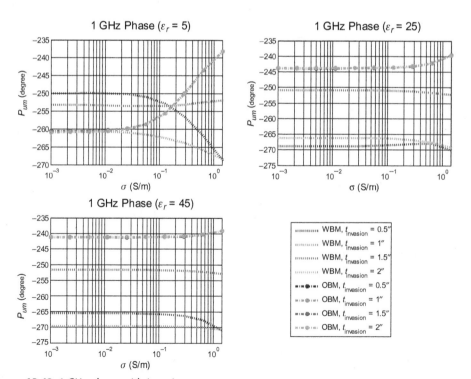

Figure 12.42 1 GHz phase with invasion zone.

10 ohm-m, which is the outer resistivity, to 1 ohm-m, the inner layer resistivity eventually. At the point where apparent resistivity is 5 ohm-m, the distance from borehole to boundary layer is defined as DOI. Since both dielectric and conductivity are measured, we redefine this concept from another point of view.

Radial models are still adopted; whereas both dielectric constant and conductivity contrast are considered. Four simplified models are shown in Fig. 12.43. Varying the inner layer radius leads to subsequent variation in tool responses. DOI is defined as the radius where tool response is the average value of results when the boundary is just at borehole wall and when it is infinitely far from borehole.

Results are presented in Fig. 12.44 for 100, 20, and 10 MHz, respectively. More significant differences in the measurements can be seen from Case 1 and Case 2, which are the conductivity contrast scenarios, than Case 3 and Case 4, for the dielectric contrast. Exchanging inner or outer layer does not affect investigation depth much; they are like flip over mirror images of each other. Another phenomenon retrieved from the plots is that phase penetrates deeper into formation in terms of dielectric constant exploration while attenuation in some degree sees farther when conductivity contrasts occur. Actually, this is consistent with previous sensitivity study. Meanwhile, greater DOI can be obtained at lower frequencies in sacrifice of signal levels. Weaker dielectric sensitivity at low frequency makes it difficult to distinguish between rocks of different dielectric constants. Thus tool response varies with inner radius gradually, resulting in relatively large DOI by definition.

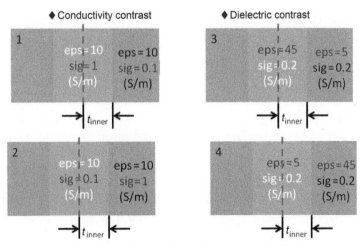

Figure 12.43 Formation structures for the DOI in a dielectric logging tool.

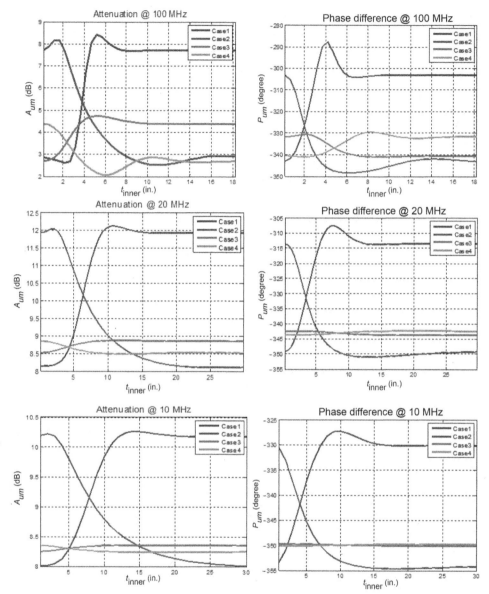

Figure 12.44 Dielectric tool responses in the formations shown in the formation models described in Fig. 12.43.

Table 12.3 lists DOI of the designed dielectric tool. For 1 GHz and 200 MHz, large dielectric constant causes oscillation in the received signal. DOI definition loses its meaning under this condition. From the table, DOI of each channel increases gradually, meaning that the tool is able to investigate into radial formation layer by layer and avoid missing information due to DOI gap.

Table 12.3 DOI of different channels in a dielectric logging tool

	Conductivity contrast		Dielectric contrast	
	DOIA (in.)	DOIP (in.)	DOIA (in.)	DOIP (in.)
1 GHz; $L = 3''/1''$	1	<1	NA	NA
200 MHz; $L = 6''/2''$	2.5	1.5	NA	NA
100 MHz; $L = 6''/2''$	4	2.5	3	5
20 MHz; $L = 8''/4''$	7	4	5	8
10 MHz; $L = 8''/4''$	8	6	>10	>10

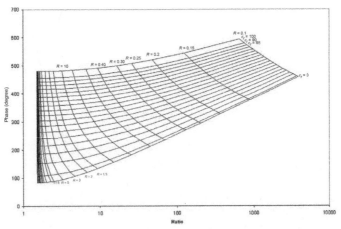

Figure 12.45 A two-dimensional plots of the dielectric constant, resistivity as functions of both amplitude ratio and phase difference of a single-frequency dielectric logging tool. The operating frequency of the tool is 1.1 GHz.

12.14 APPLICATIONS OF DIELECTRIC TOOLS

As we can see from the previous sections, both amplitude ratio and phase difference are functions of both dielectric constant and conductivity. In practice, in order to obtain both dielectric constant and resistivity simultaneously, inversion algorithm must be applied. One of the inversion method is to use a 2D chart as shown in Fig. 12.45. This chart relates four parameters: dielectric constant of the formation, resistivity of the formation, amplitude ratio, and phase difference measured by the dielectric tool at a fixed frequency. Fig. 12.45 is obtained by computation assuming the tool in a homogeneous formation and the pad is closely attached to the formation surface. It is possible to make the similar plots when the pad to formation distance and mud parameters are considered. If the measured amplitude and phase difference is obtained from the dielectric tool, a unique point in the 2D plot is obtained and the dielectric constant and resistivity of the formation is determined simultaneously. If the measured points are not found in the data set, interpolations are used between data points.

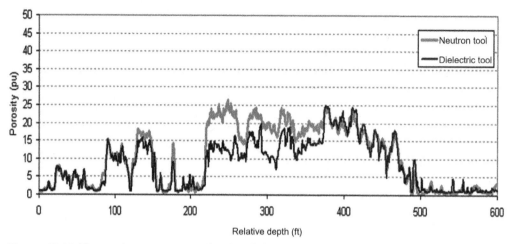

Figure 12.46 Measured water porosity by the dielectric logging tool operating at a single frequency of 1.1 GHz compared with the porosity measured by a neutron tool. The curve split reflects the hydrocarbon contents.

Fig. 12.46 shows a measured data of the dielectric tool compared with the data from a neutron porosity tool in the same formation. The dielectric constant measurement has been interpreted into water-saturated formation porosity data, which is compared with the porosity measured by a neutron tool. We know that the water porosity measured by the dielectric tool reflects the porosity occupied by the water in the formation. However the porosity measured by the neutron tool gives the total hydrogen-occupied porosity of the formation, including water, oil, and gas. In Fig. 12.46, in the sections above 220 ft, and below 375 ft, the two porosity curves overlap each other, which mean that the formation pores are full of water and no hydrocarbon can be found. However, in the section between 220 and 375 ft, the water porosity and neutron porosity split and the water porosity is less than the neutron porosity. This indicates that the volume of water takes only part of the total pore space in the formation and the rest of the pore space may be taken by oil or gas. Therefore petrophysicist would like to see the split between water porosity measured by a dielectric tool and the neutron porosity measured by a neutron logging tool.

12.15 SUMMARY

This chapter gives a comprehensive evaluation of the tool performance of a dielectric logging tool in a wide variety of environments. Influence of the pad, borehole mud, mud cake and invasion has been studied. According to tool simulation, pad correction need to be conducted if simplified mathematical models were used for analysis.

Despite of variation in field distribution pattern, mud type in the borehole does not make much difference in attenuation and phase difference from receiver groups. Existence of mud cake and invasion plays a nonnegligible role in measurements. However the influence fades away as working frequency gets lower.

DOI and vertical resolution of the tool are discussed as well. Interest on dielectric sensitivity is taken into account in both cases. Results prove that the tool is able to detect layers as thin as 1 in. either it is high conductive or high dielectric using 1 GHz channel. In addition, DOI of each array increase from 1 to 10 in. gradually, satisfying design requirements.

REFERENCES

[1] L.C. Shen, Problems in Dielectric-Constant Logging and Possible Routes to Their Solution, November–December 1985.
[2] L.C. Shen, A Laboratory Technique for Measuring Dielectric Properties of Core Samples at Ultrahigh Frequencies, 1985.
[3] T. Zhang, et al., Error Quantification of Dielectric Spectroscopy on Carbonate Core Plugs, Society of Petroleum Engineers, 2011.
[4] T. Zhang, et al., Dielectric Response of Carbonate Core-Plugs—Influence of Heterogeneous Rock Properties on Permittivity, Society of Petroleum Engineers, 2010.
[5] M. Pirrone, et al., An innovative dielectric dispersion measurement for better evaluation of thin layered reservoirs applied in a South Italy well, in: Offshore Mediterranean Conference, 2011.
[6] M. Pirrone, et al., A Novel Approach Based on Dielectric Dispersion Measurements to Evaluate the Quality of Complex Shaly-Sand Reservoirs, Society of Petroleum Engineers, 2011.
[7] D.P. Schmitt, et al., Revisiting Dielectric Logging in Saudi Arabia: Recent Experiences and Applications in Development and Exploration Wells, Society of Petroleum Engineers, 2011.
[8] N.V. Seleznev, et al., Applications of Dielectric Dispersion Logging to Oil Shale Reservoirs, Society of Petrophysicists and Well-Log Analysts, 2011.
[9] C.A.P. Suarez, et al., Dielectric Logging Uncovers New Reserves in a Reactivated Colombian Field, Society of Petrophysicists and Well-Log Analysts, 2012.
[10] M. Hizem, et al., Dielectric Dispersion: A New Wireline Petrophysical Measurement, Society of Petroleum Engineers, 2008.
[11] Y. Ramos, et al., Reservoir Evaluation and Completion Optimization in Heavy Oil Sands with Dielectric and Diffusion Measurements, Society of Petroleum Engineers, 2012.
[12] J.D. Little, et al., Dielectric Dispersion Measurements in California Heavy Oil Reservoirs, Society of Petrophysicists and Well-Log Analysts, 2010.
[13] L. Mosse, et al., Dielectric Dispersion Logging in Heavy Oil: A Case Study from The Orinoco Belt, Society of Petrophysicists and Well-Log Analysts, 2009.
[14] I. Al-Qarshubi, et al., Quantification of Remaining Oil Saturation Using a New Wireline Dielectric Dispersion Measurement—A Case Study from Dukhan Field Arab Reservoirs, Society of Petroleum Engineers, 2011.
[15] S.T. Grayson, J.L. Hemingway, A Heavy Oil Mobility Quicklook Using Dielectric Measurements at Four Depths of Investigation, Society of Petroleum Engineers, 2013.
[16] M. Han, et al., Continuous Estimate of Cation Exchange Capacity from Log Data: A New Approach Based on Dielectric Dispersion Analysis, Society of Petrophysicists and Well-Log Analysts, 2012.
[17] A. Almarzooq, et al., Characterization of Shale Gas Rocks Using Dielectric Dispersion and Nuclear Magnetic Resonance, Society of Petroleum Engineers, 2013.
[18] A.F. Abdel Aal, et al., Integration of Dielectric Dispersion and 3D NMR Characterizes the Texture and Wettability of a Cretaceous Carbonate Reservoir, Society of Petroleum Engineers, 2013.

[19] N. Heaton, et al., Novel In Situ Characterization of Heavy Oil Integrating NMR And Dielectric Logs, Society of Petrophysicists and Well-Log Analysts, 2012.
[20] H. Bachman, et al., Practical Down Hole Dielectric and Diffusion-Based Magnetic Resonance Workflow for Saturations and Viscosity of Heavy Oil Reservoirs Using a Laboratory Viscosity Calibration, Society of Petroleum Engineers, 2012.
[21] A. Al-Yaarubi, N.I.-h Al-Bulushi, R. Al Mjeni, Downhole Heavy Oil Characterization Using 3D NMR and Modern Dielectric Logs, Society of Petroleum Engineers, 2012.
[22] <http://www.halliburton.com/public/lp/contents/Data_Sheets/web/H/H07231.pdf>.

APPENDIX

Analytical derivation of homogeneous isotropic formation Jacobian matrix

$$\text{Jacobian} = \begin{bmatrix} \dfrac{\partial A}{\partial \sigma} & \dfrac{\partial A}{\partial \omega\varepsilon} \\ \dfrac{\partial P}{\partial \sigma} & \dfrac{\partial P}{\partial \omega\varepsilon} \end{bmatrix}$$

Here attenuation and phase difference are defined as following:

$$\frac{H_1}{H_2} = \frac{R_1 + jX_1}{R_2 + jX_2}$$

$$A = \text{Re}\left(\log_{10}\left(\frac{H_1}{H_2}\right)\right) = \log_{10}\left(\frac{\sqrt{R_1^2 + X_1^2}}{\sqrt{R_2^2 + X_2^2}}\right); \quad \text{unit: np/m}$$

$$P = \text{Im}\left(\log_{10}\left(\frac{H_1}{H_2}\right)\right) = \text{angle}(H_1) - \text{angle}(H_2); \quad \text{unit: rads}$$

To compute partial differentiation chain rule is applied here. For example, $\partial A/\partial \sigma$ consists of four terms representing $\partial R_1/\partial \sigma$, $\partial R_2/\partial \sigma$, $\partial X_1/\partial \sigma$, and $\partial X_2/\partial \sigma$. It is similar for $\partial A/\partial \varepsilon_r, \partial P/\partial \sigma$, and $\partial P/\partial \varepsilon_r$.

$$\frac{\partial A}{\partial \sigma} = \frac{\partial A}{\partial R_1}\frac{\partial R_1}{\partial \sigma} + \frac{\partial A}{\partial R_2}\frac{\partial R_2}{\partial \sigma} + \frac{\partial A}{\partial X_1}\frac{\partial X_1}{\partial \sigma} + \frac{\partial A}{\partial X_2}\frac{\partial X_2}{\partial \sigma} \quad \text{(A.1a)}$$

$$\frac{\partial P}{\partial \sigma} = \frac{\partial P}{\partial R_1}\frac{\partial R_1}{\partial \sigma} + \frac{\partial P}{\partial R_2}\frac{\partial R_2}{\partial \sigma} + \frac{\partial P}{\partial X_1}\frac{\partial X_1}{\partial \sigma} + \frac{\partial P}{\partial X_2}\frac{\partial X_2}{\partial \sigma} \quad \text{(A.1b)}$$

$$\frac{\partial A}{\partial \varepsilon_r} = \frac{\partial A}{\partial R_1}\frac{\partial R_1}{\partial \varepsilon_r} + \frac{\partial A}{\partial R_2}\frac{\partial R_2}{\partial \varepsilon_r} + \frac{\partial A}{\partial X_1}\frac{\partial X_1}{\partial \varepsilon_r} + \frac{\partial A}{\partial X_2}\frac{\partial X_2}{\partial \varepsilon_r} \quad \text{(A.1c)}$$

$$\frac{\partial P}{\partial \varepsilon_r} = \frac{\partial P}{\partial R_1}\frac{\partial R_1}{\partial \varepsilon_r} + \frac{\partial P}{\partial R_2}\frac{\partial R_2}{\partial \varepsilon_r} + \frac{\partial P}{\partial X_1}\frac{\partial X_1}{\partial \varepsilon_r} + \frac{\partial P}{\partial X_2}\frac{\partial X_2}{\partial \varepsilon_r} \quad \text{(A.1d)}$$

Partial differentiation of attenuation/phase difference with respect to real/imaginary part of near receiver signal can be expressed as following:

$$\frac{\partial A}{\partial R_1} = \frac{R_1}{R_1^2 + X_1^2}; \quad \frac{\partial A}{\partial X_1} = \frac{X_1}{R_1^2 + X_1^2}; \quad \text{(A.2a)}$$

$$\frac{\partial A}{\partial R_2} = \frac{-R_2}{R_2^2 + X_2^2}; \quad \frac{\partial A}{\partial X_2} = \frac{-X_2}{R_2^2 + X_2^2}; \quad \text{(A.2b)}$$

$$\frac{\partial P}{\partial R_1} = \frac{-X_1}{R_1^2 + X_1^2}; \quad \frac{\partial P}{\partial X_1} = \frac{R_1}{R_1^2 + X_1^2}; \quad \text{(A.2c)}$$

$$\frac{\partial P}{\partial R_2} = \frac{-X_2}{R_2^2 + X_2^2}; \quad \frac{\partial P}{\partial X_2} = \frac{R_2}{R_2^2 + X_2^2}; \quad \text{(A.2d)}$$

The second multiplication terms in Eq. (A.1) can be calculated from H field expression in a similar way. As we know, wavenumber in the formation can be expressed as:

$$k = \sqrt{\mu}\sqrt{\omega^2\varepsilon_0^2\varepsilon_r^2 + \sigma^2} e^{i\delta} \quad \text{(A.3)}$$

where $\delta = -\frac{1}{2}\operatorname{atan}\left(\frac{\sigma}{\omega\varepsilon_0\varepsilon_r}\right)$. Also, we can separate the real and imaginary part of wavenumber to the following parameters:

$$\alpha = \operatorname{Re}\{k\}L = (\omega\sqrt{\mu}L)\left(\varepsilon_0^2\varepsilon_r^2 + \left(\frac{\sigma}{\omega}\right)^2\right)^{1/4} \cos\delta \quad \text{(A.4a)}$$

$$\beta = \operatorname{Im}\{k\}L = (\omega\sqrt{\mu}L)\left(\varepsilon_0^2\varepsilon_r^2 + \left(\frac{\sigma}{\omega}\right)^2\right)^{1/4} \sin\delta \quad \text{(A.4b)}$$

Thus partial differentiation of real and imaginary part of wavenumber with respect to conductivity and relative dielectric constant can be calculated. This is the first stage of the chain theory.

$$\frac{\partial \delta}{\partial \sigma} = -\frac{1}{2}\frac{\omega\varepsilon_0\varepsilon_r}{\omega^2\varepsilon_0^2\varepsilon_r^2 + \sigma^2} \quad \text{(A.5a)}$$

$$\frac{\partial \delta}{\partial \varepsilon_r} = \frac{1}{2}\frac{\omega\varepsilon_0\sigma}{\omega^2\varepsilon_0^2\varepsilon_r^2 + \sigma^2} \quad \text{(A.5b)}$$

$$\frac{\partial \alpha}{\partial \sigma} = (\omega\sqrt{\mu}L)\left[\frac{\sigma}{2\omega^2}\left(\varepsilon_0^2\varepsilon_r^2 + \left(\frac{\sigma}{\omega}\right)^2\right)^{-3/4} \cos\delta - \sin\delta\left(\varepsilon_0^2\varepsilon_r^2 + \left(\frac{\sigma}{\omega}\right)^2\right)^{1/4}\left(\frac{\partial \delta}{\partial \sigma}\right)\right] \quad \text{(A.5c)}$$

$$\frac{\partial \alpha}{\partial \varepsilon_r} = (\omega\sqrt{\mu}L)\left[\frac{\varepsilon_0^2 \varepsilon_r}{2}\left(\varepsilon_0^2 \varepsilon_r^2 + \left(\frac{\sigma}{\omega}\right)^2\right)^{-3/4} \cos\delta - \sin\delta\left(\varepsilon_0^2 \varepsilon_r^2 + \left(\frac{\sigma}{\omega}\right)^2\right)^{1/4}\left(\frac{\partial\delta}{\partial\varepsilon_r}\right)\right]$$
(A.5d)

$$\frac{\partial \beta}{\partial \sigma} = (\omega\sqrt{\mu}L)\left[\frac{\sigma}{2\omega^2}\left(\varepsilon_0^2 \varepsilon_r^2 + \left(\frac{\sigma}{\omega}\right)^2\right)^{-3/4} \sin\delta + \cos\delta\left(\varepsilon_0^2 \varepsilon_r^2 + \left(\frac{\sigma}{\omega}\right)^2\right)^{1/4}\left(\frac{\partial\delta}{\partial\sigma}\right)\right]$$
(A.5e)

$$\frac{\partial \beta}{\partial \varepsilon_r} = (\omega\sqrt{\mu}L)\left[\frac{\varepsilon_0^2 \varepsilon_r}{2}\left(\varepsilon_0^2 \varepsilon_r^2 + \left(\frac{\sigma}{\omega}\right)^2\right)^{-3/4} \sin\delta + \cos\delta\left(\varepsilon_0^2 \varepsilon_r^2 + \left(\frac{\sigma}{\omega}\right)^2\right)^{1/4}\left(\frac{\partial\delta}{\partial\varepsilon_r}\right)\right]$$
(A.5f)

Similarly, H field can be expressed in real and imaginary parts as well. The second stage of chain theory is to compute partial differentiation of real and imaginary parts of H field with respect to real and imaginary part of wavenumber, and then find their differentiation to formation properties.

$$Hz = \frac{e^\beta}{2\pi L^3}[\cos\alpha(1-\beta) + \alpha\sin\alpha] + j\frac{e^\beta}{2\pi L^3}[(\beta-1)\sin\alpha + \alpha\cos\alpha] = R + jX$$
(A.6a)

$$\frac{\partial R}{\partial \alpha} = \frac{e^\beta}{2\pi L^3}(\beta\sin\alpha + \alpha\cos\alpha)$$
(A.6b)

$$\frac{\partial R}{\partial \beta} = \frac{e^\beta}{2\pi L^3}(-\beta\cos\alpha + \alpha\sin\alpha)$$
(A.6c)

$$\frac{\partial X}{\partial \alpha} = \frac{e^\beta}{2\pi L^3}(-\alpha\sin\alpha + \beta\cos\alpha) = -\frac{\partial R}{\partial \beta}$$
(A.6d)

$$\frac{\partial X}{\partial \beta} = \frac{e^\beta}{2\pi L^3}(\beta\sin\alpha + \alpha\cos\alpha) = \frac{\partial R}{\partial \alpha}$$
(A.6e)

To reach the final results, apply chain theory again:

$$\frac{\partial R_1}{\partial \varepsilon_r} = \frac{\partial R_1}{\partial \alpha_1}\frac{\partial \alpha_1}{\partial \varepsilon_r} + \frac{\partial R_1}{\partial \beta_1}\frac{\partial \beta_1}{\partial \varepsilon_r} \tag{A.7a}$$

$$\frac{\partial R_2}{\partial \varepsilon_r} = \frac{\partial R_2}{\partial \alpha_2}\frac{\partial \alpha_2}{\partial \varepsilon_r} + \frac{\partial R_2}{\partial \beta_2}\frac{\partial \beta_2}{\partial \varepsilon_r} \tag{A.7b}$$

$$\frac{\partial X_1}{\partial \varepsilon_r} = \frac{\partial X_1}{\partial \alpha_1}\frac{\partial \alpha_1}{\partial \varepsilon_r} + \frac{\partial X_1}{\partial \beta_1}\frac{\partial \beta_1}{\partial \varepsilon_r} \tag{A.7c}$$

$$\frac{\partial X_2}{\partial \varepsilon_r} = \frac{\partial X_2}{\partial \alpha_2}\frac{\partial \alpha_2}{\partial \varepsilon_r} + \frac{\partial X_2}{\partial \beta_2}\frac{\partial \beta_2}{\partial \varepsilon_r} \tag{A.7d}$$

CHAPTER 13

Finite Element Method for Solving Electrical Logging Problems in Axially Symmetrical Formations

Contents

13.1	Overview of the Numerical Simulation Methods for Well Logging Problems	504
13.2	Finite Element Method Based on Magnetic Field	505
	13.2.1 Magnetic field equations	506
13.3	Analysis of Transverse Electric Mode and Transverse Magnetic Mode	507
13.4	Vector Matrix Equation of Magnetic Field and Impedance Matrix	509
13.5	The Basis Functions	511
	13.5.1 Triangular element basis functions	511
	13.5.2 Rectangular element basis functions	514
13.6	Evaluation of Impedance Element Matrix for Rectangular Element Based on H_ϕ	515
13.7	Evaluation of Impedance Element Matrix for Rectangular Element Based on ρH_ϕ	518
13.8	Evaluation of Impedance Element Matrix for Triangular Element Based on H_ϕ	520
	13.8.1 Formulation of element matrix	520
	13.8.2 Evaluation of short terms in element matrix	522
	13.8.3 Evaluation of long term in element matrix	524
13.9	FEM Based on Electric Field	532
	13.9.1 Electric field equation	532
	13.9.2 FEM vector matrix equation of electric field	533
13.10	Evaluation of Triangular Element Matrix Based on E_ϕ (TE Mode)	535
13.11	FEM Model of Sources	537
	13.11.1 Source models	537
	13.11.2 Source model 1—electrical current loop	537
	13.11.3 Source model 2—magnetic current loop	537
	13.11.4 Source model 3—electrodes	538
	13.11.5 FEM solutions for the source existing on boundaries	540
References		545
Appendix A Vector Analysis in Cylindrical Coordinates		546
Appendix B Computation Method for Matrix Assembling Rule		547
Appendix C Computation Method of Element Matrix for Rectangular Element Based on H_ϕ (Section 13.6)		548
Appendix D Computation Method of Element Matrix for Rectangular Element Based on ρH_ϕ (Section 13.7)		550
Appendix E Term 4 (Section 13.8.3.9)		552
Appendix F Computation Method of Element Matrix for Triangular Element Based on H_ϕ (Section 2.7.4)		554
Appendix G Computation Method of Element Matrix for Triangular Element Based on E_ϕ (Section 3.3)		556

13.1 OVERVIEW OF THE NUMERICAL SIMULATION METHODS FOR WELL LOGGING PROBLEMS

In previous chapters, the methods in solving an electromagnetic (EM) logging problems are mostly analytical or semi-analytical (hybrid). Analytical methods have clear physical and mathematical meanings in each solution step. The analytical methods are usually accurate and fast in computation. Therefore, for inversion and fast view of tool performance, analytical methods are very effective. However, if complicated geometry is involved, in terms of either tool structure or formation structure, it is difficult to find analytical solutions. As computational technology evolves, numerical methods are widely used. Commercial software packages such as COMSOL [1], HFSS (ANSYS) [2], and Maxwell (ANSYS) [3] are considered as effective numerical tools in logging tool design and analysis. Strictly speaking, the numerical method is a technique for finding approximate solutions numerically. It can solve complicated one-dimensional (1D), two-dimensional (2D), and three-dimensional (3D) problems but its computational speed is generally slow and largely depends on its grid size. Different methods will be chosen with respect of different applications. Table 13.1 compares analytical, semi-analytical (hybrid), and numerical methods.

With the development of computer technique, numerical methods are becoming increasingly important in EM modeling. There are several kinds of numerical method which are widely used: finite element method (FEM), method of moment (MOM), finite difference method (FDM), and finite difference method in time domain (FDTD). In general, all of these methods are capable of solving Maxwell equations either in time or frequency domain and can be used to solve logging problems. However, each numerical method has its own characteristics and are suitable for certain applications. Table 13.2 compares different numerical methods [4]. In Table 13.2, PDE is a short for partial differential equation and IE is for integral equation. Since the logging problems are mostly single low frequency inhomogeneous volume dominated problems, FEM and FDM are the better choices. If the problem involved has a complex structure, then FEM can have more flexibility. Considering the applications

Table 13.1 Comparison between analytical, hybrid, and numerical method

	Analytical method	Hybrid method	Numerical method
Has analytical formula	Yes	Half	No
Can solve 1D problem	Yes	Yes	Yes
Can solve 2D problem	No	Yes	Yes
Can solve 3D problem	No	No	Yes
Speed	Very fast	Fast	Depends on grid size
Accuracy	Very accurate solution	Depend on each case	Approximate solution

Table 13.2 Comparison between different numerical methods

Method	FEM	FDM	FDTD	MOM
Domain	Frequency	Frequency	Time	Frequency
Multifrequency	No	No	Yes	No
Is arbitrary grid	Yes	No	No	Yes
PDE or IE	PDE	PDE	PDE	IE
Surface or volume	Volume	Volume	Volume	Surface
Is background modeled	Yes	Yes	Yes	No
High or low frequency	Both	Both	Prefer high	Both
Matrix system size	Large	Large	Large	Small
Matrix system density	Very sparse	Very sparse	Very sparse	Full
Is background need to be truncated	Yes	Yes	Yes	No
Best suited for	Inhomogeneous medium	Inhomogeneous medium	Inhomogeneous medium	Or piecewise homogeneous medium
Best suited for	Volume dominated problem	Volume dominated problem	Volume dominated problem	Surface dominated problem

to a deviated well which has more complicated geometry and properties, FEM is chosen to be used in this chapter.

FEM is a numerical technique for finding approximate solutions of partial differential equations, which is Maxwell's equation for EM modeling. FEM allows detailed and flexible visualization of complicated formation and different tool structures. The accuracy required and the associated computational time requirements can be managed simultaneously for most applications. It has become the most powerful and popular numerical method now in use. The specialties of the application of FEM to well logging problems discussed in this book include: the analysis of resistivity imaging tools, the simulation of laterolog tools, EM telemetry system and through casing resistivity tool. In this chapter, we will establish the basics of the FEM solution method to pave the way for the future analysis.

13.2 FINITE ELEMENT METHOD BASED ON MAGNETIC FIELD

Unlike analytical methods, numerical methods usually divide the space (and/or time) into small segments and establish relations between fields in each segment in the process of solving a partial differential equation with boundary conditions. By the segmentation, the solution becomes local at each element. The field in this element is usually defined at the geometric center of the element. When the number of elements is large enough for a limited space, the approximate solution approaches accurate solution. Theoretically, no matter how small the element is, the solution is not as accurate. However, for most engineering problems, as long as the accuracy is better than the manufacturing or measurement tolerance, the solution is satisfactory. In numerical simulation, there is always a contradictory story: accuracy versus the computation speed. Most cases in numerical simulation, we have to optimize the gridding scheme so that finer grids are arranged at the area where field changes rapidly, whereas course grids are assigned at the space where field changes slowly. In FEM, the Maxwell Equations are not solved directly, instead, an energy equation is derived and solved for either magnetic field (TM mode) or electric field (TE mode). Therefore the solution method is divided into two: magnetic field–based FEM and electric field–based FEM. We will discuss these two methods in details in the following sections in the frequency domain.

13.2.1 Magnetic field equations

In reference to Chapter 2, Fundamentals of Electromagnetic Fields Induction Logging Tools, if the EM fields are assumed to be time harmonic with an $e^{j\omega t}$ time variation, and be assumed to exist in a conductive medium, Faraday's and Ampere's laws become

$$\nabla \times \underline{E} = -j\omega \underline{\underline{\mu}} \cdot \underline{H} - \underline{M_s} \tag{13.1}$$

$$\nabla \times \underline{H} = j\omega \underline{\underline{\varepsilon}} \cdot \underline{E} + \underline{\underline{\sigma}} \cdot \underline{E} + \underline{J_s} \tag{13.2}$$

respectively, where $\underline{E}$ and $\underline{H}$ are the electric and magnetic fields, $\underline{M_s}$ and $\underline{J_s}$ are source electric and magnetic current densities, ω is angular frequency, $\underline{\underline{\mu}} = \mu_0 \underline{\underline{\mu_r}}$ and $\underline{\underline{\varepsilon}} = \varepsilon_0 \underline{\underline{\varepsilon_r}}$ are spatially varying permeability and permittivity, respectively and they are assumed to be dyadic functions to allow inhomogeneous and anisotropic mediums, $\underline{\underline{\sigma}} E$ describes the induced or eddy currents inside the formation and its value is determined by the formation conductivity $\underline{\underline{\sigma}}$ [4]. In well logging, formation conductivity $\underline{\underline{\sigma}}$ usually varies between 10^{-4} and 10 S/m. If the complex permittivity is defined as

$$\underline{\underline{\varepsilon_c}} = \underline{\underline{\varepsilon}} - j\frac{\underline{\underline{\sigma}}}{\omega} \tag{13.3}$$

Eq. (13.2) can be rewritten as

$$\nabla \times \underline{H} = j\omega \underline{\underline{\varepsilon_c}} \cdot \underline{E} + \underline{J_s} \tag{13.4}$$

Eq. (13.4) can also be expressed as

$$\underline{E} = \frac{1}{j\omega}\left(\underline{\underline{\varepsilon_c}}^{-1}\cdot\nabla\times\underline{H}\right) - \frac{1}{j\omega}\left(\underline{\underline{\varepsilon_c}}^{-1}\cdot\underline{J_s}\right) \qquad (13.5)$$

Take Eq. (13.5) into (13.1) and eliminate $\underline{E}$ to obtain magnetic field equation as

$$\frac{1}{j\omega}\nabla\times\left(\underline{\underline{\varepsilon_c}}^{-1}\cdot\nabla\times\underline{H}\right) + j\omega\underline{\underline{\mu}}\cdot\underline{H} = \frac{1}{j\omega}\nabla\times\left(\underline{\underline{\varepsilon_c}}^{-1}\cdot\underline{J_s}\right) - \underline{M_s} \qquad (13.6)$$

13.3 ANALYSIS OF TRANSVERSE ELECTRIC MODE AND TRANSVERSE MAGNETIC MODE

In cylindrical coordinates, both magnetic field $\underline{H}$ and electric field $\underline{E}$ have three components

$$\underline{H} = H_\rho\underline{\hat{\rho}} + H_z\underline{\hat{z}} + H_\phi\underline{\hat{\phi}} \qquad (13.7)$$

$$\underline{E} = E_\rho\underline{\hat{\rho}} + E_z\underline{\hat{z}} + E_\phi\underline{\hat{\phi}} \qquad (13.8)$$

where H_ϕ can generate E_ρ and E_z, E_ϕ can generate H_ρ and H_z, and these two set waves are denoted as transverse magnetic (TM) mode and transverse electric (TE) mode, respectively. To TM mode and TE mode, taking Eq. (A.1) to magnetic field equation (13.6) we obtain

$$\frac{1}{j\omega}\nabla\times\left(\underline{\underline{\varepsilon_c}}^{-1}\cdot\left(-\frac{\partial H_\phi}{\partial z}\underline{\hat{\rho}} + \frac{1}{\rho}\frac{\partial}{\partial\rho}(\rho H_\phi)\underline{\hat{z}}\right)\right) + j\omega\underline{\underline{\mu}}\cdot H_\phi\underline{\hat{\phi}} = \frac{1}{j\omega}\nabla\times\left(\underline{\underline{\varepsilon_c}}^{-1}\cdot\underline{J_s}\right) - \underline{M_s} \qquad (13.9)$$

$$\frac{1}{j\omega}\nabla\times\left(\underline{\underline{\varepsilon_c}}^{-1}\cdot\left(\frac{1}{\rho}\frac{\partial H_z}{\partial\phi}\underline{\hat{\rho}} + \underline{\hat{\phi}}\left(\frac{\partial H_\rho}{\partial z} - \frac{\partial H_z}{\partial\rho}\right) - \underline{\hat{z}}\frac{1}{\rho}\frac{\partial H_\rho}{\partial\phi}\right)\right) + j\omega\underline{\underline{\mu}}\cdot(H_\rho\underline{\hat{\rho}} + H_z\underline{\hat{z}})$$
$$= \frac{1}{j\omega}\nabla\times\left(\underline{\underline{\varepsilon_c}}^{-1}\cdot\underline{J_s}\right) - \underline{M_s} \qquad (13.10)$$

respectively.

If the following three assumptions are satisfied in a vertical well: (1) earth formations are axially symmetric (which means $\underline{\underline{\mu}}$ and $\underline{\underline{\varepsilon_c}}$ are axially symmetric); (2) sources are axially symmetric (which means $\underline{M_s}$ and $\underline{J_s}$ are axially symmetric); (3) the earth formations are composed of isotropic or transverse isotropic (TI) medium. The EM fields generated by the source are axially symmetric and the derivatives of $\underline{H}$ to $\hat{\phi}$ component are zero as

$$\frac{\partial H_\rho}{\partial \phi} = \frac{\partial H_\phi}{\partial \phi} = \frac{\partial H_z}{\partial \phi} = 0 \tag{13.11}$$

and the permeability and complex permittivity can be expressed as

$$\underline{\underline{\mu}} = \mu_h \hat{\rho}\hat{\rho} + \mu_h \hat{\phi}\hat{\phi} + \mu_v \hat{z}\hat{z} \tag{13.12}$$

$$\underline{\underline{\varepsilon_c}} = \varepsilon_{ch} \hat{\rho}\hat{\rho} + \varepsilon_{ch} \hat{\phi}\hat{\phi} + \varepsilon_{cv} \hat{z}\hat{z} \tag{13.13}$$

Using Eqs. (13.11)–(13.13) in (13.9) and (13.10), TM mode and TE mode of magnetic field equation (13.6) can be expressed as:

$$\left[-\frac{1}{j\omega} \frac{\partial}{\partial z} \left(\varepsilon_{ch}^{-1} \frac{\partial H_\phi}{\partial z} \right) - \frac{1}{j\omega} \frac{\partial}{\partial \rho} \left(\varepsilon_{cv}^{-1} \frac{1}{\rho} \frac{\partial}{\partial \rho} (\rho H_\phi) \right) + j\omega \mu_h \cdot H_\phi \right] \hat{\phi}$$
$$= \frac{1}{j\omega} \nabla \times \left(\underline{\underline{\varepsilon_c}}^{-1} \cdot \underline{J_s} \right) - \underline{M_s} \tag{13.14}$$

$$\left[-\frac{1}{j\omega} \frac{\partial}{\partial z} \left(\varepsilon_{ch}^{-1} \left(\frac{\partial H_\rho}{\partial z} - \frac{\partial H_z}{\partial \rho} \right) \right) + j\omega \mu_h H_\rho \right] \hat{\rho}$$
$$+ \left[\frac{1}{j\omega\rho} \frac{\partial}{\partial \rho} \left(\rho \varepsilon_{ch}^{-1} \left(\frac{\partial H_\rho}{\partial z} - \frac{\partial H_z}{\partial \rho} \right) \right) + j\omega \mu_v H_z \right] \hat{z} = \frac{1}{j\omega} \nabla \times \left(\underline{\underline{\varepsilon_c}}^{-1} \cdot \underline{J_s} \right) - \underline{M_s} \tag{13.15}$$

respectively. Eqs. (13.14) and (13.15) show the coupling between TM mode and TE mode is zero if the three assumptions listed above are satisfied. TM mode is generated if the source existing on the right side of Eq. (13.14) is in ϕ direction, and TE mode is generated if the source existing on the right side of Eq. (13.15) is in $\underline{\rho}$ direction or $\underline{z}$ direction. The relationship is shown in Table 13.3.

TM mode will be analyzed in this section using magnetic field equation and TE mode will be analyzed in Section 13.4 using electric field equation. Table 13.3 shows that for TM mode magnetic field can be expressed as

$$\underline{H} = H_\phi \hat{\phi} \tag{13.16}$$

Table 13.3 The relationship between source and the mode generated by source in vertical well

Sources	Fields
$M_{s\phi}\hat{\phi}$, $J_{s\rho}\hat{\rho}$, $J_{sz}\hat{z}$	TM mode: $H_{\phi}\hat{\phi}$, $E_{\rho}\hat{\rho}$, and $E_z\hat{z}$
$J_{s\phi}\hat{\phi}$, $M_{s\rho}\hat{\rho}$, $M_{sz}\hat{z}$	TE mode: $E_{\phi}\hat{\phi}$, $H_{\rho}\hat{\rho}$, and $H_z\hat{z}$

13.4 VECTOR MATRIX EQUATION OF MAGNETIC FIELD AND IMPEDANCE MATRIX

In a numerical solution, magnetic field $\underline{H}$ must be approximated in Eq. (13.6), and it will not be possible to satisfy the equality everywhere in the solution domain. Instead, the equality is enforced in the sense of a weighted average. This is done by requiring the equality of an inner product of both sides of Eq. (13.6) with a set of weighting or testing functions [4]. The inner product used in this text is defined as

$$\langle \underline{A}; \underline{B} \rangle_{\Omega} \equiv \int_{\Omega} \underline{A} \cdot \underline{B} d\Omega \tag{13.17}$$

This type of inner product is called a symmetric or pseudo-inner product. We first assume the availability of a suitable set of testing functions $\underline{\Omega}_m$, then multiply both sides of Eq. (13.6) by $\underline{\Omega}_m$, and integrate over whole solution domain Ω, Eq. (13.6) leads to

$$\frac{1}{j\omega}\left\langle \underline{\Omega}_m; \nabla \times \left(\underline{\underline{\varepsilon}_c}^{-1} \cdot \nabla \times \underline{H}\right)\right\rangle_{\Omega} + j\omega \left\langle \underline{\Omega}_m; \underline{\underline{\mu}} \cdot \underline{H} \right\rangle_{\Omega}$$
$$= \frac{1}{j\omega}\left\langle \underline{\Omega}_m; \nabla \times \left(\underline{\underline{\varepsilon}_c}^{-1} \cdot \underline{J_s}\right)\right\rangle_{\Omega} - \left\langle \underline{\Omega}_m; \underline{M_s} \right\rangle_{\Omega} \tag{13.18}$$

Using the identity $\nabla \cdot (\underline{A} \times \underline{B}) = \underline{B} \cdot (\nabla \times \underline{A}) - \underline{A} \cdot (\nabla \times \underline{B})$ and the divergence theorem $\iiint_V \nabla \cdot \underline{C} dV = \oiint_S \underline{C} \cdot dS$ we obtain

$$\langle \underline{B}, \nabla \times \underline{A} \rangle_{\Omega} = \langle \nabla \times \underline{B}, \underline{A} \rangle + \oiint_S (\underline{A} \times \underline{B}) \cdot \hat{n} dS \tag{13.19}$$

where S is the boundary of the solution domain Ω. If vector $\underline{A}$ is defined as

$$\underline{A} = \underline{\underline{\varepsilon}_c}^{-1} \nabla \times \underline{H} \tag{13.20}$$

Eq. (13.19) can be used to reduce the second-order derivative to a first-order derivative in Eq. (13.18). Taking Eqs. (13.19) and (13.20) into the first symmetric product of Eq. (13.18) can obtain

$$\left\langle \underline{\Omega}_m; \nabla \times \left(\underline{\underline{\varepsilon}_c}^{-1} \cdot \nabla \times \underline{H}\right)\right\rangle_{\Omega} = \left\langle \nabla \times \underline{\Omega}_m; \underline{\underline{\varepsilon}_c}^{-1} \cdot \nabla \times \underline{H} \right\rangle_{\Omega} + \oiint_S \left(\left(\underline{\underline{\varepsilon}_c}^{-1} \cdot \nabla \times \underline{H}\right) \times \underline{\Omega}_m\right) \cdot \hat{n} dS \tag{13.21}$$

Substituting Eq. (13.21) into (13.18) one can obtain the weak form of the magnetic Eq. (13.6) as

$$\frac{1}{j\omega}\left\langle \nabla\times\underline{\Omega}_m; \underline{\underline{\varepsilon}}_c^{-1}\cdot\nabla\times\underline{H}\right\rangle_\Omega + j\omega\left\langle \underline{\Omega}_m; \underline{\underline{\mu}}\cdot\underline{H}\right\rangle_\Omega$$
$$= \frac{1}{j\omega}\left\langle \underline{\Omega}_m; \nabla\times\left(\underline{\underline{\varepsilon}}_c^{-1}\cdot\underline{J}_s\right)\right\rangle_\Omega - \left\langle \underline{\Omega}_m; \underline{M}_s\right\rangle_\Omega - \frac{1}{j\omega}\oiint_S \left(\left(\underline{\underline{\varepsilon}}_c^{-1}\cdot\nabla\times\underline{H}\right)\times\underline{\Omega}_m\right)\cdot\hat{n}dS$$
(13.22)

with $m = 1, 2, \ldots, N$. In Eq. (13.22), the left side of the equality represents the relationship between magnetic field $\underline{H}$, right side represents sources existing inside the solution domain Ω or at the boundary S. Therefore by applying inner products as defined in Eq. (13.17), we converted the differential equation problem in (13.6) into an integral equation in (13.22).

If magnetic field $\underline{H}$ is expanded in the same set of basis function $\underline{\Omega}_n$ as used in testing the wave equation (this choice is known as Galerkin's method),

$$\underline{H} \approx \sum_{n=1}^{N} h_n \underline{\Omega}_n \qquad (13.23)$$

If we substitute Eq. (13.23) into the left part of Eq. (13.22),

$$\sum_{n=1}^{N}\left\{\frac{1}{j\omega}\left\langle \nabla\times\underline{\Omega}_m; \underline{\underline{\varepsilon}}_c^{-1}\cdot\nabla\times\underline{\Omega}_n\right\rangle_\Omega + j\omega\left\langle \underline{\Omega}_m; \underline{\underline{\mu}}\cdot\underline{\Omega}_n\right\rangle_\Omega\right\}h_n$$
$$= \frac{1}{j\omega}\left\langle \underline{\Omega}_m; \nabla\times\left(\underline{\underline{\varepsilon}}_c^{-1}\cdot\underline{J}_s\right)\right\rangle_\Omega - \left\langle \underline{\Omega}_m; \underline{M}_s\right\rangle_\Omega - \frac{1}{j\omega}\oiint_S \left(\left(\underline{\underline{\varepsilon}}_c^{-1}\cdot\nabla\times\underline{H}\right)\times\underline{\Omega}_m\right)\cdot\hat{n}dS$$
(13.24)

Eq. (13.24) can be expressed as

$$[Z_{m,n}][h_n] = [V_m] \qquad (13.25)$$

where

$$Z_{m,n} = \frac{1}{j\omega}\left\langle \nabla\times\underline{\Omega}_m; \underline{\underline{\varepsilon}}_c^{-1}\cdot\nabla\times\underline{\Omega}_n\right\rangle_\Omega + j\omega\left\langle \underline{\Omega}_m; \underline{\underline{\mu}}\cdot\underline{\Omega}_n\right\rangle_\Omega \qquad (13.26)$$

is defined as a global impedance matrix and

$$V_m = \frac{1}{j\omega}\left\langle \underline{\Omega}_m; \nabla\times\left(\underline{\underline{\varepsilon}}_c^{-1}\cdot\underline{J}_s\right)\right\rangle_\Omega - \left\langle \underline{\Omega}_m; \underline{M}_s\right\rangle_\Omega - \frac{1}{j\omega}\oiint_S \left(\left(\underline{\underline{\varepsilon}}_c^{-1}\cdot\nabla\times\underline{H}\right)\times\underline{\Omega}_m\right)\cdot\hat{n}dS$$
(13.27)

is a global vector. The value of global vector V_m is decided by different sources and will be discussed in the latter section of this chapter in detail.

According to the "Matrix Assembly Rule" (Appendix B), impedance global matrix $Z_{m,n}$ can be assembled by impedance element matrix $Z^e_{i,j}$. Eq. (13.26) can be expressed as the combination of reciprocal capacitance element matrix and inductance element matrix

$$Z^e_{i,j} = \frac{1}{j\omega}\left\langle \nabla \times \underline{\Omega}^e_i; \underline{\underline{\varepsilon}}_c^{-1} \cdot \nabla \times \underline{\Omega}^e_j \right\rangle_\Omega + j\omega \left\langle \underline{\Omega}^e_i; \underline{\underline{\mu}} \cdot \underline{\Omega}^e_j \right\rangle_\Omega \qquad (13.28)$$

$$Z^e_{i,j} = \frac{1}{j\omega}\Pi^e_{ij} + j\omega L^e_{ij} \qquad (13.29)$$

where reciprocal capacitance element matrix and inductance element matrix are

$$\Pi^e_{ij} = \left\langle \nabla \times \underline{\Omega}^e_i; \underline{\underline{\varepsilon}}_c^{-1} \cdot \nabla \times \underline{\Omega}^e_j \right\rangle_\Omega \qquad (13.30)$$

$$L^e_{ij} = \left\langle \underline{\Omega}^e_i; \underline{\underline{\mu}} \cdot \underline{\Omega}^e_j \right\rangle_\Omega \qquad (13.31)$$

respectively.

Eq. (13.28) can be evaluated numerically, but in this case, it may be evaluated analytically.

13.5 THE BASIS FUNCTIONS

As illustrated in Table 13.1, TM mode has only a ϕ directional magnetic field and it is axially symmetric, so the whole solution domain Ω can be considered as the revolution of ρz plane by the axis around the ϕ direction, as shown in Fig. 13.1A. Now the 3D problem can be simplified as revolution of 2D model. Fig. 13.1B shows the 2D model which is divided into triangular elements on ρz plane when $\phi = 0$ and Fig. 13.1C shows the 2D model which is divided into rectangular elements on ρz plane when $\phi = 0$. Element basis functions for triangular element and rectangular element will be introduced in Sections 13.5.1 and 13.5.2, respectively.

13.5.1 Triangular element basis functions

We first consider the element basis functions for the triangular element since it is the mostly used element in FEM analysis. Let a point in the triangle be designated by the position vector $\underline{r}$ with coordinates (ρ, z). As Fig. 13.2A shows, the point defines a subdivision of the triangle into three subtriangles [4]. The area of the subtriangle opposite

512 Theory of Electromagnetic Well Logging

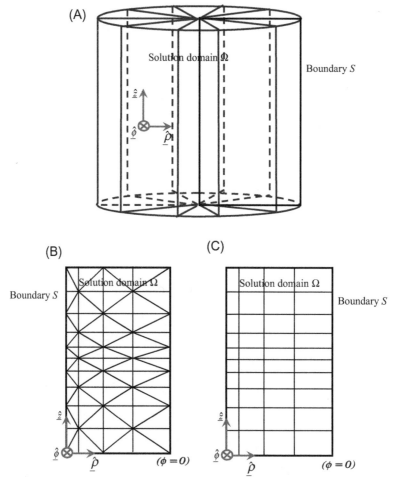

Figure 13.1 Solution domain Ω and boundary S. (A) Solution domain can be considered as the revolution of ρz plane by axis around ϕ direction. (B) Plane ρz ($\phi = 0$) is subdivided into triangular elements. (C) Plane ρz ($\phi = 0$) is subdivided into rectangular elements.

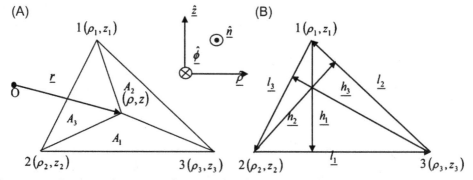

Figure 13.2 Geometrical quantities for triangular element. (A) Subdivision of a triangle into three subareas defining normalized area coordinates. (B) Edge and height vectors defined on a triangle.

vertex i has area A_i. Element basis functions for triangular element are then defined and satisfy the following conditions

$$\xi_i = \frac{A_i}{A_e}, \quad i = 1, 2, 3 \tag{13.32}$$

$$\xi_1 + \xi_2 + \xi_3 = 1 \tag{13.33}$$

where A_e is the area of element e. Points in a triangle may be represented by linearly interpolating its vertex coordinates:

$$\underline{r} = \underline{r}_1^e \xi_1 + \underline{r}_2^e \xi_2 + \underline{r}_3^e \xi_3 \tag{13.34}$$

$$\rho = \hat{\underline{\rho}} \cdot \underline{r} = \hat{\underline{\rho}} \cdot (\underline{r}_1^e \xi_1 + \underline{r}_2^e \xi_2 + \underline{r}_3^e \xi_3) = \rho_1^e \xi_1 + \rho_2^e \xi_2 + \rho_3^e \xi_3 \tag{13.35}$$

$$z = \hat{\underline{z}} \cdot \underline{r} = \hat{\underline{z}} \cdot (\underline{r}_1^e \xi_1 + \underline{r}_2^e \xi_2 + \underline{r}_3^e \xi_3) = z_1^e \xi_1 + z_2^e \xi_2 + z_3^e \xi_3 \tag{13.36}$$

Fig. 13.2B shows the position of vertex i, edge vector $\underline{l}_i$, and height vector $\underline{h}_i$. Notice that the triangular element used in this project is defined counterclockwise and exists in the $\rho - z$ plane, the triangle's unit normal is $\hat{n} = -\hat{\phi}$. Table 13.4 summarizes the computation of an element triangle's parameters [4].

The integral over the triangular element can be expressed as

$$\iint_{S_e} f(\underline{r}) dS = 2A_e \int_0^1 \int_0^{1-\xi_{i+1}} f(\underline{r}_1 \xi_1 + \underline{r}_2 \xi_2 + \underline{r}_3 \xi_3) d\xi_{i-1} d\xi_{i+1} \tag{13.37}$$

Table 13.4 Geometrical quantities defined on triangular elements

Edge vectors	$\underline{l}_i = \underline{r}_{i-1} - \underline{r}_{i+1} = (\rho_{i-1} - \rho_{i+1})\hat{\underline{\rho}} + (z_{i-1} - z_{i+1})\hat{\underline{z}}; \ l_i = \|\underline{l}_i\|; \ \hat{\underline{l}}_i = \frac{\underline{l}_i}{l_i}, \quad i = 1, 2, 3$
Area	$A_e = \frac{\|\underline{l}_{i-1} \times \underline{l}_{i+1}\|}{2} = \frac{(\rho_3 - \rho_2)(z_1 - z_3) - (z_3 - z_2)(\rho_1 - \rho_3)}{2}$
Height vectors	$\hat{\underline{h}}_i = \hat{\phi} \times \hat{\underline{l}}_i = \frac{-(\rho_{i-1} - \rho_{i+1})\hat{\underline{z}} + (z_{i-1} - z_{i+1})\hat{\underline{\rho}}}{l_i}, \quad i = 1, 2, 3;$ $h_i = \frac{2A_e}{l_i}; \ \underline{h}_i = h_i \hat{\underline{h}}_i$
Coordinates	$\xi_i = \frac{(\rho_{i-1} - \rho)(z_{i-1} - z_{i+1}) - (\rho_{i-1} - \rho_{i+1})(z_{i-1} - z)}{2A_e}$
Coordinates gradients	$\nabla \xi_i = -\frac{\hat{\underline{h}}_i}{h_i} = \frac{(\rho_{i-1} - \rho_{i+1})\hat{\underline{z}} - (z_{i-1} - z_{i+1})\hat{\underline{\rho}}}{2A_e}, \quad i = 1, 2, 3$

where

$$dS = 2A_e d\xi_{i-1} d\xi_{i+1} \tag{13.38}$$

13.5.2 Rectangular element basis functions

Compared to triangular basis function, rectangular basis function is relatively simple. Now let us discuss the element basis function for the rectangular element. Let a point with coordinates (ρ, z) divide the rectangle which has four vertexes located at coordinates (ρ_i, z_i) ($i = 1, 2, 3, 4$). As Fig. 13.3 shows, the point defines a subdivision of the rectangle into four subrectangles. The area of the subrectangle opposite vertex i has area S_i. Element basis functions are defined as and satisfy

$$\theta_i = \frac{S_i}{S_e}, \quad i = 1, 2, 3, 4 \tag{13.39}$$

$$\theta_1 + \theta_2 + \theta_3 + \theta_4 = 1 \tag{13.40}$$

$$\frac{\theta_1}{\theta_3} = \frac{\theta_2}{\theta_4} \tag{13.41}$$

where S_e is the area of rectangular element e [5].

If the center point coordinates are defined as $\rho_0 = \frac{\rho_1 + \rho_2}{2}$ and $z_0 = \frac{z_1 + z_3}{2}$, and the width and length of the element are defined as $\Delta \rho = \frac{\rho_2 - \rho_1}{2}$ and $\Delta z = \frac{z_3 - z_1}{2}$, then element basis function for the rectangular element can be expressed as

$$\theta_i = \frac{1}{A_e} \left((-1)^{\text{int}(\frac{i+1}{2})} (z - z_0) + \frac{\Delta z}{2} \right) \left((-1)^i (\rho - \rho_0) + \frac{\Delta \rho}{2} \right) \tag{13.42}$$

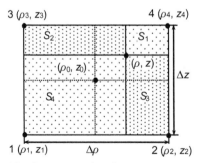

Figure 13.3 Geometrical quantities for rectangular element.

The derivative of basis function to ρ and z coordinates can be also obtained as

$$\frac{\partial \theta_i}{\partial z} = \frac{1}{A_e}(-1)^{\text{int}\left(\frac{i+1}{2}\right)}\left((-1)^i(\rho - \rho_0) + \frac{\Delta\rho}{2}\right) \quad (13.43)$$

$$\frac{\partial \theta_i}{\partial \rho} = \frac{1}{A_e}(-1)^i\left((-1)^{\text{int}\left(\frac{i+1}{2}\right)}(z - z_0) + \frac{\Delta z}{2}\right) \quad (13.44)$$

$$\frac{\partial}{\partial \rho}(\rho\theta_i) = \frac{1}{A_e}\left((-1)^{\text{int}\left(\frac{i+1}{2}\right)}(z - z_0) + \frac{\Delta z}{2}\right)\left((-1)^i(2\rho - \rho_0) + \frac{\Delta\rho}{2}\right) \quad (13.45)$$

13.6 EVALUATION OF IMPEDANCE ELEMENT MATRIX FOR RECTANGULAR ELEMENT BASED ON H_ϕ

Since $\underline{\Omega}_n$ is the testing function and also the basis function representing magnetic field $\underline{H}$, it has the same $\hat{\phi}$ direction as $\underline{H}$. The same $\underline{\Omega}_i^e$ has only $\hat{\phi}$ direction. If it is defined as

$$\underline{\Omega}_i^e = \theta_i\hat{\phi} \quad (13.46)$$

where θ_i is the basis function for rectangular element (discussed in Section 13.7), the analysis is based on H_ϕ.

From Appendix A, Eq. (A.1) can obtain the curl of $\underline{\Omega}_i^e$ as

$$\nabla \times \underline{\Omega}_i^e = \nabla \times \theta_i\hat{\phi} = \hat{\rho}\left(-\frac{\partial \theta_i}{\partial z}\right) + \hat{z}\left(\frac{1}{\rho}\frac{\partial}{\partial \rho}(\rho\theta_i)\right) \quad (13.47)$$

First define item 1, item 2, item 3, and item 4 as

$$\text{Item 1} = \int_{\rho_{\min}}^{\rho_{\max}} \rho\left((-1)^i(\rho - \rho_0) + \frac{\Delta\rho}{2}\right)\left((-1)^j(\rho - \rho_0) + \frac{\Delta\rho}{2}\right)d\rho$$

$$= \left(\frac{3 + (-1)^i(-1)^j}{12}\rho_0 + \frac{(-1)^i + (-1)^j}{24}\Delta\rho\right)(\Delta\rho)^3 \quad (13.48)$$

$$\text{Item 2} = \int_{z_{\min}}^{z_{\max}} \left((-1)^{\text{int}\left(\frac{i+1}{2}\right)}(z - z_0) + \frac{\Delta z}{2}\right)\left((-1)^{\text{int}\left(\frac{j+1}{2}\right)}(z - z_0) + \frac{\Delta z}{2}\right)dz$$

$$= \frac{(-1)^{\text{int}\left(\frac{i+1}{2}\right)}(-1)^{\text{int}\left(\frac{j+1}{2}\right)} + 3}{12}(\Delta z)^3$$

$$(13.49)$$

$$\text{Item 3} = \int_{\rho_{\min}}^{\rho_{\max}} \frac{1}{\rho}\left((-1)^i(2\rho-\rho_0)+\frac{\Delta\rho}{2}\right)\left((-1)^j(2\rho-\rho_0)+\frac{\Delta\rho}{2}\right)d\rho$$

$$= ((-1)^i+(-1)^j)(\Delta\rho)^2 + \left((-1)^i\rho_0-\frac{\Delta\rho}{2}\right)\left((-1)^j\rho_0-\frac{\Delta\rho}{2}\right)\ln\left(\frac{\rho_{\max}}{\rho_{\min}}\right) \quad (13.50)$$

$$\text{Item 4} = \int_{\rho_{\min}}^{\rho_{\max}} \frac{1}{\rho}\left((-1)^i(\rho-\rho_0)+\frac{\Delta\rho}{2}\right)\left((-1)^j(\rho-\rho_0)+\frac{\Delta\rho}{2}\right)d\rho$$

$$= ((-1)^i+(-1)^j)(\Delta\rho)^2/2 - (-1)^i(-1)^j\rho_0\Delta\rho \quad (13.51)$$

$$+ \left((-1)^i\rho_0-\frac{\Delta\rho}{2}\right)\left((-1)^j\rho_0-\frac{\Delta\rho}{2}\right)\ln\left(\frac{\rho_{\max}}{\rho_{\min}}\right)$$

Then impedance element matrix (13.29) will be analyzed. Inductance element matrix can be obtained as

$$L_{i,j}^e = \left\langle \underline{\Omega_i^e}; \underline{\underline{\mu}} \cdot \underline{\Omega_j^e} \right\rangle_\Omega$$

$$= \iiint_V \left(\theta_i\hat{\underline{\phi}}\right)\cdot\left(\underline{\underline{\mu}}\cdot\theta_j\hat{\underline{\phi}}\right)dV$$

$$= \int_{z_{\min}}^{z_{\max}}\int_{\rho_{\min}}^{\rho_{\max}} \left(\theta_i\hat{\underline{\phi}}\right)\cdot\left(\mu_h\hat{\underline{\phi}}\hat{\underline{\phi}}\bullet\theta_j\hat{\underline{\phi}}\right)2\pi\rho d\rho dz$$

$$= 2\pi\mu_h \int_{z_{\min}}^{z_{\max}}\int_{\rho_{\min}}^{\rho_{\max}} \rho\theta_i\theta_j d\rho dz$$

$$= \left\{ \begin{array}{l} \dfrac{2\pi\mu_h}{(Ae)^2}\int_{z_{\min}}^{z_{\max}} \left((-1)^{\mathrm{int}\left(\frac{i+1}{2}\right)}(z-z_0)+\dfrac{\Delta z}{2}\right)\left((-1)^{\mathrm{int}\left(\frac{j+1}{2}\right)}(z-z_0)+\dfrac{\Delta z}{2}\right)dz \cdot \\[2ex] \int_{\rho_{\min}}^{\rho_{\max}} \rho\left((-1)^i(\rho-\rho_0)+\dfrac{\Delta\rho}{2}\right)\left((-1)^j(\rho-\rho_0)+\dfrac{\Delta\rho}{2}\right)d\rho \end{array} \right\}$$

$$= \frac{2\pi\mu_h}{(Ae)^2}\frac{(-1)^{\mathrm{int}\left(\frac{i+1}{2}\right)}(-1)^{\mathrm{int}\left(\frac{j+1}{2}\right)}+3}{12}(\Delta z)^3\left(\frac{3+(-1)^i(-1)^j}{12}\rho_0+\frac{(-1)^i+(-1)^j}{24}\Delta\rho\right)(\Delta\rho)^3$$

$$= \frac{2\pi\mu_h}{(Ae)^2}*\text{item 2}*\text{item 1}$$

$$(13.52)$$

Then reciprocal capacitance element matrix is calculated as

$$\Pi_{ij}^e = \left\langle \nabla \times \underline{\Omega}_i^e; \underline{\underline{\varepsilon}}_c^{-1} \cdot \nabla \times \underline{\Omega}_j^e \right\rangle_\Omega$$

$$= \iiint_v \left(\nabla \times \underline{\Omega}_i^e \right) \cdot \left(\underline{\underline{\varepsilon}}_c^{-1} \cdot \nabla \times \underline{\Omega}_j^e \right) dV$$

$$= \iint_s \left(\hat{\underline{\rho}} \left(-\frac{\partial \Lambda_i^e}{\partial z} \right) + \hat{\underline{z}} \left(\frac{1}{\rho} \frac{\partial}{\partial \rho} (\rho \Lambda_i^e) \right) \right) \cdot \left(\underline{\underline{\varepsilon}}_c^{-1} \cdot \left(\hat{\underline{\rho}} \left(-\frac{\partial \Lambda_j^e}{\partial z} \right) + \hat{\underline{z}} \left(\frac{1}{\rho} \frac{\partial}{\partial \rho} (\rho \Lambda_j^e) \right) \right) \right) * 2\pi \rho \, dS$$

$$= \iint_s \left(\hat{\underline{\rho}} \left(-\frac{\partial \Lambda_i^e}{\partial z} \right) + \hat{\underline{z}} \left(\frac{1}{\rho} \frac{\partial}{\partial \rho} (\rho \Lambda_j^e) \right) \right) \cdot \left(\hat{\underline{\rho}} \left(-\frac{1}{\varepsilon_{ch}} \frac{\partial \Lambda_i^e}{\partial z} \right) + \hat{\underline{z}} \left(\frac{1}{\varepsilon_{cv}} \frac{1}{\rho} \frac{\partial}{\partial \rho} (\rho \Lambda_j^e) \right) \right) * 2\pi \rho \, dS$$

$$= 2\pi \int_{z_{min}}^{z_{max}} \int_{\rho_{min}}^{\rho_{max}} \left(\frac{\rho}{\varepsilon_{ch}} \frac{\partial \Lambda_i^e}{\partial z} \frac{\partial \Lambda_j^e}{\partial z} + \frac{1}{\varepsilon_{cv}} \frac{1}{\rho} \frac{\partial}{\partial \rho} (\rho \Lambda_i^e) \frac{\partial}{\partial \rho} (\rho \Lambda_j^e) \right) d\rho \, dz$$

$$= \begin{cases} \dfrac{2\pi(-1)^{\text{int}\left(\frac{i+1}{2}\right)}(-1)^{\text{int}\left(\frac{j+1}{2}\right)}}{(A_e)^2 \varepsilon_{ch}} \int_{z_{min}}^{z_{max}} dz \int_{\rho_{min}}^{\rho_{max}} \rho \left((-1)^i (\rho - \rho_0) + \dfrac{\Delta \rho}{2} \right) \\ \left((-1)^j (\rho - \rho_0) + \dfrac{\Delta \rho}{2} \right) d\rho \\ + \dfrac{2\pi}{(A_e)^2 \varepsilon_{cv}} \int_{z_{min}}^{z_{max}} \left((-1)^{\text{int}\left(\frac{i+1}{2}\right)} (z - z_0) + \dfrac{\Delta z}{2} \right) \left((-1)^{\text{int}\left(\frac{j+1}{2}\right)} (z - z_0) + \dfrac{\Delta z}{2} \right) dz \cdot \\ \int_{\rho_{min}}^{\rho_{max}} \dfrac{1}{\rho} \left((-1)^i (2\rho - \rho_0) + \dfrac{\Delta \rho}{2} \right) \left((-1)^j (2\rho - \rho_0) + \dfrac{\Delta \rho}{2} \right) d\rho \end{cases}$$

$$= \begin{cases} \dfrac{2\pi(-1)^{\text{int}\left(\frac{i+1}{2}\right)}(-1)^{\text{int}\left(\frac{j+1}{2}\right)}}{(A_e)^2 \varepsilon_{ch}} \Delta z \left(\dfrac{3 + (-1)^i(-1)^j}{12} \rho_0 + \dfrac{(-1)^i + (-1)^j}{24} \Delta \rho \right) (\Delta \rho)^3 \\ + \dfrac{2\pi}{(A_e)^2 \varepsilon_{cv}} \dfrac{(-1)^{\text{int}\left(\frac{i+1}{2}\right)}(-1)^{\text{int}\left(\frac{j+1}{2}\right)} + 3}{12} (\Delta z)^3 \cdot \\ \left(((-1)^i + (-1)^j)(\Delta \rho)^2 + \left((-1)^i \rho_0 - \dfrac{\Delta \rho}{2} \right) \left((-1)^j \rho_0 - \dfrac{\Delta \rho}{2} \right) \ln \left(\dfrac{\rho_{max}}{\rho_{min}} \right) \right) \end{cases}$$

$$= \frac{2\pi(-1)^{\text{int}\left(\frac{i+1}{2}\right)}(-1)^{\text{int}\left(\frac{j+1}{2}\right)}}{(A_e)^2 \varepsilon_{ch}} * \Delta z * \text{item 1} + \frac{2\pi}{(A_e)^2 \varepsilon_{cv}} * \text{item 2} * \text{item 3}$$

(13.53)

Fortran computer program is shown in Appendix C (Section 13.6).

13.7 EVALUATION OF IMPEDANCE ELEMENT MATRIX FOR RECTANGULAR ELEMENT BASED ON ρH_ϕ

Since $\underline{\Omega}_n$ is the testing function and also the basis function representing magnetic field $\underline{H}$, it has the same $\hat{\phi}$ direction as $\underline{H}$. The same $\underline{\Omega}_i^e$ has only $\hat{\phi}$ direction. If it is defined as

$$\rho \underline{\Omega}_i^e = \theta_i \hat{\phi} \tag{13.54}$$

where θ_i is basis function for rectangular element, the analysis is based on ρH_φ as

$$(\rho H) \approx \sum_{n=1}^{N} (\rho h)_n \theta_n \tag{13.55}$$

$$[Z_{m,n}][(\rho h)_n] = [V_m] \tag{13.56}$$

From Appendix A, Eq. (A.1) can obtain the curl of $\underline{\Omega}_i^e$ as

$$\nabla \times \underline{\Omega}_i^e = \nabla \times \left(\frac{\theta_i}{\rho}\hat{\phi}\right) = \hat{\rho}\left(-\frac{1}{\rho}\frac{\partial \theta_i}{\partial z}\right) + \hat{z}\left(\frac{1}{\rho}\frac{\partial}{\partial \rho}(\theta_i)\right) \tag{13.57}$$

Impedance element in Eq. (13.29) will be analyzed. First we find inductance element matrix as

$$\begin{aligned}
L_{i,j}^e &= \left\langle \underline{\Omega}_i^e; \underline{\underline{\mu}} \cdot \underline{\Omega}_j^e \right\rangle_\Omega \\
&= \iiint_v \left(\frac{\theta_i}{\rho}\hat{\phi}\right) \cdot \left(\underline{\underline{\mu}} \cdot \frac{\theta_i}{\rho}\hat{\phi}\right) dV \\
&= \int_{z_{min}}^{z_{max}} \int_{\rho_{min}}^{\rho_{max}} \left(\frac{\theta_i}{\rho}\hat{\phi}\right) \cdot \left(\mu_h \hat{\phi}\hat{\phi} \cdot \frac{\theta_i}{\rho}\hat{\phi}\right) 2\pi \rho d\rho dz \\
&= 2\pi \mu_h \int_{z_{min}}^{z_{max}} \int_{\rho_{min}}^{\rho_{max}} \frac{1}{\rho} \theta_i^e \theta_j^e d\rho dz \\
&= \begin{cases} \frac{2\pi \mu_h}{(A_e)^2} \int_{z_{min}}^{z_{max}} \left((-1)^{\text{int}\left(\frac{i+1}{2}\right)}(z-z_0) + \frac{\Delta z}{2}\right)\left((-1)^{\text{int}\left(\frac{j+1}{2}\right)}(z-z_0) + \frac{\Delta z}{2}\right) dz \cdot \\ \int_{\rho_{min}}^{\rho_{max}} \frac{1}{\rho}\left((-1)^i(\rho-\rho_0) + \frac{\Delta \rho}{2}\right)\left((-1)^j(\rho-\rho_0) + \frac{\Delta \rho}{2}\right) d\rho \end{cases} \\
&= \begin{cases} \frac{2\pi \mu_h (-1)^{\text{int}\left(\frac{i+1}{2}\right)}(-1)^{\text{int}\left(\frac{j+1}{2}\right)} + 3}{(A_e)^2} \cdot \frac{12}{12}(\Delta z)^3 \cdot \\ \left(((-1)^i + (-1)^j)(\Delta \rho)^2/2 - (-1)^i(-1)^j \rho_0 \Delta \rho \right. \\ \left. + \left((-1)^i \rho_0 - \frac{\Delta \rho}{2}\right)\left((-1)^j \rho_0 - \frac{\Delta \rho}{2}\right) \ln\left(\frac{\rho_{max}}{\rho_{min}}\right)\right) \end{cases} \\
&= \frac{2\pi \mu_h}{(A_e)^2} \text{item 2} * \text{item 4}
\end{aligned} \tag{13.58}$$

Then reciprocal capacitance element matrix is calculated as

$$\Pi^e_{ij} = \left\langle \nabla \times \underline{\Omega}^e_i; \underline{\underline{\varepsilon}}_c^{-1} \cdot \nabla \times \underline{\Omega}^e_j \right\rangle_\Omega$$

$$= \iiint_V (\nabla \times \underline{\Omega}^e_i) \cdot (\underline{\underline{\varepsilon}}_c^{-1} \cdot \nabla \times \underline{\Omega}^e_j) dV$$

$$= \iint_S \left(\hat{\rho}\left(-\frac{1}{\rho}\frac{\partial \theta_i}{\partial z}\right) + \hat{z}\left(\frac{1}{\rho}\frac{\partial}{\partial \rho}(\theta_i)\right) \right) \cdot \left(\underline{\underline{\varepsilon}}_c^{-1} \cdot \left(\hat{\rho}\left(-\frac{1}{\rho}\frac{\partial \theta_i}{\partial z}\right) + \hat{z}\left(\frac{1}{\rho}\frac{\partial}{\partial \rho}(\theta_i)\right) \right) \right) * 2\pi\rho\, dS$$

$$= \iint_S \left(\hat{\rho}\left(-\frac{1}{\rho}\frac{\partial \theta_i}{\partial z}\right) + \hat{z}\left(\frac{1}{\rho}\frac{\partial}{\partial \rho}(\theta_i)\right) \right) \cdot \left(\hat{\rho}\left(-\frac{1}{\varepsilon_{ch}}\frac{1}{\rho}\frac{\partial \theta_i}{\partial z}\right) + \hat{z}\left(\frac{1}{\varepsilon_{cv}}\frac{1}{\rho}\frac{\partial}{\partial \rho}(\theta_i)\right) \right) * 2\pi\rho\, dS$$

$$= 2\pi \int_{z_{min}}^{z_{max}} \int_{\rho_{min}}^{\rho_{max}} \left(\frac{1}{\varepsilon_{ch}}\frac{1}{\rho}\frac{\partial \theta^e_i}{\partial z}\frac{\partial \theta^e_j}{\partial z} + \frac{1}{\varepsilon_{cv}}\frac{1}{\rho}\frac{\partial \theta^e_i}{\partial \rho}\frac{\partial \theta^e_j}{\partial \rho} \right) d\rho dz$$

$$= \begin{cases} \dfrac{2\pi(-1)^{\text{int}\left(\frac{i+1}{2}\right)}(-1)^{\text{int}\left(\frac{j+1}{2}\right)}}{(A_e)^2 \varepsilon_{ch}} \int_{z_{min}}^{z_{max}} dz \int_{\rho_{min}}^{\rho_{max}} \frac{1}{\rho}\left((-1)^i(\rho-\rho_0) + \frac{\Delta\rho}{2}\right) \\ \left((-1)^j(\rho-\rho_0) + \frac{\Delta\rho}{2}\right) d\rho \\ + \dfrac{2\pi(-1)^i(-1)^j}{(A_e)^2 \varepsilon_{cv}} \int_{z_{min}}^{z_{max}} \left((-1)^{\text{int}\left(\frac{i+1}{2}\right)}(z-z_0) + \frac{\Delta z}{2}\right) \\ \left((-1)^{\text{int}\left(\frac{j+1}{2}\right)}(z-z_0) + \frac{\Delta z}{2}\right) dz \int_{\rho_{min}}^{\rho_{max}} \frac{1}{\rho} d\rho \end{cases}$$

$$= \begin{cases} \dfrac{2\pi(-1)^{\text{int}\left(\frac{i+1}{2}\right)}(-1)^{\text{int}\left(\frac{j+1}{2}\right)}}{(A_e)^2 \varepsilon_{ch}} \Delta z \left(((-1)^i + (-1)^j)(\Delta\rho)^2/2 - (-1)^i(-1)^j \rho_0 \Delta\rho + \left((-1)^i \rho_0 - \frac{\Delta\rho}{2}\right)\left((-1)^j \rho_0 - \frac{\Delta\rho}{2}\right) \ln\left(\frac{\rho_{max}}{\rho_{min}}\right) \right) \\ + \dfrac{2\pi(-1)^i(-1)^j(-1)^{\text{int}\left(\frac{i+1}{2}\right)}(-1)^{\text{int}\left(\frac{j+1}{2}\right)} + 3}{(A_e)^2 \varepsilon_{cv}} \dfrac{1}{12}(\Delta z)^3 \ln\left(\frac{\rho_{max}}{\rho_{min}}\right) \end{cases}$$

$$= \frac{2\pi(-1)^{\text{int}\left(\frac{i+1}{2}\right)}(-1)^{\text{int}\left(\frac{j+1}{2}\right)}}{(A_e)^2 \varepsilon_{ch}} * \Delta z * \text{item 4} + \frac{2\pi(-1)^i(-1)^j}{(A_e)^2 \varepsilon_{cv}} * \text{item 2} * \ln\left(\frac{\rho_{max}}{\rho_{min}}\right) \tag{13.59}$$

Fortran computer program is shown in Appendix D (Section 13.7).

13.8 EVALUATION OF IMPEDANCE ELEMENT MATRIX FOR TRIANGULAR ELEMENT BASED ON H_ϕ

Since $\underline{\Omega}_n$ is the testing function and also the basis function representing magnetic field $\underline{H}$, it has the same $\hat{\phi}$ direction as $\underline{H}$. The same $\underline{\Omega}_i^e$ has only $\hat{\phi}$ direction. If it is defined as

$$\underline{\Omega}_i^e = \xi_i \hat{\phi} \tag{13.60}$$

where ξ_i is basis function for triangular element, the analysis is based on H_ϕ (Fig. 13.4).

It is noted that H_ϕ is everywhere continuous in solution domain Ω since it is tangent to boundary S. Both continuity and the differentiability requirement of the model may be simultaneously realized if piecewise linear (PWL) representation is chosen for H_ϕ [4]. Scalar basis functions defined from $\underline{\Omega}_n = \Lambda_n \hat{\phi}$ are chosen to be pyramidal interpolation functions and PWL approximation may be represented as a linear combination of a set of ξ, as shown in Ref. [6].

13.8.1 Formulation of element matrix

According to Eq. (A.5), the curl of $\underline{\Omega}_i^e$ can be modeled into lower order as

$$\begin{aligned}
\nabla \times \underline{\Omega}_i^e &= \nabla \times \xi_i \hat{\phi} = (\nabla \xi_i) \times \hat{\phi} + (\nabla \times \hat{\phi}) \cdot \xi_i \\
&= \left(-\frac{\hat{h}_i}{h_i}\right) \times \hat{\phi} + \hat{\underline{z}} \frac{1}{\rho} \cdot \xi_i \\
&= -\frac{\underline{l}_i}{h_i l_i} + \hat{\underline{z}} \frac{1}{\rho} \cdot \xi_i \\
&= \hat{\underline{\rho}} \frac{\rho_{i+1} - \rho_{i-1}}{2A_e} + \hat{\underline{z}} \frac{Z_{i+1} - Z_{i-1}}{2A_e} + \hat{\underline{z}} \frac{\xi_i}{\rho}
\end{aligned} \tag{13.61}$$

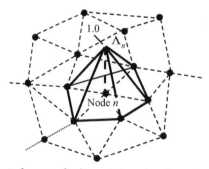

Figure 13.4 Illustration of basis function for linear triangular elements.

The impedance element (13.29) will be analyzed. First take the inductance element matrix as

$$L_{i,j}^e = \left\langle \underline{\Omega}_i^e; \underline{\underline{\mu}} \cdot \underline{\Omega}_j^e \right\rangle_\Omega$$
$$= \iiint_V \left(\Lambda_i^e \underline{\hat{\phi}} \right) \cdot \left(\underline{\underline{\mu}} \cdot \Lambda_j^e \underline{\hat{\phi}} \right) dV$$
$$= \iiint_V \left(\Lambda_i^e \underline{\hat{\phi}} \right) \cdot \left(\mu_h \underline{\hat{\phi}} \underline{\hat{\phi}} \cdot \Lambda_j^e \underline{\hat{\phi}} \right) dV$$
$$= \mu_h \iint_S \Lambda_i^e \Lambda_j^e 2\pi \rho \, dS \qquad (13.62)$$
$$= 2\pi \mu_h \iint_S \xi_i \xi_j (\rho_1 \xi_1 + \rho_2 \xi_2 + \rho_3 \xi_3) dS$$
$$= 2\pi \mu_h \iint_S (\rho_1 \xi_1 \xi_i \xi_j + \rho_2 \xi_2 \xi_i \xi_j + \rho_3 \xi_3 \xi_i \xi_j) dS$$
$$= 2\pi \mu_h \text{ Term 5}$$

Then take the reciprocal capacitance element matrix as

$$\Pi_{ij}^e = \left\langle \nabla \times \underline{\Omega}_i^e; \underline{\underline{\varepsilon}}_c^{-1} \cdot \nabla \times \underline{\Omega}_j^e \right\rangle_\Omega$$
$$= \iiint_V (\nabla \times \underline{\Omega}_i^e) \cdot (\underline{\underline{\varepsilon}}_c^{-1} \cdot \nabla \times \underline{\Omega}_j^e) dV$$
$$= \iint_S \left(\hat{\rho} \frac{\rho_{i+1} - \rho_{i-1}}{2A_e} + \hat{z} \frac{Z_{i+1} - Z_{i-1}}{2A_e} + \hat{z} \frac{\xi_i}{\rho} \right) \cdot \left(\underline{\underline{\varepsilon}}_c^{-1} \cdot \left(\hat{\rho} \frac{\rho_{j+1} - \rho_{j-1}}{2A_e} + \hat{z} \frac{Z_{j+1} - Z_{j-1}}{2A_e} + \hat{z} \frac{\xi_j}{\rho} \right) \right) * 2\pi \rho \, dS$$
$$= \iint_S \left(\hat{\rho} \frac{\rho_{i+1} - \rho_{i-1}}{2A_e} + \hat{z} \frac{Z_{i+1} - Z_{i-1}}{2A_e} + \hat{z} \frac{\xi_i}{\rho} \right) \cdot \left(\hat{\rho} \varepsilon_{ch}^{-1} \frac{\rho_{j+1} - \rho_{j-1}}{2A_e} + \hat{z} \varepsilon_{cv}^{-1} \frac{Z_{j+1} - Z_{j-1}}{2A_e} + \hat{z} \varepsilon_{cv}^{-1} \frac{\xi_j}{\rho} \right) * 2\pi \rho \, dS$$
$$= 2\pi \iint_S \left(\frac{(\rho_{i+1} - \rho_{i-1})(\rho_{j+1} - \rho_{j-1})}{4 A_e^2 \varepsilon_{ch}} \rho + \frac{(Z_{i+1} - Z_{i-1})(Z_{j+1} - Z_{j-1})}{4 A_e^2 \varepsilon_{cv}} \rho + \frac{Z_{i+1} - Z_{i-1}}{2 A_e \varepsilon_{cv}} \xi_j + \frac{Z_{j+1} - Z_{j-1}}{2 A_e \varepsilon_{cv}} \xi_i + \frac{1}{\varepsilon_{cv}} \frac{\xi_i \xi_j}{\rho} \right) dS$$
$$= 2\pi \left(\frac{1}{\varepsilon_{ch}} \text{Term 1a} + \frac{1}{\varepsilon_{cv}} \text{Term 1b} + \frac{1}{\varepsilon_{cv}} \text{Term 2} + \frac{1}{\varepsilon_{cv}} \text{Term 3} + \frac{1}{\varepsilon_{cv}} \text{Term 4} \right)$$
$$(13.63)$$

Using Eqs. (13.62) and (13.63) can get impedance element matrix

$$\left[Z_{i,j}^e\right] = \frac{2\pi}{j\omega}\left(\frac{1}{\varepsilon_{ch}}\text{Term 1a} + \frac{1}{\varepsilon_{cv}}\text{Term 1b} + \frac{1}{\varepsilon_{cv}}\text{Term 2} + \frac{1}{\varepsilon_{cv}}\text{Term 3} + \frac{1}{\varepsilon_{cv}}\text{Term 4}\right)$$
$$+ (j\omega 2\pi\mu_h)\text{Term 5}$$

(13.64)

where

$$\text{Term 1a} = \iint_s \frac{(\rho_{i+1} - \rho_{i-1})(\rho_{j+1} - \rho_{j-1})}{4A_e^2}\rho dS \qquad (13.65)$$

$$\text{Term 1b} = \iint_s \frac{(Z_{i+1} - Z_{i-1})(Z_{j+1} - Z_{j-1})}{4A_e^2}\rho dS \qquad (13.66)$$

$$\text{Term 2} = \iint_s \frac{Z_{i+1} - Z_{i-1}}{2A_e}\xi_j dS \qquad (13.67)$$

$$\text{Term 3} = \iint_s \frac{Z_{j+1} - Z_{j-1}}{2A_e}\xi_i dS \qquad (13.68)$$

$$\text{Term 4} = \iint_s \frac{\xi_i \xi_j}{\rho} dS \qquad (13.69)$$

$$\text{Term 5} = \iint_s (\rho_1 \xi_1 \xi_i \xi_j + \rho_2 \xi_2 \xi_i \xi_j + \rho_3 \xi_3 \xi_i \xi_j) dS \qquad (13.70)$$

Results of Term 1a, Term 1b, Term 2, Term 3, and Term 5 are relatively simple and are defined as short terms, which will be illustrated in Section 13.8.2. Result of Term 4 is defined as long term, which will be illustrated in Section 13.8.3.

13.8.2 Evaluation of short terms in element matrix

Term 1a, Term 1b, Term 2, Term 3, and Term 5 (Section 13.8.1) are defined as short terms because the results are simple. In Eqs. (13.62)–(13.70), ρ is the ρ coordinate of point (ρ, z) in the element as shown in Fig. 13.2A. It can be expressed as:

$$\rho = \hat{\underline{\rho}} \cdot \underline{r} = \hat{\underline{\rho}} \cdot (\underline{r}_1^e \xi_1 + \underline{r}_2^e \xi_2 + \underline{r}_3^e \xi_3) = \rho_1^e \xi_1 + \rho_2^e \xi_2 + \rho_3^e \xi_3 \qquad (13.71)$$

The integrals of ξ_i and ρ can be evaluated by:

$$\iint_S \xi_i dS = 2A_e \int_0^1 \int_0^{1-\xi_{i+1}} \xi_i d\xi_i d\xi_{i+1} = 2A_e \frac{1}{6} = \frac{A_e}{3} \quad (13.72)$$

$$\iint_S \rho dS = \iint_S (\rho_1^e \xi_1 + \rho_2^e \xi_2 + \rho_3^e \xi_3) dS = \frac{A_e}{3}(\rho_1 + \rho_2 + \rho_3) \quad (13.73)$$

Another identity which will be used to evaluate Terms is

$$\iint_S \xi_1^\alpha \xi_2^\beta \xi_3^\gamma dS = \frac{2A_e \alpha! \beta! \gamma!}{(\alpha + \beta + \gamma + 2)!} \quad (13.74)$$

Using Eq. (13.73), Term 1a can be obtained by

$$\text{Term 1a} = \iint_S \frac{(\rho_{i+1} - \rho_{i-1})(\rho_{j+1} - \rho_{j-1})}{4A_e^2} \rho dS$$

$$= \frac{(\rho_{i+1} - \rho_{i-1})(\rho_{j+1} - \rho_{j-1})}{4A_e^2} \frac{A_e}{3}(\rho_1 + \rho_2 + \rho_3) \quad (13.75)$$

$$= \frac{(\rho_1 + \rho_2 + \rho_3)}{12A_e}(\rho_{i+1} - \rho_{i-1})(\rho_{j+1} - \rho_{j-1})$$

Using Eq. (13.73), Term 1b can be obtained by

$$\text{Term 1b} = \iint_S \frac{(Z_{i+1} - Z_{i-1})(Z_{j+1} - Z_{j-1})}{4A_e^2} \rho dS$$

$$= \frac{(Z_{i+1} - Z_{i-1})(Z_{j+1} - Z_{j-1})}{4A_e^2} \frac{A_e}{3}(\rho_1 + \rho_2 + \rho_3) \quad (13.76)$$

$$= \frac{(\rho_1 + \rho_2 + \rho_3)}{12A_e}(Z_{i+1} - Z_{i-1})(Z_{j+1} - Z_{j-1})$$

Using Eq. (13.72), Term 2 can be obtained by

$$\text{Term 2} = \iint_S \frac{Z_{i+1} - Z_{i-1}}{2A_e} \xi_j dS$$

$$= (Z_{i+1} - Z_{i-1}) \frac{1}{2A_e} \frac{A_e}{3} \quad (13.77)$$

$$= \frac{1}{6}(Z_{i+1} - Z_{i-1})$$

Using Eq. (13.72), Term 3 can be obtained by

$$\text{Term 3} = \iint_S \frac{Z_{j+1} - Z_{j-1}}{2A_e} \xi_i dS$$
$$= \frac{1}{6}(Z_{j+1} - Z_{j-1})$$
(13.78)

Using Eq. (13.74), Term 5 can be obtained by

$$\text{Term 5} = \iint_S (\rho_1 \xi_1 \xi_i \xi_j + \rho_2 \xi_2 \xi_i \xi_j + \rho_3 \xi_3 \xi_i \xi_j) dS$$
$$= \frac{A_e}{60} \left\{ \rho_1 \begin{bmatrix} 6 & 2 & 2 \\ 2 & 2 & 1 \\ 2 & 1 & 2 \end{bmatrix} + \rho_2 \begin{bmatrix} 2 & 2 & 1 \\ 2 & 6 & 2 \\ 1 & 2 & 2 \end{bmatrix} + \rho_3 \begin{bmatrix} 2 & 1 & 2 \\ 1 & 2 & 2 \\ 2 & 2 & 6 \end{bmatrix} \right\}$$
(13.79)

13.8.3 Evaluation of long term in element matrix

The evaluation of Term 4 becomes complicated because the use of cylindrical coordinates causes the appearance of ρ. Taking Eq. (13.71) into Term 4 (Section 13.8.1) can obtain

$$(\text{Term 4})_{ij}^e = \iint_S \frac{\xi_i \xi_j}{\rho} dS$$
$$= \iint_S \frac{\xi_i \xi_j}{\rho_i^e \xi_i + \rho_{i+1}^e \xi_{i+1} + \rho_{i-1}^e \xi_{i-1}} dS$$
$$= 2A_e \int_0^1 \int_0^{1-\xi_{i+1}} \frac{\xi_i \xi_j}{\rho_i^e \xi_i + \rho_{i+1}^e \xi_{i+1} + \rho_{i-1}^e (1 - \xi_i - \xi_{i+1})} d\xi_i d\xi_{i+1}$$
$$= 2A_e \int_0^1 \int_0^{1-\xi_{i+1}} \frac{\xi_i \xi_j}{\xi_i(\rho_i^e - \rho_{i-1}^e) + (\rho_{i-1}^e + \xi_{i+1}(\rho_{i+1}^e - \rho_{i-1}^e))} d\xi_i d\xi_{i+1}$$
(13.80)

Eight conditions which are used to evaluate $(\text{Term 4})_{ij}^e$ are illustrated in Table 13.5 and Fig. 13.5 and will be analyzed, respectively.

13.8.3.1 Condition (1)
Taking condition (1) to Eq. (13.80) can obtain

$$(\text{Term 4})_{ij}^e \bigg|_{i=j, \rho_i \neq \rho_{i+1} \neq \rho_{i-1}} = 2A_e \int_0^1 \int_0^{1-\xi_{i+1}} \frac{\xi_i^2}{\xi_i(\rho_i - \rho_{i-1}) + (\rho_{i-1} + \xi_{i+1}(\rho_{i+1} - \rho_{i-1}))} d\xi_i d\xi_{i+1}$$
(13.81)

Table 13.5 Eight conditions used to evaluate Term 4

Condition (1)	$i = j, \rho_i \neq \rho_{i+1} \neq \rho_{i-1}$
Condition (2)	$i \neq j, \rho_i \neq \rho_j \neq \rho_k$
Condition (3)	$i = j, \rho_i = \rho_{i+1}$
Condition (4)	$i = j, \rho_i = \rho_{i-1}$
Condition (5)	$i = j, \rho_{i-1} = \rho_{i+1}$
Condition (6)	$i \neq j, \rho_i = \rho_j$
Condition (7)	$i \neq j, \rho_i = \rho_k$
Condition (8)	$i \neq j, \rho_j = \rho_k$

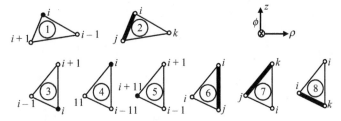

Figure 13.5 Eight conditions used to evaluate Term 4.

Using

$$\int \frac{x^2}{b+ax} dx = \frac{2\log(b+ax)b^2 + ax(ax - 2b)}{2a^3} \tag{13.82}$$

where

$$a = \rho_i - \rho_{i-1}; b = \rho_{i-1} + \xi_{i+1}(\rho_{i+1} - \rho_{i-1}) \tag{13.83}$$

can obtain

$$(\text{Term } 4)_{ij}^{e}\Big|_{i=j, \rho_i \neq \rho_{i+1} \neq \rho_{i-1}}$$

$$= 2A_e \int_0^1 \left\{ \begin{array}{l} \dfrac{(\rho_{i-1} + \xi_{i+1}(\rho_{i+1} - \rho_{i-1}))^2 \ln\left(\dfrac{\rho_i + \xi_{i+1}(\rho_{i+1} - \rho_i)}{\rho_{i-1} + \xi_{i+1}(\rho_{i+1} - \rho_{i-1})}\right)}{(\rho_i - \rho_{i-1})^3} \\ + \dfrac{(\rho_i + 2\rho_{i+1} - 3\rho_{i-1})\xi_{i+1}^2 + (-2\rho_i - 2\rho_{i+1} + 6\rho_{i-1})\xi_{i+1} + (\rho_i - 3\rho_{i-1})}{2(\rho_i - \rho_{i-1})^2} \end{array} \right\} d\xi_{i+1}$$

$$\tag{13.84}$$

Using

$$\int (f+ex)^2 \log\left(\frac{b+ax}{d+cx}\right) dx = -\frac{(ad-bc)e^2 x^2}{6ac} + \frac{(ad-bc)e(-3afc+bec+ade)x}{3a^2 c^2}$$

$$+ \frac{1}{3}(3f^2 + 3exf + e^2 x^2)\log\left(\frac{b+ax}{d+cx}\right)x + \frac{(e^2 b^3 - 3aefb^2 + 3a^2 f^2 b)\log(b+ax)}{3a^3}$$

$$+ \frac{(-e^2 d^3 + 3cefd^2 - 3c^2 f^2 d)\log(d+cx)}{3c^3}$$

(13.85)

where

$$a = \rho_{i+1} - \rho_i;\ b = \rho_i;\ c = \rho_{i+1} - \rho_{i-1};\ d = \rho_{i-1};\ e = \rho_{i+1} - \rho_{i-1};\ f = \rho_{i-1} \quad (13.86)$$

to obtain

$$(\text{Term } 4)^e_{ij}\Big|_{i=j,\rho_i \neq \rho_{i+1} \neq \rho_{i-1}} = \frac{A_e}{3(\rho_i - \rho_{i+1})^3 (\rho_i - \rho_{i-1})^3 (\rho_{i+1} - \rho_{i-1})}$$

$$\times \begin{cases} [\rho_i(\rho_i - \rho_{i-1})(\rho_{i+1} - \rho_{i-1})(\rho_{i+1} - \rho_i)(-\rho_i^2 + 3\rho_i \rho_{i+1} + 3\rho_i \rho_{i-1} - 5\rho_{i+1}\rho_{i-1})] \\ + 2\ln(\rho_i)\rho_i[3\rho_{i+1}^2 \rho_{i-1}^2 (\rho_{i+1} - \rho_{i-1}) + \rho_i^2(\rho_{i+1}^3 - \rho_{i-1}^3) + \rho_i(-3\rho_{i+1}^3 \rho_{i-1} + 3\rho_{i+1}\rho_{i-1}^3)] \\ + 2\ln(\rho_{i+1})[\rho_{i+1}^3(\rho_{i-1} - \rho_i)^3] \\ + 2\ln(\rho_{i-1})[\rho_{i-1}^3(\rho_i - \rho_{i+1})^3] \end{cases}$$

(13.87)

13.8.3.2 Condition (2)
Taking condition (2) to Eq. (13.80) can obtain

$$(\text{Term } 4)^e_{ij}\Big|_{i \neq j, \rho_i \neq \rho_j \neq \rho_k} = 2A_e \int_0^1 \int_0^{1-\xi_{j1}} \frac{\xi_i \xi_j}{\xi_i(\rho_i - \rho_k) + (\rho_k + \xi_j(\rho_j - \rho_k))} d\xi_i d\xi_j \quad (13.88)$$

Using

$$\int \frac{x}{b+ax} dx = \frac{ax - b\log(b+ax)}{a^2} \quad (13.89)$$

where

$$a = \rho_i - \rho_k; b = \rho_k + \xi_j(\rho_j - \rho_k) \tag{13.90}$$

can obtain

$$(\text{Term 4})^e_{ij}|_{i \neq j, \rho_i \neq \rho_j \neq \rho_k} = 2A_e \int_0^1 \left\{ -\frac{\xi_j(\rho_k + \xi_j(\rho_j - \rho_k)) \ln\left(\frac{\rho_i + \xi_j(\rho_j - \rho_i)}{\rho_k + \xi_j(\rho_j - \rho_k)}\right)}{(\rho_i - \rho_k)^2} + \frac{\xi_j(1 - \xi_j)}{(\rho_i - \rho_k)} \right\} d\xi_j \tag{13.91}$$

Using

$$\int x(f + ex) \log\left(\frac{b + ax}{d + cx}\right) dx = \frac{d^2 f \log(d + cx)}{2c^2} - \frac{d^3 e \log(d + cx)}{3c^3} + \frac{e \log(b + ax) b^3}{3a^3}$$

$$- \frac{f \log(b + ax) b^2}{2a^2} - \frac{(ad - bc)ex^2}{6ac} + \frac{(bc - ad)fx}{2ac}$$

$$+ \frac{(a^2 d^2 - b^2 c^2)ex}{3a^2 c^2} + \frac{1}{3} ex^3 \log\left(\frac{b + ax}{d + cx}\right) + \frac{1}{2} fx^2 \log\left(\frac{b + ax}{d + cx}\right) \tag{13.92}$$

where

$$a = \rho_j - \rho_i; b = \rho_i; c = \rho_j - \rho_k; d = \rho_k; e = \rho_j - \rho_k; f = \rho_k \tag{13.93}$$

to obtain

$$(\text{Term 4})^e_{ij}|_{i \neq j, \rho_i \neq \rho_j \neq \rho_k} = \frac{A_e}{3(\rho_i - \rho_j)^3(\rho_i - \rho_k)^2(\rho_j - \rho_k)^2}$$

$$\times \left\{ \begin{array}{l} [(\rho_i - \rho_j)(\rho_j - \rho_k)(\rho_k - \rho_i)(-\rho_i \rho_j^2 - \rho_i^2 \rho_j + \rho_j^2 \rho_k + \rho_i^2 \rho_k)] \\ + \ln(\rho_i)[\rho_i^2(\rho_j - \rho_k)^2(-2\rho_i \rho_j + 3\rho_j \rho_k - \rho_i \rho_k)] \\ + \ln(\rho_j)[\rho_j^2(\rho_i - \rho_k)^2(2\rho_i \rho_j + \rho_j \rho_k - 3\rho_i \rho_k)] \\ + \ln(\rho_k)[\rho_k^3(\rho_i - \rho_j)^3] \end{array} \right\} \tag{13.94}$$

13.8.3.3 Condition (3)

Taking condition (3) to Eq. (13.80) can obtain

$$\begin{aligned}
(\text{Term } 4)^e_{ij}|_{i=j,\rho_i=\rho_{i+1}} &= 2A_e \int_0^1 \int_0^{1-\xi_{i-1}} \frac{\xi_i^2}{\rho_i \xi_i + \rho_{i-1}\xi_{i-1} + \rho_{i+1}(1-\xi_i-\xi_{i-1})} d\xi_i d\xi_{i-1} \\
&= 2A_e \int_0^1 \int_0^{1-\xi_{i-1}} \frac{\xi_i^2}{\rho_i \xi_i + \rho_{i-1}\xi_{i-1} + \rho_i(1-\xi_i-\xi_{i-1})} d\xi_i d\xi_{i-1} \\
&= 2A_e \int_0^1 \int_0^{1-\xi_{i-1}} \frac{\xi_i^2}{\rho_i + (\rho_{i-1}-\rho_i)\xi_{i-1}} d\xi_i d\xi_{i-1} \\
&= 2A_e \int_0^1 \frac{\frac{1}{3}(1-\xi_{i-1})^3}{\rho_i + (\rho_{i-1}-\rho_i)\xi_{i-1}} d\xi_{i-1}
\end{aligned}$$

(13.95)

Using the integral formula:

$$\int \frac{(1-x)^3}{b+ax} dx = -\frac{(x-1)^3}{3a} + \frac{(a+b)(x-1)^2}{2a^2} - \frac{(a+b)^2(x-1)}{a^3}$$
$$- \frac{(-a^3 - 3ba^2 - 3b^2a - b^3)\log((x-1)a+a+b)}{a^4}$$

(13.96)

where

$$a = \rho_{i-1} - \rho_i; \quad b = \rho_i \tag{13.97}$$

can obtain

$$(\text{Term } 4)^e_{ij}|_{i=j,\rho_i=\rho_{i+1}} = A_e \times \frac{2\rho_i^3 - 9\rho_i^2 \rho_{i-1} + 18\rho_i \rho_{i-1}^2 - 11\rho_{i-1}^3 + 6\rho_{i-1}^3 \ln\left(\frac{\rho_{i-1}}{\rho_i}\right)}{9(\rho_i - \rho_{i-1})^4}$$

(13.98)

13.8.3.4 Condition (4)

Taking condition (4) to Eq. (13.80) can obtain

$$(\text{Term } 4)^e_{ij}|_{i=j,\rho_i=\rho_{i-1}} = A_e \times \frac{2\rho_i^3 - 9\rho_i^2 \rho_{i+1} + 18\rho_i \rho_{i+1}^2 - 11\rho_{i+1}^3 + 6\rho_{i+1}^3 \ln\left(\frac{\rho_{i+1}}{\rho_i}\right)}{9(\rho_i - \rho_{i+1})^4}$$

(13.99)

13.8.3.5 Condition (5)

Taking condition (5) to Eq. (13.80) can obtain

$$(\text{Term } 4)^e_{ij}|_{i=j,\rho_{i-1}=\rho_{i+1}} = 2A_e \int_0^1 \int_0^{1-\xi_{i+1}} \frac{\xi_j^2}{\rho_i \xi_i + \rho_{i-1}\xi_{i-1} + \rho_{i+1}(1 - \xi_i - \xi_{i-1})} d\xi_i d\xi_{i+1}$$

$$= 2A_e \int_0^1 \int_0^{1-\xi_{i+1}} \frac{\xi_i^2}{\rho_i \xi_i + \rho_{i-1}\xi_{i-1} + \rho_{i-1}(1 - \xi_i - \xi_{i-1})} d\xi_i d\xi_{i+1}$$

$$= 2A_e \int_0^1 \int_0^{1-\xi_{i+1}} \frac{\xi_i^2}{\rho_{i-1} + (\rho_i - \rho_{i-1})\xi_i} d\xi_i d\xi_{i+1}$$

(13.100)

Using

$$\int \frac{x^2}{b+ax} dx = \frac{2 \log(b+ax)b^2 + ax(ax - 2b)}{2a^3}$$

(13.101)

where

$$a = \rho_i - \rho_{i-1}; \; b = \rho_{i-1},$$

(13.102)

can obtain

$$(\text{Term } 4)^e_{ij}|_{i=j,\rho_{i-1}=\rho_{i+1}} = 2A_e \int_0^1 \left\{ \frac{\rho_{i-1}^2 \ln(\rho_i + \xi_{i+1}(\rho_{i-1} - \rho_i))}{(\rho_i - \rho_{i-1})^3} - \frac{\rho_{i-1}^2 \ln(\rho_{i-1})}{(\rho_i - \rho_{i-1})^3} + \frac{(\rho_i - \rho_{i-1})\xi_{i+1}^2 + (-2\rho_i + 4\rho_{i-1})\xi_{i+1} + (\rho_i - 3\rho_{i-1})}{2(\rho_i - \rho_{i-1})^2} \right\} d\xi_{i+1}$$

(13.103)

Using

$$\int \log(b+ax) dx = \log(b+ax)x - x + \frac{b \log(b+ax)}{a}$$

(13.104)

where

$$a = \rho_{i-1} - \rho_i; \; b = \rho_i,$$

(13.105)

can obtain

$$(\text{Term } 4)^e_{ij}|_{i=j,\rho_{i-1}=\rho_{i+1}} = A_e \times \frac{\rho_i^3 - 6\rho_i^2 \rho_{i-1} + 3\rho_i \rho_{i-1}^2 + 2\rho_{i-1}^3 + 6\rho_i \rho_{i-1}^2 \ln\left(\frac{\rho_i}{\rho_{i-1}}\right)}{3(\rho_i - \rho_{i-1})^4}$$

(13.106)

13.8.3.6 Condition (6)

Taking condition (6) to Eq. (13.80) can obtain

$$(\text{Term 4})^e_{ij}|_{i \neq j, \rho_i = \rho_j} = 2A_e \int_0^1 \int_0^{1-\xi_{j1}} \frac{\xi_i \xi_j}{\xi_i(\rho_i - \rho_k) + (\rho_k + \xi_j(\rho_i - \rho_k))} d\xi_i d\xi_j \quad (13.107)$$

Using

$$\int \frac{x}{b + ax} dx = \frac{ax - b \log(b + ax)}{a^2} \quad (13.108)$$

where

$$a = \rho_i - \rho_k; \quad b = \rho_k + \xi_j(\rho_i - \rho_k) \quad (13.109)$$

can obtain

$$(\text{Term 4})^e_{ij}|_{i \neq j, \rho_i = \rho_j} = 2A_e \int_0^1 \left\{ -\frac{\xi_j(\rho_k + \xi_j(\rho_i - \rho_k)) \ln\left(\frac{\rho_i}{\rho_k + \xi_j(\rho_i - \rho_k)}\right)}{(\rho_i - \rho_k)^2} + \frac{\xi_j(1 - \xi_j)}{(\rho_i - \rho_k)} \right\} d\xi_j \quad (13.110)$$

Using

$$\int x(f + ex) \log\left(\frac{b}{d + cx}\right) dx = -\frac{e \log(d + cx) d^3}{3c^3} + \frac{exd^2}{3c^2} + \frac{f \log(d + cx) d^2}{2c^2}$$
$$- \frac{ex^2 d}{6c} - \frac{fxd}{2c} + \frac{ex^3}{9} + \frac{fx^2}{4} + \frac{1}{3}ex^3 \log\left(\frac{b}{d + cx}\right) + \frac{1}{2}fx^2 \log\left(\frac{b}{d + cx}\right) \quad (13.111)$$

where

$$b = \rho_i; \quad c = \rho_i - \rho_k; \quad d = \rho_k; \quad e = \rho_i - \rho_k; \quad f = \rho_k \quad (13.112)$$

can obtain

$$(\text{Term 4})^e_{ij}|_{i \neq j, \rho_i = \rho_j} = A_e \times \frac{2\rho_i^3 - 9\rho_i^2 \rho_k + 18\rho_i \rho_k^2 - 11\rho_k^3 + 6\rho_k^3 \ln\left(\frac{\rho_k}{\rho_i}\right)}{18(\rho_i - \rho_k)^4} \quad (13.113)$$

13.8.3.7 Condition (7)
Taking condition (7) to Eq. (13.80) can obtain

$$(\text{Term 4})^e_{ij}|_{i \neq j, \rho_i = \rho_k} = 2A_e \int_0^1 \int_0^{1-\xi_j} \frac{\xi_i \xi_j}{\rho_i \xi_i + \rho_j \xi_j + \rho_k(1 - \xi_i - \xi_j)} d\xi_i d\xi_j$$

$$= 2A_e \int_0^1 \int_0^{1-\xi_j} \frac{\xi_i \xi_j}{\rho_i \xi_i + \rho_j \xi_j + \rho_i(1 - \xi_i - \xi_j)} d\xi_i d\xi_j \quad (13.114)$$

$$= 2A_e \int_0^1 \int_0^{1-\xi_j} \frac{\xi_i \xi_j}{\rho_i + (\rho_j - \rho_i)\xi_j} d\xi_i d\xi_j$$

$$= A_e \int_0^1 \frac{\xi_j (1-\xi_j)^2}{\rho_i + (\rho_j - \rho_i)\xi_j} d\xi_j$$

Using

$$\int \frac{x(1-x)^2}{b+ax} dx = \frac{x^3}{3a} - \frac{(2a+b)x^2}{2a^2} + \frac{(a+b)^2 x}{a^3} + \frac{(-b^3 - 2ab^2 - a^2 b)\log(b+ax)}{a^4} \quad (13.115)$$

where

$$a = \rho_j - \rho_i; \quad b = \rho_i \quad (13.116)$$

can obtain

$$(\text{Term 4})^e_{ij}|_{i \neq j, \rho_i = \rho_k} = A_e \times \frac{\rho_k^3 - 6\rho_k^2 \rho_j + 3\rho_k \rho_j^2 + 2\rho_j^3 + 6\rho_k \rho_j^2 \ln\left(\frac{\rho_k}{\rho_j}\right)}{6(\rho_k - \rho_j)^4} \quad (13.117)$$

13.8.3.8 Condition (8)
Taking condition (8) to Eq. (13.80) can obtain

$$(\text{Term 4})^e_{ij}|_{i \neq j, \rho_j = \rho_k} = A_e \times \frac{\rho_k^3 - 6\rho_k^2 \rho_i + 3\rho_k \rho_i^2 + 2\rho_i^3 + 6\rho_k \rho_i^2 \ln\left(\frac{\rho_k}{\rho_i}\right)}{6(\rho_k - \rho_i)^4} \quad (13.118)$$

13.8.3.9 Computation algorithm for Term 4
From Sections 13.8.3.1–13.8.3.8, there is an impression that the entire procedure of the formulation is very complicated. The final result is, however, very simple. To illustrate, Fortran computer program is shown in Appendix E (Section 13.8.3.9).

13.9 FEM BASED ON ELECTRIC FIELD

From Section 13.3, we know that an EM wave in an axially symmetric formation can be divided into two sets of modes: TM and TE modes. In Section 13.8, we discussed the solutions of the field in cylindrical coordinates based on TM mode. In this section, we will discuss the solution based on TE mode. The solution methodology is similar to the solution to TM mode.

13.9.1 Electric field equation

Eq. (13.1) can be rewritten as

$$\underline{H} = -\frac{1}{jw}(\underline{\underline{\mu}}^{-1} \cdot \nabla \times \underline{E}) - \frac{1}{jw}(\underline{\underline{\mu}}^{-1} \cdot \underline{M_s}) \qquad (13.119)$$

Take Eq. (13.119) into (13.4) and eliminate $\underline{H}$ to obtain electric field equation

$$\frac{1}{j\omega}\nabla \times (\underline{\underline{\mu}}^{-1} \cdot \nabla \times \underline{E}) + j\omega\underline{\underline{\varepsilon_c}} \cdot \underline{E} = -\underline{J_s} - \frac{1}{j\omega}\nabla \times (\underline{\underline{\mu}}^{-1} \cdot \underline{M_s}) \qquad (13.120)$$

where

$$\underline{\underline{\varepsilon_c}} = \underline{\underline{\varepsilon}} - j\frac{\underline{\underline{\sigma}}}{w} \qquad (13.121)$$

is complex permittivity, $\underline{E}$ and $\underline{H}$ are the electric and magnetic fields, $\underline{M_s}$ and $\underline{J_s}$ are source electric and magnetic current densities, ω is angular frequency, $\underline{\underline{\mu}} = \mu_0\underline{\underline{\mu_r}}$ and $\underline{\underline{\varepsilon}} = \varepsilon_0\underline{\underline{\varepsilon_r}}$ are spatially varying permeability and permittivity, respectively, and $\underline{\underline{\sigma}}$ is earth conductivity.

In this section, similar to Section 13.2, we still assume (1) earth formations are axially symmetric, (2) sources are axially symmetric, and (3) the earth formations are composed of TI medium. Therefore electric fields generated by source are axially symmetric and are free from the variation of component φ

$$\frac{\partial E_\rho}{\partial \phi} = \frac{\partial E_\phi}{\partial \phi} = \frac{\partial E_z}{\partial \phi} = 0 \qquad (13.122)$$

and permeability and complex permittivity can be expressed as

$$\underline{\underline{\mu}} = \mu_h\hat{\underline{\rho}}\hat{\underline{\rho}} + \mu_h\hat{\underline{\phi}}\hat{\underline{\phi}} + \mu_v\hat{\underline{z}}\hat{\underline{z}} \qquad (13.123)$$

$$\underline{\underline{\varepsilon_c}} = \varepsilon_{ch}\hat{\underline{\rho}}\hat{\underline{\rho}} + \varepsilon_{ch}\hat{\underline{\phi}}\hat{\underline{\phi}} + \varepsilon_{cv}\hat{\underline{z}}\hat{\underline{z}} \qquad (13.124)$$

Following the same procedure as in Section 13.2 results in the same conclusion: the coupling between TM mode and TE mode is zero if the three assumptions listed above are satisfied. TM mode has been analyzed in Section 13.8 and TE mode will be analyzed in this section. Table 13.1 shows that for TE mode only E_ϕ, H_ρ, and H_z exist, and the electric field can be expressed as

$$\underline{E} = E_\phi \underline{\hat{\phi}} \tag{13.125}$$

13.9.2 FEM vector matrix equation of electric field

Choose a set of testing functions $\underline{\Omega}_m$, and apply the product in Eq. (13.17) to electric field equation (13.120) to obtain

$$\frac{1}{j\omega}\left\langle \underline{\Omega}_m; \nabla \times \left(\underline{\underline{\mu}}^{-1} \cdot \nabla \times \underline{E}\right) \right\rangle_\Omega + j\omega \left\langle \underline{\Omega}_m; \underline{\underline{\varepsilon}}_c \cdot \underline{E} \right\rangle_\Omega$$
$$= -\left\langle \underline{\Omega}_m; \underline{J}_s \right\rangle_\Omega - \frac{1}{j\omega}\left\langle \underline{\Omega}_m; \nabla \times \left(\underline{\underline{\mu}}^{-1} \cdot \underline{M}_s\right) \right\rangle_\Omega \tag{13.126}$$

If vector $\underline{A}$ is defined as

$$\underline{A} = \underline{\underline{\mu}}^{-1} \cdot \nabla \times \underline{E} \tag{13.127}$$

taking Eqs. (13.19) and (13.127) into the first symmetric product of Eq. (13.126) can obtain

$$\left\langle \underline{\Omega}_m; \nabla \times \left(\underline{\underline{\mu}}^{-1} \cdot \nabla \times \underline{E}\right) \right\rangle_\Omega = \left\langle \nabla \times \underline{\Omega}_m; \underline{\underline{\mu}}^{-1} \cdot \nabla \times \underline{E} \right\rangle_\Omega + \oint_S \left(\left(\underline{\underline{\mu}}^{-1} \cdot \nabla \times \underline{E}\right) \times \underline{\Omega}_m\right) \cdot \hat{\underline{n}} dS \tag{13.128}$$

Taking Eq. (13.128) into (13.126) can obtain the weak form of the electric equation (13.120) as

$$\frac{1}{j\omega}\left\langle \nabla \times \underline{\Omega}_m; \underline{\underline{\mu}}^{-1} \cdot \nabla \times \underline{E} \right\rangle_\Omega + j\omega \left\langle \underline{\Omega}_m; \underline{\underline{\varepsilon}}_c \cdot \underline{E} \right\rangle_\Omega$$
$$= -\left\langle \underline{\Omega}_m; \underline{J}_s \right\rangle_\Omega - \frac{1}{j\omega}\left\langle \underline{\Omega}_m; \nabla \times (\underline{\underline{\mu}}^{-1} \cdot \underline{M}_s) \right\rangle_\Omega - \frac{1}{j\omega}\oint_S \left(\left(\underline{\underline{\mu}}^{-1} \cdot \nabla \times \underline{E}\right) \times \underline{\Omega}_m\right) \cdot \hat{\underline{n}} dS \tag{13.129}$$

with $m = 1, 2, \ldots, N$. In Eq. (13.129), the left side of the equality represents the relationship between electric field $\underline{E}$, right side represents sources existing inside the solution domain Ω or at the boundary S.

If electric field $\underline{E}$ is expended in the same set of basis function $\underline{\Omega}_n$ as used in testing the wave equation (this choice is known as Galerkin's method),

$$\underline{E} \approx \sum_{n=1}^{N} e_n \underline{\Omega}_n \qquad (13.130)$$

substituting Eq. (13.130) into (13.129) we obtain

$$\sum_{n=1}^{N} \left\{ \frac{1}{j\omega} \left\langle \nabla \times \underline{\Omega}_m; \underline{\underline{\mu}}^{-1} \cdot \nabla \times \underline{\Omega}_n \right\rangle_\Omega + j\omega \left\langle \underline{\Omega}_m; \underline{\underline{\varepsilon}}_c \cdot \underline{\Omega}_n \right\rangle_\Omega \right\} e_n$$

$$= -\left\langle \underline{\Omega}_m; \underline{J}_s \right\rangle_\Omega - \frac{1}{j\omega} \left\langle \underline{\Omega}_m; \nabla \times (\underline{\underline{\mu}}^{-1} \cdot \underline{M}_s) \right\rangle_\Omega - \frac{1}{j\omega} \oiint_S \left((\underline{\underline{\mu}}^{-1} \cdot \nabla \times \underline{E}) \times \underline{\Omega}_m \right) \cdot \hat{n} dS$$

$$(13.131)$$

Eq. (13.131) can be expressed as

$$[Y_{m,n}][e_n] = [I_m] \qquad (13.132)$$

where

$$Y_{m,n} = \frac{1}{j\omega} \left\langle \nabla \times \underline{\Omega}_m; \underline{\underline{\mu}}^{-1} \cdot \nabla \times \underline{\Omega}_n \right\rangle_\Omega + j\omega \left\langle \underline{\Omega}_m; \underline{\underline{\varepsilon}}_c \cdot \underline{\Omega}_n \right\rangle_\Omega \qquad (13.133)$$

is the admittance global matrix and

$$I_m = -\left\langle \underline{\Omega}_m; \underline{J}_s \right\rangle_\Omega - \frac{1}{j\omega} \left\langle \underline{\Omega}_m; \nabla \times (\underline{\underline{\mu}}^{-1} \cdot \underline{M}_s) \right\rangle_\Omega - \frac{1}{j\omega} \oiint_S ((\underline{\underline{\mu}}^{-1} \cdot \nabla \times \underline{E}) \times \underline{\Omega}_m) \cdot \hat{n} dS$$

$$(13.134)$$

is global vector.

According to the Matrix Assembly Rule, admittance global matrix $Y_{m,n}$ can be assembled by the admittance element matrix $Y_{i,j}^e$. Eq. (13.133) can be expressed as the combination of the reciprocal inductance element matrix and the capacitance element matrix

$$Y_{i,j}^e = \frac{1}{j\omega} \left\langle \nabla \times \underline{\Omega}_i^e; \underline{\underline{\mu}}^{-1} \cdot \nabla \times \underline{\Omega}_j^e \right\rangle_\Omega + j\omega \left\langle \underline{\Omega}_i^e; \underline{\underline{\varepsilon}}_c \cdot \underline{\Omega}_j^e \right\rangle_\Omega \qquad (13.135)$$

$$Y_{i,j}^e = \frac{1}{j\omega} \Gamma_{ij}^e + j\omega C_{ij}^e \qquad (13.136)$$

where the reciprocal inductance element matrix and the capacitance element matrix are

$$\Gamma_{ij}^e = \left\langle \nabla \times \underline{\Omega}_i^e; \underline{\underline{\mu}}^{-1} \cdot \nabla \times \underline{\Omega}_j^e \right\rangle_\Omega \qquad (13.137)$$

$$C_{ij}^e = \left\langle \underline{\Omega}_i^e; \underline{\underline{\varepsilon}}_c \cdot \underline{\Omega}_j^e \right\rangle_\Omega \qquad (13.138)$$

respectively.

Eq. (13.135) can be evaluated numerically, but in this case, it may be evaluated analytically.

13.10 EVALUATION OF TRIANGULAR ELEMENT MATRIX BASED ON E_ϕ (TE MODE)

Since $\underline{\Omega}_n$ is the testing function and also basis function to represent magnetic field $\underline{E}$, it has the same $\hat{\phi}$ direction as $\underline{E}$. The same $\underline{\Omega}_i^e$ has only $\hat{\phi}$ direction. If it is defined as

$$\underline{\Omega}_i^e = \xi_i \hat{\underline{\phi}} \qquad (13.139)$$

where ξ_i is the basis function for triangular element (Section 13.5.1), the analysis is based on E_ϕ.

Admittance element matrix (13.136) will be analyzed. Following the same steps as Eq. (13.63), we can get capacitance element matrix as

$$\begin{aligned}
C_{i,j}^e &= \left\langle \underline{\Omega}_i^e; \underline{\underline{\varepsilon}}_c \cdot \underline{\Omega}_j^e \right\rangle_\Omega \\
&= \iiint_v \left(\xi_i \hat{\underline{\phi}} \right) \cdot \left(\underline{\underline{\varepsilon}}_c \cdot \xi_j \hat{\underline{\varphi}} \right) dV \\
&= \iiint_v \left(\xi_i \hat{\underline{\phi}} \right) \cdot \left(\varepsilon_{ch} \hat{\underline{\phi}} \hat{\underline{\phi}} \cdot \xi_j \hat{\underline{\phi}} \right) dV \\
&= \varepsilon_{ch} \iint_s \xi_i \xi_j 2\pi \rho \, dS \\
&= 2\pi \varepsilon_{ch} \iint_s \xi_i \xi_j (\rho_1 \xi_1 + \rho_2 \xi_2 + \rho_3 \xi_3) dS \\
&= 2\pi \varepsilon_{ch} \iint_s (\rho_1 \xi_1 \xi_i \xi_j + \rho_2 \xi_2 \xi_i \xi_j + \rho_3 \xi_3 \xi_i \xi_j) dS \\
&= 2\pi \varepsilon_{ch} \text{ Term 5}
\end{aligned} \qquad (13.140)$$

Following the same steps as Eq. (13.62), we can get reciprocal inductance element matrix as

$$\Gamma_{i,j}^e = \left\langle \nabla \times \underline{\Omega}_i^e; \underline{\underline{\mu}}^{-1} \nabla \times \underline{\Omega}_j^e \right\rangle_\Omega$$

$$= \iiint_V (\nabla \times \underline{\Omega}_i^e) \cdot (\underline{\underline{\mu}}^{-1} \cdot \nabla \times \underline{\Omega}_j^e) dV$$

$$= \iint_S \left(-\frac{l_i}{2A_e} + \hat{z}\frac{\xi_i}{\rho} \right) \cdot \left(\underline{\underline{\mu}}^{-1} \cdot \left(-\frac{l_j}{2A_e} + \hat{z}\frac{\xi_j}{\rho} \right) \right) * 2\pi\rho dS$$

$$= \iint_S \left(\hat{\rho}\frac{\rho_{i+1} - \rho_{i-1}}{2A_e} + \hat{z}\frac{Z_{i+1} - Z_{i-1}}{2A_e} + \hat{z}\frac{\xi_i}{\rho} \right) \cdot$$
$$\left(\underline{\underline{\mu}}^{-1} \cdot \left(\hat{\rho}\frac{\rho_{j+1} - \rho_{j-1}}{2A_e} + \hat{z}\frac{Z_{j+1} - Z_{j-1}}{2A_e} + \hat{z}\frac{\xi_j}{\rho} \right) \right) * 2\pi\rho dS$$

$$= \iint_S \left(\hat{\rho}\frac{\rho_{i+1} - \rho_{i-1}}{2A_e} + \hat{z}\frac{Z_{i+1} - Z_{i-1}}{2A_e} + \hat{z}\frac{\xi_i}{\rho} \right) \cdot$$
$$\left(\hat{\rho}\mu_h^{-1}\frac{\rho_{j+1} - \rho_{j-1}}{2A_e} + \hat{z}\mu_v^{-1}\frac{Z_{j+1} - Z_{j-1}}{2A_e} + \hat{z}\mu_v^{-1}\frac{\xi_j}{\rho} \right) * 2\pi\rho dS$$

$$= 2\pi \iint_S \left(\frac{(\rho_{i+1} - \rho_{i-1})(\rho_{j+1} - \rho_{j-1})}{4A_e^2\mu_h}\rho + \frac{(Z_{i+1} - Z_{i-1})(Z_{j+1} - Z_{j-1})}{4A_e^2\mu_v}\rho \right.$$
$$\left. + \frac{Z_{i+1} - Z_{i-1}}{2A_e\mu_v}\xi_j + \frac{Z_{j+1} - Z_{j-1}}{2A_e\mu_v}\xi_i + \frac{1}{\mu_v}\frac{\xi_i\xi_j}{\rho} \right) dS$$

$$= 2\pi \left(\frac{1}{\mu_h}\text{Term 1a} + \frac{1}{\mu_v}\text{Term 1b} + \frac{1}{\mu_v}\text{Term 2} + \frac{1}{\mu_v}\text{Term 3} + \frac{1}{\mu_v}\text{Term 4} \right)$$

(13.141)

Taking Eqs. (13.141) and (13.140) into (13.136) we can get the admittance element matrix

$$[Y_{i,j}^e] = \frac{2\pi}{j\omega} \left(\frac{1}{\mu_h}\text{Term 1a} + \frac{1}{\mu_v}\text{Term 1b} + \frac{1}{\mu_v}\text{Term 2} + \frac{1}{\mu_v}\text{Term 3} + \frac{1}{\mu_v}\text{Term 4} \right)$$
$$+ (j\omega 2\pi\varepsilon_{ch})\text{Term 5}$$

(13.142)

The value of Terms has been illustrated in Sections 13.8.2 and 13.8.3.

13.11 FEM MODEL OF SOURCES

13.11.1 Source models

As we discussed in the previous sections, the FEM does not solve the Maxwell's equations directly, instead, it solves the energy equations in terms of integrals in the solution volumes. We have discussed the ways to handle TE and TM modes in terms of fields. However, sources in the logging problems must be taken care of before we can solve the problem numerically due to the special shape of the sources in the logging tools. In this section, three source models commonly used in the numerical simulation of logging problems will be discussed for magnetic field equations and electric field equations which were discussed in Sections 13.9 and 13.10.

13.11.2 Source model 1—electrical current loop

Electrical current loop is one of the mostly used sources in well logging. The application examples include induction and LWD resistivity tools. In this kind of tools, the source is an electric current loop (e.g., coil antenna), which is axially symmetric along the z axis and exists inside solution domain Ω, expressed as:

$$\underline{J_s} = J_{s\phi}\underline{\hat{\phi}} \tag{13.143}$$

Fields generated by this source are TE mode ($E_\phi\hat{\phi}$, $H_\rho\hat{\rho}$, and $H_z\hat{z}$), and they would fade to zero on the infinite boundary S_{whole}. Eq. (13.129) is simplified to

$$\frac{1}{j\omega}\left\langle \nabla\times\underline{\Omega_m}; \underline{\underline{\mu}}^{-1}\cdot\nabla\times\underline{E}\right\rangle_\Omega + j\omega\left\langle\underline{\Omega_m}; \underline{\underline{\varepsilon_c}}\cdot\underline{E}\right\rangle_\Omega = -\left\langle\underline{\Omega_m}; \underline{J_s}\right\rangle_\Omega \tag{13.144}$$

where the global vector is defined as

$$I_m = -\left\langle\underline{\Omega_m}; \underline{J_s}\right\rangle_\Omega \tag{13.145}$$

as illustrated in Fig. 13.6. Notice that Ω_{whole} is the domain of whole space, Ω is solution domain, S_{whole} is the infinite boundary, S is the boundary and here they satisfy

$$S = S_{\text{whole}} \tag{13.146}$$

$$\Omega = \Omega_{\text{whole}} \tag{13.147}$$

13.11.3 Source model 2—magnetic current loop

The example of magnetic current loop in the logging problem is the toroid antenna. The toroid antenna can be modeled as a magnetic current loop. The source is

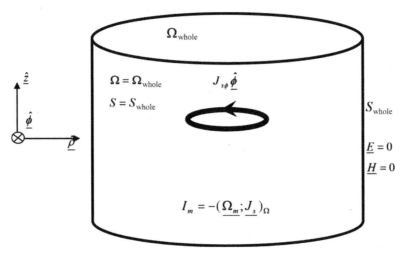

Figure 13.6 The source of electric current loop (coil antenna).

magnetic current loop, which is axially symmetric along z axis and exists inside solution domain Ω, to be expressed as:

$$\underline{M}_s = M_{s\phi}\hat{\phi} \tag{13.148}$$

Fields generated by this source are TM mode ($H_\phi\hat{\phi}$, $E_\rho\hat{\rho}$, and $E_z\hat{z}$), and they would fade to zero on the infinite boundary S_{whole}. Eq. (13.22) is simplified to

$$\frac{1}{j\omega}\left\langle \nabla\times\underline{\Omega}_m; \underline{\underline{\varepsilon}}_c^{-1}\cdot\nabla\times\underline{H}\right\rangle_\Omega + j\omega\left\langle\underline{\Omega}_m;\underline{\underline{\mu}}\cdot\underline{H}\right\rangle_\Omega = -\left\langle\underline{\Omega}_m;\underline{M}_s\right\rangle_\Omega \tag{13.149}$$

where the global vector in the equation is defined as

$$V_m = -\left\langle\underline{\Omega}_m;\underline{M}_s\right\rangle_\Omega \tag{13.150}$$

as illustrated in Fig. 13.7. They satisfy

$$S = S_{\text{whole}} \tag{13.151}$$

$$\Omega = \Omega_{\text{whole}} \tag{13.152}$$

13.11.4 Source model 3—electrodes

The source (e.g., electrode-type tool or current source or voltage source) generates current density as:

$$\underline{J}_s = J_{s\rho}\hat{\rho} + J_{sz}\hat{z} \tag{13.153}$$

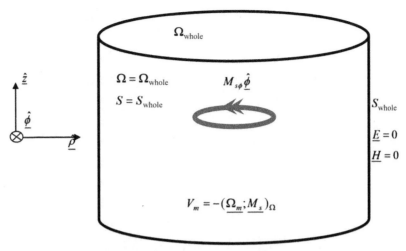

Figure 13.7 The source of magnetic current loop (toroid antenna).

Fields generated by this source are TM mode ($H_\phi \hat{\underline{\phi}}$, $E_\rho \hat{\underline{\rho}}$, and $E_z \hat{\underline{z}}$). Eq. (13.22) is simplified to

$$\frac{1}{j\omega} \left\langle \nabla \times \underline{\Omega}_m ; \underline{\underline{\varepsilon}}_c^{-1} \cdot \nabla \times \underline{H} \right\rangle_\Omega + j\omega \left\langle \underline{\Omega}_m ; \underline{\underline{\mu}} \cdot \underline{H} \right\rangle_\Omega$$
$$= \frac{1}{j\omega} \left\langle \underline{\Omega}_m ; \nabla \times (\underline{\underline{\varepsilon}}_c^{-1} \cdot \underline{J}_s) \right\rangle_\Omega - \frac{1}{j\omega} \oiint_S ((\underline{\underline{\varepsilon}}_c^{-1} \cdot \nabla \times \underline{H}) \times \underline{\Omega}_m) \cdot \hat{n} dS \quad (13.154)$$

where the first term on the right side is the electric current source, and the second term is the electromagnetic field existing on the boundary. In practice, it is hard to handle the source with the curl format, which is shown in the term of $\frac{1}{j\omega} \left\langle \underline{\Omega}_m ; \nabla \times \underline{\underline{\varepsilon}}_c^{-1} \underline{J}_s \right\rangle_\Omega$, so equivalent sources which are easier to handle will be generated to replace the actual sources. Based on the Equivalence Principle [7], the actual sources and equivalent sources produce the same field inside the solution region Ω, so it is not necessary to know the actual sources. Equivalent sources here are EM fields $\underline{E}$ and $\underline{H}$ existing on the boundary S_{tool}, which surrounds the actual sources region Ω_{tool}, as shown in Fig. 13.8, where Ω_{tool} is the region that includes electrical current source and whole tool, S_{tool} is the boundary used to cover Ω_{tool}, Ω_{whole} is the whole space region, S_{whole} is boundary of infinite and also the boundary of Ω_{whole}, Ω is solution domain, S is boundary of Ω, and they satisfy:

$$S = S_{tool} + S_{whole} \quad (13.155)$$

$$\Omega = \Omega_{whole} - \Omega_{tool} \quad (13.156)$$

By the Equivalence Principle, the EM fields in solution domain Ω are only simulated by fields existing on the boundary S_{tool} and the actual source $\underline{J_s}$ which exists outside Ω does not need to be considered. By this method and using Eq. (13.4), Eq. (13.22) can be simplified to

$$\frac{1}{j\omega}\left\langle \nabla\times\underline{\Omega_m};\underline{\underline{\varepsilon_c}}^{-1}\cdot\nabla\times\underline{H}\right\rangle_\Omega + j\omega\left\langle\underline{\Omega_m};\underline{\underline{\mu}}\cdot\underline{H}\right\rangle_\Omega = -\oiint_S (\underline{E}\times\underline{\Omega_m})\cdot\hat{n}dS \quad (13.157)$$

where the global vector is defined as

$$V_m = -\oiint_S (\underline{E}\times\underline{\Omega_m})\cdot\hat{n}dS \quad (13.158)$$

13.11.5 FEM solutions for the source existing on boundaries

From Section 13.11.4, we found that the sources can be equivalently replaced by the sources on the boundaries in a small region around the sources. Therefore as long as we can find the solution to the sources on the boundaries of a region using FEM scheme, the source issue can be solved. In this section we will discuss how to apply FEM to the source existing on boundaries illustrated in Fig. 13.8 and Eq. (13.158). The real source, which is surrounded by boundary S_{tool} can be electrode-type logging tool, voltage source, or current source. Li demonstrated the boundary-value problem for a direct current electrode-type logging tool based on Laplace's equation [8]. Fig. 13.9 demonstrates the boundary-value problem for alternating current (AC)

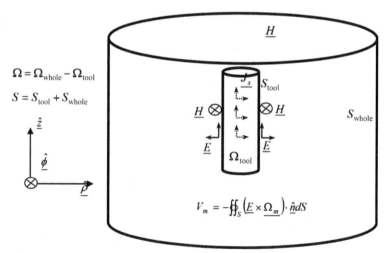

Figure 13.8 The source existing on boundaries.

electrode-type logging tool based on Maxwell's equation. It follows the same rule for other AC tools or just current sources and voltage sources.

Fig. 13.9 simplifies the same problem in Fig. 13.8 from three dimension to two dimension since the EM fields are axial symmetric. The direction of boundary S is counterclockwise and boundary S is encompassed by four parts: the infinite boundary Γ_1, the borehole axis Γ_2, the conducting surface Γ_3, and insulating surface Γ_4. Surface Γ_3 and Γ_4 together make up the boundary S_{tool} which surrounds the actual sources

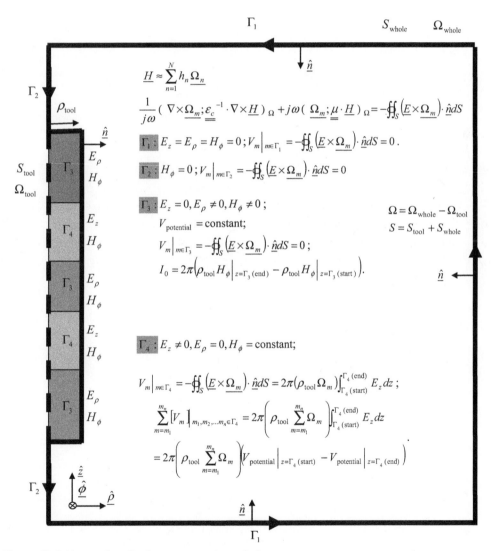

Figure 13.9 How to handle the source on boundaries.

region Ω_{tool}, where Γ_3 is the surface to allow the source current to flow in and out of the solution domain Ω and it is assumed to be perfect electric conductor (PEC) to maintain the constant potential, and Γ_4 is the surface without source current flowing through it and is assumed to be insulator and so magnetic field H_ϕ remains constant. Γ_1 is theoretically in infinite region, but in computation model, it is the place where fields attenuate to zero. Γ_2 is borehole axis where magnetic field H_ϕ is zero. Eq. (13.158) will be evaluated separately on different boundaries. The following items should be noticed:

1. On the infinite boundary Γ_1, EM fields $\underline{E}$ and $\underline{H}$ attenuate to zero. Notice $\underline{\Omega}_m$ is the testing function and also will be used as basis function to represent $\underline{H}$ on each node. It has the same $\hat{\phi}$ direction as magnetic field $\underline{H}$. Eq. (13.158) will be zero if either $\underline{n} \times \underline{E} = 0$ or $\underline{n} \times \underline{\Omega}_m = 0$ on boundary. So if node m belongs to boundary Γ_1, it satisfies:

$$E_z = E_\rho = H_\phi = 0 \quad (m \in \Gamma_1) \tag{13.159}$$

$$V_m\big|_{m \in \Gamma_1} = -\oiint_S (\underline{E} \times \underline{\Omega}_m) \cdot \hat{\underline{n}} dS = 0 \tag{13.160}$$

2. On the borehole axis boundary Γ_2, electric field H_ϕ is zero. So if node m belongs to boundary Γ_2, it satisfies:

$$H_\phi = 0 \quad (m \in \Gamma_2) \tag{13.161}$$

$$V_m\big|_{m \in \Gamma_2} = -\oiint_S (\underline{E} \times \underline{\Omega}_m) \cdot \hat{\underline{n}} dS = 0 \tag{13.162}$$

3. On the conducting surface Γ_3, total current flowing through boundary Γ_3 is I_0. Since the surface is along z direction and is assumed to be PEC to assure a constant potential, there is no E_z existing, so $\hat{\underline{n}} \times \underline{E} = \hat{\rho} \times \underline{E}$ is zero. So if node m belongs to boundary Γ_3, it satisfies:

$$E_z = 0, E_\rho \neq 0, H_\phi \neq 0 \quad (m \in \Gamma_3) \tag{13.163}$$

$$V_{\text{potential}} = \text{constant} \quad (m \in \Gamma_3) \tag{13.164}$$

$$V_m\big|_{m \in \Gamma_3} = -\oiint_S (\underline{E} \times \underline{\Omega}_m) \cdot \hat{\underline{n}} dS = 0 \tag{13.165}$$

From Maxwell's equation, the relationship between EM fields can be shown as

$$-\frac{\partial H_\phi}{\partial z} = (j\omega\varepsilon + \sigma_h)E_\rho \tag{13.166}$$

$$\frac{1}{\rho}\frac{\partial}{\partial \rho}(\rho H_\phi) = (j\omega\varepsilon + \sigma_v)E_z \tag{13.167}$$

If the frequency is low, the relationship between EM fields and current I_0 is

$$\begin{aligned}
I_0 = I_{\text{current}}\Big|_{\Gamma_3} &= 2\pi\rho_{\text{tool}}\int_{\Gamma_3(\text{start})}^{\Gamma_3(\text{end})}\sigma_h E_\rho dl \\
&= 2\pi\rho_{\text{tool}}\int_{\Gamma_3(\text{start})}^{\Gamma_3(\text{end})}\left(-\frac{\partial}{\partial z}H_\phi\right)dl \\
&= 2\pi\rho_{\text{tool}}\int_{\Gamma_3(\text{start})}^{\Gamma_3(\text{end})}\left(-\frac{\partial}{(-\partial l)}H_\phi\right)dl \\
&= 2\pi\left(\rho_{\text{tool}}H_\phi\Big|_{z=\Gamma_3(\text{end})} - \rho_{\text{tool}}H_\phi\Big|_{z=\Gamma_3(\text{start})}\right)
\end{aligned} \tag{13.168}$$

4. On the insulating surface Γ_4, since there is no current flowing through it, E_ρ is zero and H_ϕ is maintained as a constant. The potential difference on surface Γ_4 is $V_{\text{potential}}|_{\Gamma_4}$ and is the integral of E_z along the surface. If node m belongs to boundary Γ_4, it satisfies:

$$E_z \neq 0, E_\rho = 0, H_\phi = \text{constant} \quad (m \in \Gamma_4) \tag{13.169}$$

$$\begin{aligned}
V_m\Big|_{m\in\Gamma_4} &= -\oiint_S (\underline{E}\times\underline{\Omega_m})\cdot\hat{n}dS \\
&= -\int_{\Gamma_4(\text{start})}^{\Gamma_4(\text{end})}\underline{\Omega_m}\cdot(\hat{n}\times\underline{E})(2\pi\rho_{\text{tool}}dz) \\
&= -\int_{\Gamma_4(\text{start})}^{\Gamma_4(\text{end})}\Omega_m\hat{\phi}\cdot(\hat{\rho}\times\hat{\underline{z}}E_z)(2\pi\rho_{\text{tool}}dz) \\
&= 2\pi(\rho_{\text{tool}}\Omega_m)\int_{\Gamma_4(\text{start})}^{\Gamma_4(\text{end})}E_z dz
\end{aligned} \tag{13.170}$$

$$\int_{\Gamma_4(\text{start})}^{\Gamma_4(\text{end})} E_z \cdot dl = V_{\text{potential}}\Big|_{z=\Gamma_4(\text{start})} - V_{\text{potential}}\Big|_{z=\Gamma_4(\text{end})} \tag{13.171}$$

Notice that H_ϕ is unknown in Eq. (13.157), and is constant on surface Γ_4, so that all nodes on Γ_4 can share one unknown, as shown in Fig. 13.10. If a global matrix is installed in this assumption, and the nodes on Γ_4 are from m_1 to m_n, the global vector can be expressed as

$$\sum_{m=m_1}^{m_n}[V_m]|_{m_1,m_2,\ldots m_n \in \Gamma_4} = \sum_{m=m_1}^{m_n}\left[2\pi(\rho_{tool}\Omega_m)\int_{\Gamma_4(start)}^{\Gamma_4(end)}E_z dz\right]$$

$$= 2\pi\left(\rho_{tool}\sum_{m=m_1}^{m_n}\Omega_m\right)\left(V_{potential}|_{z=\Gamma_4(start)} - V_{potential}|_{z=\Gamma_4(end)}\right)$$

(13.172)

If the impedance element matrix is built based on H_ϕ, it satisfies

$$\sum_{m=m_1}^{m_n}\Omega_m = 1 \qquad (13.173)$$

$$\sum_{m=m_1}^{m_n}[V_m]|_{m_1,m_2,\ldots m_n \in \Gamma_4} = 2\pi\rho_{tool}(V_{potential}|_{z=\Gamma_4(start)} - V_{potential}|_{z=\Gamma_4(end)}) \qquad (13.174)$$

If the impedance element matrix is built based on ρH_ϕ, it satisfies

$$\sum_{m=m_1}^{m_n}\rho\Omega_m = 1 \qquad (13.175)$$

$$\sum_{m=m_1}^{m_n}[V_m]|_{m_1,m_2,\ldots m_n \in \Gamma_4} = 2\pi(V_{potential}|_{z=\Gamma_4(start)} - V_{potential}|_{z=\Gamma_4(end)}) \qquad (13.176)$$

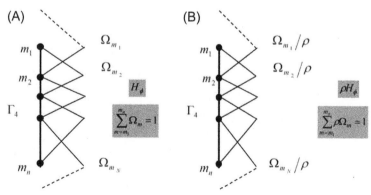

Figure 13.10 Nodes on surface Γ_4.

Similarly, consider the case when the source model is electric current loop illustrated in Fig. 13.6 and Eq. (13.145). If it is assumed that source is coil antenna with ϕ directional electric current $I_{s\phi}$, then electric current density can be written as

$$\underline{J_s} = I_{s\phi}\delta(\rho - \rho_i)\delta(z - z_i)\underline{\hat{\phi}} \tag{13.177}$$

If admittance global matrix is built based on H_ϕ, then the global vector is simplified to

$$\begin{aligned}
[I_m] &= -\left\langle \underline{\Omega_m}; \underline{J_s} \right\rangle_\Omega \\
&= \iiint_V \left(\Lambda_m \hat{\phi}\right) \cdot \left(I_{s\phi}\delta(\rho - \rho_i)\delta(z - z_i)\underline{\hat{\phi}}\right) dV \\
&= \iint_S \Lambda_m I_{s\phi}\delta(\rho - \rho_i)\delta(z - z_i) 2\pi\rho dS \\
&= 2\pi\rho_m I_{s\phi}
\end{aligned} \tag{13.178}$$

The source model is magnetic current loop illustrated in Fig. 13.7 and Eq. (13.150). If it is assumed that the source is toroidal antenna with ϕ directional magnetic current $M_{s\phi}$, then magnetic current density can be written as

$$\underline{M_s} = I_{m\phi}\delta(\rho - \rho_i)\delta(z - z_i)\underline{\hat{\phi}} \tag{13.179}$$

If the impedance global matrix is built based on H_ϕ, then the global vector is simplified to

$$[V_m] = -\left\langle \underline{\Omega_m}; \underline{M_s} \right\rangle_\Omega = 2\pi\rho_m I_{m\phi} \tag{13.180}$$

If the impedance global matrix is built based on ρH_ϕ, then the global vector is simplified to

$$[V_m] = -\left\langle \underline{\Omega_m}; \underline{M_s} \right\rangle_\Omega = 2\pi I_{m\phi} \tag{13.181}$$

REFERENCES

[1] COMSOL Multiphysics. <www.comsol.com/multiphysics>.
[2] ANSYS HFSS—High Frequency Electromagnetic Field Simulation. <http://www.ansys.com/Products/Electronics/ANSYS-HFSS>.
[3] ANSYS Maxwell—Low Frequency Electromagnetic Field Simulation. <http://www.ansys.com/Products/Electronics/ANSYS-Maxwell>.
[4] D.R. Wilton, Scattering, in: R. Pike, P.C. Sabatier (Eds.), Computational Method, Academic Press, London, 2002, pp. 316−365 (Chapter 1.5.5).
[5] S.J. Li, R.C. Liu, Simulation of AC Laterolog Response in a 2-D Formation, Well Logging Laboratory, Technical Report, No.27, University of Houston, 2006 (Chapter 1).

[6] J.M. Jin, "The Finite Element Method in Electromagnetics, second ed., Wiley, New York, NY, 2002, pp. 273–333, 339 (Chapter 8).
[7] R.F. Harrington, Time-Harmonic Electromagnetic fields, McGraw-Hill, New York, NY, 1964.
[8] J. Li, Modeling electrode-type logging tools in 2-D formations by the finite element method (Master thesis), University of Houston, 2000.

APPENDIX A VECTOR ANALYSIS IN CYLINDRICAL COORDINATES

The vector product in cylindrical coordinates is

$$\nabla \times \underline{A} = \begin{Bmatrix} \hat{\rho}\left[\dfrac{1}{\rho}\dfrac{\partial A_z}{\partial \phi} - \dfrac{\partial A_\phi}{\partial z}\right] \\ +\hat{\phi}\left[\dfrac{\partial A_\rho}{\partial z} - \dfrac{\partial A_z}{\partial \rho}\right] \\ +\hat{z}\left[\dfrac{1}{\rho}\dfrac{\partial}{\partial \rho}(\rho A_\phi) - \dfrac{1}{\rho}\dfrac{\partial A_\rho}{\partial \phi}\right] \end{Bmatrix} \qquad (A.1)$$

$$\nabla \cdot \underline{A} = \dfrac{1}{\rho}\dfrac{\partial}{\partial \rho}(\rho A_\rho) + \dfrac{1}{\rho}\dfrac{\partial A_\phi}{\partial \phi} + \dfrac{\partial A_z}{\partial z} \qquad (A.2)$$

$$\nabla U = \hat{\rho}\dfrac{\partial U}{\partial \rho} + \hat{\phi}\dfrac{1}{\rho}\dfrac{\partial U}{\partial \phi} + \hat{z}\dfrac{\partial U}{\partial z} \qquad (A.3)$$

$$\nabla^2 U = \dfrac{1}{\rho}\dfrac{\partial}{\partial \rho}\left(\rho \dfrac{\partial U}{\partial \rho}\right) + \dfrac{1}{\rho^2}\dfrac{\partial^2 U}{\partial \phi^2} + \dfrac{\partial^2 U}{\partial z^2} \qquad (A.4)$$

respectively. Notice the term $\frac{1}{\rho}$ appears in all operators and makes the application of FEM formulation more difficult than in Cartesian coordinates. Other identities which will be used in this chapter are

$$\begin{aligned} \nabla \times (U\underline{A}) &= U\nabla \times \underline{A} - \underline{A} \times \nabla U \\ \nabla \times (\underline{A} \times \underline{B}) &= \underline{A}\nabla \cdot \underline{B} - \underline{B}\nabla \times \underline{A} + (\underline{B} \cdot \nabla)\underline{A} - (\underline{A} \cdot \nabla)\underline{B} \\ \nabla \cdot (\underline{A} \times \underline{B}) &= \underline{B} \cdot \nabla \times \underline{A} - \underline{A} \cdot \nabla \times \underline{B} \\ \nabla \cdot (\nabla \times \underline{A}) &= 0 \\ \iiint_V \nabla \cdot \underline{A}\, dV &= \oiint_S \underline{A} \cdot dS \\ \iiint_V \nabla \times \underline{A}\, dV &= -\oiint_S \underline{A} \times dS \\ \underline{A} \cdot (\underline{B} \times \underline{C}) &= \underline{C} \cdot (\underline{A} \times \underline{B}) = \underline{B} \cdot (\underline{C} \times \underline{A}) \end{aligned} \qquad (A.5)$$

APPENDIX B COMPUTATION METHOD FOR MATRIX ASSEMBLING RULE

```fortran
SUBROUTINE AssembleElement(NumberOfElement)
!
!...Subroutine AssembleElement is used to assemble LocalMatrix to SS
!
!...subroutine ElementMatrix is in ElementMatrixModule, it is used to
!calculate LocalMatrix.
!
!...SS is the matrix used to assemble all elements
!...SRS is vector including all nodes's source information
!...LocalMatrix is the local(element) matirx
!...LocalVector is local vector (source information), it is forced
!to be 0, because source will be added in module NonConstantConstraintModule.
!
INTEGER, INTENT(IN) :: NumberOfElement
INTEGER :: INOD,JNOD,IDES,JDES
INTEGER :: NumberOfSS,NumberOfLocalMatrix
COMPLEX*16 :: LocalVector
COMPLEX*16 :: LocalMatrix(MaxNumberOfLocalMatrix)
!
!...ElementMatrix is used to calculate local matrix
    CALL ElementMatrix(NumberOfElement,LocalMatrix)
!
    NumberOfLocalMatrix=0
    DO 24 INOD=1,MaxNumberOfLocalNode
        NumberOfNode=IABS(NodeFromElement(NumberOfElement,INOD))
        IDES=MatrixLineOfNode(NumberOfNode)
        MatrixLineOccupancy2(IDES)=1
!
!...LocalVector is forced to be 0.
!       SRS(IDES)=SRS(IDES)+LocalVector(NumberOfElement,INOD)
        LocalVector=(0.0,0.0)
        SRS(IDES)=SRS(IDES)+LocalVector
!
        DO 23 JNOD=INOD,MaxNumberOfLocalNode
            NumberOfNode=IABS(NodeFromElement(NumberOfElement,JNOD))
            JDES=MatrixLineOfNode(NumberOfNode)
            NumberOfLocalMatrix=NumberOfLocalMatrix+1
            IF(IDES.LE.JDES) NumberOfSS=MatrixDiagonalPosition(IDES)+JDES-IDES
            IF(IDES.GT.JDES) NumberOfSS=MatrixDiagonalPosition(JDES)+IDES-JDES
            SS(NumberOfSS)=SS(NumberOfSS)+LocalMatrix(NumberOfLocalMatrix)
            IF((INOD.NE.JNOD).AND.(IDES.EQ.JDES))
SS(NumberOfSS)=SS(NumberOfSS)+LocalMatrix(NumberOfLocalMatrix)
23      CONTINUE
24  CONTINUE
!
RETURN
!
END SUBROUTINE AssembleElement
```

APPENDIX C COMPUTATION METHOD OF ELEMENT MATRIX FOR RECTANGULAR ELEMENT BASED ON H_ϕ (SECTION 13.6)

```fortran
SUBROUTINE ElementMatrix_Rectangle_NodeBase_Hphi(NumberOfElement,LocalMatrix)
!...this subroutine is used to calculate element matrix (LocalMatrix)
!...Element is triangle element.
!...Matrix is impedance matrix for formula [Zmn][Hn]=[Vm], formula is base on magnetic field.
!...Matrix is base on NumberOfNodee base function.
!...Input NumberOfElement is the number of the element
!...Output LocalMatrix is local matrix for element NumberOfElement.
!
INTEGER, INTENT(IN) :: NumberOfElement
COMPLEX*16, INTENT(OUT):: LocalMatrix(10)
!
INTEGER :: I,J,N,Index
REAL*8 :: ai,aj,bi,bj
!INTEGER :: NumberOfNode
REAL*8 :: R(4),Z(4)
REAL*8 :: RouDelta,RouCenter,RouMax,RouMin
REAL*8 :: ZDelta,ZCenter
REAL*8 :: Area
REAL*8 :: Item1,Item2,Item3
COMPLEX*16 :: Inductance(4,4),ReciprocalCapacitance(4,4) ! 3x3 element capacitance and reciprocal inductance element
COMPLEX*16 :: Matrix(4,4)
!
REAL*8 :: Mu,Epsilon
!COMPLEX*16 :: Mu,Epsilon
COMPLEX*16 :: EpsilonC_H
COMPLEX*16 :: EpsilonC_V
COMPLEX*16 :: SigmaC_H
COMPLEX*16 :: SigmaC_V
!
!...numerical integration for rectangular element
!
!              o----------------------o
!              |   0    0    0        |
!              |   0    0    0        |
!              |   0    0    0        |
!              o----------------------o
!
!...rectangular element coordinates
!
!           3(K+1,I) (R3,Z3)      4(K+1,I+1)  (R4,Z4)
!              o----------------------o
!              |                      |
!              |       0 (R0,Z0)      |    element
!              |                      |
!              o----------------------o
!           1(K,I) (R1,Z1)       2(K,I+1) (R2,Z2)

    MaxNumberOfLocalNode=4
    MaxNumberOfLocalMatrix=10
!
    DO I=1,4
        R(I)=Coordinates(ABS(NodeFromElement(NumberOfElement,I)),1)
        Z(I)=Coordinates(ABS(NodeFromElement(NumberOfElement,I)),2)
    END DO

    ZDelta=Z(3)-Z(1)
    RouDelta=R(2)-R(1)
    RouMax=R(2)
    RouMin=R(1)
    Roucenter=(R(1)+R(2))/2.
    Zcenter=(Z(1)+Z(3))/2.
    Area=ZDelta*RouDelta
    LocalMatrix=0.
!
    Epsilon=Epsilon_0*EpsilonROfElement(NumberOfElement)
```

```fortran
        Mu=Mu_0*MuROfElement(NumberOfElement)
      ! ResistivityOfElement(NumberOfElement,1) is Rh
      ! ResistivityOfElement(NumberOfElement,2) is Rv
      !
        EpsilonC_H=Epsilon-(0.0,1.0)/Omega/ResistivityOfElement(NumberOfElement,1)
        EpsilonC_V=Epsilon-(0.0,1.0)/Omega/ResistivityOfElement(NumberOfElement,2)
        SigmaC_H=(0.0,1.0)*Omega*Epsilon+1/ResistivityOfElement(NumberOfElement,1)
        SigmaC_V=(0.0,1.0)*Omega*Epsilon+1/ResistivityOfElement(NumberOfElement,2)
      !
        Index=0
          DO 2 I=1,4
              DO 3 J=I,4
      !
      !...item1,item2,item3
              ai=(-1)**(INT((I+1)/2))
              aj=(-1)**(INT((J+1)/2))
              bi=(-1)**I
              bj=(-1)**J
              item1=( (3+bi*bj)/12*RouCenter+(bi+bj)/24*RouDelta)*(RouDelta)**3;
              item2=(ai*aj+3)/12*(ZDelta)**3;
              item3=(bi+bj)*(RouDelta)**2+( bi*RouCenter-RouDelta/2 )*( bj*RouCenter-RouDelta/2
    )*log(RouMax/RouMin);
      !
      !...reciprocal capacitance arrays
      !...reciprocal capacitance arrays
              ReciprocalCapacitance(I,J)=(2*Pi*ai*aj/Area/Area/EpsilonC_H)*ZDelta*item1+ &
                                (2*Pi/Area/Area/EpsilonC_V)*item2*item3
      !         ReciprocalCapacitance(I,J)=(2*Pi*ai*aj/Area/Area/SigmaC_H)*ZDelta*item1+ &
      !                           (2*Pi/Area/Area/SigmaC_V)*item2*item3
      !
      !...inductance arrays
              Inductance(I,J)=(2*Pi*Mu/Area/Area)*item1*item2
      !
      !...Element Matrix = Impedance = jwL+(1/jwC)
              Matrix(I,J)=(0.,1.0)*Omega*Inductance(I,J)+ &
                          1./((0.,1.)*Omega)*ReciprocalCapacitance(I,J)
      !         Matrix(I,J)=(0.,1.0)*Omega*Inductance(I,J)+ReciprocalCapacitance(I,J)
      !
      !...LocalMatrix is expression of ElementMatrix in one dimension, above diagonal line
      !...LocalMatrix(1)=Matrix(1,1);LocalMatrix(2)=Matrix(1,2);LocalMatrix(3)=Matrix(1,3);
      !...LocalMatrix(4)=Matrix(2,2);LocalMatrix(5)=Matrix(2,3);LocalMatrix(6)=Matrix(3,3);
              Index=Index+1
              LocalMatrix(Index)=Matrix(I,J)
      !
    3         CONTINUE
    2     CONTINUE
      !
    RETURN
      !
    END SUBROUTINE ElementMatrix_Rectangle_NodeBase_Hphi
```

APPENDIX D COMPUTATION METHOD OF ELEMENT MATRIX FOR RECTANGULAR ELEMENT BASED ON ρH_ϕ (SECTION 13.7)

```fortran
      SUBROUTINE ElementMatrix_Rectangle_NodeBase_RouHphi(NumberOfElement,LocalMatrix)
      !...this subroutine is used to calculate element matrix (LocalMatrix)
      !...Element is triangle element.
      !...Matrix is impedance matrix for formula [Zmn][Hn]=[Vm], formula is base on magnetic field.
      !...Matrix is base on NumberOfNodee base function.
      !...Input NumberOfElement is the number of the element
      !...Output LocalMatrix is local matrix for element NumberOfElement.
      !
      INTEGER, INTENT(IN) :: NumberOfElement
      COMPLEX*16, INTENT(OUT):: LocalMatrix(10)
      !
      INTEGER :: I,J,N,Index
      REAL*8  :: ai,aj,bi,bj
      !INTEGER :: NumberOfNode
      REAL*8  :: R(4),Z(4)
      REAL*8  :: RouDelta,RouCenter,RouMax,RouMin
      REAL*8  :: ZDelta,ZCenter
      REAL*8  :: Area
      REAL*8  :: Item1,Item2,Item3,Item4
      COMPLEX*16 :: Inductance(4,4),ReciprocalCapacitance(4,4) ! 3x3 element capacitance and reciprocal inductance element
      COMPLEX*16 :: Matrix(4,4)
      !
      REAL*8  :: Mu,Epsilon
      !COMPLEX*16 :: Mu,Epsilon
      COMPLEX*16 :: EpsilonC_H
      COMPLEX*16 :: EpsilonC_V
      COMPLEX*16 :: SigmaC_H
      COMPLEX*16 :: SigmaC_V
      !
      !...numerical integration for rectangular element
      !
      !             O----------------------O
      !             |   0    0    0        |
      !             |   0    0    0        |
      !             |   0    0    0        |
      !             O----------------------O
      !
      !...rectangular element coordinates
      !
      !         3(K+1,I) (R3,Z3)       4(K+1,I+1) (R4,Z4)
      !             O----------------------O
      !             |                      |
      !             |       0 (R0,Z0)      |  element
      !             |                      |
      !             O----------------------O
      !         1(K,I) (R1,Z1)         2(K,I+1) (R2,Z2)
      !
      MaxNumberOfLocalNode=4
      MaxNumberOfLocalMatrix=10
      !
          DO I=1,4
              R(I)=Coordinates(ABS(NodeFromElement(NumberOfElement,I)),1)
              Z(I)=Coordinates(ABS(NodeFromElement(NumberOfElement,I)),2)
          END DO
      !
      ZDelta=Z(3)-Z(1)
      RouDelta=R(2)-R(1)
      RouMax=R(2)
      RouMin=R(1)
      Roucenter=(R(1)+R(2))/2.
      Zcenter=(Z(1)+Z(3))/2.
      Area=ZDelta*RouDelta
      LocalMatrix=0.
      !
          Epsilon=Epsilon_0*EpsilonROfElement(NumberOfElement)
```

```fortran
      Mu=Mu_0*MuROfElement(NumberOfElement)
!     ResistivityOfElement(NumberOfElement,1) is Rh
!     ResistivityOfElement(NumberOfElement,2) is Rv
!
      EpsilonC_H=Epsilon-(0.0,1.0)/Omega/ResistivityOfElement(NumberOfElement,1)
      EpsilonC_V=Epsilon-(0.0,1.0)/Omega/ResistivityOfElement(NumberOfElement,2)
      SigmaC_H=(0.0,1.0)*Omega*Epsilon+1/ResistivityOfElement(NumberOfElement,1)
      SigmaC_V=(0.0,1.0)*Omega*Epsilon+1/ResistivityOfElement(NumberOfElement,2)
!
      Index=0
         DO 2 I=1,4
            DO 3 J=I,4
!
!...item1,item2,item3
            ai=(-1)**(INT((I+1)/2))
            aj=(-1)**(INT((J+1)/2))
            bi=(-1)**I
            bj=(-1)**J
            item1=( (3+bi*bj)/12*RouCenter+(bi+bj)/24*RouDelta)*(RouDelta)**3;
            item2=(ai*aj+3)/12*(ZDelta)**3;
            item3=(bi+bj)*(RouDelta)**2+( bi*RouCenter-RouDelta/2 )*( bj*RouCenter-RouDelta/2 &
)*log(RouMax/RouMin);
            item4=(bi+bj)*(RouDelta)**2/2-bi*bj*RouCenter*RouDelta+ &
                  ( bi*RouCenter-RouDelta/2 )*( bj*RouCenter-RouDelta/2 )*log(RouMax/RouMin)
!
!...reciprocal capacitance arrays
!...reciprocal capacitance arrays
!
!           ReciprocalCapacitance(I,J)=(2*Pi*ai*aj/Area/Area/EpsilonC_H)*ZDelta*item1+ &
!                                       (2*Pi/Area/Area/EpsilonC_V)*item2*item3
            ReciprocalCapacitance(I,J)=(2*Pi*ai*aj/Area/Area/SigmaC_H)*ZDelta*item4+ &
                                        (2*Pi*bi*bj/Area/Area/SigmaC_V)*item2*log(RouMax/RouMin)
!
!...inductance arrays
            Inductance(I,J)=(2*Pi*Mu/Area/Area)*item2*item4
!
!...Element Matrix = Impedance = jwL+(1/jwC)
!           Matrix(I,J)=(0.,1.0)*Omega*Inductance(I,J)+ &
!                       1./((0.,1.)*Omega)*ReciprocalCapacitance(I,J)
            Matrix(I,J)=(0.,1.0)*Omega*Inductance(I,J)+ReciprocalCapacitance(I,J)
!
!...LocalMatrix is expression of ElementMatrix in one dimension, above diagonal line
!...LocalMatrix(1)=Matrix(1,1);LocalMatrix(2)=Matrix(1,2);LocalMatrix(3)=Matrix(1,3);
!...LocalMatrix(4)=Matrix(2,2);LocalMatrix(5)=Matrix(2,3);LocalMatrix(6)=Matrix(3,3);
            Index=Index+1
            LocalMatrix(Index)=Matrix(I,J)
!
3           CONTINUE
2        CONTINUE
!
RETURN
!
END SUBROUTINE ElementMatrix_Rectangle_NodeBase_RouHphi
```

APPENDIX E TERM 4 (SECTION 13.8.3.9)

```fortran
SUBROUTINE GetMatrixTerm4(M,N,R,Area,IntegralResult)
!
!...this subroutine is used to calculate ingerage of eta(m)*eta(n)/rou on triangle element area
!...R(1:3) is rou value of triangle vertices, Area is triangle area
!
!...for arbitrary triangle
!
!                 (R(1),Z(1))
!                      1
!                    / |\
!                   /  | \
!                  /   |  \
!                 /    |   \ 1_2
!                /     |    \
!               /      |     \
!               |_            \
!            2 ------------------>\ 3
!         (R(2),Z(2))           (R(2),Z(2))
INTEGER, INTENT(IN) :: M,N
REAL*8, INTENT(IN)  :: R(3)
REAL*8, INTENT(IN)  :: Area
REAL*8, INTENT(OUT) :: IntegralResult

INTEGER :: I,J,K
!
    IntegralResult=0
!
!M=1,N=1=>i=1,j=2,k=3;M=2,N=2=>i=2,j=3,k=1;M=3,N=3=>i=3,j=1,k=2;
!M=1,N=2=>i=1,j=2,k=3;M=1,N=3=>i=1,j=3,k=2;M=2,N=3=>i=2,j=3,k=1;
!M=2,N=1=>i=2,j=1,k=3;M=3,N=1=>i=3,j=1,k=2;M=3,N=2=>i=3,j=2,k=1;
    i=M
    j=M+1
    IF (j.eq.4) j=1
    IF (M.NE.N) j=N
    k=6-i-j
!
!----------
!integrate[eta(m)*eta(m)/rou]dS, when R(1),R(2) and R(3) are not equal to each other
!
!M.EQ.N
    IF ((M.EQ.N).AND.(R(i).NE.R(j)).AND.(R(i).NE.R(k)).AND.(R(i).NE.R(k))) THEN
        IntegralResult=Area/( 3*(R(i)-R(j))**3*(R(i)-R(k))**3*(R(j)-R(k)) ) &
                       *( R(i)*(R(i)-R(j))*(R(j)-R(k))*(R(k)-R(i))* &
                           (-R(i)**2+3*R(i)*R(j)+3*R(i)*R(k)-5*R(j)*R(k))* &
                          +2*LOG(R(i))*R(i)*(3*R(j)**2*R(k)**2*(R(j)-R(k))+ &
                                R(i)**2*(R(j)**3-R(k)**3)+R(i)*(- &
3*R(j)**3*R(k)+3*R(j)*R(k)**3)) &
                          +2*LOG(R(j))*R(j)**3*(R(k)-R(i))**3 &
                          +2*LOG(R(k))*R(k)**3*(R(i)-R(j))**3 )
    END IF
!M.NE.N
    IF ((M.NE.N).AND.(R(i).NE.R(j)).AND.(R(i).NE.R(k)).AND.(R(j).NE.R(k))) THEN
        IntegralResult=Area/( 3*(R(i)-R(j))**3*(R(i)-R(k))**2*(R(j)-R(k))**2 ) &
                       *( (R(i)-R(j))*(R(j)-R(k))*(R(k)-R(i))*(-R(i)*R(j)**2- &
R(i)**2*R(j)+R(j)**2*R(k)+R(i)**2*R(k)) &
                          +Log(R(i))*R(i)**2*(R(j)-R(k))**2*(-2*R(i)*R(j)+3*R(j)*R(k)-
R(i)*R(k)) &
                          +Log(R(j))*R(j)**2*(R(i)-R(k))**2*(2*R(i)*R(j)+R(j)*R(k)-3*R(i)*R(k)) &
                          +Log(R(k))*R(k)**3*(R(i)-R(j))**3 )
    END IF
!
!-------
!integrate[eta(m)*eta(m)/rou]dS, when R(i)=R(j) OR R(i)=R(K) OR R(J)=R(j)
!
!M.EQ.N and R(i)=R(j)
    IF ((M.EQ.N).AND.(R(i).EQ.R(j))) THEN
        IntegralResult=Area*(2*R(i)**3-9*R(i)**2*R(k)+18*R(i)*R(k)**2-11*R(k)**3- &
                6*R(k)**3*LOG(R(i))+6*R(k)**3*LOG(R(k)))/9/(R(i)-R(k))**4
    END IF
!M.EQ.N and R(i)=R(k)
    IF ((M.EQ.N).AND.(R(i).EQ.R(k))) THEN
```

```fortran
            IntegralResult=Area*(2*R(i)**3-9*R(i)**2*R(j)+18*R(i)*R(j)**2-11*R(j)**3- &
                6*R(j)**3*LOG(R(i))+6*R(j)**3*LOG(R(j)))/9/(R(i)-R(j))**4
        END IF
!M.EQ.N and R(j)=R(k)
        IF ((M.EQ.N).AND.(R(j).EQ.R(k))) THEN
            IntegralResult=Area*(R(i)**3-6*R(i)**2*R(k)+3*R(i)*R(k)**2+2*R(k)**3+ &
                            6*R(i)*R(k)**2*LOG(R(i)/R(k)))/3/(R(i)-R(k))**4
        END IF
!M.NE.N and R(i)=R(j)
        IF ((M.NE.N).AND.(R(i).EQ.R(j))) THEN
            IntegralResult=Area*(2*R(i)**3-9*R(i)**2*R(k)+18*R(i)*R(k)**2-11*R(k)**3+ &
                +6*R(k)**3*Log(R(k)/R(i)))/18/(R(i)-R(k))**4
        END IF
!M.NE.N and R(i)=R(k)
        IF ((M.NE.N).AND.(R(i).EQ.R(k))) THEN
            integralResult=Area*(R(k)**3-6*R(k)**2*R(j)+3*R(k)*R(j)**2+2*R(j)**3+ &
                6*R(k)*R(j)**2*Log(R(k)/R(j)))/6/(R(k)-R(j))**4
        END IF
!M.NE.N and R(j)=R(k)
        IF ((M.NE.N).AND.(R(j).EQ.R(k))) THEN
             integralResult=Area*(R(k)**3-6*R(k)**2*R(i)+3*R(k)*R(i)**2+2*R(i)**3+ &
                6*R(k)*R(i)**2*Log(R(k)/R(i)))/6/(R(k)-R(i))**4
        END IF
!
RETURN
!
END SUBROUTINE GetMatrixTerm4
```

APPENDIX F COMPUTATION METHOD OF ELEMENT MATRIX FOR TRIANGULAR ELEMENT BASED ON H_ϕ (SECTION 2.7.4)

```fortran
SUBROUTINE ElementMatrix_Triangle_NodeBase_Hphi(NumberOfElement,LocalMatrix)
!...this subroutine is used to calculate element matrix (LocalMatrix)
!...Element is triangle element.
!...Matrix is impedance matrix for formula [Zmn][Hn]=[Vm], formula is base on magnetic field.
!...Matrix is base on node base function.
!...Input NumberOfElement is the number of the element
!...Output LocalMatrix is local matrix for element NumberOfElement.
!
INTEGER, INTENT(IN) :: NumberOfElement
COMPLEX*16, INTENT(OUT):: LocalMatrix(6)
!
INTEGER :: I,J,K,M,N,Index
REAL*8 :: R(5),Z(5),l(3),h(3)   !Real R and Z only has 3 value.
REAL*8 :: Area
REAL*8 :: IntegralResult
REAL*8 :: Term1a,Term1b,Term2,Term3,Term4,Term5
COMPLEX*16 :: Inductance(3,3),ReciprocalCapacitance(3,3) ! 3x3 element capacitance and reciprocal inductance element
COMPLEX*16 :: Matrix(3,3)
!
REAL*8 :: Mu,Epsilon
!COMPLEX*16 :: Mu,Epsilon
COMPLEX*16 :: EpsilonC_H
COMPLEX*16 :: EpsilonC_V
!
!...structure of triangle element
!
!                 1
!               /|I\
!              / . |  \
!          l_3 /   |    \
!            /     |     \  l_2              Triangle geometry;
!           /   h_1|      \                  vector h_i is from vertex i to
!          /       |v      \                 edge i and perpendicular to edge i;
!         |_               \                 2*area =|l_i X l_i+1| = |l_i||h_i|
!       2 ------------------>\ 3
!                 l_1
!
!      (R(1),Z(1))    (R(4),Z(4))
!                 1
!               / |\
!              /  | \
!             /   |  \
!            /    |   \  l_2
!           /     |    \
!          /      |     \
!         |_             \
!       2 ------------------>\ 3
!      (R(2),Z(2))        (R(3),Z(3))
!      (R(5),Z(5))
!
!
      MaxNumberOfLocalNode=3
      MaxNumberOfLocalMatrix=6

!...element coordinates, area, edge and height
!...2*area =|l_i X l_i+1| = |l_i||h_i|
      DO I=1,3
          R(I)=Coordinates(ABS(NodeFromElement(NumberOfElement,I)),1)
          Z(I)=Coordinates(ABS(NodeFromElement(NumberOfElement,I)),2)
      END DO
      R(4)=R(1)  !R(4),R(5),Z(4) and Z(5) are for calculation convenience
      R(5)=R(2)
```

```fortran
      Z(4)=Z(1)
      Z(5)=Z(2)
      Area=0.5*((R(3)-R(2))*(Z(1)-Z(3))-(Z(3)-Z(2))*(R(1)-R(3)));
!     l(1)=sqrt((Z(3)-Z(2))*(Z(3)-Z(2))+(R(3)-R(2))*(R(3)-R(2)));
!     l(2)=sqrt((Z(1)-Z(3))*(Z(1)-Z(3))+(R(1)-R(3))*(R(1)-R(3)));
!     l(3)=sqrt((Z(2)-Z(1))*(Z(2)-Z(1))+(R(2)-R(1))*(R(2)-R(1)));
!     h(1)=2*Area/l(1);
!     h(2)=2*Area/l(2);
!     h(3)=2*Area/l(3);
!     eta(1)=((x(3)-x)*(y(3)-y(2))-(x(3)-x(2))*(y(3)-y))/2/Area
!     eta(2)=((x(1)-x)*(y(1)-y(3))-(x(1)-x(3))*(y(1)-y))/2/Area
!     eta(3)=((x(2)-x)*(y(2)-y(1))-(x(2)-x(1))*(y(2)-y))/2/Area
!
!...test
      IF(Area.LT.0) THEN
          WRITE(*,*)
          WRITE(*,*) 'Wrong Message: Triangle vertices should be arranged counter clockwise.'
          WRITE(*,*)
          STOP
      END IF
!---end of test .
!
!...calculate Epsilon, Mu, EpsilonC_H, EpsilonC_V
      Epsilon=Epsilon_0*EpsilonROfElement(NumberOfElement)
      Mu=Mu_0*MuOfElement(NumberOfElement)
!     ResistivityOfElement(NumberOfElement,1) is Rh
!     ResistivityOfElement(NumberOfElement,2) is Rv
      EpsilonC_H=Epsilon-(0.0,1.0)/Omega/ResistivityOfElement(NumberOfElement,1)
      EpsilonC_V=Epsilon-(0.0,1.0)/Omega/ResistivityOfElement(NumberOfElement,2)
!
!...calculate element matrix
      LocalMatrix(:)=0.
      Index=0
      DO M=1,3
          DO N=M,3
!
!...Term1a,Term1b,Term2,Term3,Term4
              Term1a=(Z(M+1)-Z(M+2))*(Z(N+1)-Z(N+2))*(R(1)+R(2)+R(3))/12/Area
              Term1b=(R(M+1)-R(M+2))*(R(N+1)-R(N+2))*(R(1)+R(2)+R(3))/12/Area
              Term2=(Z(M+1)-Z(M+2))/6.0   !(z(m+1)-z(m-1))/6
              Term3=(Z(N+1)-Z(N+2))/6.0   !(z(n+1)-z(n-1))/6
              CALL GetMatrixTerm4(M,N,R,Area,IntegralResult)
              Term4=IntegralResult
!
!...Term5=Ae/60*{R(1)[6 2 2;2 2 1;2 1 2]+
!                R(2)[2 2 1;2 6 2;1 2 2]+R(3)[2 1 2;1 2 2;2 2 6]}
              Term5=2*R(1)+2*R(2)+2*R(3)
              IF( M /= N) Term5 = Term5-R(6-M-N) ! correct off-diagonal entries
              IF( M == N) Term5 = Term5+4*R(M)   ! correct diagonal entries
              Term5=Term5*Area/60
!
!...reciprocal capacitance arrays
              ReciprocalCapacitance(M,N)=Term1a/EpsilonC_H+Term1b/EpsilonC_V+ &
                         Term2/EpsilonC_V+Term3/EpsilonC_V+Term4/EpsilonC_V
              ReciprocalCapacitance(M,N)=2*Pi*ReciprocalCapacitance(M,N)
!
!...inductance arrays
              Inductance(M,N)=2*Pi*Mu*Term5
!
!...Element Matrix = Impedance = jwL+(1/jwC)
              Matrix(M,N)=(0.,1.0)*Omega*Inductance(M,N)+ &
                         1./((0.,1.)*Omega)*ReciprocalCapacitance(M,N)
!                 Matrix(M,N)=-Omega**2*Inductance(M,N)+ReciprocalCapacitance(M,N)
!
!...LocalMatrix is expression of ElementMatrix in one dimension, above diagonal line
!...LocalMatrix(1)=Matrix(1,1);LocalMatrix(2)=Matrix(1,2);LocalMatrix(3)=Matrix(1,3);
!...LocalMatrix(4)=Matrix(2,2);LocalMatrix(5)=Matrix(2,3);LocalMatrix(6)=Matrix(3,3);
              Index=Index+1
              LocalMatrix(Index)=Matrix(M,N)
!             WRITE(*,*) 'Term4(',M,N,')=',Term4
!             WRITE(*,*) 'Matrix(',M,N,')=',Matrix(M,N)
!             WRITE(*,*) 'localMatrix(',Index,')=',LocalMatrix(Index)
          END DO
      END DO
!
RETURN
END SUBROUTINE ElementMatrix_Triangle_NodeBase_Hphi
```

APPENDIX G COMPUTATION METHOD OF ELEMENT MATRIX FOR TRIANGULAR ELEMENT BASED ON E_ϕ (SECTION 3.3)

```fortran
SUBROUTINE ElementMatrix_Triangle_NodeBase_Ephi(NumberOfElement,LocalMatrix)
INTEGER, INTENT(IN) :: NumberOfElement
COMPLEX*16, INTENT(OUT):: LocalMatrix(6)
!
INTEGER :: I,J,K,M,N,Index
REAL*8 :: R(5),Z(5),l(3),h(3)   !Real R and Z only has 3 value.
REAL*8 :: Area
REAL*8 :: IntegralResult
REAL*8 :: Term1a,Term1b,Term2,Term3,Term4,Term5
COMPLEX*16 :: ReciprocalInductance(3,3),Capacitance(3,3) ! 3x3 element capacitance and reciprocal inductance element
COMPLEX*16 :: Matrix(3,3)
!
REAL*8 :: Mu,Epsilon
!COMPLEX*16 :: Mu,Epsilon
COMPLEX*16 :: EpsilonC_H
COMPLEX*16 :: EpsilonC_V
!
!...structure of triangle element
!
!                     1
!                   /|\
!                  / | \
!           l_3  /   |  \
!              /     |   \  l_2
!             /    h_1|   \
!            |_       v    \
!           2 ----------------->\ 3
!                     l_1
!
!    (R(1),Z(1))    (R(4),Z(4))
!                     1
!                   / |\
!                  /  | \
!                 /   |  \  l_2
!                /    |   \
!               |_    |    \
!           2 ----------------->\ 3
!   (R(2),Z(2))         (R(3),Z(3))
!   (R(5),Z(5))
!
    MaxNumberOfLocalNode=3
    MaxNumberOfLocalMatrix=6
!
!...element coordinates, area, edge and height
!...2*area =|l_i X l_i+1| = |l_i||h_i|
    DO I=1,3
        R(I)=Coordinates(ABS(NodeFromElement(NumberOfElement,I)),1)
        Z(I)=Coordinates(ABS(NodeFromElement(NumberOfElement,I)),2)
    END DO
    R(4)=R(1) !R(4),R(5),Z(4) and Z(5) are for calculation convenience
    R(5)=R(2)
    Z(4)=Z(1)
    Z(5)=Z(2)
        Area=0.5*((R(3)-R(2))*(Z(1)-Z(3))-(Z(3)-Z(2))*(R(1)-R(3)));
!       l(1)=sqrt((Z(3)-Z(2))*(Z(3)-Z(2))+(R(3)-R(2))*(R(3)-R(2)));
!       l(2)=sqrt((Z(1)-Z(3))*(Z(1)-Z(3))+(R(1)-R(3))*(R(1)-R(3)));
!       l(3)=sqrt((Z(2)-Z(1))*(Z(2)-Z(1))+(R(2)-R(1))*(R(2)-R(1)));
!       h(1)=2*Area/l(1);
!       h(2)=2*Area/l(2);
!       h(3)=2*Area/l(3);
!       eta(1)=((x(3)-x)*(y(3)-y(2))-(x(3)-x(2))*(y(3)-y))/2/Area
!       eta(2)=((x(1)-x)*(y(1)-y(3))-(x(1)-x(3))*(y(1)-y))/2/Area
!       eta(3)=((x(2)-x)*(y(2)-y(1))-(x(2)-x(1))*(y(2)-y))/2/Area
```

```fortran
!
!...test
      IF(Area.LT.0) THEN
         WRITE(*,*)
         WRITE(*,*) 'Wrong Message: Triangle vertices should be arranged counter clockwise.'
         WRITE(*,*)
         STOP
      END IF
!---end of test .
!
!...calculate Epsilon, Mu, EpsilonC_H, EpsilonC_V
      Epsilon=Epsilon_0*EpsilonROfElement(NumberOfElement)
      Mu=Mu_0*MuOfElement(NumberOfElement)
    ! ResistivityOfElement(NumberOfElement,1) is Rh
    ! ResistivityOfElement(NumberOfElement,2) is Rv
      EpsilonC_H=Epsilon-(0.0,1.0)/Omega/ResistivityOfElement(NumberOfElement,1)
      EpsilonC_V=Epsilon-(0.0,1.0)/Omega/ResistivityOfElement(NumberOfElement,2)
!
!...calculate element matrix
      LocalMatrix(:)=0.
      Index=0
      DO M=1,3
         DO N=M,3
!
!...Term1a,Term1b,Term2,Term3,Term4
            Term1a=(Z(M+1)-Z(M+2))*(Z(N+1)-Z(N+2))*(R(1)+R(2)+R(3))/12/Area
            Term1b=(R(M+1)-R(M+2))*(R(N+1)-R(N+2))*(R(1)+R(2)+R(3))/12/Area
            Term2=(Z(M+1)-Z(M+2))/6.0   !(z(m+1)-z(m-1))/6
            Term3=(Z(N+1)-Z(N+2))/6.0   !(z(n+1)-z(n-1))/6
            CALL GetMatrixTerm4(M,N,R,Area,IntegralResult)
            Term4=IntegralResult
!
!...Term5=Ae/60*{R(1)[6 2 2;2 2 1;2 1 2]+
!              R(2)[2 2 1;2 6 2;1 2 2]+R(3)[2 1 2;1 2 2;2 2 6]}
            Term5=2*R(1)+2*R(2)+2*R(3)
            IF( M /= N) Term5 = Term5-R(6-M-N) ! correct off-diagonal entries
            IF( M == N) Term5 = Term5+4*R(M)   ! correct diagonal entries
            Term5=Term5*Area/60
!
!...reciprocal inductance arrays
            ReciprocalInductance(M,N)=Term1a/Mu+Term1b/Mu+ &
                        Term2/Mu+Term3/Mu+Term4/Mu
            ReciprocalInductance(M,N)=2*Pi*ReciprocalInductance(M,N)
!
!...capacitance arrays, only relate to EpsilonC_H
            Capacitance(M,N)=2*Pi*EpsilonC_H*Term5

!
!...Element Matrix = Admittance = jwC+(1/jwL)
            Matrix(M,N)=(0.,1.0)*Omega*Capacitance(M,N)+ &
                        1./((0.,1.)*Omega)*ReciprocalInductance(M,N)
!              Matrix(M,N)=-Omega**2*Capacitance(M,N)+ReciprocalInductance(M,N)
!
!...LocalMatrix is expression of ElementMatrix in one dimension, above diagonal line
!...LocalMatrix(1)=Matrix(1,1);LocalMatrix(2)=Matrix(1,2);LocalMatrix(3)=Matrix(1,3);
!...LocalMatrix(4)=Matrix(2,2);LocalMatrix(5)=Matrix(2,3);LocalMatrix(6)=Matrix(3,3).
            Index=Index+1
            LocalMatrix(Index)=Matrix(M,N)
!           WRITE(*,*) 'Term4(',M,N,')=',Term4
!           WRITE(*,*) 'Matrix(',M,N,')=',Matrix(M,N)
!           WRITE(*,*) 'localMatrix(',Index,')=',LocalMatrix(Index)
!
         END DO
      END DO
!
RETURN
!
END SUBROUTINE ElementMatrix_Triangle_NodeBase_Ephi
```

CHAPTER 14

Resistivity Imaging Tools

Contents

14.1 Introduction	559
14.2 Water-Based Mud Resistivity Imaging Tool	561
14.3 The Oil-Based Mud Resistivity Imager	562
14.3.1 Circuit model of an oil-based mud resistivity imager	564
14.3.2 Three-dimensional finite element analysis of an OBMI tool	568
14.3.3 Vertical resolution of OBMI tool	572
14.4 Conclusions	576
References	578

14.1 INTRODUCTION

Resistivity imagers are widely used in the logging industry to obtain resistivity changes in both vertical and azimuthal directions in a relatively high resolution, from which the lithology, cracks, and dip angles of the formation can be derived [1–5]. The resistivity imaging tools have relatively shallow depth of investigation (DOI) compared to other logging tools. The DOI of a resistivity imaging tool is in the range of one to a few inches. In the wireline logging process, the tool is mounted on a pad and is pushed against the borehole wall as shown in Fig. 14.1A and B. The pusher usually has 4–6 arms and each arm is mounted with a resistivity image pad. Due to the limit of the pusher, the area of the measurement does not cover the full borehole surface. The covered area is usually about 60% of the total borehole surface. Therefore the measured images are sliced in the azimuthal direction as shown in Fig. 14.1C. Due to the tool applications can be in both water-based mud (WBM) and oil-based mud, the tool configurations are quite different. In this chapter, we will discuss both cases. Logging-while-drilling (LWD) resistivity imagers are directly mounted on the mandrel as shown in Fig. 14.2A. Since the LWD tools are constantly rotating, the measured resistivity images are also continuous as shown in Fig. 14.2B.

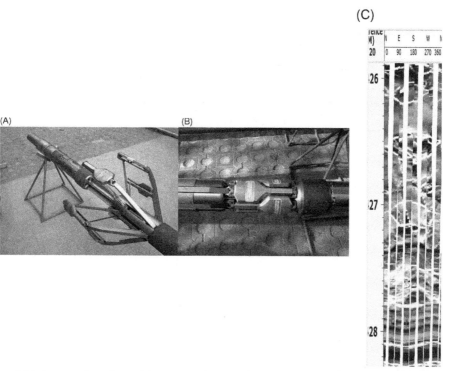

Figure 14.1 A water-based mud imager with 6 pushing arms and pads. (A) Pusher is open, (B) pusher is closed, and (C) the measured resistivity image.

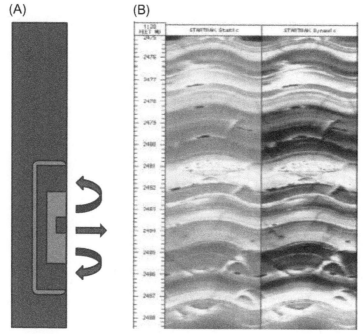

Figure 14.2 LWD resistivity imager (A) the configuration of an LWD imager and (B) the continuous resistivity image obtained by the LWD imager.

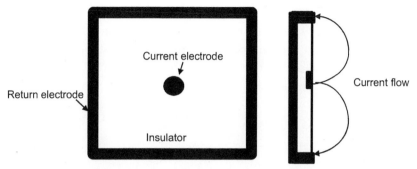

Figure 14.3 The basic resistivity imager configuration. The outside boundary is the return electrode for the current emitting electrode at the center.

14.2 WATER-BASED MUD RESISTIVITY IMAGING TOOL

Due to the fact that WBM is mostly used in drilling activities, the WBM resistivity imager is one of the most common imaging tools in the logging industry. A WBM resistivity imager is basically an array of Ohm meters as shown in Fig. 14.1. The tool is usually composed of an array of current ejection buttons and a current return electrode. A fixed voltage is applied to each of the buttons and currents flowing out from these buttons are measured. Since the voltage is fixed, the current measured at each button is proportional to the formation resistivity adjacent to the measurement point. In this sense, the resistivity imager has no focusing mechanism and therefore, DOI of this tool is relatively shallow. The operating frequency is also low and can be considered as a direct current (DC) tool. The use of alternating current (AC) instead of DC is to facilitate the signal measurement and processing. Fig. 14.3 shows a schematic drawing of a basic resistivity imager used in WBM. The basic structure of the imager is composed of one current emitting button and a return electrode a distance away from the button. The actual imager pad may have two or more rows of buttons, which are used for increased radial resolution and data correlation.

From Fig. 14.3, we can see that this configuration can be considered as a point source with the return electrode some distance away. When the voltage between the button electrode and the return electrode is fixed at V, the measured current is I, the resistivity of the formation near the current electrode can be calculated as:

$$\rho = k \frac{V}{I} \qquad (14.1)$$

where k is the tool constant. The tool constant is usually obtained when the tool is calibrated in a known formation, such as in a saline solution with known resistivity. In practical implementation, each button in the pad is applied with a slightly different frequency and the measurement is only performed at the applied frequency. Therefore the interference between buttons is minimized.

Resolution can be evaluated by using simple analytical equations. Assuming the button shown in Fig. 14.3 has a semi-spherical shape of radius a, the return electrode is very far from the button and symmetrical with respect to the button. The electric field $E(r)$ and the potential $\phi(r)$ of the button in the front side of the pad can be easily found by using Gaussian theorem:

$$\overline{E} = \frac{I}{4\pi r^2 \sigma}\overline{r} = \frac{I\rho}{4\pi r^2}\overline{r} \qquad (14.2a)$$

$$\phi = \frac{I}{4\pi r \sigma} = \frac{I\rho}{4\pi r} \qquad (14.2b)$$

where I is the current flowing out from the button, σ is the conductivity of the formation, ρ is the resistivity of the formation, and $\overline{r}$ is the unit vector pointing to the vertical direction of the button surface. From Eqs. (14.2a,b) we can see that the field is inversely proportional to the distance away from the button surface and the E field is directly proportional to the measured current at the button. Fig. 14.4 shows the current distribution near a simplified resistivity image pad as in Fig. 14.3. We can see that the E field decays very fast with distance. Therefore we can conclude that the DOI of the imagers based on the button structure is very shallow. Fig. 14.5 shows the potential distribution in the space near the button in the cross-sectional area of the WBM pad. To investigate the relationship of the distance between the electrode-ground plan and the DOI of a WBM imager, consider a numerical model shown in Fig. 14.6A where a low resistivity layer between the pad and the formation is added to represent the mud layer between the pad and the formation. Fig. 14.6B shows the potential distribution along near the electrode when the mud layer presents. Comparing the potential distributions shown in Figs. 14.4—14.6, we can have the following two conclusions: (1) the return ground plan has impacts to the DOI (Figs. 14.4 and 14.5) and (2) the mud layer due to the loose contact between the pad and the formation wall will also influence the DOI. Fig. 14.6C shows the simulated DOI as a function of the distance between the electrode and return ground plan when the mud layer presents. From the DOI plot as a function of the electrode spacing, it is clear that the spacing has impact to the DOI but not critical. The conductive mud has very little impact to the DOI since the higher resistivity of the formation has much greater attenuation to the field.

14.3 THE OIL-BASED MUD RESISTIVITY IMAGER

The resistivity imager discussed in Section 14.2 works well when the mud in the borehole is conductive. If it is used in an oil-based mud borehole, there will be a thin

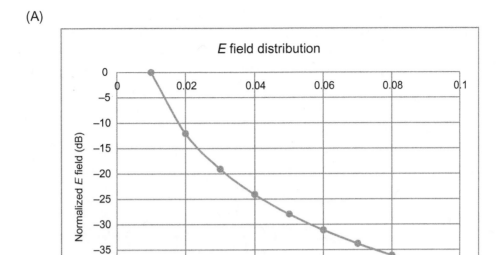

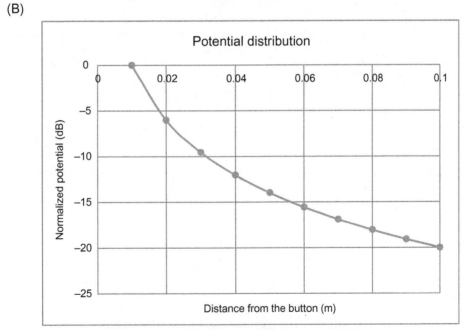

Figure 14.4 (A) The normalized potential E field and (B) potential distributions near the button in a 1 ohm-m homogeneous formation and a 1 A current source as a function of the distance from the button.

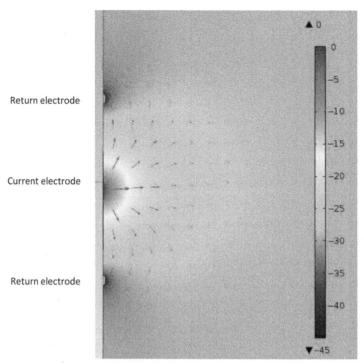

Figure 14.5 The potential field distribution into the formation of a WBM resistivity imager when the button diameter is 1 cm, the formation resistivity is 1 ohm-m, and the distance between the electrode and the return ground plan is 7.5 cm.

layer of nonconductive mud in between the pad and the borehole. Due to the low operating frequency, the coupling between the pad current electrodes and the borehole becomes very weak. The measured resistivity is completely blocked by the thin layer of insulating mud. Therefore the resistivity imager discussed in Section 14.2 will not work in the oil-based mud due to the thin layer of oil-based mud. To overcome this problem, a resistivity imager that can be used in an oil-based mud is developed. This device has two electrodes symmetrically excited with an AC voltage about 100 kHz. The measurement is done by a pair of electrode located in the middle of the two excitation electrodes as shown in Fig. 14.7.

14.3.1 Circuit model of an oil-based mud resistivity imager

The oil based mud imager (OBMI) tool is based on four-terminal measurement method. The schematic in Fig. 14.7 shows a pad applied against the borehole wall with possibly a small standoff. An AC source, I, is injected into the formation between two current-injector electrodes A and B located above and below five pairs of small button sensors. The potential difference δV is measured between the button sensors C and D [6–8].

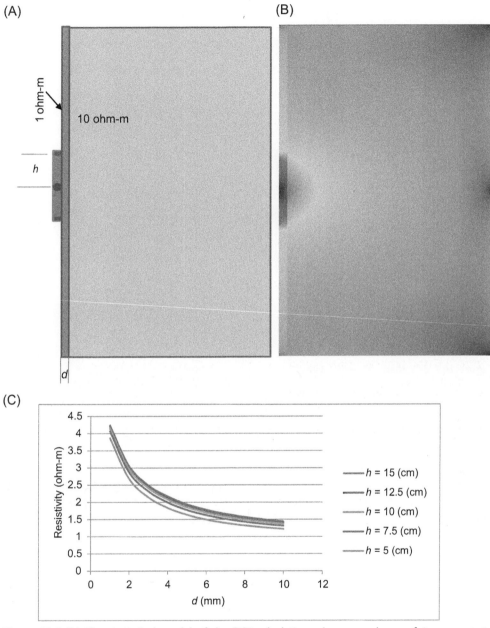

Figure 14.6 (A) The numerical model of the DOI calculation when a mud gap of 1 cm presents between the pad and the formation; (B) the potential distribution in the mud gap and the formation; and (C) the DOI as a function of the spacing between the button and the return ground obtained from the model shown in (A). The mud resistivity is 1 ohm-m, the mud gap is 1 cm, the button diameter is 1 cm, and the formation resistivity is 10 ohm-m.

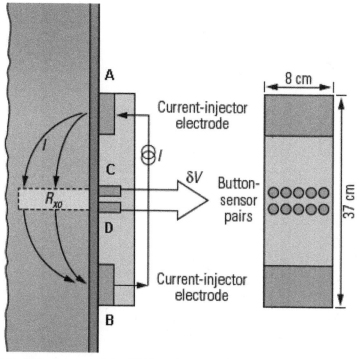

Figure 14.7 Schematic diagram of the OBMI pad against the borehole wall in side-view (left) and in front-view (right) [6].

To illustrate the operation of the OBMI tool, consider the two button model shown in Fig. 14.8A. The key assumption in this model is that the current source is an ideal current source, in other word, the current source can generate a constant current I regardless of the resistance of the load. The equivalent circuit of this model is shown in Fig. 14.8B. If we further assume the input impedance of the measurement amplifier is infinity, or much greater than the resistance of the gap between the measurement button and the borehole wall and the formation, the circuit in Fig. 14.8B can be simplified to Fig. 14.8C.

Although the operating frequency of the oil-based mud imager is not very high, due to the high resistance of the mud, the displacement currents cannot be ignored compared with the conducting current, which is small. Therefore the circuit model in Fig. 14.8 has capacitors to provide paths to the displacement currents. To understand the operating concept of the oil-based mud imager, we simplify the circuit in Fig. 14.8B to Fig. 14.8C assuming the mud has much higher resistance than the formation. Using the model in Fig. 14.8B, we can easily figure out the voltage measurement by the imager is

$$V_0 = Z_f I_0 \tag{14.3}$$

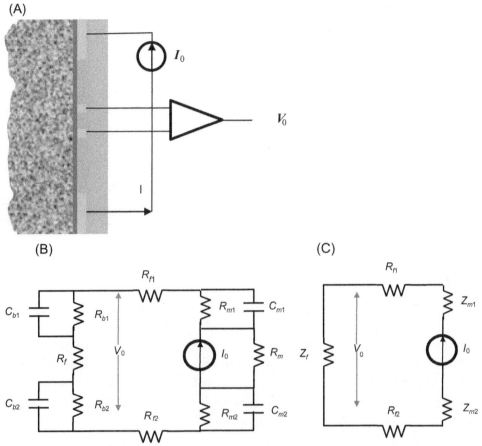

Figure 14.8 The equivalent circuits of an oil-based mud imager. Where R_{m1} and C_{m1} are the mud resistance and capacitance between the upper transmitter electrode and the formation; R_{m2} and C_{m2} are the mud resistance and capacitance between the upper transmitter electrode and the formation; R_m represents the leakage resistance between two transmitting electrodes; R_{f1} and R_{f2} are the formation resistances between the transmitter electrodes and the upper and lower measurement buttons; R_{b1}, C_{b1}, and R_{b2}, C_{b2} are mud resistance and capacitance between the formation and the measurement buttons; and R_f is the formation resistance between two measurement buttons, (A) the schematic structure of an oil-based mud imager; (B) the equivalent circuit of the oil-based mud imager shown in (A); and (C) simplified equivalent circuit of the oil-based mud imager shown in (A), where $R'_f = Z_{b1} + R_f + Z_{b2}$. Where Z_{b1} and Z_{b2} is the complex impedance of R_{b1}, C_{b1} and R_{b2} and C_{b2}.

Since

$$Z_f = R_f + \frac{R_{b1}}{1 + j\omega C_{b1} R_{b1}} + \frac{R_{b2}}{1 + j\omega C_{b2} R_{b2}} \tag{14.4}$$

Note that in the circuit model shown in Fig. 14.8, the current source is assumed to be an ideal current source with a constant current I_0 and infinitively large impedance. Or practically, the impedance of the current source is much higher than that of the mud resistance R_m. The voltage measured by the imager receiver is a function of both formation resistivity R_f and mud resistivity and the capacitance between the measurement electrode and the formation (C_{b1}, C_{b2}, R_{b1}, R_{b2}). If the mud has a dielectric constant of ε_m, and a resistivity of ρ_m, the distance between the pad and the formation surface is δ, the effective surface of the measurement electrode is S_e, the effective resistance and capacitance can be expressed as

$$R_b = \rho_m \frac{\delta}{S_e} \tag{14.5a}$$

$$C_b = \varepsilon_m \frac{S_e}{\delta} \tag{14.5b}$$

Substituting Eqs. (14.5a,b) into (14.4), we have

$$Z_f = R_f + 2\left(\frac{\delta}{S_e}\right)\frac{\rho_m}{1 + j\omega\rho_m\varepsilon_m} \tag{14.6}$$

In deriving Eq. (14.6), it is assumed that the two measurement buttons have the same electrical properties including area and distance from the formation. From Eqs. (14.6) and (14.3) we can see that the measured voltage is a function of formation resistivity but also a function of mud parameters. In most cases, mud properties do not change sharply.

14.3.2 Three-dimensional finite element analysis of an OBMI tool

The analysis in Section 14.3.1 is qualitative and can be used as a guide for the development of OBMI tools. However, to accurately understand the OBMI probe and its performances in various logging conditions, a complete numerical simulation must be conducted. To do so, a three-dimensional (3D) finite element method (FEM) mode is established to simulate the performance of the pad.

14.3.2.1 Development of OBMI numerical model

The 3D FEM model used for OBMI simulations is developed using COMSOL Multiphysics. Fig. 14.9 shows the geometry of the model. In the model, a homogenous medium represents geological formation. The cylindrical borehole is filled with oil-based mud. An OBMI pad is applied on the borehole wall to image formation resistivity. The geometry and the electrical properties of the OBMI pad can be

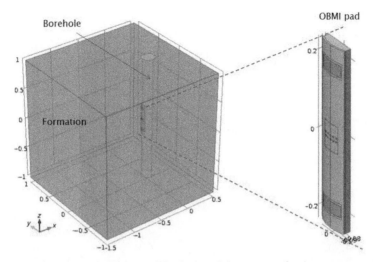

Figure 14.9 Three-dimensional OBMI model. The model consists of a homogenous formation, a cylindrical borehole filled with oil-based mud, and an OBMI pad. The geometry and electrical properties of the model can be adjusted for different simulation purposes.

Table 14.1 Electrical properties of the OBMI model

Material	Resistivity (ohm-m)	Relative permittivity
Formation	0.1–10000	5
Oil-based mud	10^6	5
OBMI pad	10^{16}	3.2
Current emitting electrodes and voltage measurement buttons	$1.67*10^{-6}$	1

adjusted for different simulation purposes. In Fig. 14.9, the diameter of the borehole is 300 mm. The length, width, and thickness of the pad are 370, 80, and 27 mm, respectively. The length and the width of the current electrodes are 80 and 50 mm, respectively. The distance between centers of the two current electrodes is 320 mm. The diameter of the measurement button is 10 mm and the distance between measurement button centers is 20 mm.

The electrical properties of the model can be adjusted for different simulation purposes. Typical values used in the OBMI numerical simulations are listed in Table 14.1.

14.3.2.2 Current distributions

Fig. 14.10 shows the current density distribution in the 3D model. The current density in the figure is in logarithm scale, i.e., $20*\log 10(J)$ where J represents the

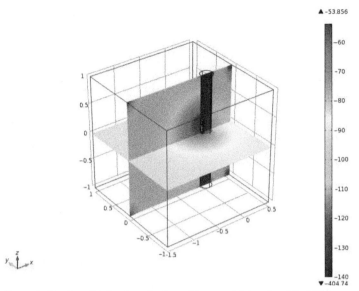

Figure 14.10 Current density distribution (in logarithm scale) in the 3D OBMI model. It can be seen that most of the current is able to be injected into formation through the highly resistive oil-based mud.

current density. Based on the OBMI model, the current density distribution in both formation and borehole have been simulated. The frequency used in the model here was 10 kHz and the injecting current was 1 mA. The material properties in Table 14.1 were used. Fig. 14.10B shows the current density (in logarithm scale) distribution on an $x-y$ plane ($z = 0$) and an $x-z$ plane ($y = 0$). The amplitude of current density has been color coded. *Red* (dark gray in print versions) color represents greater current density. It can be seen that the current in the borehole is rather small except the area near the two current pads, which indicates that most of the current is able to penetrate the resistive oil-based mud into the formation. It can be concluded that most of the transmitting current can penetrate the oil layer and pass through the formation despite the existence of the gap between the pad and the formation.

14.3.2.3 dv versus formation resistivity

Fig. 14.11 shows the voltage difference on measurement buttons dv versus formation resistivity of the model described in Section 14.2.1 (Fig. 14.8 and Table 14.1). In this section, the pad standoff is set to be 1 mm. The emitting current is set to be 1 mA. It can be seen that the dv has a linear relationship with formation resistivity within a quite wide range (0.1–10,000 ohm-m).

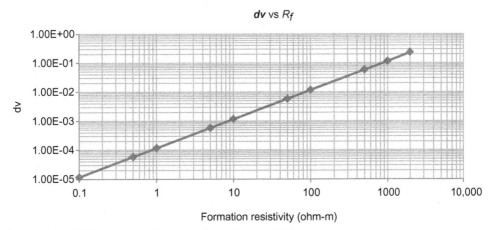

Figure 14.11 OBMI response (*dv*) versus formation resistivity.

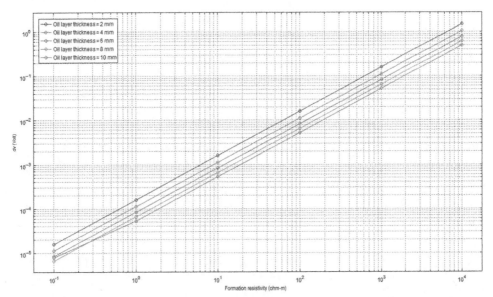

Figure 14.12 OBMI response versus formation resistivity for different standoff distance. The OBMI response decreases with increasing standoff distance. This is because the impedance between current electrodes and formation wall increases with increasing standoff distance, and accordingly the current flows through the formation decreases.

14.3.2.4 Effect of the thickness of oil-based mud layer (standoff distance)

Standoff distance usually refers to the thickness of the oil layer between the current electrodes and the formation wall. Fig. 14.12 shows the voltage difference on measurement buttons dv versus formation resistivity for different standoff distances. The standoff distance was set to be 2, 4, 6, and 8 mm, respectively. It can be seen that the

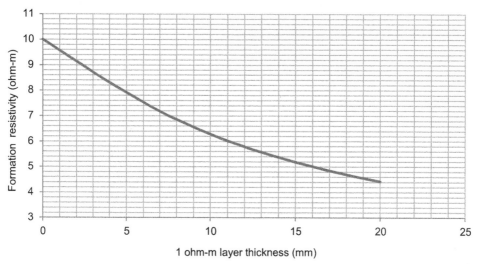

Figure 14.13 Simulated formation resistivity for a two-cylindrical layer formation model. The layer adjacent to the borehole wall has resistivity of 1 ohm-m, the other layer deeper has a resistivity of 10 ohm-m.

dv decrease with increasing standoff distance. That is because the capacitive impedance between current electrodes and formation wall increases with increasing standoff distance. Accordingly, the current flow through the formation decreases and the current leakage inside the OBMI pad increases.

14.3.2.5 Depth of investigation of the OBMI Tool

The DOI is a very important parameter of an OBMI tool. It tells us how deep the OBMI tool can see into the formation. In this section, we still use a model similar with that in Fig. 14.6 of Section 14.2. The model has two cylindrical layers: the layer adjacent to the borehole wall has resistivity of 1 ohm-m and the other layer deeper has a resistivity of 10 ohm-m. Fig. 14.13 shows the simulated formation resistivity obtained from the OBMI model as the 1 ohm-m layer thickness increases. If we define the thickness of the 1 ohm-m layer as DOI when the resistivity reads 5 ohm-m, we can see that the OBMI model described in this section has a DOI of 16 mm. Similar with WBM imager tool, the oil-based mud imager tool also has quite shallow DOI and can only depict the resistivity.

14.3.3 Vertical resolution of OBMI tool

In general, for galvanic methods (e.g., OBMI) used in well logging, the separation distance between current electrodes determines DOI and the separation distance between measurement buttons determines detection resolution. To investigate the

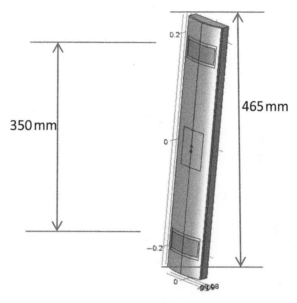

Figure 14.14 Schematic diagram of the OBMI pad model. The length, width, and thickness of the pad are 465, 80, and 25 mm, respectively. The length and the width of the current electrodes are 65 and 27 mm, respectively. The distance between measurement buttons is 9 mm.

Table 14.2 Parameters of borehole environment

Parameter	Value
Oil-based mud resistivity	10^6 ohm-m
Oil-based mud dielectric constant	5
Borehole diameter	8.5 in.
Oil film thickness (standoff distance)	2 mm

vertical resolution of this OBMI pad, we consider a numerical model as shown in Fig. 14.14 and use the formation and borehole parameters as shown in Table 14.2.

To investigate the vertical resolution of the OBMI tool, let us consider a numerical model with a homogenous medium with a resistivity of 1 and 10 ohm-m as background and a 3 cm thin bed anomaly inserted in the middle of the homogenous medium, corresponding to the position of $z = 0$ in the model as shown in Fig. 14.15. The computation of the resistivity measured by the OBMI pad uses the following linear conversion from the measured voltage between the electrodes and the current in the transmitter pad:

$$R = K*dV/I \qquad (14.7)$$

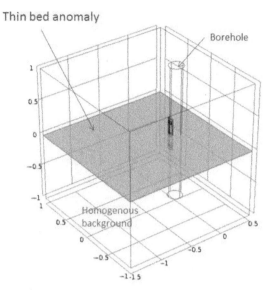

Figure 14.15 The OBMI model used in the inversion exercise. The model consists a homogenous background with resistivity of 1 and 10 ohm-m, respectively. A thin resistivity anomaly bed was added in the background with different resistivity. The thickness of the anomaly bed is 3 cm.

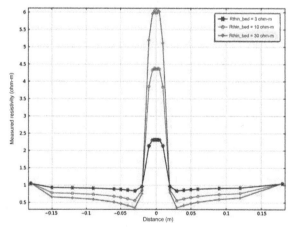

Figure 14.16 Inversion result of the OBMI model. The OBMI model has a homogenous background with resistivity of 1 ohm-m. The thickness of the bed anomaly is 3 cm. The resistivity of the bed anomaly are 3 ohm-m (*blue* (dark gray in print versions)), 10 ohm-m (*green* (light gray in print versions)), and 30 ohm-m (*red* (gray in print versions)), respectively.

The dV and I are measured voltage and current by the OBMI system and the parameter K is a constant usually called tool constant and can be obtained by a simple calibration procedure.

Figs. 14.16 and 14.17 show the inversion result for different anomaly resistivity with background resistivity of 1 and 10 ohm-m, respectively. From Figs. 14.16 and

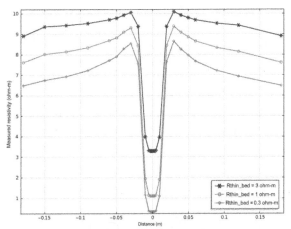

Figure 14.17 Inversion result of the OBMI model. The OBMI model has a homogenous background with resistivity of 10 ohm-m. The thickness of the bed anomaly is 3 cm. The resistivities of the bed anomaly are 3 ohm-m (*blue* (dark gray in print versions)), 1 ohm-m (*green* (light gray in print versions)), and 0.3 ohm-m (*red* (gray in print versions)), respectively.

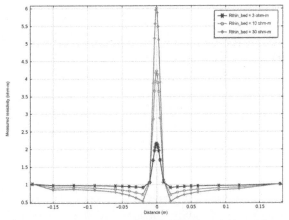

Figure 14.18 Inversion result of the OBMI model. The OBMI model has a homogenous background with resistivity of 1 ohm-m. The thickness of the bed anomaly is 1 cm. The resistivities of the bed anomaly are 3 ohm-m (*blue* (dark gray in print versions)), 10 ohm-m (*green* (light gray in print versions)), and 30 ohm-m (*red* (gray in print versions)), respectively.

14.17, it can be seen that the results can clearly reflect the position and the thickness of the anomaly. The apparent resistivity derived from the inversion result is slightly different from the true resistivity of the anomaly bed. This is because the anomaly effect is smoothed by the background and the phenomenon is commonly seen in inversion problem of geophysics. In addition, we can see that at sharp bed boundary, the OBMI results can suffer from shoulder-bed effects and distortions—as do in laterologs and conventional microresistivity imagers, but for reasons arising from different measurement principles. The severity of the distortion depends on the bed thickness,

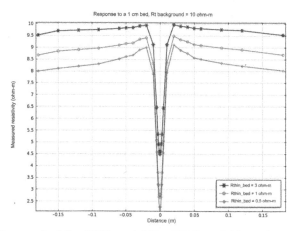

Figure 14.19 Inversion result of the OBMI model. The OBMI model has a homogenous background with resistivity of 10 ohm-m. The thickness of the bed anomaly is 1 cm. The resistivities of the bed anomaly are 3 ohm-m (*blue* (dark gray in print versions)), 1 ohm-m (*green* (light gray in print versions)), and 0.3 ohm-m (*red* (gray in print versions)), respectively.

resistivity contrast between the imaged thin bed and the shoulder beds and on whether the thin bed is more resistive or more conductive than the shoulder beds.

To investigate the vertical resolution of the OBMI pad, the thickness of the bed anomaly was changed to be 1 cm. Fig. 14.18 shows the results for different resistivities of the anomaly with background resistivity of 1 ohm-m. Fig. 14.19 shows the inversion result for different anomaly resistivities with background resistivity of 10 ohm-m. It can be seen that the inversion results still reflect the position and the thickness of the bed anomaly very well and the vertical resolution of the OBMI pad is within 1 cm.

A resistivity imaging routine has been developed in this project. Fig. 14.20 shows some examples of apparent resistivity images for the model with a different background resistivity, anomaly resistivity, and anomaly thickness.

14.4 CONCLUSIONS

The resistivity imaging tools are designed to obtain high-resolution resistivity distributions of the formation near the borehole. Based on the applications of such tools in different mud types, the resistivity imagers can be categorized into WBM imagers and oil-based mud imagers. The imager tools can be designed for wireline system and can also be used as a LWD tool. The WBM imagers are relatively simpler compared to the oil-based mud imagers. The WBM imagers use a button electrode to emit

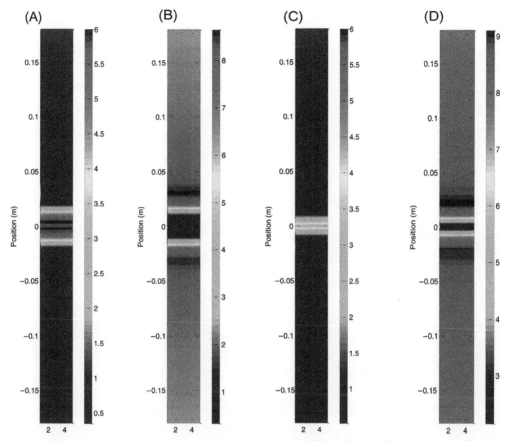

Figure 14.20 Resistivity image based on inversion results. (A) Background resistivity = 1 ohm-m, anomaly thickness = 3 cm, anomaly resistivity = 30 ohm-m; (B) background resistivity = 10 ohm-m, anomaly thickness = 3 cm, anomaly resistivity = 0.3 ohm-m; (C) background resistivity = 1 ohm-m, anomaly thickness = 1 cm, anomaly resistivity = 30 ohm-m; (D) background resistivity = 10 ohm-m, anomaly thickness = 1 cm, anomaly resistivity = 0.3 ohm-m.

currents to the formation. The return electrodes are usually the tool body. When a low-frequency voltage is applied to the button, the currents are measured and hence the resistivity is calculated using Ohm's law. Therefore the WBM imager is basically a two-terminal measurement like a multi-mater. However the oil-based mud imager has to penetrate through high-resistivity oil layer in order to measure the formation resistivity. To do so, the oil-based mud imager uses a four-terminal measurement principle with higher frequencies. There are two current emitting electrodes and two voltage measurement electrodes. The differential voltages between the measurement electrodes are directly proportional to the formation resistivity when a constant current is applied to the current electrodes. In both cases, the circuit model can be used to analyze the performance of the tool due to the low-frequency nature of the tool. For more accurate analysis, numerical modeling can provide more details in the tool design.

REFERENCES

[1] B. Anderson, I. Bryant, M. Luling, B. Spies, K. Helbig, Oilfield anisotropy: its origins and electrical characteristics, Oilfield Rev. 6 (4) (October 1994) 48–56.
[2] D. Robertson, F. Kuchuk, The value of variation, Middle East Well Eval. Rev. 18 (1997) 42–55.
[3] S. Luthi, P. Soulhaite, Fracture apertures from electrical borehole scans, Geophysics 55 (7) (July 1990) 821–833.
[4] M. Akbar, M. Petricola, M. Watfa, M. Badri, M. Charara, A. Boyd, et al., Classic interpretation problems: evaluating carbonates, Oilfield Rev. 7 (1) (January 1995) 38–57.
[5] M. Akbar, B. Vissapragada, A. Alghamdi, D. Allen, M. Herron, A. Carnegie, et al., A snapshot of carbonate reservoir evaluation, Oilfield Rev. 12 (4) (Winter 2000/2001) 20–41.
[6] P. Cheung, G. Cook, G. Flournoy, et al., A clear picture in oil based mud, Oilfield Rev. (Winter 2001/2002) 1–27.
[7] A.J. Hayman, P. Cheung, Formation Imaging While Drilling in Non-Conductive Fluids. US patent: 7,242,194 B2, 2007.
[8] P. Cheung, D. Pittman, A. Hayman, et al. Field test results of a new oil-base mud formation imager tool, in: Transactions of the SPWLA 42nd Annual Logging Symposium, Houston, Texas, USA, June 17–20, 2001, paper.

CHAPTER 15

Laterolog Tools and Array Laterolog Tools

Contents

15.1	Introduction	579
15.2	Basics of Electrode Type of Logging Tools	580
15.3	The Laterolog Focusing Principle and the Model of Dual Laterolog Tool	583
15.4	Application of Finite Element Method on Alternating Current Dual Laterolog Tool	590
	15.4.1 Choice of equation, basis function, and element matrix	590
	15.4.2 Application of source model to alternating current dual laterolog tool	591
	15.4.3 Flowchart of computational code for alternating current dual laterolog tool and finite element meshes (Fig. 15.7)	594
15.5	Validation of the Computational Method	594
15.6	Simulation Result	597
	15.6.1 Grid size and computation time	597
	15.6.2 Current pattern of LLd and LLs	598
	15.6.3 Groningen effect	598
	15.6.4 Frequency effect	601
	15.6.5 Invasion effect	602
15.7	Array Laterolog Tool	603
	15.7.1 The tool structure and the tool response	603
	15.7.2 The focusing method of an array laterolog tool	609
References		619
Appendix A Computation Method of Source Model for Alternating Current Dual Laterolog Tool		620

15.1 INTRODUCTION

In well logging, it is desired to have an image of the formations around the borehole. In other words, the information of the mud, invasion zones, and true formation in the radial direction along the borehole can provide critical evidence to the petrophysicists to characterize the formation. The radial resistivity distribution can be obtained by a laterolog tool. The investigation depth of laterolog can be designed to measure resistivities of the formation at a desired depth. Early dual laterolog measures the resistivity of the formation at two different depths of investigation: one near the borehole and the other can reach the formation a few feet away. The depth of investigation

(DOI) of an array laterolog ranges from 8 in. to about 30 in. The DOI is largely determined by the separation of electrodes.

Laterolog and Array Laterolog are basically direct current (DC) tools that have focusing capabilities. In this chapter, we will briefly explore the working principle and applications of the laterolog. There are quite a few different ways that a laterolog tool focusing the current to the formation. For different purpose of applications, the laterolog tools have different configurations. Based on the number of investigation depth desired, the tool can be designed to have multiple electrode pairs. In this chapter, we will start from basic theory of laterolog tool operation and present the focusing method used in a laterolog tool. The performance of a laterolog tool in a complicated formation is obtained by using finite element method (FEM) numerical simulation discussed in Chapter 13, Finite Element Method for Solving Electrical Logging Problems in Axially Symmetrical Formations.

Most laterolog tool operates at low frequencies in the range of tens of hertz to a few hundred hertz to facilitate the electronic design. However the principle of laterolog is based on DC analysis. Due to the low frequency, the displacement current can be ignored and the system can be analyzed with the static assumption. It is easier to describe the working principle of a laterolog tool using a resistor network. For more complicated formations, the tool performance can be obtained by using numerical method such as FEM as presented in Chapter 13, Finite Element Method for Solving Electrical Logging Problems in Axially Symmetrical Formations. We will briefly discuss the method to apply the FEM to simulate an electrode-type logging tool under alternating current (AC) conditions, where the dual laterolog tool (DLT) is chosen as an example to show the details of the application. The same method can also be applied to the simulation of the other laterolog tool.

15.2 BASICS OF ELECTRODE TYPE OF LOGGING TOOLS

Many readers who have high-frequency electromagnetic (EM) background and antenna design experience often try to relate the focusing mechanism of an array antenna with that of a laterolog since they share the same idea of "focusing" the EM field. Unfortunately, they are quite different. The phased array antenna operates at a high frequency where the antenna separation is compatible with the wavelength. The array is so designed that the field at undesired direction of radiation is canceled by another antenna element. The cancellation is based on the reversed phase in the undesired direction of radiation of both antenna elements due to physical separation.

The focusing principle of a laterolog is based on the potential balance between two electrodes. Fig. 15.1 shows the basic focus idea in a laterolog. Fig. 15.1A shows that when there are two insulating electrodes A1 and A2, each generating potentials

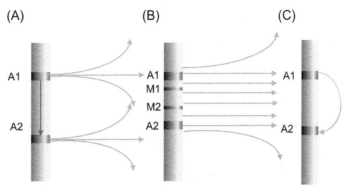

Figure 15.1 The focusing principle of a laterolog. (A) Two electrodes A1 and A2 with different potential in a homogeneous formation; (B) the current lines in a formation with the potentials adjusted so that the voltages between the measurement electrodes M1 and M2 are equal; and (C) when the return electrode of A1 is connected to A2, the detection range is close to borehole.

relative to the ground, which is far away from A1 and A2. The current will flow to the formation and will also flow from A1 to A2 via borehole mud due to the potential difference between A1 and A2. If in a homogeneous formation, the potential between A1 and A2 are adjusted so that $V_{A1} = V_{A2}$, the current flowing from A1 to A2 in the vertical direction in borehole will be zero. Therefore the current will flow to the formation and circle back to the ground which is far away, causing the current line seemingly straight near the electrodes, as shown in Fig. 15.1B. However, when the formation is nonhomogeneous, the current flow between A1 and A2 is affected by the formation inhomogeneity. The potential of A1 and A2 may not be equal in order to eliminate the borehole current. In this case, two measuring electrodes M1 and M2 are added between A1 and A2. The tool constantly adjusting the potential difference between A1 and A2 so that the voltage V_{M1}, V_{M2} between M1 and M2 are equal to zero. Therefore the current flow is maintained to be in the formation. This is the focusing method used in far detection mode of a laterolog.

On the other hand, if we would like to just sense the formation resistivity near the borehole, we can set the electrode A2 as a return electrode of A1, the current lines will be curved and the detecting distance will be near the electrode as shown in Fig. 15.1C.

At this point, we can see that the laterolog tool is an electrode-type tool. The basic structure of the tool has at least two current emitting electrodes and two measurement or voltage monitoring electrodes shown in Fig. 15.1B. To understand the operation principle in detail, we need to first know the characteristics of an electrode in a conducting formation.

Let us first consider a point source of current I, located in a homogenous media with conductivity of σ. We can use the analogy between static charge in homogeneous space and the constant current case [1]. The current density in the formation due to a point source is

$$J = \frac{J_0}{4\pi R^2} \tag{15.1a}$$

where R is the distance from the source point to the field point. J_0 is the current density on the electrode cylinder as depicted in Fig. 15.2, which is

$$J_0 = \frac{I}{2\pi a l} \tag{15.1b}$$

Therefore we can find the electric field in the formation due to the current electrode

$$E = \frac{J}{\sigma} = \frac{J_0}{4\pi R^2 \sigma} \tag{15.2}$$

The potential of the point at R, with assumptions that the zero potential is at infinity, is

$$d\phi = \int_R^\infty E dR = \int_R^\infty \frac{J_0}{4\pi R^2 \sigma} dR = \frac{J_0}{4\pi R \sigma} \tag{15.3}$$

From Eq. (15.3), for any point (x_s, y_s, z_s) on the electrode, if we assume that $r \gg r_s$, the contribution to the potential of any point in the space (x, y, z) can be found to be

$$\Phi(x, y, z) = \frac{J_0}{4\pi\sigma\sqrt{(z-z_s)^2 + r^2}} \tag{15.4}$$

where $r = \sqrt{x^2 + y^2 + z^2}$ and $r_s = \sqrt{x_s^2 + y_s^2 + z_s^2}$.

Note that in Eq. (15.4), the assumption made is based on a long and thin electrode, which is close to real case.

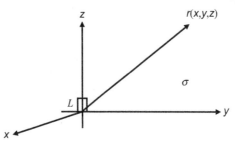

Figure 15.2 A cylindrical electrode is in a homogeneous formation with conductivity of σ. The length of the electrode is L and radius is a.

Now, let us consider the electrode of a laterolog. If the electrode has a diameter of a and a length of L, the surface potential due to the current I can be found by integration of the potential due to a point source as shown in Eq. (15.4) in the z direction

$$V = \frac{J_0}{4\pi\sigma} \int_0^L \Phi(x, y, z) dz = \frac{I}{4\pi^2 \sigma a L} \ln\left(\frac{L}{a}\right) \qquad (15.5)$$

From Eq. (15.5), we can also obtain the impedance of the electrode in a homogeneous formation

$$Z = \frac{1}{4\pi^2 \sigma a L} \ln\left(\frac{L}{a}\right) \qquad (15.6)$$

15.3 THE LATEROLOG FOCUSING PRINCIPLE AND THE MODEL OF DUAL LATEROLOG TOOL

The most commonly used laterolog tool is the dual laterolog. The DLT is designed to have dual DOIs so that we can obtain information of the formation at the desired depth radially. In the logging industry, the DLT is considered to be one of the golden standard of resistivity measurements. To understand the focusing mechanism, we can use a simple circuit to simulate a laterolog. Consider the simplified structure of a laterolog shown in Fig. 15.1B. There are two current emitting electrodes and two measurement electrodes. A circuit model of this simplified tool is shown in Fig. 15.3.

In Fig. 15.3, source V_1 is from the main current electrode. Source V_2 is a focusing electrode. V_{M1} and V_{M3} are the measurement electrodes. R_1 and R_2 represent the leakage resistance between the current electrodes and the measurement electrodes via both tool internal leakage and the borehole leakage. Note that these resistances may not be the same due to current leakage and formation inhomogeneity. R_{c1} and R_{c2} are resistances of the formation. R_{b1} and R_{b2} are resistances representing current leakage of borehole and tool body between source electrodes. These resistances can be estimated using the formula given in Eq. (15.6). From Fig. 15.3, the focusing method

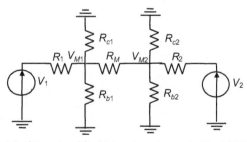

Figure 15.3 A circuit model of the simplified laterolog shown in Fig. 15.1B.

can be interpreted as follows. For a given V_1, the value of V_2 is adjusted so that the voltage difference between two measuring electrodes are zero. When $V_{M1} - V_{M2} = 0$, there will be no current flow in R_M. Therefore the current will have to flow through R_1 and R_2, and then terminated by R_{b1} and R_{c1}, and R_{b2} and R_{c2}. When the focusing condition $V_{M1} = V_{M2}$ is reached, the two current paths by V_{M1}, R_1, R_{b1}, and R_{c1}, and V_{M2}, R_2, R_{b2}, and R_{c2} become independent. Let us consider the tool performance in a homogeneous formation. From Fig. 15.3, we can obtain the voltages at the measurement electrodes:

$$V_{Mi} = V_0 \frac{R_{bi} R_{ci}}{R_i(R_{ci} + R_{bi}) + R_{bi} R_{ci}} \quad (i = 1, 2) \tag{15.7}$$

In a homogeneous formation, $V_{M1} = V_{M2}$ when the tool is focused, we can have the following identity,

$$\frac{V_1 R_{c1}}{R_1 \left(\frac{R_{c1}}{R_{b1}} + 1\right) + R_{c1}} = \frac{V_2 R_{c2}}{R_2 \left(\frac{R_{c2}}{R_{b2}} + 1\right) + R_{c2}} \tag{15.8}$$

Solutions to Eq. (15.8) will make the laterolog tool focused. The solutions also depend on the structure of the tool. We know that in the homogeneous formation, the resistance of the formation should be uniform, therefore, $R_{c1} = R_{c2} = R_c$. If the electrodes are well insulated from each other, e.g., $R_b \gg R_c$, Eq. (15.8) becomes

$$\frac{V_1}{\frac{R_1}{R_{c1}} + 1} = \frac{V_2}{\frac{R_2}{R_{c2}} + 1} \tag{15.9}$$

Eq. (15.9) is the simplified focusing condition for the laterolog tool. If the formation is homogeneous, $R_{c1} = R_{c2}$, and the internal leakage resistances are the same for both current electrodes, which means $R_1 = R_2$, then the focusing condition becomes

$$V_1 = V_2 \tag{15.10}$$

which is the ideal focusing condition.

In most cases, the borehole resistivity and the internal impedance of the voltage sources must be considered, a more detailed circuit for the simplified laterolog is shown in Fig. 15.4. Compared with the simplified circuit model, six more resistors are added. R_U and R_D are the resistances to model the upper and lower current path from the current emitting electrodes to the tool body. R_{s1} and R_{s2} are internal resistance of the voltage sources, and R_{m1} and R_{m2} are resistance of the mud looking from the measurement electrodes in radial direction. R_{f1} and R_{f2} are formation resistance.

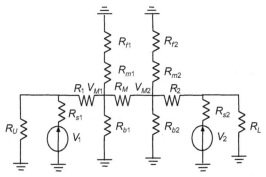

Figure 15.4 A more complete circuit model for the simplified laterolog tool. R_U and R_D are the resistances to model the upper and lower current path from the current emitting electrodes to the tool body. R_{s1} and R_{s2} are internal resistance of the voltage sources, and R_{m1} and R_{m2} are resistance of the mud looking from the measurement electrodes in radial direction. R_{f1} and R_{f2} are formation resistance.

It is not hard to solve the circuit in Fig. 15.4. Resistances R_U and R_D can be balanced out if there are more focusing electrodes added to the upper and lower side of the electrodes V_1 and V_2. R_{s1} and R_{s2} are internal resistance of the voltage sources V_1 and V_2, which are usually small and they are not a function of formation and borehole parameters.

After understanding the working principle of a laterolog tool, we now can analyze the focusing conditions mathematically. Let us use most commonly used dual laterolog as an example. In practice, sometimes, the DLT is also combined with a microspherical focused tool to detect even shallow regions such as invasion resistivity. With different measurement depth one could then solve for the formation and invasion resistivities (R_t and R_{xo}), and the invasion diameter (d_i), assuming negligible or easily correctable shoulder-bed effect [2].

The tool is split for illustrative purposes only; laterolog-shallow (LLs) and laterolog-deep (LLd) currents are axisymmetric. I_o denotes survey current and I_a is bucking current. The MicroSFL is located on the lower (A2′) electrode.

The MicroSFL was used to give an accurate estimate of R_{xo} and to delineate bed boundaries. With R_{xo} known, a dual DOI laterolog tool is then optimized to determine the remaining two unknowns, d_i and R_t. The electrode configuration for the two DLT arrays is shown in Fig. 15.5 [3]: a shallow DOI measurement, LLs and a deep measurement, LLd.

Both LLs and LLd use the same electrodes and same survey current beam, but different focusing methods to provide two different depths of investigation. DLT tool used in this chapter has 13 electrodes, including 11 electrodes on the tool body and 2 remote electrodes. On the tool body there are five current electrodes, A0, A1, A1′, A2, A2′, and six voltage electrodes, M1, M1′, M2, M2′, M3, M3′, which are also called monitoring electrodes. The returned electrode B is located at the surface to

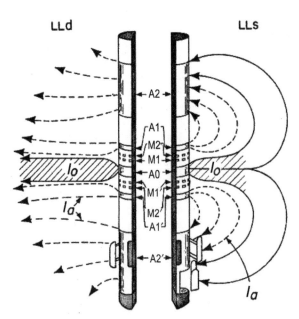

Figure 15.5 Schematic diagram of the dual laterolog tool configuration and current patterns.

collect all emitted currents. The reference potential electrode, N, is located about 80 ft above the tool at the top of the insulated bridle which is used to support the weight of the tool and provide electrical communication between the downhole and surface electronics. The circuit is designed so that current is only emitted from or return to electrodes A0, A1, A2, A1′, A2′, B, and there is no current flowing through electrodes M1, M1′, M2, M2′, M3, M3′, and N. The frequencies used for LLs and LLd are 280 and 35 Hz, respectively. Table 15.1 shows DLT electrode positions, where the number which is not underlined is from Anderson [2], and the number underlined is generated for simulation purposes.

In LLd mode, a constant survey current I_0 is emitted by main current electrode A0. The current electrodes A1, A1′, A2, and A2′ are set at almost the same potential and emit bucking currents into the formation surrounding the tool. Current is controlled to ensure that the potentials of the monitoring electrodes M1, M1′, M2, and M2′ are the same; this condition guarantees there is no vertical direction current flowing in the vicinity of the monitoring electrodes M1, M1′, M2, and M2′, and forces the survey current beam to be well focused into any bed adjacent to the A0 electrode. Current is controlled to ensure that the potentials of electrodes A2, A2′, M3, and M3′ are the same; this condition guarantees there is no vertical direction current flowing through monitoring electrodes M3 and M3′, and forces the currents deep into the formation. The electrode pair M3 and M3′ is also used to further ensure that a uniform potential gradient is maintained between the bucking electrodes when there are high resistivity contrasts between beds.

Table 15.1 DLT electrode position (in in.)

	Position (in.)		
B	$\overline{5020}$	to	$\overline{5000}$
N	$\overline{960}$	to	$\overline{955}$
A2	174.5	to	69.5
M3	$\overline{39.3}$	to	$\overline{39.25}$
A1	38.8	to	24.5
M2	$\overline{16.0}$	to	$\overline{15.7}$
M1	$\overline{10.0}$	to	$\overline{9.7}$
A0	4.5	to	−4.5
M1′	$\overline{-9.7}$	to	$\overline{-10}$
M2′	$\overline{-15.7}$	to	$\overline{-16.0}$
A1′	−24.5	to	−38.8
M3′	$\overline{-39.25}$	to	$\overline{-39.3}$
A2′	−69.5	to	−174.5

The monitoring electrodes are thin rings. The number which is not underlined is from Anderson [2], and the number underlined is generated for simulation purposes.

All currents from the current electrodes are collected by returned electrode B located at the surface. This focusing method ensures that the LLd mode has a large DOI. The nominal thickness of the survey current beam is about 2 ft. Apparent resistivity R_a for the LLd is computed from the ohmic drop to the current emitted from A0 between equipotential surfaces passing through the monitoring electrodes and the reference potential electrode N,

$$R_a = K_{LLd} \frac{(V_{M1} + V_{M1'})/2 - V_N}{I_{A0}} \tag{15.11}$$

where K_{LLd} is the tool constant in the LLd mode.

The 13 equations for focusing conditions above the 13 LLd monitoring electrodes are:

$$I_{A2'} + I_{A1'} + I_{A0} + I_{A1} + I_{A2} + I_B = 0 \tag{15.12a}$$

$$I_{M3} + I_{M3'} = 0 \tag{15.12b}$$

$$I_{M2} + I_{M2'} = 0 \tag{15.12c}$$

$$I_{M1} + I_{M1'} = 0 \tag{15.12d}$$

$$I_N = 0 \tag{15.12e}$$

$$U_{A2} = U_{A2'} \tag{15.12f}$$

$$U_{M3} = U_{M3'} \qquad (15.12\text{g})$$

$$U_{A1} = U_{A1'} \qquad (15.12\text{h})$$

$$U_{M2} = U_{M2'} \qquad (15.12\text{i})$$

$$U_{M1} = U_{M1'} \qquad (15.12\text{j})$$

$$U_{M1} = U_{M2} \qquad (15.12\text{k})$$

$$U_{A2} = U_{M3} \qquad (15.12\text{l})$$

$$I_{A0} = I0 \qquad (15.12\text{m})$$

where I is the current flowing into or from formation through the electrodes and V is the potential of the electrodes. The linear system of equations describing focusing conditions for LLd can be written in matrix form as

(F_{LLd}) Focusing condition LLd

$$\begin{bmatrix}
1 & 0 & 1 & 0 & 1 & 0 & 0 & 1 & 0 & 0 & 1 & 0 & 1 & 0 & 0 & 0 & 0 & 0 & 0 & 0 & 0 & 0 & 0 & 0 \\
0 & 0 & 0 & 1 & 0 & 0 & 0 & 0 & 0 & 0 & 1 & 0 & 0 & 0 & 0 & 0 & 0 & 0 & 0 & 0 & 0 & 0 & 0 & 0 \\
0 & 0 & 0 & 0 & 0 & 1 & 0 & 0 & 0 & 1 & 0 & 0 & 0 & 0 & 0 & 0 & 0 & 0 & 0 & 0 & 0 & 0 & 0 & 0 \\
0 & 0 & 0 & 0 & 0 & 0 & 1 & 0 & 1 & 0 & 0 & 0 & 0 & 0 & 0 & 0 & 0 & 0 & 0 & 0 & 0 & 0 & 0 & 0 \\
0 & 1 & 0 & 0 & 0 & 0 & 0 & 0 & 0 & 0 & 0 & 0 & 0 & 0 & 0 & 0 & 0 & 0 & 0 & 0 & 0 & 0 & 0 & 0 \\
0 & 0 & 0 & 0 & 0 & 0 & 0 & 0 & 0 & 0 & 0 & 0 & 0 & 0 & 1 & 0 & 0 & 0 & 0 & 0 & 0 & 0 & 0 & -1 \\
0 & 0 & 0 & 0 & 0 & 0 & 0 & 0 & 0 & 0 & 0 & 0 & 0 & 0 & 0 & 1 & 0 & 0 & 0 & 0 & 0 & 0 & -1 & 0 \\
0 & 0 & 0 & 0 & 0 & 0 & 0 & 0 & 0 & 0 & 0 & 0 & 0 & 0 & 0 & 0 & 1 & 0 & 0 & 0 & 0 & -1 & 0 & 0 \\
0 & 0 & 0 & 0 & 0 & 0 & 0 & 0 & 0 & 0 & 0 & 0 & 0 & 0 & 0 & 0 & 1 & 0 & 0 & 0 & -1 & 0 & 0 & 0 \\
0 & 0 & 0 & 0 & 0 & 0 & 0 & 0 & 0 & 0 & 0 & 0 & 0 & 0 & 0 & 0 & 0 & 1 & 0 & -1 & 0 & 0 & 0 & 0 \\
0 & 0 & 0 & 0 & 0 & 0 & 0 & 0 & 0 & 0 & 0 & 0 & 0 & 0 & 0 & 1 & -1 & 0 & 0 & 0 & 0 & 0 & 0 & 0 \\
0 & 0 & 0 & 0 & 0 & 0 & 0 & 0 & 0 & 0 & 0 & 0 & 0 & 1 & -1 & 0 & 0 & 0 & 0 & 0 & 0 & 0 & 0 & 0 \\
0 & 0 & 0 & 0 & 0 & 0 & 1 & 0 & 0 & 0 & 0 & 0 & 0 & 0 & 0 & 0 & 0 & 0 & 0 & 0 & 0 & 0 & 0 & 0
\end{bmatrix} \cdot$$

$[I_B; I_N; I_{A2}; I_{M3}; I_{A1}; I_{M2}; I_{M1}; I_{A0}; I_{M1'}; I_{M2'}; I_{A1'}; I_{M3'}; I_{A2'}; V_B; V_N; V_{A2}; V_{M3}; V_{A1}; V_{M2}; V_{M1};$
$V_{A0}; V_{M1'}; V_{M2'}; V_{A1'}; V_{M3'}; V_{A2'}]' = \begin{bmatrix} 0 & 0 & 0 & 0 & 0 & 0 & 0 & 0 & 0 & 0 & 0 & 0 & I0 \end{bmatrix}'$

(15.13)

where the left side of Eq. (15.13) is defined as focusing condition LLd F_{LLd}.

In LLs mode, the only difference is that current electrodes A2 and A2′ are set to be returned electrodes to collect the current emitted from electrodes A0, A1, and A1′. Since the current returns are so close, the survey current diverges quickly when it enters the formation, resulting in a shallow DOI for LLs mode. As in the LLd mode, current is controlled to ensure that the potentials of the monitoring electrodes M1, M1′, M2, and M2′ are the same; this condition guarantees there is no vertical direction current flowing in the vicinity of the monitoring electrodes M1, M1′, M2, and M2′, and forces the survey current beam to be well focused into any bed adjacent to the A0 electrode. The nominal thickness of the survey current beam is 2 ft. The equation used to compute apparent resistivity R_a for LLs is

$$R_a = K_{LLs} \frac{(V_{M1} + V_{M1'})/2 - V_N}{I_{A0}} \qquad (15.14)$$

where K_{LLs} is the tool coefficient for LLs mode.

The 13 equations for focusing conditions above 13 monitoring electrodes are:

$$I_{A2'} + I_{A1'} + I_{A0} + I_{A1} + I_{A2} = 0 \qquad (15.15a)$$

$$I_{M3} + I_{M3'} = 0 \qquad (15.15b)$$

$$I_{M2} + I_{M2'} = 0 \qquad (15.15c)$$

$$I_{M1} + I_{M1'} = 0 \qquad (15.15d)$$

$$I_N = 0 \qquad (15.15e)$$

$$I_B = 0 \qquad (15.15f)$$

$$U_{A2} = U_{A2'} \qquad (15.15g)$$

$$U_{M3} = U_{M3'} \qquad (15.15h)$$

$$U_{A1} = U_{A1'} \qquad (15.15i)$$

$$U_{M2} = U_{M2'} \qquad (15.15j)$$

$$U_{M1} = U_{M1'} \qquad (15.15k)$$

$$U_{M1} = U_{M2} \qquad (15.15l)$$

$$I_{A0} = I0 \qquad (15.15m)$$

where I is the current flowing into or from formation through the electrodes and V is the potential of the electrodes. The linear system of equations describing focusing condition for LLs can be written in matrix form as

$$(F_{LLs}) \text{ Focusing condition LLs}$$

$$\begin{bmatrix} 0 & 0 & 1 & 0 & 1 & 0 & 0 & 1 & 0 & 0 & 1 & 0 & 1 & 0 & 0 & 0 & 0 & 0 & 0 & 0 & 0 & 0 & 0 & 0 \\ 0 & 0 & 0 & 1 & 0 & 0 & 0 & 0 & 0 & 0 & 1 & 0 & 0 & 0 & 0 & 0 & 0 & 0 & 0 & 0 & 0 & 0 & 0 & 0 \\ 0 & 0 & 0 & 0 & 0 & 1 & 0 & 0 & 0 & 1 & 0 & 0 & 0 & 0 & 0 & 0 & 0 & 0 & 0 & 0 & 0 & 0 & 0 & 0 \\ 0 & 0 & 0 & 0 & 0 & 0 & 1 & 0 & 1 & 0 & 0 & 0 & 0 & 0 & 0 & 0 & 0 & 0 & 0 & 0 & 0 & 0 & 0 & 0 \\ 0 & 1 & 0 & 0 & 0 & 0 & 0 & 0 & 0 & 0 & 0 & 0 & 0 & 0 & 0 & 0 & 0 & 0 & 0 & 0 & 0 & 0 & 0 & 0 \\ 1 & 0 & 0 & 0 & 0 & 0 & 0 & 0 & 0 & 0 & 0 & 0 & 0 & 0 & 0 & 0 & 0 & 0 & 0 & 0 & 0 & 0 & 0 & 0 \\ 0 & 0 & 0 & 0 & 0 & 0 & 0 & 0 & 0 & 0 & 0 & 0 & 0 & 0 & 1 & 0 & 0 & 0 & 0 & 0 & 0 & 0 & 0 & -1 \\ 0 & 0 & 0 & 0 & 0 & 0 & 0 & 0 & 0 & 0 & 0 & 0 & 0 & 0 & 0 & 1 & 0 & 0 & 0 & 0 & 0 & 0 & -1 & 0 \\ 0 & 0 & 0 & 0 & 0 & 0 & 0 & 0 & 0 & 0 & 0 & 0 & 0 & 0 & 0 & 0 & 1 & 0 & 0 & 0 & 0 & -1 & 0 & 0 \\ 0 & 0 & 0 & 0 & 0 & 0 & 0 & 0 & 0 & 0 & 0 & 0 & 0 & 0 & 0 & 0 & 0 & 1 & 0 & 0 & 0 & -1 & 0 & 0 & 0 \\ 0 & 0 & 0 & 0 & 0 & 0 & 0 & 0 & 0 & 0 & 0 & 0 & 0 & 0 & 0 & 0 & 0 & 0 & 1 & 0 & -1 & 0 & 0 & 0 & 0 \\ 0 & 0 & 0 & 0 & 0 & 0 & 0 & 0 & 0 & 0 & 0 & 0 & 0 & 0 & 0 & 0 & 0 & 1 & -1 & 0 & 0 & 0 & 0 & 0 \\ 0 & 0 & 0 & 0 & 0 & 0 & 1 & 0 & 0 & 0 & 0 & 0 & 0 & 0 & 0 & 0 & 0 & 0 & 0 & 0 & 0 & 0 & 0 & 0 \end{bmatrix} \cdot$$

$[I_B; I_N; I_{A2}; I_{M3}; I_{A1}; I_{M2}; I_{M1}; I_{A0}; I_{M1'}; I_{M2'}; I_{A1'}; I_{M3'}; I_{A2'}; V_B; V_N; V_{A2}; V_{M3}; V_{A1};$
$V_{M2}; V_{M1}; V_{A0}; V_{M1'}; V_{M2'}; V_{A1'}; V_{M3'}; V_{A2'}]' = \begin{bmatrix} 0 & 0 & 0 & 0 & 0 & 0 & 0 & 0 & 0 & 0 & 0 & I0 \end{bmatrix}'$

(15.16)

where the left part of Eq. (15.16) is defined as focusing condition LLs F_{LLs}.

15.4 APPLICATION OF FINITE ELEMENT METHOD ON ALTERNATING CURRENT DUAL LATEROLOG TOOL

After understanding the basics of the laterolog tool, it is necessary to study the details of the tool performance. Although analytic solutions may be obtained, the numerical solutions are more convenient in understanding the tool physics. In this section, we will use FEM simulations discussed in Chapter 14, Resistivity Imaging Tools, to solve the field distributions and tool performance in complicated formations. We will also discuss issues related to the tool such as Groningen effects.

15.4.1 Choice of equation, basis function, and element matrix

As illustrated in Section 13.11, the electric current source exists only in ρ direction for the AC DLT tool. From the analysis of Section 13.3, Table 13.3, it is known that only transverse magnetic mode (E_ρ, E_z, and H_φ) exists. H_φ term will be considered to build the FEM equation. The magnetic field equation and the weak form of the magnetic equation were discussed in Sections 13.1 and 13.3, and expressed in Eqs. (13.6) and (13.22) as

$$\frac{1}{j\omega}\nabla\times\left(\underline{\underline{\varepsilon_c}}^{-1}\cdot\nabla\times\underline{H}\right)+j\omega\underline{\underline{\mu}}\cdot\underline{H}=\frac{1}{j\omega}\nabla\times\left(\underline{\underline{\varepsilon_c}}^{-1}\cdot\underline{J_s}\right)-\underline{M_s} \quad (13.6)$$

$$\frac{1}{j\omega}\left\langle\nabla\times\underline{\Omega_m};\underline{\underline{\varepsilon_c}}^{-1}\cdot\nabla\times\underline{H}\right\rangle_\Omega+j\omega\left\langle\underline{\Omega_m};\underline{\underline{\mu}}\cdot\underline{H}\right\rangle_\Omega$$
$$=\frac{1}{j\omega}\left\langle\underline{\Omega_m};\nabla\times(\underline{\underline{\varepsilon_c}}^{-1}\cdot\underline{J_s})\right\rangle_\Omega-\left\langle\underline{\Omega_m};\underline{M_s}\right\rangle_\Omega-\frac{1}{j\omega}\oiint_S\left(\left(\underline{\underline{\varepsilon_c}}^{-1}\cdot\nabla\times\underline{H}\right)\times\underline{\Omega_m}\right)\cdot\hat{n}dS$$
$$(13.22)$$

Basis functions were discussed in Section 13.5.1 for triangular element basis functions and Section 13.5.2 for rectangular element basis functions. Element matrix based on ρH_ϕ is used and matrix system $[Z_{m,n}][(\rho h)_n]=[V_m]$ is built for the simulation of the AC DLT tool. Formulas for the element matrix are illustrated in Sections 13.7 and 13.9 if the solution domain is divided into rectangular elements and triangular elements, respectively.

15.4.2 Application of source model to alternating current dual laterolog tool

The source model used for AC DLT tool is source model 3: sources existing on boundaries, which is illustrated in Section 13.11.5 and Fig. 13.9. In this model, the whole tool Ω_{tool} is taken from the solution domain Ω and current sources inside the tool do not need to be considered. The sources now are EM fields existing on the tool boundary Γ_{tool}. As shown in Fig. 15.6, if the number of electrodes is $N_{electrode}$ and number of insulators is $N_{insulator}$, then the number of unknowns on the tool boundary Γ_{tool} can be expressed as

$$N_{unknowns\_of\_tool}=N_{insulator}+2*N_{electrode} \quad (15.17)$$

The unknowns on the tool boundary are $\rho H_\phi(i)$ ($i=1$, $N_{insulator}$), $V(i)$ ($i=1$, $N_{electrode}$), and $I(i)$ ($i=1$, $N_{electrode}$), separately. For the DLT tool shown in Fig. 15.6, $N_{insulator}$ is 14, $N_{electrode}$ is 13, and $N_{unknowns\_of\_tool}$ is 40. The next

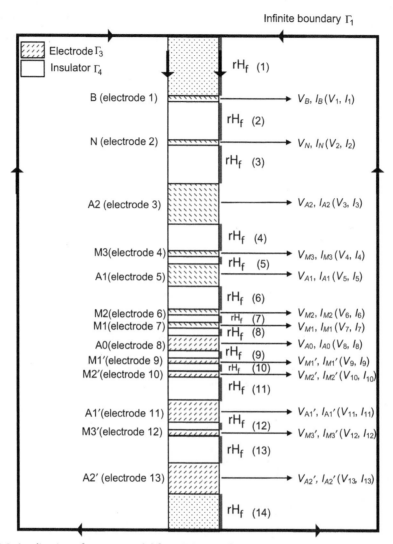

Figure 15.6 Application of source model for AC DLT tool.

paragraphs will illustrate how to build $N_{\text{unknowns\_of\_tool}}$ equations to obtain the solution for sources.

1. The first $N_{\text{insulator}}$ equation

As mentioned before, the source model decides the value of the global vector. For source model 3, the element of the global vector is nonzero only when the nodes exist on boundary Γ_4. If impedance matrix is built based on ρH_ϕ, then this element of the global vector can be expressed by Eq. (13.176). For the DLT tool shown in Fig. 15.6, there are 14 ($N_{\text{insulator}}$) boundaries of Γ_4, so there are 14 nonzero elements in the global vector. In computation code, these 14 elements are put at the front of the global vector, and the matrix system can be written as

$$[Z_{i,j}]_{(i=1,N;j=1,N)} \cdot \begin{bmatrix} \rho H_\phi(1) \\ \rho H_\phi(2) \\ \rho H_\phi(3) \\ \cdots \\ \rho H_\phi(12) \\ \rho H_\phi(13) \\ \rho H_\phi(14) \\ \rho H_\phi(15) \\ \rho H_\phi(16) \\ \cdots \\ \rho H_\phi(N-1) \\ \rho H_\phi(N) \end{bmatrix} = \begin{bmatrix} 2\pi(0 - V_1) \\ 2\pi(V_1 - V_2) \\ 2\pi(V_2 - V_3) \\ \cdots \\ 2\pi(V_{11} - V_{12}) \\ 2\pi(V_{12} - V_{13}) \\ 2\pi(V_{13} - 0) \\ 0 \\ 0 \\ \cdots \\ 0 \\ 0 \end{bmatrix} \quad (15.18)$$

where N is the number of total unknown ρH_ϕ in solution domain Ω.

After applying Gaussian elimination, all other nodes are removed except the nodes existing on boundary Γ_4; the matrix can be expressed as

$$[A_{i,j}]_{(i=1,14;j=1,14)} \cdot \begin{bmatrix} \rho H_\phi(1) \\ \rho H_\phi(2) \\ \rho H_\phi(3) \\ \cdots \\ \rho H_\phi(12) \\ \rho H_\phi(13) \\ \rho H_\phi(14) \end{bmatrix} - \begin{bmatrix} 2\pi(0 - V_1) \\ 2\pi(V_1 - V_2) \\ 2\pi(V_2 - V_3) \\ \cdots \\ 2\pi(V_{11} - V_{12}) \\ 2\pi(V_{12} - V_{13}) \\ 2\pi(V_{13} - 0) \end{bmatrix} = 0 \quad (15.19)$$

2. The second $N_{\text{electrode}}$ equations

The second $N_{\text{electrode}}$ equations come from $I_0 = 2\pi(\rho_{\text{tool}} H_\phi|_{z=\Gamma_3(\text{end})} - \rho_{\text{tool}} H_\phi|_{z=\Gamma_3(\text{start})})$ (Eq. 4.26) and can be expressed as

$$\rho H_\phi(i+1) - \rho H_\phi(i) - \frac{I(i)}{2\pi} = 0, \quad (i = 1, N_{\text{electrode}}) \quad (15.20)$$

3. The third $N_{electrode}$ equations

The third $N_{electrode}$ equations come from focusing condition and can be expressed as

$$F_{LLd} \cdot \begin{bmatrix} I \\ V \end{bmatrix} = \begin{bmatrix} 0 & 0 & 0 & 0 & 0 & 0 & 0 & 0 & 0 & 0 & 0 & 0 & I0 \end{bmatrix}' \quad (15.21)$$

$$F_{LLs} \cdot \begin{bmatrix} I \\ V \end{bmatrix} = \begin{bmatrix} 0 & 0 & 0 & 0 & 0 & 0 & 0 & 0 & 0 & 0 & 0 & 0 & I0 \end{bmatrix}' \quad (15.22)$$

where F_{LLd} and F_{LLs} are focusing condition for LLd and LLs separately (Eqs. 15.13 and 15.16).

Combining Eq. (15.19) with (15.22), the equations used to handle the source of the DLT tools can be expressed as

$$\begin{bmatrix} B_{11} & B_{12} & B_{13} \\ B_{21} & B_{22} & B_{23} \\ B_{31} & B_{32} & B_{33} \end{bmatrix}_{40 \times 40} \cdot \begin{bmatrix} \rho H_\phi \\ I \\ V \end{bmatrix}_{40 \times 1} = \begin{bmatrix} C_1 \\ C_2 \\ C_3 \end{bmatrix}_{40 \times 1} \quad (15.23)$$

B_{11} and B_{13} are obtained from Eq. (15.19). B_{21} and B_{22} come from Eq. (15.20). B_{32} and B_{33} come from Eq. (15.21) for the LLd tool and from Eq. (15.22) for the LLs tools. B_{12}, B_{23}, B_{31} are zero.

The computation method of source model for the AC DLT tool is shown in Appendix A.

15.4.3 Flowchart of computational code for alternating current dual laterolog tool and finite element meshes (Fig. 15.7)

Fig. 15.8 lists different two-dimensional finite element meshes [4].

15.5 VALIDATION OF THE COMPUTATIONAL METHOD

The numerical method discussed in this chapter is validated by the results from Anderson [2], where the frequency of the LLd is set to 35 Hz and the frequency of the LLs is set to 280 Hz. First is the comparison of tool coefficients. Tool coefficients are the most important variable to calculate apparent resistivity as $R_a = K_{LLd} \frac{(V_{M1} + V_{M1'})/2 - V_N}{I_{A0}}$ (Eq. 15.11) and $R_a = K_{LLs} \frac{(V_{M1} + V_{M1'})/2 - V_N}{I_{A0}}$ (Eq. 15.14). Tool coefficients are first calculated by the computational code to guarantee that the apparent resistivity calculated is equal to the true formation resistivity.

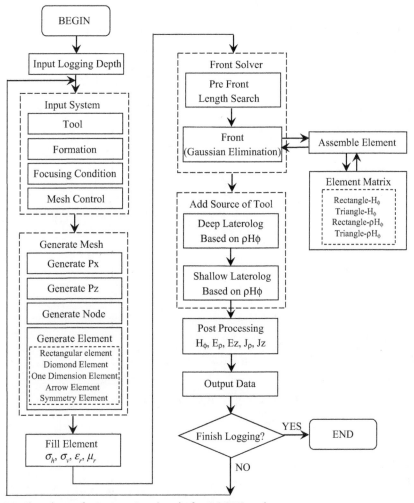

Figure 15.7 Flowchart of computational code for AC DLT tool.

Table 15.2 lists the tool coefficients calculated by Anderson and this chapter. K_{LLs} is exactly the same and K_{LLd} is very close to each other. The slight difference is caused by the different setting of voltage electrodes M, where Anderson assumed that the electrodes M are thin rings and in the model of electrodes M in this chapter it still has actual geometry.

Fig. 15.8 shows the comparison of dual laterologs computed from a benchmark formation which has 30 in. of invasion in some beds. The borehole has a radius of 4 in. and a resistivity of 0.5 ohm-m. The resistivity of the formation is 0.5, 5, and 50 ohm-m, and the resistivity of the invasion is 2.5 and 10 ohm-m. The LLs and LLd logs were not borehole corrected since correction is only necessary for large holes. In

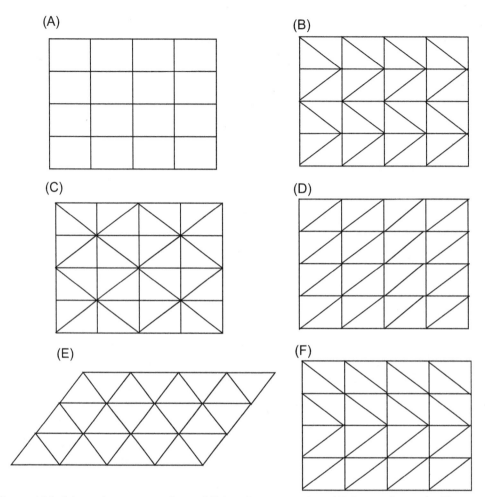

Figure 15.8 Schematic representations of finite element meshes. (A) Rectangle mesh, (B) arrow mesh, (C) diamond mesh, (D) one-directional mesh, (E) hexagonal mesh, and (F) symmetric mesh.

Table 15.2 Validation from tool coefficient

	K_{LLs}	K_{LLd}
Anderson	1.45	0.89
Author	1.45	0.87

the uninvaded bed between 47 and 57 ft, LLd departs from R_t because the survey current flows preferentially in the conductive bed (squeeze effect), while much of the bucking current remains in the resistive shoulders as it flows to the remote return. Fig. 15.9 shows the result for an artificial testing formation [2].

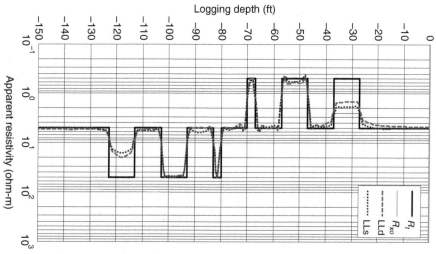

Figure 15.9 Validation from dual laterolog computed logs [2].

Table 15.3 Grid size for AC DLT tool

RgridNumberOfBorehole	20 (unit)
RgridDeltaR	0.05 (in.)
RgridRatio	1.25
RgridNumberOfRatio	50 (unit)
ZgridNumberOfEachElectrode	10 (unit)
ZgridDeltaZ1	0.188 (in.)
ZgridRatio1	1.0
ZgridNumberOfRatio1	12 (unit)
ZgridDeltaZ2	0.2 (in.)
ZgridRatio2	1.04 (unit)
ZgridNumberOfRatio2	200 (unit)

15.6 SIMULATION RESULT

15.6.1 Grid size and computation time

Table 15.3 shows the grid size for AC DLT tool. The range of solution domain Ω which is set for simulation is: R directional range is from 0 to 1000 ft, Z-directional range is from -1000 to 1000 ft. Table 15.4 shows the computer system and computation time. From the table, it can be seen that the simulation time for one logging point is only 4 seconds.

Table 15.4 Computer system and computation time

System	Microsoft Windows XP, Version 2002
CPU	Intel Core 2 Duo
Processor	2.00 GHz
RAM	3.00 GB
Time	4 S

15.6.2 Current pattern of LLd and LLs

Fig. 15.10 shows the current density image of the deep laterolog tool, when the tool is running in a homogeneous medium with the resistivity of 1 ohm-m and frequency of 35 Hz. In Fig. 15.10, (A) is the ρ-directional current density, (B) is z-directional current density, and (C) is the amplitude of current density, where (A1), (B1), (C1) are the big view of the whole solution region with the axis range [0 1000 −1000 1000] ft and (A2), (B2), (C2) are the small view concentrate the region around the tool with the axis range [0 50 −50 50] ft. From the big view, it can be seen that the currents flow out of current electrodes, deep into the earth formation and then be collected by the returned electrode B which is put in the far distance of 418 ft in the simulation. Electrodes A series and B are drawn to scale in (C1) and (C2). Reference potential electrode N cannot be located by (A1); it is between electrodes B and A2 in the place where there is zero current density. Voltage electrodes M series can be judged from the distribution of Jz which is shown in (B2).

Fig. 15.11 shows the current density image of the shallow laterolog tool. From the big view, it can be seen that there is no current collected by electrode B; currents flow out from electrode A0, A1, and A1′ and be collected by electrode A2 and A2′, as shown in Fig. 15.11C2. By comparing Figs. 15.10 and 15.11, it shows currents of LLd flow deeper into the formation than currents of LLs, giving a better DOI.

15.6.3 Groningen effect

LLd logs have increasing resistivity gradient when the tool is in a conductive bed which is below a resistive bed as shown in Fig. 15.12. The measured resistivity in this situation will not reflect the true resistivity of the formation. This phenomena is called Groningen effect. The Groningen effect can be simulated by using the FEM numerical model. As shown in Fig. 15.12, three different formation conditions are simulated when the resistivity of the resistive bed is 100, 1000, and 10,000 ohm-m separately. The bed below resistive bed is a conductive bed and normally has 1 ohm-m resistivity. Groningen effect is more severe when the resistivity of the resistive bed is high, as shown in the logging plots. The main reason to cause Groningen effect is the skin effect. With return electrode B at the surface, the LLd currents often had to travel 1−2 miles to reach the current return. At 35 Hz and 1 ohm-m, the skin depth, δ, is

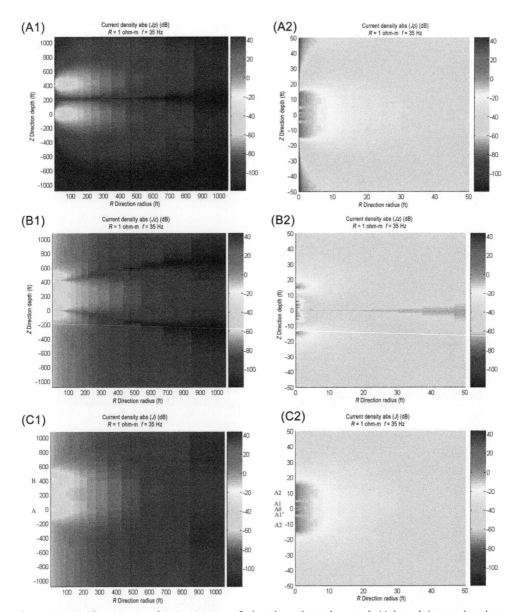

Figure 15.10 The current density image of the deep laterolog tool. LLd tool is running in a homogeneous medium with the resistivity of 1 ohm-m and frequency of 35 Hz. In the figure, there are (A) abs($J\rho$), (B) abs(Jz), and (C) abs(J) which are all plotted in dBA/m^2, with the range of (1) axis = [0 1000 −1000 1000] ft and (2) axis = [0 50 −50 50]. Currents flow out from electrode A0, A1, A1′, A2, A2′ and are collected by electrode B. Electrodes locations are drawn in (C1) and (C2).

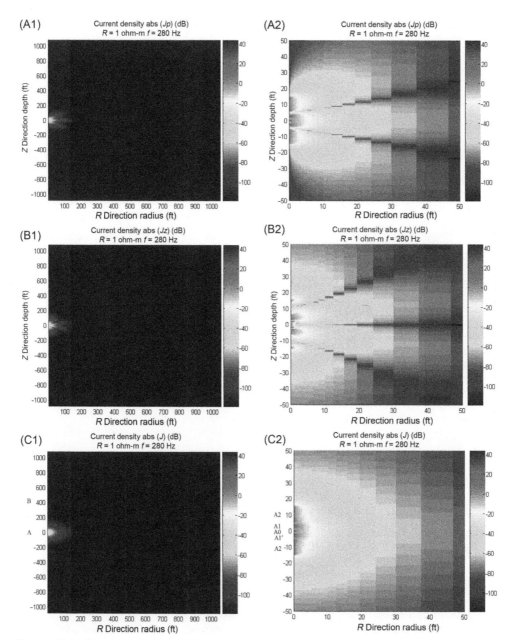

Figure 15.11 The current density image of the shallow laterolog tool. LLs tool is running in a homogeneous medium with the resistivity of 1 ohm-m and frequency of 35 Hz. In the figure, there are (A) abs($J\rho$), (B) abs(Jz), and (C) abs(J) which are all plotted in dBA/m^2, with the range of (1) axis = [0 1000 −1000 1000] ft and (2) axis = [0 50 −50 50]. Currents flow out from electrode A0, A1, and A1′ and are collected by electrode A2, A2′. Electrodes locations are drawn in (C1) and (C2).

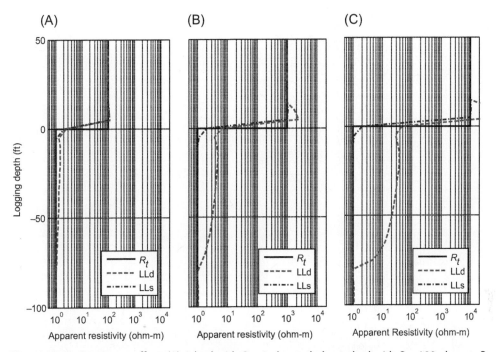

Figure 15.12 Groningen effect. (A) A bed with $R = 1$ ohm-m below a bed with $R = 100$ ohm-m, R_{LLd} reach the high value of 1.5 ohm-m in the depth of -15 ft. (B) A bed with $R = 1$ ohm-m below a bed with $R = 1000$ ohm-m; R_{LLd} reach the high value of 4.8 ohm-m in the depth of -15 ft. (C) A bed with $R = 1$ ohm-m below a bed with $R = 10,000$ ohm-m; R_{LLd} reach the high value of 37.9 ohm-m in the depth of -15 ft.

about 280 ft, which is small in comparison. Because of skin effect, the current returning to B is constrained to remain within a cylinder of radius of δ around the cable carrying current down to the tool, effectively forming a coaxial current beam. This confinement of the current around the cable creates an additional AC impedance, which in turn generates a negative potential at N and distorts the apparent resistivity readings [2].

15.6.4 Frequency effect

Fig. 15.13 shows the frequency effect on DLT tools obtained by using FEM simulation, where (A) is frequency effect on LLs tool and (B) is on LLd tool. The tools are run in a homogeneous formation, the resistivity of the formation is from 0.1 to 10,000 ohm-m. It shows that LLs is not sensitive to the frequencies, so it can run well in relatively high frequency, e.g., 280 Hz. LLd is very sensitive to the frequencies when formation resistivity is low. For example, when the formation resistivity is 0.1 ohm-m, a LLd with 1000 Hz frequency can over read the resistivity

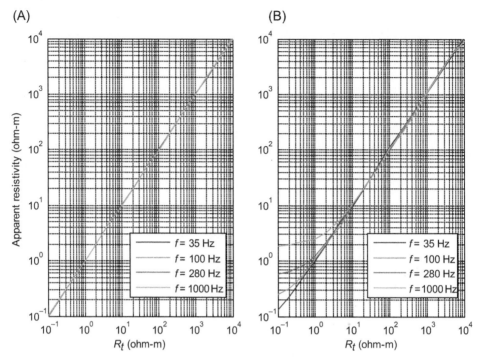

Figure 15.13 Frequency effect. (A) Frequency effect on LLs and (B) frequency effect on LLd.

as 1.9 ohm-m, a LLd with 280 Hz frequency can over read the resistivity as 0.6 ohm-m and a LLd with 100 Hz frequency can over read the resistivity as 0.25 ohm-m. LLd can only work well in low frequencies and that is why it is set to work on 35 Hz.

15.6.5 Invasion effect

Fig. 15.14 shows the invasion effect on DLT tools obtained by using FEM simulation, where (A) is invasion effect on LLs tool and (B) is on LLd tool. The tools are run in a five-layer formation with resistivity of 5, 0.5, 5, 50, 5 ohm-m. The invasions exist on the second and the forth layer with the invasion resistivity of 2.5 and 10 ohm-m. The invasion radius changed from 10 to 50 in. The borehole is a water-based mud borehole with a resistivity of 0.5 ohm-m and radius of 4 in. It shows that tools' response are very sensitive to the invasion resistivity and invasion radius. With the invasion radius increasing from 10 to 50 in., the response of LLs and LLd are shifting from the true formation resistivity R_t to the invasion resistivity R_{xo}. When the invasion radius is about or more than 40 in., it is hard for both LLs and LLd to test the true formation resistivity.

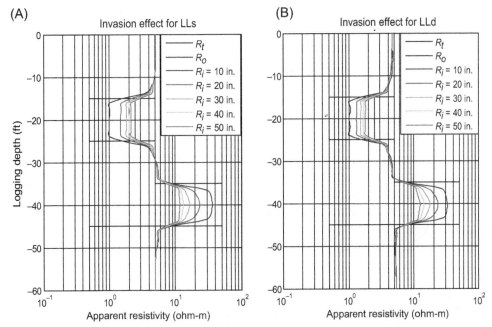

Figure 15.14 Invasion effect. (A) Invasion effect on LLs. (B) Frequency effect on LLd.

15.7 ARRAY LATEROLOG TOOL

15.7.1 The tool structure and the tool response

From the previous sections, we learnt that by correctly control the current in the current emitting electrodes with voltage feedback from the monitoring electrodes, the combined current flow emitted from the electrodes can be "steered" to flow into different depth of the formation, providing the resistivity at the different DOI. To have more detailed radial formation information, it is natural to increase the number of electrodes. Array laterolog tool was developed [5] to provide more resistivity curves at different DOI in the formation. Fig. 15.15 shows a typical array laterolog tool electrode arrangements. In the tool shown in Fig. 15.15, there are 11 current electrodes (A0, A1 ... A6, and A1', A2' ... A6') and 20 measurement electrodes (M1, M2, ... M10, M1', M2' ... M10'). Note that the array laterolog does not have a return electrode at the surface, the return electrode is A6 and A6'. Typical dimensions of the electrode system are shown in Table 15.5. Therefore the Groningen effect is minimized since A6 and A6' are much closer to the tool compared to the surface in DLT case. Although the array laterolog tool is more complicated in operation, the focusing mechanism is very similar to that of a DLT.

There has been discussions of the laterolog data processing mechanism in terms of software focusing and hardware focusing. Theoretically, if we have enough

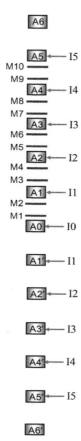

Figure 15.15 An array laterolog tool: A1–A6 and A1'–A6' are current emission electrodes and M1–M10 are measurement electrodes. The dimensions of the tool are shown in Table 15.5.

information, along the borehole (measured potentials and currents), we can always solve for the same amount of unknowns if these measurements are independent. The focusing conditions can be mathematically imposed onto these equations as conditions for the solutions. In this sense, we can always arbitrarily emit current into the formation, e.g., applying the same current or voltages to the transmitting electrodes and measure the voltages on the measurement electrodes. By applying appropriate focusing conditions mathematically, or, soft focusing conditions, different DOI can be achieved. This idea is named as software focusing since no hardware focusing is necessary. However, in practice, there may be difficulties. For example, the current flowing into the formation may be very small compared with the current flowing in the borehole. This case happens when the borehole mud is very conductive and the formation is very resistive. Therefore without appropriate hardware focusing mechanism, which

Table 15.5 The electrode arrangement of an array laterolog tool
Electrode distances from the center of the A0 electrode

Current electrodes	Bottom (in.)	Top (in.)
A0		0.6
A1	7	18
A2	24	30
A3	36	43
A4	51	60
A5	75	133.5
A6	141	261

Monitoring electrodes	Bottom (in.)	Top (in.)
M1	2	3.2
M2	5	6.2
M3	18.3	19.5
M4	22.5	23.7
M5	30.3	31.5
M6	34.5	35.7
M7	43.3	44.5
M8	49.5	50.7
M9	60.3	61.5
M10	73.5	74.7

physically "push" the current into the formation and prevent the azimuthal current flow, pure software focusing is very difficult. A more practical focusing method is to use both hardware and software focusing. Due to the complexity of the electrode system, the focusing method is rather complicated. We will discuss the focusing method for the array laterolog tool in Section 15.7.2.

From the DLT tool, we understand that if we move the current emitting electrodes away from each other, the DOI of the tool will increase as the distance of the electrodes separation increases. Array laterolog can be considered as a combination of many two-electrode tools. Fig. 15.16 shows the focusing mechanism of the array laterolog tool. There are six measurement modes in the tool shown in Fig. 15.16. The voltage curves in the figure show the voltage distribution on the current emitting electrodes for each mode. We can clearly see that mode 0 has only A0 with high voltage and the rest of the electrodes are shorted to the ground. However, in Mode 5, all current emitting electrodes are energized to maximize the DOI. Fig. 15.17 shows the current distribution inside a homogeneous formation for Mode 1 to Mode 5. Mode 0 has the same current distribution as a point source and is not plotted in Fig. 15.17. The numerical simulation used COMSOL commercial software package. From these

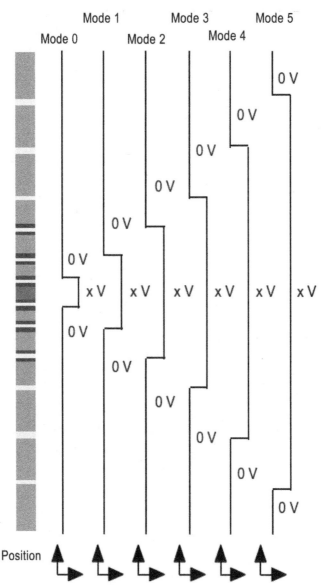

Figure 15.16 The array laterolog tool focusing mechanism. The *green area* (the light colored are, centered at about 1/4 and 3/4 in the vertical direction along the tool (left of the figure)) of the tool represents the current emitting electrodes. *Yellow areas* are insulators. *Blues* are measurement electrodes and the center electrode A0 is shown in *red*.

plots, we can see that as the mode number increases, the focusing effects of the tool are enhanced. For Mode 5, the current at the tool center is almost a straight line, which can flow in the formation further and therefore, generate greatest DOI for the tool. Use the same 10 and1 ohm-m cylindrically layered structure as discussed in

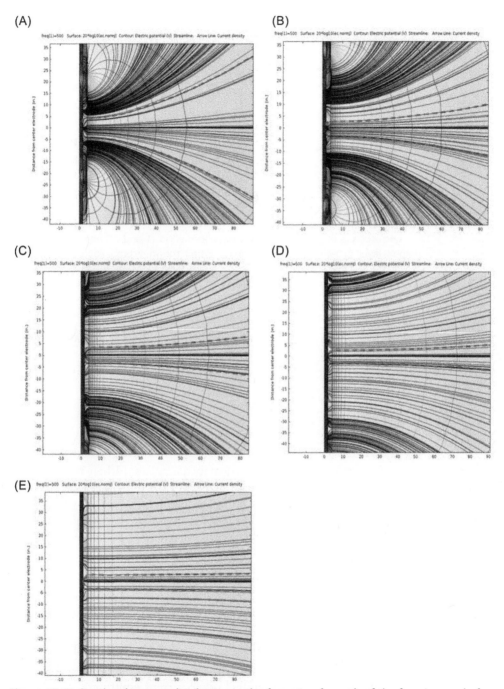

Figure 15.17 Simulated current distribution in the formation for each of the focusing mode from Mode 1 to Mode 5. (A) Mode 1; (B) Mode 2; (C) Mode 3; (D) Mode 4; and (E) Mode 5.

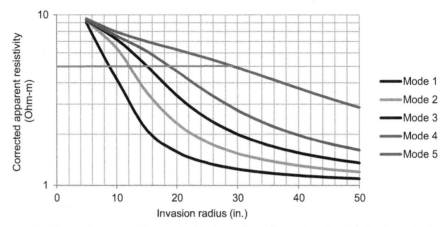

Figure 15.18 The tool response in a cylindrical two-layer formation (1 and 10 ohm-m) when the radius of the 1-ohm-m layer increases (invasion radius). When the measured apparent resistivity reaches 5 ohm-m, the radius of the 1-ohm-m layer is defined as the DOI of the mode.

Table 15.6 The DOI of the array laterolog tool shown in Fig. 15.15

Mode	1	2	3	4	5
DOI (in.)	9.5	12	15	19	29.5

Chapter 7, Induction and LWD Tool Response in a Cylindrically Layered Isotropic Formation, to evaluate the DOI of the tool, we can obtain the DOI curves for each mode. Fig. 15.18 shows that for the tool in Fig. 15.15, as the radius of the 1-ohm-m formation increases the measured apparent resistivity decreases. When the apparent resistivity reaches 5 ohm-m, the radius of the 1-ohm-m formation is defined as DOI of the mode. Table 15.6 is the DOI of the tool at different mode. Note that Mode 0 is usually used for the measurement of the resistivity of the mud and the DOI is very limited.

If different DOI is desired, the electrode spacing must be adjusted. Next, let us discuss the vertical resolution of the laterolog tool. Use the same numerical model, let us consider a 1-ft thin layer with a resistivity of 10 ohm-m sandwiched in a two shoulder beds of 1 ohm-m as shown in Fig. 15.19A. Fig. 15.19B shows the tool response computed by using COMSOL software package. An interesting observation can be drawn from Fig. 15.19 that the vertical response of the tool for each mode has little difference. This is due to the fact that the measurement of the apparent resistivity is based on the current emitted from A0 and all other current emitting electrodes are basically "guard" electrodes to keep the current in A0 electrode flow as desired and they are not counted for evaluation of the apparent resistivity.

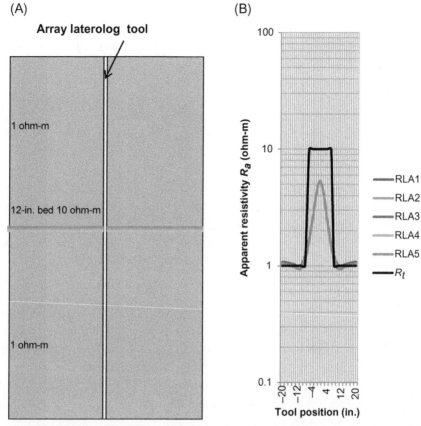

Figure 15.19 The vertical resolution of the array laterolog tool shown in Fig. 15.15. (A) a 1-ft, 10-ohm-m formation sandwiched by two 1-ohm-m shoulder beds used for evaluation of the tool vertical resolution and (B) the computed tool response in the formation in (A).

15.7.2 The focusing method of an array laterolog tool

There are many different ways to discuss the focusing mechanism of the array laterolog tool. Due to the low frequency (or DC) nature of the tool operation, the analysis of the array laterolog is rather straight forward and much simpler than that for induction tools since only static equations are solved.

The practical focusing method of array laterolog is a combination of hardware and software focusing. The hardware injects the currents in a way that is as close as focused as possible. Hardware focusing alone, however, is subject to physical limitations which, in a dynamic environment such as well logging, result in slight voltage imbalances on the array. Software focusing, on the other hand, uses mathematical superposition of signals to ensure that the focusing conditions are satisfied to rectify any imperfections.

Other issue to be considered in practical application is the logging speed. As we discussed in the previous sections, the laterolog has six measurement modes for different DOI. Each mode will have different electrode configurations and accomplish the measurements. If the tool electronics switch to each electrode configuration in a series fashion, the processing speed of the electronics may not be fast enough to handle the six measurement modes. If we do the measurement once using DC or a single frequency, we may not be able to distinguish from which electrode the current is from since the location of the current source is necessary for the focusing and measurement as we will see in the later part of this chapter. Consider the practical logging speed is about 360 ft/h, which is 0.1 ft/s, we expect the processing speed is fast enough so that each mode of the laterolog tool measures the same point and will not be affected by the motion of the logging tool. The practical solution to this problem is to use the frequency division multiplexing (FDM) method commonly used in communication systems. The idea is to use slightly different frequency combinations at each electrode and use signal analysis method such as fast Fourier transform (FFT) to obtain the information of the source locations.

Fig. 15.20 shows the frequency assignment configuration of the array laterolog tool shown in Fig. 15.15, where A0 is survey current electrode, A1−A5 and A1′−A5′ are guard current electrode, A6 and A6′ are current return electrodes, M1−M10 and M1′−M10′ are voltage monitoring electrodes. Different frequencies are assigned to different current electrodes. The central current electrode A0 carries a maximum of six frequencies ($f_0, f_1, f_2, f_3, f_4, f_5$) whereas the rest current electrodes carry less frequencies. In practice, the symmetrical guard electrodes (i.e., A1 and A1′; A2 and A2′; A3 and A3′; A4 and A4′; A5 and A5′) and voltage electrodes (i.e., M1 and M1′; M2 and M2′, ... M10 and M10′) around the central current electrode A0 are short circuited for simplifying electrical system of the tool.

To make things easier to understand, we use a resistor network to represent the formations between electrodes and therefore, the problem becomes the solution of a circuits. Due to the fact that the tool is symmetrical, only half of the tool is considered in a homogenous formation. The idea is to model the formation between electrodes by using a lumped resistor or conductor. The pros of this method is that the physics of the focusing method can be clearly described. However the cons are the measurement electrodes cannot be simulated in the model. The array laterolog shown in Fig. 15.20 can be modeled by a resistive network as shown in Fig. 15.21 in a homogeneous formation. The current and voltage relations are represented by a conductor network with the electrodes as sources.

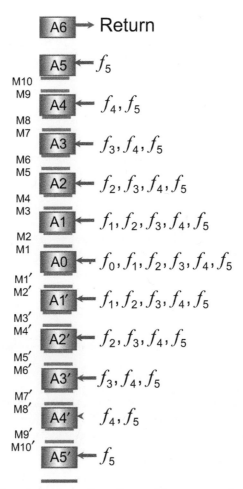

Figure 15.20 Array laterolog electrode configuration and frequency assignment.

The voltages and currents in the network can be expressed as

$$\begin{bmatrix} I_0 \\ I_1 \\ I_2 \\ I_3 \\ I_4 \\ I_5 \\ I_6 \end{bmatrix} = \begin{bmatrix} Y_{0,0} & Y_{0,1} & Y_{0,2} & Y_{0,3} & Y_{0,4} & Y_{0,5} & Y_{0,6} \\ Y_{1,0} & Y_{1,1} & Y_{1,2} & Y_{1,3} & Y_{1,4} & Y_{1,5} & Y_{1,6} \\ Y_{2,0} & Y_{2,1} & Y_{2,2} & Y_{2,3} & Y_{2,4} & Y_{2,5} & Y_{2,6} \\ Y_{3,0} & Y_{3,1} & Y_{3,2} & Y_{3,3} & Y_{3,4} & Y_{3,5} & Y_{3,6} \\ Y_{4,0} & Y_{4,1} & Y_{4,2} & Y_{4,3} & Y_{4,4} & Y_{4,5} & Y_{4,6} \\ Y_{5,0} & Y_{5,1} & Y_{5,2} & Y_{5,3} & Y_{5,4} & Y_{5,5} & Y_{5,6} \\ Y_{6,0} & Y_{6,1} & Y_{6,2} & Y_{6,3} & Y_{6,4} & Y_{6,5} & Y_{6,6} \end{bmatrix} \begin{bmatrix} V_0 \\ V_1 \\ V_2 \\ V_3 \\ V_4 \\ V_5 \\ V_6 \end{bmatrix} \quad (15.24)$$

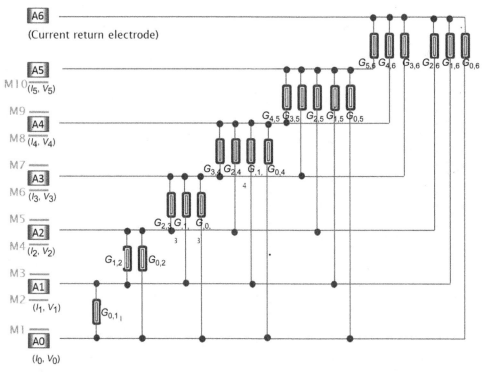

Figure 15.21 The resistor network of the array laterolog tool. This model only has half of the tool and the other half is symmetrical.

where Y_{ij} is the admittance between electrode i and j ($i = 0,1,2,\ldots 6, j = 0,1,2\ldots 6$), and

$$Y_{ij} = -G_{ij} \tag{15.25}$$

For the reason of reciprocal,

$$Y_{ij} = Y_{ji} \tag{15.26}$$

Use the law of current preservation, we have,

$$Y_{ii} = -\sum_{j=0}^{6} Y_{ij} (i \neq j) \tag{15.27}$$

Note that V_6 is a return electrode and

$$V_6 = 0 \tag{15.28}$$

$$I_6 = -\sum_{k=0}^{5} I_k \tag{15.29}$$

If the following voltages are applied to the electrodes:

$$A_0: \overline{V_{A_0}} = V_{A0}(\cos \omega_0 t + \cos \omega_1 t + \cos \omega_2 t + \cos \omega_3 t + \cos \omega_4 t + \cos \omega_5 t)$$
$$A_1: \overline{V_{A_1}} = V_{A1}(\cos \omega_1 t + \cos \omega_2 t + \cos \omega_3 t + \cos \omega_4 t + \cos \omega_5 t)$$
$$A_2: \overline{V_{A_2}} = V_{A2}(\cos \omega_2 t + \cos \omega_3 t + \cos \omega_4 t + \cos \omega_5 t) \quad (15.30)$$
$$A_3: \overline{V_{A_3}} = V_{A3}(\cos \omega_3 t + \cos \omega_4 t + \cos \omega_5 t)$$
$$A_4: \overline{V_{A_4}} = V_{A4}(\cos \omega_4 t + \cos \omega_5 t)$$
$$A_5: \overline{V_{A_5}} = V_{A5}\cos \omega_5 t$$

where $\omega_0, \omega_1, \omega_2, \omega_3, \omega_4, \omega_5$ are angular frequencies corresponding to $f_0, f_1, f_2, f_3, f_4, f_5$.

Using the superposition theorem, at electrode A0, we can obtain the relations between currents and voltages using the relation given in Eq. (15.24):

$$\begin{aligned}
I_{A_0}^{(f_0)} &= Y_{0,0} V_{A_0}^{(f_0)} \\
I_{A_0}^{(f_1)} &= Y_{0,0} V_{A_0}^{(f_1)} + Y_{0,1} V_{A_2}^{(f_1)} \\
I_{A_0}^{(f_2)} &= Y_{0,0} V_{A_0}^{(f_2)} + Y_{0,1} V_{A_1}^{(f_2)} + Y_{0,2} V_{A_2}^{(f_2)} \\
I_{A_0}^{(f_3)} &= Y_{0,0} V_{A_0}^{(f_3)} + Y_{0,1} V_{A_1}^{(f_3)} + Y_{0,2} V_{A_2}^{(f_3)} + Y_{0,3} V_{A_3}^{(f_3)} \\
I_{A_0}^{(f_4)} &= Y_{0,0} V_{A_0}^{(f_4)} + Y_{0,1} V_{A_1}^{(f_4)} + Y_{0,2} V_{A_2}^{(f_4)} + Y_{0,3} V_{A_3}^{(f_4)} + Y_{0,4} V_{A_4}^{(f_4)} \\
I_{A_0}^{(f_5)} &= Y_{0,0} V_{A_0}^{(f_5)} + Y_{0,1} V_{A_1}^{(f_5)} + Y_{0,2} V_{A_2}^{(f_5)} + Y_{0,3} V_{A_3}^{(f_5)} + + Y_{0,4} V_{A_4}^{(f_5)} + Y_{0,5} V_{A_0}^{(f_5)}
\end{aligned} \quad (15.31)$$

The matrix form of Eq. (15.31) is

$$\begin{bmatrix} V_{A_0}^{(f_0)} & & & & & \\ V_{A_0}^{(f_1)} & V_{A_1}^{(f_1)} & & & & \\ V_{A_0}^{(f_2)} & V_{A_1}^{(f_2)} & V_{A_2}^{(f_2)} & & & \\ V_{A_0}^{(f_3)} & V_{A_1}^{(f_3)} & V_{A_2}^{(f_3)} & V_{A_3}^{(f_3)} & & \\ V_{A_0}^{(f_4)} & V_{A_1}^{(f_4)} & V_{A_2}^{(f_4)} & V_{A_3}^{(f_4)} & V_{A_4}^{(f_4)} & \\ V_{A_0}^{(f_5)} & V_{A_1}^{(f_5)} & V_{A_2}^{(f_5)} & V_{A_3}^{(f_5)} & V_{A_4}^{(f_5)} & V_{A_5}^{(f_5)} \end{bmatrix} \begin{bmatrix} Y_{0,0} \\ Y_{0,1} \\ Y_{0,2} \\ Y_{0,3} \\ Y_{0,4} \\ Y_{0,5} \end{bmatrix} = \begin{bmatrix} I_{A_0}^{(f_0)} \\ I_{A_0}^{(f_1)} \\ I_{A_0}^{(f_2)} \\ I_{A_0}^{(f_3)} \\ I_{A_0}^{(f_4)} \\ I_{A_0}^{(f_5)} \end{bmatrix} \quad (15.32)$$

Similarly, we can obtain the current and voltage relations at all other electrodes:
For electrode A1 to A5:

$$\begin{bmatrix} V_{A_1}^{(f_1)} & & & & \\ V_{A_1}^{(f_2)} & V_{A_2}^{(f_2)} & & & \\ V_{A_1}^{(f_3)} & V_{A_2}^{(f_3)} & V_{A_3}^{(f_3)} & & \\ V_{A_1}^{(f_4)} & V_{A_2}^{(f_4)} & V_{A_3}^{(f_4)} & V_{A_4}^{(f_4)} & \\ V_{A_1}^{(f_5)} & V_{A_2}^{(f_5)} & V_{A_3}^{(f_5)} & V_{A_4}^{(f_5)} & V_{A_5}^{(f_5)} \end{bmatrix} \begin{bmatrix} Y_{1,1} \\ Y_{1,2} \\ Y_{1,3} \\ Y_{1,4} \\ Y_{1,5} \end{bmatrix} = \begin{bmatrix} I_{A_1}^{(f_1)} - V_{A_0}^{(f_1)} Y_{1,0} \\ I_{A_1}^{(f_2)} - V_{A_0}^{(f_2)} Y_{1,0} \\ I_{A_1}^{(f_3)} - V_{A_0}^{(f_3)} Y_{1,0} \\ I_{A_1}^{(f_4)} - V_{A_0}^{(f_4)} Y_{1,0} \\ I_{A_1}^{(f_5)} - V_{A_0}^{(f_5)} Y_{1,0} \end{bmatrix} \quad (15.33)$$

$$\begin{bmatrix} V_{A_2}^{(f_2)} & 0 & 0 & 0 \\ V_{A_2}^{(f_3)} & V_{A_3}^{(f_3)} & 0 & 0 \\ V_{A_2}^{(f_4)} & V_{A_3}^{(f_4)} & V_{A_4}^{(f_4)} & 0 \\ V_{A_2}^{(f_5)} & V_{A_3}^{(f_5)} & V_{A_4}^{(f_5)} & V_{A_5}^{(f_5)} \end{bmatrix} \begin{bmatrix} Y_{2,2} \\ Y_{2,3} \\ Y_{2,4} \\ Y_{2,5} \end{bmatrix} = \begin{bmatrix} I_{A_2}^{(f_2)} - V_{A_0}^{(f_2)} Y_{2,0} - V_{A_1}^{(f_2)} Y_{2,1} \\ I_{A_2}^{(f_3)} - V_{A_0}^{(f_3)} Y_{2,0} - V_{A_1}^{(f_3)} Y_{2,1} \\ I_{A_2}^{(f_4)} - V_{A_0}^{(f_4)} Y_{2,0} - V_{A_1}^{(f_4)} Y_{2,1} \\ I_{A_2}^{(f_5)} - V_{A_0}^{(f_5)} Y_{2,0} - V_{A_1}^{(f_5)} Y_{2,1} \end{bmatrix} \quad (15.34)$$

$$\begin{bmatrix} V_{A_3}^{(f_3)} & 0 & 0 \\ V_{A_3}^{(f_4)} & V_{A_4}^{(f_4)} & 0 \\ V_{A_3}^{(f_5)} & V_{A_4}^{(f_5)} & V_{A_5}^{(f_5)} \end{bmatrix} \begin{bmatrix} Y_{3,3} \\ Y_{3,4} \\ Y_{3,5} \end{bmatrix} = \begin{bmatrix} I_{A_3}^{(f_3)} - V_{A_0}^{(f_3)} Y_{3,0} - V_{A_1}^{(f_3)} Y_{3,1} - V_{A_2}^{(f_3)} Y_{3,2} \\ I_{A_3}^{(f_4)} - V_{A_0}^{(f_4)} Y_{3,0} - V_{A_1}^{(f_4)} Y_{3,1} - V_{A_2}^{(f_4)} Y_{3,2} \\ I_{A_3}^{(f_5)} - V_{A_0}^{(f_5)} Y_{3,0} - V_{A_1}^{(f_5)} Y_{3,1} - V_{A_2}^{(f_5)} Y_{3,2} \end{bmatrix} \quad (15.35)$$

$$\begin{bmatrix} V_{A_4}^{(f_4)} & 0 \\ V_{A_4}^{(f_5)} & V_{A_5}^{(f_5)} \end{bmatrix} \begin{bmatrix} Y_{4,4} \\ Y_{4,5} \end{bmatrix} = \begin{bmatrix} I_{A_4}^{(f_4)} - V_{A_0}^{(f_4)} Y_{4,0} - V_{A_1}^{(f_4)} Y_{4,1} - V_{A_2}^{(f_4)} Y_{4,2} - V_{A_3}^{(f_4)} Y_{4,3} \\ I_{A_4}^{(f_5)} - V_{A_0}^{(f_5)} Y_{4,0} - V_{A_1}^{(f_5)} Y_{4,1} - V_{A_2}^{(f_5)} Y_{4,2} - V_{A_3}^{(f_5)} Y_{4,3} \end{bmatrix} \quad (15.36)$$

$$[V_{A_5}^{(f_5)}][Y_{5,5}] = \left[I_{A_5}^{(f_5)} - V_{A_0}^{(f_5)} Y_{5,0} - V_{A_1}^{(f_5)} Y_{5,1} - V_{A_2}^{(f_5)} Y_{5,2} - V_{A_3}^{(f_5)} Y_{5,3} - V_{A_4}^{(f_5)} Y_{5,4} \right] \quad (15.37)$$

With these equations, we can discuss the focusing algorithm of the laterolog tool. Assuming the measurement electrodes are very close to the current emitting electrodes, and the voltages measured at the measurement electrodes are the same as the current electrodes, for Mode 0, which is a nonfocusing mode of the laterolog tool, we have:

$$I_{A_0}^{f_0} = Y_{00} V_{A_0}^{f_0} \quad (15.38)$$

However, Mode 1 requires the potential at A1 is equal to that at A2, and the current emitted from the focusing electrode A1 can be obtained by solving the voltage−current relation in Eq. (15.33),

$$V^{(f_1)}_{A_0\_m} = V^{(f_1)}_{A_1\_m} \tag{15.39}$$

$$I^{(f_1)}_{A_0}/(G_{0,2} + G_{0,3} + G_{0,4} + G_{0,5} + G_{0,6}) = I^{(f_1)}_{A_1}/(G_{1,2} + G_{1,3} + G_{1,4} + G_{1,5} + G_{1,6}) \tag{15.40}$$

The current at the Guard electrode A1 is then expressed as

$$I^{(f_1)}_{A_1} = \frac{V^{(f_1)}_{A_0}(G_{1,2} + G_{1,3} + G_{1,4} + G_{1,5} + G_{1,6})}{G_{0,2} + G_{0,3} + G_{0,4} + G_{0,5} + G_{0,6}} \tag{15.41}$$

For Mode 2, the focusing conditions are:

$$\begin{cases} V^{(f_2)}_{A_0\_m} = V^{(f_2)}_{A_1\_m} \\ V^{(f_2)}_{A_0\_m} = V^{(f_2)}_{A_2\_m} \end{cases} \tag{15.42}$$

Use the current and voltage conditions in Eq. (15.34), we have,

$$\begin{cases} I^{(f_2)}_{A_0}/(G_{0,3} + G_{0,4} + G_{0,5} + G_{0,6}) = I^{(f_2)}_{A_1}/(G_{1,3} + G_{1,4} + G_{1,5} + G_{1,6}) \\ I^{(f_2)}_{A_0}/(G_{0,3} + G_{0,4} + G_{0,5} + G_{0,6}) = I^{(f_2)}_{A_2}/(G_{2,3} + G_{2,4} + G_{2,5} + G_{2,6}) \end{cases} \tag{15.43}$$

Solving for the focusing currents on A1 and A2,

$$\begin{cases} I^{(f_2)}_{A_1} = \dfrac{V^{(f_2)}_{A_0}(G_{1,3} + G_{1,4} + G_{1,5} + G_{1,6})}{(G_{0,3} + G_{0,4} + G_{0,5} + G_{0,6})} \\ I^{(f_2)}_{A_2} = \dfrac{V^{(f_2)}_{A_0}(G_{2,3} + G_{2,4} + G_{2,5} + G_{2,6})}{(G_{0,3} + G_{0,4} + G_{0,5} + G_{0,6})} \end{cases} \tag{15.44}$$

For Mode 3, the focusing conditions are:

$$\begin{cases} V^{(f_3)}_{A_0\_m} = V^{(f_3)}_{A_1\_m} \\ V^{(f_3)}_{A_0\_m} = V^{(f_3)}_{A_2\_m} \\ V^{(f_3)}_{A_0\_m} = V^{(f_3)}_{A_3\_m} \end{cases} \tag{15.45}$$

$$\begin{cases} I^{(f_3)}_{A_0}/(G_{0,4} + G_{0,5} + G_{0,6}) = I^{(f_3)}_{A_1}/(G_{1,4} + G_{1,5} + G_{1,6}) \\ I^{(f_3)}_{A_0}/(G_{0,4} + G_{0,5} + G_{0,6}) = I^{(f_3)}_{A_2}/(G_{2,4} + G_{2,5} + G_{2,6}) \\ I^{(f_3)}_{A_0}/(G_{0,4} + G_{0,5} + G_{0,6}) = I^{(f_3)}_{A_3}/(G_{3,4} + G_{3,5} + G_{3,6}) \end{cases} \tag{15.46}$$

And the focusing currents are:

$$\begin{cases} I_{A_1}^{(f_3)} = \dfrac{I_{A_0}^{(f_3)}(G_{1,4} + G_{1,5} + G_{1,6})}{(G_{0,4} + G_{0,5} + G_{0,6})} \\[6pt] I_{A_2}^{(f_3)} = \dfrac{I_{A_0}^{(f_3)}(G_{2,4} + G_{2,5} + G_{2,6})}{(G_{0,4} + G_{0,5} + G_{0,6})} \\[6pt] I_{A_3}^{(f_3)} = \dfrac{I_{A_0}^{(f_3)}(G_{3,4} + G_{3,5} + G_{3,6})}{(G_{0,4} + G_{0,5} + G_{0,6})} \end{cases} \quad (15.47)$$

For Mode 4, four focusing conditions are given:

$$\begin{cases} V_{A_0\_m}^{(f_4)} = V_{A_1\_m}^{(f_4)} \\ V_{A_0\_m}^{(f_4)} = V_{A_2\_m}^{(f_4)} \\ V_{A_0\_m}^{(f_4)} = V_{A_3\_m}^{(f_4)} \\ V_{A_0\_m}^{(f_4)} = V_{A_4\_m}^{(f_4)} \end{cases} \quad (15.48)$$

$$\begin{cases} I_{A_0}^{(f_4)}/(G_{0,5} + G_{0,6}) = I_{A_1}^{(f_4)}/(G_{1,5} + G_{1,6}) \\ I_{A_0}^{(f_4)}/(G_{0,5} + G_{0,6}) = I_{A_2}^{(f_4)}/(G_{2,5} + G_{2,6}) \\ I_{A_0}^{(f_4)}/(G_{0,5} + G_{0,6}) = I_{A_3}^{(f_4)}/(G_{3,5} + G_{3,6}) \\ I_{A_0}^{(f_4)}/(G_{0,5} + G_{0,6}) = I_{A_4}^{(f_4)}/(G_{4,5} + G_{4,6}) \end{cases} \quad (15.49)$$

And the four focusing currents can be solved as:

$$\begin{cases} I_{A_1}^{(f_4)} = \dfrac{I_{A_0}^{(f_4)}(G_{1,5} + G_{1,6})}{(G_{0,5} + G_{0,6})} \\[6pt] I_{A_2}^{(f_4)} = \dfrac{I_{A_0}^{(f_4)}(G_{2,5} + G_{2,6})}{(G_{0,5} + G_{0,6})} \\[6pt] I_{A_3}^{(f_4)} = \dfrac{I_{A_0}^{(f_4)}(G_{3,5} + G_{3,6})}{(G_{0,5} + G_{0,6})} \\[6pt] I_{A_4}^{(f_4)} = \dfrac{I_{A_0}^{(f_4)}(G_{4,5} + G_{4,6})}{(G_{0,5} + G_{0,6})} \end{cases} \quad (15.50)$$

Lastly, for Mode 5 the focusing conditions are:

$$\begin{cases} V_{A_0\_m}^{(f_5)} = V_{A_1\_m}^{(f_5)} \\ V_{A_0\_m}^{(f_5)} = V_{A_2\_m}^{(f_5)} \\ V_{A_0\_m}^{(f_5)} = V_{A_3\_m}^{(f_5)} \\ V_{A_0\_m}^{(f_5)} = V_{A_4\_m}^{(f_5)} \\ V_{A_0\_m}^{(f_5)} = V_{A_5\_m}^{(f_5)} \end{cases} \qquad (15.51)$$

$$\begin{cases} I_{A_0}^{(f_5)}/G_{0,6} = I_{A_1}^{(f_5)}/G_{1,6} \\ I_{A_0}^{(f_5)}/G_{0,6} = I_{A_2}^{(f_5)}/G_{2,6} \\ I_{A_0}^{(f_5)}/G_{0,6} = I_{A_3}^{(f_5)}/G_{3,6} \\ I_{A_0}^{(f_5)}/G_{0,6} = I_{A_4}^{(f_5)}/G_{4,6} \\ I_{A_0}^{(f_5)}/G_{0,6} = I_{A_5}^{(f_5)}/G_{5,6} \end{cases} \qquad (15.52)$$

The currents from the five electrodes under focusing conditions are

$$\begin{cases} I_{A_1}^{(f_5)} = \dfrac{I_{A_0}^{(f_5)} G_{1,6}}{G_{0,6}} \\ I_{A_2}^{(f_5)} = \dfrac{I_{A_0}^{(f_5)} G_{2,6}}{G_{0,6}} \\ I_{A_3}^{(f_5)} = \dfrac{I_{A_0}^{(f_5)} G_{3,6}}{G_{0,6}} \\ I_{A_4}^{(f_5)} = \dfrac{I_{A_0}^{(f_5)} G_{4,6}}{G_{0,6}} \\ I_{A_5}^{(f_5)} = \dfrac{I_{A_0}^{(f_5)} G_{5,6}}{G_{0,6}} \end{cases} \qquad (15.53)$$

Under the focusing condition, the current near the center electrode A0 flows deeper into the formation as shown in Figs. 15.17 and 15.18. We define the apparent resistivity based on the current flow from the center electrodes, which is voltage and current ratio in a specific focusing mode. Therefore, for each mode, the apparent resistivity measured by the tool may be different depending on the invasion profile of the formation. For each mode, the apparent resistivity is defined as:

$$R_a^{f_n} = k_n \frac{V_0^{f_n}}{I_0^{f_n}} \quad (n = 0, 1, \ldots 5) \tag{15.54}$$

where $R_a^{f_n}$ is the apparent resistivity of Mode n ($n = 0, 1, \ldots 5$); f_n is the nth frequency used in the tool; k_n is the tool constant of Mode n, which can be found during tool calibration process; $V_0^{f_n}$ is the voltage on the center electrode A0 at frequency n; and $I_0^{f_n}$ is the current on the center electrode A0 at frequency f_n. Using the definition in Eq. (15.50) and the resistor network discussed in this section, the apparent resistivity can be solved.

Apparent resistivity R_a:

$$\text{Mode 0:} R_a^{(f_0)} = K_0/[G_{0,1} + G_{0,2} + G_{0,3} + G_{0,4} + G_{0,5} + G_{0,6}]$$

$$\text{Mode 1:} R_a^{(f_1)} = K_1/[G_{0,2} + G_{0,3} + G_{0,4} + G_{0,5} + G_{0,6}]$$

$$\text{Mode 2:} R_a^{(f_2)} = K_2/[G_{0,3} + G_{0,4} + G_{0,5} + G_{0,6}]$$

$$\text{Mode 3:} R_a^{(f_3)} = K_3/[G_{0,4} + G_{0,5} + G_{0,6}]$$

$$\text{Mode 4:} R_a^{(f_4)} = K_4/[G_{0,5} + G_{0,6}]$$

$$\text{Mode 5:} R_a^{(f_5)} = K_5/G_{0,6}$$

Fig. 15.22 shows the simulated apparent resistivity of a formation with different bed thickness when the focusing method discussed in this section is applied. Since the formation model used for the simulation has an invasion profile, which is a normal low resistivity mud filtrate invasion into a sandstone zone, the apparent resistivity from each mode "sees" different due to the invasion profile. The deeper detection mode (e.g., Mode 5) sees higher resistivity whereas the shallow detection mode (e.g., Mode 1) measures lower resistivity. Using the different resistivity measurements, the petrophysicists can determine the invasion profile of the formation.

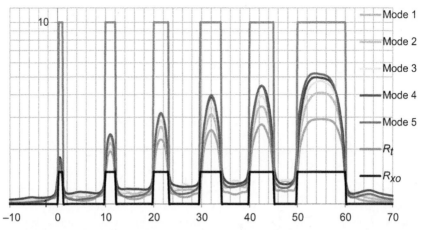

Figure 15.22 The simulated apparent resistivity of a layered formation with the invasion. The focusing conditions used in this simulation is shown in this section. The invasion profile of the formation is a step of 1 and 15 ohm-m as shown in *dotted line* and expressed as R_{xo} in the figure. The mud resistivity is 0.1 ohm-m, borehole diameter is 8 in., and invasion radius is 15 in.

In Fig. 15.22, we notice that the maximum reading of the apparent resistivity is less than that of the true formation. This is because the influence of the mud is 0.1 ohm-m.

REFERENCES

[1] L.C. Shen, J.A. Kong, Applied electromagnetism, in: PWS Series in Engineering, third ed., pp. 365, 1995, ISBN-13: 978-0534947224 (Chapter 12).
[2] B.I. Anderson, Modeling and inversion methods for the interpretation of resistivity logging tool response (Doctoral thesis), Delft University Press, Dundee, 2001, pp. 116–160.
[3] J. Suau, P. Grimaldi, A. Pupon, P. Pouhaite, The dual laterolog–Rxo tool, in: Proceedings 1972 SPE Annual Technical Conference and Exhibition, Society of Petroleum Engineers, Paper SPE 4018, 1972.
[4] J.M. Jin, The Finite Element Method in Electromagnetics, second ed., Wiley, New York, NY, 2002, pp. 273–333, 339 (Chapter 8).
[5] J.W. Smits, I. Dubourg, M.G. Luling, et al. Improved resistivity interpretation utilizing a new array laterolog tool and associated inversion processing, in: Presented at the SPE Annual Technical Conference and Exhibition, New Orleans, Louisiana, 27-30 September 1998. SPE-49328-MS, 1998. http://dx.doi.org/10.2118/49328-MS.

APPENDIX A COMPUTATION METHOD OF SOURCE MODEL FOR ALTERNATING CURRENT DUAL LATEROLOG TOOL

```
SUBROUTINE NonConstantConstraint_AC_Laterolog_Deep_RouHPhi
!
INTEGER :: L,M,N,MatrixSize
!
COMPLEX*16 :: Is,Vs !current source or voltage source
COMPLEX*16,ALLOCATABLE :: RouHphi(:)
COMPLEX*16,ALLOCATABLE :: I(:)
COMPLEX*16,ALLOCATABLE :: V(:)
COMPLEX*16,ALLOCATABLE :: X(:) !B*X = Vector
COMPLEX*16,ALLOCATABLE :: Vector(:),VectorCopy(:)
COMPLEX*16,ALLOCATABLE :: SourceMatrix(:,:) !matrix information for source nodes
COMPLEX*16,ALLOCATABLE :: A(:,:) !matrix information for source nodes on 4 insulators
COMPLEX*16,ALLOCATABLE :: B(:,:),Bcopy(:,:) !matrix information for 3 electrodes and 4 insulators
!
              NumberOfInsulator = NonConstantConstraintNumber
              NumberOfElectrode = NumberOfInsulator-1
  MatrixSize = NumberOfInsulator + 2*NumberOfElectrode
!
  IF (.NOT.ALLOCATED(NonConstantConstraintValue)) &
    ALLOCATE(NonConstantConstraintValue(NonConstantConstraintNumber))
              NonConstantConstraintValue = (0.0,0.0)
!
              ALLOCATE(SourceMatrix(NonConstantConstraintNumber,NonConstantConstraintNumber))
              ALLOCATE(A(NonConstantConstraintNumber,NonConstantConstraintNumber))
              ALLOCATE(B(MatrixSize,MatrixSize))
              ALLOCATE(Bcopy(MatrixSize,MatrixSize))
              SourceMatrix = (0.0,0.0)
              A = (0.0,0.0)
              B = (0.0,0.0)
Bcopy = (0.0,0.0)
!
ALLOCATE(RouHphi(NumberOfInsulator))
ALLOCATE(I(NumberOfElectrode))
ALLOCATE(V(NumberOfElectrode))
ALLOCATE(X(MatrixSize))
ALLOCATE(Vector(MatrixSize))
ALLOCATE(VectorCopy(MatrixSize))
              RouHphi = (0.0,0.0)
              I = (0.0,0.0)
              V = (0.0,0.0)
              X = (0.0,0.0)
              Vector = (0.0,0.0)
              VectorCopy = (0.0,0.0)
!
```

```fortran
!...read source matrix information from file 'SourceMatirx.dat'
                OPEN(1,FILE = 'SourceMatirx.dat',STATUS = 'UNKNOWN')
                DO M = 1,NonConstantConstraintNumber
                DO N = 1,NonConstantConstraintNumber
                                READ(1,*) SourceMatrix(M,N)
                                A(M,N) = SourceMatrix(M,N)
                END DO
                END DO
                CLOSE(1)
!
!...form matrix B: Line 1 to NumberOfInsulator
                DO M = 1,NumberOfInsulator
                DO N = 1,NumberOfInsulator
                                B(M,N) = A(M,N)
                END DO
                END DO
                DO M = 1,NumberOfInsulator
                DO N = NumberOfInsulator + 1,NumberOfInsulator +
                    NumberOfElectrode
                                B(M,N) = 0
                END DO
                END DO
                DO M = 1,NumberOfInsulator
                                L = 0
                DO N = NumberOfInsulator + NumberOfElectrode + 1,
                    NumberOfInsulator + NumberOfElectrode + NumberOfElectrode
                                L = L + 1
!                               IF (L.EQ.M) B(M,N) = -1*2*Pi*RadiusOfElectrode
!                               IF (L.EQ.(M-1)) B(M,N) = 1*2*Pi*RadiusOfElectrode
                                IF (L.EQ.M) B(M,N) = -1*2*Pi
                                IF (L.EQ.(M-1)) B(M,N) = 1*2*Pi
                END DO
                END DO
!
!...form matrix B: Line NumberOfInsulator + 1 to
NumberOfInsulator + NumberOfElectrode
                L = 0
                DO M = NumberOfInsulator + 1,NumberOfInsulator + NumberOfElectrode
                                L = L + 1
                DO N = 1,NumberOfInsulator + NumberOfElectrode + NumberOfElectrode
                                IF (N.EQ.L) B(M,N) = 1
                                IF (N.EQ.(L + 1)) B(M,N) = -1
!                               IF (N.EQ.(L + NumberOfInsulator)) B(M,N) = -
                                    (1./2./Pi/RadiusOfElectrode)
                                IF (N.EQ.(L + NumberOfInsulator)) B(M,N) = -
                                    (1./2./Pi)
                END DO
                END DO
```

```
!
!...form matrix B: Line NumberOfInsulator + NumberOfElectrode + 1 to
! NumberOfInsulator + 2*NumberOfElectrode

  DO M = NumberOfInsulator + NumberOfElectrode + 1, NumberOfInsulator +
    2*NumberOfElectrode

    DO N = NumberOfInsulator + 1, NumberOfInsulator + 2*NumberOfElectrode
      B(M,N) = FocusingConditionDeep(M-NumberOfInsulator-NumberOfElectrode,
        N-NumberOfInsulator)
    END DO
  END DO
!
!Line NumberOfInsulator + 2*NumberOfElectrode
!last line for CurrentOfSource or VoltageOfSource
!last line is: I(IndexOfElectrodeA0) = CurrentOfSource
! or V(IndexOfElectrodeA0) = VoltageOfSource
                  M = NumberOfInsulator + 2*NumberOfElectrode
                  IF ((CurrentOfSource.NE.0).AND.(VoltageOfSource.EQ.0)) THEN
                                  B(M,NumberOfInsulator + IndexOfElectrodeA0) = 1
                  ELSE IF ((CurrentOfSource.EQ.0).AND.(VoltageOfSource.NE.0)) THEN
                                  B(M,NumberOfInsulator + NumberOfElectrode +
                                    IndexOfElectrodeA0) = 1
                  ELSE
  WRITE(*,*)
                                  WRITE(*,*) 'Wrong Message: Please check input
                                    current or voltage source value.'
                                  WRITE(*,*)
  STOP
                  END IF
!
!...form vector
                  IF ((CurrentOfSource.NE.0).AND.(VoltageOfSource.EQ.0)) THEN
                                  Vector(NumberOfInsulator +
                                    2*NumberOfElectrode) = CurrentOfSource
                  ELSE IF ((CurrentOfSource.EQ.0).AND.(VoltageOfSource.NE.0)) THEN
                                  Vector(NumberOfInsulator +
                                    2*NumberOfElectrode) = VoltageOfSource
                  ELSE
  WRITE(*,*)
                                  WRITE(*,*) 'Wrong Message: Please check input
                                    current or voltage source value.'
                                  WRITE(*,*)
  STOP
                  END IF
!
!...slove B*X = Vector
Bcopy = B
VectorCopy = Vector
CALL SolveMatrixB(B,Vector,X,MatrixSize)
```

```fortran
              B = Bcopy
              Vector = VectorCopy
              !
              !...get RouHphi,I and V
                          DO M = 1,NumberOfInsulator
                                        RouHphi(M) = X(M)
                                        NonConstantConstraintValue(M) = RouHphi(M)
                          END DO
                          L = 0
                          DO M = NumberOfInsulator + 1,NumberOfInsulator +
                             NumberOfElectrode
                                        L = L + 1
                                        I(L) = X(M)
                          END DO
                          L = 0
                          DO M = NumberOfInsulator + NumberOfElectrode + 1,
                             NumberOfInsulator + 2*NumberOfElectrode
                                        L = L + 1
                                        V(L) = X(M)
                          END DO
              !
              !...impedanceofsystem
                          Vs = V(IndexOfElectrodeA0)
                          Is = I(IndexOfElectrodeA0)
                          VoltageOfSourceAterCalculation = Vs
                          CurrentOfSourceAterCalculation = Is
                          ImpedanceOfSystemDLL = Vs/Is
              !
              !...apprent resistivity of deep laterolog is K_DLL*(U(M1)-U(N))/Is
              ApparentResistivityOfDeepLaterolog = K_DLL*(V(IndexOfElectrodeM1)-V
              (IndexOfElectrodeN))/Is
              !
              !...test: Output matrix B and Vector for testing purpose
              OPEN(1,FILE = 'MatrixB.dat',STATUS = 'UNKNOWN')
              DO M = 1,MatrixSize
                DO N = 1,MatrixSize
                  WRITE(1,20) B(M,N)
                END DO
                END DO
              CLOSE(1)
              OPEN(1,FILE = 'Vector.DAT',STATUS = 'UNKNOWN')
              DO M = 1,MatrixSize
                WRITE(1,20) Vector(M)
              END DO
              CLOSE(1)
              20                FORMAT(2F30.20)
```

```fortran
!
!...test: testing if B*X is equal to Vector
  DO M=1,MatrixSize
    VectorCopy(M)=0
  DO N=1,MatrixSize
    VectorCopy(M)=VectorCopy(M)+B(M,N)*X(N)
END DO
END DO
DO M=1,MatrixSize
    VectorCopy(M)=VectorCopy(M)-Vector(M)
END DO
!
                 OPEN(1,FILE='VectorError.DAT',STATUS='UNKNOWN')
DO M=1,MatrixSize
  WRITE(1,*) VectorCopy(M)
END DO
CLOSE(1)
!
!...test: output value of X
                 OPEN(1,FILE='ValueOfX.DAT',STATUS='UNKNOWN')
DO M=1,MatrixSize
WRITE(1,*) 'X(',M,') = ',X(M)
END DO
CLOSE(1)
!
!...deallocate
                 DEALLOCATE(RouHphi)
                 DEALLOCATE(I)
                 DEALLOCATE(V)
                 DEALLOCATE(X)
                 DEALLOCATE(Vector)
                 DEALLOCATE(VectorCopy)
                 DEALLOCATE(A)
                 DEALLOCATE(B)
  DEALLOCATE(Bcopy)
                 DEALLOCATE(SourceMatrix)
!
                 RETURN
!
END SUBROUTINE NonConstantConstraint_AC_Laterolog_Deep_RouHPhi
```

CHAPTER 16

Theory of the Through-Casing Resistivity Logging Tool

Contents

16.1	Introduction	625
16.2	Through-Casing Resistivity Measurement Procedure	626
16.3	Circuit Model of the Through-Casing Resistivity Tool	628
16.4	Finite Element Method Simulation of the Through-Casing Resistivity Logging Tool	632
	16.4.1 Numerical simulation of the through-casing resistivity tool with electrodes	633
16.5	Through Casing Logs from a Toroidal Antenna	637
	16.5.1 Choice of element matrix and source model	638
	16.5.2 Comparison with published literature	638
	16.5.3 Simulation results	640
References		643

16.1 INTRODUCTION

Conventional laterolog, logging while drilling, or induction resistivity tools log a formation before the borehole is cased. As the oil and gas prices increase and the improvements of the drilling technology, many oil fields are reevaluated and thinner layers of reservoir that were considered less profitable are reopened for production. Many oil fields are now in the process of second or third production period. To locate the oil- or gas-bearing zones accurately, it becomes necessary to evaluate the formation in the previously cased holes. It is relatively easy to use radiation tools or acoustic tools through-casing metal. However, using electrical ways is rather difficult due to the shielding effect of the metal casing. When a metal casing is applied, conventional resistivity logging tools are not operative because the electrically conductive casing practically shields electromagnetic (EM) signals from the tools and prevents any electrical or electromagnetic energy from entering the formation. However, there are many instances when resistivity logging through the casing is needed. For example, measuring formation resistivity profile change during the years of production is critical to assess the oil or gas reservoir for further exploration in an old oil field. Therefore the application of through-casing resistivity (TCR) tool becomes important for secondary production of old reservoirs.

In this chapter, a tool that is designed to measure the resistivity of the formation through a metal casing is analyzed based on two different methods: circuit model and numerical method using finite element method (FEM). The tool consists of a current-emitting electrode or loop and several potential electrodes located inside a metal casing. In both circuit model and numerical model, we use point electrodes and ring electrode as current-emitting method. In numerical method, we will also discuss the possibility of using a toroid antenna for signal emission since it is possible to have a TCR system without contacting the casing if toroid coils are used. A cylindrical cement layer may be present between the casing and the outside formation. The potential electrodes and the electronic circuit inside the tool record a signal that is proportional to the second derivative of the electric potential in the axial direction. Theoretical formulas of this signal have been derived and numerical results have been obtained. The chapter will show that the tool signal is a function of the resistivity of the formation and is affected by other parameters such as diameter, thickness, conductivity of the metal casing, and thickness and conductivity of the cement. The tool response has less dependence on the conductivity of the mud as most other resistivity tools due to the existence of the metal casing and the mud is inside the casing.

Several patents were filed for the TCR tools [1–5]. The basic idea of these patents is to use a very low-frequency (approximately 1 Hz) electric source to reduce the shielding effect of the metal casing and use the second-order derivatives of the received signal to obtain the information of the formation resistivity. Theoretical analyses of these ideas are given by Kaufman [6,7] and Schenkel [8]. Kaufman modeled the conducting casing approximately as a sheet of current without thickness and computed the potential and its derivatives along the axis of the casing. The source is modeled as a point electrode. Schenkel used an integral equation approach and then solved the problem numerically. We first use a simple circuit model to explain the concept of the TCR tool in Section 16.2. Following the circuit model, numerical results using FEM are used to simulate more complicated cases including the TCR tool using toroid antennas instead of electrodes and ring as emitters. Applications and tool performance of the TCR tool are studied by Tabarrovsky et al. [9].

16.2 THROUGH-CASING RESISTIVITY MEASUREMENT PROCEDURE

A model of a TCR tool using electrodes as both current emitting and potential detecting method in a homogeneous formation is illustrated in Fig. 16.1. To make the discussions easier, we assume that the spacing between all the electrodes is the same. Although low-frequency alternating current signals are used to make measurement easier, direct current (DC) analysis can be used without loss of generality. Assume a DC is injected into the casing through electrode I_0. By open circuiting the return electrode F, the current is forced to return to the source through the formation; this is called the measurement mode. The current is returned through the F electrode when

Theory of the Through-Casing Resistivity Logging Tool

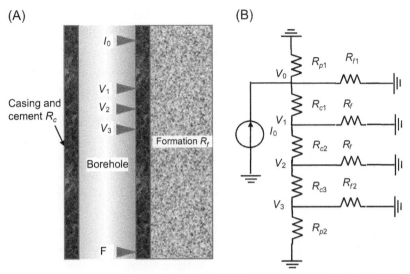

Figure 16.1 A circuit representation of the TCR tool. (A) The physical model of the TCR tool structure and (B) the equivalent circuit of the TCR tool.

it is closed; this condition is known as either the calibration or null mode. The electronic circuit, consisting of switches and differential amplifiers, is designed to measure the electrostatic potential, and the differences in potentials at electrodes V_1, V_2, and V_3, in reference to an electrode that is far away from the tool, which is the measurement ground. V_1, V_2, and V_3 are connected to a cascade of differential amplifiers, the output of the circuit is approximately proportional to the second derivative of the electrostatic potential at the inner surface of the casing. That is the assembly of differential amplifiers measures the voltage U'', and calculate

$$U'' = \Delta V_1 - \Delta V_2 \tag{16.1a}$$

where

$$\Delta V_1 = V_1 - V_2 \tag{16.1b}$$

$$\Delta V_2 = V_2 - V_3 \tag{16.1c}$$

the voltages, V_1, V_2, and V_3, and the spacing are defined in Fig. 16.1. In the null mode, the current flows from A to F with very little current leaking into the formation. Under this condition, the voltage is measured and is denoted as U_{cal}. Without changing the current flow, the voltage difference between V_1 and V_2 (or V_2 and V_3) and I_0 electrodes is measured and is denoted as,

$$U_{cal} = V_1 - V_2 \tag{16.2a}$$

$$V_0 = V_0 \tag{16.2b}$$

The electrode F is then opened, forcing the current to flow through the casing and into the formation. In this mode, known as the measurement mode, the voltages V_1, V_2, and V_3 are measured and the formation resistivity can be calculated.

16.3 CIRCUIT MODEL OF THE THROUGH-CASING RESISTIVITY TOOL

To have a general idea of the TCR tool, we can use a simple resistive network model. Consider the measurement mode. If we carefully think about the TCR tool, the system can be represented by a simple circuit. Consider the schematic of the TCR tool shown in Fig. 16.1. As discussed in Section 16.2, it has a DC or near DC current source, which injects current through electrode I_0, a return electrode F that is far away from the rest of the tool, three voltage measurement electrodes V_1, V_2, and V_3. The measurements are done using the second-order difference of the measured voltage Δ^2. We hope the measured Δ^2 is directly related to the formation resistivity.

The TCR tool can be modeled by a resistor network as shown in Fig. 16.1B. In Fig. 16.1B, R_{p1} and R_{p2} are resistance of the casing seeing from the tool, usually very small. R_{c1}, R_{c2}, and R_{c3} are resistance of casing between electrodes. R_{f1}, R_f, and R_{f2} are the resistance of the formation viewed from the current source, voltage electrodes V_1, V_2, and V_3, respectively. To solve the circuit, we can simplify the drawing in Fig. 16.1B to Fig. 16.2. In Fig. 16.2, we assume the casing resistance between voltage probes is equal and

$$R'_{p1} = \frac{R_{p1} R_{f1}}{R_{p1} + R_{f1}} \quad \text{and} \quad R'_{p2} = \frac{R_{p2} R_{f2}}{R_{p2} + R_{f2}}$$

Apply mesh current method to solve the circuit problem in Fig. 16.2, we have

$$- I_0 R'_{p1} + I_1 \left(R'_{p1} + R_{c1} + R_f \right) + I_2 R_f = 0 \tag{16.3a}$$

$$- I_1 R_f + I_2 (R_{c2} + 2R_f) - I_3 R_f = 0 \tag{16.3b}$$

$$- I_2 R_f + I_3 (R_{c2} + R'_{p2} + R_f) = 0 \tag{16.3c}$$

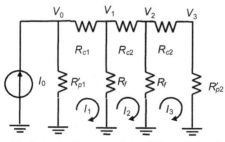

Figure 16.2 Equivalent simplified circuit diagram of the TCR tool showing in Fig. 16.1.

To simplify the solution, we can assume that $R'_{p1} = R'_{p2} = R_p$, $R_{c1} = R_{c2}$. This assumption is usually true when the measurement electrode spacing is identical. Solve for I_1, I_2, and I_3 we have,

$$I_1 = \frac{I_0 R_p R_f}{R_f + R_p + R_c} + I_3 \quad (16.4a)$$

$$I_2 = \frac{I_0 R_p R_f}{(R_f + R_p + R_c)(R_{c2} + 2R_f) - 2R_f^2} \quad (16.4b)$$

$$I_3 = \frac{I_0 R_p R_f^2}{[(R_f + R_p + R_c)(R_{c2} + 2R_f) - 2R_f^2](R_f + R_p + R_c) - (R_f + R_p + R_c)R_f^2} \quad (16.4c)$$

The voltages in the circuit showing in Fig. 16.2 can be obtained:

$$\Delta V_1 = V_1 - V_2 = I_2 R_c \quad (16.5a)$$

$$\Delta V_2 = V_2 - V_3 = I_3 R_c \quad (16.5b)$$

$$\Delta^2 = \Delta V_1 - \Delta V_2 = (I_2 - I_3) R_c \quad (16.5c)$$

$$U'' = \frac{I_0 R_c R_p R_f (R_c + R_p)}{[(R_f + R_p + R_c)^2 (R_c + 2R_f) - 2R_f^2](R_f + R_p + R_c)} \quad (16.6)$$

Considering the fact that $R_c \ll R_p$, and $R_c \ll R_f$, when F is open (considered F is very far from the tool in this case, e.g., on the surface), Eq. (16.6) can be simplified:

$$U'' \approx \frac{R_c}{2} \frac{I_0}{2 + \frac{R_f}{R_p}} \quad (16.7)$$

From Eq. (16.7), it is seen that the second-order difference of the voltage measured from inside of the casing is inversely proportional to the resistivity of the formation. We also note that the value of the measured voltage is directly proportional to the resistivity of the casing, which is usually very small, which makes the measured voltage also small. Longer distance between probes will increase the measured voltage value. R_p is largely determined by the resistivity of the casing and the distance between the return electrode and the current injection electrode. In practice, R_p is about 5–10 times greater than R_c.

If we assume a tool geometry as shown in Fig. 16.3A, we can calculate the parameters required in Eq. (16.7). The distance between current injection electrode and the return electrode is 2 m, the distance between voltage electrodes is 0.5 m. $R_p = 5 \times 10^{-4}$ ohm, $R_c = 5 \times 10^{-5}$ ohm. We assume the current injected into the casing is 1 A. The response of the tool can be computed. Fig. 16.3B and C shows the

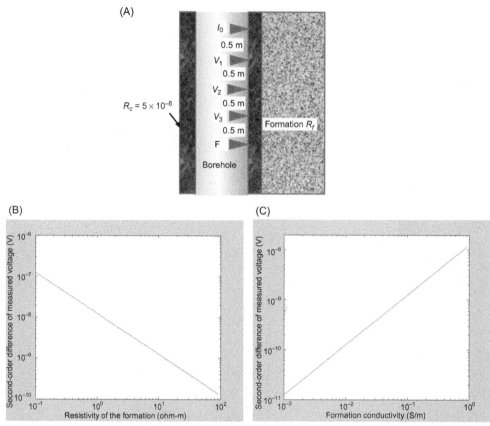

Figure 16.3 The simulated TCR double difference voltage versus resistance in the homogeneous formation computed using the approximate equation (16.7). The distance between current injection electrode and the return electrode is 2 m, the distance between voltage electrodes is 0.5 m. $R_p = 5 \times 10^{-4}$ ohm, $R_c = 5 \times 10^{-5}$ ohm and the current injected into the casing is 1 A. (A) The model of the TCR tool; (B) the computed TCR tool performance as a function of formation resistance; and (C) the computed tool performance as a function of the formation conductivity.

computed double difference voltage as a function of formation resistivity and formation conductivity in double logarithm scale based on the values given above.

From Fig. 16.3 we can see that the measured double difference voltage is a linear function of the conductivity and resistivity of the formation in the double logarithm scale. We also noticed that the measured signal is relatively small in the range of nanovolts. In Fig. 16.3, the resistivity and conductivity of the formation is calculated by multiplying or dividing the probe spacing.

In the discussions above, we assume the values of R_p and R_c are known. Actually, R_p and R_c values can be found in the calibration procedure (null mode). In the calibration mode, the F electrode is closed and the current is returned from the

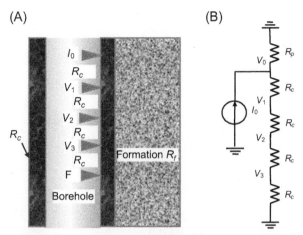

Figure 16.4 The equivalent circuit of the TCR in calibration mode when electrode F is closed assuming the distance between all electrodes are equal. (A) The physical model of the TCR tool in calibration mode and (B) the equivalent circuit of the TCR tool in the calibration mode.

F electrode which is now, the same distance from electrode V_3, as the distance between V_1 and V_2 (or V_2 to V_3). If we consider the equivalent circuit shown in Fig. 16.1B, the circuit is simplified to the one shown in Fig. 16.4.

From Fig. 16.4, we can find the value of R_c and R_p,

$$R_c = \frac{V_1 - V_2}{I_0} \tag{16.8a}$$

and

$$R_p = \frac{4V_0 R_c}{4I_0 R_c - V_0} \tag{16.8b}$$

Combining Eqs. (16.8a) and (16.8b), we have

$$R_p = 4R_0 \frac{(V_1 - V_2)/V_0}{4(V_1 - V_2)/V_0 - 1} \tag{16.8c}$$

where $R_0 = \frac{V_0}{I_0}$ is the resistance of the system looking from the source in the calibration mode. In practice, R_0 can be found by measuring the voltage and current output at the source terminal. From Eq. (16.8c), we can see that the formation resistivity can be found by two steps each time the tool is moved to a new position. The first measurement is a calibration process when electrode F is engaged and the second measurement is done when the electrode F is retracted and the current return electrode is placed far away from the tool. Once these parameters are found, the formation resistivity can be obtained by inverting Eq. (16.7):

$$R_f = \left(\frac{R_c I_0}{2U''} - 2\right) R_p \qquad (16.9)$$

$$\rho_f = \frac{R_f}{d} \qquad (16.10)$$

where d is the distance between the measurement electrodes, and ρ_f is the resistivity of the formation.

16.4 FINITE ELEMENT METHOD SIMULATION OF THE THROUGH-CASING RESISTIVITY LOGGING TOOL

Section 16.3 gives us an idea of the TCR tool working principle—when the conductivity of the casing is present, there is a possible solution to measure the resistivity of the formation from inside the casing. In the homogenous formation, the formation conductivity can be obtained by using Eqs. (16.7) and (16.8). However, to further investigate the performance of the TCR tool in a rather complicated formation, such as layered formation with cement between the casing and the formation, it is difficult to use a simple circuit model. However, as we discussed in Chapter 13, Finite Element Method for Solving Electrical Logging Problems in Axially Symmetrical Formations, it is possible to use FEM numerical modeling to study the performance of the TCR tool in a complicated environment. In this section, we will use FEM to model the TCR tool and study the performance of the tool. We will also explore the possibility to use a toroid antenna as the transmitter for the TCR tool. The idea of using a toroidal antenna was studied by Pardo et al. [10,11].

The EM field of the electrode type TCR tool was first studied by Kaufman [6,7]. According to Kaufman's study, the borehole region is composed of three zones: the near zone, the intermediate zone, and the far zone. Within each zone, the field behaves in very specific ways: (1) in the near zone, the behavior of the field changes quickly with the distance from the source; (2) in the far zone, the field is similar to that in a uniform medium with the formation resistivity; and (3) in the intermediate zone, the second-order derivative of the potential is almost constant and is proportional to the formation conductivity,

$$U'' \propto \sigma_f \quad \text{if:} 10 < (d/a) < 10^3; \quad z < \sqrt{\frac{2\pi a \Delta a \sigma_c}{\sigma_f}} \qquad (16.11)$$

where σ_f is the conductivity of the formation, d is the distance from transmitter to receiver, a is the radius of the borehole, Δa is the thickness of casing, and σ_c is the conductivity of the casing. The behavior of the fields in the intermediate zone creates a possibility of detecting the formation conductivity outside the casing from the tool inside the borehole. Refer to Fig. 16.1, the schematic of the through-casing logging tool, where the source is located at point I_0, and the potentials at points V_1, V_2, and V_3 are measured.

Generally, as we assumed in Section 16.3, the distance between the measurements electrodes are set equal. To use the FEM simulation, we rewrite the measurement quantities using field definition. The second derivative of potential can be expressed as

$$\frac{\partial^2 U}{\partial z^2} = \frac{\frac{V_1 - V_2}{d} - \frac{V_2 - V_3}{d}}{d/2} \tag{16.12}$$

If we use difference to approximate the derivatives,

$$E_{z1} = \frac{V_1 - V_2}{d} \tag{16.13}$$

$$E_{z2} = \frac{V_2 - V_3}{d} \tag{16.14}$$

the second-order derivative of the potential can also be expressed as the first difference of the vertical component of the electric field

$$\frac{\partial^2 U}{\partial z^2} = \frac{E_{z1} - E_{z2}}{d} \tag{16.15}$$

16.4.1 Numerical simulation of the through-casing resistivity tool with electrodes

As we discussed in the previous sections, the TCR tool inject a constant survey current I_0 through the current electrode, then flows through the casing, and the whole formation, and finally is collected by returned electrode F. This condition leads to $I_0 + IF = 0$. There are two focusing conditions for the electrode tool:

$$I_0 + I_F = 0 \tag{16.16}$$

$$I_0 = \text{constant\_} I_F \tag{16.17}$$

Points V_1, V_2, and V_3 are chosen to measure the potential differences. The potential differences between these measurement electrodes can be calculated by the integration of electric field

$$V_{12} = \int_{V_1}^{V_2} E_z \cdot dl \tag{16.18}$$

$$V_{23} = \int_{V_2}^{V_3} E_z \cdot dl \tag{16.19}$$

The final logs are the second derivative of potential, which is expressed in Eq. (16.12). The reference point of the logging depth is the position V_2.

If the tool has electrodes that contact the casing electrically, it is called a contact-casing TCR tool. The potential differences are measured along the casing. Otherwise, it is a noncontact-casing TCR tool and the potential differences are measured inside the casing without contact electrodes.

16.4.1.1 Choice of element matrix and source model

The element matrix is based on ρH_ϕ, which was discussed in Chapter 13, Finite Element Method for Solving Electrical Logging Problems in Axially Symmetrical Formations, and matrix system $[Z_{m,n}][(\rho h)_n] = [V_m]$ is built for the simulation. The source model used for the through-casing logging tool is source model 3: sources existing on boundaries, which is illustrated in Section 13.11.5 and Fig. 13.9. The method of handling the source here is similar to that used to handle the source of the dual laterolog tool, which is illustrated in Section 15.4. There are $N_{\text{insulator}} + 2*N_{\text{electrode}}$ equations needed to solve all unknown source conditions; the first $N_{\text{insulator}}$ equations are shown as $[A_{i,j}] \cdot [\rho H_\phi(j)] - [2\pi(V_{i-1} - V_i)] = 0$, ($i = 1, N_{\text{insulator}}$) in Eq. (15.19) and the second $N_{\text{electrode}}$ equations are shown as $\rho H_\phi(i+1) - \rho H_\phi(i) - \frac{I(i)}{2\pi} = 0$, ($i = 1, N_{\text{electrode}}$) in Eq. (15.20). The only difference is the third $N_{\text{electrode}}$ equations, which is shown in Eq. (15.23).

16.4.1.2 Simulation results

In this section, the through-casing measurements of the second derivative of potential in Kaufman's tool will be discussed. In the simulation, source electrode is placed in the depth of $z = 0$ ft, and the returned electrode is placed in the depth of $z = 1200$ ft. The radius of the electrode is 2 in., and the two electrodes are contacting casing using a connection line with resistivity equal to $2.3*10^{-7}$ ohm-m and relative permeability equal to 85. Point V_1, V_2, and V_3 are placed in the depth of $z = 60$ ft, $z = 66$ ft, and $z = 72$ ft, respectively along the casing. The geometry of the cased well is shown in Fig. 16.1. The current of the source is set to be 1 A. Fig. 16.5 shows through-casing measurements of the second derivative of potential against frequency. In the low frequencies, which is less than 5 Hz in this case, the measured values are almost kept in a constant. When frequency increases, the values decreases.

Fig. 16.6 shows the through-casing measurements of the second derivative of potential as a function of the formation conductivity in an electrode type tool in a homogeneous formation with conductivity changing from 0.01 to 10 S/m. It shows that amplitude of the second derivative of potential is proportional to the formation conductivity as $Abs(dU^2/dz^2) \propto \sigma$. The result is not linear when the conductivity is higher than 2 S/m.

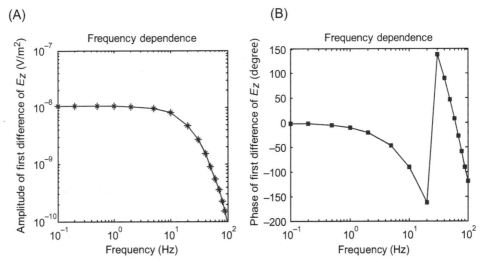

Figure 16.5 The second derivative of the potential versus frequency of a TCR tool. (A) amplitude; (B) phase.

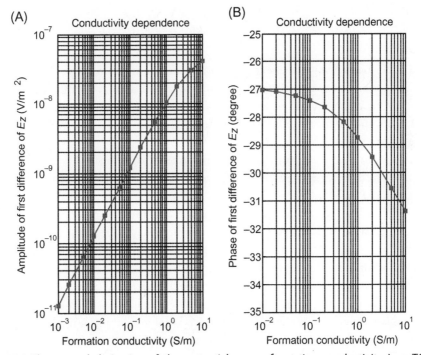

Figure 16.6 The second derivative of the potential versus formation conductivity in a TCR tool. (A) amplitude; (B) phase.

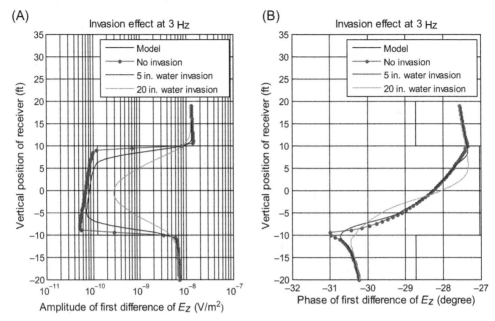

Figure 16.7 The second derivative of potential in invaded formation measured by a TCR tool. (A) Amplitude; (B) phase.

Fig. 16.7 shows invasion effect on through-casing measurements of the electrode type TCR tool. It shows that through-casing measurements are sensitive to formation resistivity; it has higher value in low-resistivity layer and lower value in high-resistivity layer. It also shows that this sensitivity decreases with the increasing of the target zone's invasion radius.

Fig. 16.8 shows a cased well with an anisotropic formation. The top and bottom layers have the horizontal resistivity of 1 ohm-m and vertical resistivity of 10 ohm-m. The middle layer has horizontal resistivity of 50 ohm-m and vertical resistivity of 200 ohm-m. Fig. 16.8 shows the through-casing measurements of the formation without and with anisotropy: the first curve is without anisotropy with both horizontal and vertical resistivities being equal to Rh; the second one is without anisotropy with both horizontal and vertical resistivities being equal to Rv; and the third one is with anisotropy with horizontal resistivity being equal to Rh and vertical resistivity being equal to Rv. It shows that in the situation of anisotropy, the through-casing logs are affected by horizontal resistivity much more than the vertical resistivity. That means through-casing logs are not sensitive to vertical resistivity.

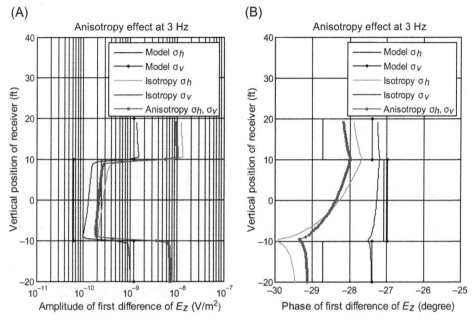

Figure 16.8 The second derivative of potential in an anisotropic formation measured by a TCR tool. (A) amplitude; (B) phase.

16.5 THROUGH CASING LOGS FROM A TOROIDAL ANTENNA

As we discussed in the beginning of this chapter, there are two possible implementations of the TCR tool. We discussed the contact version in the previous sections. The contact electrodes method is mostly used in the real applications. Note that in practice, most casing are rusted and maintaining a good electric contact from the electrode to the casing is not an easy task. To avoid high contact resistance, many mechanical ways are developed such as the use of sharp tips of the electrodes. At each logging point, the tool must be moved and contact electrodes have to be retracted during the move. Due to the mechanical motion, the logging speed is limited and the tool reliability is compromised. To improve the tool performance, noncontact method is preferred. In this section, we will discuss another possible implementation method of the TCR tool, which is noncontact TCR tool using toroid antennas. So far, there are no commercially available toroid antenna—based TCR tool available. The discussions in this section are intended to be more exploratory than practical applications. Using FEM, this tool can be analyzed.

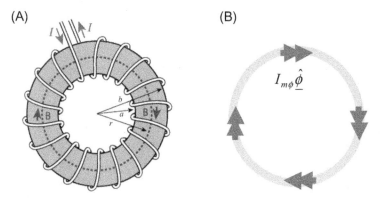

Figure 16.9 Toroidal antenna used in a noncontact TCR tool. (A) Schematic of the toroidal antenna. (B) Source model of the antenna is a magnetic current loop.

16.5.1 Choice of element matrix and source model

Fig. 16.9 shows the schematic of a toroidal antenna; its source model is a magnetic current loop. The element matrix based on ρH_ϕ, which was discussed in Chapter 13, Finite Element Method for Solving Electrical Logging Problems in Axially Symmetrical Formations, is used and matrix system $[Z_{m,n}][(\rho h)_n] = [V_m]$ is built for the simulation. The source model for a toroidal antenna is a magnetic current loop, which was discussed in Section 13.11.3. Since element matrix is built based on ρH_ϕ, the impedance global matrix can be expressed as $[V_m] = -\langle \underline{\Omega_m}; \underline{M_s} \rangle_\Omega = 2\pi I_{m\phi}$, as shown in Eq. (13.150).

16.5.2 Comparison with published literature

To make sure the computation method is effective and correct, any algorithm must be verified. One of the most effective way to do the verification is to use the data in a published literature. In this section, we will compare the data obtained in this chapter and compare it with the result by Pardo [10]. Fig. 16.10 shows the geometry of a cased well with a homogeneous formation. The thickness of the casing is 1.27 cm with resistivity equal to $2.3*10^{-7}$ ohm-m and relative permeability equal to 85. The casing is surrounded by a 5-cm layer of cement with a resistivity of 2 ohm-m. The radius of the borehole is 10 cm, and it is filled with mud with a resistivity of 1 ohm-m. Measurements are based on the use of one toroidal transmitter and two toroidal receiver antennas located 1.25 and 1.5 m above the transmitter. The transmitter is modeled by prescribing an impressed volume magnetic current $M_{imp} = \delta(z)\delta(\rho - a)I_{m\phi}\hat{\phi}$, where $I_{m\phi}$ is set at 50 in the simulation and $a = 10$ cm, just contacting the casing. The first difference of the vertical component of the electric field is calculated by $\frac{\partial^2 U}{\partial z^2} = \frac{E_{z1} - E_{z2}}{d_{12}}$ in Eq. (16.15). Fig. 16.11 shows the through-casing measurement of the first vertical difference of E_z as a function of frequency,

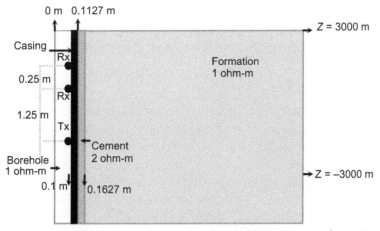

Figure 16.10 Geometry of the tool and the cased well with a homogeneous formation.

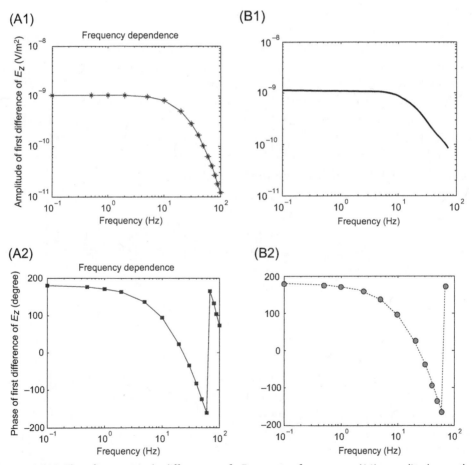

Figure 16.11 The first vertical difference of E_z versus frequency. (A1) amplitude—author, (B1) amplitude—Pardo; (A2) phase—author, (B2) phase—Pardo.

where (A1) is the amplitude calculated by the author, (B1) is the amplitude from Pardo, (A2) is the phase from the author, and (B2) is the phase from Pardo. It shows that the results from the author and the results from Pardo match very well.

16.5.3 Simulation results

Fig. 16.12 shows through-casing measurements of the first vertical difference of E_z against conductivity in a homogeneous formation with conductivity changing from 0.001 to 10 S/m. The geometry of the tool and the cased well is shown in Fig. 7.3. It shows that amplitude of the first vertical difference of E_z is proportional to the formation conductivity as $Abs(dE_z/dz) \propto \sigma$, but the result is not linear when the conductivity is higher than 2 S/m. The reason is that the receivers are not located in the intermediate zone for high-conductivity formation. If the receivers are put in a farther position, the relationship between first vertical difference of E_z and conductivity would be linear even in a high-conductivity formation. The distance from intermediate zone to transmitter increases with increasing of formation conductivity and frequency.

Fig. 16.13 shows a three-layer cased well with an invasion zone existing in the middle layer. The resistivities of the layers are 1, 200, and 1 ohm-m, respectively and

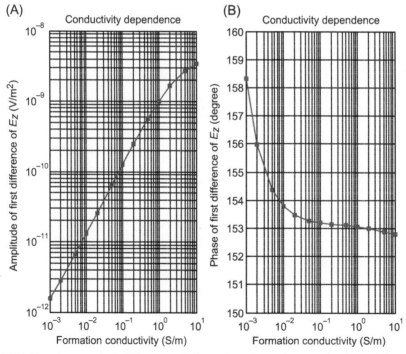

Figure 16.12 The first vertical difference of E_z against formation conductivity. (A) Amplitude; (B) phase.

Theory of the Through-Casing Resistivity Logging Tool   641

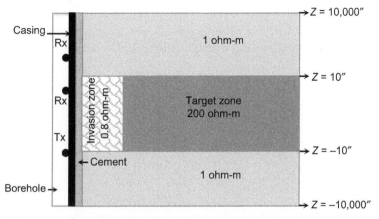

Figure 16.13 Geometry of a cased well with invasion zone.

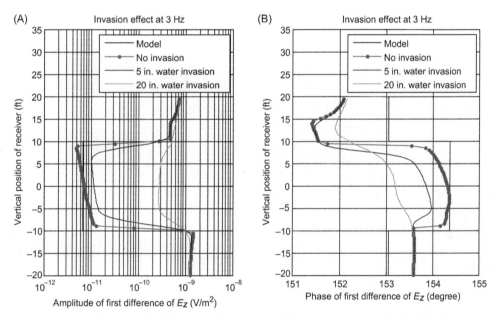

Figure 16.14 The first vertical difference of E_z in invaded formation. (A) Amplitude; (B) phase.

the resistivity of the invasion zone is 0.8 ohm-m. Fig. 16.14 shows the through-casing measurements when the radius of the invasion zone is 0, 5, and 20 in., respectively. It shows that through-casing measurements are sensitive to formation resistivity; it has higher value in low-resistivity layer and lower value in high-resistivity layer. It also shows that this sensitivity decreases with the increasing of the target zone's invasion radius. If the invasion radius reaches 20 in. or above, this sensitivity is very small for a toroidal transmitter.

Fig. 16.15 shows a cased well with anisotropy formation. The top and bottom layers have the horizontal resistivity of 1 ohm-m and vertical resistivity of 10 ohm-m. The middle layer has horizontal resistivity of 50 ohm-m and vertical resistivity of 200 ohm-m. Fig. 16.16 shows the through-casing measurements of the formation without and with anisotropy: the first curve is without anisotropy with both horizontal and vertical resistivities being equal to Rh; the second one is without anisotropy with both

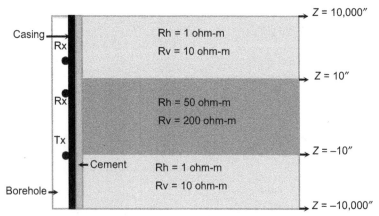

Figure 16.15 Geometry of a cased well with anisotropy formation.

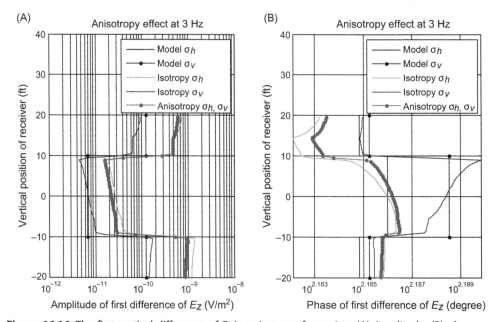

Figure 16.16 The first vertical difference of E_z in anisotropy formation. (A) Amplitude; (B) phase.

horizontal and vertical resistivities being equal to Rv; and the third one is with anisotropy with horizontal resistivity being equal to Rh and vertical resistivity being equal to Rv. It shows that in the situation of anisotropy, the through-casing logs are affected by horizontal resistivity much more than the vertical resistivity. That means through-casing logs are not sensitive to vertical resistivity and cannot measure it.

REFERENCES

[1] R.E. Fearon, Method and Apparatus for Electric Well Logging, U.S. Patent No. 2,729,784, January 3, 1956.
[2] A.A. Kaufman, Conductivity Determination in a Formation Having a Cased Well, U.S. Patent No. 4,796,186, January 3, 1989.
[3] B. Vail, Methods and Apparatus for Measurement of the Resistivity of Geological Formations From Within Cased Boreholes, U.S. Patent No. 4,820,989, April 11, 1989.
[4] M.F. Gard, J.E. Kingman, J.D. Klein, Methods and Apparatus for Measuring the Electrical Resistivity of Geologic Formations Through Metal Drill Pipe for Casing, Patent No. 4,837,518, June 6, 1989.
[5] B.T. Vail, Methods and Apparatus for Measurement of Electronic Properties of Geological Formations Through Borehole Casing, U.S. Patent No. 4,882,542, 1989.
[6] A.A. Kaufman, W.E. Wightman, A transmission line method for electrical logging through casing, Geophysics 58 (12) (1993) 1739–1747.
[7] A. Kaufman, The electric field in a borehole with a casing, Geophysics 55 (1) (1990) 29–38.
[8] C.J. Schenkel, The electrical resistivity method in cased boreholes (Ph.D. thesis), University of California, Berkeley, May 1991.
[9] L.A. Tabarrovsky, M.E. Cram, T.V. Tamarchenko, K.M. Strack, B.S. Zinger, Through-casing resistivity (TCR): physics, resolution, and 3-D effects, in: 35th Annual Logging Symposium Transactions: Society of Professional Well Log Analysts, Paper TT, June 1994.
[10] D. Pardo, C. Torres-Verdin, L. Demkowicz, Simulation of multifrequency borehole resistivity measurements through metal casing using a goal-oriented hp-finite element method, IEEE Trans. Geosci. Remote Sens. 44 (2006) 2125–2135.
[11] D. Pardo, C. Torres-Verdin, L. Demkowicz, Feasibility study for 2D frequency-dependent electromagnetic sensing through casing, Geophysics 72 (3) (2007) F111–F118.

CHAPTER 17

Electromagnetic Telemetry System and Electromagnetic Short Hop Telemetry in a Logging-While-Drilling/Measuring-While-Drilling Tool

Contents

17.1 Introduction to Logging-While-Drilling/Measuring-While-Drilling Uplink and Downlink Technologies	645
17.2 The Numerical Model of Electromagnetic Telemetry System	649
17.3 Application of Finite Element Method on Electromagnetic Telemetry Systems	652
17.3.1 Choice of equation, basis function, and element matrix	652
17.3.2 Application of source model for Electromagnetic Telemetry System	653
17.3.3 Flowchart of computational code for EM telemetry system	656
17.4 Validation of the Computation Algorithm in a Cased Borehole	659
17.5 Simulation Result Without Casing	661
17.5.1 Grid size and computation time for different borehole depth	661
17.5.2 Current distribution pattern in a 9000-ft-depth borehole	663
17.5.3 Voltage on receiver	667
17.5.4 System impedance	672
17.6 Short Hop Electromagnetic Telemetry Used in a Near Bit Logging-While-Drilling Sensor	680
17.6.1 Simulation of the short hop communication system	681
17.6.2 Input current and impedance of a toroidal transmitter versus formation resistivity	685
17.7 Conclusions	687
References	688
Appendix A Computation Method of Source Model for EM Telemetry System (Section 17.3.2)	688

17.1 INTRODUCTION TO LOGGING-WHILE-DRILLING/MEASURING-WHILE-DRILLING UPLINK AND DOWNLINK TECHNOLOGIES

The logging-while-drilling (LWD)/measuring-while-drilling (MWD) tools send signals from the downhole to the surface or vice versa wirelessly due to the real-time data transmission requirements and the difficulties in wiring the drill pipes. For many years the LWD/MWD data transmission has been relying on the mud pulse technology and the data rate is limited by the nature of the mechanical waves and sources to generate the mechanical waves. So far, the reported high-speed mud pulser can reach close to 50 bps in favorite conditions [1]. The data rate of the most commonly used mud pulser is around 1−5 bps. Comparing to any communication system we use nowadays, this data rate is way too low. However, due to the slow drilling speed, this data rate is still tolerable. Fig. 17.1 is a schematic of a mud circulation system in a drilling pit.

Mud pulsers are basically an encoded "blocker" of the mud flow in the downhole causing the mud pressure change in the mud system. A pressure detector on the surface can pick up the attenuated mud pressure change and decode the data as shown in Fig. 17.2. There are several ways in implementing a mud pulser. Most popular way is to use a plunger which generates a pressure increase when actuated and therefore, a positive mud pulse signal is produced as shown in Fig. 17.2B. The other one would be opposite as shown in Fig. 17.2C by releasing mud from the drill pipe to the annulus area. Fig. 17.2D shows a rotating wheel in the plane perpendicular to the mud flow, which generates positive continuous mud pulses. Since the motion of the wheel is in the sheer direction of the mud flow, it may produce a faster mud pulse [2]. Similar idea can be used to transmit data downwards. The uplink and downlink system can be frequency divided. The downlink data rate is much lower in the range of 0.1 bps. Therefore the data transmission can be duplex by frequency division multiplexing. Fig. 17.3 shows a downlink data transmission method.

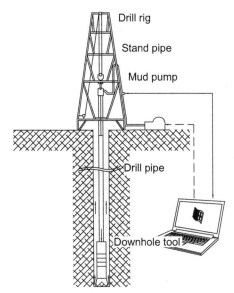

Figure 17.1 A mud circulation system in a drill pit.

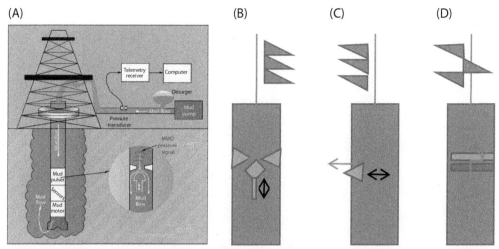

Figure 17.2 A mud pulse telemetry in a drilling system for transmitting data from downhole to the surface. (A) The mud pulse telemetry system; (B) positive pulser; (C) negative pulser; and (D) positive rotary pulser.

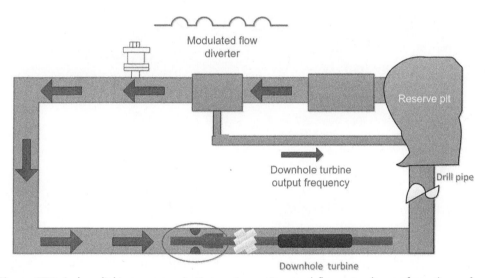

Figure 17.3 A downlinking communication system using mud flow rate change from the surface by bypassing part of the mud flow and detect the rotation speed of the downhole turbine.

The detection of the downlink is usually at the downhole turbine generator by measuring the turbine rotation speed. When the flow rate of the mud changes, the turbine generator rotation speed will change, which cause the frequency of the turbine output change. Therefore the turbine frequency (or RPM) is directly proportional to the mud flow.

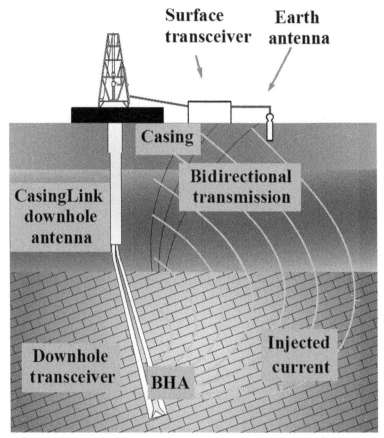

Figure 17.4 The electromagnetic telemetry system. The downhole antenna is placed near the drill bit in the MWD/LWD tool string and the surface electrodes are between the drill pipe and a point on the surface about 100 m away in the ground.

Mud pulsers are mechanical actuators, which are effective and reliable. However, due to the nature of mechanical devices, they suffer from limited data rate, life span, wear, blockage, and corrosion. What is more, they completely rely on the continuation of the mud in the drilling system. In unbalanced drilling or air drilling, the mud pulsers will not work. One of the other natural choices of downhole data transmission is the use of EM method as we do in the air. We can transmit gigabits per second in air using microwave frequencies. Unfortunately, due to the loss of the formation, downhole EM transmission is difficult even at a few bit per second at a distance of a few thousand meters. The basic idea of EM telemetry is shown in Fig. 17.4. A transmitter/receiver antenna is placed inside the borehole and a receiving/transmitting antenna is placed on the surface. For the surface receiver, the drill pipe is used as one electrode and the other electrode is placed about 100 m away. The potential difference is measured and recoded as receiver/transmitter. In this chapter, we will use numerical method discussed in Chapter 13, Finite Element Method for Solving Electrical Logging Problems in Axially Symmetrical Formations, to analyze the EM telemetry. The important parameters that interest us include the transmission distance, field distributions, and impedance of the downhole antennas. For downhole antennas, we will discuss two different possible implementation antenna structures: gap antenna and the coaxial antenna.

17.2 THE NUMERICAL MODEL OF ELECTROMAGNETIC TELEMETRY SYSTEM

In this section, we will discuss numerical modeling method using finite element method (FEM) to analyze EM telemetry system. The details of FEM analysis have been presented in Chapter 13, Finite Element Method for Solving Electrical Logging Problems in Axially Symmetrical Formations. In this section, we will apply the FEM to the EM telemetry analysis. Fig. 17.5 shows a numerical model of EM telemetry system. There is an insulating gap between upper collar and lower collar. A current source or voltage source (transmitter) is applied between the gap. The inner and outer boundaries of the source are assumed to be electrical insulator. Here the top and bottom boundaries of the source are assumed to be perfect electric conductors. Fig. 17.5B and C plots the models of the current source and the voltage source.

Therefore, in the model of Fig. 17.5, there is no transverse directional flow of current from the inner and outer boundaries of the source, and the source current will flow out vertically from top boundary of the source to the upper pipe and through the earth formation, then flow back to the lower pipe, and finally flow into the bottom boundary of the source. Assuming a source current I_s exits only in z direction, the source can be understood as: I_s is equal to the total amount of current passing vertically through a disk of radius R_B (the outer boundary of the insulator) minus the total amount of current passing vertically through a disk of radius R_A (the inner boundary of the insulator). Assuming a voltage source V_s, the potential difference on source is the integral of E field from lower boundary of the insulator to its upper boundary. These two kinds of source can convert to each other.

An electromagnetic signal will be emitted into the formation from the source of the EM telemetry system. The signal can be detected at the surface as a very small voltage drop between the drill pipe and an electrode placed in the ground about 100 m away. The signaling system depends on the voltage drop being detectable in the presence of the ambient electromagnetic noise. The prediction of this voltage drop at different depths, formation resistivity, and frequency is therefore crucial and involves the solution of an electromagnetic field problem in two dimensions (vertical well) or three dimensions (deviated well). Only a vertical well will be analyzed in this chapter. For an axis-symmetric vertical well, where the source exists only in

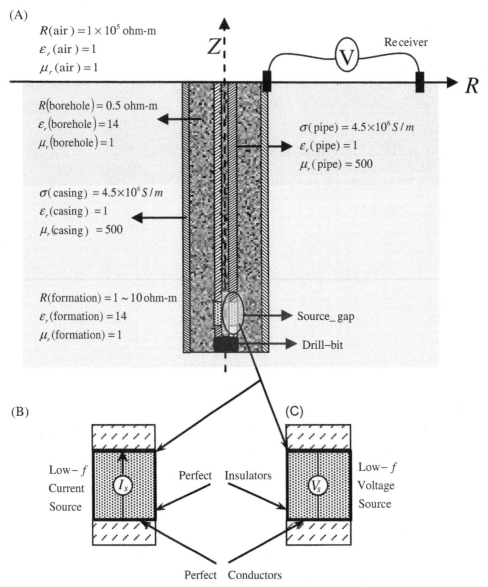

Figure 17.5 The numerical model of the EM telemetry system. (A) Model of the EM telemetry system, (B) model of current source, and (C) model of voltage source.

the z direction, only the transverse magnetic (TM) mode exists from the analysis of Chapter 13, Finite Element Method for Solving Electrical Logging Problems in Axially Symmetrical Formations, and the H_φ term will be considered to build the FEM equation. A schematic of the EM telemetry system is shown in Fig. 17.6 [3].

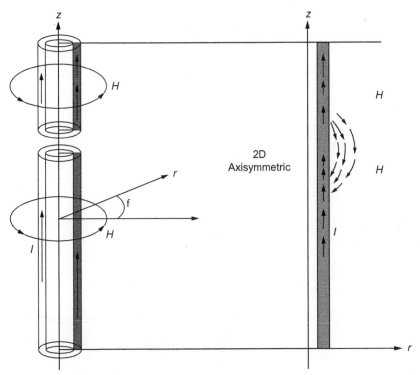

Figure 17.6 Schematic of the EM telemetry system.

Table 17.1 Material properties used in the simulation of an EM telemetry

Material	Conductivity, σ_h	Conductivity, σ_v	Permeability, μ_r	Permittivity, ε_r
Drill-pipe	4.5×10^6 S/m	4.5×10^6 S/m	500.0	1.0
Casing	4.5×10^6 S/m	4.5×10^6 S/m	500.0	1.0
Drill-bit	4.5×10^6 S/m	4.5×10^6 S/m	500.0	1.0
WBM wells	2 S/m	2 S/m	1	14.0
OBM wells	0.5×10^{-3} S/m	0.5×10^{-3} S/m	1	14.0
Earth	1.0×10^{-4} to 10 S/m	1.0×10^{-4} to 10 S/m	1	14.0
Air	1.0×10^{-5} S/m	1.0×10^{-5} S/m	1	1
Sea water	4	4	1	80.0

Material properties of the drill-pipe, drill-bit, casing, oil-based mud (OBM) and water-based mud (WBM) boreholes, earth formation, air, sea water are shown in Table 17.1 [4]. Example of geometry of the EM telemetry system is shown in Table 17.2.

Table 17.2 Example of geometry of EM telemetry system

Geometry	Value
Drill pipe: inside radius	2.135 in.
Drill pipe: outside radius	2.5 in.
Drill bit depth	Variable
Borehole: radius	5.625 in.
Borehole depth	= drill bit depth
Transmitter: gap thickness	1 in.
Transmitter: gap-bit distance	30 ft
Transmitter: frequency	1–100 Hz
Receiver: start radius	5.625 in.
Receiver: end radius	446 ft
Receiver: depth	−0.1 ft

17.3 APPLICATION OF FINITE ELEMENT METHOD ON ELECTROMAGNETIC TELEMETRY SYSTEMS

17.3.1 Choice of equation, basis function, and element matrix

As illustrated in Section 17.2, electric current source exists only in z direction for EM telemetry system in a vertical borehole. From the analysis in Section 13.3, Table 13.3, only the TM mode (E_ρ, E_z, and H_ϕ) exists and the H_ϕ term will be considered to establish the FEM equation. The magnetic field equation and the weak form of the magnetic equation were discussed in Sections 13.2 and 13.3, and expressed in Eqs. (13.6) and (13.22) as

$$\frac{1}{j\omega}\nabla\times\left(\underline{\underline{\varepsilon_c}}^{-1}\cdot\nabla\times\underline{H}\right)+j\omega\underline{\underline{\mu}}\cdot\underline{H}=\frac{1}{j\omega}\nabla\times\left(\underline{\underline{\varepsilon_c}}^{-1}\cdot\underline{J_s}\right)-\underline{M_s} \qquad (13.6)$$

$$\frac{1}{j\omega}\left\langle\nabla\times\underline{\Omega_m};\underline{\underline{\varepsilon_c}}^{-1}\cdot\nabla\times\underline{H}\right\rangle_\Omega+j\omega\left\langle\underline{\Omega_m};\underline{\underline{\mu}}\cdot\underline{H}\right\rangle_\Omega$$

$$=\frac{1}{j\omega}\left\langle\underline{\Omega_m};\nabla\times\left(\underline{\underline{\varepsilon_c}}^{-1}\cdot\underline{J_s}\right)\right\rangle_\Omega-\left\langle\underline{\Omega_m};\underline{M_s}\right\rangle_\Omega-\frac{1}{j\omega}\oiint_S\left(\left(\underline{\underline{\varepsilon_c}}^{-1}\cdot\nabla\times\underline{H}\right)\times\underline{\Omega_m}\right)\cdot\hat{n}dS$$

$$(13.22)$$

Basis functions were discussed in Section 13.5.1 for triangular elements and Section 13.5.2 for rectangular elements. The element matrix based on ρH_φ is used and matrix system $[Z_{m,n}]\,[(\rho h)_n]=[V_m]$ is established for the simulation of EM telemetry system. Formulas for element matrix are illustrated in Sections 13.7 and 13.8 if the solution domain is divided into rectangular elements and triangular elements, respectively.

17.3.2 Application of source model for Electromagnetic Telemetry System

Considering the source models we discussed in Chapter 13, Finite Element Method for Solving Electrical Logging Problems in Axially Symmetrical Formations, we can determine that the Source model 3 is appropriate since we can consider the two sections of the collar are simply two electrodes (Section 13.11.4). Therefore Source model 3 will be applied to the source model of EM telemetry system, as shown in Fig. 17.7. In Fig. 17.7, we notice that the left part is the image of the right part so as to provide more space to illustrate the problem. The EM telemetry system has two kinds of boundaries: one is a whole space boundary which is the outside loop and is surrounded by infinite boundary Γ_1 and borehole axis Γ_2; the other is a tool boundary which is the inside loop and is surrounded by electrode boundary Γ_3 (top and bottom of the source) and insulator boundary Γ_4 (inside and outside of source). The solution domain is the whole space minus tool space $\Omega = \Omega_{\text{whole}} - \Omega_{\text{tool}}$ and the boundary is the outside boundary plus the tool boundary $S = S_{\text{tool}} + S_{\text{whole}}$. The *blue lines* (the outer boundaries) show the directions

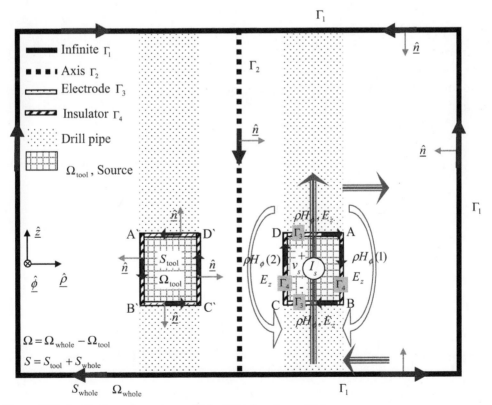

Figure 17.7 Application of source model for EM telemetry system.

of the boundaries, with the unit normal $\hat{n}$ pointed into the solution domain Ω. The source exists inside the tool space Ω_{tool} and is surrounded by tool boundary S_{tool}; the z directional current I_s flows through electrode boundary Γ_3 and the potential difference existing on boundary Γ_4 is V_s, which is the voltage of source. The *red lines* show how current flows in the source, the region around source and the region of earth formation.

As we discussed in Section 17.3.1, the matrix system is built upon $[Z_{m,n}][(\rho h)_n] = [V_m]$, where $[V_m]$ is the global vector, the following relationships exist for different boundaries as:

1. Boundary Γ_1 and Γ_2 and Γ_3. From Section 13.11.4 we obtain

$$V_m\Big|_{m \in \Gamma_1, \Gamma_2, \Gamma_3} = -\oiint_S (\underline{E} \times \underline{\Omega}_m) \cdot \hat{n} dS = 0 \tag{17.1}$$

2. Boundary DA is the top boundary of the source and is electrode boundary Γ_3. From Eq. (13.167) we obtain

$$I_s = 2\pi \left(\rho H_\phi \Big|_{\rho = \Gamma_3(\text{end})} - \rho H_\phi \Big|_{z = \Gamma_3(\text{start})} \right) = 2\pi \left((\rho H_\phi)_{\text{point\_}A} - (\rho H_\phi)_{\text{point\_}D} \right) \tag{17.2}$$

3. Boundary BC is the bottom boundary of the source and is electrode boundary Γ_3. From Eq. (13.167) we obtain

$$-I_s = 2\pi \left(\rho H_\phi \Big|_{\rho = \Gamma_3(\text{end})} - \rho H_\phi \Big|_{z = \Gamma_3(\text{start})} \right) = 2\pi \left((\rho H_\phi)_{\text{point\_}C} - (\rho H_\phi)_{\text{point\_}B} \right) \tag{17.3}$$

4. Boundary CD is the inside boundary of the source and is insulator boundary Γ_4. From Eq. (13.173) we obtain

$$\sum_{m=m_1}^{m_n} [V_m]\Big|_{m_1, m_2, \ldots m_n \in \Gamma_4 = CD} = 2\pi \left(V_{\text{potential}}\Big|_{z = \Gamma_4(\text{start})} - V_{\text{potential}}\Big|_{z = \Gamma_4(\text{end})} \right)$$
$$= 2\pi \left(V_{\text{potential\_point\_}C} - V_{\text{potential\_point\_}D} \right) = -2\pi V_s \tag{17.4}$$

5. Boundary AB is the outside boundary of the source and is insulator boundary Γ_4. From Eq. (13.173) we obtain

$$\sum_{m=m_1}^{m_n} [V_m]\Big|_{m_1, m_2, \ldots m_n \in \Gamma_4} = 2\pi \left(V_{\text{potential}}\Big|_{z = \Gamma_4(\text{start})} - V_{\text{potential}}\Big|_{z = \Gamma_4(\text{end})} \right) \tag{17.5}$$
$$= 2\pi \left(V_{\text{potential\_point\_}A} - V_{\text{potential\_point\_}B} \right) = 2\pi V_s$$

Notice on boundary CD and AB, ρH_ϕ is a constant as

$$(\rho H_\phi)_{point\_A} = (\rho H_\phi)_{point\_B} = \rho H_\phi(1) \tag{17.6}$$

$$(\rho H_\phi)_{point\_C} = (\rho H_\phi)_{point\_D} = \rho H_\phi(2) \tag{17.7}$$

respectively. From Eqs. (17.1)–(17.7), the matrix system for EM telemetry system can be expressed as:

$$Z_{1,1}\rho H_\phi(1) + Z_{1,2}\rho H_\phi(2) + \ldots + Z_{1,n}\rho H_\phi(n) = 2\pi V_s \tag{17.8}$$

$$Z_{2,1}\rho H_\phi(1) + Z_{2,2}\rho H_\phi(2) + \ldots + Z_{2,n}\rho H_\phi(n) = -2\pi V_s \tag{17.9}$$

$$Z_{n,1}\rho H_\phi(1) + Z_{n,2}\rho H_\phi(2) + \ldots + Z_{n,n}\rho H_\phi(n) = 0 \tag{17.10}$$

$$2\pi \rho H_\phi(1) - 2\pi \rho H_\phi(2) = I_s \tag{17.11}$$

After applying Gaussian Elimination to eliminate all other nodes except nodes existing on boundary Γ_4, the final source mode on the EM telemetry system can be expressed as:

$$A_{1,1}\rho H_\phi(1) + A_{1,2}\rho H_\phi(2) = 2\pi V_s \tag{17.12}$$

$$A_{2,1}\rho H_\phi(1) + A_{2,2}\rho H_\phi(2) = -2\pi V_s \tag{17.13}$$

$$2\pi \rho H_\phi(1) - 2\pi \rho H_\phi(2) = I_s \tag{17.14}$$

If the source is voltage source and its value is constant, other variables can be calculated as

$$\rho H_\phi(1) = \frac{A_{1,2} + A_{2,2}}{A_{1,1}A_{2,2} - A_{1,2}A_{2,1}}(2\pi)V_s \tag{17.15}$$

$$\rho H_\phi(2) = -\frac{A_{1,1} + A_{2,1}}{A_{1,1}A_{2,2} - A_{1,2}A_{2,1}}(2\pi)V_s \tag{17.16}$$

$$I_s = \frac{A_{1,1} + A_{2,1} + A_{1,2} + A_{2,2}}{A_{1,1}A_{2,2} - A_{1,2}A_{2,1}}(2\pi)^2 V_s \tag{17.17}$$

If the source is current source and its value is constant, other variables can be calculated as

$$\rho H_\phi(1) = \frac{A_{1,2} + A_{2,2}}{A_{1,1} + A_{2,1} + A_{1,2} + A_{2,2}} \frac{I_s}{2\pi} \tag{17.18}$$

$$\rho H_\phi(2) = -\frac{A_{1,1} + A_{2,1}}{A_{1,1} + A_{2,1} + A_{1,2} + A_{2,2}} \frac{I_s}{2\pi} \tag{17.19}$$

$$V_s = \frac{A_{1,1}A_{2,2} - A_{1,2}A_{2,1}}{A_{1,1} + A_{2,1} + A_{1,2} + A_{2,2}} \frac{I_s}{(2\pi)^2} \tag{17.20}$$

The impedance of the EM telemetry system is calculated by

$$Z_s = \frac{V_s}{I_s} \tag{17.21}$$

Understanding the changing of system impedance resulting from different EM telemetry systems will help to make the decision on how to choose a source. Simulation results will be shown in Section 17.4.

The computation method of the source model for the EM telemetry system is shown in Appendix A.

17.3.3 Flowchart of computational code for EM telemetry system
17.3.3.1 Computation of current density, EM fields and receiver voltage
In low frequency and the TM mode assumption, Ampere's laws can be simplified as (Fig. 17.8)

$$\nabla \times \underline{H} = \underline{\sigma} \cdot \underline{E} = \sigma_h E_\rho \hat{\rho} + \sigma_v E_z \hat{z} = J_\rho \hat{\rho} + J_z \hat{z} \tag{17.22}$$

where E_ρ and E_z are ρ directional and z directional electric fields separately, and J_ρ and J_z are ρ directional and z directional eddy currents separately. In cylindrical coordinates and the TM mode assumption, the curl of magnetic fields is simply

$$\nabla \times \underline{H} = -\frac{\partial H_\phi}{\partial z}\hat{\rho} + \frac{1}{\rho}\frac{\partial}{\partial \rho}(\rho H_\phi)\hat{z} \tag{17.23}$$

The relationship between magnetic fields and eddy currents can be obtained as

$$J_\rho = -\frac{\partial H_\phi}{\partial z} = -\frac{1}{\rho}\frac{\partial(\rho H_\phi)}{\partial z} \tag{17.24}$$

$$J_z = \frac{1}{\rho}\frac{\partial}{\partial \rho}(\rho H_\phi) \tag{17.25}$$

as shown in Fig. 17.9.

Electromagnetic Telemetry System and Electromagnetic Short Hop Telemetry 657

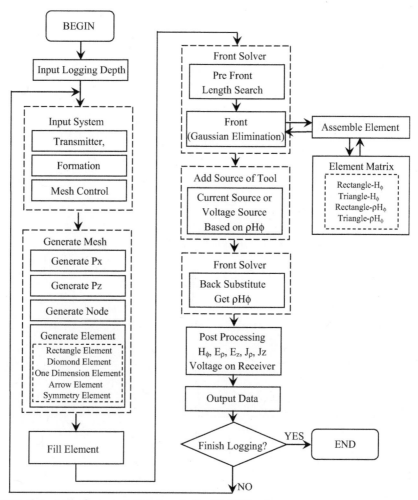

Figure 17.8 Flowchart of computational code for AC DLT tool.

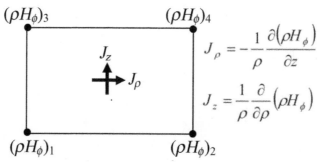

Figure 17.9 Relationship between magnetic field and eddy current.

If the eddy currents in the center point of the element are calculated, they can be simply

$$J_\rho = -\frac{1}{\rho_0} \frac{1}{2\Delta z} \left((\rho H_\phi)_3 + (\rho H_\phi)_4 - (\rho H_\phi)_1 - (\rho H_\phi)_2 \right) \quad (17.26)$$

$$J_z = \frac{1}{\rho_0} \frac{1}{2\Delta \rho} \left((\rho H_\phi)_2 + (\rho H_\phi)_4 - (\rho H_\phi)_1 - (\rho H_\phi)_3 \right) \quad (17.27)$$

which is shown in Fig. 17.9. If the eddy currents which are not limited in the center point of the element are calculated, they can be obtained by

$$J_\rho = -\frac{1}{\rho} \frac{\partial (\rho H_\phi)}{\partial z} = -\frac{1}{\rho} \frac{\partial \left(\sum_{i=1}^{4} (\rho H_\phi)_i \theta_i \right)}{\partial z} = -\frac{1}{\rho} \sum_{i=1}^{4} \left((\rho H_\phi)_i \frac{\partial (\theta_i)}{\partial z} \right)$$

$$= -\frac{1}{\rho} \sum_{i=1}^{4} \left((\rho H_\phi)_i (-1)^{\text{int}(\frac{i+1}{2})} \left((-1)^i (\rho - \rho_0) + \frac{\Delta \rho}{2} \right) \frac{1}{\Delta \rho \Delta z} \right) \quad (17.28)$$

$$J_z = \frac{1}{\rho} \frac{\partial}{\partial \rho} (\rho H_\phi) = \frac{1}{\rho} \frac{\partial \left(\sum_{i=1}^{4} (\rho H_\phi)_i \theta_i \right)}{\partial \rho} = \frac{1}{\rho} \sum_{i=1}^{4} \left((\rho H_\phi)_i \frac{\partial (\theta_i)}{\partial \rho} \right)$$

$$= \frac{1}{\rho} \sum_{i=1}^{4} \left((\rho H_\phi)_i (-1)^i \left((-1)^{\text{int}(\frac{i+1}{2})} (z - z_0) + \frac{\Delta z}{2} \right) \frac{1}{\Delta \rho \Delta z} \right) \quad (17.29)$$

where θ_i is rectangular element basis functions illustrated in Section 13.5.2. The electric field can be obtained from eddy current by

$$E_\rho = \frac{J_\rho}{\sigma_h} \quad (17.30)$$

$$E_z = \frac{J_z}{\sigma_v} \quad (17.31)$$

The eddy currents calculated from the two methods are equal to each other. The voltage on receiver can be obtained by

$$V_{\text{receiver}} = \int_{\rho_1}^{\rho} E_\rho d\rho \quad (17.32)$$

17.4 VALIDATION OF THE COMPUTATION ALGORITHM IN A CASED BOREHOLE

The numerical computation algorithm must be validated before we can use it. To do so, we usually compare the computed results with published literature. In Ref. [3], Vong has both simulation and field test data of a telemetry system when the borehole is cased. In Vong's paper, the carrier frequency is 4.89 Hz in order to improve the skin depth. A fully cased borehole was used for the experiment. The well was drilled to a depth of 610 m with casing placed up to 579 m downhole. The borehole mud has a higher salinity with a measured conductivity of 11.97 S/m, which is considered to be salty mud. The earth formation has an average conductivity of 0.03 S/m. The receiving antenna is from the casing to a position on the surface about 110 m away from the well. The properties of this system are shown in Table 17.3.

The calculated and measured results are shown in Fig. 17.10, where "measured" is the measured result, "FE model 1 (Vong, Lagrange)" is the result from Ref. [4], "FE model 2 (Vong, Normal)" is the result from Ref. [3], and "FE model 3 (self)" is the result from the model illustrated in this chapter. Table 17.4 shows the simulation time of the different models. It shows that the model used in this chapter and the 2006

Table 17.3 Properties of EM telemetry system with casing

Property	Value
Drill pipe: inside radius	2.135 in.
Drill pipe: outside radius	2.5 in.
Drill pipe: conductivity	$4.5*10^6$ S/m
Drill bit depth	Variable
Casing: inside radius	5.625 in.
Casing: outside radius	6.125 in.
Casing depth	−579 m
Casing: conductivity	$4.5*10^6$ S/m
Borehole: radius	6.125 in.
Borehole depth	−610 m
Borehole: conductivity	11.97 S/m
Transmitter: gap thickness	1 in.
Transmitter: gap-bit distance	30 ft
Transmitter: current of source	10 A
Transmitter: frequency	4.89 Hz
Receiver: start radius	6.125 in.
Receiver: end radius	110 m
Receiver: depth	−0.03 m
Earth formation conductivity	0.03 S/m
Ground depth	0 m
Air resistivity	$1*10^5$ ohm-m

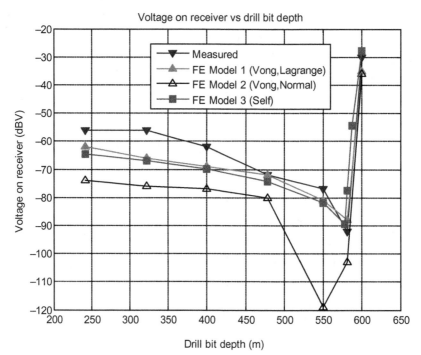

Figure 17.10 Calculated and measured EM signals for cased well.

Table 17.4 Comparison of simulation time

Scheme	Equations	Time (s)
Vong (2006), coarse grid	493,848	360
Vong (2006) fine grid	980,514	600
Current method	183,111	16

model used by Vong achieved very similar results, but the model in this chapter is much faster.

The results also show that with the casing acting as a shield, the signal level decreases significantly. One distinctive trend of the detected electromagnetic signal is that it exhibits a rapid diminishing effect as the EM telemetry system makes an exit at the end of the casing into open ground. This is clearly illustrated in Fig. 17.10. This phenomenon is due to the cancellation effect of the opposing eddy currents generated by the metal casing and the conducting drill pipe.

In Table 17.4, Vong's results are from a Quad AMD Opteron computer with 1.8-GHz processors and a total of 20 GB of RAM; the result discussed in this chapter is from a personal computer with 2.0-GHz processor and 3.0 GB of RAM.

17.5 SIMULATION RESULT WITHOUT CASING

In this section, we will test the stability of the FEM method by comparing the results from different grid size and lists computation times for different borehole depth. This section will also study the relationship between current distribution, voltage on the receiver and impedance of the system with borehole depth, borehole type, formation resistivity, and gap spacing. The properties of the EM telemetry system are shown in Table 17.5, where range shows the range of the parameter, and value shows the value of parameter if it is not specially mentioned in the examples.

17.5.1 Grid size and computation time for different borehole depth

Table 17.6 shows the results of a 2000-ft-depth and 5000-ft-depth EM telemetry system calculated using different grid: coarse grid, fine grid, and very fine grid. The range of solution domain Ω is: r-direction range is from 0 to 1000 ft, z-direction range is from $(DrillBitDepth - 400 * ZgridDeltaZ2)$ to $(GroundDepth + 200 * ZgridDeltaZ2)$. Comparing the results from different grid size, it can be seen that the stability of the algorithm is satisfactory. Fine grid is chosen to be used in the future simulation considering both computation time and accuracy.

Table 17.7 shows the system and properties of a personal computer which is used to calculate different EM telemetry systems discussed in this chapter. Table 17.8 and

Table 17.5 Properties of EM telemetry system without considering casing

	Value	Range
Drill pipe: inside radius	2.135 in.	
Drill pipe: outside radius	2.5 in.	
Drill pipe: conductivity	$4.5*10^6$ S/m	
Drill bit depth	Variable	Variable
Borehole: radius	5.625 in.	
Borehole depth	= drill bit depth	100–9000 ft
Borehole: resistivity	WBM 0.5 ohm-m	OBM 2000 ohm-m
		WBM 0.5 ohm-m
Transmitter: gap thickness	1 in.	1–10 in.
Transmitter: gap-bit distance	30 ft	
Transmitter: current of source	1 A	
Transmitter: frequency	10 Hz	1–100 Hz
Receiver: start radius	5.625 in.	
Receiver: end radius	446 ft	
Receiver: depth	−0.1 ft	
Earth formation resistivity	10 ohm-m	$1e^{-1}$–$1e^{+4}$ ohm
Ground depth	0 ft	
Air resistivity	$1*10^5$ ohm-m	

Table 17.6 Comparison of different grid size

	Coarse grid	Find grid	Very fine grid
RgridNumberOfInsidePipe	4 (unit)	8 (unit)	12 (unit)
RgridNumberOfDrillPipe	4 (unit)	8 (unit)	12 (unit)
RgridNumberOfOutsidePipe	4 (unit)	8 (unit)	12 (unit)
RgridDeltaR	0.05 (in.)	0.05 (in.)	0.05 (in.)
RgridRatio	1.48	1.25	1.17
RgridNumberOfRatio	30 (unit)	50 (unit)	70 (unit)
ZgridNumberOfGap	4 (unit)	8 (unit)	12 (unit)
ZgridDeltaZ1	1.0 (in.)	1.0 (in.)	1.0 (in.)
ZgridRatio1	1.20	1.16	1.11
ZgridNumberOfRatio1	18 (unit)	18 (unit)	18 (unit)
ZgridDeltaZ2	24 (in.)	12 (in.)	6 (in.)
ZgridNumberAboveGround	200 (unit)	200 (unit)	200 (unit)
ZgridNumberBelowDrillbit	400 (unit)	400 (unit)	400 (unit)
	2000-ft-depth borehole		
Impedance of system (ohm-m)	$0.407559633 + 0.005964321j$	$0.408763006 + 0.005919551j$	$0.408579252 + 0.006049483j$
Voltage on receiver (dBV)	-50.7518984	-50.6635042	-50.5105116
	5000-ft-depth borehole		
Impedance of system (ohm-m)	$0.407638656 + 0.006011884j$	$0.408841430 + 0.005973304j$	$0.408664989 + 0.006109718j$
Voltage on receiver (dBV)	-85.7917715	-85.5982587	-84.9611987

Table 17.7 Computer system and property

System	Microsoft Windows XP, Version 2002
CPU	Intel Core 2 Duo
Processor	2.00 GHz
RAM	3.00 GB

Table 17.8 Grid size and computation time in different borehole depth

Borehole depth (ft)	R Directional grid number	Z Directional grid number	Global matrix size	Simulation time (second)
500	76	910	69,111*69,111	6
1000	76	1410	107,111*107,111	9
1500	76	1910	145,111*145,111	12
2000	76	2410	183,111*183,111	16
2500	76	2910	221,111*221,111	19
3000	76	3410	259,111*259,111	22
3500	76	3910	297,111*297,111	26
4000	76	4410	335,111*335,111	29
4500	76	4910	373,111*373,111	33
5000	76	5410	411,111*411,111	36
5500	76	5910	449,111*449,111	39
6000	76	6410	487,111*487,111	43
6500	76	6910	525,111*525,111	45
7000	76	7410	563,111*563,111	48
7500	76	7910	601,111*601,111	52
8000	76	8410	639,111*639,111	56
8500	76	8910	677,111*677,111	59
9000	76	9410	715,111*715,111	63

Fig. 17.11 show the relationship of grid size and computer time for different borehole depth. It shows that the computation time for a 2000-ft-depth borehole is only 16 seconds with global matrix size of 183,111*183,111 and the computation time for a 9000-ft-depth borehole is 63 seconds with global matrix size of 715,111*715,111. Fine grid is used in the simulation shown. If coarse grid is chosen to be used, computation time will be significantly shorter.

17.5.2 Current distribution pattern in a 9000-ft-depth borehole

Fig. 17.12 shows the current density image of an EM telemetry system in a 9000-ft-depth borehole with different frequencies ($f = 10$ Hz and $f = 100$ Hz) and different formation resistivities ($R = 0.1$, 1, and 10 ohm-m). The resistivity of the borehole mud is set to be equal to the resistivity of the formation. The horizontal axis is the grid number in the r direction and vertical axis in Z direction. Current density values

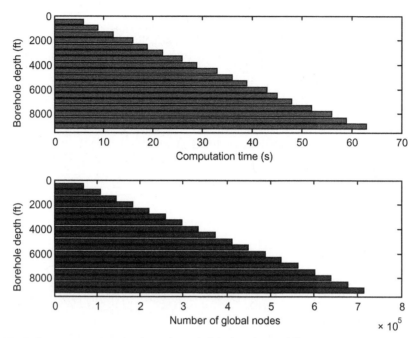

Figure 17.11 Computation time and number of global nodes in different borehole depth.

are expressed in color. Note that the drill pipe is located between the grid number 9 and 16. Fig. 17.12 shows that with the decreasing of the frequency and increasing of the formation resistivity, current can flow further along the pipe and into the earth formation, thus the voltage received by the receiver which is located on the surface will be higher. In Fig. 17.12, let us consider the current distribution at the depth where the current strength is colored in *green* (the value of current density is around -70 dBA/m^2) on the outside boundary of the drill pipe when the source is located at the depth of -9000 ft: in Fig. 17.12C ($R = 10$ ohm-m, $f = 10$ Hz), *green* region appear up to the ground; in Fig. 17.12B ($R = 1$ ohm-m, $f = 10$ Hz), the *green* region appear around the depth of -5000 ft; in Fig. 17.12A ($R = 0.1$ ohm-m, $f = 10$ Hz), *green* region appear around the depth of -7500 ft; in Fig. 17.12F ($R = 10$ ohm-m, $f = 100$ Hz), *green* region appear around the depth of -4000 ft; in Fig. 17.12E ($R = 1$ ohm-m, $f = 100$ Hz), *green* region appear around the depth of -7000 ft; in Fig. 17.12D ($R = 0.1$ ohm-m, $f = 100$ Hz), *green* region appear around the depth of -8500 ft. In the later discussion of the received voltage at the receiver on the surface, Section 17.5.3 shows that the voltage received by receiver for the system shown in Fig. 17.12C is around -135 dBV if assuming a WBM wells, -240 dBV for Fig. 17.12F, -265 dBV for Fig. 17.12B, and -560 dBV for Fig. 17.12E.

Fig. 17.13 is a vector plot showing how current flows from the source to pipe, and then into the earth formation, where Fig. 17.13A shows the region around source

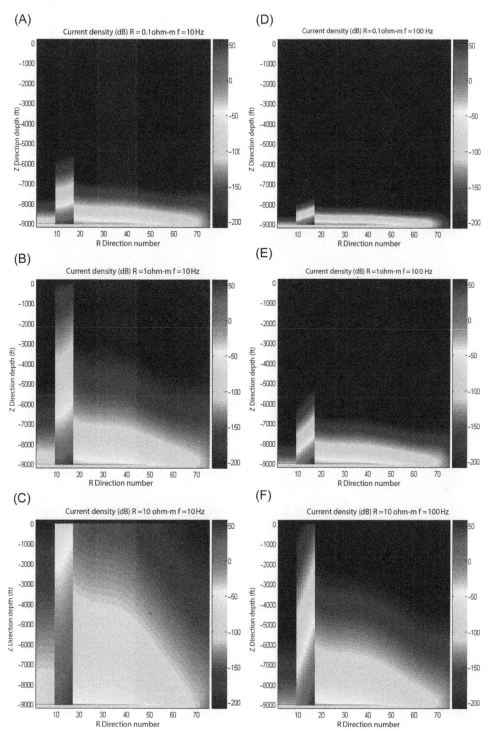

Figure 17.12 Current density image in a 9000-ft-depth borehole. (A) $R = 0.1$ ohm-m, $f = 10$ Hz; (B) $R = 1$ ohm-m, $f = 10$ Hz; (C) $R = 10$ ohm-m, $f = 10$ Hz; (D) $R = 0.1$ ohm-m, $f = 100$ Hz; (E) $R = 1$ ohm-m, $f = 100$ Hz; (F) $R = 10$ ohm-m, $f = 100$ Hz.

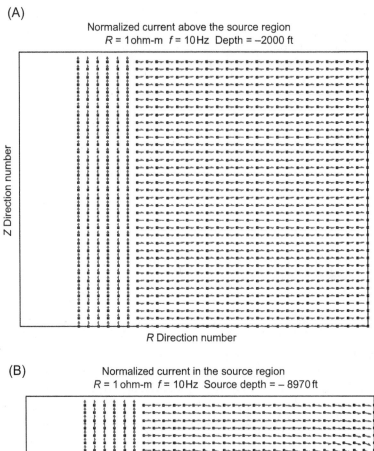

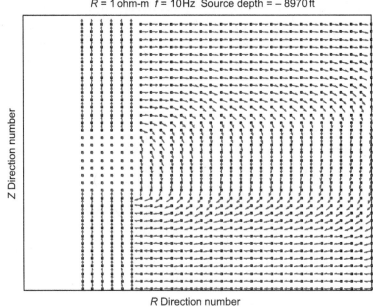

Figure 17.13 Current injected into earth formation from source. (A) Region around source, (B) region above source.

and Fig. 17.13B shows the region above the source. In the region around source, current flows up vertically from the source to the upper part of pipe, then turns back to the lower part of the pipe, finally flows back vertically to the source, which was also explained in Fig. 17.11. In this figure, the horizontal axis is the grid number in the r direction and vertical axis in Z direction. Current density directions are expressed in vector arrows.

17.5.3 Voltage on receiver

Potential difference V is received by receiver on the surface. The potential difference is caused by the current flow from the drill pipe to a point relative far from the pipe, usually about 100 m. Fig. 17.14 shows the received voltage in dBV as a function of transmitter depth and operating frequency. The transmitter depth is from -100 to -9000 ft and frequency is set from 1 to 100 Hz when the formation resistivity is 10 ohm-m and gap spacing for the source is 1 in. Fig. 17.14A is the result for a WBM borehole and Fig. 17.14B is the result from an OBM borehole. The results show that the signal can propagate much further in the OBM wells than in the WBM wells. For example, when borehole depth is -9000 ft and frequency is 1 Hz, voltage on receiver is -95 dBV in the WBM wells and -50 dBV in the OBM wells. This can be easily understood that when the borehole mud is not conductive in the OBM case, the current flowing along the drill pipe will have much less dissipation to the formation than that when the borehole mud is conductive in the WBM case. In other words, the conductive mud in the WBM case "short circuits" the current flowing along the drill pipe making the transmission loss much higher. We can predict that when the conductivity of the borehole mud increases, the transmission loss will also increase.

Fig. 17.14 also shows that the voltage on receiver increases when the borehole depth decreases and when frequency decreases. Shorter transmission distance apparently suffers less attenuation and the lower frequency has less signal attenuation due to longer skin depth than higher frequencies. Fig. 17.15 shows the same relationship as Fig. 17.14 when the formation resistivity is 1 ohm-m. This figure shows lower formation resistivity results, lower voltage received by receiver, which is also the effect of skin depth since lower formation resistivity makes the skin depth shorter for the same frequency. This conclusion is emphasized more in Fig. 17.16, which shows the relationship of receiver voltage with borehole depth (from -100 to -9000 ft) and formation resistivity (from 0.1 to 10,000 ohm-m) when the frequency is 10 Hz and gap spacing for the source is 1 in. Fig. 17.17 shows the effect of gap spacing on receiver voltage when the frequency is 10 Hz and formation resistivity is 10 ohm-m. From Fig. 17.17, we can see that the received signal strength is not very sensitive to the gap spacing, especially, in the OBM case.

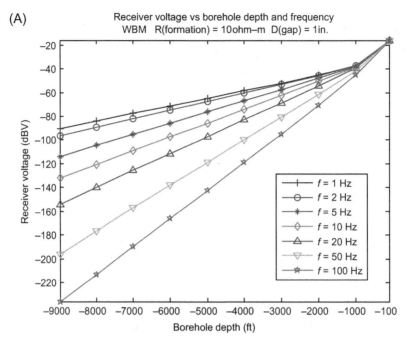

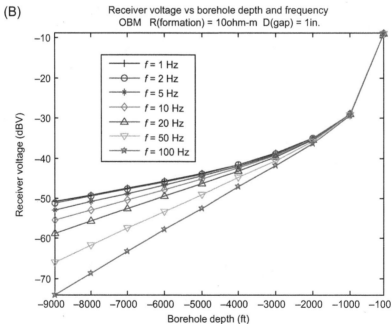

Figure 17.14 Receiver voltage in different borehole depth and frequency 1 when R(formation) = 10 ohm-m and D(gap) = 1 in. (A) WBM wells, (B) OBM wells.

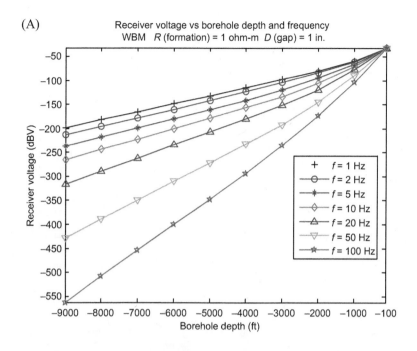

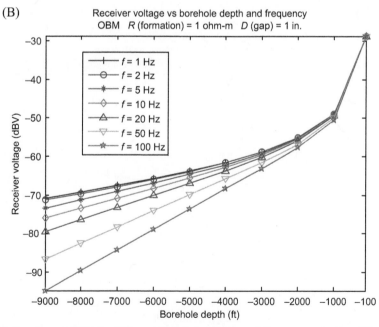

Figure 17.15 Receiver voltage in different borehole depth and frequency 2 when R(formation) = 1 ohm-m and D(gap) = 1 in. (A) WBM wells, (B) OBM wells.

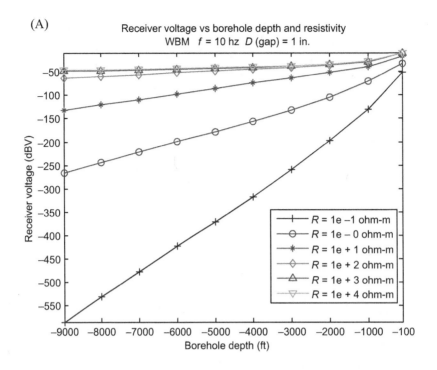

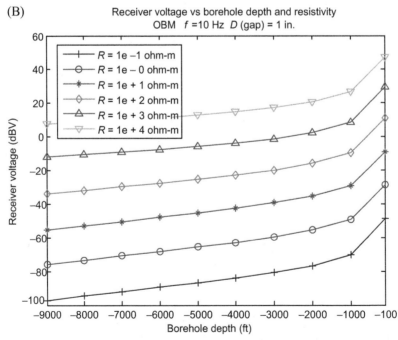

Figure 17.16 Receiver voltage in different borehole depth and formation resistivity when $f = 10$ Hz and D(gap) = 1 in. (A) WBM wells, (B) OBM wells.

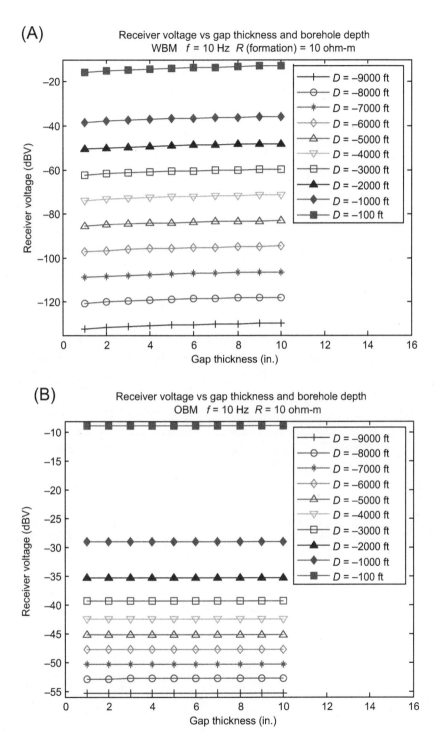

Figure 17.17 Receiver voltage in different gap thickness and borehole depth when $f = 10$ Hz and $R = 10$ ohm-m. (A) WBM wells, (B) OBM wells.

17.5.4 System impedance

System impedance is defined as V_s/I_s, where V_s is the voltage applied across the gap and I_s is source current. System impedance is an important parameter when designing the EM telemetry system. Fig. 17.18 shows the relationship between the impedance and the depth of the transmitter and frequency (from 1 to 100 Hz) when the formation resistivity is 10 ohm-m and a gap spacing of 1 in. Fig. 17.18A is the case when a WBM is used and Fig. 17.18B shows the result when an OBM is employed. From Fig. 17.18, we can see that the system impedance is not sensitive to the frequency in the OBM case and has slight sensitivity in the WBM case when the formation resistivity is 10 ohm-m. The system impedance is sensitive to the borehole depth when the borehole has a shallow depth (less than 2000 ft). Fig. 17.18 also shows that system impedance is very sensitive to the borehole mud. The system impedance for WBM well is around 0.4 ohm-m, and is around 24 ohm-m for OBM wells when the well is deep.

Fig. 17.19 shows the same relationship as Fig. 17.18, whereas the formation resistivity is 1 ohm-m. From Fig. 17.19 we can see that now the system impedance is sensitive to the frequency in WBM wells because the borehole resistivity (0.5 ohm-m) and formation resistivity (1 ohm-m) are similar. The system impedance changes from 0.71 to 0.85 ohm-m when the frequency is from 1 to 100 Hz in a deep WBM wells. The system impedance is around 24 ohm-m in a deep OBM wells; it is the same value as Fig. 17.18B when the formation resistivity is 10 ohm-m.

Fig. 17.20 is the telemetry impedance as a function of well depth (from −100 to −9000 ft) and formation resistivity (from 0.1 to 10,000 ohm-m) when the frequency is 10 Hz and source gap spacing of 1 in. It is shown that the formation resistivity has the big impact on system impedance, no matter for the WBM wells or the OBM wells. The impact is more obvious in OBM well. System impedance increases when the formation resistivity increases.

Fig. 17.21 shows the effect of gap spacing on the system impedance when the frequency is 10 Hz and formation resistivity is 10 ohm-m. It shows that system impedance is not sensitive to the gap spacing in OBM wells, its value is around 24 ohm-m when the well is deep. It also shows that the system impedance is a little more sensitive to the gap spacing in WBM well than the OBM wells. The system impedance increases from 0.4 to 0.55 ohm-m when the gap spacing increases from 1 to 10 in. in a deep WBM wells.

The impedance characterization of the EM telemetry system can be easily understood by using a simplified circuit model. Fig. 17.22 shows a circuit model of the EM telemetry system.

From Fig. 17.22B, we can easily understand the physics in this section quantitatively presented in Figs. 17.18−17.21. Considering the collar as a cylindrical

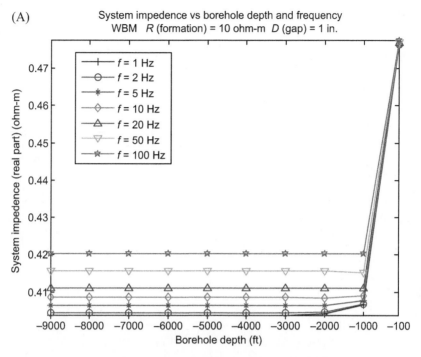

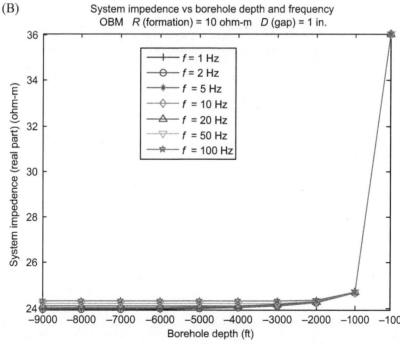

Figure 17.18 System impedance in different borehole depth and frequency when R(formation) = 10 ohm-m and D(gap) = 1 in. (A) WBM wells, (B) OBM wells.

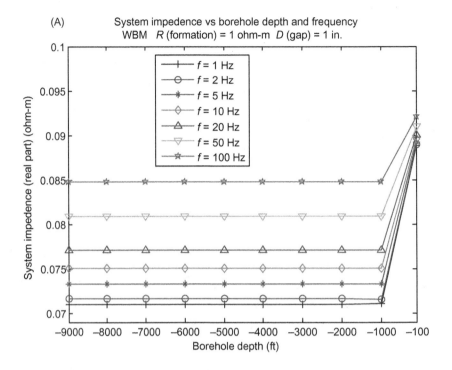
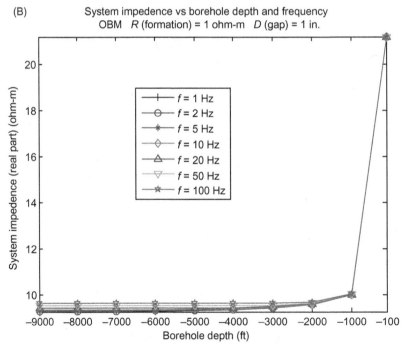

Figure 17.19 System impedence in different borehole depth and frequency when R(formation) = 1 ohm-m and D(gap) = 1 in. (A) WBM wells, (B) OBM wells.

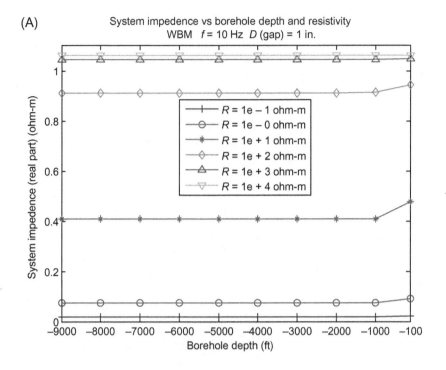

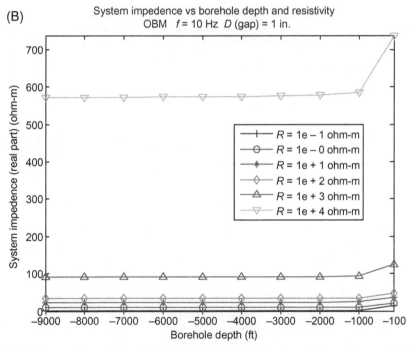

Figure 17.20 System impedance in different borehole depth and formation resistivity when $f = 10$ Hz and D(gap) $= 1$ in. (A) WBM wells, (B) OBM wells.

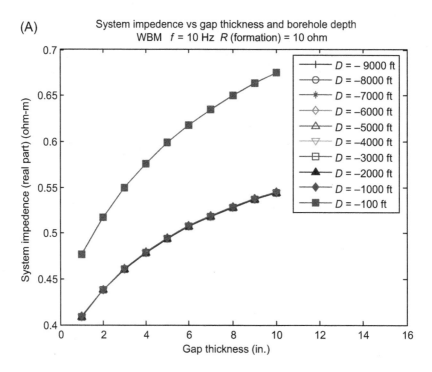

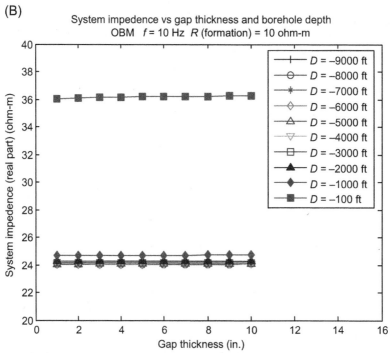

Figure 17.21 System impedance as a function of different gap spacing and borehole depth when $f = 10$ Hz and R(formation) = 10 ohm-m. (A) WBM wells, (B) OBM wells.

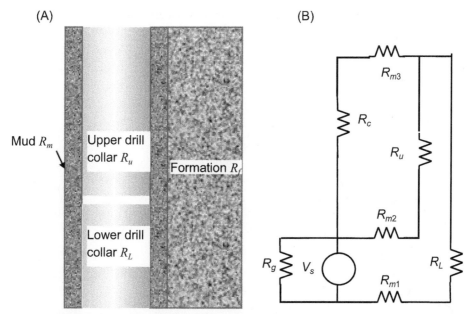

Figure 17.22 The simplified equivalent circuit of the EM telemetry system used to analyze the input impedance at the gap between upper and lower sections of the drill collar. (A) the schematic of the EM telemetry system and (B) the simplified circuit, where V_s is the voltage source applied to the gap; R_g is the equivalent resistance of the gap due to mud; R_{m1}, R_{m2}, and R_{m3} are the resistance from sections of collar below the gap, above the gap, and from the top part of the collar, respectively. R_c is the resistance of the drill collar; R_L and R_u are the resistance of the formation presented to the lower section and upper section of the drill collar, respectively.

electrode, from the direct current (DC) resistance of the cylindrical electrode in a homogeneous media given in Eq. (15.6), we know that the longer the collar, the lower the DC resistance as plotted in Fig. 17.23 for a given formation conductivity and collar diameter.

$$Z = \frac{1}{4\pi^2 \sigma a L} \ln\left(\frac{L}{a}\right) \tag{15.6}$$

Although the resistance curve given in Fig. 17.23 is for DC case and the impedance in Figs. 17.18–17.21 are obtained using AC computation, at low frequencies, these two values should be close to each other. We should notice that the resistance showing in Fig. 17.23 is very low compared with the impedance given in Fig. 17.18–17.21 for a long drill pipe. This is because the impedance given in Fig. 17.23 is only one part of the two series resistances given in Fig. 17.18–17.21 as explained in Fig. 17.22. Considering the WBM case, since the borehole resistance is relatively low, the

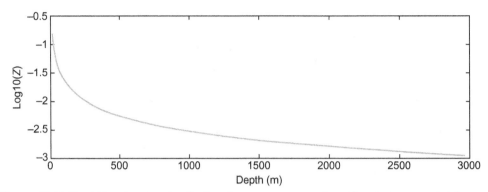

Figure 17.23 The DC resistance of collar in a homogeneous media with a 1 ohm-m resistivity as a function of the depth calculated using Eq. (15.6). Note that the impedance of the gap is a series of upper and lower section of the gap.

impedance showing in Figs. 17.18–17.21 can be calculated using the two-series resistance of the upper and lower section of the drill pipe:

$$Z_d = \frac{1}{4\pi^2 \sigma a L_L} \ln\left(\frac{L_L}{a}\right) - \frac{1}{4\pi^2 \sigma a L_U} \ln\left(\frac{L_U}{a}\right) \qquad (17.33)$$

where Z_d is the differential impedance shown in Figs. 17.18–17.21, L_U and L_L are the length of the drill pipes above and below the excitation gap. Since L_U is usually much greater than L_L, the impedance is largely determined by the first term in Eq. (17.33):

$$Z_d \approx \frac{1}{4\pi^2 \sigma a L_L} \ln\left(\frac{L_L}{a}\right) \quad \text{when } L_U \gg L_L \qquad (17.34)$$

From Table 17.5, we learn that L_L in the computation of Figs. 17.18–17.21 is 30 ft. We can calculate Z_d using Eq. (17.34) and find that $Z_d = 0.15$ ohm when the formation resistivity is 1 ohm-m, which is close to the values shown in Figs. 17.18–17.21. This also explains the impedance shown in Figs. 17.18–17.21 does not change as a function of depth when the drill pipe is longer than 100 m. Based on Eq. (17.33), the telemetry impedance is determined by the shorter sections of the drill pipe below the excitation gap, which is fixed at 30 ft, and therefore the impedance will not change much when the drill pipe above the gap is much longer than 30 ft.

Fig. 17.24 shows another possible implementation of the EM telemetry system, where a coaxial structure is used and a voltage excitation is applied between the inner and outer conductor instead of a gap as discussed in the previous sections. The advantages of this structure are that the mechanical implementation is rather easier compared with the gap structure, which is mechanically challenging due to

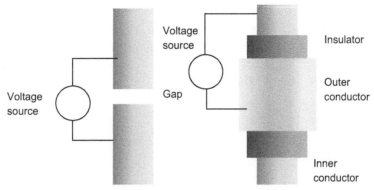

Figure 17.24 The gap EM telemetry device and the coaxial EM telemetry device.

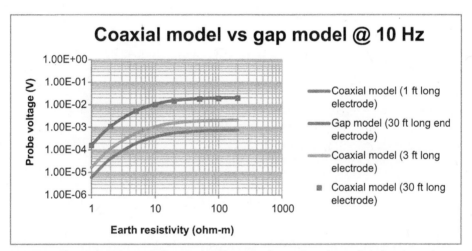

Figure 17.25 The comparison of the simulated received signal at the surface between the coaxial EM telemetry system and the gap EM telemetry system under the same conditions.

the torque required at the gap. Fig. 17.25 shows the comparison of the coax telemetry and the gap telemetry designs under the same formation conditions. From Fig. 17.25, we can see an interesting phenomenon that when the length of the lower section of the drill pipe is the same of that of the coax section, the two system will be identical in terms of received signal strength. This means that these two different implementation will have the same transmission distance if the length of the coax section is identical with the bottom part of the drilling section from the lower end of the gap insulation. In practice, the bottom of the gap is usually composed of a section of the EM telemetry collar, the mud motor or a rotary steerable drilling tool, and the drill bit, which is usually more than 30 ft long. Therefore the gap structure naturally uses this section of the drill pipe whereas the coax structure does not. For an EM telemetry tool design, we should consider both pros and cons of these two systems.

17.6 SHORT HOP ELECTROMAGNETIC TELEMETRY USED IN A NEAR BIT LOGGING-WHILE-DRILLING SENSOR

Short hop telemetry is usually used for the communications between tool sections or between different tools in an LWD system when the wiring is difficult or impossible. The typical use of the short hop device is in the near bit tool and in a rotary steerable tool to hop over the mud motor sections where wiring is difficult. Figs. 17.26 and 17.27 show a schematic of a near bit LWD system where a short hop wireless communication system is employed to send the measured signals to the receiver bypassing the mud motor. Near bit LWD device is mostly used for geosteering. The near bit device is directly mounted right behind the drill bit in front of a rotary

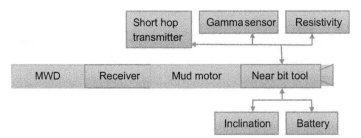

Figure 17.26 A block diagram of a near bit LWD system. The mud motor section is usually 30 ft long and the data measured by the near bit tool is sent wirelessly from the short hop antenna to the short hop receiver over the length of the mud motor.

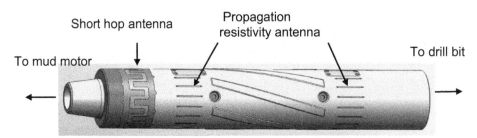

Figure 17.27 A three-dimensional model of a near bit LWD system. The total length of the near bit device is about 40 in.

steerable drilling system or a mud motor. The advantages of the near bit device is its vicinity to the drill bit. It detects the formation and drilling information at the drill bit in real time. In most cases, the near bit device detects the near bit inclination, natural (directional or nondirectional) gamma, sometimes resistivity sensors are also integrated. The near bit device is battery powered and the data measured by the device is wirelessly sent to a receiving device mounted over the rotary steerable system or over a mud motor. The communication distance is in the range of 30–40 ft. Due to the short transmission distance, this downhole wireless communication system is called a short hop communication system. In this section, we will discuss the performance of the short hop system used in a near bit LWD device as an example of the EM telemetry.

17.6.1 Simulation of the short hop communication system

In Fig. 17.27, we can see that there is a gap between the upper and lower section of the collar at the short hop antenna area. Note that this gap only appears on the outer surface of the collar while the body of the collar is solid. There is a toroid antenna installed under the gap area. The toroidal antenna generates a current on the outer surface of the collar and the current flows along the collar and leaks to the formation or mud and loops back to the other side of the gap as shown in Fig. 17.28.

Note that the short hop system would also work in OBM due to the connection to the formation through the drill bit and displacement current coupled through the OBM. In Fig. 17.27, zig-zag shaped short hop antenna slots are designed. However, any small gap between upper and lower collar surface would work equally.

As an example, consider the short hop communication system shown in Fig. 17.29. A toroid antenna of 300 turns over a m-metal sheets is used for both transmitter and receiver. The excitation voltage applied to the transmitter antenna is

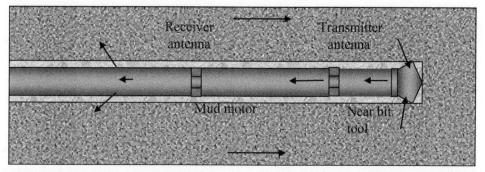

Figure 17.28 The current flow paths of the short hop communication system in a near bit tool. The *red arrows* (black in print versions) show the current paths. The current flowing in the formation also includes the current flow in the borehole. The operating frequency is usually in the range of tens of kilohertz.

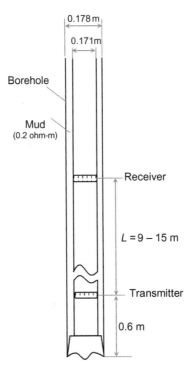

Figure 17.29 Geometry of the at-bit short hop telemetry system.

1 V. The distance between the transmitter and the receiver is assumed to be 9−15 m, which covers the length of a mud motor and the LWD device such as a resistivity LWD tool. Numerical modeling of the short hop communication system can be done using FEM simulation software as we discussed in this chapter. The simulation is a three-dimensional problem since we have to consider the limited length of the drilling system. In the following discussions, the simulation is done using the commercial software HFSS by ANSYS.

The first thing we need to consider in a communication system is the channel characteristics such as bandwidth and signal attenuations. However the channel characteristics are a complex function of many parameters such as resistivity of borehole mud, antenna slot dimensions, transmission distance, and formation resistivity. For a given formation resistivity of 0.35 ohm-m, and a toroid core cross section of $0.25'' \times 1''$, and a relative magnetic permeability of 2000, Fig. 17.30 shows the simulated received voltage as a function of frequency at different transmission distance. From Fig. 17.30, we can clearly see that the transmission channel has a bandpass behavior. The high loss at low frequencies is caused by the efficiency of the toroid since the toroid antenna has higher efficiency at higher frequencies. The increased loss at higher frequencies are caused by the low pass nature of the transmission media,

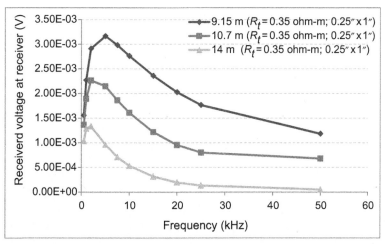

Figure 17.30 Modeled frequency response of the signal received above mud motor ($R_t = 0.35$ ohm-m).

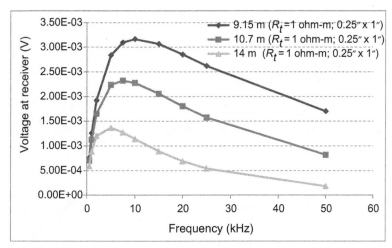

Figure 17.31 Modeled frequency response of the signal received above mud motor ($R_t = 1$ ohm-m).

e.g., the formations, and the mud. At higher frequencies, the skin depth becomes smaller and the loss increases. Similar channel characteristics can be seen from Fig. 17.31 when the formation resistivity is increased to 1 ohm-m. Comparing Figs. 17.30 and 17.27, we can see that as the formation resistivity increases, the path loss will increase due to less current in the formation.

The modeling data show that the signal strength is a function of the transmitter—receiver spacing, formation resistivity, and signal frequency. Under a fixed excitation (e.g., 1 V), the maximum signal may occur at 2 kHz, 10–15-m T-R spacing, 0.35-ohm-m formation resistivity. The signal can also reach maximum value at 10 kHz, 9.15-m T-R spacing, in 1-ohm-m formation.

The modeling data show that the frequency band of 2–20 kHz offers fair signals to be measured. For the reason of high-speed data transmission, higher carrier frequency is preferred. However, at lower frequencies, the path loss is lower resulting in the increase in the reliability and robustness of the data communications. To investigate the details of the influence of the path loss by the formation resistivity, let us consider the two end of the preferred frequency band and change the formation resistivity. Fig. 17.32 compares the signal strengths at 4 and 10 kHz in various formations and at different transmitter–receiver spacing.

Observing the curves of the received signals at different frequencies as a function of formation resistivity as shown in Fig. 17.32, it is clear that when the resistivity of the formation is higher, the current return path will have less current flow and the attenuation will be higher. However, at the frequency near 10 kHz, the signal attenuation

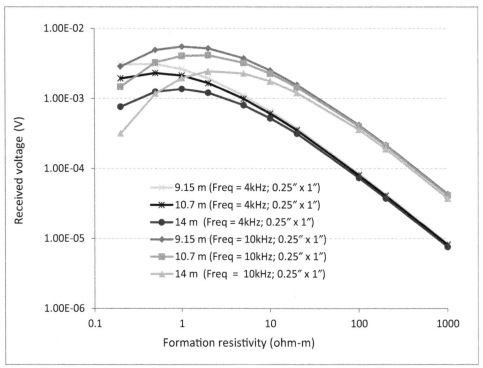

Figure 17.32 Modeled signal strength received above mud motor in different formations. In high-resistivity formation (>1), 10 kHz frequency generates greater signal magnitude than 4 kHz.

seems to be tolerable throughout the formation resistivity up to 1000 ohm-m when the attenuation is about 50 dB. Therefore 10 kHz is recommended for the short hop telemetry system. It would be even better if the signal frequency is selectable from 2 to 20 kHz for different situation depending on the formation resistivity.

17.6.2 Input current and impedance of a toroidal transmitter versus formation resistivity

Using the results from Section 17.6.1, we determined the operating frequency range and the signal strength of the received signal, which can help in designing the hardware for the short hop communication system. In this section, we are interested in understanding the transmitter current necessary to drive the transmitter antenna and the factors that impact to the current. Due to the fact that the near bit system is operated by downhole batteries, the understanding and optimization of the power consumption is critical in designing the short hop hardware.

If a 1-V excitation voltage is applied to the transmitter toroid of 100 turns over the m-metal with a relative magnetic permeability of 2000 and a cross section of 0.25″ × 1″, over a collar as shown in Fig. 17.29, assuming the formation resistivity is 0.35 and 1 ohm-m, without a borehole, the computed current flowing into the toroid antenna is plotted in Figs. 17.33 and 17.34 as a function of operating frequency. It is seen that at higher frequency, the toroid has higher efficiency and the current has a smaller real part compared to the current at lower frequencies (<1 kHz). Note that the imaginary part represents the inductive energy storage and can be canceled by

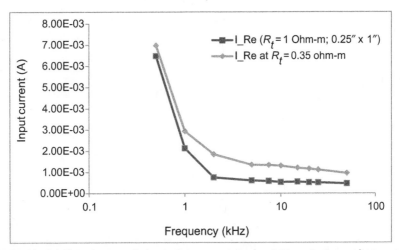

Figure 17.33 Modeled real part of the input current under 1-V excitation voltage at different frequencies; *blue line* (dark gray in print versions) for 0.35-ohm-m formation and *red line* (light gray in print versions) for 1-ohm-m formation.

using a serial capacitor in the tuning process once the center frequency is determined. We can clearly see that for a fixed excitation voltage, the real part of input current decreases with the increase of formation resistivity, and the imaginary part of the input current looks insensitive to formation resistivity, but either real component or imaginary component of the input current decreases with frequency. The lower the frequency is, the faster the current decreases.

To find out the impedance change of the transmitter toroid with the formation resistivity, we consider two fixed frequency points: 4 and 10 kHz and two different toroid core sizes: $0.25'' \times 2''$ and $0.25'' \times 1''$. Note that in practice, a resonant capacitor is always used to make the input impedance of the toroid a real value, the impedance of the toroid will be determined only by the real part of the current since the voltage applied to the toroid antenna is a 1 V real value:

$$Z = V/\text{Re}(I) \qquad (17.35)$$

Therefore the impedance of the toroid antenna can be computed using Eq. (17.35). Due to the fact that the toroid antenna impedance is also a function of the gaps between the figures of the antenna cover (Fig. 17.27), the figures can be included in the simulation to obtain the antenna current with 1-V voltage excitation. Fig. 17.35 shows the computed real part of the impedance of the toroid antenna as a function of the frequency. It is seen that the real part of the antenna impedance after cancellation of the imaginary part of the impedance by adding a series capacitor increases with the increase in formation resistivity. We can also find out that the impedance is also greatly impacted by the metal fingers due to the eddy current consumption by the metal collar.

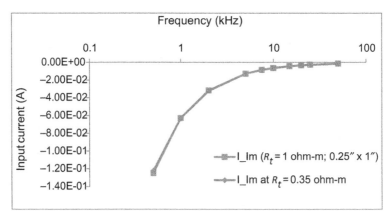

Figure 17.34 Modeled imaginary part of the input current under 1-V excitation voltage at different frequencies; *blue line* (dark gray in print versions) for 0.35-ohm-m formation and *red line* (light gray in print versions) for 1-ohm-m formation.

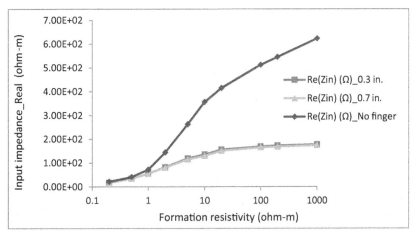

Figure 17.35 Modeled real part of the input impedance under 1-V excitation voltage as a function of the formation resistivity and metal finger thickness.

17.7 CONCLUSIONS

In this chapter, we discussed the theoretical and practical sides of the electromagnetic telemetry system used in the LWD/MWD downhole communications. There are two EM telemetry systems discussed: the EM telemetry to link the downhole device to the surface device and the EM telemetry to link two separated downhole devices, which is also called short hop EM telemetry. The antennas used in the EM telemetry can be coils or toroid. The effective communication distance for a downlink EM telemetry is in the range of 3 km with a data rate less than 10 bps. Increased depth will have to compromise in the data rate. Compared to the mud pulser, EM telemetry is relatively more reliable for shallow wells whereas the mud pulser has longer transmission distance. Using FEM simulation, we can obtain various parameters of the EM telemetry system such as transmission distance in various formations resistivity background, impedance of the antennas, received voltage range, and frequency response of the EM telemetry system. The short hop EM telemetry was discussed using the example of the communication system in a near bit LWD device. The short hop EM telemetry is widely used in downhole environment. Interconnections among downhole devices are not easy due to the harsh environments. Wireless communications among downhole devices become a common practice in the downhole tool design. In most cases, the downhole devices requiring short hop telemetry are battery operated and the power consumption of the short hop system is an important factor to be considered. In this chapter, we discussed the optimized operating frequency, the impedance of the toroid antenna, and the received signal strength. These parameters are critical in the design of a short hop communication system hardware.

REFERENCES

[1] I. Dowell, A. Mills, M. Lora, Chapter 15—Drilling-data acquisition, in: R.F. Mitchell (Ed.), *Petroleum Engineering Handbook*. II—Drilling Engineering, Society of Petroleum Engineers, 2006, pp. 647–685. ISBN 978-1-55563-114-7.
[2] C. Klotz, P. Bond, I. Wasserman, S. Priegnitz, A new mud pulse telemetry system for enhanced MWD/LWD applications, in: IADC/SPE 112683, Presented at the IADC/SPE Drilling Conference Held in Orlando, Florida, U.S.A., March 4–6, 2008.
[3] P.K. Vong, D. Rodger, A. Marshall, Modeling an electromagnetic telemetry system for signal transmission in oil fields, IEEE Trans. Magn. 41 (May 2005) 2008–2011.
[4] P.K. Vong, H.C. Lai, D. Rodger, Modeling electromagnetic field propagation in eddy-current regions of low conductivity, IEEE Trans. Magn. 42 (April 2006) 1267–1270.

APPENDIX A COMPUTATION METHOD OF SOURCE MODEL FOR EM TELEMETRY SYSTEM (SECTION 17.3.2)

```fortran
SUBROUTINE NonConstantConstraint_Telemetry_RouHphi
!
INTEGER :: I,J
COMPLEX*16 :: Is,Vs !current source or voltage source
COMPLEX*16 :: IsTest,VsTest1,VsTest2 !current source or voltage source
COMPLEX*16,ALLOCATABLE :: SourceMatrix(:,:) !matrix information for source nodes
COMPLEX*16,ALLOCATABLE :: A(:,:) !matrix information for source nodes
COMPLEX*16,ALLOCATABLE :: X(:) !X = RouHphi
!
!...read source matrix information from file 'SourceMatirx.dat'
      ALLOCATE(SourceMatrix(NonConstantConstraintNumber,
         NonConstantConstraintNumber))
      ALLOCATE(A(NonConstantConstraintNumber,NonConstantConstraintNumber))
      ALLOCATE(X(NonConstantConstraintNumber))
      OPEN(1,FILE='SourceMatirx.dat',STATUS='UNKNOWN')
      DO I=1,NonConstantConstraintNumber
      DO J=1,NonConstantConstraintNumber
         READ(1,*) SourceMatrix(I,J)
         A(I,J) = SourceMatrix(I,J)
      END DO
      END DO
      CLOSE(1)
!
!...relationship between source nodes and source
!
      IF (((CurrentOfSource.NE.0).AND.(VoltageOfSource.EQ.0)) THEN
         Is = CurrentOfSource
         Vs = Is/2/pi/2/pi*( A(1,1)*A(2,2)-A(1,2)*A(2,1) )/( A(1,1) +
            A(1,2) + A(2,1) + A(2,2) )
         X(1) = Is/2/pi*( A(1,2) + A(2,2) )/( A(1,1) + A(1,2) + A(2,1) + A(2,2) )
         X(2) = -Is/2/pi*( A(1,1) + A(2,1) )/( A(1,1) + A(1,2) + A(2,1) + A(2,2) )
```

```
        ELSE IF ((CurrentOfSource.EQ.0).AND.(VoltageOfSource.NE.0)) THEN
            Vs = VoltageOfSource
            Is = Vs*2*Pi*2*Pi*( A(1,1)+A(1,2)+A(2,1)+A(2,2) )/( A(1,1)*
               A(2,2)-A(1,2)*A(2,1) )
            X(1) = Vs*2*pi*( A(1,2)+A(2,2) )/( A(1,1)*A(2,2)-A(1,2)*A(2,1) )
            X(2) = -Vs*2*pi*( A(1,1)+A(2,1) )/( A(1,1)*A(2,2)-A(1,2)*A(2,1) )
        ELSE
            WRITE(*,*)
            WRITE(*,*) 'Wrong Message: Please check CurrentOfSource and
               VoltageOfSource.'
            WRITE(*,*)
            STOP
        END IF
!
!...test if Vs and Is is in right value
        VsTest1 = ( A(1,1)*X(1)+A(1,2)*X(2) )/(2*Pi)
        VsTest2 = ( A(2,1)*X(1)+A(2,2)*X(2) )/(2*Pi)
        IsTest = 2*pi*X(1)-2*Pi*X(2)
!
!...resistivity of gap
        VoltageOfSource = Vs
        CurrentOfSource = Is
        ImpedanceOfSystem = Vs/Is
!
!...give X(RouHphi) to NonConstantConstraintValue
ALLOCATE(NonConstantConstraintValue(NonConstantConstraintNumber))
        NonConstantConstraintValue = (0.0,0.0)
        DO I = 1,NonConstantConstraintNumber
            NonConstantConstraintValue(I) = X(I)
        END DO
!
        RETURN
!
END SUBROUTINE NonConstantConstraint_Telemetry_RouHphi
```

APPENDIX A

LogSimulator User Manual
Theory of Electromagnetic Well Logging

A.1 INTRODUCTION

This software package is a companion of Liu's book "Theory of Electromagnetic Well Logging." It is intended for the reader and students to practice some of the theory described in the book. However, the software package can also be used to help the data interpretation of the logging data. The logging tools used in this software are limited to electrical logging tools only, not including tools based on other physics.

LogSimulator simulates tool responses in a formation. In this program, three types of logging tools are included:
1. Induction
2. Laterolog
3. Logging while drilling resistivity

All these three tools measure formation resistivity or its reciprocal conductivity.

A basic *Induction* tool has transmitter coils and receiver coils. The transmitter coil is fed by an alternating current source and induces eddy current loops by generating magnetic field in the formations. These current loops in turn induce electromotive force (EMF) in the receiver coils in the same way, and the induced EMF is directly proportional to the formation conductivity. Induction tool is best used in resistive drilling fluids, e.g., oil-based mud. Detailed induction tool theory used in this software is given in Chapter 2, Fundamentals of Electromagnetic Fields Induction Logging Tools, and Chapter 7, Induction and Logging-While-Drilling Tool Response in a Cylindrically Layered Isotropic Formation, of the book.

A *Laterolog* tool uses electrodes to send currents into the formation. By controlling the currents from these nodes and keeping them in a focus condition, the tool will have greater depth of investigations. Laterolog tool is best used with low-resistive drilling fluids. The Laterolog tools are described in Chapter 15, Laterolog Tools and Array Laterolog Tools, of the book.

Logging-while-drilling (LWD) resistivity tools usually use three coils, one transmitter and two receivers. It uses high-frequency signals propagating in the formation to measure the resistivity of the formation. Two receivers are used to compensate the measurements and cancel out uncertainty from electrical devices or environments. The LWD principle is given in Chapter 2, Fundamentals of Electromagnetic Fields Induction Logging Tools, and Chapter 4, Triaxial Induction and Logging-While-Drilling Resistivity Tool Response in Homogeneous Anisotropic Formations, in the book.

Geology structures of the Earth can be very complicated. In this program, three types of formations are considered:

1. Isotropic 1D
2. Cylindrical 1D
3. Anisotropic 2D

An *Isotropic 1D* formation is a vertically layered formation, with each layer has different electromagnetic properties, e.g., resistivity, permittivity (epsilon), and permeability (mu).

A *Cylinder 1D* formation is a zoned formation, each zone is a cylinder around the borehole with different radius. Each zone has its electromagnetic properties, e.g., resistivity, permittivity (epsilon), and permeability (mu).

An *Anisotropic 2D* formation has two levels of structures. The first level is a layered formation, each layer has both horizontal (R_h) and vertical (R_v) resistivity, and a number of zones. Each zone has different radius and may have different horizontal (R_h) and vertical (R_v) resistivity.

A.2 USE OF THE PROGRAM

The software interface is shown in Fig. A.1. Use of the program is very straightforward, simply choose a tool and a formation, and then click Run.

On the left-hand side the software interface displays tool and formation. The results of the simulation are displayed on the right, together with the formation resistivity for comparison. The names of the tool and formation selected are displayed on the status bar at the bottom of the interface.

A.2.1 Default tools and formations

For user's convenience, program has implemented a few default tools and formations.

There are six default tools, each type has two tools, as shown in Fig. A.2.

There are also three default formations, as shown in Fig. A.3:

6STAIRS: Isotropic 1D
CYL3: Cylinder 1D
2D: Anisotropic 2D

A.2.2 Edit tools and formations

To edit a built-in tool or formation, first click Select Tool/Formation from menu. On Select Tool/Formation dialog, click Edit button, Edit Tool/Formation dialog opens (see below). Make any changes, click OK to save changes and back to selection dialog. You may also click Save to File to save edited tool/formation as shown in Fig. A.4. Fig. A.4 is an MWD tool Edit dialog.

Appendix A 693

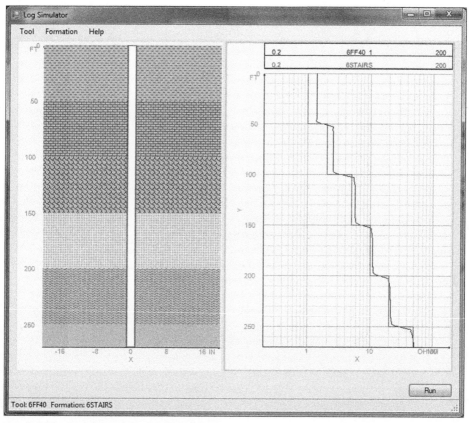

Figure A.1 The interface of the LogSimulator.

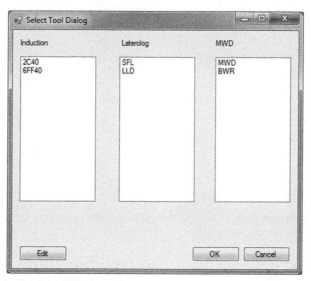

Figure A.2 The tool selection dialog box.

Figure A.3 Built-in formations.

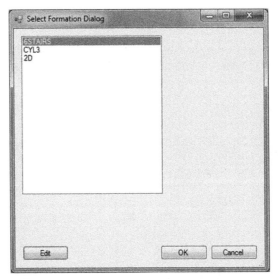

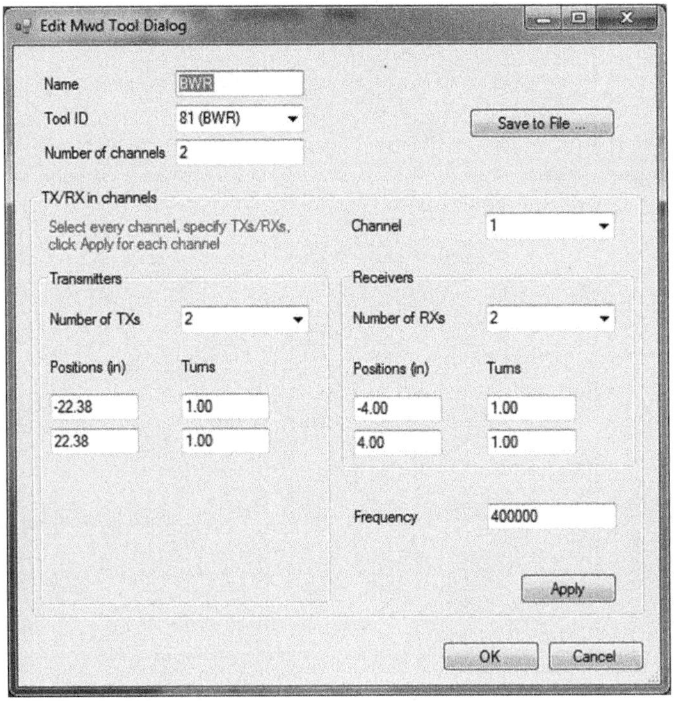

Figure A.4 Example of tool parameter editing.

On the top left are three parameters: Name (Tool), Tool ID, and Number of channels, to which user may make changes. Every time the Number of channels is changed, the list items in the Channel dropdown list below will change accordingly. For example, if user changes the Number of channels to 3, the list items in the Channel dropdown list will be 1, 2, and 3.

In the middle are the TX/RX parameters in every channel. To make changes to these parameters in any channel, user must first select that channel, make any changes, and then click Apply button to save changes to that channel. If you forget clicking Apply button, all changes to that channel will be lost. Clicking OK saves the top left three parameters, and will not save TX/RX parameters specified in each channel.

Fig. A.5 is an example of editing an Anisotropic 2D formation.

On the top left, there are four parameters: Name (tool), Borehole radius, Mud resistivity, and Number of layers, to which user may make changes. Every time the Number of layers are changed, the list items in the Layer dropdown list below will change accordingly. For example, if user changes the Number of layers to 3, the list items in the Layer dropdown list will be 1, 2, and 3.

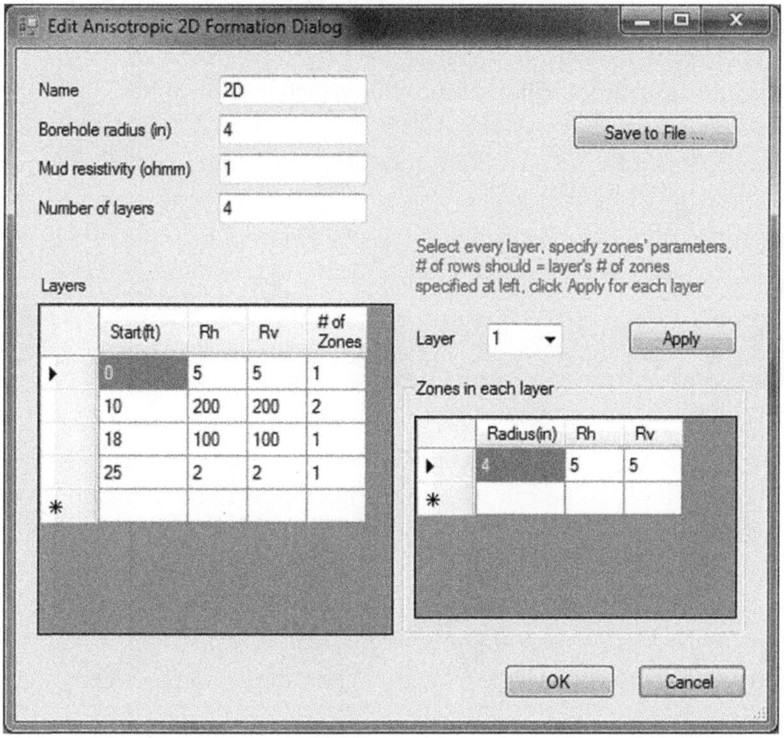

Figure A.5 An example of editing a formation.

The middle left are the parameters for all layers. Here are three rules must be followed:
1. The number of rows in this Layer table must match the Number of layers specified above.
2. The Start depth on the first row must be 0.
3. The Start depths in the first column must be increasing.

In the middle right are the zone parameters in every layer. To make changes to the zone parameters in any layer, user must first select that layer from the Layer dropdown list, make any changes, and then click Apply button to save changes to zone parameters in that layer. If you forget clicking Apply button, all changes to zone parameters in that layer will be lost. Clicking OK only saves the top left four parameters. The zone table rule must be followed in the Zone table:

The number of rows in this Zone table must match the number of Zones of that layer specified in the left Layer table.

A.2.3 Create tools/formations

From Tool/Formation menu, select New and the type of Tool/Formation to open Create specific Tool/Formation dialog. Because using the same dialog as in editing, all functions in Edit Tools/Formations apply here. There are two differences between creating and editing modes:
1. There is an Open from File button, with which user may load the corresponding tool/formation previously saved.
2. The tool/formation created will be appended in the corresponding tool/formation list as user clicks OK button. It is suggested that the name of created be different from those already exist in the list.

A.3 RUN SIMULATION

A.3.1 Tool and formation combination

As described previously, there are three types of tools and three types of formations. Not all types of tools can be run simulation with all types of formations.

An Induction tool can be run with either Isotropic 1D or Cylinder 1D formations, but cannot run with any Anisotropic 2D formation;

A Laterolog tool can be only run with Anisotropic 2D formations;

An MWD tool can be only run with Isotropic 1D formations.

A.3.2 Simulation results

As an Induction tool run with an Isotropic 1D formation, it usually turns out one resistivity curve. The default depth range is calculated based on the formation

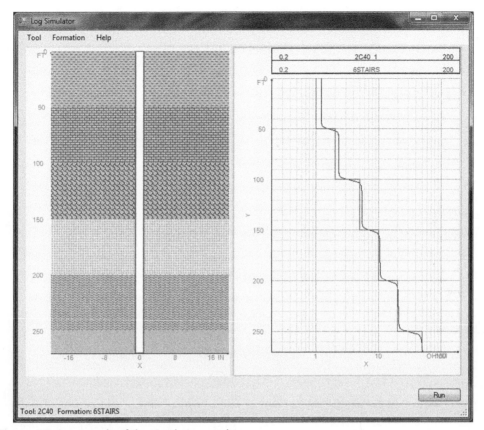

Figure A.6 An example of the simulation result.

thickness plus 20 ft in the last layer. The curve with the formation's name is the resistivity curve of the formation. Fig. A.6 is an example of the simulation result.

When an Induction tool runs with a Cylinder 1D formation, the result will be a single number shown on the dialog in Fig. A.7.

When a Laterolog tool runs with an Anisotropic 2D formation, the result is a single resistivity curve. The default depth range is calculated based on the formation thickness plus 20 ft in the last layer. The curve with the formation's name is the R_v curve of the formation, as shown in Fig. A.8.

When an MWD tool runs with an Isotropic 1D formation, the result is $2N$ curves, where N is the number of channels, and each channel has two curves: apparent phase and attenuation resistivity. The default depth range is calculated based on the formation thickness plus 20 ft in the last layer. The curve with the formation's name is the apparent resistivity curve of the formation, as shown in Fig. A.9.

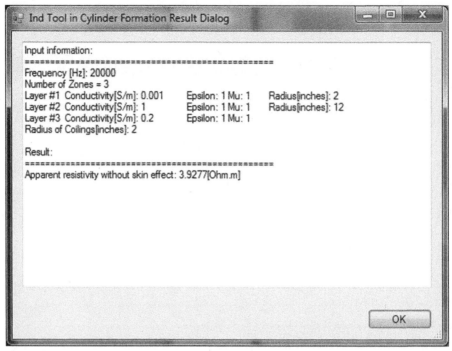

Figure A.7 The dialog displaying results of the induction tool response in a cylindrically layered formation.

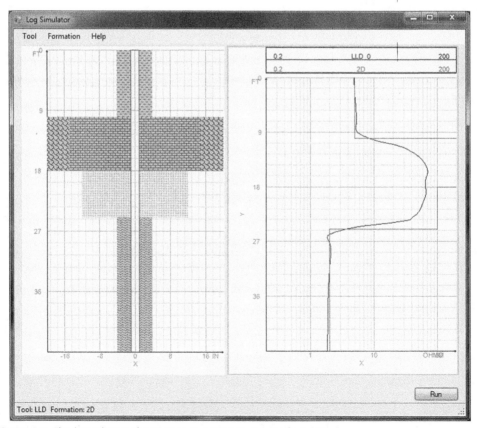

Figure A.8 The laterolog tool response in a 2D anisotropic formation.

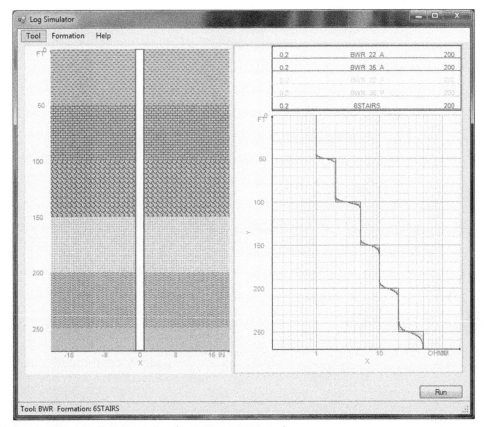

Figure A.9 The simulation results of a LWD resistivity tool.

A.3.3 Display settings

To make changes to the display settings, right click mouse on the plots, and then click Plot Property..., the Plot Property dialog opens (Fig. A.10).

On the General panel (Fig. A.10), user may change the display ranges, fonts, and major and minor grids, etc. On the Curve panel (Fig. A.11), which applies to the result plot only, user may change the line color and width of a curve selected.

To change the display ranges of the depth, user may also use pan and zoom feature. To pan, press, and hold the CTRL key, use the left mouse button to move the depth ranges; to zoom, press, and hold the SHIFT key, use the mouse wheel to zoom in/out the depth ranges.

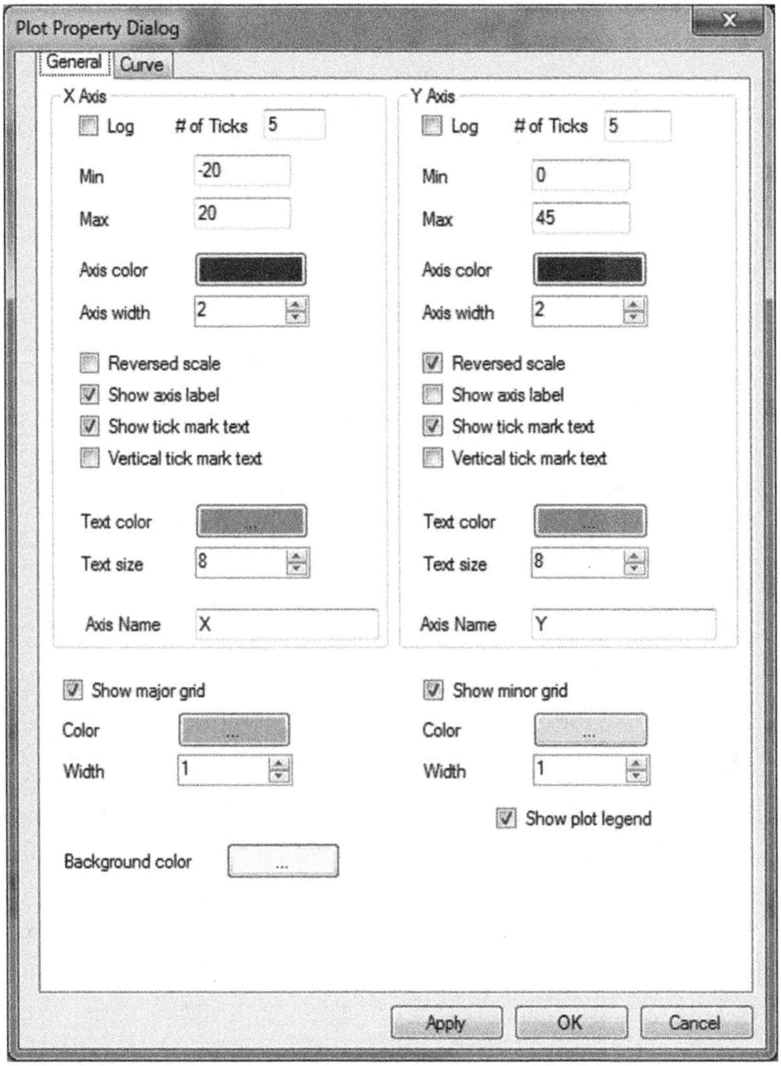

Figure A.10 The display window.

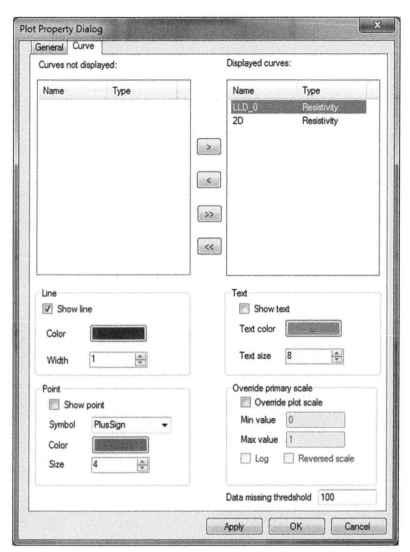

Figure A.11 Curve display options.

INDEX

Note: Page numbers followed by "*f*" and "*t*" refer to figures and tables, respectively.

A

AC. *See* Alternating current (AC)
AC DLT. *See* Alternating current dual laterolog tool (AC DLT)
Acoustic logging. *See* Sonic logging
Admittance element matrix, 535–536
ADR. *See* Azimuthal Deep Resistivity (ADR)
Ahead-of-the-bit boundaries, 409
Ahead-of-the-bit field distribution of LWD tools, 410–414
Alternating current (AC), 561
Alternating current dual laterolog tool (AC DLT), 590–591
  application of source model to, 591–594, 620–624
  computational code for, 594, 595*f*
  grid size for, 597*t*
Amos's subroutine, 212
Amplitude-based apparent conductivity, 287–290, 293
Analog Circuit Test, 117
Anisotropic formation, sensitivity analysis in, 459–463
Anisotropic model, three-layer, 310*f*
Anisotropy impact, analysis of
  to the resistivity LWD tool, 175–178
Antenna equivalent dipole, 449*f*
Antenna spacing, 451–454
Apparent conductivity, 34–35, 37, 293, 335–336
ARC measurements. *See* Array Resistivity Compensated (ARC) measurements
ARC475 tool, 133, 134*f*, 136
Archie's Law, 55
arcVISION LWD, 133
Array induction logs, 270–273
Array laterolog tool, 579–580, 603–619, 604*f*
  electrode configuration and frequency assignment, 611*f*
  focusing method of, 606*f*, 609–619
  hardware focusing, 603–605
  software focusing, 603–605
  tool structure and tool response, 603–608
Array Resistivity Compensated (ARC) measurements, 133
Array Wave Resistivity (AWR) tool, 131
Asphalt and mortar with different moisture contents, experimental data of, 106–108
At-bit short hop telemetry system, geometry of, 682*f*
"At-the-bit" measurement, 409
Automatic measurement system using parallel-disk technique, 80–83, 80*f*
AWR tool. *See* Array Wave Resistivity (AWR) tool
Axial coil transmitter, 422–423
Azimuthal Deep Resistivity (ADR), 130–131
Azimuthal direction, 425, 559
AziTrack, 128
  tool configuration, 362–363, 362*f*
AziTrak Deep Azimuthal Resistivity measurement tool, 129, 129*f*

B

Bedding system, 165
Bergman–Milton simple pole theory, 56
Bessel functions, 40, 212
BHC system. *See* Borehole compensation (BHC) system
BHS formula. *See* Bruggeman–Hanai–Sen (BHS) formula
Biaxial anisotropic homogeneous formation, triaxial induction logging tool response in, 136–160
  full magnetic field response, 143–147
    with arbitrary tool axis, 148–149
  numerical examples, 152–160
  spectrum-domain solution to Maxwell's equations, 140–143
  triple integrals, computation of, 149–152

Biaxial anisotropic-layered formation
  triaxial induction and LWD logging tool response in. *See* Triaxial induction and LWD logging tool response in biaxial anisotropic-layered formation
"Biaxial" anisotropy, 137
BIM. *See* Born iterative method (BIM)
Bisection method, 393
Bolzano bisection method, 393
Borehole compensation (BHC) system, 129–130, 132–133
Borehole conductivity, 414, 444
Borehole mud, influence on dielectric logging tools, 475–478
Born Approximation, 333–334, 348
Born inversion, 333–334
Born iterative method (BIM), 334
Boundary detection
  processing flow of, in geosteering, 392–393
  using orthogonal antennas, 420–424
Boundary distance inversion, 391–403, 392f
  Bolzano bisection method, 393
  processing flow of boundary detection in geosteering, 392–393
  simulation results, 393–400
  simulation results with noise added, 400–403
  theory of inversion, 391
  workflow of inversion problem, 392
Bruggeman–Hanai–Sen (BHS) formula, 57–58
Bulk material, resistivity of, 52–53
Button electrodes, 415

C

Cavity-backed slot antenna, 467–470, 467f
  electrical field |E| distribution of, 468f
  radiation pattern, 469f
CDR tool. *See* Compensated Dual Resistivity (CDR) tool
Centerfire system, 129–130, 130f
Cholesky factorization, 298–301
Coaxial EM telemetry device, 679f
Coil tools, 418–420
Coils, 410–411
Cole–Cole model, 65
Compact Propagation Resistivity (CPR) Tool, 129–130
Compensated Dual Resistivity (CDR) tool, 132–133
Compensated LWD configuration, 362–363
Compensated Wave Resistivity (CWR) tool, 132
Complex image theory, in nonperfect medium, 352, 354–361
  dipole in lossless half-space, 355–358
  dipole in the dissipative media, 358–361
  horizontal dipole in half-space, 355
Complex permittivity, 20
Complex refractive index method (CRIM), 60–62, 61f
Compton scattering effect, 8
COMSOL, 421, 472–473, 504, 605–608
COMSOL Multiphysics, 463–467, 568–569
Conductivity, apparent, 34–35
Conductivity, frequency dispersion of, 67–69
Conductivity contrast, effects of, 382
Contact-casing TCR tool, 634
Convergence algorithm, 172
CPMG sequence, 10, 10f
CPR Tool. *See* Compact Propagation Resistivity (CPR) Tool
CRIM. *See* Complex refractive index method (CRIM)
Cross-component measurements, 407–408
CWR tool. *See* Compensated Wave Resistivity (CWR) tool
Cylindrical coordinates, vector analysis in, 546
Cylindrically layered formation, analytical solutions in, 205
  arbitrary cylindrical layered formations in spectral domain
    derivation for expression of electrical field for, 247–249
  arbitrary cylindrically layered media, response of induction and LWD tools in, 217–240
    discussions of convergence, accuracy, and numerical computation, 230–233
    geometrical configuration, 224–225, 224f, 225f
    induction and LWD tool response with complex invasion profile, 233–235
    magnetic mud influence to an induction and LWD tool response, 236–240
    methodology, 225–230
  four-layer cylindrical medium, induction and LWD tool response in, 206–217
    borehole effects to induction logging tool, 217

geometrical configuration, 207–208, 207f, 208f
influence of mandrel conductivity to LWD tool performance, 217
LWD resistivity tool response with borehole mud and mandrel, 214–216
solution method, 208–213
homogeneous formation in spectral domain, derivation for the expression of electrical field for, 245–246
spectral domain, derivation for the magnetic fields in, 243–244

D

Dash lines, 6
DBIM. *See* Distorted Born iterative method (DBIM)
DC analysis. *See* Direct current (DC) analysis
DCtool. *See* Direct current (DC) tool
Debye model, 65
Deep Propagation Resistivity (DPR) tool, 128
Deep propagation tool (DPT), 450
Deep-looking directional resistivity tool, 424–434
   forward modeling of, 429–434
      1-ohm-m boundary, 434, 434f
      physics of directional resistivity tool, 425–429
DeepTrak, 128
Density log, 8
Depth of Investigation (DOI), 32, 424, 451–452, 491–494
   for LWD tools, 409–410
Dielectric antenna pad, 449f
Dielectric constant
   frequency dispersion of, 64–67
   of rocks, 52–54
Dielectric constant model
   and conversion charts, 136, 136t
   for different LWD tools, 137f
Dielectric logging tools, 11–12, 447
   antenna spacing, 451–454
   applications of, 495–496
   borehole mud influence, 475–478
   cavity-backed slot antenna, 467–470
   Depth of Investigation, 491–494
   design and modeling, using 3D numerical modeling software package, 463–467
   dielectric constant, 447–448, 451, 454, 489
   effects of the pad, 470–474

frequency selection of, 451
history, 449–451
mud cake and invasion, 483–491
sensitivity analysis, 454–458
   in anisotropic formation, 459–463
   isotropic formation, 454–458
   multicomponent dielectric tool responses at 1 GHz, 459f
   sensitivity contour plot, 457f
   signal and sensitivity plot at 1 GHz, 456f
tool pad configuration, 453f
vertical resolution, 479–482
Dirac delta function, 252–254
Direct current (DC) analysis, 626–628
Direct current (DC) tool, 561
Directional electromagnetic (EM) method, 407
Directional LWD resistivity tools, 407
   deep-looking, 424–434
   physics of, 425–429
Distance inversion based on Gauss–Newton algorithm, 435–443
Distorted Born iterative method (DBIM), 334
DLT. *See* Dual laterolog tool (DLT)
DOI. *See* Depth of Investigation (DOI)
Downhole processors, 391
Downlink data transmission method, 646–647, 647f
DPR tool. *See* Deep Propagation Resistivity (DPR) tool
DPT. *See* Deep propagation tool (DPT)
Dual laterolog tool (DLT), 583–590
   application of FEM on AC DLT, 590–594
      application of source model to AC DLT, 591–594, 620–624
      choice of equation, basis function, and element matrix, 590–591
      computational code for AC DLT, 594, 595f
   computational method, validation of, 594–596
   configuration and current patterns, 586f
   electrode position, 587t
   laterolog-deep (LLd) mode, 586–587
   laterolog-shallow (LL) mode, 589
   simulation result, 597–602
      current pattern of LLd and LLs, 598
      frequency effect, 601–602
      grid size and computation time, 597
      Groningen effect, 598–601
      invasion effect, 602

E

E field analysis of the circuit model of the parallel-disk sample holder, 112–116
E field distribution inside the sample holder
  at 1 GHz, 78
  at 2 MHz, 78–80
Eigenmode expansion method, 252, 292–293, 348
Electric field equation, 532–533
Electrical current loop, 537
Electrical logging, 2–3
Electrical resistivity, 52–53
Electrical survey, 2–3
Electrode device, 3, 3f
Electrode type of logging tools, basics of, 580–583
Electrode type TCR tool, EM field of, 632–633
Electrodes, source model, 538–540
Electromagnetic fields, 164
  due to a magnetic dipole in a homogeneous medium, 23–26
  of electrode type TCR tool, 632–633
Electromagnetic motive force (EMF), 17–18
Electromagnetic signals, 448
Electromagnetic telemetry system, 645, 648f, 651f
  application of FEM on EM telemetry systems, 652–658
    choice of equation, basis function, and element matrix, 652
    flowchart of computational code for EM telemetry system, 656–658
    source model, application of, 653–656, 653f, 688–689
  logging-while-drilling (LWD)/measuring-while-drilling (MWD) tools, 645–649
  material properties used in the simulation of, 651t
  numerical model of, 649–651, 650f
  simulation result without casing, 661–680
    current distribution pattern in a 9000-ft-depth borehole, 663–667
    grid size and computation time for different borehole depth, 661–663
    system impedance, 672–680
    voltage on receiver, 667–671
  validation of computation algorithm in a cased borehole, 659–660
Electromagnetic theory, 410
Electromagnetic Wave Resistivity (EWR), 130–131
Electromagnetic waves, sources of, 22f
Electromotive force (EMF), 18–19
  in the receiving coil and the use of bucking coil, 26–31
EM fields. See Electromagnetic fields
EM signals. See Electromagnetic signals
EM theory.Electromagnetic theory
EMF. See Electromagnetic motive force (EMF); Electromotive force (EMF)
Equipment calibration synopsis, 117–119
  HP4191A RF Impedance analyzer, 82–83, 118
  HP4275A LCR Meter, 82, 95, 117
  HP8510C Network analyzer, 104–105, 119
Equivalence Principle, 539–540
Equivalent two-layer model, 357–358
  by applying the image theory, 357f
EWR. See Electromagnetic Wave Resistivity (EWR)

F

Fast forward modeling method, 353
Fast Hankel transform, 172
FDM. See Finite difference method (FDM)
FDM method. See Frequency division multiplexing (FDM) method
FDTD. See Finite difference method in time domain (FDTD)
FEM. See Finite element method (FEM)
15-layer synthetic formation, inversion in, 324–325
Finite coil in homogeneous formation, 127
Finite difference method (FDM), 504–505, 505t
Finite difference method in time domain (FDTD), 504–505, 505t
Finite element method (FEM), 503, 626, 649
  application of, on EM telemetry systems, 652–658
    application of source model for electromagnetic telemetry system, 653–656
    choice of equation, basis function, and element matrix, 652
    flowchart of computational code for EM telemetry system, 656–658

application on AC DLT, 590–594
  application of source model to AC DLT, 591–594, 620–624
  choice of equation, basis function, and element matrix, 590–591
  computational code for AC DLT, 594, 595f
based on electric field, 532–535
  electric field equation, 532–533
  FEM vector matrix equation of electric field, 533–535
based on magnetic field, 505–507
  magnetic field equations, 506–507
basis functions, 511–515
  rectangular element, 514–515
  triangular element, 511–514
comparison with other numerical methods, 505t
computation method of element matrix for rectangular element based on $H\varphi$, 548–549
computation method of element matrix for triangular element
  based on $E\varphi$, 556
  based on $H\varphi$, 554–555
evaluation of impedance element matrix for rectangular element
  based on $H\varphi$, 515–517
  based on $\rho H\varphi$, 518–519
evaluation of impedance element matrix for triangular element, 520–531
  evaluation of long term in element matrix, 524–531
  evaluation of short terms in element matrix, 522–524
  formulation of element matrix, 520–522
evaluation of triangular element matrix based on $E\varphi$, 535–536
FEM solutions for the source existing on boundaries, 540–545
simulation of through-casing resistivity logging tool, 632–636
source models, 537–545
  electrical current loop, 537
  electrodes, 538–540
  magnetic current loop, 537–538
vector analysis in cylindrical coordinates, 546

Five-layer synthetic formation, inversion of, 319–324
Forward modeling, 333–334, 351–352
  and image theory, 353–361
  of deep-looking tool with tilted antennas, 429–434
Fourier transform, 230, 241
Frequency dispersion
  of conductivity, 67–69
  of dielectric constant, 64–67
Frequency division multiplexing (FDM) method, 610
Full magnetic field response
  of triaxial induction sonde in a biaxial anisotropic medium, 143–147
  with arbitrary tool axis, 148–149

G

Galerkin's method, 510
Gamma ray log, 7
Gap EM telemetry device, 679f
Gauss law, 112
Gaussian theorem, 562
Gauss–Laguerre quadrature, 149
Gauss–Legendre quadrature method, 197, 197
Gauss–Newton algorithm, 297–299, 301–302, 407–408, 435–438
  distance inversion based on, 435–443
  inversion in three-layer formations, 441–443
  inversion in two-layer formations, 438–441
Gauss-Quadrature algorithm, 172
Geiger–Mueller (G-M) counter, 7
Geometrical quantities defined on triangular elements, 513t
Geosteering, 13–15, 352, 352f
  application of image theory in. *See* Image theory
Gerschgorin circle theorem, 301
Golden section search, 308, 309f
Green functions, 164
Groningen effect, 598–601
Groove effects
  in formation with borehole, 284–286
  in homogeneous medium, 283
Guard electrode tool, 6

H

Hertz potential, 21–23, 38–41, 167–168, 171, 355–356
Hertz vector potential, 122, 124–126
   derivation of, in multiple layer formation, 179–186
      x-direction magnetic dipole, 179–185
      z-direction magnetic dipole, 185–186
Hessian singular value decomposition analysis, 459–461
HFDT. *See* High frequency dielectric tool (HFDT)
HFSS. *See* High Frequency Electromagnetic Field (HFSS)
High frequency dielectric tool (HFDT), 450
High Frequency Electromagnetic Field (HFSS), 78, 102, 504
High-speed mud pulser, 645–646
Homogeneous anisotropic formation, LWD tool response in, 127–136
   commercial LWD/MWD tools, 127–136
      APS WPR Wave Propagation Resistivity (WPR) Sub, 127–128, 128f
      Array Resistivity Compensated (ARC) measurements, 133
      Array Wave Resistivity (AWR) tool, 131
      AziTrak Deep Azimuthal Resistivity, 129, 129f
      Centerfire system and Compact Propagation Resistivity (CPR) Tool, 129–130
      Compensated Dual Resistivity (CDR) tool, 132–133
      Compensated Wave Resistivity (CWR) tool, 132
      Electromagnetic Wave Resistivity (EWR) and Azimuthal Deep Resistivity (ADR), 130–131
      Multifrequency Resistivity High-Temperature (MFR HT) sensor, 134–136
      Multiple Propagation Resistivity, 128, 128f
      PeriScope, 133
      Slim Compensated Wave Resistivity (SCWR), 132
   dielectric constant model and conversion charts, 136, 136t
Homogeneous biaxial anisotropic medium
   spectral-domain solution to Maxwell's equations in, 188–192
Homogeneous formation, finite coil in, 127
Homogeneous inversion, 35–36
Homogeneous isotropic lossy media, magnetic dipole in, 122–123
Homogeneous transverse isotropic lossy media, magnetic dipole in, 124–126
HP4191A RF Impedance analyzer, 82–83, 118
HP4275A LCR Meter, 82, 95, 117
HP8510C Network Analyzer, 104–105, 119

I

IFFT (Inverse Fourier Transform), 230–231, 233
Image theory, 351, 354f
   boundary distance inversion, 391–403
      Bolzano bisection method, 393
      processing flow of boundary detection in geosteering, 392–393
      simulation results, 393–400
      simulation results with noise added, 400–403
      theory of inversion, 391
      workflow of inversion problem, 392
   of PEC interface, 354f
   relative error, 382, 398–399, 402
   simulation results and discussions, 361–391
      calculation speed, 387
      effects of conductivity contrast, 382
      frequency, 382–384
      logging with high deviated angle, 387–391
      one-dimensional formation model, 361–362
      spacing, 384–387
      tool configuration, 362–363
   theory of forward modeling using, 353–361
      complex image theory in nonperfect medium, 354–361
      review of traditional image theory, 353–354
Impedance element matrix, evaluation of
   for rectangular element, 515–519
   for triangular element, 520–531
INDTRI, 361–362
Induced electromotive force in the receiving coil and the use of bucking coil, 26–31
Induction and LWD resistivity tool response in 2D isotropic formation, 251
   array induction logs, 270–273
   formulations, 252–263
   measurement-while-drilling logs, 273–281
   numerical consideration, 263–265

simulation of effects of mandrel grooves on MWD conductivity logs, 281–291
  effects of conversion table, 287–291
  groove effects in formation with borehole, 284–286
  groove effects in homogeneous medium, 283
  theoretical MWD models, 281–283
verifications, 265–269
Induction and LWD tool response in cylindrically layered isotropic formation, 205
  in arbitrary cylindrically layered media, 217–240
    discussions of convergence, accuracy, and numerical computation, 230–233
    geometrical configuration, 224–225, 224f, 225f
    induction and LWD tool response with complex invasion profile, 233–235
    magnetic mud influence to an induction and LWD tool response, 236–240
    methodology, 225–230
  in four-layer cylindrical medium, 206–217
    borehole effects to induction logging tool, 217
    geometrical configuration, 207–208, 207f, 208f
    influence of the mandrel conductivity to LWD tool performance, 217
    LWD resistivity tool response with borehole mud and mandrel, 214–216
    solution method, 208–213
  spectral domain
    derivation for expression of electrical field for arbitrary cylindrical layered formations in, 247–249
    derivation for expression of electrical field for homogeneous formation in, 245–246
    derivation for magnetic fields in, 243–244
Induction arrays, 45–48
Induction logging data, direct inversion of, 37
Induction logging tool, basic, 4–5, 4f
Induction logs, inversion of
  in two-dimensional formation, 330–346
    results from least squares inversion, 338–346
    theory of 2D induction log inversion, 334–338

Induction tool, 28–29, 37–47, 429
InSite ADR Azimuthal Deep Resistivity Sensor, 131
Inversion method for triaxial induction and LWD logging data in 1D and 2D formations, 295
  Cholesky factorization, 299–301
  constraints, 303–304
  Gauss–Newton algorithm, 297–299
  initial values, 304–310
    boundary merge, 308–309
    initial boundary locations, 306–307
    inverting for, 304–306
    noise analysis, 309–310
  inversion of induction logs in a two-dimensional formation, 330–346
    results from the least squares inversion, 338–346
    theory of 2D induction log inversion, 334–338
  inversion results and analysis, 310–330
    15-layer synthetic formation, inversion in, 324–325
    five-layer synthetic formation, inversion of, 319–324
    real, isotropic formation with synthetic data, inversion of, 325–330
    real log data, inversion of, 330
    synthetic log inversion, 315–319
    synthetic log inversion using all nine components of magnetic field, 310–315
  Jacobian matrix, 302–303
  line search, 301–302
Inversion methods, 391
Inversion problem, workflow of, 392
Isotropic formation (2D), induction and LWD resistivity tool response in, 251
  array induction logs, 270–273
  formulations, 252–263
  measurement-while-drilling logs, 273–281
  numerical consideration, 263–265
  simulation of effects of mandrel grooves on MWD conductivity logs, 281–291
    conversion table, effects of, 287–291
    groove effects in formation with borehole, 284–286
    groove effects in homogeneous medium, 283
    theoretical MWD models, 281–283
  verifications, 265–269

J

Jacobian matrix, 302–303, 337, 438, 459

L

Lateral device, 3–4, 3f
Laterolog, 579–580
  basic, 6, 6f
  circuit model, 583f, 585f
  dual laterolog. See Dual laterolog tool (DLT)
  electrode type of, 580–583
  focusing principle of, 580–581, 583–590
Laterolog-deep (LLd) current, 585–586, 598
Laterolog-shallow (LLs) current, 585–586, 598
Least squares inversion, results from, 338–346
Lichtnecker–Rother (LR) formula, 63
Line search, 301–302
Linear triangular elements, basis function for, 520f
LiveLink, 463–467
LLCM formula. See Lorentz–Lorenz, Clausius–Mossotti (LLCM) formula
LLd current. See Laterolog-deep (LLd) current
LLs current. See Laterolog-shallow (LLs) current
Logging problems, 206
  in vertical borehole, 205–206
Logging tool, 411
  dielectric, 452f, 483–486, 495t
  electrode type of, 580–583
  induction, 4–5, 4f
  neutron, 7–8
  pad-type dielectric, 12f
  propagation, 5–6, 5f
  through-casing, 632–634
  triaxial induction, 136–137
Logging while drilling (LWD), 2, 13, 15, 447–448
Logging-while-drilling (LWD) resistivity imagers, 559, 560f
Logging-while-drilling (LWD) tools, 351–352, 407–408, 424, 645–649
  ahead-of-the-bit field distribution of, 410–414
  analysis of anisotropy impact to, 175–178
  in biaxial anisotropic-layered formation. See Triaxial induction and LWD logging tool response in biaxial anisotropic-layered formation
  in cylindrically layered isotropic formation. See Cylindrically layered formation, analytical solutions in
  depth of investigation (DOI) for, 409–410
  in homogeneous anisotropic formation. See Homogeneous anisotropic formation, LWD tool response in
  in 1D and 2D formations. See Inversion method for triaxial induction and LWD logging data in 1D and 2D formations
  in TI formation. See Triaxial induction tool and LWD tool response in TI formation
  in 2D isotropic formation. See Induction and LWD resistivity tool response in 2D isotropic formation
  short hop telemetry used in, 680–686
Lorentz–Lorenz, Clausius–Mossotti (LLCM) formula, 59–60, 60f
Lossy media
  homogeneous isotropic, 122–123
  homogeneous transverse, 124–126
Low-frequency tools, 410
LR formula. See Lichtnecker–Rother (LR) formula
LWD. See Logging while drilling (LWD)

M

Magnetic current loop, 412, 537–538
Magnetic dipole
  electromagnetic fields due to, in a homogeneous medium, 23–26
  in homogeneous isotropic lossy media, 122–123
  in homogeneous transverse isotropic lossy media, 124–126
  in layered formation, 166–171
    magnitude of reflection and refraction magnetic fields, 171
    x-directed, 168–169, 179–185
    y-directed, 169–171
    z-directed, 167, 185–186
  moment, 23
  source, in transverse isotropic homogeneous formation, 165–166
Magnetic field
  finite element method based on, 505–507
  synthetic log inversion using all nine components of, 310–315
  synthetic log inversion using diagonal components of, 315–319
  vector matrix equation of, 509–511

Magnetic field equations, 506–507
MATLAB, 463–467
Matrix
   admittance element matrix, 535–536
   derivation of, 202–204
   impedance element matrix, 515–517
   Jacobian matrix, 302–303, 337, 438, 459
   triangular element matrix, evaluation of, 535–536
Matrix assembling rule, computation method for, 547
Matrix Assembly Rule, 511, 534–535
Maxwell (ANSYS), 504
Maxwell's equations, 17–20, 54, 140–141, 164, 210, 241, 412, 504–505
   spectral-domain solution to, 140–143
      in homogeneous biaxial anisotropic medium, 188–192
Measurement-while-drilling (MWD) conductivity logs, 273–291
   conversion table, effects of, 287–291
   groove effects in formation with borehole, 284–286
   groove effects in homogeneous medium, 283
   theoretical MWD models, 281–283
Measuring while drilling (MWD), 2, 13–15, 127
Measuring-while-drilling (MWD) tools, 275–277, 293, 645–649
Method of moment (MOM), 504–505, 505t
MFR HT sensor. See Multifrequency Resistivity High-Temperature (MFR HT) sensor
MicroSFL, 585
Mixing formulas, 57–63
   Bruggeman–Hanai–Sen (BHS) formula, 57–58
   complex refractive index method (CRIM), 60–62
   Lichtnecker–Rother (LR) formula, 63
   Lorentz–Lorenz, Clausius–Mossotti (LLCM) formula, 59–60
MOM. See Method of moment (MOM)
Mud circulation system in a drill pit, 646f
Mud pulse telemetry in a drilling system, 647f
Mud pulsers, 646–649
Multifrequency Resistivity High-Temperature (MFR HT) sensor, 134–136
Multiinvasion zones, 252–254, 272–273, 277, 293

Multiple Propagation Resistivity, 128, 128f
MWD. See Measuring while drilling (MWD)

N

NaviGator, 128
Near bit logging-while-drilling sensor, 680f
   short hop electromagnetic telemetry used in, 680–686
   three-dimensional model of, 680f
Neutron log, 7–8
New-generation dielectric tools, 448–450
NMR logging. See Nuclear Magnetic Resonance (NMR) logging
NMR tools. See Nuclear Magnetic Resonant (NMR) tools
Nuclear logging, 6–8
   density log, 8
   gamma ray log, 7
   neutron log, 7–8
Nuclear Magnetic Resonance (NMR) logging, 9–11
Nuclear Magnetic Resonant (NMR) tools, 449
Numerical model of electromagnetic telemetry system, 649–651, 650f
Numerical simulation methods, 504–505
   of through-casing resistivity tool, 633–636
      choice of element matrix and source model, 634
      simulation results, 634–636

O

OBM. See Oil-based mud (OBM)
OBMI tools
   current distributions, 569–570
   depth of investigation of, 572
   development of, 568–569
   dv versus formation resistivity, 570
   effect of the thickness of oil-based mud layer, 571–572
   pad, 573–574, 573f
   vertical resolution of, 572–576
Ohm's Law, 2–3, 576–578
Oil and gas exploration, 1–2
Oil-based mud (OBM), 413–414, 475–476, 672
Oil-based mud resistivity imager, 562–576
   circuit model of, 564–568, 567f
   three-dimensional FEM of OBMI tool, 568–572
   vertical resolution of OBMI tool, 572–576

Oklahoma model, 319, 324, 325f
One-dimensional formation model, 361–362
1D-layered model, 173f
1-V excitation voltage, 681–682, 685–686
Orthogonal antennas, 407–408
    boundary detection using, 420–424
Oscillating magnetic dipole, 22f, 23f

P

Pad-type tools, 12, 12f, 470–472
Parallel-disk capacitor, 74, 75f, 112, 112f
Parallel-disk measurement method, 71–72
Parallel-disk sample holder, 70–71
    circuit model of, 72–74
    circuit parameters of, 74–75, 75f
    E field analysis of the circuit model of, 112–116
Parallel-disk technique, 71, 78–80
    analysis of dynamic range of, 78–80
    automatic measurement system using, 80–83
PathFinder AWR array, 131
PathFinder Compensated Wave Resistivity (CWR) tool, 132
PathFinder Slim Compensated Wave Resistivity (SCWR), 132
PEC. See Perfect electric conductor (PEC)
Perfect electric conductor (PEC), 353
Perfect magnetic conductor (PMC) interface, 353
PeriScope, 133
    tool configuration of, 133f
Permittivity, 20
Phase-based apparent conductivity, 287–290
Photoelectric log. See Density log
π network circuit model, 72, 73f
PMC interface. See Perfect magnetic conductor (PMC) interface
Polarization, 10, 10f
Pore spaces in rocks, 52, 55
Porosity of a rock, 52
Potential difference, 564, 648–649, 667
Propagation logging method, 5–6, 5f
Pseudo-inner product, 509

Q

Quasistatic approximations and skin depth, 31–33
Quikrete Mortar Mix No. 1102, 106–107

R

Radiation logging. See Nuclear logging
Radiofrequency Module, 463–467
Real, isotropic formation with synthetic data inversion of, 325–330
Real log data, inversion of, 330
Real-time inversion, 351–352
Rectangular element basis functions, 514–515
Resistivity and dielectric constant of rocks, 52–54
Resistivity imagers. See Resistivity imaging tools
Resistivity imaging tools, 559
    depth of investigation (DOI) of, 559, 562
    oil-based mud resistivity imager, 562–576
        circuit model of, 564–568
        3D finite element analysis of OBMI tool, 568–572
        vertical resolution of OBMI tool, 572–576
    water-based mud resistivity imaging tool, 561–562
Resistivity logging methods, 2–4
Resistivity of a bulk material, 52–53
Resistivity-at-the-Bit tool, 409
Rocks, measurement methods of electrical properties of, 69–95
    automatic measurement system using parallel-disk technique, 80–83
    computation of dielectric constant and conductivity of test samples, 77–78
    differences between the LF and the HF measurements, 92–94
    error analysis, 95
    experimental data and discussions, 83–91
    parallel-disk measurement method, 71–72
    parallel-disk sample holder
        circuit model of, 72–74
        circuit parameters of, 74–75, 75f
    parallel-disk technique, analysis of dynamic range of, 78–80
        E field distribution inside the sample holder at 1 GHz, 78
        E field distribution inside the sample holder at 2 MHz, 78–80
    performance analysis at high frequencies, 76–77
    rock measurements, background of, 70–71

S

Saline solutions with different salinities, experimental data of, 105–106, 106t
Schlunberger's ARC tool, physical parameters of, 134t
Scintillation counter, 7
SCWR. See Slim Compensated Wave Resistivity (SCWR)
Sediment rocks, electrical properties of
  Archie's Law, 55
  background review, 55–57
  E field analysis of the circuit model of parallel-disk sample holder, 112–116
  equipment calibration synopsis, 117–119
    HP4191A RF Impedance analyzer, 82–83, 118
    HP4275A LCR Meter, 95, 117
    HP8510C Network analyzer, 104–105, 119
  frequency dispersion of the conductivity, 67–69
  frequency dispersion of dielectric constant, 64–67
  measurement methods of electrical properties of rocks, 69–95
    analysis of dynamic range of the parallel-disk technique, 78–80
    automatic measurement system using the parallel-disk technique, 80–83
    background of rock measurements, 70–71
    circuit model of the parallel-disk sample holder, 72–74
    circuit parameters of the parallel-disk sample holder, 74–75, 75f
    computation of the dielectric constant and the conductivity of test samples, 77–78
    differences between the LF and the HF measurements, 92–94
    error analysis, 95
    experimental data and discussions, 83–91
    parallel-disk measurement method, 71–72
    performance analysis at high frequencies, 76–77
  mixing formulas, 57–63
    Bruggeman–Hanai–Sen (BHS) formula, 57–58
    complex refractive index method (CRIM), 60–62
    Lichtnecker–Rother (LR) formula, 63
    Lorentz–Lorenz, Clausius–Mossotti (LLCM) formula, 59–60
  resistivity and dielectric constant of rocks, 52–54
  TM_{010} resonant cavity technique, 95–110
    dynamic range of the resonant cavity technique, 98–104
    measurement system and experimental data of the resonant cavity technique, 104–110
    theory of the TM_{010} resonant cavity technique, 95–98
Shale resistivity, 174–175
Short hop EM telemetry, 680–686
  current flow paths of, 681f
  input current and impedance of a toroidal transmitter versus formation resistivity, 685–686
  simulation of, 681–685
Signal analysis method, 610
Slim Compensated Wave Resistivity (SCWR), 132
Sommerfeld integral, 356, 360
Sonic logging, 8–9, 9f
Source models, used in the numerical simulation, 537–538
  electrical current loop, 537
  electrodes, 538–540
  magnetic current loop, 537–538
Sources, 21
Spectral domain
  derivation for electrical field for arbitrary cylindrical layered formations in, 247–249
  derivation for electrical field for homogeneous formation in, 245–246
  derivation for magnetic fields in, 243–244
Spectral-domain solution to Maxwell's equations in homogeneous biaxial anisotropic medium, 188–192
Spectrum domain solutions and two-coil induction tools in layered media, 37–45
Spectrum-domain solution to Maxwell's equations, 140–143
Symmetric product, 509
Synthetic log inversion, 315–319
  using all nine components of magnetic field, 310–315
System impedance, 672–680

T

TCR logging tool. See Through-casing resistivity (TCR) logging tool
Telemetry impedance, 672, 678

Theory of inversion, 391
  See also Inversion method for triaxial induction and LWD logging data in 1D and 2D formations
3D electromagnetic simulation software, 463–467
3D model, 421, 422f
  with a z-direction toroidal transmitter and a y-direction coil receiver, 421f
Three-layer 1D model, 388–391
Three-layer equivalent model, 358, 362f, 373
  by applying the image theory, 358f
Three-layer formations, inversion in, 441–443
Threshold of length (TOL), 393
Through-casing resistivity (TCR) logging tool, 625
  circuit model of, 627f, 628–632
  finite element method simulation of, 632–636
    numerical simulation, 633–636
  geometry of a cased well with anisotropy formation, 642f
  measurement procedure, 626–628
  reservoirs, 625
  toroidal antenna, through casing logs from, 637–643
    choice of element matrix and source model, 638
    comparison with published literature, 638–640
    simulation results, 640–643
TM_{010} resonant cavity technique, 95–110
  asphalt and mortar with different moisture contents, experimental data of, 106–108
  automatic measurement system of, 104–105
  dynamic range of, 98–104
    error analysis of simulation data, 103–104
    simulation data, 98–103
  error analysis, 109
  saline solutions with different salinities, experimental data of, 105–106
  theory of, 95–98
TOL. *See* Threshold of length (TOL)
Tool constant and skin-effect correction, 35–37
Toroid antenna impedance, 686
Toroidal antenna, 411–412, 681
  ahead-of-the-bit field of, 414f, 415f
  configuration of, 416f
  current map of, 416f
  electric field distribution of, 414f
  implementation of, 411f
  through casing logs from, 637–643
    choice of element matrix and source model, 638
    comparison with published literature, 638–640
    simulation results, 640–643
Toroidal tools, 411–412, 423, 444
Toroidal transmitter, 413f, 415–420, 685–686
  modeling, 413f
Transmitter–receiver spacing, 416–417
Transverse electric mode, analysis of, 507–508
Transverse isotropic homogeneous formation, magnetic dipole source in, 165–166
Transverse magnetic mode, analysis of, 507–508
Transverse relaxation time, 11
Triangular element basis functions, 511–514
Triaxial induction and LWD logging data, inversion method for. *See* Inversion method for triaxial induction and LWD logging data in 1D and 2D formations
Triaxial induction and LWD logging tool response in biaxial anisotropic-layered formation, 187
  double integrals, computation of, 196–197
  layered medium, propagation in, 193–196
  matrix, derivation of, 202–204
  numerical examples, 197–201
  spectral-domain solution to Maxwell's equations in homogeneous biaxial anisotropic medium, 188–192
  unbounded medium, propagation in, 193
Triaxial induction logging tool response in biaxial anisotropic homogeneous formation, 136–160
  full magnetic field response, 143–147
    with arbitrary tool axis, 148–149
  numerical examples, 152–160
  spectrum-domain solution to Maxwell's equations, 140–143
  triple integrals, computation of, 149–152
Triaxial induction tool and LWD tool response in TI formation, 163
  analysis of anisotropy impact to the resistivity LWD tool, 175–178
  convergence algorithm, 172
  derivation of Hertz vector potential in multiple layer formation, 179–186

x-direction magnetic dipole, 179–185
z-direction magnetic dipole, 185–186
magnetic dipole source in transverse isotropic homogeneous formation, 165–166
magnitude of reflection and refraction magnetic fields, 171
simulation results and analysis, 172–175
x-directed magnetic dipole, 168–169
y-directed magnetic dipole, 169–171
z-directed magnetic dipole, 167
Triple integrals, computation of, 149–152
TRITI2011_series, 429
2C40 tool, 263–269, 310
2D axial symmetrical system, 483–486
2D induction log inversion
 flowchart of, 335f
 theory of, 334–338
Two-layer equivalent model, 357–358
 by applying the image theory, 357f
Two-layer formation model, 407–408, 417–418, 423f, 438, 444
 inversion in, 438–441, 439f

V

Vector matrix equation of magnetic field and impedance matrix, 509–511

W

Water-based mud (WBM), 413–414, 448, 475–476
 boreholes, 672
 -filled borehole, 420
 resistivity imaging tool, 560f, 561–562, 576–578
Wave number, 252–254
Wave Propagation Resistivity (WPR) Sub, 127–128, 128f
WBM. *See* Water-based mud (WBM)
Weatherford MFR tool, 135f
 physical parameters of, 135t
Well Logging Lab, 361–362
Well logging methods, 2–6
 basic induction logging tool, 4–5, 4f
 basic laterolog, 6
 basic propagation logging method, 5–6, 5f
 basic resistivity logging methods, 2–4
Wireless communications among downhole devices, 687
Wireline logging and logging while drilling, 12–13
WPR Sub. *See* Wave Propagation Resistivity (WPR) Sub

X

x-directed magnetic dipole, 168–169, 179–185

Y

y-directed magnetic dipole, 169–171

Z

z-directed magnetic dipole, 167, 185–186
Zero-dimensional inversion, 304–305, 308, 313, 328
 flowchart of, 305f